SECOND EDITION 2019

PHYSICS

WITH HEALTH SCIENCE APPLICATIONS

PAUL PETER URONE
R.E. TREMBLAY

Kendall Hunt
publishing company

Cover image © Shutterstock, Inc.

www.kendallhunt.com
Send all inquiries to:
4050 Westmark Drive
Dubuque, IA 52004-1840

4th Printing

PAK ISBN 978-1-5249-5062-0
Text Alone ISBN 978-1-5249-5061-3

Published in the United States of America

CONTENTS

PREFACE

To the student:

The purpose of this book is to transmit an understanding of basic concepts in physics and to prepare students to apply them both professionally and in everyday life. It is intended for the courses in physics with health science applications commonly taken by students majoring in allied health, nursing, physical therapy, kinesiology, and related disciplines.

The general approach in the text is to introduce topics with familiar everyday examples before proceeding to general principles. Numerous examples with step-by-step solutions and applications related to the health sciences and everyday situations are used to illustrate physical principles. This enables the student to understand the general principles and their relevance; the unifying aspect of physical laws and the basic simplicity of nature form the underlying theme of the text.

We have developed a website that includes step-by-step solutions to numerous homework problems. The website also contains approximately 30 audiovisual lessons on fundamental topics. Additionally we have included 11 labs that enhance and expand topics ranging from vector addition to radioactive half-life. We hope that you enjoy your journey through the wonderful world of physics and its applications to the health care field.

To the instructor:

The content, organization, graphics, and learning tools were specifically developed and written to provide instructional flexibility in course length, difficulty, and topical emphasis.

This textbook is intended for use in one-semester or two-quarter courses. To allow greater flexibility in its use, the book contains more material than can be covered in a one-semester course. However, it is structured so that many topics can be omitted without loss of continuity. No material is specifically denoted as optional since that decision should be left to the instructor and based on the type of course involved. Chapters 7, 9, 12, 13, and 15, and the last section or sections of Chapters 4, 5, 6, 11, 14, 16, and 18, are primarily applications. These chapters and sections can be emphasized, covered briefly or omitted without loss of continuity and without sacrificing the study of basic physical principles.

Certain topics are presented with greater than normal emphasis.

Both Chapters 6 and 7 are devoted to fluids because of their great biological and medical importance. Chapter 18, on radioactivity and nuclear physics, also contains more than the normal amount of information, including numerous tables, calculations of radiation doses, and diagnostic and therapeutic uses of radiation. Special attention is also devoted to other interesting subjects, with sections or entire chapters on energy and power in humans, heat and the human body, medical and biological applications of fluids, hearing, vision, electrical safety, bioelectricity, and medical applications of EM radiation. Numerous other health science applications are an integral part of the text. The manner in which subjects are presented, with worked examples, review questions, and problems, also allows students to learn about many topics outside the classroom and makes the book a more valuable resource.

The ordering of topics is largely traditional, the exception being that heat is studied before fluids, allowing easier introduction of such concepts as diffusion and osmosis. The ordering of some topics can be changed without difficulty. The chapters on optics and vision,(Chapters 14 and 15) follow those on electricity and magnetism (Chapters 10-13) but could be covered right after Chapter 8, on sound. Chapters 5-7 on heat and fluids can be covered anytime after energy is introduced in Chapter 4.

More than 1000 questions and problems are included throughout the book at the ends of chapters. In addition, there are mathematics tutorials in Appendix A as part of a review for students whose math background is weak or distant. The 450-plus chapter review questions requiring descriptive answers are structured to have students identify, analyze, and generalize physical principles. Every effort has been made to make the questions a learning device as well as a testing device. The 550-plus problems are similarly designed. Problems are of three levels of difficulty, and both their level and the section to which they are most closely related are identified. These features allow maximum pedagogical benefit and are convenient for the instructor if material is omitted.

Over 100 numerical examples illustrate physical principle by posing a question or situation. The solution is then developed to illustrate the physical principle and the problem-solving process. Each topic is introduced in enough depth to be understood at an introductory level and to provide students with tools to answer questions and solve problems on their own.

Over 300 drawings and photographs illustrate the text material.

Self-explanatory figure captions allow the student to understand the concept of each figure without having to read the text in detail.

The International System of Units (SI units) is used throughout the text but not to the exclusion of other units, since in many applications non-SI units are traditional. These units are freely introduced where appropriate and related to SI units through conversions. This allows students to understand the interconnection of units while still seeing the familiar units most commonly used in specific applications.

Students and faculty, please feel free to reach out to us with questions, comments and suggestions.

Sincerely,
Paul Peter Urone and R.E. Tremblay

ACKNOWLEDGEMENTS

My deepest gratitude to Paul Peter Urone for writing an inspired and comprehensive first edition, upon its foundation this second edition resides. His encouragement and insight have been invaluable. Thank you to the team at Kendall Hunt, in particular, Bob Largent, acquisitions manager, Torrie Johnson, project coordinator and Katie Benzer, web support and managing editor Angela Lampe. I could not have asked for a more supportive team. Thank you to my high school physics teacher, Dr. John Karajanis and undergraduate physics teacher, Dr. William S. Porter, for unwavering support and guidance along this journey of exploration and enlightenment. This work has been inspired by countless students whose dedication to learning, demanded comprehensive explanations with expanded examples and solutions. This work is for them and those that follow. Thank you to my wife Cathy whose love, unwavering support and devotion, has allowed me to pursue my passion. Thank you to our daughter Maddie who allowed me to see through the eyes of a child once again as she grew to be the woman that any parent would be proud of.

Sincerely,
R.E. Tremblay

1 Introduction

"I learned very early the difference between knowing the name of something and knowing something."

Richard P. Feynman

Physicists are fond of saying that physics is the most basic of all sciences. There is truth in this statement, for physics concerns itself with questions that are basic to all natural sciences. The word physics comes from Greek, meaning "of nature" or "natural philosophy." Until recent times natural philosophy encompassed many fields, including astronomy, chemistry, biology, mathematics, medicine, philosophy, and religion. Physics is the remaining core of natural philosophy and concerns itself with questions of what underlies the interactions of matter, energy, space, and time, and even with what constitutes realty.

It is not surprising that a field that has produced the theories of relativity and quantum mechanics and has drastically altered our concept of the universe has an aura of mystery—of being remote from everyday experience and impossible to understand. Nothing could be farther from the truth. Once one becomes accustomed to looking for an explanation of various phenomena in terms of underlying scientific principles, it is possible to see physics everywhere. Physics is all around us. The flight of birds, the operation of a microwave oven, the color of the sunset, and the pitch of one's voice all have basic explanations in physics. Those explanations can be understood by anyone, not just professional scientists. Welcome to the wonderful world of physics!

The purpose of this text is to aid in the study of the principles of physics and their applications in the health sciences. Those principles are easier to digest and are more meaningful if their application in familiar settings is made apparent. The following section describes a few situations in the health sciences in which physics is important.

1.1 PHYSICS IN THE HEALTH SCIENCES

Physics is encountered in many situations—recreational, and even social. The situations described in this section are but a few of the many, particularly in the health sciences, which will be studied in depth in later sections of the text.

Athletics: We do not all have an intuitive feeling for how to use our bodies most effectively. Even the greatest athletes learn from their coaches. One important area of study in modern athletics is kinesiology, literally the study of motion. It is based on the relationships between distance, time, velocity, and acceleration (covered in Chapter 2), as well as on the concepts of force, work, energy, and power (developed in Chapters 3 and 4). Studying those chapters will allow a deeper understanding of the body, its muscles, and its utilization of energy in terms of underlying physics that may already be intuitively familiar.

For example, it will be clear why it is harder to carry an object at arm's length than close to the body. Experience makes it obvious, but physics tells why. (It has to do with where the muscles are attached to bones in relationship to the joints and something called torque.)

Traction Systems: Some traction systems seem to have wires, pulleys, and weights going every which way and performing altogether mysterious tasks. The study of forces in Chapter 3 helps explain traction systems. It will show the importance of not only the strength of a force, but also the direction of the force and the point where it is applied. The strength of the force in traction will obviously depend on how large a weight is used. The direction of the force will be the same as the direction of the wire attached to the subject. The point of application is the place where the wire is attached and effects the torque as much as the force does. Analysis of the system using Newton's laws in Chapter 3 will show the effect of these three factors.

Nutrition and Exercise: Few things have caught the attention of the public as have nutrition and exercise over the last several years. It turns out that work is the manifestation of energy, changing forms. In humans, work changes stored food energy into heat, motion, and other forms of energy. When work, energy, power, and, efficiency are studied in Chapter 4, they will be related to food energy and human exercise as well as many other examples. Energy is one of the central concepts of physics and will be important in almost every chapter following its introduction in Chapter 4.

Body Temperature: Humans and other warm-blooded animals maintain a constant body temperature by converting food energy to heat energy. However, the body continues to produce heat even when surrounding temperatures are higher than body temperature. That excess heat is dissipated by perspiring. Heat as a special form of energy is studied in Chapter 5. Various methods of heat transfer are presented and it will become clear why perspiration is the body's only possible method of dissipating heat when surrounding temperatures are high. It will also be seen why an alcohol rub reduces body temperature, as might be necessary with a high fever. The concept of efficiency makes it evident that the body creates even more heat than normal during exercise since a large fraction of the food energy used in producing muscle contractions ends up as heat instead (efficiency is less than 100%). As a consequence, the body requires more cooling and perspires more during exercise than when at rest.

Physical Therapy: Patients undergoing physical therapy usually have weakened or damaged muscles or suffer from nerve disorders that make it difficult for them to move their muscles effectively. A great deal of physical therapy takes place in water because the water helps to support the weight of the person. The buoyant force supplied by the water, greatly reduces the effective weight of the person and of his limbs, making it possible for him to perform exercises that would be impossible out of the water. The buoyant force is one aspect of the physics of fluids covered in Chapter 6.

The physics of fluids has a tremendous number of applications in biological systems—so many that Chapter 7 is entirely devoted to them.

Blood Flow and Respiration: Liquids and gases can be made to move by the application of pressure. Pressure is studied in Chapters 6 and 7, as are methods for transmitting and measuring pressure. The flow of liquids and gases in general as well as in biological systems will also be studied in depth. For example, the heart creates blood pressure differences by exerting forces on the blood with a muscular contraction. The subsequent blood flow is regulated by blood vessels changing diameter and thereby changing their resistance to flow. Other examples of the body's use of pressure include breathing, maintenance of reduced pressure in the chest cavity to keep the lungs from collapsing, and pressure in the eye to maintain its shape.

Hearing: Hearing is the perception of sound. Sound will be the first example of a wave phenomenon. Hearing and sound will be studied in Chapters 8 and 9, where it will be shown how hearing does not simply reproduce the actual physical properties of sound. For example, loudness is the perceived intensity of sound waves. However, humans do not perceive ultrasound at all, so loudness is not a perfect indication of intensity and hence differs from that physical characteristic.

Ultrasonic Scanners: Ultrasound is any sound that is above the frequency that an average person can hear. Ultrasound still behaves in a fashion similar to audible sound waves. For example, it scatters from boundaries between substances and so can be used to probe the inside of the body noninvasively, much as boaters use sonar to determine depth and image of fish in dark waters. Ultrasonic waves can be made perfectly safe by keeping their intensity low enough. If this is done, the ultrasound cannot cause injury because it lacks the energy to do so. In Chapter 8, ultrasonic scanners are explained and compared with other tools for probing the interior of the body, such as x-rays. (X-rays are a type of light that is yet another type of wave.)

Electrical Safety: Certain medical procedures make hospital patients extremely sensitive to electrical shock. In the discussion of electrical safety in Chapter 12, the reasons for that sensitivity will be explained and some of the major methods of protection will be presented. These include the three-wire system, proper grounding of appliances, and the use of circuit breakers. As usual, the study of electrical safety will be based on the principles of physics (learned from the study of electricity and magnetism in Chapter 10 and of simple electric circuits in Chapter 11).

Nervous System: The nervous system is a complex of biological electric circuits that controls the muscles, among other things. Chapter 13 is entirely devoted to bioelectricity. In that chapter it will be seen that bioelectricity can be recorded and interpreted to yield a great deal of information on the functioning of certain body organs. The most common such recording is the electrocardiogram, literally a recording of the electrical impulses that control the beating of the heart. Electrocardiograms give detailed information about the condition of that organ. Similarly, information can be obtained about brain functions by recording its electrical impulses in an electroencephalogram. (This is possible even though the brain is extremely complex, even when compared with the most advanced modern computers.)

Vision: Most people consider vision to be their most important sense. Vision is covered in Chapter 15 as one application of the general physics of optics. Among the aspects of vision that have their explanation in the laws of physics are how the eye forms an image on the retina and the correction of common vision defects. The laws of optics, which are applied in Chapter 14 to lenses, mirrors, and microscopes, can also describe the near miracle of vision.

From Microwave Deep Heating to Sunlamps: These are but two examples of applications of electromagnetic (EM) waves, studied in Chapter 16. EM waves take many forms, including radio waves, microwaves, visible light, ultraviolet light (as from a sunlamp), x-rays, and gamma rays. The behavior of all these EM waves is analogous to that of sound waves. Learning the essential physics behind EM waves will make it possible to understand why they exhibit so many different properties. For example, microwaves can be used for deep heating, while ultraviolet waves cannot; ultraviolet waves cause both tanning and sunburn and can be used to sterilize objects even if the waves are very dim.

Spectroanalysis: It is a useful tool in the detection of trace amounts of toxic substances. It is based on the fact that all elements and compounds emit EM spectra that are uniquely characteristic of the particular substance. The uniqueness of atomic spectra is explained by atomic physics, the subject of Chapter 17. Spectroanalysis is used in medicine and a host of other disciplines, from chemistry to astronomy. It was used as a tool long before atomic physics was understood, but now we, unlike our ancestors, can understand not only how it works but why.

X-Rays: X-rays are a part of the EM spectrum, which will be studied in Chapters 16–18. As in the immediately preceding discussion of microwaves and ultraviolet waves, it will be possible to understand why x-rays have the properties they do and why they are so useful as a diagnostic tool in medicine. The study of the effects of radiation on biological organisms in Chapter 18 will show that x-rays are hazardous and cannot be made perfectly safe—that is, their use involves a calculated risk.

Radiotherapy, Radiation Diagnostics, and Radiation Protection: Chapter 18 starts with radioactivity and nuclear physics. It then applies nuclear physics and principles of physics studied earlier to such things as radiotherapy, radiation diagnostics, and radiation protection. It will be possible to understand the uses as well as the hazards of radiation. The energy and other characteristics of radiation and the physical laws governing it give insight into these problems and help one gain the ability to assess for oneself the risk versus the benefit.

These are but a few of the applications of physics presented in this text. In addition to the health science applications presented here, there will be numerous other applications in many areas. One subject is built upon another in physics, and as one nears the end of this text it will be possible to see how interconnected the field of physics is. The underlying unity of physics is perhaps its most beautiful aspect. There exist a relatively small number of basic principles that govern all of nature. An appreciation of that beauty is one of the goals of this text.

1.2 MODELS, THEORIES, AND LAWS

The laws of physics were developed over many centuries. Periods of stagnation were interspersed with eras of explosive and revolutionary discoveries. How were these laws discovered? Can we take advantage of the methods used by scientists to understand and give order to the myriad of phenomena observed in nature?

The study and understanding of physics is made easier by using the same techniques that scientists use to make discoveries. These include building models, theories, and laws and performing experiments. We can take advantage of the known organization, and unity of nature to understand, and apply the laws of physics.

Much of what physics, especially modern physics, concerns itself with is not accessible to ordinary senses of sight, smell, touch, or hearing. Radiation, for example, is impossible to sense but is very real, the nucleus of an atom cannot be seen with the most powerful microscopes, and some of the most important information in medicine comes from x-rays that aren't sensed by the eye. Yet, by building mental images called models, it is possible to understand phenomena that must be observed with instruments and not with our innate senses.

A model is an analogy to objects or phenomena that are generally familiar and can be experienced directly. Models serve as very useful mental images to help picture what is going on in a system that cannot be sensed directly. One example is the planetary model of the atom, which pictures electrons orbiting the nucleus just as planets orbit the sun. The electrons don't actually revolve around the nucleus, but many of the observed characteristics of atoms correspond very well to this mental image. Another example of a model is the wave theory of light, which pictures light having characteristics similar to water waves and sound waves. Note that the word theory is used instead of model in this case. A model becomes a full-fledged theory if it is widely successful in its applications and is not contradicted by experiment. When ideas reach the point of broadly successful application, it is tempting to think of them as being literally true and to lose sight of the fact that we are dealing with an analogy. At the beginning of the twentieth century, Albert Einstein, Max Planck, and others realized that light has both wave and particle properties.

A scientific **law** describes what happens in nature in a general way.

An example is Newton's laws of motion, which accurately describe the relationship between motion and force. Laws have been observed to describe nature accurately in so many circumstances that they are thought of as being absolutely true and immutable. Models can help visualize why certain laws work. Laws themselves say what happens but may not say why.

1.2.1 The Importance of Experiments

The validity of models, theories, and laws is judged on their success in describing the results of experiments. Since it is impossible to perform all imaginable experiments, it is impossible to be absolutely certain that even the most basic scientific laws are always valid. We must be willing to let go of a law, no matter how dear it is, if experiments show it to be invalid. Obviously, experiments that show violations of scientific laws must be carefully checked. Usually, thousands of other experiments have had results consistent with the law, so the chance that the experiment reveals a violation of the law is wrong cannot be neglected. The one thing in which scientists do have absolute faith is experimental observation. If something is observed in a good-quality, reproducible experiment, then it is true no matter how bizarre it may seem.

Many of the results of modern physics do in fact seem bizarre and fantastic. The reason for this is that physicists have used sophisticated equipment to observe objects and phenomena that are not observable with ordinary senses. These objects generally fall into one or both of two categories: They are objects that are very small or objects that are moving near the speed of light. Nature is significantly different for very small and very fast-moving objects than it is in everyday life. Consequently, the models devised to aid in understanding and describing these objects and phenomena contradict everyday experience.

An example of a seemingly bizarre effect that has real consequences is the slowing of time as objects approach the speed of light. This effect is part of the theory of relativity, but nothing like it is experienced in everyday life. Time seems to go by at the same rate whether a person is traveling at

Table 1.1 The Realms of Physics

	Slow	Fast (>1% speed of light)
Large	Classical	Relativity
Small (submicroscopic)	Quantum mechanics	Relativistic quantum mechanics

60 or 600 mi/hr. Time actually does slow down for a person or object traveling in a jet at 600 mi/hr, but it is such a small effect that it is difficult to observe. However, the effect is very large and has significant consequences for very fast-moving objects, such as certain types of radiation coming to earth from outer space. Some types of this cosmic radiation should die out before they reach the surface of the earth, but since time slows down for them they do reach earth. They then form part of background radiation that affects us all.

Table 1.1 gives the realms of the major theories of physics. Classical physics is an extremely good approximation and can be used to a very high degree of accuracy for objects larger than about molecular size and moving slower than about 1% of the speed of light. (Even space probes travel at less than 0.005% of the speed of light.) Classical physics therefore describes almost all everyday experience. Consequently, classical physics fits common intuition very well and seems more logical than some aspects of modern physics.

When dealing with small objects it is necessary to use the theory of quantum mechanics. Similarly, when dealing with objects that are moving at very high speeds it is necessary to use the theory of relativity. Finally, when dealing with small fast-moving objects it is necessary to use a combination of the two, ***relativistic quantum mechanics***. In fact, relativistic quantum mechanics works for everything from electrons to athletes to clusters of galaxies and could be used in all situations. However, the mathematical difficulties are such that relativity, quantum mechanics, and relativistic quantum mechanics are only used when absolutely necessary (i.e., for small and/or fast moving objects).

Chapters 16–18 deal with topics in relativity and quantum mechanics (together also called modern physics). At that time it will be seen that some of the most exciting discoveries of this century have applications in the health sciences and everyday life. The subjects preceding the last three chapters, which belong largely to classical physics, not only have many applications of their own, but also are crucial to the understanding of later material. Physics is a discipline that builds new concepts upon a foundation of earlier concepts. It cannot be effectively studied without starting from some very basic definitions. In the following section, the process begins with definitions of the units upon which all measurable quantities are based.

1.3 LENGTH, MASS, AND TIME: THE BASIC UNITS

Almost every measurable quantity presented in this book can be expressed as a combination of three basic entities: length, mass, and time. A fourth basic entity, charge, will be added when electricity and magnetism are introduced in Chapter 10. The basic entities are measured in various units. For example, meters are units for measuring length, kilograms for mass, and seconds for time. All other physically measurable entities, such as energy, are measured in units that are various combinations of length, mass, time, and charge, implying that those are the most basic entities.

Several systems of units are in common use. Among them are the familiar British system of feet, pounds, and seconds (which the British have abandoned but the Americans have not) and two major metric systems.

This text emphasizes metric SI units but does not use them exclusively. The abbreviation SI stands for the French Système International. The SI units of length, mass, and time are the meter, kilogram, and second. Units other than SI are used in some applications; in applications this text will use the most common units, whether SI or not. For example, units of millimeters of mercury are commonly used for measuring blood pressure and will be used in this text even though they are not SI units. *Appendix B* gives conversion factors among common systems of units.

The SI unit of length, the meter, was originally devised to be one ten-millionth of the distance from the equator to either pole. The meter is now defined as 1/299,792,458th of the distance light travels in a vacuum in 1 second. (The second is defined below.) There is nothing profound about this; measuring the distance light travels in a second is simply the most accurate method of defining distance. Since the meter is accepted as the international standard, other units of length are based on the meter. For example, 1 centimeter is equal to one-hundredth of a meter, and 1 in. is equal to 2.54 cm.

Metric systems of units, such as SI units, use Greek ***metric prefixes*** to denote powers of 10. (See Appendix A3 for an explanation of scientific notation and powers of 10.) Table 1.2 gives the most

Table 1.2 Metric Prefixes

Prefix	Abbreviation	Value
tera	T	10^{12}
giga	G	10^{9}
mega	M	10^{6}
kilo	k	10^{3}
hecto	h	10^{2}
deka	da	10^{1}
—	—	$10^{0}(=1)$
deci	d	10^{-1}
centi	c	10^{-2}
milli	m	10^{-3}
micro	μ	10^{-6}
nano	n	10^{-9}
pico	p	10^{-12}
femto	f	10^{-15}
atto	a	10^{-18}

common of these, one of which has already been mentioned: One centimeter equals one-hundredth of a meter. The major advantage of metric systems is that most unit conversions within the systems involve only factors of 10.

The SI unit of ***mass***, the kilogram (1000 g), is based on a standard platinum–iridium cylinder that is kept near Paris. All other mass standards are based on a comparison to this standard kilogram. The mass of an object is directly related to the number of atoms and molecules in it. As Chapter 18 will clarify, an object made of a pure substance, and having a mass in grams equal to the molecular weight (defined in Chapter 18) of the substance, contains 6.022×10^{23} molecules of that substance (6.02×10^{23} is Avogadro's number in SI units). Mass is thus an indicator of just how much substance is in an object—the more mass, the more atoms and molecules. You will soon see that mass is a measure of an object's inertia.

The SI unit of ***time*** is the same as in many other systems—the second. One second was historically defined as 1/86,400th of a mean solar day. One second is now defined in terms of certain vibrations of a cesium atom—as the time required for 9,192,631,770 of those vibrations. There is nothing particularly basic about the cesium atom; again, it is just easier to measure its vibrations accurately than to measure accurately the mean length of a day (which actually changes slightly over time because of shifts of material in the Earth, tidal effects, melting and freezing of glaciers, etc.).

An example of a measurable quantity expressed in terms of two of the three basic units, length and time, is speed, which in SI units is expressed in meters per second. To convert a speed of 1 m/s to km/hr,

$$1\frac{\text{m}}{\text{s}} \cdot \frac{1\text{km}}{1000\text{m}} \cdot \frac{3600\text{s}}{1\text{hr}} = 3.6\frac{\text{km}}{\text{hr}}.$$

This conversion of meters per second to kilometers per hour is an example of unit analysis and illustrates a commonsense approach that will be used when appropriate throughout the text. Following the discussion of problem-solving the next chapter, however, a more formal approach will usually be preferred.

As new entities, such as energy, are presented, new units will be required to measure those entities. Those units can always be expressed in terms of length, mass, and time. Appendix B is a convenient summary of units and how to convert between them. It also lists the abbreviations for units that will be used throughout the text.

2 Motion

"I know not how I seem to others, but to myself I am but a small child wandering upon the vast shores of knowledge, every now and then finding a small bright pebble to content myself with while the vast ocean of undiscovered truth lay before me."

Sir Isaac Newton

Motion is apparent in widely ranging phenomena, from blood cells squeezing through capillaries to planets moving across the sky. Historically, motion was one of the first phenomena to be studied carefully. Some progress was made in the understanding of motion in ancient times, particularly by the philosophers of classical Greece, but it was not until the Renaissance that the basic laws of motion were discovered. Many individuals made important contributions, but two stand above the rest: **Galileo Galilei** (1564–1642) and **Isaac Newton** (1642–1727).

If Galileo's predecessors had placed a greater value on experimentation, they might have made more progress than they did. Instead, most natural philosophy was based on logical argument and the constraining influence of a particular school of thought. The transition that Galileo and others made from dogma to experimentation was not without pain; Galileo himself was forced by the **Inquisition** to recant his work and lived the last years of his life under a form of house arrest.

The central ideas regarding motion developed by Galileo and Newton remained essentially intact until 1905, when Albert Einstein (1879–1955) published his first paper on the theory of relativity. Even today the classical theory of Galileo, Newton, and others describes motion with extremely good precision as long as the object being described moves slower than about 1% of the speed of light.

The formal name for the study of motion exclusive of the causes of motion is *kinematics*, which comes from a Greek word meaning "motion." Motion itself is important, but there is another reason for studying it before any other topic. An understanding of motion makes it easier to understand the topics of force and energy covered in later chapters.

2.1 TIME, DISPLACEMENT, (lesson on speed, velocity and acceleration)

In order to describe motion it is necessary to specify *time, displacement, speed, velocity,* and *acceleration.* We will discuss each of these topics in detail.

Time: Motion always takes place over a period of time. Even light, which travels at the fastest possible speed, takes time to get from one place to another. It will therefore be necessary to specify time in order to describe motion. As stated in Chapter 1, time is one of the basic measurable entities in physics. Time is normally measured in seconds or in other related units, such as minutes or hours. Philosophers still puzzle over the nature of time, but the physicist's perception of time is simple: Time is measured in terms of change. If nothing changes, then it is impossible to tell that time has passed. All devices that measure time measure change. For example, one day is the amount of time for the Earth to rotate on its axis once. The symbol "*t*" is usually used to represent time.

Displacement, ΔX, by definition, is an object's distance and direction from its starting position. Don't let the Greek letter delta (Δ) scare you. It simply means "a change in." If we let "*X*" represent an object's position, then ΔX means "a change in position." Consider the following example.

EXAMPLE 2.1

If a physics student drives 12 km north, to her class, and then drives 14 km south to her grandma's house, what is her displacement for the day?

Ans. Since displacement is an object's distance and direction from its starting position, her displacement is 2 km south. As this example implies, there is a difference between distance and displacement. A distance has no given direction; for example, 6.0 km is a distance. Distance is an example of a measurement that is a scalar—because we don't care about direction. A displacement, on the other hand, has both magnitude and a direction; 6.0 km east is an example of displacement. The magnitude is 6.0 km and the direction is east. Any quantity that has both a magnitude and a direction is called a *vector*. When we add vectors, we must pay attention to both the magnitude and direction. Two plus two is not always four. More about this later.

Speed is the rate at which an object changes its position. Velocity is an object's speed and direction. For example, 60 mi./hr *north* is an object's velocity because both its speed and direction are given. We see that velocity as well as displacement, is a vector.

An object's *average* velocity can be determined by dividing its displacement by the elapsed time.

Average velocity $\equiv \dfrac{\Delta X}{\Delta t}$ The $\equiv$ symbol is secret code for "by definition." An object's average velocity, by definition, is its displacement divided by the elapsed time. Another method of writing average velocity is to place a bar above the *V*.

$$\text{Average velocity } \overline{V} \equiv \frac{\Delta X}{\Delta t} \quad \textbf{\textit{Know}} \tag{2.1}$$

EXAMPLE 2.2

In the case of the student driving to her physics class, let's determine her average velocity, in km/hr, if it took her 15.0 minutes to travel 12.0 km north to class.

$$\textbf{Answer. } \overline{V} \equiv \frac{\Delta X}{\Delta t} = \frac{12.0\,\text{km}}{15.0\,\text{min}} \cdot \frac{60.0\,\text{min}}{1.00\,\text{hr}} \text{north} = 48.0 \frac{\text{km}}{\text{hr}} \text{north}$$

Please notice that we converted minutes to hours and that we also had to provide the direction, north, in order to complete the answer. Answers must be in the correct units and vectors always have a direction.

Acceleration: The acceleration of an object is the rate at which its velocity changes. We can write the definition of acceleration with our shorthand notation as follows:

$$a \equiv \frac{\Delta V}{\Delta t} \quad \textit{Know} \tag{2.2}$$

When calculating the change in velocity, ΔV, always subtract the first velocity from the second. $\Delta V \equiv V_2 - V_1$

Please note that an object's acceleration is always in the same direction as its change in velocity.

Typical units of acceleration are $\frac{\frac{m}{s}}{s} = \frac{m}{s^2}$ ***Know***

Sometimes, an object's "change in velocity" is given in the problem. However, there will be problems in which an object's change in velocity is not directly given. In order to determine the change in velocity of an object, we always subtract an earlier velocity from a later velocity.

EXAMPLE 2.3

Calculate the acceleration of a car that changes its velocity from 10.0 km/hr west to 90.0 km/hr due west in 12.0 seconds.

$$\textbf{Ans. } a \equiv \frac{\Delta V}{\Delta t} = \frac{V_2 - V_1}{\Delta t} = \frac{90.0 \frac{\text{km}}{\text{hr}} \text{west} - 10.0 \frac{\text{km}}{\text{hr}} \text{west}}{12.0\,\text{s}} = 6.67 \frac{\text{km/hr}}{\text{s}} \text{west}$$

If we are asked to convert the acceleration units to the more typical $\frac{m}{s^2}$, here is how it is done:

$$6.67 \frac{\frac{\text{km}}{\text{hr}} \cdot \frac{1000\,\text{m}}{\text{km}} \cdot \frac{1.00\,\text{hr}}{3600\,\text{s}}}{\text{s}} = 1.85 \frac{\text{m}}{\text{s}^2} \text{west}$$

*When converting units, it is useful to think of multiplying by a special fraction whose numerator and denominator are equal to each other. Consider it as a special version of one. You can always multiply by one, without changing the value of something.

For example, we converted from hours to seconds by multiplying by $\frac{1.00\text{hr}}{3600\text{s}}$. Because 1 hour equals 3600 seconds, we were able to convert from hours to seconds without changing the value of the original velocity. Similarly, we converted from kilometer to meters by multiplying by $\frac{1000\,\text{m}}{1\,\text{km}}$.

Things to know for Section 2.1: *Vectors* have magnitude and direction. *Displacement*, *velocity*, and *acceleration* are vectors and therefore have direction.

ΔX, *Displacement* is an object's distance and direction from its starting position

Average velocity $\overline{V} \equiv \frac{\Delta X}{\Delta t}$ Units are $\frac{\text{m}}{\text{s}}$

Average acceleration $a \equiv \frac{\Delta V}{\Delta t} = \frac{V_2 - V_1}{\Delta t}$ Units are $\frac{\text{m}}{\text{s}^2}$

Now go try a few homework problems that relate to this material.

2.2 LINEAR MOTION AND PROBLEM SOLVING

Linear motion is motion along a straight line. It is one dimensional. That means that in a particular problem, you can go forward or backward—but not up or left or in any other direction other than forward or backward. In a different linear motion problem, the object might be able to go up or down, but not in any other direction. Remember that a vector has both magnitude and direction. Because vectors can be mathematically challenging, it is often instructive to begin with linear problems that involve objects that can only move in one direction or the opposite direction. We can designate one direction as positive and the opposite direction as negative. That's a bit easier than jumping right in with sine, cosine, tangent, and the Pythagorean theorem.

2.2.1 Instantaneous Velocity

What if we want to know the velocity of an object at some moment in time? As long as its acceleration is constant, we can use the formula:

$$V = V_o + at \qquad \textbf{\textit{Know}} \tag{2.3}$$

V_o is the velocity of the object at the beginning of the problem
"a" is the acceleration of the object
"t" is an instant in time. In this equation, "t" is the moment in time at which we want to know the object's velocity

EXAMPLE 2.4

A sprinter is able to produce an acceleration of 8.00 m/s^2 for the first 0.750 seconds of a race. Calculate the velocity of the sprinter at 0.75 seconds if she (a) starts from rest (b) has a running start of 1.00 m/s when the race begins.

a. $V = V_o + at$ From rest means that $V_o = 0\frac{\text{m}}{\text{s}}$

$$V = V_o + at = 0\frac{\text{m}}{\text{s}} + 8.00\frac{\text{m}}{\text{s}^2}\cdot 0.750\,\text{s} = 6.00\frac{\text{m}}{\text{s}}\text{ forward.}$$

b. Running start, V_o = 1.00 m/s

$$V = V_o + at = 1.00\frac{\text{m}}{\text{s}} + 8.00\frac{\text{m}}{\text{s}^2}\cdot 0.750\text{s} = 7.00\frac{\text{m}}{\text{s}}\text{ forward}$$

It helps to have a running start.

EXAMPLE 2.5

Now suppose that the sprinter continues to run at 6.00 m/s until the end of the race. After crossing the finish line the sprinter *accelerates backward* at 2.00 m/s². How long will it take her to stop?

$V_o = 6.00\ \frac{\text{m}}{\text{s}}$ →

$a = 2.00\ \frac{\text{m}}{\text{s}^2}$ backward ←

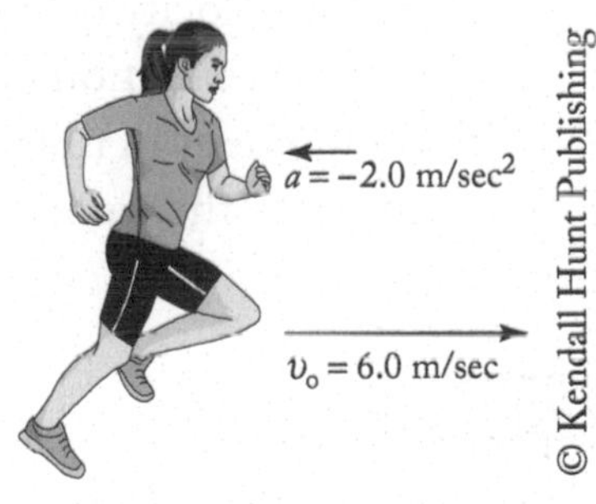

Solution: The initial velocity in this problem is 6.0 m/s and the final velocity is 0 m/s. In general, you should never guess or assume that the initial or final velocity of a problem is zero. Examine the question carefully, in order to determine these values.

We will solve Equation 2.3 for time. If your algebra is too rusty to follow this solution, please seek help.

$$V = V_o + at$$

Step by step 1: Subtract V_o, the initial velocity, from both sides of the equals sign.

$$V - V_o = at$$

Step 2: Divide both sides of the equation by the acceleration, *a*.

$$\frac{V - V_o}{a} = t$$

Step 3: Rewrite the equation with the time "*t*" on the left side of the equals sign and fill in the given values. Because the acceleration was backward, we enter its value as a negative.

$$t = \frac{V - V_o}{a} = \frac{0\frac{\text{m}}{\text{s}} - 6.00\frac{\text{m}}{\text{s}}}{-2.00\frac{\text{m}}{\text{s}^2}} = 3\,\text{seconds}$$

By convention, we usually let forward be the positive direction and backward be the negative direction. In problems that involve up or down, we usually let up be positive and down be negative.

Displacement—a more complete picture.

As we mentioned earlier, the displacement of an object is its distance and direction from its starting position. If you are asked to determine an object's position at a particular time, the displacement equation is the first rule that should come to mind. It allows us to include information concerning the initial velocity of an object and the object's acceleration. It may look scary, but it is useful and powerful.

Things to know: Before proceeding, please memorize the following:

The displacement equation: $\Delta X = V_o t + \frac{1}{2}at^2$ (2.4)

ΔX is displacement

"V_o" is the velocity of the object at the beginning of the problem.

"a" is the acceleration of the object

"t" is a moment in time. In the following example, "t" is the moment in time at which we want to know the woman's displacement.

EXAMPLE 2.6

How far does the splinter in example 2.5 travel in stopping? In example 2.5, the splinter was moving forward at 6.00 m/s and began accelerating backward at 2.00 m/s². We determined in Example 2.4 that it would require 3.00 seconds to stop.

In order to answer a "how far" question, we will use the displacement equation. Just think of it as a template to be followed. Don't forget to keep track of direction. In front of the initial velocity and acceleration vectors, we use the plus sign for forward and the minus sign for backward.

Step by step 1: Write down the displacement equation.

$$\Delta X = V_o t + \frac{1}{2}at^2$$

Where V_o = 6.00 m/s time is 3.00 s a = −2.00 m/s²

Step 2: Fill in the values. Make sure that you keep track of direction using plus and minus signs. Because the *acceleration was backward we will enter its value as a negative.*

$$\Delta X = V_o t + \frac{1}{2}at^2 = 6.00\frac{\text{m}}{\text{s}} \cdot 3.00\,\text{s} + \frac{1}{2}\left(-2.00\frac{\text{m}}{\text{s}^2}\right)(3.00\,\text{s})^2$$

Step 3: Do the math.

ΔX = 18 m − 9.0 m = 9.0 m

Step 4: Give the complete answer.

Because the answer came out positive, we can say that the *sprinter's displacement was 9.0 m forward while stopping.*

Things to know for Section 2.2: Before proceeding, if you haven't already done so, please memorize the following:

$$V = V_o + at \tag{2.3}$$

$$\Delta X = V_o t + \frac{1}{2}at^2 \tag{2.4}$$

ΔX is displacement
"V_o" is the velocity of the object at the beginning of the problem.
"a" is the acceleration of the object
"t" is a moment in time. In this equation, "t" is the moment in time at which we want to know the object's displacement.

Now go try a few **questions** and **homework problems** that relate to this material.

2.3 GRAPHICAL REPRESENTATION OF MOTION

2.3.1 Position versus Time

The motion equations represent the relationships between time, distance, velocity and acceleration mathematically. Graphs display these same relationships visually. Graphing an object's position versus time, can be a great way to determine what is happening to the object's position and velocity as time goes by. We can also determine if the object is accelerating. The *slope* of a *position versus time* graph indicates an object's velocity. If the slope does not change, then we know the object's velocity didn't change and therefore it did not accelerate. If the slope gets steeper, then the object's velocity has increased and the object accelerated in the same direction that it was moving in. For linear motion, a steeper slope on a *position versus time graph* indicates that an object is speeding up. If on the other hand, the slope decreases we know that the object slowed down. When an object is slowing down, it is accelerating backward.

In the *position versus time* graph of Figure 2.3, notice that the slope of line 1 is decreasing. The object is slowing down until time t_1. At time t_1, the slope is zero, indicating that the object is not moving at that instant. After time t_1 the slope is negative, indicating that the object is moving back toward its starting point. The slope on line 2 is constant, indicating that the object is moving away from its starting position at constant velocity. Line 3 has an increasing slope, indicating that the object is moving faster and faster away from its starting position. The changing slope of lines 1 and 3 indicate that the object's velocity was always changing. When an object's velocity has changed, it has accelerated.

Velocity versus time graphs for the same motions described in Figure 2.3.

In the *velocity versus time* graph of Figure 2.4, notice that *line 1* has a positive velocity and therefore was moving forward until time t_1. At time t_1 its velocity is zero and from time t_1 onward, its velocity is negative—the object was now moving backward. The slope of the line in a velocity versus time graph is its acceleration.

Figure 2.3 Graphs of *position* versus time for three different motions. The slope of the line in a position vs time graph indicates the velocity of the object.

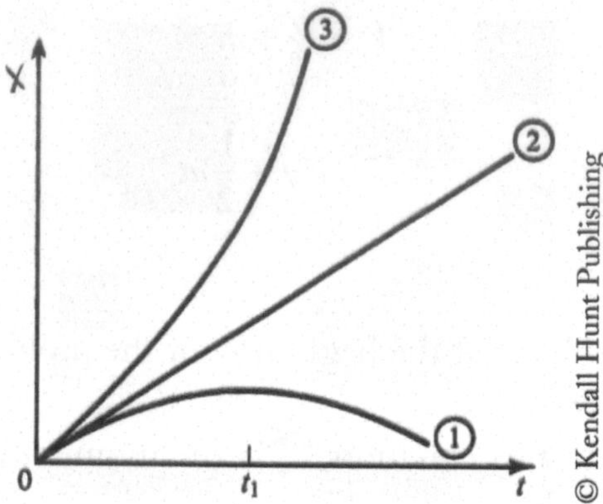

Figure 2.4 Graphs of velocity versus time for three different motions—the same ones described in Figure 2.3.

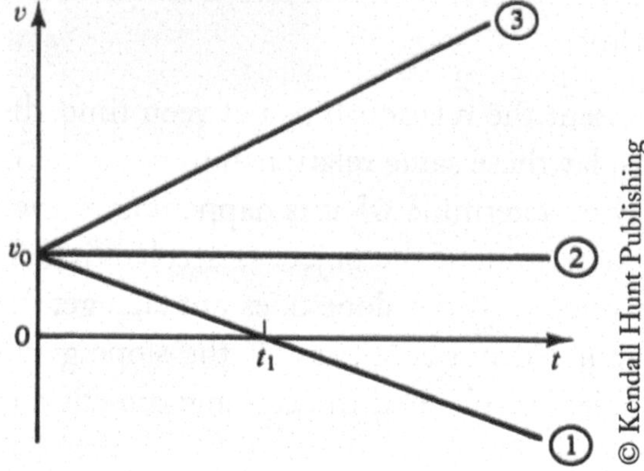

Since line 1 has a negative slope, the object was always accelerating backward even while it was moving backward.

Line 2 represents a constant velocity. Line 2 has a slope of zero, indicating that it has zero acceleration—the object is not accelerating, therefore, its velocity never changes.

Line 3 represents increasing velocity and its positive slope indicates that the object is always accelerating forward.

Now go try questions 2.13 and 2.14. You will gain more experience with graphing position, velocity and acceleration vs. time in Lab. 1.

2.4 MOTION EQUATIONS AND PROBLEM-SOLVING

It is critical to your success that you make your best effort to *memorize the key equations and fundamental units*. We have listed them in the "Things to know" sections. After they are memorized, you must practice using them by doing the homework problems. If the problems are hard for you, do them more than once.

Here are some general guidelines that have been useful to past students:

1. Examine the problem to determine which physical principles are involved. As you practice, and begin to understand the underlying physics, it will become easier and easier, to identify these principles.
2. Identify exactly what is asked. Ask yourself, "what is the question?".
3. Make a list of given and inferred information.
4. Write down the equation that relates to the question or problem. If you have memorized your key equations, this step gets a lot easier. Then solve the equation for the unknown value. We will refer to this version of the equation as your *working equation*. If you have trouble with some of the homework problems, do them two or three times. You will be surprised at how easy they become with a little practice.
5. Substitute the given and inferred information into your working equation. Include all of the appropriate units. Units are your friends. Spend time with them.
6. Interpret your answer. If the velocity comes out negative, does that mean the object is moving up or down, forward or backward? Check your answer to see if it is reasonable. Does it make sense? As you progress through this course, you will develop a foundation of knowledge that will assist you in determining whether or not an answer is reasonable.

Now go back and try some of the homework problems that gave you trouble the last time.

2.5 GRAVITY AND FALLING BODIES

As far as we know, there are only four fundamental forces in nature. Gravity was the first of these forces to be studied. Galileo was the first to demonstrate that all objects, in free-fall, accelerate down, at the same rate—if air resistance is negligible. Objects are in free-fall, when the only force acting on them is gravity. The mass of the object in free-fall, does not affect its acceleration. Interestingly, it does not even matter whether the object is moving up, down, left, right, or horizontally. If it is in free fall, near the earth's surface, it will accelerate *down*, at 9.80 m/s^2 as long as air friction can be neglected (Figure 2.5).

Galileo's recorded experiments dealt a death blow to the 1000-year-old notion that heavier objects in free fall, must always accelerate more than lighter objects. Galileo was the first scientist to rely on experiments in order to gather information concerning the physics of the topic that he was investigating. Galileo is often credited with being the father of modern science because of his forceful demonstration of the value of observation and the discoveries he made through his ingenious experiments.

Figure 2.5 A lead ball and a feather will fall with the same acceleration if the effect of air resistance is made small enough. Astronaut David R. Scott demonstrated that this is true, during the Apollo 15 mission to the moon (July 26–August 7, 1971).

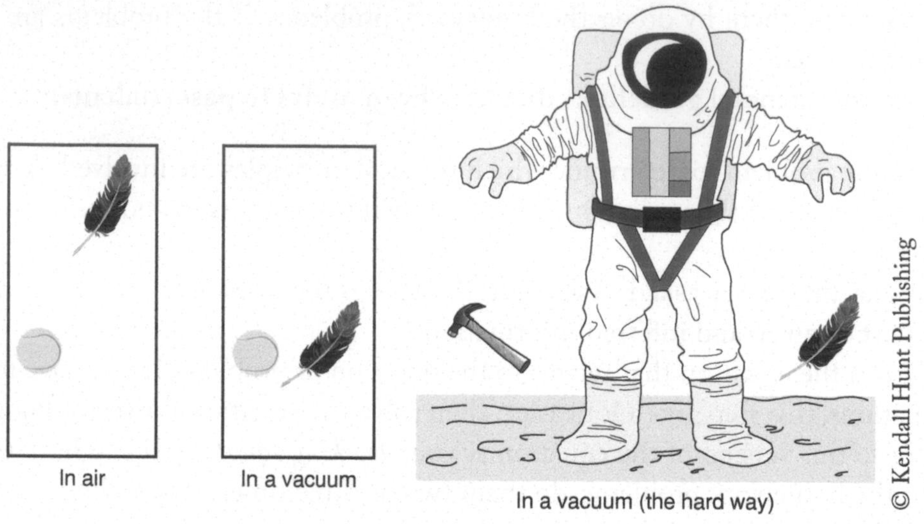

https://commons.wikimedia.org/wiki/File:Apollo_15_feather_and_hammer_drop.ogv

The value for the acceleration of an object, near the earth's surface is 9.80 m/s^2 down. It is given the symbol *g*, because it is caused by the force of gravity. It is important to note that "*g*" does not stand for gravity. It represents the acceleration of the object, not the force applied to it. By convention, the down direction is usually given a negative sign and up is positive. However, in problems where only the down direction is involved, you are free to let the down direction be positive.

Because all objects in free fall accelerate down at the same rate, 9.80 m/s^2, we can substitute the value of "*g*," for the value of the objects acceleration in problems involving objects in free fall.

EXAMPLE 2.7

A tennis ball is hit straight up with an initial speed of 30 m/s. (a) Where is the tennis ball relative to the tennis racket 2 seconds after it is hit? (b) What is its velocity 2 seconds after it is hit?

Solution: (a) If we want to know where the ball is, we should first consider the displacement equation that you memorized earlier. It will give us the ball's distance and direction from its starting point. The starting point and time occur the instant that the ball leaves the racket.

a. Where is the tennis balls position, relative to the tennis racket, 2 seconds after it is hit?

Step by step 1: List the known and unknown values.

V_o = 30.0 m/s up; t = 2.00 seconds; a = 9.80 m/s^2 down; ΔX = ?

Step 2: Write down the appropriate equation and solve it for the unknown.

$$\Delta X = V_o t + \frac{1}{2}at^2$$

Since the unknown is ΔX, our equation is already solved for the unknown. We now have our working equation.

Step 3: Mathematically account for the fact that vectors have direction information. We will do this by letting the up direction be positive and down be negative.

Step 4: Replace the variable letters with the correct values. Notice that we enter minus 9.80 m/s^2 into the equation for the acceleration of the ball.

$$\Delta X = V_o t + \frac{1}{2}at^2$$

$$\Delta X = V_o t + \frac{1}{2}at^2 = 30.0\frac{\text{m}}{\text{s}}\cdot 2\text{s} + \frac{1}{2}\left(-9.80\frac{\text{m}}{\text{s}^2}\right)\cdot(2.00\,\text{s})^2$$

$$\Delta X = 60.0\,\text{m} - 19.6\,\text{m} = 40.4\,\text{m}$$

Answer: Because ΔX came out positive, we say that *the ball is 40.4 m above the racket*
Example 2.7 (b) What is the tennis ball's velocity 2.00 seconds after it is hit?

Step by step 1: List the known and unknown values.
V_o = 30.0 m/s up; t = 2.00 seconds; a = 9.80 m/s^2 downward;
V = ?

Step 2: Write down the appropriate equation and solve it for the unknown.

$V = V_o + at$

Since the unknown is V, our equation is already solved for the unknown and we have our working equation.

Step 3: Mathematically account for the fact that vectors have direction information. Usually, we will do this by letting the up direction be positive and down be negative.

Step 4: Replace the variable letters with the correct values. Notice that we enter minus 9.80 m/s^2 into the equation for the acceleration of the ball.

$$V = V_o + at$$

$$V = V_o + at = 30.0\frac{\text{m}}{\text{s}} + \left(-9.80\frac{\text{m}}{\text{s}^2}\right)\cdot 2.00\,\text{s} = 10.4\frac{\text{m}}{\text{s}}$$

Step 5 Interpret the Answer: Because the velocity came out positive, we say that the ball was moving *10.4 m/s upward.*

EXAMPLE 2.8

A coin is dropped into a wishing well. The surface of the well water is 60 meter below the point of release. Neglecting air friction: (a) Calculate the position of the coin 3 seconds after it is released. (b) How long does it take the coin to reach the water?

a. Calculate the position of the coin 3 seconds after it is released.

Step by step 1: List the known and unknown values.

$V_o = 0$ m/s; $t = 3.00$ seconds; $a = 9.80$ m/s^2 downward; ΔX=?

Step 2: Write down the appropriate equation and solve it for the unknown.

$$\Delta X = V_o t + \frac{1}{2}at^2$$

Since the unknown is ΔX, our equation is already solved for the unknown.

Step 3: Mathematically account for the fact that vectors have direction information. Usually, we will do this by letting the up direction be positive and down be negative.

Step 4: Replace the variable letters with the correct values. Notice that we enter minus 9.80 m/s^2 into the equation for the acceleration of the ball.

$$\Delta X = V_o t + \frac{1}{2}at^2$$

$$\Delta X = 0\frac{\text{m}}{\text{s}}(3.00\,\text{s}) + \frac{1}{2}\left(-9.80\frac{\text{m}}{\text{s}^2}\right)(3.00\,\text{s})^2 = -44.1\,\text{m}$$

Step 5 Interpret the Answer: Because the displacement came out negative, we say that the coin is *44.1 m* *<u>below</u>* *the point of release.*

b. **Ex. 2.8** How long does it take the coin to reach the water?

Step by step 1: List the known and unknown values.

$\Delta X = 60.0$ m down; $V_o = 0$ m/s; $a = 9.80$ m/s^2 downward;
t = ?

Step 2: Write down the appropriate equation and solve it for the unknown.

$$\Delta X = V_o t + \frac{1}{2}at^2$$

Because the initial velocity is 0 m/s, the equation reduces to:

$$\Delta X = \frac{1}{2}at^2$$

In order to get our working equation, we must solve for "t":

$$t = \sqrt{\frac{2\Delta X}{a}}$$

if you couldn't do this step, seek help.

Step 3: Mathematically account for the fact that vectors have direction information. Usually, we will do this by letting the up direction be positive and down be negative.

Step 4: Replace the variable letters with the correct values. Notice that we enter minus 9.80 m/s^2 into the equation for the acceleration of the coin and we enter minus 60 m for its displacement.

$$t = \sqrt{\frac{2(-60.0\,\text{m})}{-9.80\frac{\text{m}}{\text{s}^2}}} = 3.50\,\text{s}$$

Step 5 Interpret the Answer: The coin hits the water *3.50 seconds after it is released.*

Review: Things to know for Chapter 2

Vectors have magnitude and direction. *Displacement, velocity, and acceleration* are vectors.

Average velocity $\bar{V} \equiv \frac{\Delta X}{\Delta t}$ Units are $\frac{\text{m}}{\text{s}}$

Average acceleration $a \equiv \frac{\Delta V}{\Delta t} = \frac{V_2 - V_1}{t}$ Units are $\frac{\text{m}}{\text{s}^2}$

g is the special acceleration of objects in free fall, near the earth's surface. When we disregard air friction, they accelerate down at 9.80 m/s^2.

ΔX, *Displacement* is an object's distance and direction from its starting position

$$\Delta X = V_o t + \frac{1}{2}at^2$$

ΔX is displacement
"V_o" is the velocity of the object at the beginning of the problem.
"a" is the acceleration of the object
"t" is a moment in time. In the equation, "t" is the moment in time at which we want to know the object's displacement.

Connections:

Equation 2.3 $V = V_o + at$ is just an expanded version of Equation 2.2 $\left(a \equiv \frac{\Delta V}{\Delta t}\right)$

$$a \equiv \frac{\Delta V}{\Delta t} = \frac{V_2 - V_1}{t}$$

Solving for V_2 yields: $V_2 = V_1 + at$ This is the same as Equation 2.3.
Homework **questions, problems** and **solutions.**

Questions

2.1 How does displacement differ from distance? Give an example of each.

2.2 Explain the difference between displacement and total distance traveled. Give an example of each.

2.3 How does velocity differ from speed? Give an example of each.

2.4 Name at least three distinct methods or devices for measuring time and explain how each measures time by the observation of a change in some measurable quantity.

2.5 Return to Example 2.1. What will be the average velocity of the students car for the day once the car has been returned home? Explain in terms of the definitions of displacement and velocity.

2.6 How are the directions of displacement, velocity, and acceleration specified in one-dimensional motion (motion along a straight line)?

2.7 Does time have a direction associated with it? Explain. (If it does, can that direction be changed?)

2.8 It is possible for the speed of an object to be zero at the same time that its acceleration is not zero. Give an example of such a situation.

2.9 An object can be accelerated even though its speed remains constant. Give an example of such a motion.

2.10 Draw a graph of displacement versus time for a person walking to the mailbox, taking a moment to mail a letter, and then running home.

2.11 Draw a graph of velocity versus time for a person walking to the mailbox, taking a moment to mail a letter, and then running home.

2.12 Draw two graphs, one of displacement versus time and the other of velocity versus time for a motion in which the initial velocity is positive and the acceleration is negative.

2.13 Draw two graphs, one of displacement versus time and the other of velocity versus time for a motion in which the initial velocity is positive and the acceleration is positive.

√2.14 Answer the following for the graphs of position versus time in Figure 2.8. There may be more than one answer for each part. (a) Which motion has the largest velocity? (b) Which has the largest initial velocity? (c) Which graph indicates accelerated motion?

√2.15 Answer the following for the graphs of velocity versus time in Figure 2.9. There may be more than one answer for each part. (a) Which motion has the largest acceleration? (b) Which has the smallest initial velocity? (c) Which has a constant acceleration? (d) Which has an acceleration of zero?

Figure 2.8

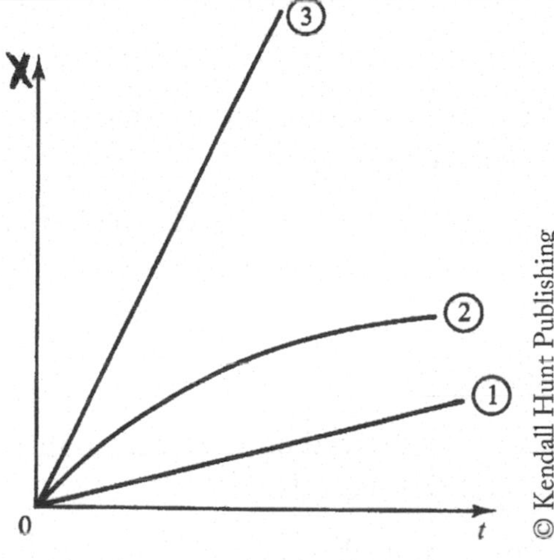

Figure 2.9

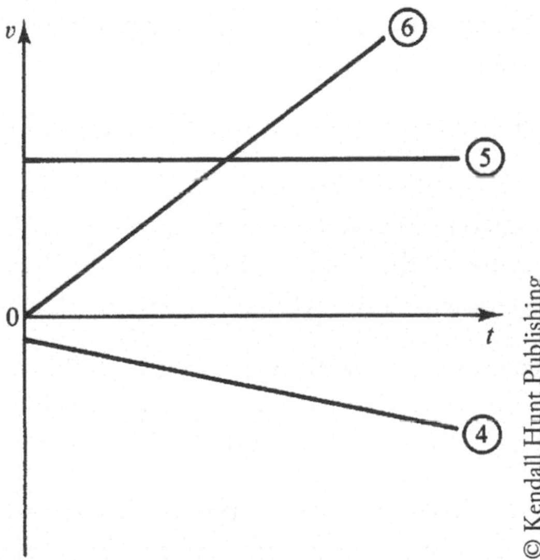

2.16 Give an example of a physical law having to do with motion due to gravity.

2.17 In the solution to Example 2.10(b) a negative value was used for displacement. What does the negative value mean and why was it used in that example?

Problems

SECTION 2.1

√1 (I) What is the displacement of a shopper in a mall who first walks 100 m straight east and then walks 125 m straight west?

√2.2 (1) On a long distance car trip, a person travels 800 km north (straight line distance on a map) in 12 hours. (a) What was the average speed? (b) What was the average velocity?

√2.3 (I) A soccer player runs 75 m in a straight line down a soccer field in 11 seconds. What was her average speed?

√2.4 (I) A cheetah can reach a final speed of 30 m/s from rest in 6.0 seconds. Calculate the average acceleration of the cheetah.

√2.5 (I) An elephant can run at a steady speed of 10 m/s and can stop in 2.5 seconds. What is the average acceleration of the elephant while stopping?

√2.6 (II) A Grand Prix racer completes a 500 km race in exactly 2 hour and 15 minute. What is the average speed of the car in kilometers per hour and meters per second?

2.7 (II) How long will it take a runner to complete a marathon race of 42.2 km if that runner can maintain an average speed of 4.1 m/s?

2.8 (II) What distance in meters will an ant cover in 1 hour if it maintains an average speed of 1.0 cm/s?

2.9 (II) (a) What is the average speed in kilometers per second of the earth around the sun given that the radius of the earth's orbit is 1.51×10^8 km? (b) What is the average velocity of the earth over a year's time?

2.10 (II) An airplane accelerates at takeoff to a speed of 200 km/hr in 18 seconds. What is its acceleration compared to that of gravity? That is, find the acceleration in meters per second squared and express it as a multiple of *g*.

2.11 (II) When the brakes of a certain car are applied, an acceleration of 6.0 m/s^2 backward is produced. (a) How much time will it take to bring the car to a halt from an initial speed of 110 km/hr? (b) How far will it travel in that time?

2.12 (II) A football player running at a speed of 9.0 m/s hits the goalpost and comes to a stop in 0.050 seconds. Calculate his acceleration and compare it with that of gravity.

2.13 (II) How fast will a motorcycle be going after accelerating from rest for 3.5 seconds if that motorcycle produces an acceleration of 7.5 m/s?

2.14 (II) An airplane making an approach for landing accelerates at a rate of 1.2 m/s^2 backward. How long will it take the airplane to reduce its speed by 300 km/hr?

√2.15 (I) What is the final speed of a bicyclist initially moving at 10 m/s who accelerates at a rate of 1.5 m/s^2 for 3.0 seconds?

√2.16 (I) A dolphin accelerates backward for 2.0 seconds from an initial speed of 7.5 m/s in order to take a look at a dolphin of the opposite sex. Calculate the dolphin's final speed if the backward acceleration had a magnitude of 1.5 m/s^2.

√2.17 (I) How far will a car travel from a standing start if it accelerates at 4.0 m/s^2 for 9.0 seconds?

2.18 (I) How fast will a runner be going if he accelerates from rest at 3.0 m/s^2 for a distance of 2.5 m?

2.19 (II) How long will it take a car to accelerate from 50 to 90 km/hr if it produces an acceleration of 3.0 m/s^2?

2.20 (II) An ice skater accelerates backward for 5.0 seconds to a final speed of 12 m/s. If the acceleration backward was at a rate of 1.5 m/s^2, what was the skater's initial speed?

2.21 (II) (a) If a freight train can accelerate at 0.050 m/s^2, how far must it travel to increase its speed from 2.0 to 8.0 m/s? (b) How long will it take the train to accomplish this?

2.22 (II) How long will it take a car to cover a distance of 100 m if it accelerates from rest at a rate of 4.0 m/s^2?

2.23 (III) A freight train moving at an initial speed of 40 m/sec puts on its brakes, producing a deceleration of 0.50 m/sec^2 (a) How long will it take the train to travel the next 100 m? (b) At what speed will it be traveling at the end of this 100 m?

2.24 (III) A 35-m-long airplane accelerates down a runway from rest. If the nose of the airplane passes over the end of the runway 1000 m from its starting point at a speed of 90 m/sec, what will be the speed of the airplane when its tail passes over the end of the runway?

2.25 (III) A car is 100 m from an intersection and is moving at 20 m/sec when the driver observes the traffic light ahead turn yellow. The yellow light will turn red in 4.0 sec. (a) Will the car be able to stop if the driver's reaction time is 0.25 sec and the brakes can produce a deceleration of 5.0 m/sec^2? (b) If the driver decides to try to make the light, the car must accelerate. Taking a reaction time of 0.25 sec into account, what acceleration must the car be able to produce to make the light?

2.26 (III) Show the missing steps in the derivation of equation 2.6.

2.27 (III) Show the missing steps in the derivation of equation 2.7.

2.28 (III) Show the missing steps in the derivation of equation 2.8.

SECTION 2.3

2.29 (I) (a) If you drop a syringe and it takes 0.18 seconds to hit the table, how high above the table was the syringe when it was released? (b) How fast was the syringe moving when it hit the the table?

2.30 (I) A child throws a ball straight down off the balcony of an apartment building at a speed of 7.5 m/s. (a) Calculate the position and velocity of the ball 2.0 seconds later. (b) Make the same calculations if the ball is thrown straight up.

2.31 (II) Calculate the displacement and velocity of a rock that is dropped from the Royal Gorge Bridge 1.0, 2.0, 3.0, and 4.0 seconds after its release. Make graphs of the position versus time and velocity versus time, drawing a line through the calculated points, to represent the motion.

2.32 (II) (a) If a rock is dropped from a high cliff, how fast will it be going when it has fallen 100 m? (b) How long will it take to fall this distance?

3

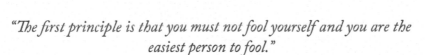

Force

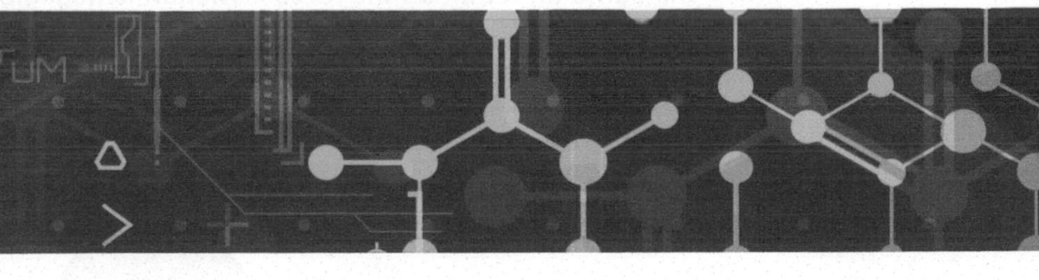

"The first principle is that you must not fool yourself and you are the easiest person to fool."

Richard P. Feynman

What causes a body to accelerate? Why do objects accelerate toward the earth when dropped? Is it natural for objects to slow down and stop, or is there something causing them to slow down? How do we prevent objects form accelerating?

3.1 FORCE, THE CAUSE OF ACCELERATION; NEWTON'S LAWS OF MOTION

Every acceleration (equivalently, every change in velocity) is caused by forces acting on a body. Conversely, if a body does not accelerate, then the total force acting on it is zero—even if several forces are present. Forces are vectors. One must remember to include direction information when adding forces.

The apparently simple idea of cause and effect that forces cause acceleration, didn't come easily. It was and still is tempting to think of common phenomena as having no cause and simply being "the nature of things." For example, a question like "Why does water flow downhill?" seems silly. Yet such questions have serious answers; in this case the force of gravity causes water to flow downhill. The genius of Galileo, Newton and others was not only in providing answers to basic questions, but also in simply being curious enough to ask basic questions.

Force is defined intuitively as a push or pull in a particular direction. If an applied force is the only one acting on a body, then the body will accelerate in the same direction as the force. The magnitude and the direction of the force, and the mass of the object, determine the acceleration of the object (see Figure 3.1). If several forces act on a body, then its acceleration is in the same direction as the total force. Newton's second law of motion describes this completely and will be discussed in detail.

Figure 3.1 (a) Any unbalanced force applied to a hockey puck will produce an acceleration in the same direction as the force if friction is negligible. (b) The magnitude of the acceleration depends on the magnitude of the force and the mass of the puck. A smaller force produces a smaller acceleration. A smaller mass results in a larger acceleration.

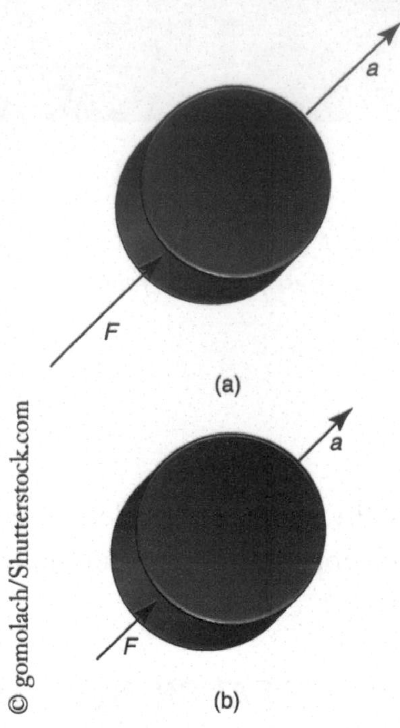

3.1.1 Newton's Laws of Motion and the Concept of a System

Sir Isaac Newton was mentioned in Chapter 2, along with Galileo, as having had a major influence on the study of motion. What Newton did was to write down the relationships between force and motion in a form that could be used to predict and describe motion. Those relationships were found to apply in every circumstance where an experiment could be performed to test them and came to be known as Newton's laws of motion. They are three in number. If you want to understand motion, you must become familiar with all three laws.

Newton's First Law: Inertia (Mass) Newton's first law is a nonmathematical statement that a net force is needed to change the motion of any body, whether it is moving or at rest.

Newton's first law of motion:

An object at rest will remain at rest, an object in motion will remain in motion, at a constant velocity, unless an unbalanced force is applied to it. Know

The phrase "at a constant velocity" means that the object cannot change its speed or direction, unless an unbalanced force is applied to it. The property of a body that causes it to remain at rest or to maintain a constant velocity is called its inertia. *Mass* is a measure of an object's inertia. It is more difficult to accelerate some bodies than others; those which have more mass are harder to accelerate. The mass of a body is proportional to the number of atoms and molecules in it. This means that the mass of a body doesn't depend on its location. For example, you would have the exact same mass on the earth, moon, or in outer space. An object's mass does not depend on its location. We will usually use kilograms as units of mass in this text.

Newton's Second Law of motion: The second law of motion gives the mathematical relationship between force, mass, and acceleration.

The acceleration produced by forces acting on a body is directly proportional to and in the same direction as the net external force and inversely proportional to the mass of the body.

Mathematically, the second law is written as:

$$a = \frac{\Sigma F}{m} \quad \textit{Know} \tag{3.1}$$

where ΣF stands for the *net force* or the sum of all external forces acting on the body and "m" is the body's mass. Newton's second law of motion tells us that the acceleration of an object is directly proportional to the net force applied to the object and is inversely proportional to the object's mass. A large force produces a large acceleration, a large mass requires a large force to make it accelerate at the same rate as a small mass. A body will accelerate in the same direction as the net force that is applied to it.

All the internal forces in a body, such as the forces between atoms and molecules in it, can be completely ignored. This is a tremendous simplification. All the parts that make up an object or set of objects under study, together with the forces and relations among them, can be described as a system. To use Newton's second law, all that need be considered are the forces acting on the system from the outside. To determine which forces are external and which are internal, one need only carefully define what the system of interest is.

Newton's second law is a much more profound statement of physical principle than anything previously presented in this text. The motion equations in Chapter 2, come straight from the basic definitions of velocity and acceleration and tell nothing more about nature than is contained in those definitions. Newton's second law, however, is based on the results of experiments and relates mass and acceleration, two previously unrelated quantities. Furthermore, Newton's second law gives a precise definition of force that is consistent with our intuitive notions of a force as a push or pull.

Newton's Third Law:

For every force applied to an object, there is another force, equal in magnitude and opposite in direction, applied to the other object. Know

In order to apply the third law correctly it is again important to define carefully the system of interest. In Figure 3.2, a swimmer pushes against the side of a pool and as a result moves away from the edge. If we define the swimmer to be the system of interest, then it becomes clear what is happening here. She exerts a force on the side of the pool, which is external to the system of interest. By Newton's third law, the side of the pool exerts a force back on the swimmer—an external force. If there is relatively little friction between the swimmer and the water, she will then move in a direction opposite to the force she exerted on the side of the pool with an acceleration proportional to the force she exerted and inversely proportional to her mass.

The pair of forces is always equal in magnitude and opposite in direction and are applied to different objects.

Units of Force: In this text, we will use the metric units of force called newtons.

Figure 3.2 The swimmer exerts a force, F_{action}, on the wall of a pool and consequently experiences a reaction force, $F_{reaction}$, of equal magnitude and opposite direction as required by Newton's third law, $F_{reaction}$ is an external force, which causes the swimmer to accelerate in the opposite to the wall. The two forces do not cancel, because they do not act on the same system.

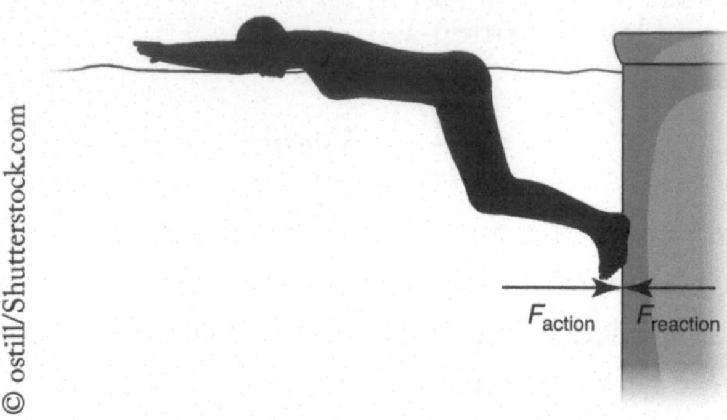

If we solve Newton's second law for force, we get: $F = ma$
If we substitute 1 kg for the mass and 1 m/s^2 for acceleration, we get

$$\text{Force units} = 1\frac{\text{kg}\cdot\text{m}}{\text{s}^2} \equiv 1\,\text{N}$$

One newton is the force required to give a mass of 1 kg an acceleration of 1 m/s^2.

$$1\,\text{Newton} \equiv 1\frac{\text{kg}\cdot\text{m}}{\text{s}^2} \quad \textit{Know}$$

EXAMPLE 3.1

A 440-g can of food is given a shove on a frictionless level surface and is observed to accelerate at a rate of 1.50 m/s^2. How much is the force, in newtons, was applied by the shove?

In order to have the force in Newtons, we must convert grams to kilograms. There are 1000 g in 1.00 kg.

$$440\,\text{g}\left(\frac{1\,\text{kg}}{1000\,\text{g}}\right) = 0.44\,\text{kg}$$

Step by step 1: List the known and unknown values.
mass = 0.44 kg; acceleration = 1.50 m/s^2; F = ?

Step 2: Write down the appropriate equation and solve it for the unknown.

$$a = \frac{\sum F}{m}$$

Solving for the net force, $\sum F$;
We get our working equation: $\sum F = ma$

Step 3: Mathematically account for the fact that vectors have direction information. Usually, we will do this by letting the forward direction be positive and backward be negative.

Step 4: Replace the variable letters with the correct values.

$$\sum F = ma = 0.44\,\text{kg}\left(1.5\frac{\text{m}}{\text{s}^2}\right) = 0.66\frac{\text{kg}\cdot\text{m}}{\text{s}^2}$$

$$\sum \text{F} = 0.66\,\text{Newtons}$$

Step 5: With a vector, you must give a magnitude and direction:

Answer. The net force applied to the can is *0.66 N forward.*

A somewhat less obvious use of Newton's laws is required in problems where several bodies and forces are involved, as in the following two examples, illustrated in Figure 3.3.

EXAMPLE 3.2

An orderly is pushing a gurney with a patient on it. The mass of the orderly is 85.0 kg, the mass of the gurney is 20.0 kg, and the mass of the patient is 50.0 kg. The orderly exerts a backward force of 100 N on the floor. What acceleration is produced, assuming the friction in the wheels is negligible?

Step by step 1: List the known and unknown values.
Total mass = 155 kg; Net force $\sum F = 100\,\text{N}$; acceleration a = ?

Step 2: Write down the appropriate equation and solve it for the unknown.

$a = \dfrac{\sum F}{m}$ We now have our working equation.

Step 3: Mathematically account for the fact that vectors have direction information. Usually, we will do this by letting the forward direction be positive and backward be negative.

Figure 3.3 Orderly pushing a gurney with a patient on it.

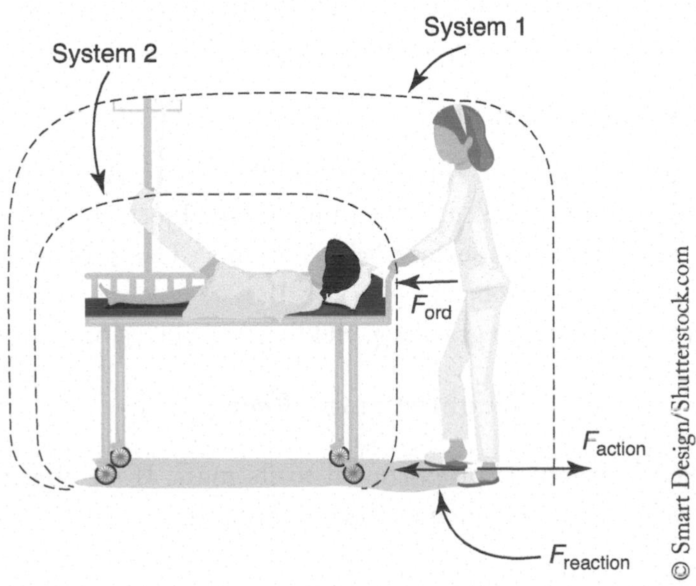

Step 4: Replace the variable letters with the correct values.

$$a = \frac{\Sigma F}{m} = \frac{100\dfrac{\text{kg}\cdot\text{m}}{\text{s}^2}}{155\,\text{kg}} = 0.645\frac{\text{m}}{\text{s}^2}$$

Step 5: With a vector, you must give a magnitude and direction:
Answer. The acceleration of the patient, orderly and gurney is *0.645 m/s² forward.*

Things to know for Section 3.1

Newton's first law of motion: *An object at rest will remain at rest, an object in motion will remain in motion, at a constant velocity, unless an unbalanced force is applied to it.*

Newton's second law of motion: $a = \frac{\Sigma F}{m}$

The acceleration produced by forces acting on a body is directly proportional to and in the same direction as the net external force and inversely proportional to the mass of the body.

Newton's Third Law: *For every force applied to an object, there is another force, equal in magnitude and opposite in direction, applied to the other object.*

Mass units are kilograms, abbreviated kg.

$$1\,\text{N} \equiv \frac{1\,\text{kg}\cdot\text{m}}{\text{s}^2}$$

You should now be ready to try the first five homework questions at the end of this chapter.

3.2 WEIGHT, FRICTION, TENSION, AND OTHER CLASSES OF FORCES

Traditionally, forces have been grouped into several classes and given names relating to their source, the nature in which they are transmitted, or describing their effects. A few of these classes of forces are of sufficient importance to warrant separate description.

3.2.1 Weight and the Force of Gravity

Weight is the force of gravity between the planet and the object. As was discussed in Section 2.3, the acceleration of an object in free-fall, g, is the same for all objects, provided that forces other than gravity are negligible. Near the earth's surface, "g" = 9.80 m/s^2. Using Newton's second law, we have:

Weight = Force of gravity on an object = ma

Here "a" is the acceleration due to gravity, g, yielding an expression for the weight of an object:

$$\text{Weight wt} = \text{mg} \quad \textit{Know} \tag{3.2}$$

Notice that an object's weight is proportional to its mass. If an object is twice as massive as another object, it will weigh twice as much as the other object.

A 141 pound woman, has a mass of approximately 64.0 kg. Let's calculate the metric weight of a 64.0 kg woman on the earth.

$$\text{Metric earth weight} = \text{mg} = 64.0\,\text{kg}\left(9.80\frac{\text{m}}{\text{s}^2}\right) = 627\frac{\text{kg}\cdot\text{m}}{\text{s}^2} = 627\,\text{N}$$

Let's calculate the weight of the same woman if she is standing in a pressurized dome, on the moon's surface. The acceleration due to gravity on the moon's surface, is 1.67 m/s^2.

$$\text{Metric moon weight} = \text{mg} = 64.0\,\text{kg}\left(1.67\frac{\text{m}}{\text{s}^2}\right) = 107\,\text{N}$$

Notice that although her mass is still 64.0 kg, her metric weight, near the moon's surface is only 107N. We see that although her mass remains constant, her weight depends on her location.

If you are within several miles of sea level, you can easily determine your mass and therefore your metric weight. *In order to determine your mass*, divide your weight (in pounds) by 2.2. For example, if you weigh 175 pounds on the earth, your mass is:

$$\text{Mass} = \frac{\text{wt}}{2.2} = \frac{175}{2.2} = 79.5\,\text{kg}$$

$$\text{Metric earth weight} = \text{mg} = 79.5\,\text{kg}\left(9.8\frac{\text{m}}{\text{s}^2}\right) = 779\frac{\text{kg}\cdot\text{m}}{\text{s}^2} = 779\,\text{N}$$

Please take a moment to determine your mass and metric weight. You will then be able to use yourself as a frame of reference when dealing with problems concerning mass and weight.

Center of Gravity: The force of gravity on a solid body can be considered to act at a single point, called the center of gravity (c.o.g.). If the body mass is symmetrically distributed, such as a baseball, then its center of gravity is at its geometrical center. An asymmetrical object, such as a person, will have a center of gravity closer to the more massive parts of the body. A number of the objects illustrated in this chapter have their centers of gravity marked in them.

3.2.2 Newton's Universal Law of Gravitation

Equation 3.2 is not the most general expression for the force of gravity. Equation 3.2 is used when we consider the force of gravity of the earth on small objects, such as ourselves. This will be the case in almost all of our experiences with the force of gravity. A more general expression for the force of gravity between any two bodies is used in astrophysics, for example. It was also discovered by Newton and is named Newton's universal law of gravitation in his honor. *Newton's universal law of gravitation* states that there is a force of attraction between any two masses. The force is proportional to the product of the masses, m_1 and m_2, and inversely proportional to the square of the distance, "d" between their centers of mass. In equation form this is:

$$F_G = \frac{\gamma m_1 m_2}{d^2}$$

Newton began to realize that earth's gravity might extend to the moon, and beyond. When Newton demonstrated that his law of gravity could be used to explain not only the weight of a person but also the motion of the planets around the sun, he proved that the universe was knowable. This is arguably the greatest scientific achievement of all time!

© Georgios Kollidas/Shutterstock.com

In 1798, Henry Cavendish assembled an experimental apparatus that was used to determine the value of the Universal Gravitational Constant (γ). Once the value of the constant was known, using the period of the Moon's revolution around the Earth, scientists were able to determine the mass of the Earth. The experiment consisted of a torsion balance with a pair of 2-in. 1.61-pound lead spheres suspended from the arm of a torsion balance and two much larger stationary lead balls (350 pounds). His work led others to accurately determine the value of the universal gravitational constant γ, gamma. It has a value $6.67 \times 10^{-11} \frac{\text{N} \cdot \text{m}^2}{\text{kg}^2}$. Move the decimal 11 places to the left and notice how tiny this number is. The fact that gamma is just a smidgen above zero indicates that gravity is fundamentally a weak force. At least one of the objects must have a very large mass in order to have a noticeable gravitational attraction between them.

Connections: All objects, in free fall, near the earth's surface, will accelerate toward the center of the earth at approximately *9.80 m/s²*. Why?

Using Newton's second law of motion, we get:

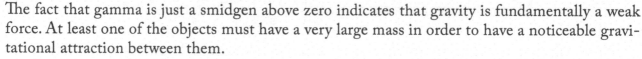

$$a = \frac{\Sigma F}{m} = \frac{F_G}{m} = \frac{\frac{\gamma m M_{\text{earth}}}{d^2}}{m} = \frac{\gamma M_{\text{earth}}}{d^2}$$

The first thing that you should notice is that the mass, "m" of the object divides into itself. We see that when we disregard air friction, the acceleration of an object in free fall does not depend on the mass of the object. Additionally, gamma γ, M_{earth}, and "d," the distance to the center of the earth from sea level, are known constants. If we substitute their values into the equation, $\frac{\gamma M_{\text{earth}}}{d^2}$, we find that $a = 9.80 \text{ m/s}^2$. That is where the value of "g" comes from!

So once again, we see that: *When air friction is negligible, all objects in free fall, near the earth's surface will accelerate down at 9.80 m/s².*

Galileo demonstrated experimentally that when air friction is neglected, all objects in free fall, near the earth's surface, accelerate down at the same rate. Newton explained why this is true, with his laws of motion and gravity. Albert Einstein developed the general theory of relativity and further clarified our understanding of gravity. Present day physicists are attempting to unify all of the laws of nature. The quest goes on.

3.2.3 Friction and the Normal Force

Friction is the familiar force that always opposes motion. Frictional forces result from physical contact between substances. The details of how friction occurs and its exact dependence on materials, speed, and so on are complex. In most cases, however, friction has one simplifying characteristic. Friction is proportional to the force exerted by one substance on another perpendicular to the surface between them—that is, the *normal force*, F_N, (so named because it is perpendicular, or normal, to the surface). *The mathematical expressions for friction are:*

Kinetic friction ≡ Friction between sliding or slipping surfaces.

$$f_K = \mu_k F_N \quad \textbf{\textit{Know}} \tag{3.4a}$$

Static friction ≡ Friction between stationary surfaces.

$$f_S = \mu_S F_N \quad \textbf{\textit{Know}} \tag{3.4b}$$

where μ_k is the *coefficient of kinetic friction*, and μ_s is the *coefficient of static friction*, and F_N is the normal force. The first expression is used for the friction between moving substances and the second for the friction between stationary substances. Table 3.1 gives typical values of coefficients of friction. Notice that the smallest values in the table are for bone lubricated with synovial fluid. Mother Nature proves to be a good engineer.

Table 3.1 Coefficients of Kinetic and Static Friction[a]

	μ_k Kinetic Friction	μ_s Static Friction
Rubber on dry concrete	0.7	1.0
Shoes on wood	0.7	0.9
Rubber on wet concrete	0.5	0.7
Wood on wood	0.3	0.5
Waxed wood on wet snow	0.100	0.140
Metal on wood	0.3	0.5
Steel on steel (dry)	0.3	0.6
Steel on steel (oiled)	0.03	0.05
Teflon on steel	0.04	0.04
Shoes on ice	0.05	0.1
Ice on ice	0.03	0.1
Steel on ice	0.02	0.4
Bone lubricated with synovial fluid.	0.015	0.016

[a]All values are approximate.

Frictional forces are generally less for sliding objects than for stationary ones, as an inspection of Table 3.1 reveals. Static friction (between two objects which are not moving relative to one another) is never larger than the outside force applied to it. Friction grows to that limit to oppose whatever force is trying to move the objects relative to one another. If no such force is acting, then there is no friction since it would be an unbalanced force and would cause the block to accelerate.

Typical of the behavior of friction is what happens when a furniture mover attempts to move a heavy box by pushing it along the floor. When he pushes on the box with a small force, the box doesn't move; evidently the force of friction has grown to be equal to the force applied. He pushes harder and still the box doesn't move; the force of friction has grown to equal the larger force he is exerting. He rests a moment; meanwhile the box doesn't move, so friction must have decreased to zero. He now exerts a mighty shove and the force applied exceeds the static friction, $f_s = \mu_s F_N$. Once the box breaks and begins to slide, kinetic, or moving friction is given by $f_k = \mu_k F_N$, which is smaller than static friction. Because the coefficient of kinetic friction is less that the coefficient of static friction, once the box is moving, less force is required to keep it going than was necessary to get it started. If you have ever used a sled with metal runners on snow, you may have noticed the difference between static and kinetic friction.

The force at right angles to the surface is called the *normal forces*. The normal force F_N, can be due to many things. During a massage, the normal force increases if the massage therapist supports more of her weight with her hands. This causes more friction because F_N is larger. To compensate, a lubricating oil with a low coefficient of friction is used.

In the simplest case of an object on a level surface, the normal force F_N, is just the force necessary to support the weight of the object and is in fact equal to the weight of the object. Because the object's acceleration is zero, the sum of all the forces on it is zero. *Therefore, on a horizontal surface, in the absence of external forces other than gravity, the normal force is equal in magnitude and opposite in direction to the object's weight.*

EXAMPLE 3.3

As shown in Figure 3.4, a 70-kg cross-country skier is on level ground in wet snow wearing waxed wooden skis. (a) What is the maximum force that can be exerted on the skier in a horizontal direction

Figure 3.4 Skier on a horizontal surface.

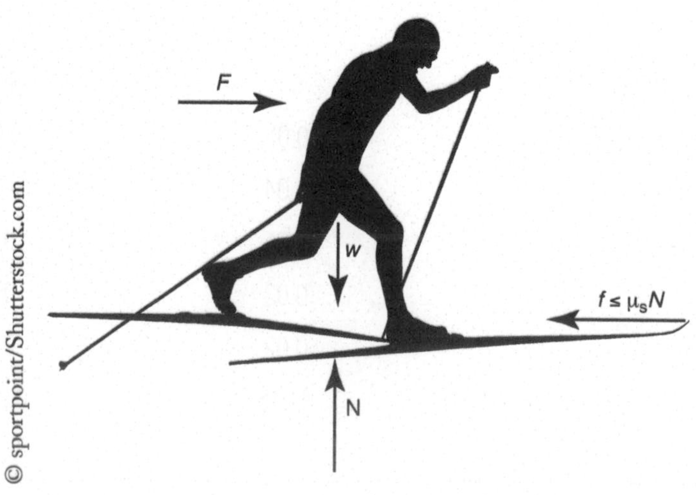

without causing him to move? (b) Once the skier is moving, what horizontal force is necessary to keep him moving at a steady velocity?

Solution:

a. The maximum force that can be exerted without moving him is equal to the maximum static friction:

Step by step 1: Static friction: $f_s = \mu_s F_N$
From Table 3.1 we know that the coefficient of static friction μ_s, is 0.140
The beginning skier is on flat, level ground, so the normal force is just the weight of the skier:

Step 2: The weight of the skier is determined by using the weight equation:

$$wt = \text{mg} = 70.0\,\text{kg} \cdot 9.80\frac{\text{m}}{\text{s}^2} = 686\ \text{N}$$

Step 3: Write down the equation for static friction and substitute the values for coefficient of static friction and the force normal. Remember, on a flat horizontal surface, as long as no other forces are being applied, the object's weight will equal the force normal.

$$f_s = \mu_s F_N = \mu_s wt = 0.140 \cdot 686\,\text{N} = 96.0\,\text{N}$$

The maximum static friction on the skier is 96.0 N.

b. What horizontal force is necessary to keep her moving at a constant velocity?

Step by step 1: The skier will be sliding, so we will use the kinetic coefficient of friction to solve our problem.

$$f_k = \mu_k F_N$$

We know from Table 3.1 that the kinetic coefficient of friction is, $\mu_k = 0.100$
The force normal will equal the skier's weight:

Step 2: We already know that the skier weighs 686 N. We substitute the kinetic coefficient of friction into our equation as follows:

$$f_k = \mu_k F_N = 0.100 \cdot 686\,\text{N} = 68.6\ \text{N}$$

The kinetic friction on his ski is 68.6 N as she moves down the slope.

3.2.4 Tension

A tension is any force carried by a flexible string, rope, cable, chain, and so on. Because the medium carrying the force is flexible, it can only pull and only exerts a force along its length. The word tension comes from a Latin word meaning "to stretch thin." In muscle systems the fibrous cords that carry forces exerted by muscles to other parts of the body are called tendons.

Tensions provide an interesting illustration of Newton's third law. Consider the following example.

EXAMPLE 3.4

A child and basket with a total mass of 10.0 kg are suspended from a scale by a cord as shown in Figure 3.5. Calculate the tension in the cord.

Solution: Because the baby is not accelerating, according to Newton's first and second laws of motion, the total force on the baby must equal zero. We can say this mathematically, following Newton's second law, as follows:

$$\sum F = Tension + weight = 0 \text{ N}$$

The baby and basket have a total mass of 10.0 kg. Remember that *mass is not the same thing as weight*. The weight, which is the force of gravity on the baby and basket is 98.0 N.

$wt = \text{mg} = 10.0\,\text{kg} \cdot 9.80 \frac{\text{m}}{\text{s}^2} = 98.0 \text{ N down}$

The only way that the net force applied to the child and basket can equal 0 N, is if the tension in the cord is 98.0 N up. The two vectors will cancel each other out and result in a net force of 0 N.

You should now be ready to try **homework problems** 6–9.

Other Classes of Forces: Are Some More Basic Than Others?

Figure 3.5

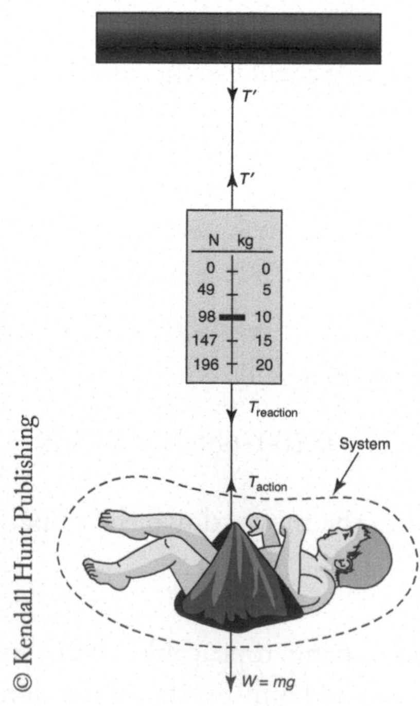

There are many classes of forces, many of them to some extent familiar—such as cohesion, adhesion, thrust, drag, lift, and shearing force. Jargon evolves in most fields and usually results in specialized names being given to entities that can be more loosely classified. Physicists by their very nature ask whether certain types of forces are more basic than others or if all forces are really the same in some way, perhaps being different manifestations of the same thing. It has been found that there are only four basic forces in nature and that two of these suffice to describe everything but nuclear phenomena. Those two forces are gravity and the electromagnetic force.

Almost every phenomena observed in everyday life can be ascribed to gravity and electromagnetic forces. Except for weight, all the forces mentioned so far in this book are electromagnetic in origin. Tension, for example, is due to the cohesive atomic and molecular electromagnetic forces acting in a string. Friction is due to electromagnetic interactions between the atoms and molecules of materials in close contact. Gravity is certainly pervasive and with us at all times, but it only accounts for the phenomenon of weight. It is not necessary to worry about the details of gravity or the electromagnetic force to solve problems involving weight, tension, and so on, but it is interesting to know that a myriad of phenomena can be explained as manifestations of just two basic forces.

Things to know for Section 3.2

Static friction: $f_S = \mu_S F_N$

Kinetic friction: $\mathrm{f}_K = \mu_k \mathrm{F}_N$

Weight wt = mg

When air friction is negligible, all objects in free fall, near the earth's surface will accelerate down at 9.80 m/s^2

3.3 VECTORS: METHODS OF ADDING AND PROJECTING FORCES

Forces are vectors, since they have both *magnitude* and *direction*. (See Section 2.1 for the definitions of vectors and scalars.) In Sections 3.1 and 3.2, care was taken to choose examples of forces whose directions were along a straight line. A technique called vector analysis is needed to handle forces that are not simply parallel or perpendicular to one another but may be at any imaginable angle with regard to each other. The techniques of vector analysis presented in this section are applicable to any type of vector, such as acceleration or velocity, but we shall apply them mostly to forces—forces are relatively easy to visualize, and they are very important.

Where it is important to take direction into account, vectors are usually represented in bold type. (Of course, if we only are interested in the magnitudes, or if all directions in a problem are the same, we can omit the boldface.) The **bold** letters are a reminder that we must apply vector analysis.

3.3.1 Graphical Method of Vector Addition

The graphical method of vector addition presented in this section is performed with pencil and paper using a ruler and a protractor. An arrow is used to represent a vector, as seen in many of this

Figure 3.6 Graphical method of vector addition, (a) Two forces F_1 and F_2 are applied to the same system, Arrows are drawn to represent the forces; their lengths are proportional to the magnitude of the force. The arrows are moved head to tail. (b) *The total force is the arrow drawn from the tail of the first to the head of the second,* The order of addition is unimportant. (c) Numerical values for the direction and magnitude of the total force can be found by drawing the vectors "to scale" and then measuring them with a protractor and ruler.

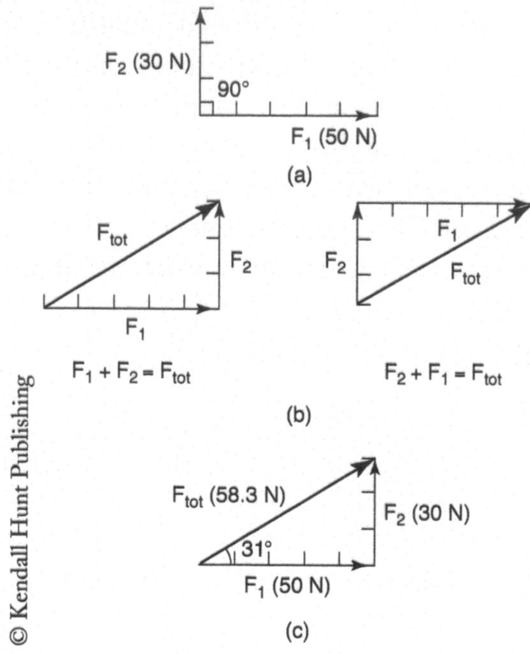

chapter's figures. The direction of the arrow represents the direction of the vector, and the length of the arrow is proportional to the magnitude of the vector. Using a ruler and a protractor, one can add and subtract vectors with an accuracy limited only by how precisely the lengths and directions of the arrows are drawn. The vector produced by adding or subtracting two or more other vectors is called the *resultant.*

Figure 3.6 illustrates how this is done. Two forces, $\boldsymbol{F_1}$ and $\boldsymbol{F_2}$, having magnitudes of 50 and 30 N, respectively, are applied to an object at right angles (90°) to one another. (The lengths of the arrows in the figure are proportional to the magnitudes of the forces.) It is clear that the direction of the resultant total force will be to the upper right, but its exact direction and magnitude are not immediately obvious. Notice that the magnitude of the total force, would only be 80 N if both forces were applied in the same direction. In this example, 50-N East plus 30-N North equals 58.3-N, 59° clockwise from North. We warned you in Chapter 2 that 2 + 2 does not always equal 4. With some practice you can learn to add vectors easily. We have placed detailed instructions for adding vectors graphically and mathematically in Lab 2.

3.3.2 Mathematical method of Vector Addition

The mathematical method of vector addition will require the use of a calculator that has sine, cosine, and tangent functions. We will use these trigonometric functions to help us break each vector into

Figure 3.7 The forefinger. Note that the tendons carrying the forces exerted by the muscles go over joints that change the direction of the force, extensor muscle.

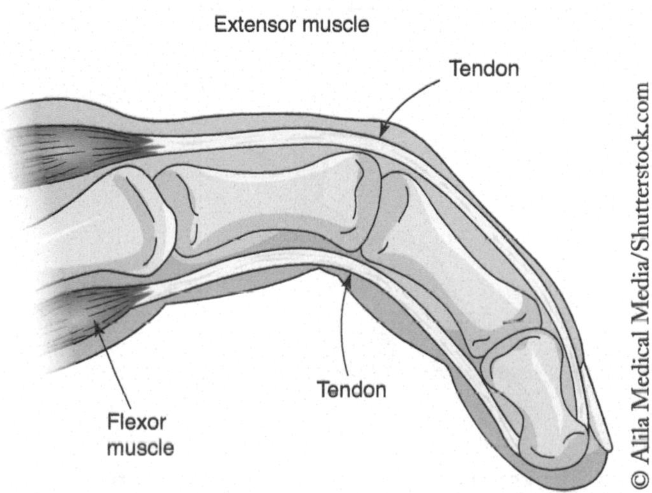

two perpendicular parts, called components. Once all of the vectors are broken up into their right angle components, the components in one direction or the opposite direction are combined. For example, if we are working with compass directions, we can combine all of the north and south components by letting north components be in the positive direction and south components be in the negative direction. Then we can total the north and south components. If the answer comes out positive, the final north–south vector is north. We can also combine all of the east and west components in a similar manner. For example, if the total of all the N–S, E–W components is 50 N east and 30 N north, we would draw a diagram like part (b) in Figure 3.6. Using the inverse tangent function, we would determine that the angle between the two vectors is 31°. Using the Pythagorean theorem, we would get a magnitude of 58.3 N. Detailed instructions for the mathematical and graphical addition of vectors are in Lab 2.

3.3.3 Forces Exerted by Muscles: An Introduction

One of the more interesting applications of force vectors is to muscle systems. Muscles exert forces directed along their lengths by contracting. In many cases a muscle is attached to two bones, often with a joint between the bones to allow movement. Those muscles that cause bones to move closer together are called *flexor* muscles, and those that cause bones to move apart are called *extensor* muscles.

An example is shown in Figure 3.7. Tendons, such as those in the finger or knee, sometimes carry force exerted by a muscle to another point and may even change its direction. There are hundreds of muscles in the body, allowing forces to be exerted in almost any direction. Several muscles act simultaneously in the shoulder to produce a net force, as shown in Figure 3.8. (The graphical method of vector addition works for any number of vectors.)

Other types of muscles may join back on themselves and cause the constriction of an opening when they contract. Such muscles, called *sphincters*, serve several functions. The sphincter at the lower end of the esophagus prevents the back flow of stomach fluids. Another sphincter muscle in the eye changes the curvature of the lens of the eye to allow clear vision of near and distant objects.

Figure 3.8 Several muscles act at once in the shoulder to produce the total force exerted on the arm. The graphical method of adding forces yields the expected result: The upward and downward components of the forces cancel, leaving a large horizontal force.

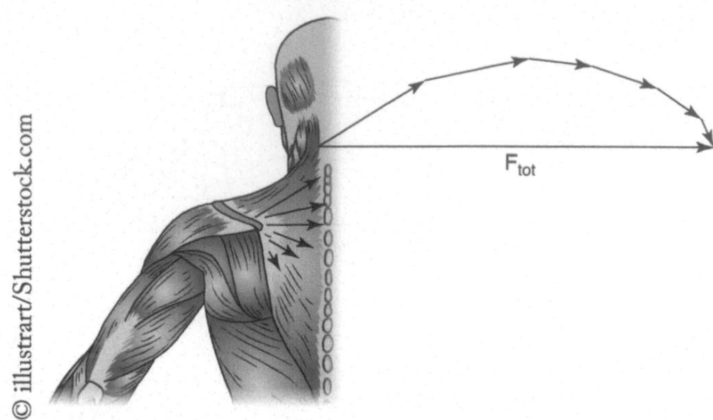

Figure 3.9 Patient in a Stryker frame. Traction is applied by hanging a weight on a wire attached to pins in the patient's temples. The pulley changes the direction of the force so that it is along the length of the spinal column.

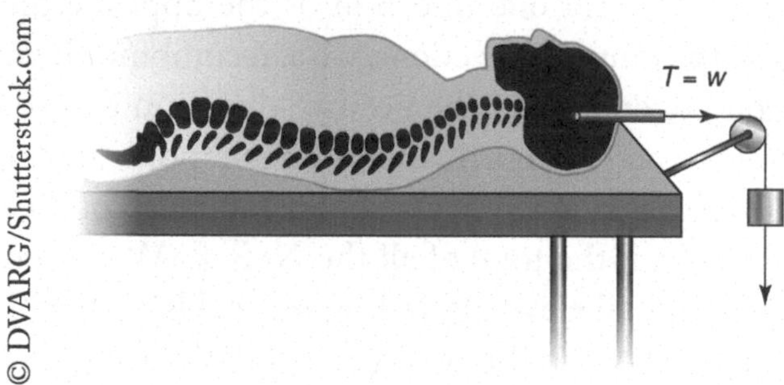

It is impossible not to admire the beauty and sophistication of the muscle systems in the body. Those in the face and hands are particularly impressive, for they allow not only complex movement but also the display of emotion. We shall return to muscle systems later in this chapter.

3.3.4 Traction Systems

Patients are placed in traction because of broken bones and spinal injuries, for example. At first glance many traction systems appear hopelessly complex, with wires going in every direction, performing unknown functions. Actually, traction systems are rather simple and easy to understand when analyzed using just a few physics principles. Figure 3.9 shows a patient in traction in a Stryker frame for an injury to the spinal vertebrae. It is necessary to stretch the spinal column along its length for the vertebrae to heal and to prevent broken vertebrae from damaging the spinal cord. The force applied is equal to the weight hanging on the right. The wire transmits the force, a tension, to the patient; the frictionless pulley simply changes the direction of the wire. Because a wire is used, the force is in the same direction as the wire at the point where it is attached to the patient. In the Stryker frame the wire exerts force parallel to the spine, as desired. The wire is attached to two pins in the patient's

Figure 3.10 Typical traction for a broken femur. The weight w_1 supports the weight of the lower leg, w_{leg}. The weight w_2 counteracts muscle contractions in the upper leg. The positions of the pulleys are adjusted so that the forces are in the proper direction.

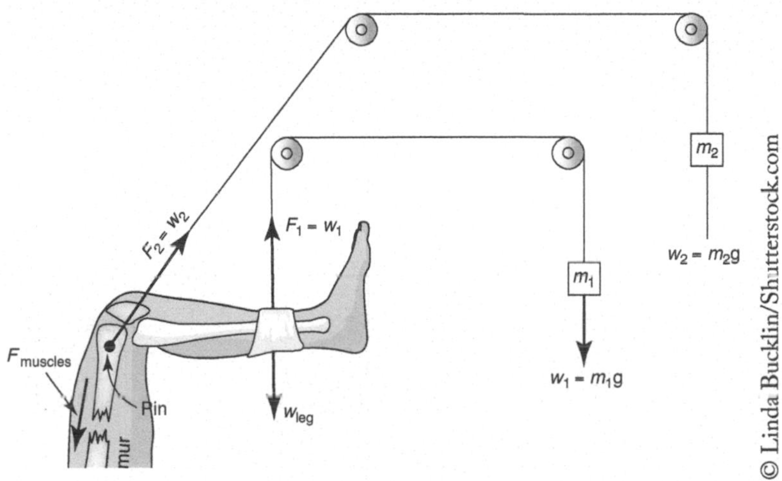

temples to allow some head movement without disturbing the direction and magnitude of the traction force. Friction between the patient and the bed prevents the patient from moving to the right.

wt = mg

For some patients a multiple-function traction system may be used, as illustrated in Figure 3.10 for a patient with a broken femur.

Two purposes are served by the system shown: first, to support the weight of the lower leg and, second, to prevent shortening and bowing of the upper leg. Because wires can only apply force along their lengths, the one attached to the lower leg supplies an upward force equal to the weight W_1 (again assuming no friction in the pulleys). The weight $\boldsymbol{W_1}$ is adjusted until it just supports the lower leg without effort on the part of the patient. The muscles in the upper leg are quite strong, and forces exerted by them tend to shorten the femur, causing permanent shortening and possible bowing if the second system isn't employed. Pins are inserted in the femur below the break, and the pulleys are adjusted until the wire attached to the pins is in the same direction as the femur. Once this is accomplished, the force applied is along the same direction as the femur so that it can counteract any muscle contractions in the upper leg. The weight $\boldsymbol{W_2}$ in figure 3.10 is adjusted to be large enough to counteract average muscle contractions in the upper leg.

It is possible to analyze any traction system by keeping two things in mind. First, the force applied is in the direction of the wire at the point where it is attached to the patient; second, the force is equal to the weight hung on the wire. If several forces act at one point, the graphical method of vector addition can be used to analyze the system. (See the diagram of Russell traction in Figure 3.11.)

3.3.5 Projection of Forces

It is sometimes convenient to break up a single force into two forces that produce an equivalent effect. These two forces give the original force when added together. This technique is called *projection*, and

Figure 3.11 Russell traction accomplishes the same purpose as the system in Figure 3.10, The forces add to give a total force that supports the lower leg and has a direction along the femur to counteract muscle contractions.

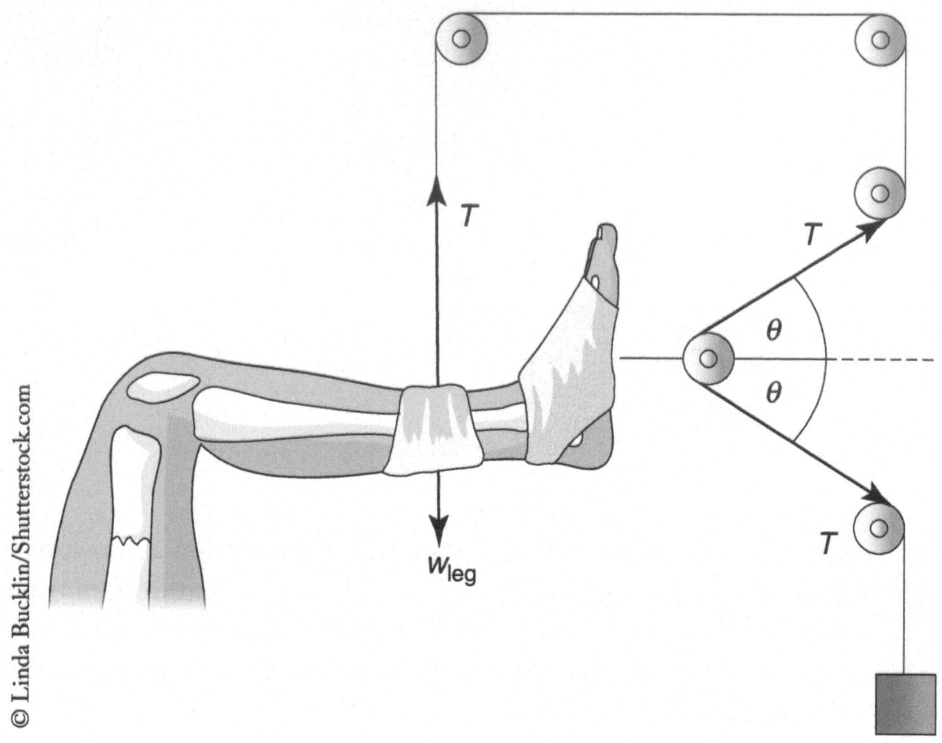

such forces are called *components* of the original force. The utility of breaking up or *projecting* forces into components is best illustrated by an example.

Figure 3.12(a) shows a shopper pushing a loaded grocery cart by exerting a *20-N* force in a direction 30° below the horizontal. What part of the *20-N* force is effective in causing the cart to move to the right? The force is not as effective as if it were horizontal since part of the force is directed downward and is taken up by the floor. (In the extreme case where the shopper pushes straight downward, the cart doesn't move forward at all.) Only the horizontal component of the force results in forward motion.

In order to determine the component of the *20-N* force that is effective in causing the cart to move to the right, we break the *20-N* force into two forces, one vertical and one horizontal. This can be done graphically or mathematically, as shown in Figure 3.12(a). An arrow is drawn to represent the force, its length being proportional to its magnitude and its direction being 30° below the horizontal. The 20-N force is broken into horizontal and vertical components as follows: First an arrow is drawn forward, parallel to the ground and represents the forward component. Then another arrow, perpendicular to the forward component is drawn down to the tip of the 20-N vector. These two arrows represent the horizontal and vertical components of the 20-N force. Because the horizontal component is adjacent to the 30° angle, the cosine function is used to calculate a value of *17.3 N forward.* The vertical component is opposite the 30° angle and therefore the Sine function is used to determine that it's value of *10.0 N down.* Note again that unless vectors are parallel, their magnitudes don't add like ordinary numbers. Once again we remind you that 2 plus 2 does not always equal 4. *When adding vectors, you must include direction.* See example 3.5 for a more detailed explanation.

Figure 3.12 (a) A 20-N force is exerted by a shopper on a grocery cart at the angle shown. Only the horizontal component of the force is effective in producing an acceleration; the vertical component is counteracted by the floor. The force is broken down into two right angle forces that produce the same effect—that is, they add to give the original force.

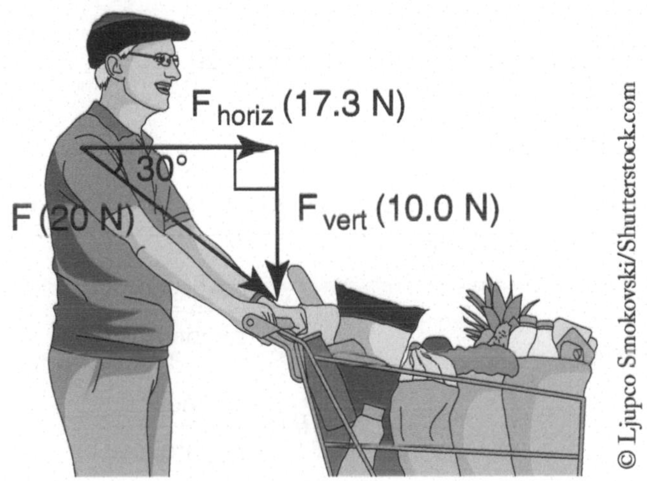

EXAMPLE 3.5

Calculate (a) the net horizontal force applied to the cart in Figure 3.12 and (b) the acceleration of the cart shown in Figure 3.12, given that the total mass of the cart plus groceries is 25.0 kg and that a *2.00-N* frictional force opposes the motion. *You should follow along with your calculator.* First, make sure that your calculator is in *degree* mode.

Solution:

a. Because the horizontal component of the force applied by the woman is adjacent to the 30° angle, we will use the cosine function to determine its value.

$$\text{Cos}(\theta) \equiv \frac{\text{adjacent}}{\text{hypotenuse}} \quad \textbf{\textit{Know}}$$

Step by step 1: Solve the cosine equation for the adjacent side. This will be our working equation.

$$\text{Cos}(\theta) \equiv \frac{\text{adjacent}}{\text{hypotenuse}}$$

$$\text{adjacent} = \text{Cos}(\theta) \cdot \text{hypotenuse}$$

Step 2: Substitute the known values and do the math. $\text{adjacent} = \text{Cos}(30°) \cdot 20.0\,\text{N} = 17.3\,\text{N}$

Step 3: Determine the *net horizontal force* applied to the cart, by getting a total of the horizontal forces. If we let the forward direction be positive and backward be negative, we get:

$$\Sigma F = 17.3\,\text{N} + -2.00\,\text{N} = 15.3\,\text{N forward.}$$

b. **Solution.** In order to determine the acceleration of the cart, we will use Newton's second law of motion.

$$a = \frac{\Sigma F}{m}$$

According to this rule, we will need to divide the *net force*, $\sum F$, applied to the cart, by the mass of the cart. The mass of the cart is given in the problem statement. It is 25.0-kg.

$$a = \frac{\sum F}{m} = \frac{15.3\frac{\text{kg}\cdot\text{m}}{\text{s}^2}}{25.0\text{kg}} = 0.612\frac{\text{m}}{\text{s}^2}\text{ forward.}$$

If you are feeling lucky, you might try **problems** 10 and 11.

Things to know for Section 3.3

The following rules are essential to learning how to add vectors mathematically. Please make an effort to know and understand them.

$$\text{Cos}(\theta) \equiv \frac{\text{adjacent}}{\text{hypotenuse}}$$

$$\text{Sin}(\theta) \equiv \frac{\text{opposite}}{\text{hypotenuse}}$$

$$\text{Tangent}(\theta) \equiv \frac{\text{opposite}}{\text{adjacent}}$$

$$\text{hypotenuse} \equiv \sqrt{opp^2 + adj^2}$$

3.4 STATICS; TORQUE AND ROTATION

3.4.1 Statics: A Special Case of Newton's Laws

Statics is the study of bodies in equilibrium, such as buildings, bridges, or patients in traction. When a body is in equilibrium, its acceleration is zero. Many times the body is at rest, but bodies in motion with a constant velocity are also in equilibrium and belong in the study of statics.

Two conditions must be satisfied for a body to be in equilibrium:

1. The net external force on the body must be zero so that its acceleration is zero.
2. The net external torque applied to the object must be zero.

This second condition brings up something quite new. It is possible to change the rate at which a body rotates, even when the net external force on it is zero, as Figure 3.13 illustrates. The figure shows two different situations in which two forces of equal magnitude and opposite direction are applied to a body resting on a frictionless surface. In both situations the net external force is zero, but the object rotates whenever the forces are not applied directly opposite to one another. Evidently, the point at which a force is applied is crucial to whether or not an object is caused to rotate.

In some situations we shall want to cause rotation rather than prevent it, such as in opening a door or flexing a limb. Whether we want to cause or prevent rotation, the point at which force is applied is important. In order to change the rate at which an object is rotating, without changing its mass distribution, we must apply an unbalanced *torque*.

Figure 3.13 It is important to consider the points at which forces are applied as well as their magnitudes and directions, In both cases shown. the net external force is zero, (a) Two forces of equal magnitude and opposite direction are applied to an object resting on a frictionless surface, equilibrium is achieved; the object remains stationary, (b) If the same two forces are applied at different points, the object rotates; equilibrium is not achieved.

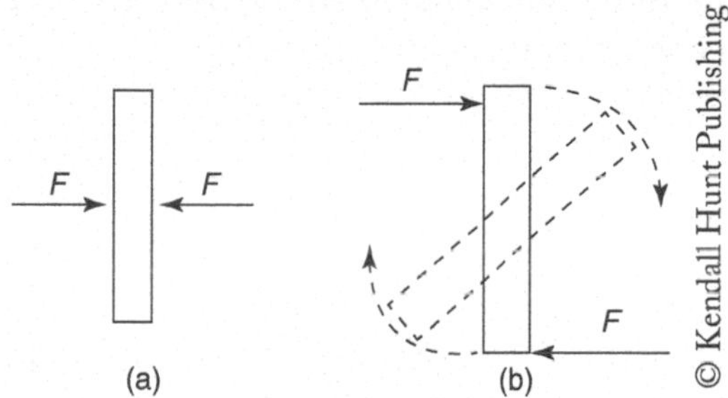

3.4.2 Torque

The effectiveness of a force in producing a rotation is called torque. It is useful to think of torque as the turning effect of a force. As illustrated in Figure 3.14, the three factors involved are the magnitude and direction of the force and the point at which it is applied. Let's consider each of those factors in turn. Clearly, the greater the magnitude of the force, the more effective it will be in causing the door to rotate about its hinges. The direction of the force is also important. If the force were tangent to the door, as in Figure 3.14(b), there would be no rotation; the person would simply be pulling or pushing on the hinges (torque would be zero). If the direction of the force were reversed, as in Figure 3.14(c), the door would rotate in the opposite sense—clockwise when viewed from above rather than counterclockwise. The third factor is the point at which the force is applied. Imagine what happens when you push on a door too close to the hinges, as in Figure 3.14(d); The distance from the pivot to the applied force is small resulting in a smaller torque than if the force is applied further from the pivot, as in Figure 3.14(e).

The two factors upon which torque depends are:

1. The magnitude of the perpendicular component of the force applied.
2. The *distance from the pivot* point that the perpendicular component of the force is applied.

The formal definition of torque in equation form takes both factors into account: *Torque is the product of the perpendicular force applied to an object and the distance "d" from the pivot.* We can state this mathematically with the following equation:

$$\text{Torque} \equiv F_{\perp} \cdot d \quad \textbf{\textit{know}} \qquad (3.5)$$

$F_{\perp}$ is the perpendicular component of the applied force.
"d" is the distance from the pivot to the place where the force is applied.

Figure 3.14 The torque produced by a force acting on a door as viewed from above. (a) Counterclockwise torque, if unopposed, will cause the door to rotate counterclockwise. (b) Torque is zero, so the force pulls on the hinges producing no rotation. (c) Clockwise torque equal and opposite to that in (α). (d) The torque is smaller because the force is applied closer to the hinges. (e) The perpendicular lever arm $l_{\perp}$ is equal to the distance down the door to the point at which the force is applied *when* the force is perpendicular to the door.

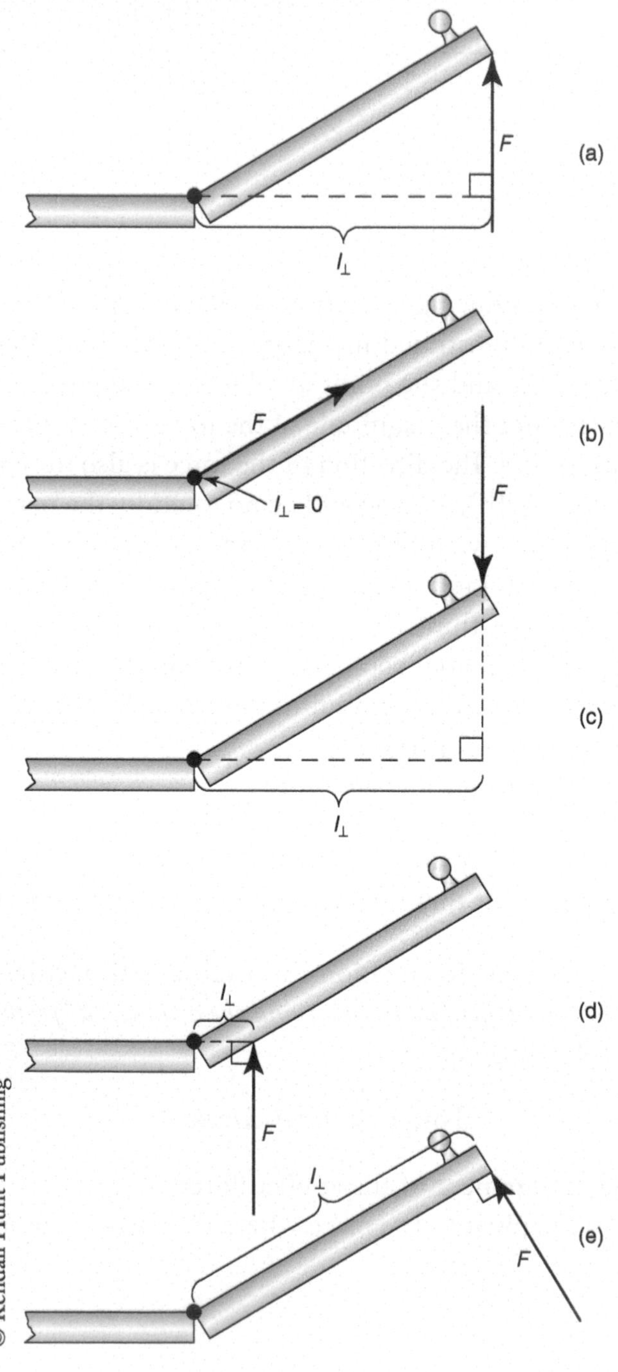

Figure 3.14 (f)

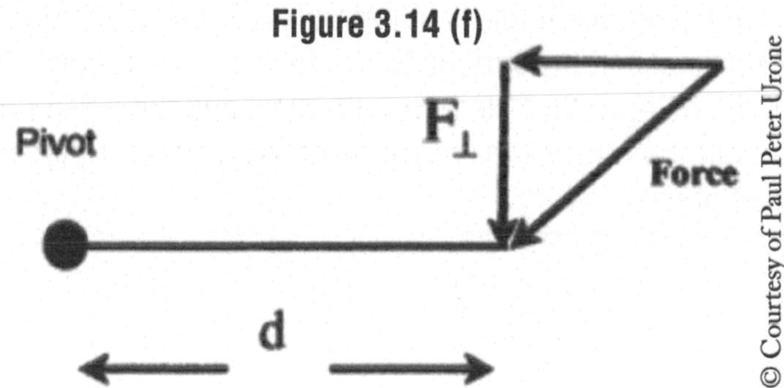

It is important to note that torque is a vector since it has direction as well as magnitude. The direction is specified as being either clockwise or counterclockwise about a pivot point as seen from your point of view.

Now that torque has been defined mathematically, the two conditions for static equilibrium can be stated in equation form. No linear acceleration means that the net external force must be zero. No angular acceleration means that the net torque must be zero—that the sum of the clockwise torques must equal the sum of the counterclockwise torques. Angular acceleration, ω, is the rate at which a system rotates.

For a body to be in equilibrium it must satisfy both of the following:

1. The net external force applied to an object must equal zero:

$$\sum F_{\text{external}} = 0\,\text{N}$$

2. The net external torque applied to the object must equal zero:

$$\sum \tau_{\text{external}} = 0\,\text{m} \cdot \text{N}$$

where τ_{external} is the external torque. An equivalent statement is:

Clockwise torque = Counterclockwise torque

The following example illustrates the use of torques in a static equilibrium situation.

EXAMPLE 3.6

(a) Calculate the force that the bicep muscle exerts to hold up the forearm and the book shown in Figure 3.15. (b) Compare the force exerted by the bicep muscle with the weight of the forearm and book.

Discussion (a) The best way to approach this problem is to make the forearm the system of interest. Four external forces act on the forearm: the weight of the forearm, the weight of the book, the force the bicep muscle exerts, and the force exerted at the elbow joint by the humerus. The second of the two conditions for equilibrium, ***Equation 3.6b***, is sufficient to solve this particular problem. One simplification can be made by taking the pivot point to be the elbow joint. The torque exerted by F_H is then zero since it is applied at the pivot point. Values of perpendicular component of the forces are easy to determine in this case because the forearm is horizontal.

a. **Solution.** Determine the F_{bicep} required to hold the book.

Figure 3.15 Forearm holding textbook as in Example 3.9, and a schematic for easier visualization of the distances of the various forces from the pivot point. The weight of the arm and that of the text produce clockwise torques, and the bicep produces a counterclockwise torque, but the force of the humerus produces no torque, because it acts directly on the pivot point.

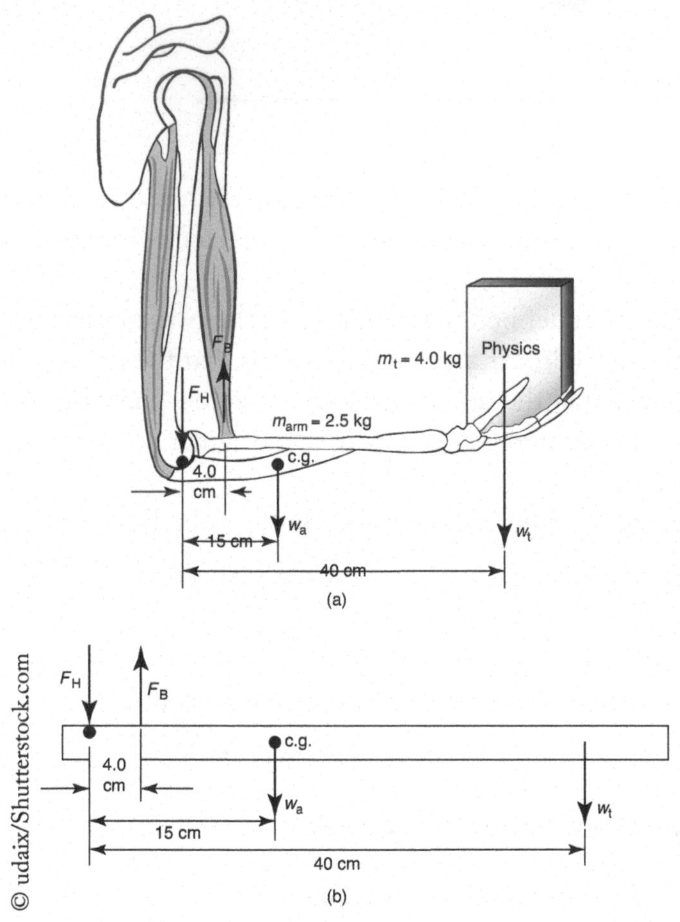

The weights of the forearm and textbook produce clockwise torques, while F_B exerted by the bicep muscle produces a counterclockwise torque.

Step by step 1: Determine the weight of the objects.
Remember that weight = mg. The weight of the arm and textbook are:

$$wt_{\text{arm}} = \text{mg} = 2.50\,\text{kg} \cdot 9.80\frac{\text{m}}{\text{s}^2} = 24.5\,\text{N}$$

$$wt_{\text{book}} = \text{mg} = 4.0\,\text{kg} \cdot 9.80\frac{\text{m}}{\text{s}^2} = 39.2\,\text{N}$$

Step 2: Because the book and arm are in rotational equilibrium, we can state that:
Clockwise torque = Counterclockwise torque

$$24.5\,\text{N} \cdot 15.0\,\text{cm} + 39.2\,\text{N} \cdot 40\,\text{cm} = F_{\text{bicep}} \cdot 4.00\,\text{cm}$$

Step 3: Solve for the force exerted by the bicep, F_{bicep}.

$$F_{\text{bicep}} = \frac{367\,\text{N} \cdot \text{cm} + 1568\,\text{N} \cdot \text{cm}}{4.00\,\text{cm}} = 484\,\text{N}$$

b. **Solution.** Compare the force exerted by the bicep muscle with the combined weight of the arm and book.

c. The combined weight of the arm and book is only 63.7 N, and yet the bicep has to apply a force of 484 N in order to maintain rotational equilibrium. Remember that:

$$\text{Torque} = F_{\perp} \cdot d \text{ Know}$$

d. Torque depends as much on distance from the pivot as it does on force.

The weight of the book and arm are applied at a distance of 40 cm and 15 cm, respectively. The bicep applies its force at only 4 cm from the elbow. Because the bicep applies its force so close to the pivot, in order for clockwise torques to equal counterclockwise torques, its force must be greater than if it was applied further out from the elbow (pivot). The bicep is said to be at a mechanical disadvantage, as are most of the muscles in the body.

It is well known that poor posture can produce back strain. Figure 3.16 illustrates why. When one stands erect, as in Figure 3.16(a), the weight of the upper body is directly over the legs and little force is exerted by back and leg muscles. Most of one's weight is then supported by the skeletal system, and not by muscle action. If one slouches or leans forward, as in Figure 3.16(b), then the center of gravity of the upper body is no longer directly over the pivot point. Now the back muscles must exert a torque about the pivot at the base of the spine to counteract the torque due to the weight of the upper body. Continual effort by the back muscles produces back strain. One of the most common complaints during pregnancy is back pain caused by carrying the weight of the unborn child with the back muscles. This can also be partially alleviated by correct posture.

Everyone says, "Lift with your legs, not your back." The reasons are exactly the same as those for maintaining good posture.

As Figure 3.17(a) illustrates, when lifting bent over, the back muscles must supply enough torque to lift both the load and the upper body. If one lifts with the legs, as shown in Figure 3.17(b), torque is supplied by the extensor muscles in the upper legs. The upper leg extensor muscles have no greater mechanical advantage than the back muscles ($F_{\perp} \cdot$ d is small for both systems); but the leg muscles are larger and the knee joint is stronger and better able to cope with the forces involved than are the lower spinal joints.

Figure 3.16 (a) A person standing erect places the center of gravity of his upper body directly above the pivot in his lower back. No torque is produced and the back muscles therefore don't have to exert torques to maintain balance. An equivalent mechanical system is shown on the right. (b) A person bending over or slouching allows the weight of his upper body to create a torque that must be compensated by torque created by the back (erector spinae) muscles. This can produce considerable muscle stain if prolonged. Note that F_B is large in the equivalent mechanical system.

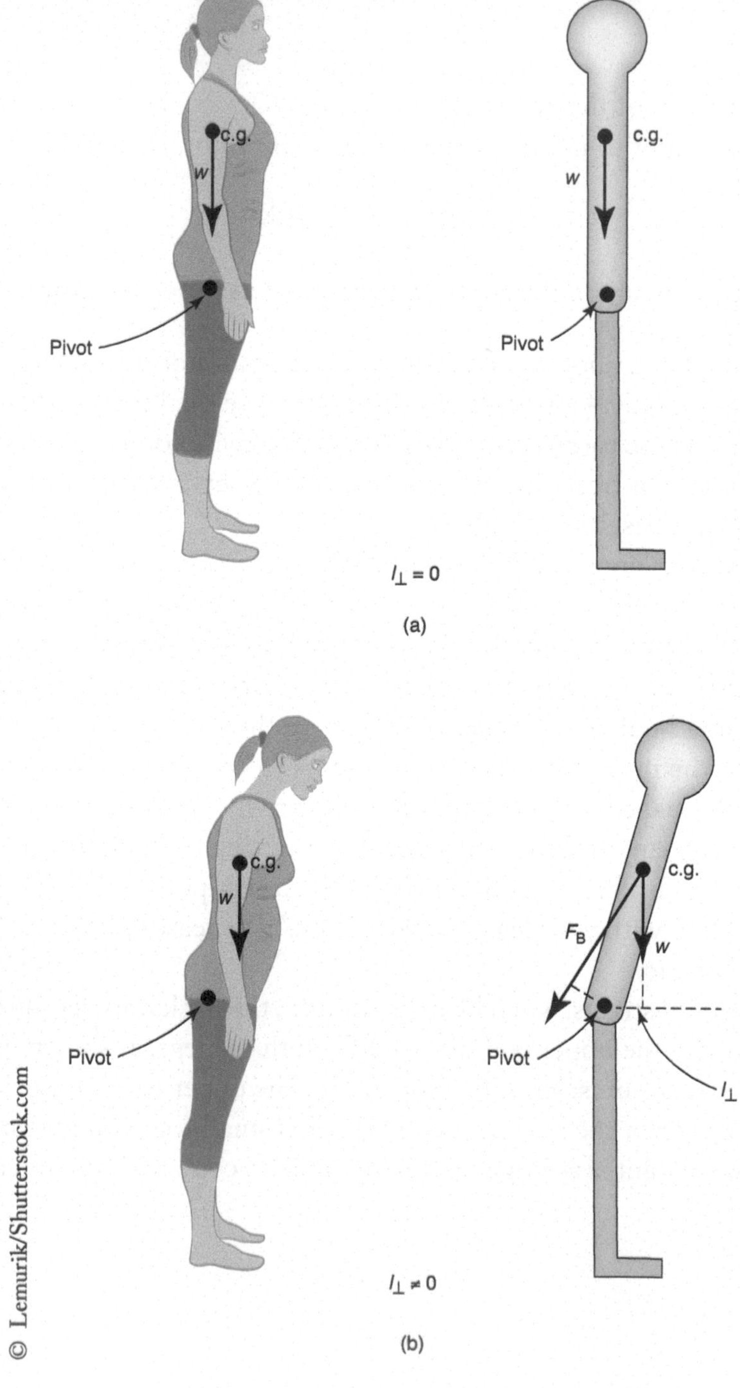

Figure 3.17 (a) If a person lifts with his back, very large forces must be generated by the back (erector spinae) muscles and be withstood by the joints in the lower back. (b) When a person lifts with his legs, the upper leg muscles (quadriceps) exert large forces that must be withstood by the knee joints. Both are better able to withstand these forces than the muscles and joints in the back. Equivalent mechanical systems are shown on the right.

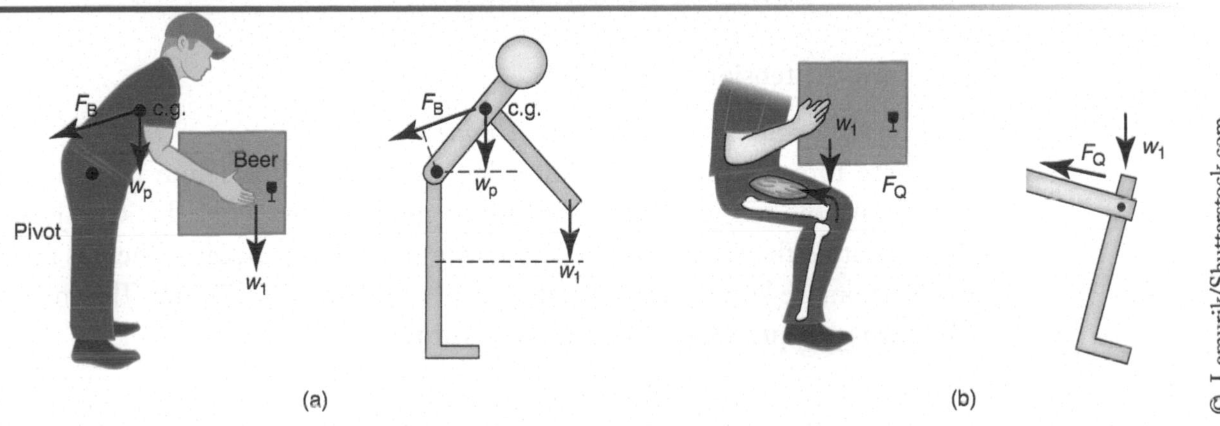

Figure 3.18 The exercise device discussed in Example 3.10. Because of its mechanical disadvantage, the Upper leg muscle must exert a considerably larger effort than it would if it could lift the load directly, F_M is shown to scale only in the equivalent mechanical system and is much larger than T.

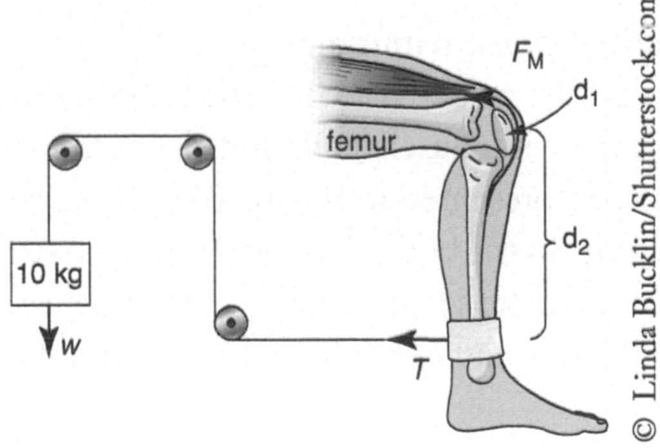

EXAMPLE 3.7

Calculate the force exerted by the extensor muscle in the upper leg when lifting a weight as shown in Figure 3.18.

Discussion: This is a statics problem since lifting the weight at a constant velocity or holding it stationary means that acceleration is zero. Torque is involved because there is a rotation about the knee joint. The tension "T" in the line is equal to the weight of the block.

Solution:

Step by step 1: Determine the tension in the extensor muscle. Remember that weight = mg. The weight of the block is:

$$Wt = mg = 10.0\text{kg} \cdot 9.80\frac{\text{m}}{\text{s}^2} = 98.0\text{N}$$

Therefore, the tension "T" is 98.0 N.

Step 2: Next recall the definition of torque: $Torque \equiv F_{\perp} \cdot d$
The three forces acting on the lower leg are those of the femur and the extensor muscle and the tension transmitted to the ankle by the wire. The knee joint is the pivot point, so the force from the femur exerts no torque. Because there is no angular acceleration, we know that this system is in rotational equilibrium. That means that,
Clockwise torque = Counterclockwise torque

$$T \cdot d_2 = F_{\text{muscle}} \cdot d_1$$

Step 3: Identify and substitute the values for T, d_1, and d_2. The distance from the knee joint and the ankle is 35 cm. Therefore, d_2 = 35.0 cm. The distance from the knee joint to the place where the force from the femur muscle is applied is 2.0 cm. Therefore, d_1 =2.00 cm.

$$F_{\text{muscle}} = \frac{98.0\,\text{N} \cdot 35.0\,\text{cm}}{2.00\,\text{cm}} = 1720\,\text{N}$$

Notice that the force exerted by the femur muscle (1720 N), is considerably larger than the weight it lifts (98.0 N). It is no wonder that people who push their bodies to extremes, such as Olympic weight lifters, must take precautions to avoid muscle pulls and torn ligaments that are a result of torque.

The ideas concerning torque presented in this section are applicable to a wide variety of systems in addition to muscles and joints in the body. Among these are martial arts, buildings, bridges, ladders against walls, simple machines (such as levers, pulleys, wheels, gears, and screws), airplane propellers, cam shafts in engines, fishing poles, and so forth. **Lab 3** should enhance your understanding of torque.

EXAMPLE 3.8

Using torque and the information in the diagram below, determine the mass of the meter stick.

Solution:

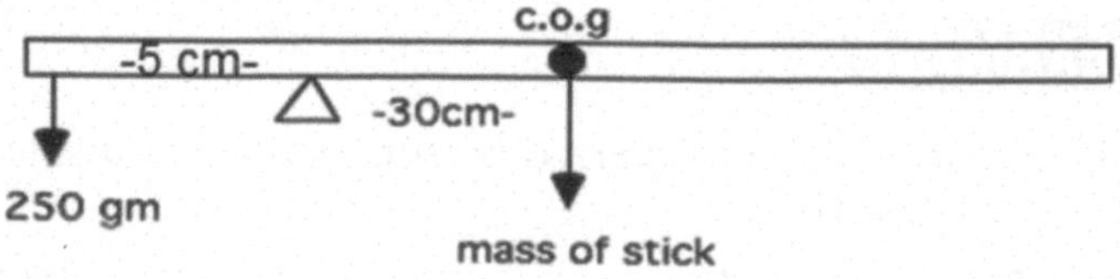

Step by step 1: Clockwise torque = Counterclockwise torque

Step 2: Mass of stick (30.0 cm) = 250 gm (5.00 cm)

Step 3: Solve for mass of stick

$$M_{\text{stick}} = \frac{250\,\text{gm} \cdot 5.00\,\text{cm}}{30.0\,\text{cm}} = 41.8\,\text{g}$$

Things to know for Section 3.4

Torque depends on the magnitude, direction and distance from the pivot that the force is applied.

Clockwise torque = Counterclockwise torque

$$Torque \equiv F_{\perp} \cdot d$$

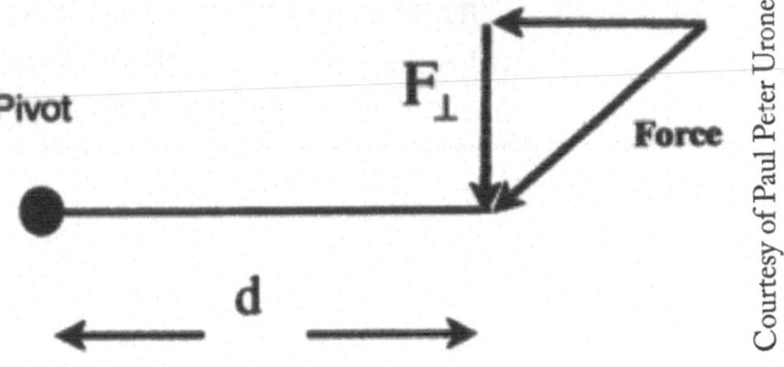

3.5 ROTATIONAL MOTION AND CENTRIPETAL FORCE

What does a person in a car feel while turning a corner? Is there any relation between what the person in the car feels and what blood in a centrifuge experiences? According to Newton's first law, bodies move in straight lines at constant speeds unless acted upon by outside forces. A force is therefore necessary to make a body follow a curved path. That force is called the *centripetal force*, and it is what the person in the car, the blood in the centrifuge, or any body moving in a curved path experiences.

First note that whenever a body moves in a curved path a rotation is involved. Figure 3.19 shows blood being centrifuged and a car rounding a curve. The center of rotation may be physical, as it is for the centrifuge, or just a point in space, as it is for the car rounding a curve. In either case the distance from the center of rotation to the object, is the radius of curvature of the path followed. Any object moving in a perfect circle at constant speed, accelerates toward the center of that circle. Because the *acceleration is directed toward the center of the circle*, it is called *centripetal acceleration*. The equation for the centripetal acceleration of an object moving in a perfect circle, at constant speed is:

$$a_c = \frac{v^2}{r} \quad \textbf{\textit{Know}}$$

From Newton's second law of motion, we know that $F = ma$. In order to calculate the *centripetal force* that produces the centripetal acceleration on the object, we substitute the centripetal acceleration equation for "a."

Therefore, the centripetal force equation is:

$$F_c = ma_c = \frac{mv^2}{r} \quad \textbf{\textit{Know}} \tag{3.7}$$

Equation 3.7 will allow us to calculate the force needed to make a body move in a curved path. where F_c is the centripetal force, m is the mass of the body, v is its speed, and r is the radius of the curve being followed. Consider the car rounding the curve. The larger the mass m of the car, the larger the force that is necessary to make it go in a curve rather than in a straight line. It is more difficult to turn corners at high speeds; this is evidenced in the equation by having v^2 in the numerator. Sharp, small radius curves are more difficult to maneuver; a small value of r in the denominator also means a large centripetal force is required to navigate the turn.

Figure 3.19 Whenever an object moves in a curved path, rotation about some point is involved. (a) A centrifuge rotates about an easily recognized pivot point. (b) A car rounding a curve rotates about a point in space to which it is not physically attached.

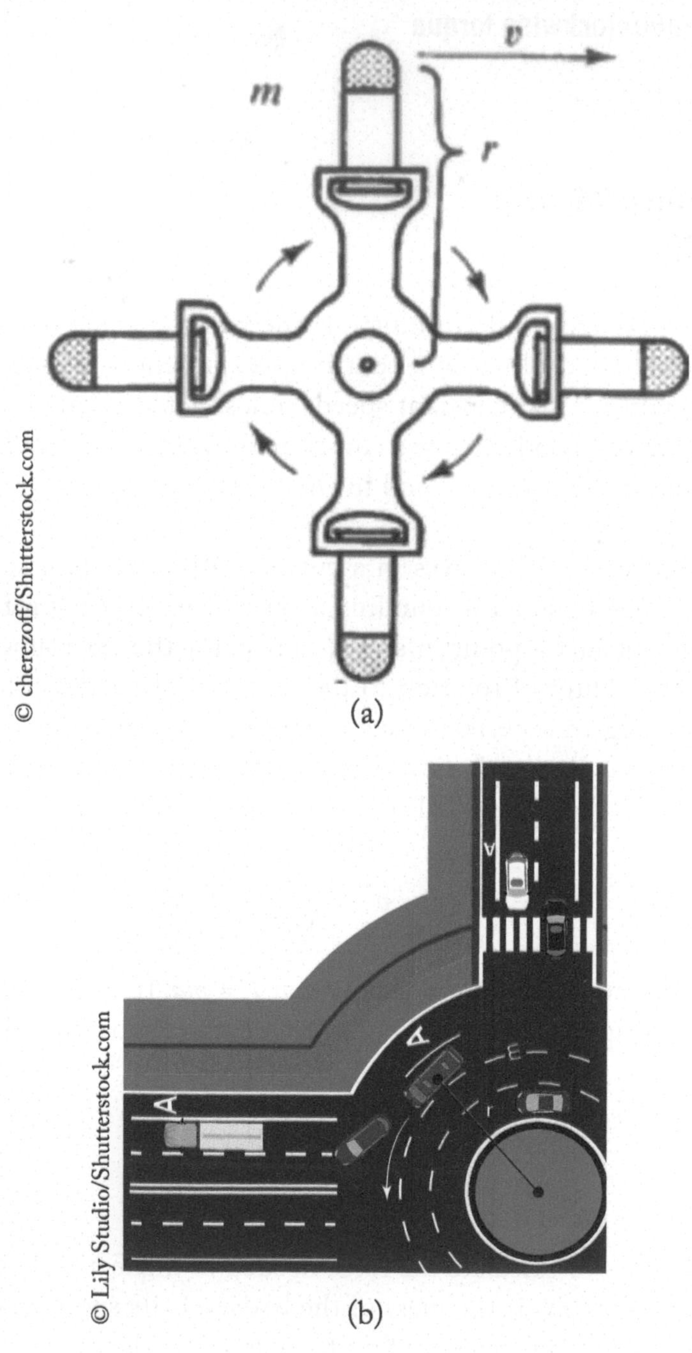

Figure 3.20 (a) When a ball is twirled at the end of a string, the centripetal force is supplied by the string in a direction perpendicular to the velocity of the ball. (b) Similarly, the door of the car supplies the centripetal force to cause the driver to move in a circular path.

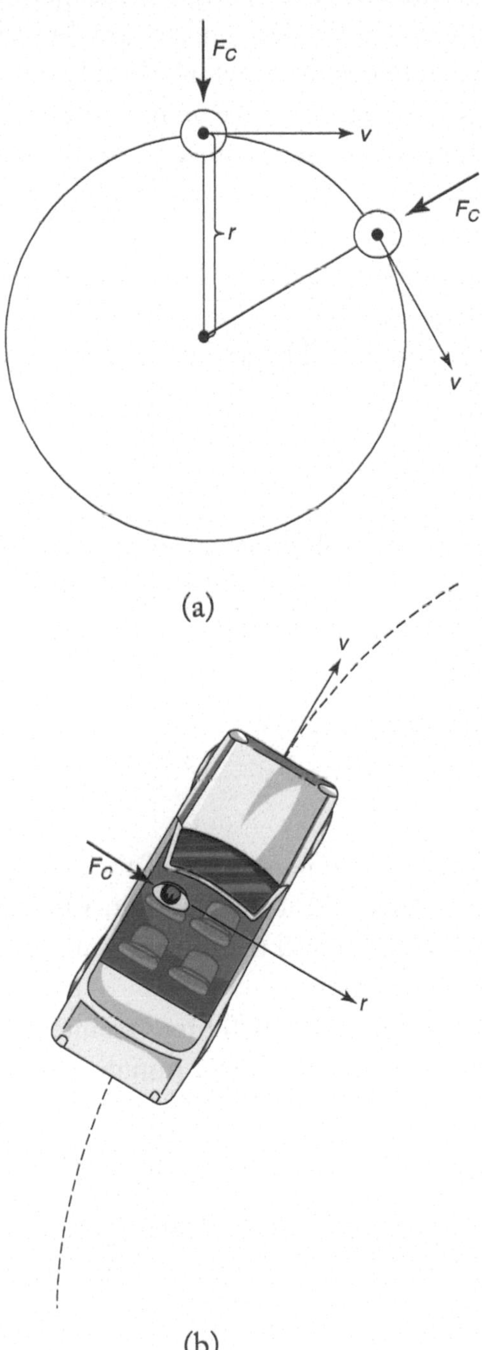

To make a body move in a circular path, a centripetal force of the magnitude given by Equation 3.7 must be exerted perpendicular to the velocity, as shown in Figure 3.20. If you have ever twirled a ball on a string, you have supplied the centripetal force through the string. If you let go, the ball flies off in a straight line tangent to the circle. Similarly, a passenger in a car feels the centripetal force acting on her when rounding a curve. If the driver makes a right turn, the door on her left pushes her toward the center of the circle, on her right. Newton's third law dictates that she will push outward on the door as the door applies an unbalance force on her *toward the center of the circle*. If the door or some part of the car does not apply an unbalance force on her, toward the center of the circle, she will not stay in the car as it rounds the turn. That would be a bad thing. All objects must follow Newton's laws of motion. They are not just suggestions.

It is perhaps easier to comprehend just how large a centripetal force is by considering how large an acceleration the centripetal force produces. In the next example, we calculate and compare the centripetal acceleration produced by an ultracentrifuge with "*g*," the acceleration caused by gravity.

EXAMPLE 3.9

An ultracentrifuge spins at 150,000 revolutions per minute and the material being centrifuged is 5.0 cm from the center of rotation. (a) Calculate the centripetal acceleration and (b) express it as a multiple of the acceleration due to gravity (in g's).

Solution:

a. The equation for centripetal acceleration is:

$$a_c = \frac{v^2}{r}$$

You may have noticed that we are not given "*v*," the linear velocity of the contents. Instead, we are told that the angular velocity is 150, 000 rev/min. Our first task is to convert 150,000 rev/min, to meters per second. You may recall that the circumference of a circle is:

Circumference of a circle = 2πr know

150,000 rev/min is the same thing as 150,000 times around a circle in 1 minute. Also note that 5.00 cm = 0.0500 m. There are 60.0 seconds in 1 minute. We will use these facts to calculate the tangential velocity, "*v*."

Step by step 1: Convert angular velocity to linear or tangential velocity.

$$v = 150{,}000\frac{rev}{1\,\text{min}}\left(\frac{2\pi \cdot 0.0500\,\text{m}}{1\,\text{rev}}\right)\left(\frac{1\,\text{min}}{60\,\text{s}}\right) = 785\frac{\text{m}}{\text{s}}$$

Step 2: Substitute the values for velocity and radius into the centripetal acceleration equation.

$$a_c = \frac{v^2}{r} = \frac{\left(785\frac{\text{m}}{\text{s}}\right)^2}{0.0500\,\text{m}} = 1.23\times 10^7\,\frac{\text{m}}{\text{s}^2}$$

b. Express as a multiple of g:

View this as a unit conversion problem where we convert m/s^2 to g's.

$$1.23\times10^{7}\,\frac{\text{m}}{\text{s}^{2}}\left(\frac{1.00\,\text{g}}{9.80\,\frac{\text{m}}{\text{s}^{2}}}\right)=1.26\times10^{6}\,\text{g}$$

This centrifuge creates an acceleration greater than 1 million times the acceleration due to gravity, meaning that the body being centrifuged feels a force more than 1 million times its weight!

Centrifuges are used to hasten settling in fluids. Ultracentrifuges, as the name implies, are the most advanced of these devices. Centripetal forces are felt by people in a car, on an amusement ride, or in an airplane when rounding curves. The centripetal accelerations to which people are subjected are considerably less than any centrifuge can exert. Race car drivers may experience accelerations as large as g when taking curves at high speeds. Fighter pilots may experience accelerations as large as 10 g, when turning their aircraft. So-called g-suits are required to keep fighter pilots from blacking out from loss of blood to the brain and to protect blood vessels in the legs from being stressed by too much blood. Astronauts are put into a large centrifuge during training so that they can experience accelerations as large as those during a launch (up to 10 g).

Objects that spin, such as wheels or centrifuges, must be designed to withstand the large forces produced. The spokes of a centrifuge, for example, must be strong enough to carry a large centripetal force to the sample holder. Flywheels in dragsters sometimes explode because the driver revs the engine too fast. This happens if the materials in the flywheel are too weak to supply the inward force needed to keep the outer rim of the flywheel moving in a circle.

Rotational motion and centripetal force are also important in the study of planetary motion or the paths of satellites. In such circumstances, the force of gravity supplies the centripetal force. Historically, this understanding was crucial to the discovery that the earth is not the center of the universe and that the planets rotate about the sun, their curved paths being caused by gravity.

Things to know for Section 3.5

Circumference of a circle: circumference $= 2\pi r$

Centripetal acceleration: $a_c = \frac{v^2}{r}$; **Centripetal force:** $F_c = ma_c = \frac{mv^2}{r}$

Things to know for Chapter 3:

$$1\,\text{N} = 1\,\frac{\text{kg}\cdot\text{m}}{\text{s}^2}$$

$$\text{weight} = \text{mg};\quad g = 9.80\ \text{m/s}^2$$

Kinetic friction: $f_k = \mu_k F_N$ Static friction: $f_s = \mu_s F_N$

Torque $\equiv F_{\perp}\cdot d$

For a body to be in equilibrium it must satisfy both of the following:

The net external force applied to the body must equal zero:
$\sum F_{\text{external}} = 0\,\text{N}$

The net external torque applied to the object must equal zero:
$\sum Torque_{\text{external}} = 0\,\text{m} \cdot \text{N}$

An equivalent statement is: Clockwise torque = Counterclockwise torque

Circumference of a circle: circumference = $2\pi r$

Newton's first law of motion: *An object at rest will remain at rest, an object in motion will remain in motion, at a constant velocity, unless an unbalanced force is applied to it.*

Newton's second law: $a = \dfrac{\sum F}{m}$

Newton's Third Law: *Whenever one body exerts a force on a second body, the second body exerts a force back on the first that is equal in magnitude and opposite in direction.*

Centripetal acceleration: $a_c = \dfrac{v^2}{r}$; **Centripetal force:** $F_c = \dfrac{mv^2}{r}$

Questions

3.1 Is it more accurate to say that forces cause motion or that forces cause changes in motion? Explain in terms of Newton's laws of motion.

3.2 Explain in view of Newton's first law why most moving objects slow down and come to rest. What condition would have to be satisfied in order to achieve perpetual motion?

3.3 What is the meaning of the word "external" in Newton's second law? How does this relate to the appropriate definition of the system of interest when using the law?

3.4 Why do you feel yourself being pushed back into the seat of an airplane on takeoff? What is actually happening in terms of Newton's laws of motion?

3.5 What are the differences between definitions and physical laws, such as Newton's laws of motion?

3.6 Give an example of a situation in which Newton's third law (action–reaction) is important.

3.7 What is the force of the gurney on the orderly in Figure 3.3? What force does the floor exert on the gurney? Which of Newton's laws of motion did you use to answer each part of this question?

3.8 How do mass and weight differ? Which is a constant property of a body and which depends on location? Is it technically correct to say that you weigh so many kilograms?

3.9 The screeching noise made by chalk on a blackboard or by car tires in a panic stop comes from a "slip-grab" motion in which the surfaces alternately slip and catch. Explain this motion using the fact that kinetic friction is less than static friction.

3.10 What is normal force and how does it affect the frictional force?

3.11 Weight, friction, and tension are all forces. Describe the distinctive characteristics of each.

3.12 A vector is any entity that has both magnitude and direction. Is torque a vector? Explain briefly.

3.13 It is sometimes very important to consider where a force is applied to a body. Give an example in which the motion of a body depends on the point at which force is applied to it.

3.14 Give a definition for torque. What are the variables on which torque depends?

3.15 Why does a person lean forward when carrying a heavy backpack?

3.16 The torques exerted by most muscles in the body are relatively small because muscles are attached close to the joints. What is an advantage to having the muscles attached close to joints? Can you think of one exceptional case in which the body exerts a very large force on an object and the muscle is attached relatively far from the joint?

3.17 Give the conditions for static equilibrium. Can a moving body be in equilibrium?

3.18 Why are forces exerted by muscles usually much larger than the forces exerted by the body on some object?

Figure 3.21 The standard technique for casting breaks in the lower leg is to immobilize the ankle and knee and to hold the knee in a flexed position (see question 3.19).

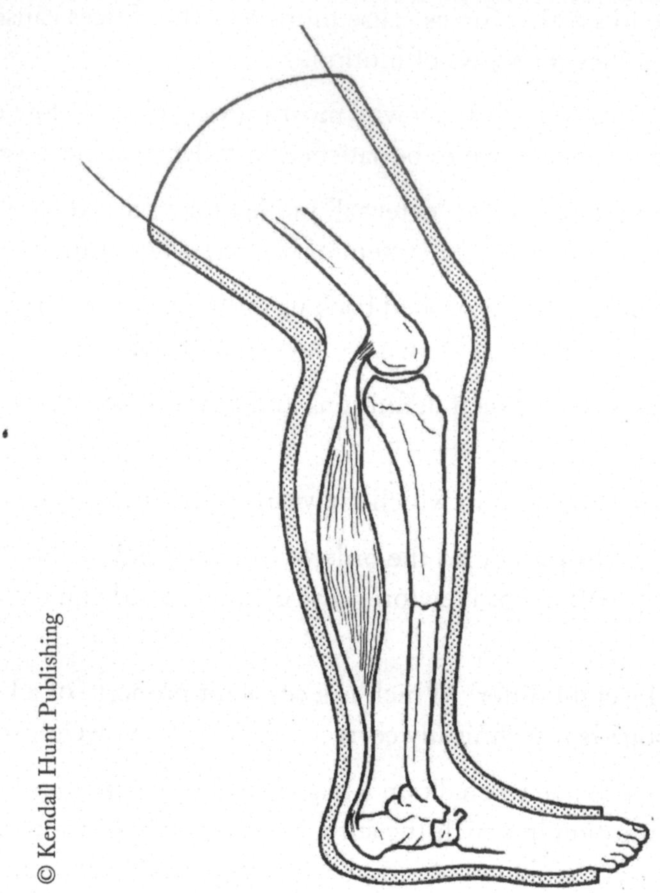

3.19 A cast placed on a broken lower leg is extended above the knee, holding the leg in a flexed position as shown in Figure 3.21. The extension above the knee prevents bowing and rotation at the break. Use the concept of torque and the fact that muscles exert more force when stretched to explain why the cast must extend above the knee and hold it in a flexed position. (Note that the same casting technique works for breaks in the forearm.)

3.20 Why must a force be exerted to cause an object to move in a curved path?

3.21 Car tires moving at 30 m/s experience forces four times greater than when moving at 15 m/s. Explain. Similarly, if a car and a truck are moving at the same speed, the truck's larger tires experience a smaller acceleration at the rim. Why is this?

3.22 Explain how a washing machine gets water out of clothes by spinning them. Use Newton's laws of motion and the concept of centripetal force in your explanation.

PROBLEMS

SECTIONS 3.1 AND 3.2

√3.1(I) What net force must be exerted on a 7.0-kg sack of potatoes to give it an acceleration of 3.5 m/s^2?

√3.2(I) A net force of 30 N is applied to an object, which is then observed to accelerate at 0.25 m/s^2. Calculate the mass of the object.

√3.3(I) Calculate the acceleration of a rocket of mass 1.2×10^6 kg if the net force on it is 2.0×10^6 N. Express in meters per second-squared and as a multiple of the acceleration of gravity, g.

√3.4(I) (a) What is the weight in newtons of a 50-kg person on earth? (b) The acceleration due to gravity on the moon is 1.67 m/s^2. What would this person's weight be on the moon?

√3.5(I) Calculate the mass in kilograms of a flea that has a weight of 5×10^{-6} N.

√3.6(I) How much kinetic friction will there be in a knee joint if the weight supported by the joint is 500 N. (Use Table 3.1 for coefficient of friction).

√3.7(I) If a steel spatula experiences a frictional force of 0.20 N when scraping against a Teflon frying pan, what is the normal force between the spatula and the pan? (Use Table 3.1 for coefficient of friction).

√3.8(I) A frictional force of 300 N is observed between two moving pieces of steel when a normal force of 1000 N exists between them. Please refer to Table 3.1 in order to determine if the steel oiled or dry?

SECTIONS 3.1 AND 3.2

√3.9(I) What is the tension in a strand of spider thread when a spider of mass 1.0×10^{-4} kg hangs motionlessly from it? Hint: The tension will have the same magnitude as the spider's weight.

3.10(II) Two children pull in opposite directions on a toy wagon of mass 8.0 kg. One exerts a force of 30 N, the other a force of 45 N. Both pull horizontally and friction is negligible. (a) Draw a diagram of the system using arrows to represent all external forces acting on it, including the force of gravity. (b) Calculate the acceleration of the wagon.

√3.11(II) An orderly exerts a horizontal force of 50.0 N on a gurney with a patient on it. The gurney and patient have a total mass of 90.0 kg. If the gurney and patient accelerate at 0.350 m/s^2, what is the magnitude and direction of the frictional force acting on the gurney?

3.12(II) A 70-kg gymnast climbs a rope. Calculate the tension in the rope (a) If the gymnast climbs at a steady speed; (b) if the gymnast accelerates upward at 0.20 m/s^2.

3.13(II) A car pulling a 1500-kg trailer produces an acceleration of 0.50 m/s^2. (a) The frictional force in the wheels of the trailer is 125 N. What force is the car exerting on the trailer? (b) What force is the trailer exerting on the car?

3.14(II) Two workers push horizontally in the same direction on a 250-kg wooden crate that is on a wooden floor. (a) What total force must they exert to start the crate moving? (b) Once moving, one worker exerts a force of 500 N and the other exerts a force of 600 N. Calculate the acceleration of the crate.

3.15(II) A large rocket has a mass at takeoff of 2.0×10^6 kg. What is its initial acceleration if its engines produce 3.5×10^7 N of thrust?

SECTIONS 3.1 AND 3.2

3.16(III) An 1100-kg car accelerates from 0 to 30 m/s in 12 seconds. If a frictional force of 200 N opposes the motion, what forces must the wheels exert backward on the pavement to cause this acceleration?

3.17(III) A basketball player crouches while waiting to jump for the ball. He lowers his center of gravity 0.30 m. When he jumps for the ball, his center of gravity reaches a height of 0.90 m above its normal position. (a) Calculate the velocity of the player when his feet left the ground. (b) Calculate the acceleration he produced to achieve this velocity. (c) What force did he exert on the floor if his mass is 110 kg?

3.18(III) (a) If half the weight of a car is supported by its drive wheels, what maximum acceleration can the car produce? Refer to Table 3.1. You may assume that the tires are rubber and the road is dry concrete. (b) What maximum backward acceleration is possible in a panic stop (tires sliding)? (c) Some cars are equipped with a device that prevents the tires from sliding while stopping. What is the maximum backward acceleration of a car equipped with this device? (d) Calculate the distance required to stop the cars in parts (b) and (c) from an initial speed of 30 m/s.

3.19(III) A freight train has two engines, each with a mass of 50,000 kg, pulling 15 freight cars, each having a mass of 40,000 kg. The first engine exerts a force of 80,000 N backward on the tracks, and the second engine exerts a force of 90,000 N backward on the tracks. (a) Calculate the acceleration of the train if the coefficient of rolling friction is 0.02. (b) What is the force in the coupling between the first and second engine? (c) What is the force between the ninth and tenth freight cars?

√3.20(I) Two movers push horizontally on a refrigerator. One pushes due north (bearing 0°°) with a force of 150 N and the other pushes due east (bearing 90°°), with a force of 200 N. Using trigonometric functions, determine the direction and magnitude of the resultant force on the refrigerator.

√3.21(I) Show that the result of the previous problem is independent of the order in which the forces are added.

√3.22(I) Using trigonometric functions, determine the magnitudes of the two forces, F_1 and F_2, which add up to the total force shown in Figure 3.25.

Figure 3.22

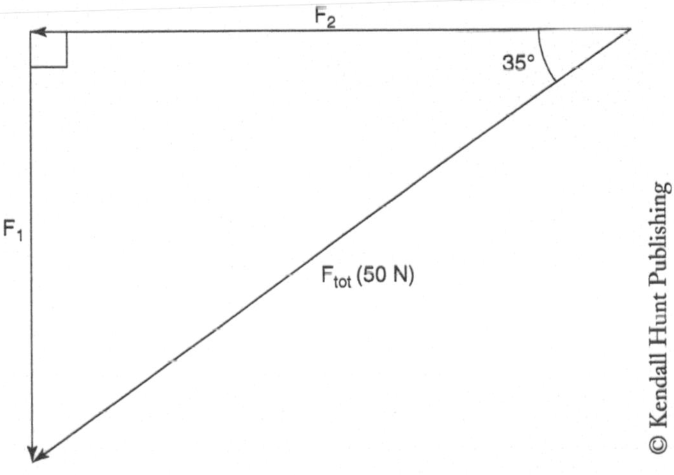

Figure 3.23

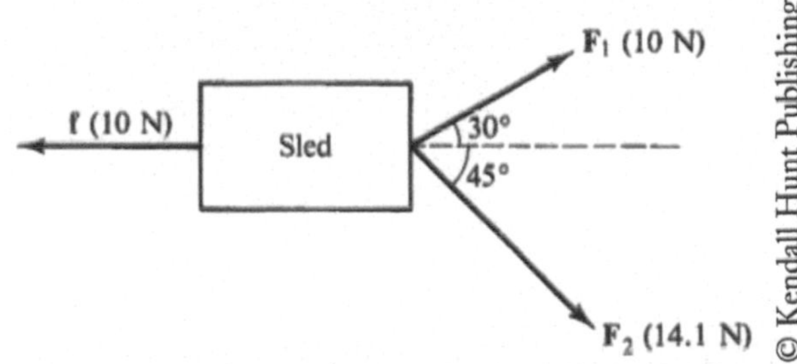

√3.23(II) Figure 3.23 represents a sled viewed from above being pulled by two children and experiencing a frictional force. Find the acceleration of the 9.0-kg sled.

3.24(II) A flea jumps by exerting a force of 1.0×10^{-5} N straight downward on the ground. A breeze is blowing into the face of the flea parallel to the ground, exerting a force of 0.50×10^{-6} N on the flea. Find the direction and magnitude of the total force on the flea given its mass to be 6.0×10^{-7} kg. (Don't neglect gravity.)

3.25(II) An extremely large tension can be created in a rope by pulling on it perpendicular to its length. This technique can be used to pull a stuck car out of the mud by tying a rope to the bumper of the car and a nearby tree and pulling on the rope as shown in Figure 3.24. Find the tension in the rope using the information in the figure. (Consider what happens when there is no motion; the applied force, T_1, and T_2 must add to zero as shown.)

3.26(II) What force is exerted on the tooth by the braces in Figure 3.25 if the tension in the wire is 25 N?

3.27(II) Two muscles in the back of the leg pull upward on the Achilles tendon, as shown in Figure 3.26. These muscles are called the medial and lateral heads of the gastrocnemius muscle. Find the magnitude and direction of the total force on the Achilles tendon. What type of movement could be caused by this force?

Figure 3.24

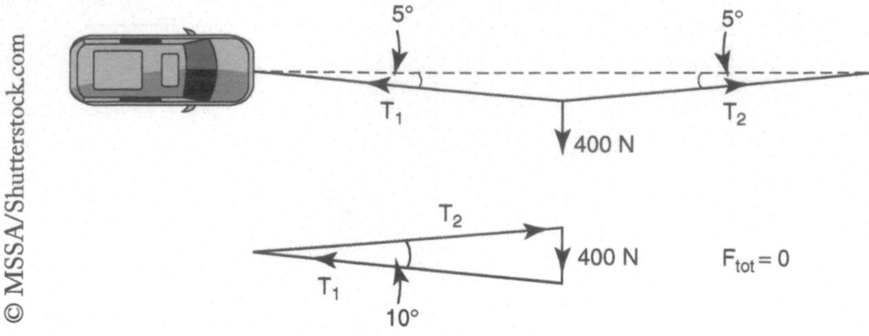

Figure 3.25

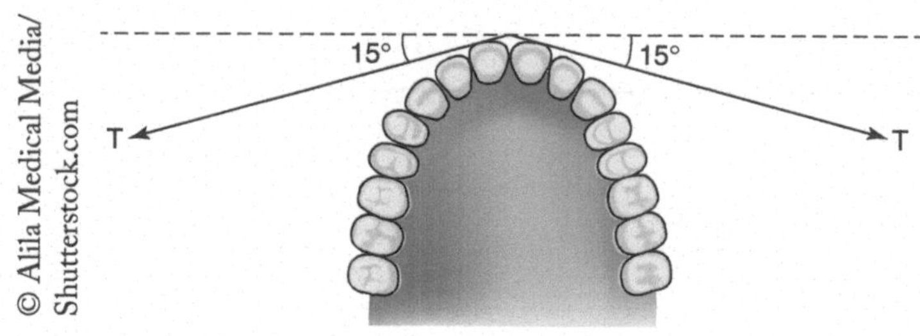

Figure 3.26

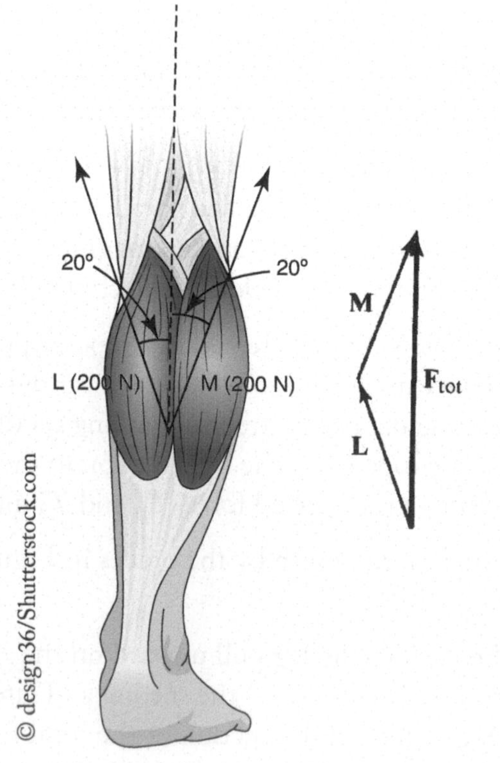

3.28(II) The upper leg muscle (quadriceps) exerts a force of 1200 N, which is carried by a tendon over the kneecap (the patella) at the angles shown in Figure 3.27. What is the direction and magnitude of the force exerted by the kneecap on the upper leg bone (the femur)?

3.29(III) When reading, the head is normally held as shown in Figure 3.28. The forces acting are the weight of the head, W, the force of the neck muscles, F_M, and the force supplied by the upper vertebrae, F_V. These forces add to zero when the head is stationary. Calculate the magnitude and direction of F_V needed to accomplish this. Does the direction of the force seem less than ideal for producing minimum strain in the upper vertebrae?

3.30(III) Three independent muscles, collectively called the hip abduction muscle, connect the upper leg bone (the femur) to the hip. Find the magnitude and direction of the total force exerted by these muscles on the upper leg using the magnitudes and directions shown in Figure 3.29.

Figure 3.27

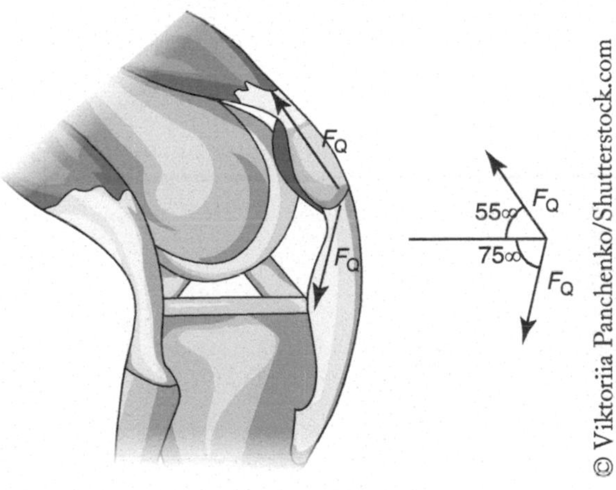

Figure 3.28

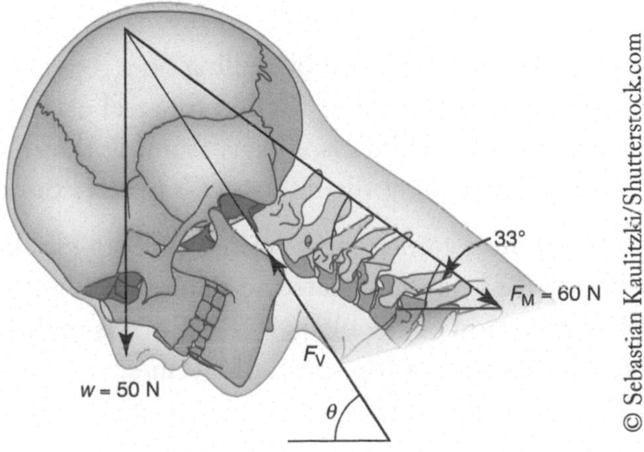

Figure 3.29

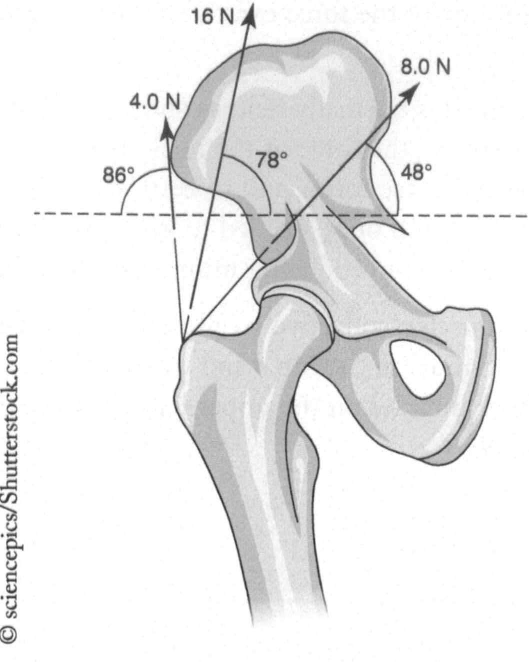

Figure 3.30

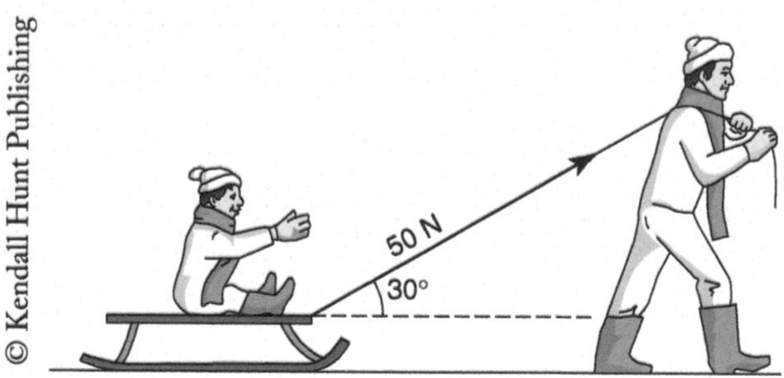

3.31(III) **(a)** A father pulls his child on a sled by pulling on a rope over his shoulder as shown in Figure 3.30. Using the coefficient of friction for steel on ice, calculate the acceleration of the sled if its mass plus the child's is 20 kg, assuming that the sled is already in motion. (b) What minimum force must the father exert to start the sled moving from rest?

3.32(III) A participant in a winter games event pushes a 50-kg block of ice across a frozen lake, as shown in Figure 3.31. (a) Calculate the minimum force to get the block moving. (b) Calculate the acceleration of the block if that force is maintained. (c) How long does it take the contestant to reach a top speed of 5.0 m/s?

Figure 3.31

SECTION 3.4

3.33(I) What force does each of four legs on a uniform bed of mass 45 kg exert on the floor?

3.34(I) A 75-kg person stands on a 5.0-kg bathroom scale. (a) How much force does the person exert on the scale? (b) What force does the scale exert on the person? (c) What force does the scale exert on the floor?

3.35(I) How much torque do you exert if you push perpendicularly on a door with a force of 30 N at a distance of 0.85 m from its hinges?

3.36(I) If a torque of 50 N-m is needed to tighten a nut on a bolt properly, what force must be exerted perpendicularly on a wrench at a point 0.20 m from the center of the bolt?

3.37(I) What force is applied by each traction set up shown in Figure 3.32?

√3.38(II) Even when the head is held erect, its center of gravity is not directly over its major point of support, the atlanto-occipital joint. The splenius muscles in the back of the neck must therefore exert a force to keep the head erect. Calculate the force they must exert, using the information in Figure 3.33, if the/mass of the head is 5.0 kg.

√3.39(II) A man stands on his toes by exerting an upward force through the Achilles tendon, as in Figure 3.34. Calculate the force in the Achilles tendon if he stands on one foot and has a mass of 80 kg. (Yes, this is another torque problem.)

3.40(II) Two children sit on a seesaw. One has a mass of 20 kg and sits 1.0 m from the pivot point. If the other child has a mass of 23 kg, how far must she sit from the pivot to balance the seesaw?

3.41(II) What force must the woman in Figure 3.35 exert on the floor with her hands in order to do a pushup?

Figure 3.32

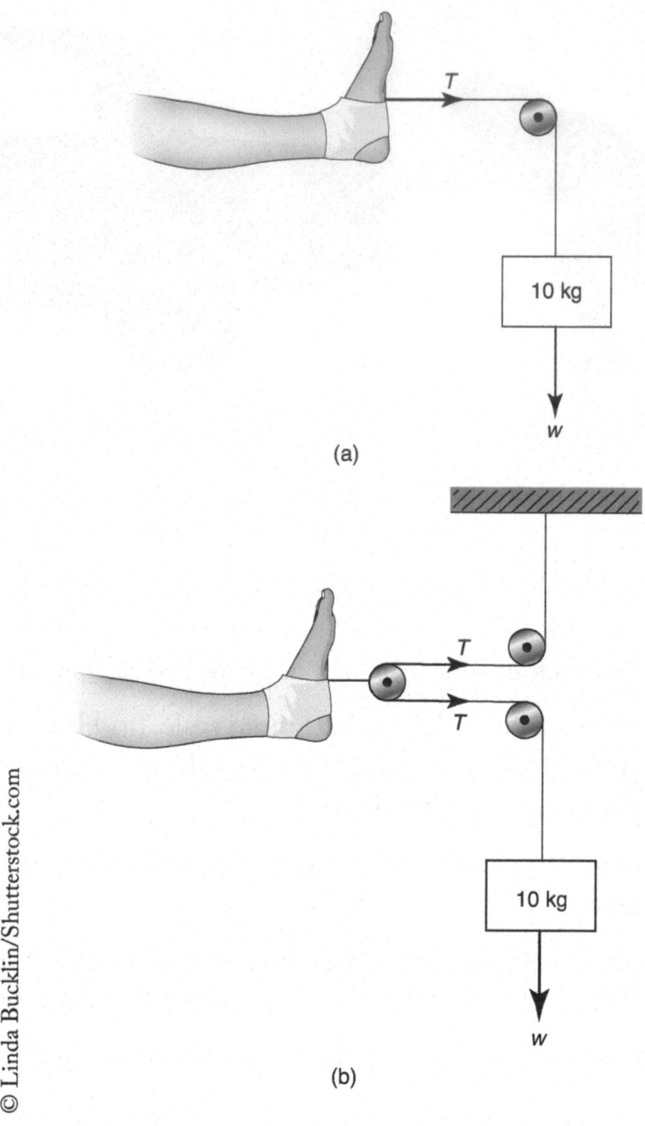

3.42(III) In Figure 3.36 a pole vaulter carries a 6.0-m long pole by pushing down with his right hand and up with his left hand. The pole is uniform and has a mass of 10 kg. Find the forces F_R and F_L.

3.43(III) One way of exercising the upper leg muscles is to lift a weight placed on the foot, as shown in Figure 3.37. Calculate the force that the muscles must exert to hold the leg in the position shown.

3.44(III) A muscle in the jaw called the masseter muscle is attached relatively far from the joint. Large forces can therefore be exerted by the back teeth. What force is exerted by the molars on the hard candy in Figure 3.38?

Figure 3.33 After B. LeVeau, M. Williams, and H. Lissner, *Biomechanics of Human Motion*, 2nd ed. (Saunders, Philadelphia, 1977), By Permission.

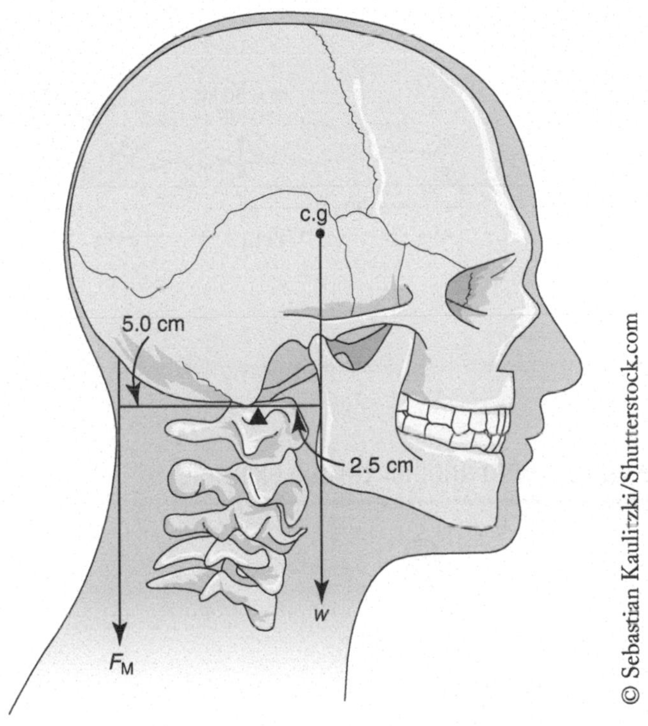

Figure 3.34 Because of torque, someone standing on their toes, applies a large force to their Achilles tendon.

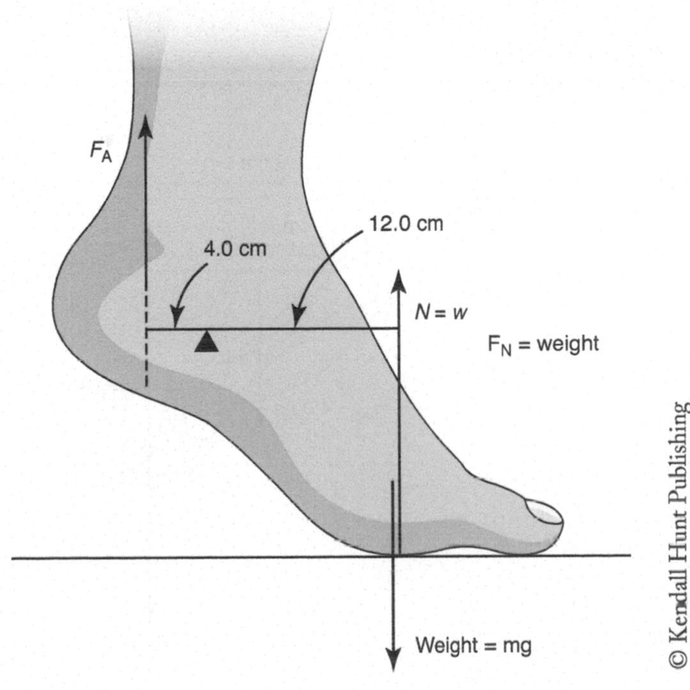

Figure 3.35 Torque is used to calculate the force that the woman must apply to the floor in order to do a push-up.

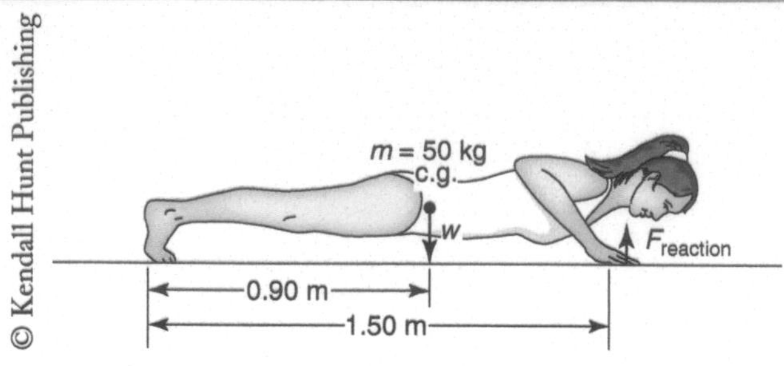

Figure 3.36 A pole vaulter with another torque example.

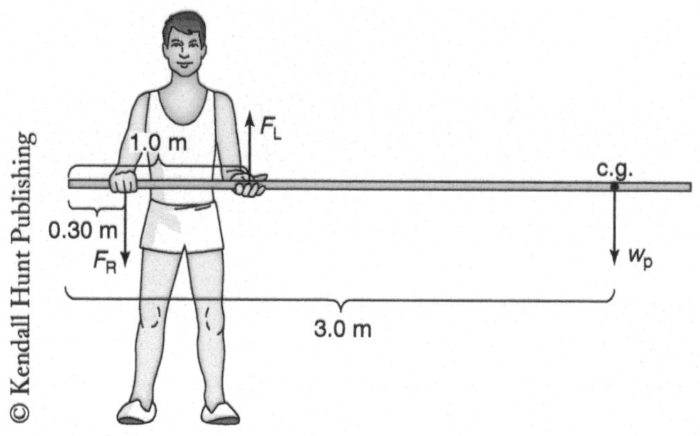

Figure 3.37 Torque is everywhere.

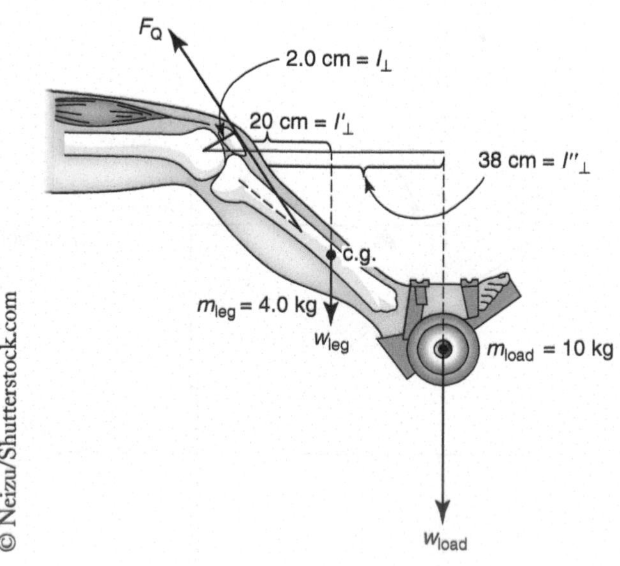

Figure 3.38 For example 3.44, another torque problem.

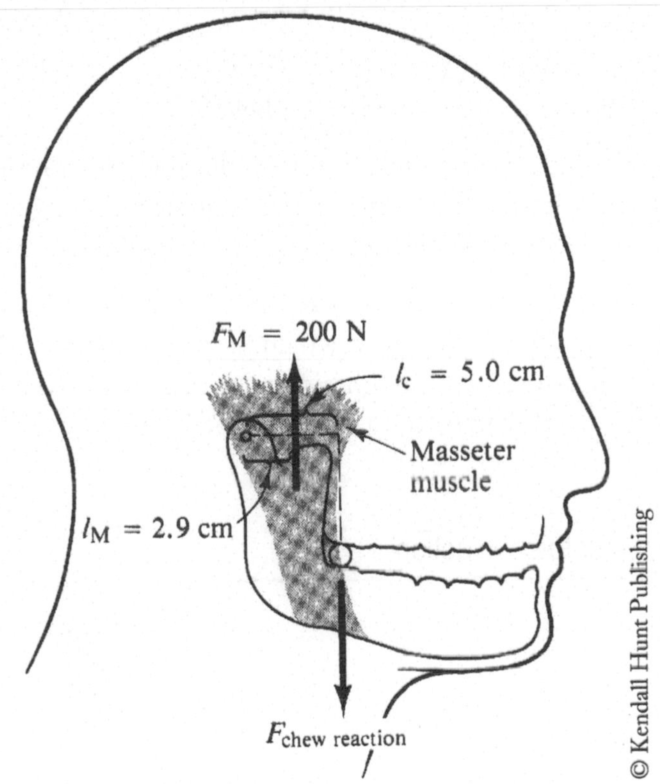

SECTION 3.5

√3.45(II) What acceleration in m/s^2 is experienced by materials 10.0 cm from the center of rotation of a centrifuge that spins at 4000 revolutions per minute?

√3.46(I) (a) How much sideways force must the wheels of a 950-kg car exert to cause it to round a corner of radius 200 m at a speed of 35 m/s? (b) What acceleration does the driver experience? Express your answer in meters per second-squared and as a multiple of *g*.

√3.47(I) (a) Calculate the acceleration of a 0.55-m-diameter car tire while the car is traveling at a constant speed of 25 m/s. (b) Compare this with the acceleration experienced by a jet car tire of diameter 1.0 m when setting a land speed record of 339 m/s.

3.48(II) What acceleration (in meters per square second) is experienced by the passengers in a jet airplane making a level turn of radius 1.0 km at a speed of 400 km/hr?

3.49(II) What centripetal force is exerted by the rope on a 1.2-kg tether ball swung in a 2.0-m-diameter circle at 45 revolutions per minute?

3.50(II) At how many revolutions per minute must an astronaut-training centrifuge rotate to produce an acceleration of 10 g if the radius of rotation is 25 m?

3.51(II) Friction between tires and pavement supplies the centripetal acceleration necessary for a car to turn. Using the coefficients of friction for rubber on concrete calculate the maximum speed at which a car can round a turn of radius 30 m (a) when the road is dry; (b) when the road is wet. (You may assume the road is level, but we know that in many instances it is banked. How can banking help?)

3.52(III) One blade on a helicopter rotates at 100 revolutions per minute and is 7.0 -m long. If the blade is uniform in construction and has mass of 200 kg, how much force must be supplied by the mechanism that attaches it to the helicopter?

3.53(III) When cyclists or runners round a curve, they instinctively lean into the turn to keep from falling over. The ideal angle at which to lean is one for which the total force of the ground on the person goes through his center of gravity, as shown in Figure 3.39. Find the angle θ for a curve of 10-m radius taken at 5.0 m/s.

Figure 3.39 Centripetal force in action.

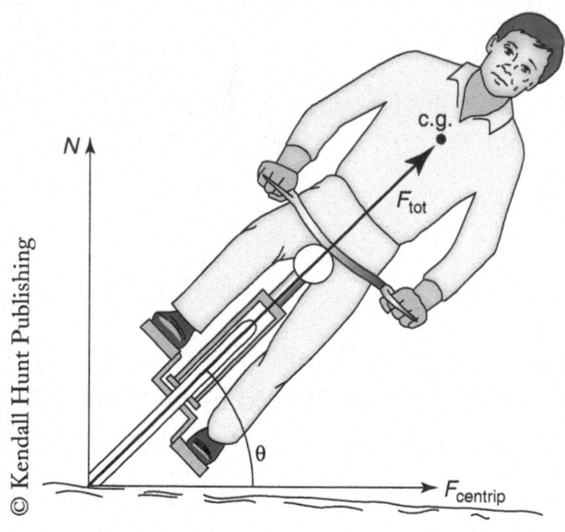

4 Work, Energy, and Power

"I think it's much more interesting to live not knowing than to have answers which might be wrong."

Richard P. Feynman

4.1 WORK: THE PHYSICIST'S DEFINITION

Energy is one of the most important concepts in physics. Energy can take many forms, all of which can be transformed from one type to another.

Sunlight is converted to chemical energy by plants by photosynthesis. The chemical energy stored in a match is converted to both visible and invisible light when it is struck.

What does a physicist mean by energy?

Energy, as we shall see, is the ability to do work. Furthermore, energy has many different forms, each of which can be converted into other forms. Sunlight is an example of electromagnetic energy. It can be converted into chemically stored energy, as in coal, oil, or food. Energy obtained from food is converted to other forms by the body, such as thermal energy to maintain body temperature, and into work done by the body. Sunlight can also be converted into electric energy by photovoltaic cells, which can in turn do work for us. All other forms of energy, such as nuclear energy, can also do work and can be converted into other types of energy (see Figures 4.1(a) and 4.1(b)).

What is most important about energy is that its total amount in a closed system is constant regardless of the types of energy involved. This experimentally determined fact, called the principle of conservation of energy, implies that energy is very basic. Moreover, the conservation of energy principle is a powerful tool in solving problems. Many of the problems in the preceding chapters, for example, can be solved relatively easily using physical principles based on energy. Say it with me, *I believe in the conservation of energy.*

Work is another concept for which the physicist's definition is not exactly the same as the one used in everyday life. The physicist means something very specific when speaking of work; some forms of human endeavor, such as studying, do not fit that definition, while others do.

The *work* done on an object by a force F is defined as the product of force in the direction of motion multiplied by the displacement ΔX of the object.

Figure 4.1a Sunlight can be converted to chemical potential energy by plants during photosynthesis.

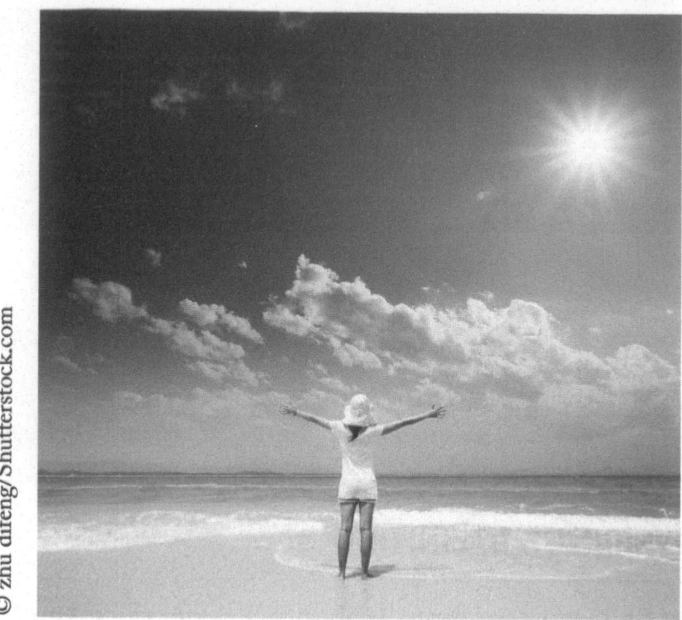

© zhu difeng/Shutterstock.com

Figure 4.1b Chemical potential energy is converted to visible and invisible light when a match is strtuck.

© Serhii Kalaba/Shutterstock.com

It is important to remember that only the component of the force in the direction of motion is used in the equation. Also note that if the object is not displaced, no work is done to it. Consider the three situations shown in Figure 4.2. In two of them no work is done since $\Delta X = 0$, even though the person involved would get tired and think of himself as having done some work. When work is done to an object, that object must gain energy. Work and energy are intimately related to one another. Work is done to or by a system only when energy is transferred into or out of that system.

(a) ΔX = 2.0 meters.

$$Work \equiv F_{\text{in the direction of motion}} \bullet \Delta X \qquad \textit{Know} \qquad \text{eq. 4.1}$$

Figure 4.2a (a) Force and motion are in the same direction, and therefore, *work is done to the box*. (b) The box is stationary, and hence, $\Delta X = 0$ m. No work is done on the box (holding, not lifting). (c) Force and motion are perpendicular; there is no force in the direction of the box's motion. No work is done on the box.

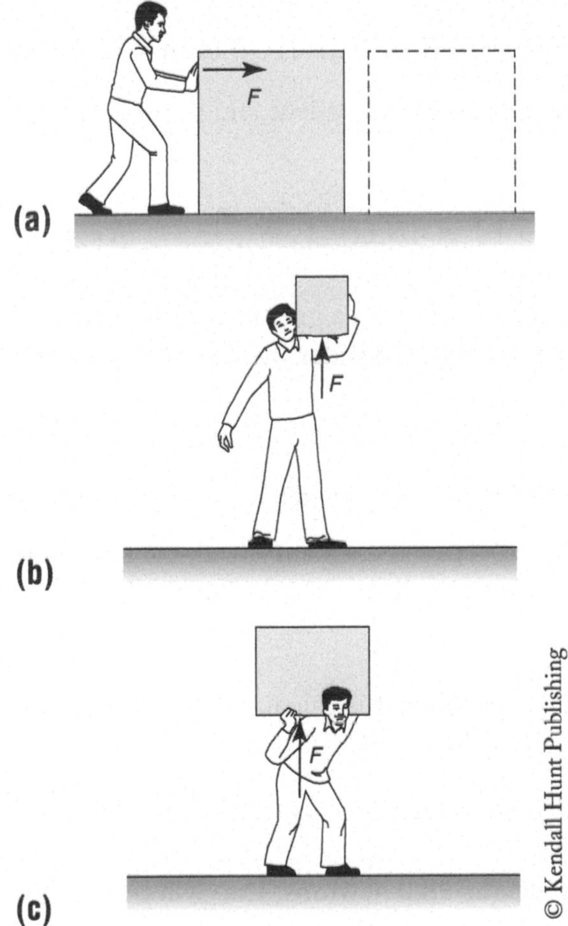

In Figure 4.2, the system is the box; only in Figure 4.2(a) is energy given to the box, in the form of energy of motion. In Figures 4.2(b) and 4.2(c), the force applied by the person merely supports the weight of the box and imparts no energy to it. The crucial point is that although the person expends energy and gets tired, none of that energy is transferred to the system of interest and hence no work is done to that system in the last two situations shown. The physicist's definition of work is designed to keep track of energy being transferred into and out of systems.

The SI unit for work is the newton-meter, which is called the joule (J) after the English physicist James Joule (1818–1889). Joule did many of the pioneering experiments establishing the equivalence of work and energy.

James Joule is credited with developing the first law of thermodynamics.

EXAMPLE 4.1

Calculate the work done by an ambulance attendant who pushes a patient on a gurney with a net horizontal force of 25 N for a distance of 3.0 meters.

Solution: The force applied is in the same direction as the object's displacement.

$$Work \equiv F\Delta X = 25.0\,\text{N} \cdot 3.00\,\text{m} = 75.0\,\text{N.m}$$

75.0 joules (J) of work were done to the patient and gurney. The patient and gurney gained 75.0 J of kinetic energy.

$$1\,\text{J} \equiv 1\text{N.m} = 1\,\frac{\text{kg}\cdot\text{m}^2}{\text{s}^2} \qquad \textit{Know}$$

Work is also done in lifting an object to a greater height because the force exerted and the distance moved are in the same direction. Consider the following example.

EXAMPLE 4.2

How much work is done in lifting a 10-kg bucket 1.2 m from the floor to a table, as in Figure 4.3?

Solution: In order to lift the groceries at constant speed, the force applied must equal to the weight of the object. It is important that you do not think that the bucket weighs 10 kg. The *mass* of the bucket is 10 kg.

Step by step 1: The weight of the bucket is calculated by multiplying its mass, by the acceleration due to gravity, "*g*."

$$Weight = \text{mg} = 10.0\,\text{kg} \cdot 9.80\,\frac{\text{m}}{\text{s}^2} = 98.0\,\frac{\text{kg}\cdot\text{m}}{\text{s}^2}$$

The bucket weighs 98.0 newtons. That will be the force required to lift the bucket.

Figure 4.3 Work is done to the bucket by the shopper in Example 4.2.

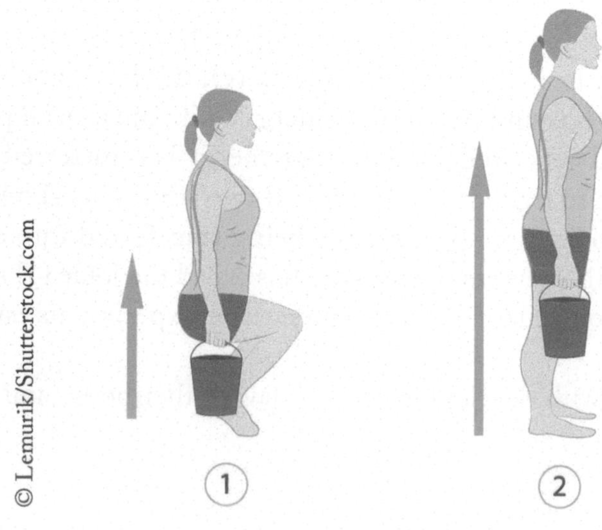

Step 2: Ignoring the slightly larger force needed to start the bucket moving, the work done to the bucket is

$$Work \equiv F\Delta X = 98.0\,\text{N} \cdot 1.20\,\text{m} = 118\,\text{N.m}$$

We see that *118 joules of work* was done **to** the bucket. Whenever work is done to an object, that object must gain some form of energy. In this example, the bucket gained 118 joules gravitational potential energy (GPE).

EXAMPLE 4.3

If the proud father in Figure 4.4 exerts a force of 10.0 newtons at an angle of 30° below the horizontal, how much work does he do in pushing the baby carriage a distance of 5.0 m?

Remember that F is the component of the force, in the direction of motion. In this example, the horizontal component of the force is labeled as F_H.

The horizontal component of the force is F_H = cos (30°) 10.0 N = 8.66 N.

Therefore, the work done to the carriage is

$$Work \equiv F\Delta X = 8.66\,\text{N} \cdot 5.00\,\text{m} = 43.3\,\text{N.m}$$

Figure 4.4 Only the component of force in the direction of the motion, F_H, produces work. See Example 4.3.

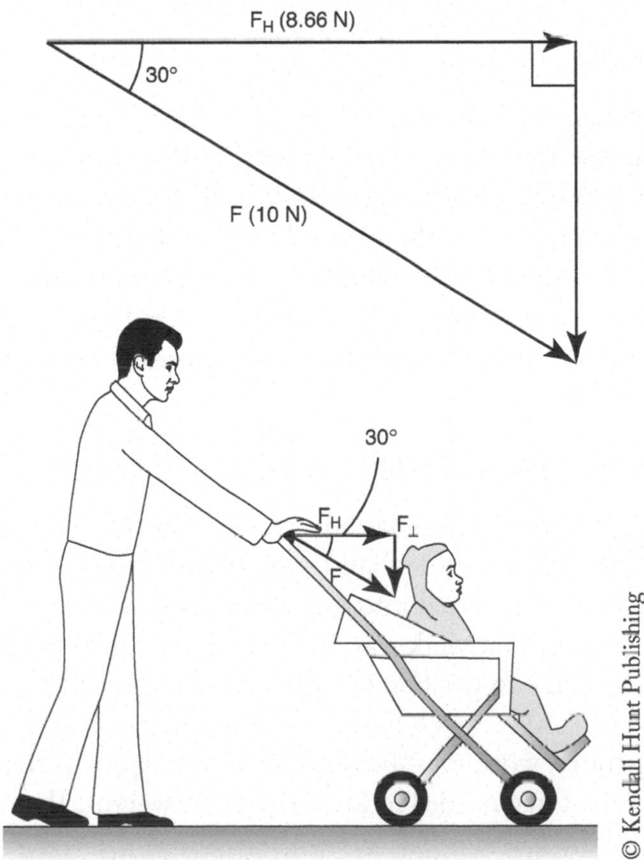

43.3 joules of work were done to the carriage and it would gain 43.3 joules of kinetic energy

****

Things to know in Section 4.1

$$Work \equiv F_{\text{in the direction of motion}} \bullet \Delta X$$

Joules are units of energy in the metric system.

$$1\text{ J} \equiv 1\text{Nm} = 1\ \frac{\text{kg}\cdot\text{m}^2}{\text{s}^2}$$

4.2 ENERGY

It's all about energy

These last two examples can be solved using principles of motion and force from Chapters 2 and 3, without considering energy. However, one of the advantages of considering energy is that calculations are often made simpler. Examples 4.4 and 4.5 require about twice as many algebraic steps if solved without considering energy. Another more important reason for considering energy is the physical insight it gives. Every imaginable physical process involves energy changing form, moving from one place to another, being turned into work, and so on. It is interesting to ask what the original source of energy is in each of the preceding examples. This brings us to the topics of gravitational potential energy (GPE) and other types of stored energy.

4.2.1 Potential Energy

An object can have energy due to its shape or position. Such energy is called potential energy (PE). A watch spring changes shape when wound up, storing the work done by the person winding it as elastic potential energy. When the spring unwinds, it does work to move the watch hands and the potential energy is transferred out of the spring. Water at the top of a hydroelectric dam has energy due to its position. As the water runs downhill through a turbine, the gravitational potential energy of the water is converted to electric energy. There are other forms of potential energy, all of them having the potential to do work. We shall be most concerned with gravitational potential energy, potential energy due to position.

4.2.2 Potential Energy: Energy Due to Position

Water at the top of a hydroelectric dam has potential energy due to its position because gravity pulls downward on it. When water moves downhill, gravity does work on it, transferring energy into the water, which can in turn be converted into any other form of energy (such as electrical). Another way of thinking about this is that work was done on the water to get it to the top of the dam; this work put potential energy due to position into the water, energy that can be extracted by letting the water go downhill.

As Figure 4.5 illustrates, work must be done to lift an object. If the box in the figure is lifted at a constant speed, then the force needed to lift it equals its weight. The distance that the box is lifted away from the center of the Earth is "*h*."

Figure 4.5 When any object is lifted to a greater height work is done since the force exerted *($F = mg$)* is in the same direction as the distance moved ($\Delta X = h$). The work done on the box is thus equal to *its gain in gravitational potential energy.*

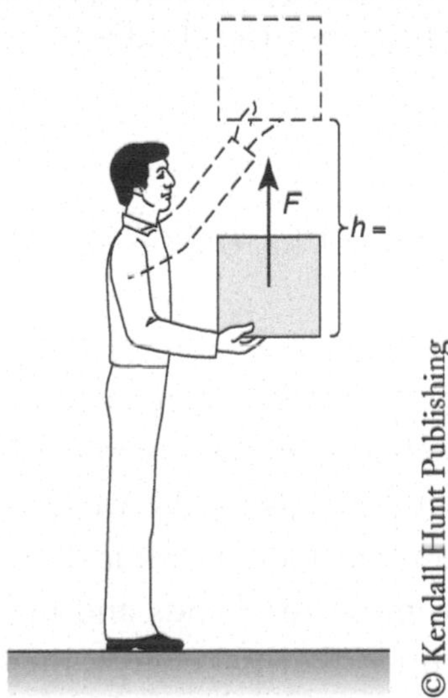

The work done to the box is $Work \equiv F\Delta X$. In this case, the force will equal the weight of the object. Remember that the weight of an object equals its mass multiplied by the acceleration due to gravity.

$$Work = F\Delta X = weight \cdot \Delta X = mgh$$

Since we are lifting the object a height "*h*," we will use the letter "h" instead of ΔX.

Therefore, in this example, the work done to the object as it was lifted can be determined from the product of the objects mass, *g*, and the vertical height that it was lifted. Because work done to an object by moving it away from the center of the earth comes up so frequently, it is given its own name. It is called *gravitational potential energy (GPE).*

$$\text{Gravitational Potential Energy } (\text{G.P.E.}) \equiv mgh \qquad \textit{Know} \qquad \text{eq. 4.2}$$

Since work is done to the box by the person, energy is given to the box in the form of energy due to position (the box is lifted so that there is no motion in the final state, and hence, no kinetic energy in that state).

If the box is dropped from its higher position, its GPE will first be converted into another form of energy, kinetic energy. As we shall see, when it strikes the floor, the kinetic energy is converted to what is known as thermal energy by frictional forces during impact. Thermal energy is the sum of the kinetic energies of the atoms and molecules of a substance. One of its effects is to increase temperature. More about kinetic energy later.

The expression for gravitational potential energy (GPE) given in Equation 4.2 is a general result, valid whenever a mass m is raised by a height, h, away from the center of the Earth. Even if energy is present in other forms, such as motion, the GPE given to an object in lifting it to a height, h, is always the product of its mass, the acceleration due to gravity (g), and the vertical distance that the object is lifted.

EXAMPLE 4.6

Calculate the work done by a 50.0-kg nurse in lifting a 25.0-kg child and herself, 0.500 m onto a table. The nurse's center of mass also rises 0.500 m (Figure 4.6).

Figure 4.6 shows the nurse lifting both the child and her own weight. Her center of gravity also rises 0.500 m.

Solution: work = gain in GPE.

$$GPE \equiv mgh = 75.0\,\text{kg} \cdot 9.80\,\frac{\text{m}}{\text{s}^2} \cdot 0.500\,\text{m} = 368\,\frac{\text{kg} \cdot \text{m}^2}{\text{s}^2}$$

The work done by the nurse lifting herself and the baby was $W = 368$ Joules. If the question had been to find only the work done on the child, then a mass of 25 kg would have been used.

Another example of gravitational potential energy (GPE) is a cuckoo clock run by weights hanging from it. The clock is wound by raising the weights, thus giving them gravitation potential energy, which is slowly used to run the clock. Work is done in running the clock because forces are exerted through some distance to move the hands and make the cuckoo sing. Most of the energy is used against friction and ends up as thermal energy, increasing the average kinetic energy of the atoms and molecules of a substance.

It can be instructive to trace energy as it changes from one form to another in various situations, whether simple or complex.

Kinetic Energy: Energy of Motion

The kinetic energy of an object is defined as follows:

$$K.E. = \frac{1}{2}mv^2 \qquad \textit{Know}$$

Any time an object is moving, it has kinetic energy. Just like gravitational potential energy, its units are

$$\frac{\text{kg} \cdot \text{m}^2}{\text{s}^2} = \text{J}$$

Other Forms of Energy

Thermal or Internal Energy. So far only two (admittedly important) forms of energy have been expressed in equation form: kinetic energy and gravitational potential energy. Even so, it is possible to calculate the amount of work or energy going into a form other than kinetic or gravitational potential energy.

Consider, for example, what happens when friction is present.

Figure 4.6

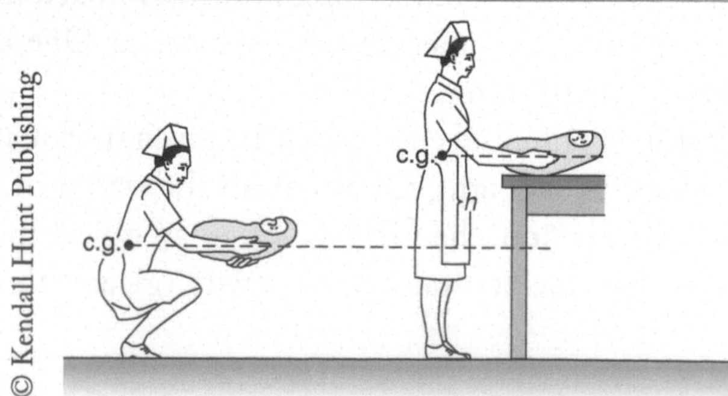

Friction always opposes motion. If someone pushes horizontally on two identical objects on level surfaces, and only one surface is frictionless, then the frictionless object ends up with a larger acceleration. For comparison's sake, let us say that the same force is exerted on each through the same distance; yet one ends up with more kinetic energy than the other. Where did the work done on the object with friction go? Part went into increasing the object's kinetic energy, by changing the speed at which it is traveling and because of the friction, part goes into the individual atoms and molecules (thermal energy) that make up the object. This causes a temperature increase of the object and the surface it is traveling on. A temperature increase corresponds to an increase in the random kinetic energies of the atoms and molecules in the object and surface. This is referred to as internal energy or *thermal energy*. Hence, the thermal energy mentioned previously is actually random internal kinetic energy, a submicroscopic phenomenon that is different from macroscopic kinetic energy, such as the overall motion of an object.

EXAMPLE 4.7 CHALLENGE EXAMPLE

An ambulance attendant pushes horizontally on a gurney with a force of 100 N, as shown in Figure 4.7. He starts from rest and pushes for a distance of 5.00 m. There is a significant but unknown amount of friction in the wheels, so that the final speed is less than if there were no friction. The final speed is 1.0 m/s and the mass of the gurney plus patient is 80.0 kg. (a) Calculate the amount of work done against friction. (b) Calculate the force of friction.

The following is a conservation of energy statement:

The total energy supplied will equal the gain in the kinetic energy of the gurney plus the work done against friction.

(a) Calculate the amount of work done against friction.

Step by step 1: $Energy_{total} = \Delta \text{K.E.} + \text{work done against friction}$

Step 2: $Work_{friction} = Energy_{total} - \Delta K.E.$

Step 3: The energy supplied equals the work done by the ambulance attendant.

$$W = F\Delta X = 100\,\text{N} \cdot 5\,\text{m} = 500\,\text{J}$$

Step 4: The gain in kinetic energy of the gurney and patient is

$$K.E. \equiv \frac{1}{2}mv^2 = \frac{1}{2} \cdot 80\,\text{kg} \cdot \left(1.00\frac{\text{m}}{\text{s}}\right)^2 = 40\frac{\text{kg} \cdot \text{m}^2}{\text{s}^2} = 40\,\text{J}$$

Step 5: $Work_{friction} = 500\text{ J} - 40\text{ J} = 460\text{ J}$

(b) Calculate the force of friction.

$$Work_{friction} = f \cdot \Delta X$$

$$friction = \frac{Work_{friction}}{\Delta X} = \frac{460\dfrac{\text{kg} \cdot \text{m}^2}{\text{s}^2}}{5.00\,\text{m}} = 92\frac{\text{kg} \cdot \text{m}}{\text{s}^2}$$

Ans. There was 92 N of frictional force.

Figure 4.7 The work done by the ambulance attendant on the gurney goes partially into kinetic energy and partially into thermal energy generated by friction in the wheels. See Example 4.5.

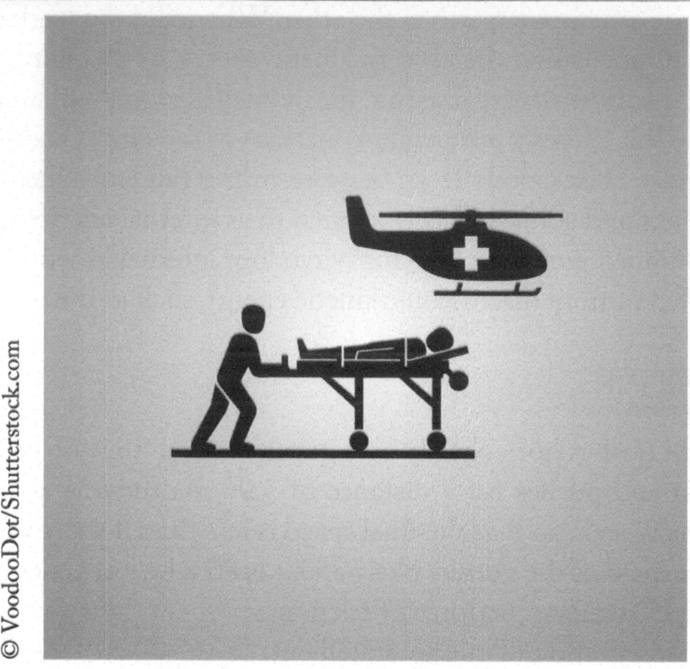

In the next chapter, it will be possible to calculate the effects of thermal energy, such as how much the temperature of the wheels in the previous example increases. At present, we shall keep track of how much work or energy went into thermal energy and not be concerned with its precise effects.

Chemically Stored Energy: Foods and Other Fuels. Many of the most common energy sources are chemical in nature. Food is but one type of chemically stored energy (a particularly unromantic way to think of a thick steak); gasoline and natural gas are others. The energy content of foods is given in units called *kilocalories* (also called dietary calories): 1 kcal equals 4186 J. Table 4.1 lists the number of kilocalories per gram for some common foods and fuels.

For foods and most fuels the process by which stored chemical energy is released is basically oxidation. In machines, the oxidation process produces thermal energy which is partially converted to work or other forms of energy. In animals, the oxidation process is complex but also can result both in thermal energy and in work being performed by the animal. If an animal consumes more food than it needs to do work or keep warm, it will convert the excess to fat, another form of stored chemical energy. Chemical energy stored in fat is used if and when there is a food deficit.

Dieting to lose weight means that the food energy intake must be less than the energy expended during ones daily activities. Increased exercise aids dieting in part because more food energy is required than normal. The body must use energy stored in fat to do work and replace food energy not consumed. Table 4.2 lists the caloric requirements of certain types of human activities.

Table 4.1 Caloric Content (kcal/g) of Common Foods and Fuels

Average carbohydrate	4.1	Ice cream, chocolate	2.22
Average protein	4.1	Lard (fat)	9.30
Average fat	9.3	Lobster, raw	0.91
		Milk, whole	0.64
Common Foods	**kcal/g**	Milk, low-fat	0.42
Apples	0.58	Oranges	0.49
Avocado	1.67	Peanuts, roasted	5.73
Baby formula	0.67	Peas	Peas
Beans, kidney	1.18	Potato, baked	0.93
Beer	0.42	Raisins	2.90
Big Mac	2.89	Shrimp, snails, raw	0.91
Butter	7.20	Sirloin, lean	1.66
Carrots	0.42	Sugar	4.00
Celery	0.15	Tomato	0.22
Cheese, cheddar	4.00	Tuna, in oil	1.97
Cheese, cottage	1.06	Wine	0.85
Chicken, roasted	1.60		
Chocolate	5.28	**Common Fuels**	**kcal/g**
Coffee, black	0.008	Coal	8.0
Cola, carbonated	0.36	Gasoline	11.4
Corn flakes	3.93	Furnace oil	10.5
Eggs	1.63	Methanol	5.2
Grapes	0.69	Natural gas	13.0
Ham, cooked	2.23	Wood (average)	4.0
Hamburger, lean	1.63		

EXAMPLE 4.8

If a person who normally requires an average of 3000 kcal of food energy per day instead consumes 4000 kcal per day, he will steadily gain weight. How many minutes of bicycling per day is required to work off this extra 1000 kcal? Use the information in Table 4.2 to help you.

Solution: Table 4.2 states that 5.7 kcal/min are used when cycling at a moderate speed.

The energy consumption rate is actually the power. Solving the power equation for time:

$$Power \equiv \frac{E}{t} \text{ therefore } t = \frac{E}{Power} = \frac{1000\,\text{kcal}}{5.7\,\frac{\text{kcal}}{\text{min}}} = 175 \text{ minutes}$$

The person in Example 4.6 will continue to use energy at a higher than normal rate for several hours after exercise. Nevertheless, the 175 minutes of exercise time is large, so to lose weight it is advisable to cut down on food as well as to exercise.

The energy requirements listed in Table 4.2 include both the energy to produce thermal energy (body heat) and the energy used to do work. No animal or machine can convert 100% of its food or fuel energy into useful work. Some energy is always wasted because no animal or machine is 100% efficient. Efficiency will be discussed further in Section 4.4.

Electric Energy. There are many examples of energy being stored in the form of electric energy. One device that stores pure electric energy is called a capacitor. Many electronic instruments, such as heart defibrillators, use capacitors to store energy. Fibrillation is a potentially fatal malfunction of the beating of the heart. Electric energy stored in a large capacitor in the defibrillator is used to cause an electric current to pass through the patient's heart to stop fibrillation, that is, to defibrillate the heart. Ironically, electric current through the heart can also cause fibrillation, depending on the size of the current.

Natural electrical activity in the body also involves capacitors. This will be discussed in Chapter 13.

Some forms of electric energy are hidden more subtly than in capacitors. Although the energy stored in the wound mainspring of a watch is usually referred to as elastic potential energy, it is actually electrical in character, as surprising as that may seem. The spring is distorted from its natural shape when it is wound, causing the atoms and molecules to be either pushed closer together

Table 4.2 Energy Consumption Rate, Power, for Various Activities[a]

Activity	Energy Consumption Rate (kcal/min)	Watts
Sleeping	1.2	83
Sitting at rest	1.7	120
Standing relaxed	1.8	125
Sitting in class	3.0	210
Walking slowly (4.8 km/hr)	3.8	265
Cycling (13–18 km/hr)	5.7	400
Playing tennis	6.3	440
Swimming breaststroke	6.8	475
Ice skating (14.5 km/hr)	7.8	545
Climbing stairs (116/min)	9.8	685
Cycling (21 km/hr)	10.0	700
Playing basketball	11.4	800
Cycling, professional racer	26.5	1855

Source: Adapted form J. Cameron and J. Skofronick, ***Medical physics*** (Wiley-Interscience, New York, 1978), by permission.

[a] Normal 76-kg male.

or pulled farther apart than they would normally be. Strong electric forces act between the atoms and molecules trying to restore them to their normal positions. The spring thus has the ability to do work—it contains energy. Therefore, the spring's energy is electrical in nature because the forces between atoms and molecules are electrical.

When any elastic medium is distorted from its natural shape, energy is stored in it in just the same manner as described for watch mainsprings. The elastic potential energy due to shape mentioned above is really electrical in nature. Rubber bands, guitar strings, and crossbows are a few more devices in which energy is stored in a distorted medium. When examined in greater depth, many but not all forms of stored energy are actually electrical in nature.

Light: Energy in a Hurry to Get Somewhere Light is a very pure form of energy that travels extremely fast, as fast as anything can. As will be explained in Chapter 16, light is an electromagnetic wave that travels as bundles of energy, called photons. Even though light is moving, its energy is not the same as the kinetic energy defined earlier. We will discuss the energy contained in photons of light, later in this text.

Sunlight is the original source of almost all energy on earth. (Nuclear energy is a major exception.) Coal, for example, contains stored chemical energy obtained from sunlight by photosynthesis in plants. Hydroelectric energy is possible because sunlight provides energy to evaporate water into the atmosphere, allowing rain to fall and be collected in reservoirs. Sunlight is used directly in a growing number of applications, from heating buildings to cooking food. Conversion of sunlight into electricity by photovoltaic cells is becoming more common as the price of the cells makes their installation more cost effective. We will study the photoelectric effect later in this text.

Visible light is only one small part of the electromagnetic spectrum, which runs from radio to x rays and beyond. Many parts of the electromagnetic spectrum are used to transmit energy, such as microwaves for cooking and deep heating and lasers for welding detached retinas. The electromagnetic spectrum will be studied in depth in later chapters.

Conversion of Mass to Energy

In 1905, Albert Einstein published, among other things, his theory of special relativity. In it, he concluded that mass and energy are interchangeable. The quantitative mass energy relationship is given in his famous equation:

$$E = mc^2 \qquad \text{eq. 4.3}$$

E is the amount of energy that is obtained by converting an amount of mass "m" to energy, and "c" is the speed of light (3.00×10^8 m/s). c^2 is a large number, and so a large amount of energy can be obtained from the conversion of a small amount of mass. (A closer approximation to the speed of light in a vacuum is 2.99792×10^8 m/s. We will usually round this number to 3.00×10^8 m/s).

The large-scale conversion of mass into energy during a process known as nuclear fusion, is the "secret" behind the sun's ability to produce the tremendous amount of energy for the 4.5 billion years that it has been shining. Every second, the sun converts 4 million tons of matter to sunlight and radiates it out to the universe.

The conversion of mass into energy is also the source of energy from nuclear power plants and atomic bombs. Not only can mass be converted to energy but energy can also be converted into matter. Mass and energy are actually the same thing in different forms.

EXAMPLE 4.9

Using $E = mc^2$,

a. Calculate the amount of energy obtained from converting 1.00 g of mass into energy. Note that 1.00 g = 1×10^{-3} kg.

$$E = mc^2 = 1.00\times10^{-3}\text{kg}\cdot\left(3.00\times10^{8}\frac{\text{m}}{\text{s}}\right)^2 = 9.00\times10^{13}\,\text{J}$$

b. An adult requires approximately 2000 kcal of energy from food per day in order to maintain their body weight. Convert 2000 kcal to joules.

Step by step 1: 1 kcal = 1000 calories. Therefore,

$$2000\text{ kcal}\cdot\frac{1000\text{ cal}}{1\text{ kcal}} = 2.00\times10^{6}\text{cal}$$

step 2: Convert from calories to joules. Note that 4.19 J = 1 calorie. Therefore,

$$2.00\times10^{6}\text{cal}\cdot\frac{4.19\text{J}}{1\text{cal}} = 8.38\times10^{6}\,\text{J}$$

c. Compare the nuclear energy contained in 1 g to the daily energy requirements of an average adult.

$$\frac{\text{Nuclear energy from 1 gram}}{\text{daily adult energy requirement}} = \frac{9.00\times10^{13}\,\text{J}}{8.38\times10^{6}\,\text{J}} = 1.07\times10^{7}$$

If humans ran on nuclear fusion, we could feed approximately 10.7 million people with a 1 g mass.

It is useful at this point to sort the various forms of energy into broad categories. The way in which these categories are defined depends on the depth to which energy is studied. For our level of treatment, the following is a reasonable classification scheme:

Kinetic energy: $K.E. \equiv \frac{1}{2}mv^2$

The energy of motion of macroscopic bodies, where m is the mass of the object and v is its speed.

Thermal energy: The kinetic energy of atoms and molecules in a body due to their random motion; also called internal energy. (Thermal energy is related to temperature, as discussed in more detail in the next chapter.)

Potential energy: The energy due to position, to gravity, or to shape or deformation. Also energy stored chemically, thermally, or electrically.

Light: Energy contained in electromagnetic photons.

Mass: It can be converted to energy and energy can be converted to mass.

Things to know for Section 4.2

Metric units of work are named joules.

$$\frac{kg \cdot m^2}{s^2} = J$$

Units of energy: joules and calories.

$$4.19 \text{ J} = 1 \text{ cal}$$

$$Work \equiv F\Delta X \quad G.P.E. \equiv mgh \quad K.E. \equiv \frac{1}{2}mv^2$$

4.3 CONSERVATION OF TOTAL ENERGY: THE CONCEPT OF A SYSTEM REVISITED

Imagine that nine-year-old Cathy has a set of 200 building blocks called Legos. Cathy's friend Karen, comes over to her house with her own set of 200 Legos. At the end of the day, Karen goes home. While picking up, Cathy and her mom can only find 196 Legos. There are four Legos missing. Cathy and her mom eventually find 3 of the missing Legos and they wonder where the fourth one is. Maybe Karen took it home by mistake or maybe it is still in the house—they are not sure. But the thing that Cathy and her mom are sure of is that the Lego is someplace. I want you to take a moment to think about how they know that the Lego is someplace. Cathy and her mom believe in the conservation of Legos. Sometimes the Legos are hard to keep track of. Sometimes they might even get out of the house, but one thing that Cathy and her mom know for sure is that the Lego has to be somewhere. Energy is like that. Energy is conserved. Sometimes it goes from one system to another, but we know that it went somewhere.

Total energy is defined as the sum of all forms of energy in a system: kinetic, potential, chemical, and so on. Experiments have shown that the total energy in a closed system is always conserved. This experimental fact is known as the principle of conservation of energy (sometimes called conservation of mass energy, since mass and energy are interconvertible). Because this principle is found to apply universally, it is of fundamental importance. Conservation of energy is a central theme of this text and allows us to solve problems that otherwise would be very difficult; it is not impossible to solve. It implies that energy can be neither created nor destroyed; it can only change form.

In order to understand fully the conservation of energy principle and to apply it correctly, we must again consider what is meant by a system, in particular a closed system. As was seen in the previous sections, energy can be transferred from one system to another if one system does work on the other. Simply put, a system is closed if no energy enters or leaves it by any method. This means, for example, that no work is done on or by the system and no energy enters or leaves it.

Consider the two systems shown in Figure 4.8(a). Neither of these systems is closed because energy is entering or leaving them. The person does work on the box to lift it; some of his energy is transferred to the box. If the two systems are combined into a single system, as in Figure 4.8(b), a closed system is formed, assuming that the thermal energy generated by the person does not leave the system.

Figure 4.8 (a) Energy is transferred from system A to system B when the box is lifted. Neither system is closed. (b) A larger single system containing both the person and the box is closed. Energy is transferred between components inside the system; none enters or leaves the system.

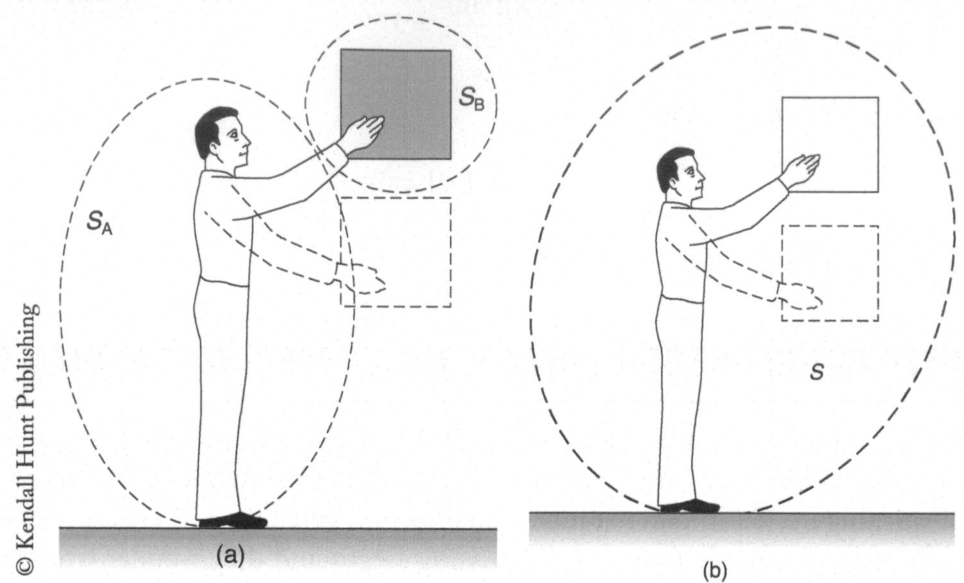

If energy cannot get out of a system, it is said to be closed. Energy is then transferred only between components inside the system. In general, energy is conserved if a large enough system is considered.

The *conservation of energy principle* can be written in equation form as

$$\text{K.E.} + \text{P.E.} + \text{all other forms of energy} = \text{Constant} \qquad \textit{Know} \qquad \text{eq. 4.4}$$

Say it with me: *I believe in the conservation of energy!*

The conservation of energy principle is very useful in solving problems. Only initial and final conditions need be considered; all intermediate events can be ignored.

This is a tremendous simplification. Consider the following example.

EXAMPLE 4.10

Figure 4.9 shows a 20-kg sled that has gotten away. The sled starts from rest and goes straight down the hill. How fast will the sled be going at the bottom of the hill if it loses 100 m in altitude and friction is negligible?

Solution: The total energy of the sled is the same at the top and bottom of the hill no matter how many bumps it goes over. When the sled gets to the bottom of the hill, it will not have any GPE remaining. Since there is no friction, energy cannot "leak" out of the system. That is, other forms of energy, OE, is 0 joules. Because we believe in the conservation of energy, we know that the total energy at the top of the hill, must equal the total energy at the bottom of the hill. So where did the energy that was stored as gravitational go to?

Equation 4.5 states that,

Figure 4.9 When the sled goes downhill its potential energy due to gravity is converted to kinetic energy and thermal energy. See Examples 4.10 and 4.11.

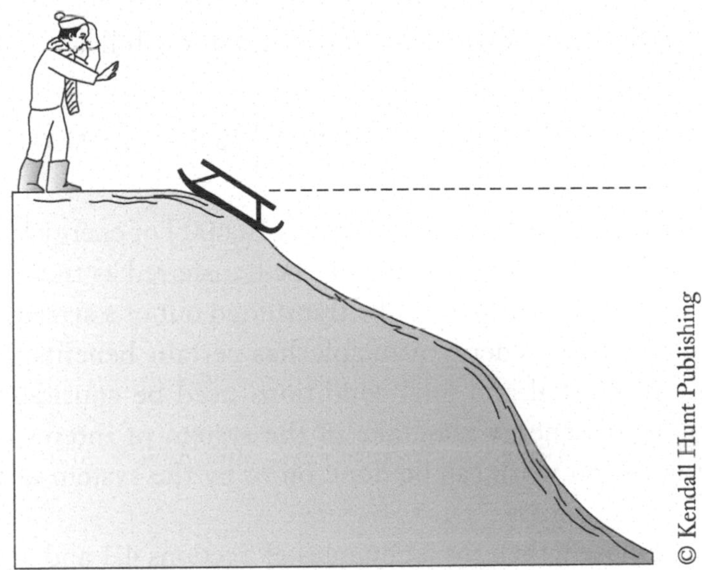

$KE + PE + OE = \text{constant}$, where OE is other forms of energy.

Using conservation of energy and the fact that there is no friction or other forms of energy involved, the GPE at the top of the hill must equal the kinetic energy at the bottom of the hill.

Step by step 1: GPE at the top of the hill = K.E. at the bottom of the hill

$$mgh_{Top} = \frac{1}{2} mv^2{}_{bottom}$$

Step 2: Solve for v, the velocity at the bottom of the hill and then substitute the values for g and h.

$$v = \sqrt{2gh} = \sqrt{2 \cdot 9.8 \frac{m}{s^2} \cdot 100m} = 44.3 \frac{m}{s}$$

EXAMPLE 4.11

Let's suppose that the sled in the previous problem is observed to have a speed of only 30.0 m/s at the bottom of the hill. How much energy is missing and where do you think that it went?

In Example 4.10, the sled had a mass of 20.0 kilograms and was 100 meters in altitude above the bottom of the hill. The total energy of the sled at the top of the hill is in the form of GPE.

Step by step 1: Calculate the GPE of the sled at the top of the hill.

$$G.P.E. \equiv mgh = 20.0\,\text{kg} \cdot 9.80 \frac{\text{m}}{\text{s}^2} \cdot 100\,\text{m} = 19{,}600\,\text{J}$$

Step 2: Given that the velocity of the sled at the bottom of the hill is 30 m/s and know that its mass is 20 kg, we can calculate the kinetic energy of the sled at the bottom of the hill.

The *kinetic energy of the sled at the bottom* of the hill is

$$K.E. \equiv \frac{1}{2}mv^2 = \frac{1}{2}\cdot 20.0\,\text{kg}\cdot\left(30.0\frac{\text{m}^2}{\text{s}}\right) = 9000\,\text{J}$$

At the bottom of the hill, the sled apparently has 10,600 J of energy less than it had at the top of the hill. Friction caused the 10,600 J of energy to be transferred as thermal energy. Thermal energy is one of the ways that energy can "hide" or be transferred out of a system.

Using the conservation of energy principle has certain benefits. It can be applied to any closed system. Only the initial and final conditions need be considered. One must, of course, examine the various forms energy may take in the system of interest. It is important that the system be closed; that is, no work can be done on or by the system and no energy can enter or leave the system.

If the system is not closed, then the techniques of Sections 4.1 and 4.2 are applied. The amount of energy transferred from one system to another and the forms the energy takes are again considered. Work done to or by the system may also need to be considered explicitly if the system is not closed, depending on what information is wanted (see the examples in Section 4.2).

Things to know for Section 4.3

The conservation of energy principle:

$$\text{K.E.} + \text{P.E.} + \text{all other forms of energy} = \text{Constant}$$

I believe in the conservation of energy.

4.4 POWER AND EFFICIENCY

4.4.1 Power: The Time Rate of Doing Work

Power is a word used in many contexts, usually with a meaning that is the same as or analogous to that used in physics. The power P is defined as the amount of work done per unit of time:

$$Power \equiv \frac{Energy}{time} \qquad Know \qquad \text{eq. 4.5}$$

Power is the rate at which energy is used or expended.

$$\text{Units: } 1\,\text{W} \equiv 1\frac{\text{J}}{\text{s}} \qquad Know$$

A car with a large engine is a good example of the meaning of power. A large engine can expend energy at a rapid rate, and it uses gasoline at a faster rate than a less powerful engine. The engine can accelerate the car to a high speed in a short time. It does a large amount of work and imparts kinetic energy to the car very quickly.

It is also able to climb hills quickly, thereby increasing GPE at a rapid rate. All this is logical and consistent with the definition of power.

The SI unit for power is the joule per second, which is called the watt (W) after James Watt (1736–1819).

Watt himself defined a unit for power, the horsepower, which is still in use today. He was interested in describing the rate at which his steam engines could do work and defined his unit in terms of the most common source of power in that era, the horse. One horsepower (hp) is now defined to equal 746 W.

$$1\,\text{hp} \equiv 746\,\text{W}$$

The watt is a familiar unit. The power of all light bulbs and many other electrical devices are rated in watts. People, as well as light bulbs, cars, and horses, are capable of consuming energy and doing work. In the next example, we see how the power output of a person compares with that of a horse.

EXAMPLE 4.12

(a) Calculate the amount of power generated by a 60.0-kg person climbing a 2.00-m-high flight of stairs in 7.00 seconds. Give your answer in watts and horsepower. (b) Repeat the calculation for the same person running up the stairs in 2.00 seconds.

Solution: (a) Work is done to increase the GPE of the person, as shown in Figure 4.10. The power generated for that purpose alone is then

$$Power \equiv \frac{Energy}{time} = \frac{\Delta G.P.E.}{\text{time}} = \frac{mgh}{t}$$

$$Power = \frac{60.0\,\text{kg}\cdot 9.8\,\frac{\text{m}}{\text{s}^2}\cdot 2.00\,\text{m}}{7.00\,\text{s}} = 168\,\frac{\text{J}}{\text{s}} = 168\,\text{W}$$

Converting to horsepower:

$$168\,\text{W}\cdot\frac{1\,\text{hp}}{746\,\text{W}} = 0.225\,\text{hp}$$

Figure 4.10 If a person has an initial horizontal velocity, only a vertical force will be needed to climb stairs. The work done is equal to the gain in gravitational potential energy, *mgh,* whenever the *vertical* height is increased. The amount of work done is independent of the path and the time it takes to climb the stairs. See Example 4.10.

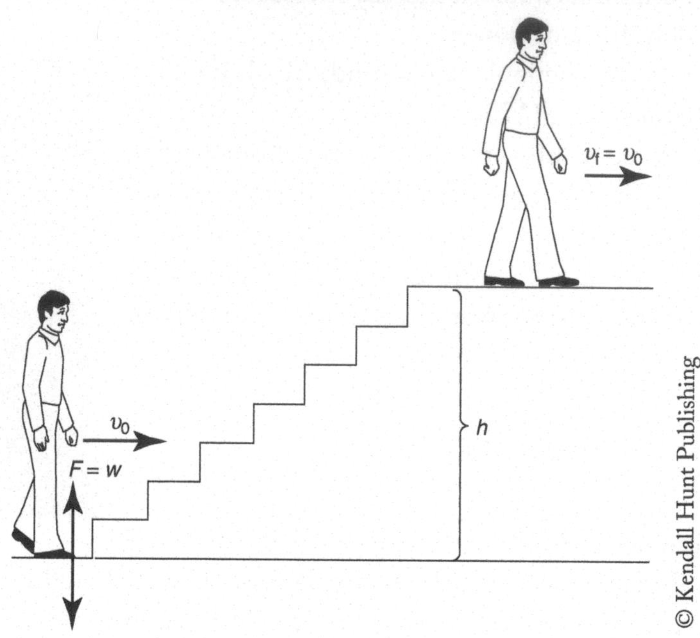

4.12(b) When the person runs up the stairs, the work done is exactly the same, but it is done in a less time.

$$Power = \frac{60.0\,\text{kg} \cdot 9.8\,\dfrac{\text{m}}{\text{s}^2} \cdot 2.00\,\text{m}}{2.00\,\text{s}} = 588\,\frac{\text{J}}{\text{s}} = 588\,\text{W}$$

Converting the power from watts to horsepower,

$$588\,\text{W} \cdot \frac{1\,\text{hp}}{746\,\text{W}} = 0.788\,\text{hp}$$

which is a considerably larger power output—in fact, a significant fraction of a horsepower—78.8% of a horsepower! It is no wonder that running upstairs is so stressful and causes the body to utilize its available energy very quickly. People with heart problems are warned that climbing stairs is one of the most stressful acts that they can perform.

Efficiency: the Real World

A certain amount of energy is always wasted when people, animals, machines, or electrical devices, for example, do work. The word "wasted" implies a value judgment regarding the various forms energy may take. Energy is never lost, but some energy always goes into a form that does not do any useful work. Light bulbs, for example, turn much of the electric energy they consume into thermal energy rather than light. Similarly, car engines produce considerable thermal energy and do not convert all the chemical energy they consume into useful work. The concept of efficiency relates to these facts. The more efficient a device is, the more work it will do with a given energy input. Efficiency is defined as

$$Efficiency \equiv \left(\frac{\text{Work or useful Energy out}}{\text{Work or Energy input}}\right)100\% \qquad \textit{Know} \qquad \text{eq. 4.6a}$$

We can equivalently define efficiency as the ratio of the useful power output to the power input of a device.

$$Efficiency \equiv \left(\frac{\text{Useful Power out}}{\text{Power input}}\right)100\% \qquad \textit{Know} \qquad \text{eq. 4.6b}$$

An efficiency of 100% indicates a perfect device. An efficiency greater than 100% is impossible because that would violate conservation of energy. In practice, an efficiency of 100% cannot be achieved, although some devices come close.

EXAMPLE 4.13

The efficiency of the human body in converting food energy into work varies with activity. When climbing stairs the efficiency of the body is 20%. How much food energy in kilocalories does the person in Example 4.12 consume in climbing one flight of stairs?

Solution.

Step by step 1: We saw in Example 4.12 that the person had a gain in their GPE of

$$\Delta G.P.E. \equiv mgh = 60.0\,\text{kg}\cdot 9.80\,\frac{\text{m}}{\text{s}^2}\cdot 2.00\,\text{m} = 1180\,\frac{\text{kg}\cdot\text{m}^2}{\text{s}^2}$$

Step 2: We are told that climbing stairs is 20% efficient. Equation 4.7 gives us the relationship between energy out, energy in, and efficiency.

$$Efficiency\left(\frac{\text{Work or useful Energy out}}{\text{Work or Energy input}}\right)100\%$$

Step 3: Solving for energy in,

$$\text{Energy input} = \frac{E_{out}\cdot 100\%}{Efficiency} = \frac{1180\,\text{J}\cdot 100\%}{20\%} = 5880\,\text{J}$$

Step 4: The person had to consume 5880 J of energy in the form of food. Because food energy is usually given in calories, which are actually kilocalories, we are asked to convert our answer to kilocalories. 1.00 kcal = 4190 J.

$$5880\,\text{J}\cdot\frac{1.00\,\text{kcal}}{4190\,\text{J}} = 1.40\text{kcal} = 1.40 \text{ diet calories}$$

Much more food energy is used than would be needed if the body were 100% efficient (five times more food energy in the last example). Even so, the amount of food energy required to climb the stairs is small compared with that in an average daily diet, or even compared with the kilocalories contained in a single gram of sugar (see Table 4.1).

What happens to "wasted food energy" in a human? It goes mostly into thermal energy. (This thermal energy may be useful in keeping warm, but it is still called wasted energy.) The amount of thermal energy generated by a person can be significant. Consider again the person climbing the stairs in Example 4.13, Figure 4.10.

EXAMPLE 4.14

(a) How much thermal energy is generated by the person climbing the stairs in figure problem 4.13? (b) What is the rate of thermal energy production in watts when the person walks up the stairs? (c) What is the rate when he runs up the stairs?

Solution (a):

Step by step 1: Identifying the various forms of energy involved and using the conservation of energy principle yields

$$\text{Total Energy in} = \text{Total Energy out}$$

Step 2: In Example 4.13, we determined that the person consumed 5880 J of energy in the form of food. Therefore, the total energy input was 5580 J.

Step 3: The total energy output resulted in an increase in GPE and thermal energy. We calculated that he had a gain of GPE of 1176 J. The "missing energy" was given off as thermal energy.

Mathematically, our reasoning looks as follows:

$$KE_{in} + PE_{in} + \text{Other Energy}_{in} = KE_{out} + PE_{out} + \text{Other Energy}_{out}$$

Note that:

Other Energy_{in} is the food energy and Other Energy_{out} is the thermal energy.

Step 4: Substitute the known values into our conservation of energy equation and solve for Thermal Energy.

$$0 \text{ joules} + 0 \text{ joules} + \text{food energy} = 0 \text{ joules} + mgh + \text{Thermal Energy}$$

$$5580 \text{ joules} = 1176 \text{ joules} + \text{Thermal Energy}$$

Therefore,

Thermal Energy = 5580 J – 1176 J = 4704 J.

Solution (b): What is the rate of thermal energy production in watts when the person is walking up the stairs in 7.00 seconds?

When you read the words "the rate of energy," from now on, realize that we are talking about energy per unit time. Power, is energy per unit time.

$$Power \equiv \frac{Energy}{time} = \frac{4704\,\text{J}}{7.00\,\text{s}} = 672\,\text{W}$$

Solution (c): What is the rate of energy production when he runs up the stairs in 2.00 seconds?

$$Power \equiv \frac{Energy}{time} = \frac{4704\,\text{J}}{2.00\,\text{s}} = 2350\,\text{W}$$

In most circumstances, the thermal energy produced by the body when doing physical work is much more than is needed to keep warm and must be dissipated. One method of getting rid of this thermal energy is the evaporation of perspiration. This and other methods of thermal-energy dissipation will be discussed in some detail in Chapter 5.

Table 4.3 gives the efficiencies of the body and other systems in converting energy into mechanical work. It is an unfortunate fact of life that all the systems listed have efficiencies of considerably less than 100%. In most cases, the efficiency cannot be made much better, even with perfect technology. The root causes of inefficiencies are discussed in some of the references at the end of this chapter.

Table 4.3 Efficiency (%) of the Body and of Mechanical Devices

Body	
Cycling	20
Swimming, surface	2
Swimming, submerged	4
Shoveling	3
Steam engine	17
Gasoline engine	38
Nuclear power plant	35
Coal power plant	42

EXAMPLE 4.15

A large nuclear power plant has an efficiency of 35.0% and generates 1000 MW of electric power. What is the plant's rate of production of waste thermal energy in megawatts?

Solution.

Step by step 1: Write the efficiency equation.

$$Efficiency \equiv \left(\frac{\text{Useful Power out}}{\text{Power input}}\right)100\% \qquad \text{eq. 4.6b}$$

Step 2: We are told that the power out is 1000 Megawatts and that efficiency is 35%. We can determine the "power in" by solving Equation 4.7b for "Power in" and substituting the given values.

$$\text{Power input} \equiv \left(\frac{\text{Useful Power out}}{35.0\%}\right)100\%$$

$$\text{Power input} \equiv \left(\frac{1000\,\text{MW}}{35.0\%}\right)100\% = 2860\,\text{MW}$$

Step 3: Because the plant is 35% efficient, we know that 65% of the power and energy is given off as waste thermal energy.

$0.65 \cdot 2860\,\text{MW} = 1860\,\text{MW}$ of waste thermal power.

Therefore, 1860 MJ of thermal energy are given off to the environment every second in order to produce 1000 Megajoules per second of electrical energy.

The amount of waste thermal energy produced by nuclear and conventional power plants is so large that cooling towers are sometimes constructed to dissipate it into the atmosphere, as seen in Figure 4.11. Alternatively, waste thermal energy may be dumped into a river, often causing problems in the local environment. In some countries, "waste" thermal energy is used to heat homes and industries. Tens of thousands of homes can be heated with the waste thermal energy of even a moderate-sized electric power plant. However, it is most common to throwaway this thermal energy since energy has historically been very cheap. It may be time to rethink this strategy.

Energy is sometimes expressed in units of power multiplied by time, such as watt-seconds or kilowatt-hours (kWh). Electric utility bills almost always list the amount of energy used in kilowatt-hours rather than joules or kilocalories. Starting with the definition of power and solving for energy gives

$$Power \equiv \frac{Energy}{time}, \text{ therefore } Energy = Power \times time$$

This equations shows that the amount of energy a device uses depends on its power rating and how long it is on. A 1000-W toaster costs less to operate 10 minute a day than a 75.0-W porch light that is on for 6.00 hours a day. The cost of using an electrical appliance for a given amount of time is easy to calculate if the power consumption of the device is known.

Figure 4.11 The most prominent feature of this nuclear power plant is its cooling towers. Huge amounts of waste thermal energy must be given off to the local environment. A form of thermal pollution.

EXAMPLE 4.16

The cost of electric energy varies around the country, but 12.0 cent per kWh is typical. What is the cost of running a 50.0-W LED TV set 6.00 hours per day for 30 days?

Solution.

Step by step 1: Solve the definition of power for energy.

$$Power \equiv \frac{Energy}{time} \text{ Therefore } Energy = Power \times time$$

Step 2: Substitute the given values. Since we want the energy in terms of Watt-hour, we convert the days into hours.

$$Energy = Power \times time = 50.0\,\text{W} \cdot \frac{6.00\,\text{hr}}{\text{day}} \cdot 30.0\,\text{day} = 9000\,\text{W} \cdot \text{hr}$$

Step 3: The cost is given to us in terms of kilowatt-hours; therefore, we convert the energy to kilowatt-hours.

$$9000\,\mathrm{W}\cdot\mathrm{hr}\cdot\frac{1\,\mathrm{kW}}{1000\,\mathrm{W}}=9.00\,\mathrm{KWh}$$

$$Energy = 9.00\ \mathrm{kWh}$$

Step 4: The cost is given to us as 12.0 cent per kWh. We will use this to determine the amount of our electric bill.

$$\mathrm{Cost}=9.00\,\mathrm{kW}\cdot\mathrm{hr}\cdot\left(\frac{12.0\ \mathrm{cent}}{\mathrm{kW}\cdot\mathrm{hr}}\right)=108\ \mathrm{cents}=\$1.08$$

The cost of running an LED TV is much less than the old 500-W CRT (cathode ray tubes) based TVs. An appliance like a toaster, which is on only about 10 minutes per day, is very cheap to operate even though it consumes 1000 W or more while on. Electric clocks use only about 1 W and are also cheap to operate even though they are on 24 hours per day. Only large appliances, such as air conditioners, water heaters, refrigerators, and TVs, that are on for several hours per day, or small appliances, such as incandescent light bulbs, that are on for many hours per day, affect electrical costs significantly. The newer 12.5-W LED light bulbs are just as bright as a 60-W incandescent light bulb and last up to 25 years, resulting in significant cost savings over their life time.

Energy units in terms of power times time are also used on heart defibrillators, many of which have meters labeled in watt-seconds instead of Joules. The use of watt-seconds relates to the operation of the defibrillator, which passes several hundred watts of electric power through the heart for a fraction of a second to stimulate it into a normal beating pattern.

Things to Know for Section 4.4

$$Power \equiv \frac{Energy}{time} \qquad 1\,\mathrm{watt} \equiv 1\frac{\mathrm{J}}{\mathrm{s}}$$

A kilowatt-hour is a unit of energy

$$Efficiency \equiv \left(\frac{\text{Work or useful Energy out}}{\text{Work or Energy input}}\right)100\%$$

4.5 WORK, ENERGY, POWER, AND EFFICIENCY IN HUMANS

Every bodily process requires energy, not only the performance of physical work, but also blood circulation, digestion, sleeping, and even thinking. Whenever energy is involved, the related concepts of work, power, and efficiency usually are important too, as many examples in previous sections indicated. The body can be thought of as an energy converter, with the source of energy being stored chemical energy (food) that the body converts into mechanical work and thermal energy and back into stored chemical energy (fat).

4.5.1 Metabolic Rates

The rate of conversion of food energy to some other form is called the metabolic rate. The total energy conversion rate of a person at rest is called the basal metabolic rate (BMR) and is divided among various systems in the body, as shown in Table 4.4. At first glance, the percentages of total BMR are a bit surprising. The largest fraction goes to the liver and spleen, with the brain coming next, followed by the skeletal muscles, which are a close third to the brain. (Of course, during vigorous exercise, the energy consumption of the skeletal muscles and heart increase markedly.) The kidneys use more energy in a resting person than does the heart! The remainder of the BMR is used in all the other organs of the body, including smooth muscles, the intestines, the rest of the lymphatic system, and bone marrow.

Table 4.4 Basal Metabolic Rate and Oxygen Consumption rates[a]

Organ	Power Consumed at Rest (kcal/min)	Power Consumed at Rest (W)	Oxygen Consumption (ml O_2/min)	Percent of BMR
Liver and spleen	0.33	23	67	27
Brain	0.23	16	47	19
Skeletal muscle	0.22	15	45	18
Kidney	0.13	9	26	10
Heart	0.08	6	17	7
Other[b]	0.23	16	48	19
Totals	1.22	85	250	100

[a] In resting 65-kg male.
[b] Includes the digestive tract, smooth muscle, skin, etc.
Adapted from J. Cameron and J. Skofronick, *Medical Physics* (Wiley-Interscience, New York, 1978), by permission.

The human brain, with its 86 billion neurons is the most complex, energy consuming brain on the planet. Approximately, $6\frac{\text{kcal of energy}}{\text{day}}$ are required to sustain approximately 1 billion neurons. With the brains 86 billion neurons, 516 Kcal of energy per day are needed by the brain. If one ate raw food, you would have to eat for 9 hours every day to supply enough energy to maintain a healthy human. That is not feasible. Cooked food is partially digested and provides the body with three times the energy that raw food does. It was the invention of cooking, 1.5 million years ago, that has allowed the human brain to evolve to its present degree of complexity. No other primate cooks.*

The BMR of an individual is related to thyroid activity. An overactive thyroid results in a high BMR and hyperactivity in severe cases. Conversely, an underactive thyroid leads to a low BMR and lethargy. The BMR also is directly related to the mass of a person, with larger people having greater BMRs. This is partly due to the need for more thermal energy to maintain body temperature in a large person and partly because big people have large organs and thus, for example, can do more work.

The dependence of BMR on mass applies both to humans and to other warm blooded animals.

Energy consumption is directly proportional to oxygen consumption, since the digestive process is basically one of oxidizing food. Approximately 4.9 kcal of energy are produced for each liter of

*Suzana Herculano-Houzel; PNAS June 26, 2012 Vol.109 Supplement 1.

oxygen consumed, independent of the type food. Because of this, some physiological measurements use the oxygen consumption rate as a measure of an individual's energy production rate.

The digestive process is quite effective in metabolizing food; only about 5% of the caloric value of foods is excreted in the feces and urine without being utilized by the body. The body stores excess food energy by producing fatty tissue. (The excess is the food energy that is not required to produce thermal energy or do work.) During rest, most of the energy consumption of the body ends up as thermal energy. Only a small amount of work is done on the outside world. Work done by the heart on the blood is eventually converted to thermal energy by friction in the circulatory system, and most skeletal muscle activity is in the form of small motions during rest.

EXAMPLE 4.17

How many grams of fat will an idle person (person at rest) gain in a day by consuming 2500 kcal of food? A person at rest consumes 1.22 kcal/min. Assume that all extra energy is converted to fat. 1.00 g of fat contains 9.30 kcal of energy

Solution

Step by step 1: First calculate the amount of energy that a person at rest requires for an entire day. A person at rest consumes 1.22 kcal/min. You should recognize the units as energy over time: Power.

$$Power \equiv \frac{Energy}{time}\text{; therefore, } Energy = Power \times time$$

$$Energy = Power \times time = \frac{1.22\,\text{kcal}}{1\,\text{min}} \times \frac{60.0\,\text{min}}{1\,\text{hr}} \times \frac{24.0\,\text{hr}}{1\,\text{day}} = 1757\,\frac{\text{kcal}}{\text{day}}$$

Step 2: Because the person consumes 2500 kcal of food in a day, the amount of extra energy will be the difference between the energy consumed and the energy expended.

$$2500\text{ kcal} - 1757\text{ kcal} = 743\text{ kcal}$$

Step 3: Finally, we get to calculate the amount of fat that is stored. The idle person is taking in 743 more kcal than he is using. We are assuming that energy is stored as fat. Using the energy value for fat we see that 1.00 g of fat contains 9.30 kcal of energy. Assuming that the same amount of energy is used to store a gram of fat as is obtained from digesting a gram of fat, we convert the energy to grams of fat as follows:

$$743\,\text{kcal} \cdot \frac{1.00\text{ g of fat}}{9.30\,\text{kcal}} = 79.8\text{ g of fat}$$

The person is gaining approximately 80.0 g of fat per day. In one year, the person would gain approximately 29 kg of fat. At 2.2 lb per kilogram, the person would gain about 64 pounds in one year. Get off the couch now!

The mass gain in the last example is a bit exaggerated since normal daily activities require more food than is necessary to maintain the BMR. But excess caloric intake is still to be avoided because several hours of exercise are required to compensate for even moderate overeating (see Example 4.11).

$$Efficiency \equiv \frac{Power_{ow}}{Power_{In}} \times 100\%$$

Efficiency and Human Activity

As seen earlier in table 4.3, fully 80%–98% of the energy consumed in various activities goes into producing thermal energy, since the efficiency of the body ranges from 2% to 20%, depending on the type of activity. Swimming on the surface is approximately 2% efficient whereas swimming underwater is approximately 4% efficient.

Why does body efficiency vary with activity? It basically depends on how many muscles are used cooperatively to produce work and on the energy used by other organs. The more muscles used to produce useful work compared with the activity of other organs, the less efficient the body is. A single muscle is only about 25% efficient in converting food energy to work; the rest goes to thermal energy. The body can thus never be more than 25% efficient, since other organs consume energy and produce no work. Leg muscles are the largest in the body; hence cycling and climbing stairs are relatively efficient processes (20%). Shoveling uses arm, shoulder, leg and back muscles and has an efficiency of only about 3%.

The body tends to be more efficient in vigorous activity, maximum efficiency being attained by well-trained athletes. Although the energy consumption of many organs increases during vigorous exercise, the energy consumption of muscles doing work increases by a larger factor. Two individuals performing the same physical activity often have different efficiencies. The person who has the fewest muscles exerting forces in opposition to one another will have the greatest efficiency and will frequently be able to accomplish the most work. Athletic training and exercise result in increased efficiency as well as increased energy consumption.

The generally low efficiency of the body means that significant amounts of thermal energy are generated. That portion not needed to maintain body temperature must be dissipated into the environment, as was mentioned previously. On the other hand, if insufficient thermal energy is generated to stay warm, the body resorts to shivering to generate more.

Table 4.5 Energy and Oxygen Consumption rates[a]

Activity	Power: Rate of Energy Consumption (kcal/min)	Power: Rate of Energy Consumption (W)	Oxygen Consumption (litres O_2/min)
Sleeping	1.2	83	0.24
Sitting at rest	1.7	120	0.34
Standing relaxed	1.8	125	0.36
Sitting in class	3.0	210	0.60
Walking slowly (4.8 km/hr)	3.8	265	0.76
Cycling (13–18 km/hr)	5.7	400	1.14
Shivering	6.1	425	1.21
Playing tennis	6.3	440	1.26
Swimming breaststroke	6.8	475	1.36
Ice skating (14.5 km/hr)	7.8	545	1.56
Climbing stairs (116/min)	9.8	685	1.96
Cycling (21 km/hr)	10.0	700	2.00
Playing basketball	11.4	800	2.28
Cycling, professional racer	26.5	1855	5.30

[a] Normal 76-kg male.

Adapted from J. Cameron and J. Skofronick, *Medical Physics* (Wiley-Interscience, New York, 1978), by permission.

EXAMPLE 4.18

A cyclist traveling at 21 km/hr consumes 2.00 L of oxygen per minute. A sleeping person only consumes 0.24 L of oxygen per minute. Compare the volume of oxygen consumed by the cyclist with that used by a person sleeping. What would you expect the ratio of the respiratory rates to be in these two activities?

Solution: Comparing Oxygen consumption rates:

$$\frac{O_2 \text{ cycling consumption rate}}{O_2 \text{ sleeping consumption rate}} = \frac{2.00\frac{\text{liter}}{\text{min}}}{0.240\frac{\text{liter}}{\text{min}}} = 8.33$$

Comparing respiration rates:

Step by step 1: First calculate the amount of oxygen consumed while cycling.

The rate for cycling is 2.00 L of O_2 per minute. Therefore, the volume consumed in 1 hour (60 minutes) is

$$\frac{2.00 \text{ L of } O_2}{1 \text{ min}} \cdot 60.0 \text{ minutes} = 120 \text{ L of } O_2 \text{ consumed while cycling for 1 hour}$$

Step 2: Calculate the amount of oxygen consumed while sleeping for 1 hour. The rate of oxygen consumption while sleeping is only 0.240 L/min.

Therefore, the volume of O_2 consumed in 1 hour, while ***sleeping*** is

$$\frac{0.240 \text{ L of } O_2}{1 \text{ min}} \cdot 60.0 \text{ minutes} = 14.4 \text{ L of oxygen consumed in 1 hour}$$

Step 3: The ratio of Oxygen consumed while cycling compared with that consumed while sleeping is

$$\frac{120 \text{ L}}{14.4 \text{ L}} = \frac{8.33}{1}$$

The cyclist is consuming 8.33 time more oxygen and energy than the sleeper.

Remember, energy consumption is directly proportional to oxygen consumption, since the digestive process is basically one of oxidizing food. Approximately, 4.9 kcal of energy is produced for each liter of oxygen consumed, independent of the type food.

A similar calculation shows that a person sitting at rest consumes 20 L of oxygen per hour. The average respiratory rate for such a person is 12 breaths of 0.5 L each per minute, or 360 L of air per hour. Air is 20% oxygen, so this would be an inhalation of 72 L of oxygen. But only 28% of the oxygen is used, (20.4 L of the 72 L). The other 72% of O_2 is exhaled and can be used for mouth-to-mouth resuscitation, blowing on charcoals, and so on.

The Anaerobic and Aerobic Phases

Anaerobic: Humans and other animals are capable of doing work for short periods of time at much higher rates than can be maintained for long periods of time. There is no doubt that this ability is useful in emergencies, allowing an animal to climb a tree very rapidly, run away, or fight fiercely when in danger. To do this, the body rapidly oxidizes available blood sugar with oxygen already present in the body and makes the resultant energy available to vital organs and skeletal muscles. The heart rate increases to circulate substances more rapidly to areas where they are needed. Blood-flow patterns are altered to send more blood to organs active in stress response, such as skeletal muscles, and less to those not immediately needed, such as the skin. The spleen contracts to put stored blood into general circulation, making more sugar available immediately. The liver transforms large amounts of stored glycogen into glucose and releases it into the circulatory system. (Glucose can be rapidly oxidized to obtain energy.) All these actions utilize oxygen already in the body, and the higher power output cannot last more than about a minute. This stage is called the *anaerobic phase* because no ingestion of oxygen is required.

Aerobic: The body quickly becomes oxygen deficient during the anaerobic phase and requires oxygen from the lungs to maintain its activity. It thus enters a new stage, called the aerobic phase. The series of actions that initiate the anaerobic phase are accompanied by other actions that make possible the long-term production of energy in the *aerobic phase*. These include sweat production to dissipate thermal energy, rapid breathing, and dilation of air passageways.

Although the body is an almost bewilderingly complex system, its utilization of energy is simple in principle. Chemical energy stored in food is converted either to work, to thermal energy, or to stored chemical energy in fatty tissue. Nothing more than the conservation of energy principle is involved.

Things to Know for Chapter 4

$$Work \equiv F_{\text{in the direction of motion}} \bullet \Delta X$$

Joules are units of energy in the metric system.

$$1\text{ joule} \equiv 1\text{Nm} = 1\,\frac{\text{kg}\cdot\text{m}^2}{\text{s}^2}$$

A kilowatt-hour is a unit of energy

Gravitational Potential Energy: $G.P.E. \equiv mgh$

Kinetic Energy: $K.E. = \frac{1}{2}mv^2$

The conservation of energy principle:

$$\text{K.E.} + \text{P.E.} + \text{all other forms of energy} = \text{Constant}$$

I believe in the conservation of energy.

$$Power \equiv \frac{Energy}{time} \quad 1\text{ watt} \equiv 1\frac{\text{joule}}{\text{sec}}$$

$$Efficiency \equiv \left(\frac{\text{Work or useful Energy out}}{\text{Work or Energy input}}\right)100\%$$

Useful information:

746 W = 1 hp Approximately 4.19 J = 1 cal

Questions

4.1 What happens to the energy content of a system when work is done on or by it?

4.2 Describe a situation in which a force is exerted for a long time but does no work. Explain in terms of the definition of work.

4.3 Give an example of a force that moves but does no work. Explain in terms of the definition of work.

4.4 In Example 4.2, the work done in lifting a sack of groceries was calculated. More work would be done if the shopper stood up while lifting the groceries. Explain.

4.5 The various forms of energy can be separated into categories as at the end of Section 4.2. For each category describe a situation in which energy belonging to that category is involved.

4.6 Describe the various forms of energy involved in a car trip taken at night. Also identify any work done on or by the car.

4.7 Describe the forms of energy involved in pole-vaulting. Also identify any work done by the pole-vaulter.

4.8 Give an example not identical to any in the chapter in which kinetic energy is converted into another form of energy.

4.9 To what form is most of the energy expended by a swimmer ultimately converted when she swims at a constant velocity?

4.10 Explain why it might take more energy to sit in a lecture than it does to sit relaxed (refer to Table 4.2). Describe what activities might require more energy and where that energy goes.

4.11 What is a closed system?

4.12 Energy can neither be created nor destroyed. How is it then that total energy is not constant in an open system?

4.13 Give an example of a closed system in which there are changes in the forms of energy but energy is conserved.

4.14 Do devices with an efficiency of less than 100% violate the conservation of energy principle? Explain.

4.15 Electric bills list energy consumed in units of kilowatt-hours. What is the relationship between kilowatt-hours and joules?

4.16 Most electrical appliances are rated in watts or horsepower. Does the rating depend on whether the appliance is operated for long or short periods of time? Explain in terms of the definition of power.

4.17 The energy content of food, as listed in Table 4.1, is determined by burning it and measuring the amount of heat released. Discuss why this is a reasonable evaluation of the amount of energy the food will release when digested.

4.18 Why is work done by using the arms less efficient than work done by using the legs, as Table 4.3 indicates?

4.19 The energy consumption rate (metabolic rate) of a person can be measured by monitoring his or her oxygen consumption. Explain the relationship between oxygen consumption and energy consumption.

4.20 Do you think the efficiency of the body is high or low when shivering? Explain.

4.21 Isometric exercises are those in which muscles are exerted against one another, such as locking the hands together and moving them up and down with one pushing against the other. Is work done? Is body heat generated? Explain.

4.22 The efficiency of the body in performing sustained work can approach but never reach 25%. Explain the limit.

4.23 What are the anaerobic and aerobic phases of metabolism?

4.24 Marathon runners speak of "hitting the wall" after 20 miles of running and never run more than about 15 miles at one time while training. What do you suppose is the source of energy after 20 miles of running, and why would it be inadvisable to run longer distances while training?

PROBLEMS

SECTION 4.1

√4.1 (I) Determine the amount of work in joules that a car that is driven 150 km must do, if an average force of 500 N is applied to overcome friction?

√4.2 (II) Calculate the work done by a 55-kg person in climbing a flight of stairs 2.0 m high.

√4.3 (III) How much work is done to the wagon by the girl pulling her brother shown in Figure 4.12, if she travels 25 m?

Figure 4.12

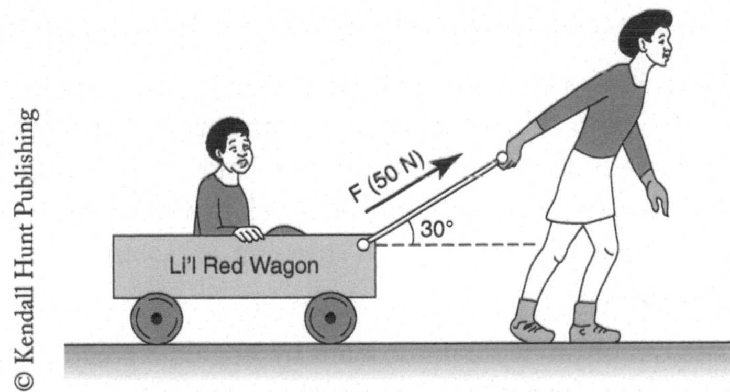

SECTION 4.2

√4.4 (I) Calculate (a) the kinetic energy of an 85.0-kg athlete running at a speed of 10.0 m/s and (b) the kinetic energy of a 60.0-kg cheetah running at 31.0 m/s. (c) Compare the kinetic energy of the cheetah to the kinetic energy of the athlete.

√4.5 (I) (a) How much energy in kilocalories is expended against gravity alone by a 75-kg man in climbing 1500 m up a mountain? (b) Convert the energy to joules.

√4.6 (I) How long can you play tennis on the energy from a 10-g pat of butter (see Tables 4.1 and 4.2)?

√4.7 (I) How many minutes of lecture can you sit through on the energy supplied by 32 g of peanuts? This is the average amount in a vending machine pack (see Tables 4.1 and 4.2).

√4.8 (II) If a 0.5 kg baseball is caught by a person whose hand recoils 0.30 m, calculate the average force on the person's hand while stopping the ball. (The initial speed of the baseball is 44.7 m/s). Assume that all of the kinetic energy of the baseball is used in doing work, moving the recoiling hand.

√4.9 (II) Calculate the gravitational potential energy of the water in a lake of volume 100 km^3 (mass = 1.0×10^{14} kg). The lake has an average height of 50 m above the hydroelectric generators at the base of the dam holding the water back.

4.10 (II) How much thermal energy is generated by the brakes of a 1.0×10^7-kg train in causing the train to slow from a speed of 30 m/s to a speed of 10 m/s? (Neglect any other forces that might also slow the train.)

√4.11 (II) (a) Given that the energy consumption rate for climbing stairs is 9.8 kcal/min, calculate the number of kilocalories needed to climb 232 stair steps in 2 minutes. (b) Calculate the gain in GPE for a 50-kg person if each step is 0.18 m high. Convert your answer to kcal and compare your result with the answer to part (a). Explain why the energy found in part (a) is larger than that found in part (b).

√4.12 (II) (a) How many kilocalories are required to walk 8.0 hours per day, as many nurses do on their jobs? (b) If the remainder of the day is spent sitting at rest for 8.0 hours and sleeping for 8.0 hours, what total number of kilocalories would be needed for the entire day?

4.13 (III) When a person jumps from some height to the ground, very large stresses are produced in the joints. If a 60-kg person jumps out of a tree from a height of 3.0 m, calculate the force in the knee joints (a) if he lands stiffly and compresses the joint materials 0.50 cm; (b) if he cushions the shock by flexing his legs upon striking the ground and his body moves 0.80 m in stopping. (c) Compare these forces with the weight of this person.

4.14 (III) Severe stresses can be produced in the joints by jogging on hard surfaces or with insufficiently padded shoes. If the downward velocity of the leg is 7.0 m/s when the jogger's foot hits the ground, and if the leg stops in a distance of 2.0 cm, calculate the force in the ankle joint. The mass of the jogger's leg is 12 kg. (You may neglect the weight of the rest of the jogger's body.) Compare this force with the weight of a 75-kg jogger.

4.15 (III) A person in a wheelchair with a total mass of 100 kg is pushed from rest to a speed of 2.0 m/s. If the effective coefficient of friction is 0.05 and the distance moved is 15 m, (a) calculate the work done. (b) Calculate the average force that must be exerted on the wheelchair to accomplish this.

SECTION 4.3

4.16 (II) A San Francisco cable car loses its brakes and runs away down a hill 30 m high. (a) Calculate its speed at the bottom of the hill if there is no friction and its initial speed is zero. (b) Suppose instead that the brakes fail only partially and the final speed is observed to be 5.0 m/s. Calculate the thermal energy generated if the cable car plus its load of passengers has a mass of 8000 kg.

4.17 (II) Using conservation of energy, compare the loss of kinetic energy and the corresponding gain in GPE in order to calculate how far a rock will rise above its point of release. Assume that it is tossed straight up with an initial speed of 10 m/s.

Please neglect air resistance. Note: This is the same problem as Example 2.8(a), but it can be solved using energy considerations alone.

4.18 (II) Surprisingly little advantage is gained by getting a running start in a downhill race. To demonstrate this, use conservation of energy to calculate the final speed of a skier who skis down a hill 80 m high with negligible friction (a) if her initial speed is zero; (b) if her initial speed is 3.0 m/s. (The final speed found in part (b) is larger than in part (a), but by far less than 3.0 m/s!)

4.19 (II) (a) How high a hill can a car coast up (engine disengaged and with negligible friction) if it has an initial speed of 90 km/h? (b) If a 1000-kg car with an initial speed of 90 km/h is observed to coast up a hill to a height of 20 m above its starting point, how much energy was converted to thermal energy by friction?

4.20 (III) (a) A person in a wheelchair approaches a 0.50-m-high ramp with a speed of 3.5 m/s. Calculate the final speed of the wheelchair at the top of the ramp if the person coasts up and friction is negligible. (b) What initial speed would be necessary for the final speed to be 1.0 m/s?

4.21 (III) Using energy considerations and neglecting air resistance, show that a rock thrown from a bridge 50 m above water with an initial speed of 15 m/s will strike the water with a speed of 34.7 m/s, regardless of the direction in which it is thrown. Note: This is the same situation posed in Example 2.10. See the discussion following that example. Does it make more sense now?

SECTION 4.4

√4.22 (I) What is the useful power output in (a) watts and (b) horsepower of a person who does 2.5×10^6 J of useful work in 4.0 hours?

√4.23 (I) What is the average power consumption in watts of an appliance that uses 0.75 kWh of energy in one day? kWh is short for kilowatt hours.

√4.24 (I) What is the efficiency of an athlete who consumes 3000 kcal of food and does 2.5×10^6 J of useful work?

√4.25 (I) Calculate the efficiency of a light bulb that consumes 60 W of electric power and puts out 10 W of visible light.

4.26 (II) How much does it cost to operate a 1.0-W electric clock for a year if electricity costs 11 cent/kWh?

4.27 (II) Suppose the swimmer uses mostly arm motion to pull himself along, exerting an average force of 80 N through a distance of 1.8 m in each stroke. The swimmer does 120 strokes per minute. (a) Calculate the work done in each stroke. (b) Calculate the useful power generated by the swimmer's arms.

4.28 (II) (a) Calculate the horsepower needed for a 1200-kg car to climb to an altitude of 1000 m from sea level in 1 hour. (b) If the useful output of the engine is actually 40 hp, where does the rest of the power go? (c) If the car engine is 20% efficient, calculate the total power produced by the car in horsepower.

4.29 (III) (a) Calculate the energy in kilocalories used by a person who does 50 chin-ups. This person has a mass of 75 kg and raises his center of gravity 0.50 m during each chin-up. (The work done in lowering the body each time doubles the amount of energy necessary.) (b) What is the total power consumed in watts if the chin-ups are done in 2.0 minutes? (c) What is the useful power output?

4.30 (III) The useful power output of Bryan Allen, who flew a human-powered airplane across the English Channel on June 12, 1979, was about 350 W. The propeller of the airplane was driven by the pilot's legs, using a bicycle-type drive mechanism. Using 20% efficiency for legs, calculate the food energy he needed for the 2.5-hour trip.

4.31 (III) (a) A 0.75-m-long crank is attached to a generator used to recharge a battery. A 10-N force is required to turn the crank and the system is 95% efficient in converting the work done by turning the crank into stored electric energy. How many turns of the crank are necessary to store 100,000 J of energy in the battery? (b) If the battery is 90% efficient in giving up its stored energy, how long can it run a 100-W light bulb?

SECTION 4.5

√4.32 (I) The BMR of a 1.5 kg premature baby is 100 kcal/day. Calculate the number of liters of oxygen consumed per day. Approximately 4.9 kcal of energy are produced for each liter of oxygen consumed. (Note: In practice, the oxygen supplied to premature babies is regulated on the basis of periodic sampling of the oxygen content of the baby's blood. Brain damage can be caused to the baby by either too little or too much oxygen.)

√4.33 (I) (a) The heart uses 7% of the human body's BMR while resting. The skeletal muscles use 18% of the body's BMR while resting. What percentage of the body's energy is used by the heart and skeletal muscles when resting? (b) The total energy consumption rate while resting is 1.22 kcal/min. How many kilocalories per minute are required for the heart and skeletal muscles?

√4.34 (II) Calculate the energy in kilocalories needed to play tennis for 2.0 hours. (b) What percentage of this energy would be supplied by a soft drink containing 50 g of sugar?

4.35 (II) It is reasonable to assume that the body obtains as much energy from breaking down its own fatty tissue as it does from digesting fat. Swimming consumes 6.8 kcal/min, sleeping consumes 1.2 kcal/min and sitting at rest consumes 1.7 kcal/min. How many grams of fat will be lost by a person who swims 2.0 hours, sleeps 8.0 hours, and sits at rest the remainder of the day, if he consumes 2000 kcal?

5

Temperature and Heat

"Religion is a culture of faith; science is a culture of doubt."

—Richard P. Feynman

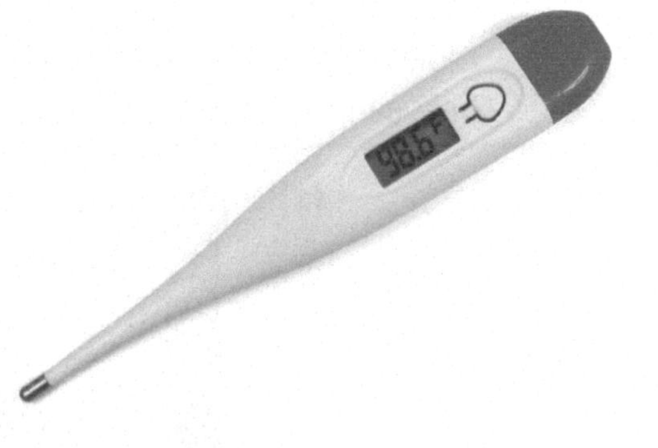

© denniro/Shutterstock.com

The words *temperature and heat* are used together so often that they may appear to mean the same thing. Although temperature and heat are closely related, they are definitely not the same thing. *Temperature* is a property of an object related to the average kinetic energy of atoms and molecules in that object; heat is a form of energy and not a property of an object. *Heat* is the amount of energy that flows from one substance to another because of a temperature difference between the two substances. One of the effects heat can have when it enters an object is to increase its temperature. Some of the other effects of heat are to melt solids or boil liquids. Under controlled circumstances, these phase changes occur at specific temperatures. The precise definitions of temperature and heat and a large number of applications of these and related topics are the subjects of this chapter.

5.1 TEMPERATURE AND PHASES OF MATTER

Temperature is an indicator of the average kinetic energy of the atoms and molecules of a substance. The temperature of an object is a number that uniquely determines whether thermal energy flows into or out of a substance. Heat is conducted from higher temperature objects to lower temperature objects. It is useful to examine the structure of matter on an atomic scale before defining temperature more precisely.

5.1.1 Atoms, Molecules, and the Phases of Matter

An atom is the smallest unit of an element. A molecule, which is a bound combination of atoms, is the smallest unit of a compound. Water molecules, for example, are composed of two hydrogen atoms and one oxygen atom. (Rather than constantly repeat the phrase "atoms or molecules" when talking about the smallest units of substances, we shall speak of molecules, considering an atom to be a molecule with only one constituent.)

Matter exists in various phases, most commonly as a solid, liquid, or gas. Molecules in a solid are bound to one another as if connected by springs, as shown in Figure 5.1(a). Some movement of the molecules is possible, but if one molecule gets too close or too far away from a neighbor, the spring-like forces return the molecule toward its average home location. A solid has rigidity and retains its shape because the average positions of the molecules in it are fixed. Molecules in a liquid are freer to move, acting like sticky ball bearings, as shown in Figure 5.1(b). Strong forces keep the molecules in a liquid from getting too close together or too far apart, but they can slide over one another with ease. A liquid will flow and cannot retain its shape unless it is in a container. In a gas, as seen in Figure 5.1(c), molecules are much farther apart than in either solids or liquids. Because of the large average separation, the forces between molecules in a gas are weak, so weak, in fact that the molecules of a gas act almost independently of one another and will escape if not in a closed container.

Figure 5.1 (a) Molecules in a solid always have the same neighbors, and their motion is limited by spring-like forces. (b) In a liquid, molecules act like sticky ball bearings that can slide over one another but cannot get too far apart or too close together. (c) In a gas, molecules can move relatively freely. Average separations are large, and the forces between molecules are weak.

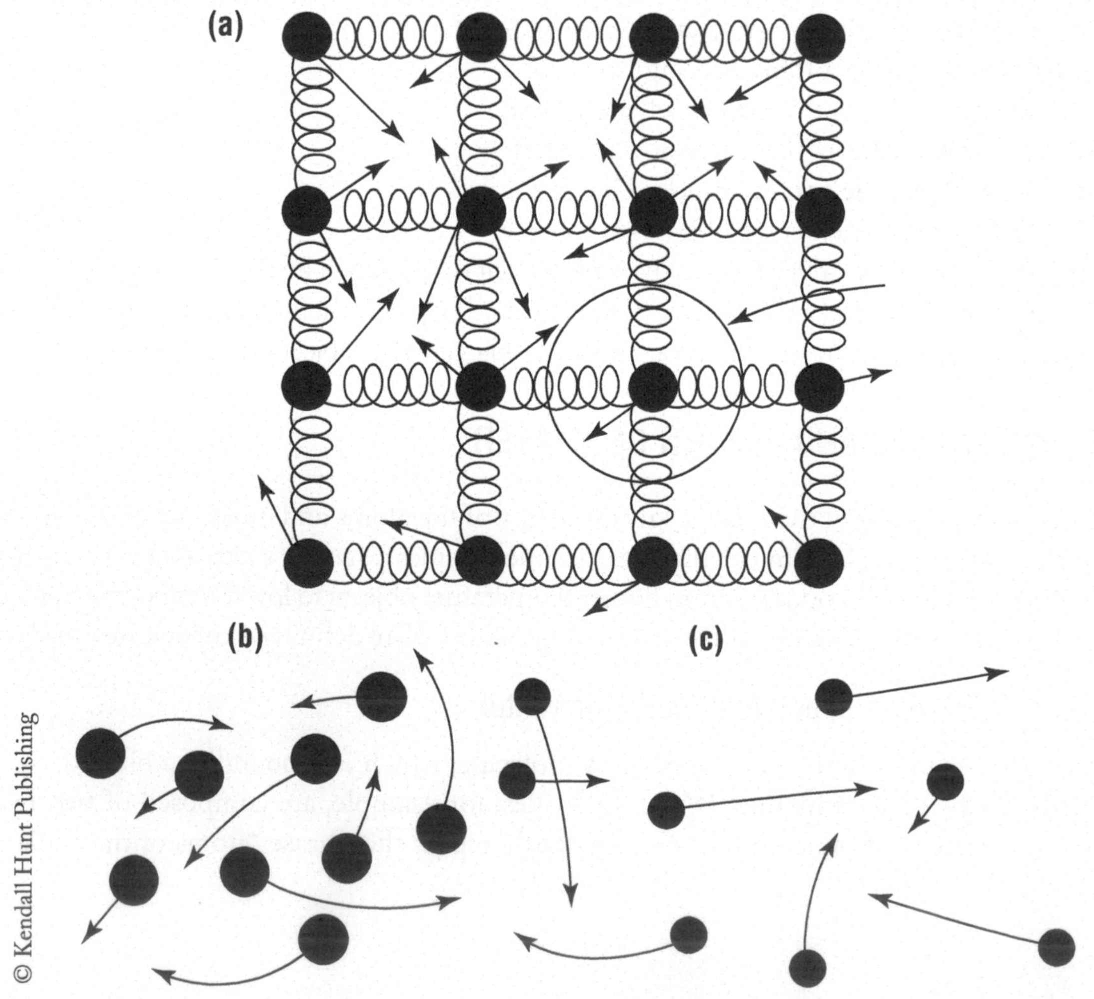

The molecules in Figure 5.1 are shown in motion. This motion exists at all times and is disordered. That is, the velocities of the molecules are distributed over a wide range, with the average speed depending on the *temperature* of the object. More precisely, *the temperature of a substance is an indicator of the average kinetic energy of its molecules*: the higher the temperature, the greater the average speed of the molecules.

5.1.2 Temperature Scales and Thermometers

The most familiar temperature scale in the United States is the Fahrenheit scale. Worldwide, the Celsius scale (once called centigrade) is by far the most common and is slowly coming into use in the United States. Another scale, important mostly in scientific and technical work, is the Kelvin scale. An obvious difference between the temperature scales is the size of their basic unit (the degree) and the temperature that they refer to as zero. At 0° Celsius and 0° Fahrenheit, there is still thermal energy that can be removed from the substance. It can get colder. On the other hand, at 0 kelvins, there is no more thermal energy that can be removed. The substance cannot get any colder than 0 K.

Both the Fahrenheit and Celsius scales are based on the freezing and boiling points of water. These two events occur at unique temperatures provided the pressure is specified (usually as standard atmospheric pressure). On the Fahrenheit scale, 32°F is the freezing point of water, and 212°F is the boiling point of water at 1 atm pressure. The Celsius scale has 0°C as the freezing point and 100°C as the boiling point at 1 atm pressure. Temperatures on the two scales can be converted to one another by using the following equation:

$$T_f = \frac{9}{5}T_C + 32 \qquad \textit{Know} \qquad \text{eq. 5.1}$$

Figure 5.2 Comparison of the Fahrenheit, Celsius, and Kelvin scales, showing the values of certain important temperatures on each scale. The maximum temperature is thought to be determined by the shortest possible wavelength, the Planck length, 1.616×10^{-27} nm. That corresponds to a temperature of 1.417×10^{32} K that occurred during the creation of the universe, the Big Bang.

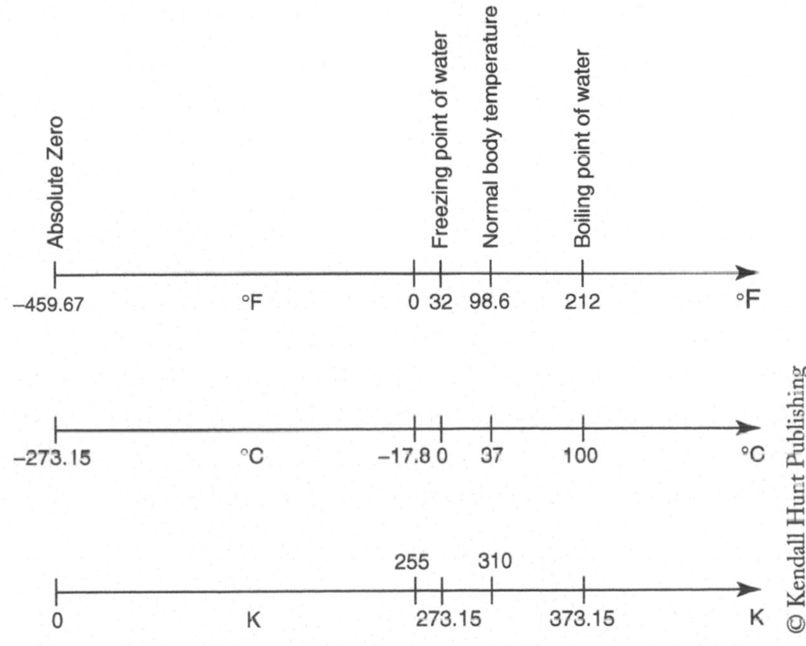

Figure 5.3 Two thermometers based on thermal expansion. (a) Mercury thermometer. The small-diameter tube connected to the reservoir is usually magnified for easier viewing. (b) Bimetallic strip made of two metals with different thermal expansions. As the temperature rises, the metal on the right expands more than the metal on the left, causing the strip to bend to the left.

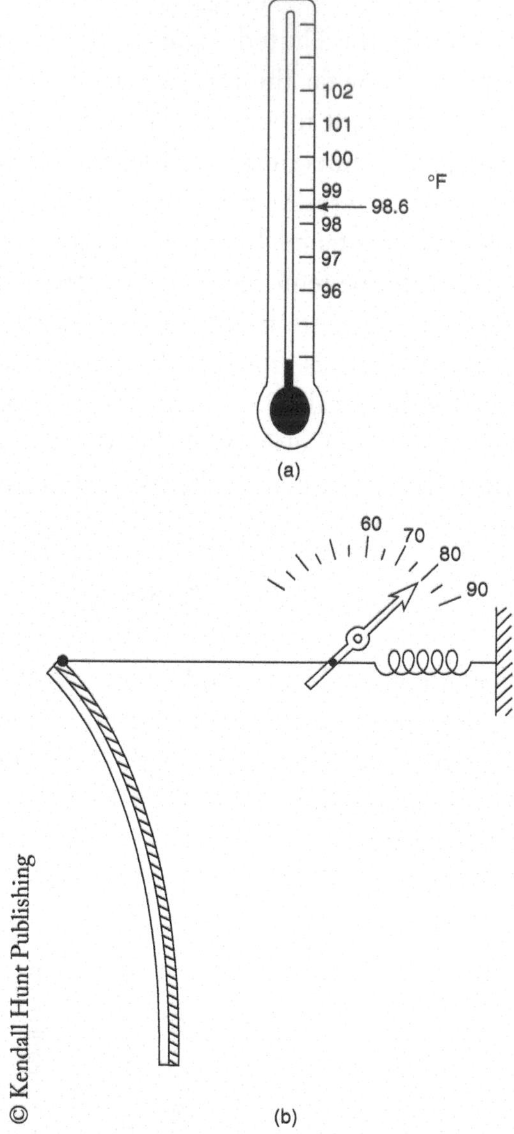

When the temperature of an object is lowered, some of the kinetic energy of its molecules is removed. There is a lower limit to temperature; at that limit, all the energy that can be removed from the molecules has been removed. This lower limit, called absolute zero, occurs at −273.15°C. On the Kelvin scale, absolute zero is taken to be 0 kelvins (abbreviated K, with no degree symbol). A 1° change in kelvins is the same as 1° temperature change in Celsius. The relationship between temperature in Celsius and temperature in kelvins is given as follows:

$$T_{\text{Kelvins}} = T_C + 273 \qquad \textit{Know} \qquad \text{eq. 5.2}$$

(See Figure 5.2)

The Kelvin scale is an absolute temperature scale. The average kinetic energy of molecules in a gas is directly proportional to the temperature in kelvins. This is also reasonably accurate in a liquid or solid.

A thermometer is any device that measures temperature. A large variety of such devices exists, including mercury thermometers and bimetallic strips, examples of which are illustrated in Figure 5.3. Both of these devices depend on the fact that most materials expand when their temperature is increased. (This effect is discussed in more detail next.) However, a thermometer may make use of any temperature-dependent property of an object; it not only changes in length or volume but also the character of electrical changes, or changes in color, for example (see Figure 5.4). *Fig. 5.4 Digital thermometers rely on electrical properties that respond to changing temperature.*

5.1.3 Thermal Expansion

Most materials expand when their temperature is increased and contract when it is decreased. This property is in part due to the increase in average kinetic energy of molecules with increasing temperature. The increased average speed of the molecules causes an increase in the average separation of the molecules. The amount that an object expands depends on its size, the material of which it is made, and the amount of the temperature change. We can calculate the change in length due to the thermal expansion of an object with the following equation:

$$\Delta L = \alpha L \Delta T \quad \textit{Know} \qquad \text{eq. 5.3}$$

where ΔL is the change in the length L of the object, ΔT is the change in temperature of the object, and α is its coefficient of linear expansion. Table 5.1 gives coefficients of linear expansion for various substances. Equation 5.3 is valid when the temperature change does not cause a phase change.

EXAMPLE 5.1

What is the change in length of a column of mercury 3.0 cm long if its temperature increases from 37°C to 40°C? Use Table 5.1 for the value of the coefficient of linear expansion of mercury.

Solution: Equation 5.3 allows us to easily calculate the change in length of the column of mercury.

Step by step 1: Write down our equation for linear expansion.

$$\Delta L = \alpha L \Delta T$$

Step 2: Determine the values of the variables in the equation.

The original length "L" is given as 3.0 cm and the temperature change ΔT is 3°C. Table 5.1 gives us the coefficient of linear expansion of mercury as $\frac{60 \times 10^{-6}}{°\text{C}}$.

Step 3: Substitute the values into our equation and do the math.

The change in length is then found directly from Equation 5.3:

$$\Delta L = \alpha L \Delta T = \frac{60 \times 10^{-6}}{°\text{C}} \cdot 3.00\,\text{cm} \cdot 3.00°\text{C} = 5.40 \times 10^{-4}\,\text{cm}$$

Table 5.1 Coefficients of Linear Expansion[a] at 20°C

Material	α (1/°C)
Solids	
Aluminum	25×10^{-6}
Brass	19×10^{-6}
Gold	14×10^{-6}
Iron or steel	12×10^{-6}
Lead	29×10^{-6}
Silver	18×10^{-6}
Glass (ordinary)	9×10^{-6}
Glass (Pyrex)	3×10^{-6}
Quartz	0.4×10^{-6}
Concrete, brick	12×10^{-6}
Marble (average)	2.5×10^{-6}
Liquids[b]	
Ether	550×10^{-6}
Ethyl alcohol	370×10^{-6}
Gasoline	320×10^{-6}
Glycerin	170×10^{-6}
Mercury	60×10^{-6}
Water	70×10^{-6}
Gases[b]	
Air and most others at atmospheric pressure	1100×10^{-6}

[a]Also called thermal expansion coefficients.
[b]Values for liquids and gases are approximate.

That change in length is much, much too small to account for the visible rise of the mercury in the thermometer. It is the expansion and contraction of the mercury reservoir in the bulb that sends the column up and down the narrow tube in the thermometer.

Bridges are designed with sliding joints so that they may expand and contract when their temperature changes, without buckling. Standard railroads are constructed with gaps where rails are joined together to allow for expansion and contraction. The creaking and pinging sounds made by a stove when it heats up are due to its catching and slipping as it expands.

Water is an important exception to the general rule that substances expand with increasing temperature and contract with decreasing temperature. Water behaves "normally" at most temperatures, but if it is cooled to 3.96°C, it will expand with further cooling as submicroscopic ice crystals begin rapidly forming, until it reaches 0°C. As water continues to cool, the cold ice/water mixture is pushed to the surface by the sinking, denser, 3.96°C water. The warmest water in a frozen lake is 3.96°C and is at its bottom. At 0°C water freezes to ice, in the process of expanding even further. Therefore, the lake freezes at the top instead of the bottom. The effects of the expansion of water with decreasing temperature, especially the expansion of water when it freezes, are far ranging. Water freezing and expanding in pipes or engine cooling systems can damage them irreparably. Damage done to foods when they are frozen is due primarily to cells burst by the expansion of water turned to ice. This is why "frost bite" is dangerous. Some foods tolerate being frozen better than others, but none tolerate repeated freezing and thawing.

5.1.4 Density

Density is defined as the mass per unit volume occupied by an object or substance:

$$density \equiv \frac{mass}{volume} \qquad Know \qquad \text{eq. 5.4}$$

We will use the Greek letter rho ρ to represent density. So, in secret code shorthand, we can write

$$\rho \equiv \frac{m}{V}$$

If something is heavy for its size, then it is dense.

Typical density units in the metric system are $\frac{\text{kg}}{\text{m}^3}$ and $\frac{\text{g}}{\text{cm}^3}$.

At 4°C, the density of water is $1000\frac{\text{kg}}{\text{m}^3}$, which is the same as $1.00\frac{\text{g}}{\text{cm}^3}$.

The density of a substance can be an important physical property.

For example, if an object is less dense than water, it will float; if it is more dense, it will sink. Density can also be used to identify a substance, for example, to determine whether a piece of jewelry is pure gold or just gold plated.

Not as exciting as the golden crown story, but of significant importance, the density of urine can give medical information. A urine density above 1.01 g/cm^3 indicates possible dehydration. Density above 1.03 g/cm^3 indicates possible problems with kidney function.

What does density have to do with temperature? The density of an object depends on its temperature because the volume of an object changes with temperature. If the volume of an object increases when its temperature is increased, then the same amount of mass occupies a larger volume. The density of the object therefore decreases with increasing temperature. This is why hot air rises. Hot air is less dense than cold air. The denser, cold air is pulled down by the force of gravity. The sinking cold air pushes the less dense warm air up and away. The behavior of water at temperatures near freezing is an exception. When water freezes, it expands. The density of ice is thus less than that of water (0.917 versus 1.00 gm/cm^3), which is why ice floats. This is crucial to life as we know it, for it means that lakes and oceans freeze from the top down and melt quickly in warm weather. For most of the substances, the solid phase is more dense than the liquid phase.

The density of a substance is also related to its molecular configuration. Looking back at Figure 5.1, one would suspect that the density of gases would be smaller compared with the densities of liquids and solids. The values given for densities in Table 5.2 confirm this suspicion. Furthermore,

Table 5.2 Densities of various substances

Substance	Density ρ, in $\frac{g}{cm^3}$
Solids	
Aluminum	2.70
Brass	8.44
Copper (average)	8.8
Gold	19.3
Iron or steel	7.8
Lead	11.3
Silver	10.1
Uranium	18.7
Concrete	2.3
Cork	0.24
Glass	2.6
Granite	2.7
Wood	0.3–0.9
Ice (0°C)	0.917
Bone	1.7
Liquids	
Water (4°C)	1.000
Blood, plasma	1.03
Blood, whole	1.05
Seawater	1.025
Mercury	13.6
Ethyl alcohol	0.79
Gasoline	0.68
Glycerin	1.26
Olive oil	0.92
Gases[a]	
Dry air	1.29×10^{-3}
Carbon dioxide	1.98×10^{-3}

Substance	Density ρ, in $\frac{g}{cm^3}$
Carbon monoxide	1.25×10^{-3}
Hydrogen	0.090×10^{-3}
Helium	0.18×10^{-3}
Methane	0.72×10^{-3}
Nitrogen	1.25×10^{-3}
Nitrous oxide	1.98×10^{-3}
Oxygen	1.43×10^{-3}
Water (100°C steam)	0.60×10^{-3}

[a]Unless otherwise specified, at 0°C and 1 atm pressure.

the densities of liquids and solids are roughly of the same order of magnitude. This is consistent with the picture that molecules in liquids and solids are essentially in contact, whereas those in gases are widely separated by empty spaces.

EXAMPLE 5.2

What is the mass of a liter of whole blood? One liter of blood is the same volume as 1000 cubic centimeters of blood. $1.00\,\text{L} \equiv 1000\,\text{cm}^3$

Solution.

Step by step 1: Recall that the definition of density.

$$\rho \equiv \frac{m}{V} \qquad \textit{Know}$$

Step 2: Look up the density of whole blood in Table 5.2.

$$\rho_{\text{blood}} = 1.05\frac{\text{g}}{\text{cm}^3}$$

Step 3: Solve the density equation for mass.

$$\rho \equiv \frac{m}{V}; \quad m = \rho \cdot V$$

Step 4: We now can determine the mass of a liter of blood by substituting the values for density and volume as follows:

$$m = \rho \cdot V = 1.05\frac{\text{g}}{\text{cm}^3}1000\,\text{cm}^3 = 1050\,\text{g} = 1.05\,\text{kg}$$

Things to know for Section 5.1

$$T_f = \frac{9}{5}T_C + 32 \qquad T_{Kelvins} = T_C + 273$$

$$\Delta L = \alpha L \Delta T$$

$$density \equiv \frac{mass}{volume} \text{ Units } \frac{\text{kg}}{\text{m}^3} \text{ or } \frac{\text{g}}{\text{cm}^3}$$

Using symbols, we write the definition of density as follows:

$$\rho \equiv \frac{m}{V}$$

5.2 HEAT

If a piece of hot metal is dropped into a bucket of water, the metal will cool down and the water will warm up until their temperatures become equal. Since temperature is related to the average kinetic energy of molecules in a system, it is apparent that energy has moved from the hot metal to the cold water. *Heat is defined as energy that flows as a result of temperature differences.* In nature, heat always flows from higher temperature objects to colder temperature objects, until a common temperature is reached.

Heat is only one of many forms that energy may take. It is a very special type of energy. *Heat is the name given only to the energy that is transferred because of the temperature difference between two locations.* The words heat and heat transfer are synonymous. It is tempting, but incorrect, to think of heat as moving into an object and being stored as heat in that object. Actually, when energy flows into an object, it often changes from one form of energy to another. As an example, when light is absorbed it is no longer light but some other form of energy. One obvious effect of heat input is to increase temperature, in which case it is converted to thermal (internal) energy.

One of the effects of heat is to change temperature; heat gained by a substance can increase temperature, and heat lost can decrease temperature. However, sometimes heat can flow into or out of a substance, without the substance undergoing a temperature change. Heat transfer can also cause phase changes, such as melting, boiling, freezing, and condensation. (Phase changes are discussed in Section 5.3.) In all instances, once heat has been transferred, it no longer exists as heat; it has been converted to some other form of energy.

5.2.1 Specific Heat Capacity "*c*"

The amount of heat, Q, that is needed to cause a temperature change depends on the mass "m" of the object, the amount of the temperature change, ΔT, and the specific heat capacity, "c," of the substance. All of this information is summarized in the following equation.

$$Q = mc\,\Delta T \qquad \textit{Know} \qquad \text{eq. 5.5}$$

Q heat is often measured in calories or joules

4.19 joules = 1 calorie

m mass, measured in grams or kilograms
c *specific heat capacity* is the amount of energy required to change the temperature of 1 g of a substance by 1°C. Units are $\frac{\text{calories}}{\text{g} \cdot {}^\circ\text{C}}$

ΔT temperature change. Unit is °C.

It takes considerably more heat to increase the temperature of some substances than others. *The amount of heat required to change the temperature of 1 g of a substance by 1° C is the property of a substance known as its specific heat capacity "c"* and is included in Equation 5.5. Specific heat depends on temperature but varies slowly enough that the values at 20°C can be used with reasonable accuracy. The specific heat capacity of water is exactly 1.0 cal/g°C at 15°C because the calorie is defined as the amount of heat that it takes to increase the temperature of 1 g of water by one degree Celsius, at 15°C. The calorie is equivalent to 4.186 J, that is, 1.0 cal = 4.186 J. Food calories are really kilocalories, that is, one diet calorie is actually 1000 cal, or 4186 J.

Rounding these values,

$$1 \text{ calorie} = 4.19 \text{ J}$$

$$1 \text{ food calorie} = 1000 \text{ calories} = 4190 \text{ J}$$

You will learn more about specific heat capacity in Lab. 4

EXAMPLE 5.3

(a) How much heat is needed to increase the temperature of 200 g of water 5.00°C? (b) How much heat is needed to increase the temperature of 200 g of lead 5.00°C?

Refer to Table 5.3 for the specific heats water and lead.

Solution (a): Liquid water has one of the highest specific heat capacities of any naturally occurring substance. $c_{\text{water}} = 1\frac{\text{cal}}{\text{g} \cdot {}^\circ\text{C}}$

$$Q = mc\Delta T = 200\,\text{g} \cdot 1\frac{\text{cal}}{\text{g} \cdot {}^\circ\text{C}} 5.00^\circ\text{C} = 1000 \text{ cal}$$

Solution (b): $Q = mc\Delta T = 200\,g \cdot 0.030\frac{\text{cal}}{\text{g} \cdot {}^\circ\text{C}} 5.00^\circ\text{C} = 30.0 \text{ cal}$

Significantly different amounts of heat are required to cause equal changes of temperature in equal masses of different substances. It is lot easier to change the temperature of lead than it is to change the temperature of water! Lead has a low specific heat capacity and water has a high specific heat capacity.

The same substance in different phases has different specific heats. Water, for example, has a specific heat of 1.0 cal/g°C, ice has 0.50 cal/g°C, and steam has 0.48 cal/g°C. Specific Heats of Various Substances at 20°C

The specific heat of a substance depends on the way in which that substance is able to distribute energy put into it. Specific heats are still a subject of research because they can yield information about the submicroscopic structure of substances. For our purposes, it will be sufficient to be aware that specific heats are characteristic of a substance and are related to complex submicroscopic phenomena.

Table 5.3 Specific Heats of Various Substances at 20°C

Substance	C (calc/g°C, kcal/kg°C)
Aluminum	0.217
Brass	0.090
Copper	0.092
Gold	0.031
Iron or Steel	0.11
Lead	0.030
Silver	0.056
Glass	0.20
Ice (–5°C)	0.50
Porcelain	0.26
Wood	0.4
Human Body (average)	0.83
Protien	0.4
Ethyl alcohol	0.58
Glycerin	0.60
Mercury	0.033
Water (15°C)	1.000
Gases at Contant Pressure	
Air	0.25
Carbon dioxide	0.199
Helium	1.24
Nitrogen	0.248
Oxygen	0.218
Water (100°C steam)	0.482

Heat is not the only cause of temperature change. A bicycle pump becomes hot with use; heat did not cause the temperature increase of the pump. Remember, heat is defined as energy transferred because of a temperature difference. The pump and its surroundings were originally at the same temperature, so heat would have no reason to flow into the pump. What, then, caused the temperature increase? Work was done to the gas as it was compressed. When work is done to a substance, it must gain energy. When a gas gains energy, its temperature increases; therefore, the air in the pump got hotter. Work can be converted into any form of energy. In this case, work was done by the person and energy went from the person into the pump, air, and tire. Some of the work became thermal energy.

Another situation in which temperature is changed by something other than heat is a bullet striking a block of wood. Both the bullet and the wood around the bullet hole experience an immediate temperature increase. Work is done by friction to stop the bullet; most of this work goes into thermal energy in the bullet and wood. The entire block of wood eventually rises in temperature; the bullet and wood around the hole transfer heat to the block because they are at a higher temperature.

EXAMPLE 5.4

Above and beyond—Although very instructive, this may be hard for you to do without help. You may want to skip this example.

If a lead ball is dropped, its temperature increases when it strikes the ground. The original gravitational potential energy of the ball is first converted to kinetic energy during the fall and then to thermal energy on impact. Calculate the temperature increase of a lead ball dropped from a height of 20 m, assuming that half the thermal energy generated on impact goes into the ball.

The gravitational potential energy of the ball can be easily calculated using the formula, $G.P.E. \equiv mgh$

Since one-half of this energy is converted to thermal energy, the amount converted to thermal energy is $\frac{1}{2}G.P.E. \equiv \frac{1}{2}mgh$

$$\text{Thermal energy} = \frac{1}{2}mgh$$

We are asked for the increase in temperature of the lead ball. The thermal energy is Q. and $Q = mc\Delta T$.

Here we go with our substitutions. Brace yourself.

$$Q = mc\,\Delta T = \frac{1}{2}mgh$$

Dividing both sides of the equation by m and solving for ΔT yields,

$$\Delta T = \frac{gh}{2c}$$

The specific heat capacity of lead is $c = 0.030\frac{\text{kcal}}{\text{kg}\cdot{}^\circ\text{C}}$

Now, we substitute the values into our working equation.

$$\Delta T = \frac{gh}{2c} = \frac{9.8\frac{\text{m}}{\text{s}^2}\cdot 20.0\,\text{m}}{2\cdot 0.030\frac{\text{kcal}}{\text{kg}\cdot{}^\circ\text{C}}} = 3213\frac{\frac{\text{kg}\times\text{m}^2}{\text{s}^2}}{\text{kcal}}{}^\circ\text{C}$$

If you have been studying really hard, you might recognize that $\frac{\text{kg}\cdot\text{m}^2}{\text{s}^2} \equiv \text{joule}$

$\Delta T = \frac{3213\,\text{J}}{\text{kcal}}\,°\text{C}$ *We have a problem.* This solution contains units of energy in both joules and kilocalories. Let us convert kilocalories to joules.

$$\Delta T = \frac{3213\,\text{J}}{\text{kcal}}\,°\text{C} \cdot \frac{1\,\text{kcal}}{4190\text{ joules}}\,°\text{C} = 0.780°\text{C}$$

We have determined that the lead ball will increase in temperature by 0.780°C.

Wow, that was intense!

Any type of work done on a system can be converted completely or partially to thermal energy, increasing the temperature of the system. The final form of the energy transferred into the system is the same whether work is done on the system or heat enters the system. This means that if one is presented with a hot object, it is impossible to know whether its high temperature was caused by heat or work. This is another reason it does not make sense to talk of the heat content of an object; it does make sense to talk of the thermal energy content and the effects that heat may have.

One person who made a very important contribution to proving that heat and work have the same thermal effects was *James Joule* 1818–1889. He did this by carefully measuring the amount of work needed to produce the same effect as a certain amount of heat. Joule's experiments and those of other scientists and engineers were performed in the early and mid-1800s, long before it was established that molecules exist and that temperature is a measure of the average kinetic energy of molecules. But it was recognized that two objects of different temperatures would reach the same temperature if put in contact. It was also recognized that hot gases can do work, as in steam engines, and that the gases cool in the process.

The experiments of Joule and others are considered to be among the most important of the era because they led to the establishment of the principle of conservation of energy and to the development of the first law of thermodynamics. Not only are there many different forms of energy, but also they are interconvertible. Energy is neither lost nor created from nothing. Say it with me. "I believe in the conservation of energy."

Figure 5.4 James Joule 1818–1889

Things to know for Section 5.2

$$Q = mc\Delta T$$

Specific heat capacity ≡ *The amount of heat required to change the temperature of 1 gram of a substance by 1°C is the property of a substance known as its specific heat capacity "c."*

5.3 PHASE CHANGES AND LATENT HEAT

5.3.1 Heat Of Fusion

Ice cubes will keep a drink cold for many minutes while they slowly melt away. Ice from the freezer is usually colder than 0°C. When it is dropped into a drink, its temperature rises to 0°C and it begins to melt. If there is enough ice, the temperature of the drink falls to 0°C and stays there until all the ice has melted. *The temperature of a mixture of ice and water that has been mixed together and has come to thermal equilibrium is defined as 0°C.* Meanwhile, heat must be entering the drink because its surroundings are warmer than it is. Where is the heat going? It is not increasing temperature because the temperature of the drink cannot rise above 0°C until all of the ice has melted. (Ice cannot exist above 0°C.) Apparently, a great deal of heat is required just to melt the ice. Experimentally, it has been found that 80 cal are required to melt a single gram of ice, the same amount of energy that would raise the temperature of 1 g of water from 0°C to 80°C!

5.3.2 Latent Heat; Heats of Fusion, and Vaporization

Most *phase changes*, or changes of a substance from one phase of matter to another, require large amounts of energy compared to the energy needed for temperature changes. Energy must be put into a substance to cause it to melt or boil. Energy must be taken out of a substance to cause it to freeze or condense (gas to liquid). The energy can be heat transfer or can be due to work done on or by the system. *Energy used to cause a phase change does not cause a temperature change.* When ice melts at 0°C, it becomes liquid water at 0°C; when water boils at 100°C, it becomes steam at 100°C. The same is true in reverse. When water at 0°C freezes, it becomes ice at 0°C; when steam at 100°C condenses, it becomes water at 100°C.

To understand where energy goes during a phase change, first consider melting and boiling; both require energy input. There are attractive forces between molecules that must be overcome during melting and boiling. For a solid to melt, the spring-like forces that hold molecules in place (see Figure 5.1) must be broken, and a certain amount of energy is required to break each "spring." For a liquid to boil, attractive forces between molecules must be overcome, and work must be done to move the molecules to the larger separations found in gases (also shown in Figure 5.1). The amount of energy required is thus proportional to the number of molecules in the object and also to the strength of the forces acting between molecules. The amount of energy required to melt a substance is expressed as

$Q = H_f m$ *Know* The melting (or freezing) equation

where H_f represents the *heat of fusion* of the substance. The *heat of fusion* is the amount of energy required to melt or freeze 1 g of a solid.

The heat of fusion for water is $80\frac{\text{cal}}{\text{g}}$.

The amount of energy required to boil (vaporize) a substance can be determined using

$Q = H_V m$. The boiling Equation 5.6b,

H_V represents the *heat of vaporization* of the substance.

The *heat of vaporization* is the amount of energy required to vaporize or condense 1 g of a substance.

The heat of vaporization for water is $540\frac{\text{cal}}{\text{g}}$. That is a lot of energy. It is the energy source for hurricanes and steam burns.

37°C is the average internal temperature of a human body. Water evaporating from a person's skin requires 580 cal/g.

The values of both H_f and H_V depend on the strength of the forces between molecules and hence on the substance (see Table 5.4). Because the energy input required to melt or boil a substance does not produce a temperature change, it is, in a way, hidden or latent. For this reason, the energy associated with a phase change is sometimes referred to as *latent heat*.

Energy must be removed to reverse the above phase changes that is, to cause freezing or condensation. The process is exactly reversible; for example, the same amount of energy must be transferred to freeze as to melt. Equations 5.6a and 5.6b are valid for phase changes in either direction. When a

Table 5.4 Latent Heats of Various Substances

Substance	Melting Point (°C)	Heat of Fusion, H_f (cal/g, kcal/kg)	Boiling Point (°C)	Heat of Vaporization, H_v (cal/g, kcal/kg)
Oxygen	−218.8	3.3	−183	51
Ethyl Alcohol	−114	25	78	204
Ammonia	−75	108	−33	327
Mercury	−39	2.8	357	70
Water	0	80	100	540[a]
Lead	327	5.9	1750	208
Aluminum	660	90	2450	2720
Silver	960	21	2193	558
Copper	1083	32	2300	1211
Uranium	1133	20	3900	454
Tungsten	3410	44	5900	1150

[a]580 at 37°C.

substance freezes or condenses, attractive forces pull molecules into place and do work in the process. If the molecules are to stay close together, then this work energy must be removed from the system, often in the form of heat. A large amount of energy must be transferred in order to make ice cubes or to melt ice cubes. Once water has cooled to 0°C, $80\frac{\text{cal}}{\text{g}}$ must still be removed to freeze the water.

EXAMPLE 5.5

How much energy is required to raise the temperature of 10 g of ice that is at –30°C to water vapor at 140°C? We will break this problem into five separate parts.

Step by step 1: We will use $Q = mc\Delta T$ to calculate the amount of energy required to warm the ice from –30°C to its melting temperature of 0°C.

$$Q = mc\Delta T = 10\,\text{g} \cdot 0.50\frac{\text{cal}}{\text{g°C}} \cdot 30.0°\text{C} = 150\ \text{cal}$$

Step 2: We will use $Q = H_f m$ to determine the amount of energy required to melt the ice.

$$Q = H_f m = 80\frac{\text{cal}}{\text{g}} \times 10\,\text{g} = 800\ \text{cal}$$

Step 3: We will use $Q = mc\Delta T$ to calculate the amount of energy required to warm the water from 0°C up to its boiling temperature of 100°C.

$$Q = mc\Delta T = 10\,\text{g} \cdot 1.00\frac{\text{cal}}{\text{g} \cdot °\text{C}} \cdot 100°\text{C} = 1000\ \text{cal}$$

Step 4: We will use $Q = H_V m$ to determine the amount of energy required to change the water into a gas (boil).

$$Q = H_V m = 540\frac{\text{cal}}{\text{g}} \cdot 10\,\text{g} = 5400\ \text{cal}$$

Step 5: We will use $Q = mc\Delta T$ to calculate the amount of energy required to warm the water from 100°C to its final temperature of 140°C.

$$Q = mc\Delta T = 10.0\,\text{g} \cdot 0.5\frac{\text{cal}}{\text{g} \cdot °\text{C}} \cdot 40°\text{C} = 200\ \text{cal}$$

Final Answer = 150 cal + 800 cal + 1000 cal + 5400 cal + 200 cal = 7550 calories.

EXAMPLE 5.6

An aluminum ice cube tray containing 800 g of water has reached 0°C. If heat is being removed from it at the rate of 15 cal/s, how much time, in minutes, will it take for all of the water to freeze?

Step by step 1: We will use the freezing equation in order to determine the amount of energy that must be removed in order to freeze 800 g of water that is already at 0°C.

$$Q = H_f m = 80\frac{\text{cal}}{\text{g}} \cdot 800\,\text{g} = 64{,}000\ \text{cal}$$

Step 2: We are told that the energy is removed at a rate of 15 cal/s. We can use this information to determine the time. Notice that 15 cal/s has the units of energy over time. That is power. $Power \equiv \frac{E}{t}$

Solving for time yields,

$$t = \frac{E}{Power} = \frac{64{,}000\ \text{cal}}{15\frac{\text{cal}}{\text{s}}} = 4270\ \text{seconds}$$

Step 3: We are asked to give the time in minutes.

$$4270\,\text{s} \cdot \frac{1\,\text{minute}}{60\,\text{s}} = 71.0\ \text{minutes}$$

Note that the aluminum tray does not affect the amount of energy that must be removed since its temperature remains constant during the freezing process.

Figure 5.5 is a graph of temperature versus time for 1 g of ice as heat is put into it at a steady rate of 1 cal/s. The temperature rises steadily to 0°C, at which temperature the ice begins to melt. The ice water mixture stays at 0°C until all the ice has melted. Once this has happened, the temperature of the water produced rises steadily but at a slower rate than for ice. This is because water has a larger specific heat capacity than ice, so more energy (and hence more time) is required to raise its temperature each Celsius degree. Once the water has reached 100°C, it begins to boil. The temperature

Figure 5.5 A graph of temperature versus time for 1 g of water to which heat is added at a steady rate. The temperature remains constant during phase changes, showing that energy is required just to change phase. The steepness, or slope, of the graph (at times other than during phase changes) is inversely proportional to the specific heats of ice, water, and steam.

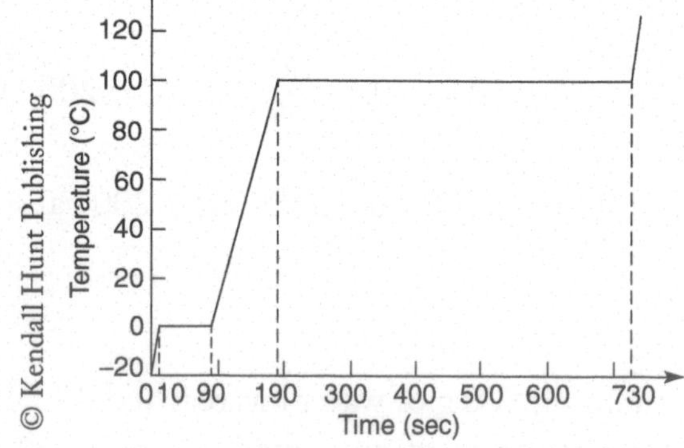

remains at 100°C until all the water has become steam. Because the heat of vaporization is larger than the heat of fusion, a longer time is required to boil the water than to melt the ice. After all the water has been converted to steam, the temperature again begins to rise steadily, in fact more rapidly than for water because the specific heat capacity of steam is less than that of water. Note that the graph in Figure 5.5 is valid only if energy is supplied at a constant rate.

5.3.3 Evaporation and Relative Humidity

Water and other substances can evaporate at temperatures far below their boiling points. A glass of water will slowly evaporate away in a few days. Ice even **sublimates** directly to water vapor. Ice cubes left in a freezer slowly disappear as water goes from its solid state directly to its gaseous state. What makes evaporation and sublimation possible at temperatures below the boiling point is random molecular motion. *Temperature is an indicator of the average kinetic energy of the molecules of a substance.* There are always a few molecules with kinetic energies far above average. Those with large enough kinetic energies are able to overcome the attractive forces that hold them in a solid or liquid, as shown in Figure 5.6(a), and escape as a gas.

Humidity has a definite effect on the net evaporation rate of water: the higher the humidity, the lower the net evaporation rate. A brief examination of humidity is necessary in order to understand its effect on evaporation.

At any given temperature, air has a certain capacity to hold water vapor. This capacity increases with temperature, as Table 5.5 shows. Relative humidity is defined as the ratio of actual vapor density to saturation vapor density, where saturation vapor density is the maximum amount that air can hold at a given temperature.

Figure 5.6 (a) Evaporation from an open container of water. More molecules escape than are recaptured. (b) A closed container prevents evaporation. The air above the water becomes saturated; more molecules are recaptured, and condensation forms.

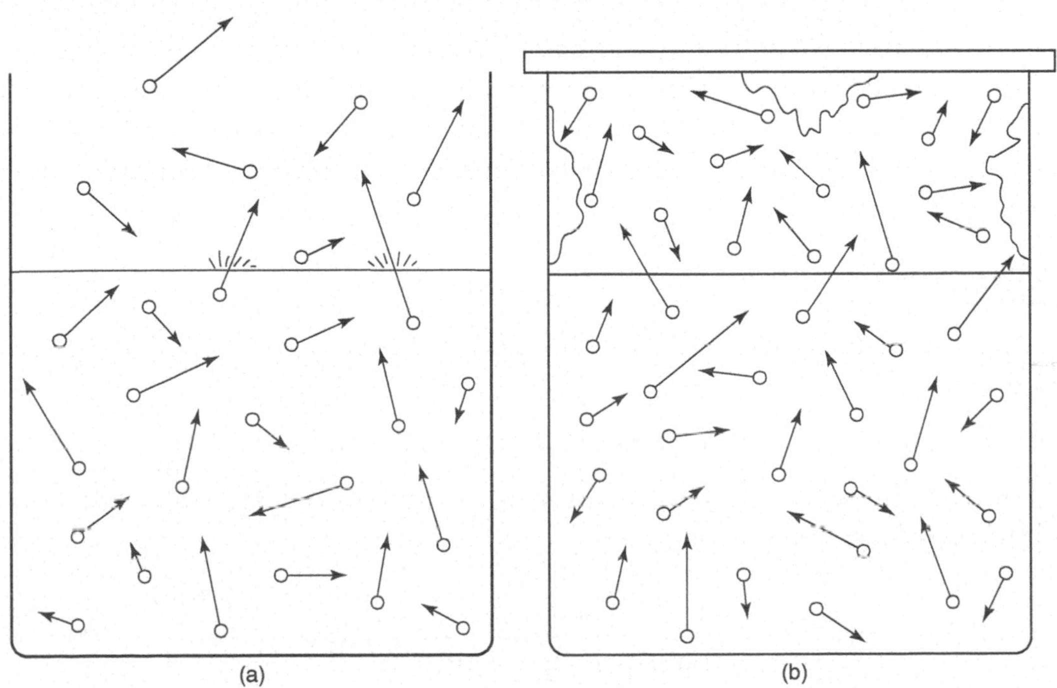

Table 5.5 Saturation Density of Water Vapor in Air

Temperature (°C)	Water Vapor Density (g/m³)
−10	2.36
0	4.85
5	6.80
10	9.40
15	12.83
20	17.30
25	23.0
30	30.4
37	44.0
40	51.1
60	130.5
80	293.8
95	505
100	598
200	7840

The quantitative expression for relative humidity is

$$\%\ \text{Relative humidity} = \left(\frac{\text{vapor density}}{\text{saturation vapor density}}\right)\cdot 100\%\ \textit{Know} \qquad \text{eq. 5.7}$$

A relative humidity of 100% means the air is totally saturated and can hold no more water vapor. The relative humidity is dependent on temperature. For example, if air with relative humidity of 100% at 20°C is heated to 30°C, its relative humidity falls to 57% (*see Example 5.7*). If this same air is cooled to less than 20°C, condensation will form, removing water vapor and preventing the relative humidity from rising above 100%.

EXAMPLE 5.7

If air with a relative humidity of 100% at 20°C is heated to 30°C, use Table 5.5 to determine its new relative humidity.

Solution: From Table 5.5, we see that the vapor pressure density is $17.3\frac{\text{g}}{\text{m}^3}$ at 20°C and that the saturation vapor pressure density at 30°C is $30.4\frac{\text{g}}{\text{m}^3}$. Substituting the values into Equation 5.7, we see that the % relative humidity is

$$\%\ \text{Relative humidity} = \left(\frac{\text{vapor density}}{\text{saturation vapor density}}\right)\cdot 100\% = \left(\frac{17.3\frac{\text{g}}{\text{m}^3}}{30.4\frac{\text{g}}{\text{m}^3}}\right)\cdot 100 = 57.0\%$$

Now, there are two reasons that relative humidity affects the net evaporation rate. First, if evaporation increases relative humidity to 100%, then water vapor simply condenses out of the air at the same rate as evaporation puts it in. Second, water molecules in the air may strike the water (or ice) and stick. The greater the humidity, the more likely this is to occur. Both these processes are happening in the closed container in Figure 5.6(b).

One example of the use of low humidity is found in self-defrosting freezers, where the air inside is kept very dry to speed sublimation and prevent frost. This can also dehydrate foods that are not properly sealed (so called "freezer burn"). People find that they are most comfortable when air has a relative humidity of 40%–50%. Lower humidity dries out skin and mucous membranes and can cause sinus trouble and other respiratory problems. Higher humidity suppresses the net evaporation rate of perspiration and may cause you to feel uncomfortably warm even at moderate temperatures. This brings up a new topic: the interplay of phase changes and the temperature of surroundings.

5.3.4 Situations Involving Both Phase and Temperature Changes

Why do people perspire when they are hot? This is the only method of cooling off that works when surrounding temperatures are warmer than body temperature. The energy needed to evaporate perspiration comes out of the body of the person. This loss of energy to evaporation can carry away waste thermal energy and may even reduce body temperature. If relative humidity is high, evaporation will be inhibited, leading to the tired but true cliché, "It isn't the heat, it's the humidity." On a hot humid day a person may seem to sweat more profusely than on a hot dry day. What is more likely is that on a dry day perspiration is less evident since it evaporates so quickly.

The cooling effect of evaporation can also be understood by considering random molecular motion. Evaporation occurs below the boiling point because some molecules have large enough speeds to break away. It follows that those with the highest speeds are the most likely to break away and that those left behind will have a lower average speed; that is, the residue will be at a lower temperature [refer to Figure 5.6(a)]. This submicroscopic explanation is entirely consistent with the large-scale explanation that the energy needed to cause the phase change comes from cooling the residue.

EXAMPLE 5.8

One method of reducing a dangerously high body temperature very quickly is to give a patient an alcohol rub. Evaporation of alcohol takes place very rapidly because air is normally devoid of alcohol vapor. (Relative humidity does not affect the evaporation of alcohol because relative humidity refers to the saturation of air with water vapor.) How many grams of alcohol must be evaporated from the surface of a 70-kg person to reduce his temperature by 1.5°C? The specific heat capacity of tissue is $0.83\frac{\text{cal}}{\text{g}\cdot{}^\circ\text{C}}$.

Table 5.4 lists the heat of vaporization for alcohol as $204\frac{\text{cal}}{\text{g}}$.

Solution: Assuming all the energy needed to evaporate the alcohol comes from the person and that his body produces negligible amounts of thermal energy from food digestion during the process.

If you believe in the conservation of energy, you know that in a closed environment: Heat gained by one system = Heat lost by another system.

More specifically, *heat gained by the evaporating alcohol will equal the heat lost by the person.*

$$Q_{\text{alcohol}} = Q_{\text{person}}$$

Since the alcohol evaporates, we will use $Q_{\text{alcohol}} = H_V m$ to represent the heat gained by the alcohol. The person's tissue will undergo a temperature change, so we will use $Q = mc\Delta T$ to calculate the heat lost by the person.

$$Q_{\text{alcohol}} = Q_{\text{person}}$$
$$H_V m_{\text{alcohol}} = mc\Delta T_{\text{Person}}$$

Solving for the mass of alcohol,

$$m_{\text{alcohol}} = \frac{mc\Delta T_{\text{Person}}}{H_V} = 70.0\times10^3\,\text{g}\cdot 0.830\frac{\text{cal}}{\text{g}\cdot{}^\circ\text{C}}\cdot 1.50^\circ\text{C} = 427\ \text{g of alcohol.}$$

Note that we converted the person's 70.0 kg of mass into grams of mass, in order to have units that are consistent with the constants.

This next example involves both a phase change and a temperature change. Energy is required for each, and these energies are equated because of the assumption that all the energy needed to evaporate the alcohol comes from the person. The basic principle involved is conservation of energy.

EXAMPLE 5.9

Suppose that 100 g of ice at an initial temperature of −10°C are put into 400 g of water at an initial temperature of 20°C. How many grams of ice will melt? The final temperature will be 0°C and there will be some ice left over. For simplicity, ignore the effects of the container.

Solution: Because the container can be ignored (not a bad assumption, since some cups are made of very good insulating materials), using our belief in the conservation of energy, we state that

Concept: Conservation of energy.

Heat gained by warming ice + heat gained by the melting ice = heat lost by the surrounding water.

Solution: Heat gained = Heat lost

Step by step 1: $Q_{\text{warming ice}} + Q_{\text{melting ice}} = Q_{\text{water}}$

$$Q = mc\Delta T_{\text{ice}} + H_f m_{\text{ice}} = mc\Delta T_{\text{water}}$$

Step 2: Factor out the mass of ice, m_{ice}.

$$m_{\text{ice}}\left(c\Delta T + H_f\right) = mc\Delta T_{\text{water}}$$

Step 3: Solve for the mass of ice, $m_{\text{ice}} = \dfrac{mc\Delta T_{\text{water}}}{c\Delta T_{\text{ice}} + H_{\text{fusion of ice}}}$

Step 4: Substitute the values.

$$m_{\text{ice}} = \frac{400\,\text{g}\cdot 1.00\frac{\text{cal}}{\text{g}\cdot{}^\circ\text{C}}\cdot 20.0^\circ\text{C}}{0.50\frac{\text{cal}}{\text{g}\cdot{}^\circ\text{C}}\cdot 10.0^\circ\text{C} + 80\frac{\text{cal}}{\text{g}}} = 94.1\ \text{g}$$

Note that 94.1 g of ice were melted. We were left with 5.9 g of ice, so we know that the final temperature was 0°C. If there had not been enough ice to cool the water to 0°C, then the final temperature would not have been 0°C.

In situations involving both phase and temperature changes the energy required for each temperature change and each phase change is identified and treated separately. All energies are considered to be positive, but a careful analysis is made of their origins and destinations.

Another interesting situation occurs when condensation forms. A great deal of energy is released when water condenses. What will its effect be on a cold drink? Water condenses on a cold drinking glass because the relative humidity of the cold air near the glass reaches 100%. Condensation causes the drink to warm up faster than it would if the air in the room were very dry and no condensation formed. Similarly, burns produced by the condensation of 100°C live steam on skin are more severe than those produced by the same mass of 100°C water. This is due to the extra energy released when the steam changes state from gas to liquid.

Things to know for Section 5.3

H_f is called the heat of fusion. The heat of fusion is the amount of energy required to melt or freeze 1 g of a solid.

$$Q = H_f m$$

The *heat of vaporization* is the amount of energy required to vaporize or condense 1 g of a substance.

$$Q = H_V m$$

$$\% \text{ Relative humidity} = \left(\frac{\text{vapor density}}{\text{saturation vapor density}} \right) \cdot 100\%$$

5.4 METHODS OF HEAT TRANSFER

Heat transfer is one of the most common of all physical phenomena. The rising sun takes the chill off of a frosty morning, while a furnace blows warm air through a home. Barefooted early risers feel the cold floor against their feet. Whenever there is a temperature difference, heat will flow. How does heat get from one place to another? There are only three methods of heat transfer: conduction, convection, and radiation. These are defined as follows:

1. *Conduction* is the transfer of heat through stationary matter by physical contact, such as between bare feet and a cold floor. Conduction is the result of kinetic energy transfered from molecule to molecule as they collide with one another.
2. *Convection* is the transfer of heat by the movement of a fluid, such as wind or ocean currents. Convection depends on gravity pulling the denser portions of a fluid beneath the less dense portions.
3. Heat transfer by *radiation* occurs when visible light, infrared (IR) radiation, or another form of *electromagnetic radiation* is emitted or absorbed. One example of this is sunlight warming the earth.

Each of these methods will now be discussed separately in some detail. Note that in real situations various combinations of all three methods may occur simultaneously.

5.4.1 Conduction

When you stand barefooted on a cold tile floor, your feet soon become cold, too. Heat is transferred by *conduction* to the floor from your feet. Because the feet are initially warmer, the molecules in the soles of the feet have a larger average kinetic energy than those in the floor. That kinetic energy is transferred to the floor by collisions between the molecules in the feet and those in the floor. When objects of different energies collide, energy is transferred from the more energetic to the less energetic object. Thus, the average kinetic energy of the molecules in the feet decreases, and your feet get cold.

Heat transfer by conduction occurs both at the surface between two systems and also within each system. Those barefoot, for example, get cold throughout, not just on the soles. The floor beneath warms up slightly, but the heat transferred to the floor is conducted away into the colder surroundings. Heat conduction within a system as well as between systems takes place by molecular collisions, which transfer heat from higher to lower temperature regions.

Why is it that a tile floor feels cold while a rug at the same temperature feels warm? Heat is conducted more rapidly by the tile than the rug. The tile floor has a higher conductivity than the rug. When you stand on cold tile, your feet get cold because heat flows out of the feet faster than it can be replaced by the body. Cold is not coming into your feet. Household insulation materials, for example, are chosen specifically for their ability to slow the rate of heat transfer by conduction because they have a low conductivity.

The *rate of heat transfer by conduction* depends on the conductivity of the material, the surface area in contact, the temperature difference between the two surfaces, and the thickness of the material through which the energy flows. As you may have guessed, we can summarize all this information in an equation.

The equation for the rate of heat transfer by conduction through a slab of material, such as that shown in Figure 5.7 is

$$\frac{Q}{t} = \frac{kA\Delta T}{L} \quad \textit{Know} \qquad \text{eq. 5.8}$$

When dealing with units, *let your constants be your guide.*

The phrase, rate of heat transfer, has been replaced with $\frac{Q}{t}$ and has units of joules per second, watts.

The coefficient of thermal conductivity of the substance, k, will have units of $\frac{\text{watts}}{\text{m}\cdot{}^\circ\text{C}}$.

"A" is the surface area in contact with another substance. Its units will be m^2.

"ΔT" is the temperature difference between the two surfaces.

"L" is the thickness of the material through which the energy is being conducted.

Looking at our equation for the rate of heat transfer by conduction, $\frac{Q}{t} = \frac{kA\Delta T}{L}$. We see that it is inversely proportional to the thickness "L" of the material. Thick layers of clothing are worn in the winter to slow the rate of heat loss by conduction. Another example is the thick layer of fat found in Arctic animals.

Figure 5.7 Heat will be conducted through a slab of material at a rate given by Equation 5.8. The slab could be anything from a window pane to a layer of body fat.

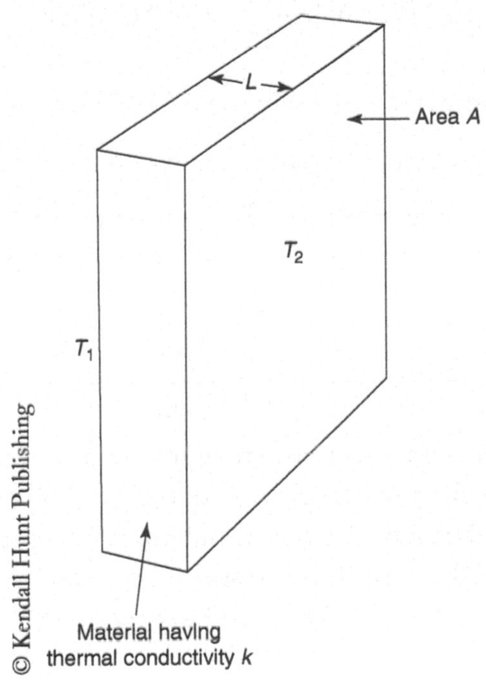

On the other hand, the rate of heat transfer by conduction is directly proportional to the area of contact. If you double the area "*A*," the rate of heat transfer by conduction will also be doubled.

This is because the number of molecules in physical contact between regions of different temperature is proportional to the area *A*. Conduction is heat transfer by molecular collision, so the greater the number of molecules in contact, the faster the rate of conduction will be. Pots and pans are usually flat-bottomed to have the largest possible area in contact with the burner on a stove. Windows are a major source of heat loss in the winter because they are very thin. To minimize this heat loss, the total window area can be limited by reducing the number and size of the windows in a building. Alternatively, the thickness of the windows can be increased in order to increase the thickness "*L*." Double-paned windows with a gas-filled gap between them can be used in order to decrease the conductivity "*k*," of the window.

EXAMPLE 5.10

Calculate the *rate of heat transfer by conduction* out of a human body, making the following assumptions: the interior temperature is 37.0°C, skin temperature is 33.0°C, the area of the person is 1.50 m^2, and the thickness of the tissues averages 1.00 cm.

Solution

Step by step 1: Examine the equation for the rate of heat transfer by conduction. $\frac{Q}{t} = \frac{kA\Delta T}{L}$

The value of *k* for human tissue is found in Table 5.6 and is $0.20\frac{\text{watts}}{\text{m}\cdot°\text{C}}$. Its units indicate that we should measure thickness in meters. The thickness of 1.00 cm must be converted to meters.

Step 2: Convert 1.00 cm to meters.

$$1.00\ \text{cm} \cdot \left(\frac{1.00\,\text{m}}{100\,\text{cm}}\right) = 0.0100\ \text{m}$$

We see that 1.00 cmentimeter is the same as 0.0100 meters.

The area "*A*" of 1.5 m^2 is given. The temperature difference ΔT is 4°C.

Step 3: Just substitute these values into Equation 5.8. The phrase "KAT lovers" helps me to remember the equation for the rate of heat loss by conduction.

$$\frac{Q}{t} = \frac{kA\Delta T}{L} = \frac{0.200\dfrac{\text{W}}{\text{m}\cdot{}^\circ\text{C}}\cdot 1.50\,\text{m}^2\cdot 4^\circ\text{C}}{0.0100\,\text{m}} = 120\ \text{W}$$

Although this rate of heat loss appears to be enough to remove waste thermal energy from a resting person (see Table 4.5), in reality the heat loss by conduction will not be this high. If that person is clothed, this rate of heat loss will be greatly reduced, but even if the person were naked, the rate of heat loss would be less than calculated because still air is such a poor conductor. In reality, convective cooling by blood circulation, convective cooling by the air, and the evaporation of perspiration are also needed.

5.4.2 Convection

Energy transfer by convection requires a fluid. A fluid is either a gas or a liquid. If it flows, it is a fluid. Convection can be natural or forced. Natural convection occurs when the force of gravity pulls the denser portions of a fluid beneath the less dense regions of the fluid. Hot air rising above a fire is actually pushed upward by sinking colder, denser air. Natural convection often occurs because materials expand when heated and contract when cooled. When a pan of water is heated on a stove, as shown in Figure 5.8, the water on the bottom is heated by conduction and expands. When the water expands, its density decreases and it is pushed to the top by sinking, denser water. These convection currents can be quite strong, and they are easily visible when a pan of water is heated vigorously.

The same method can be used to heat a home, as Figure 5.9 shows.

The forced-air furnace, as its name implies, creates forced convection. Although gases and liquids are not good conductors (see Table 5.6), they can transfer heat very rapidly by convection. *Heat transfer by convection* can be rapid because large numbers of molecules move large distances, whereas in conduction, molecules move only very small distances, transferring energy by collision.

Convection is very important in weather and climate. The Gulf Stream, for example, carries large amounts of warm water northward and is responsible for the moderate climate of Europe. Weather systems often involve large amounts of heat transfer by convection. For example, a cold front can move south out of Canada and rapidly lower temperatures in its path. One interesting example of forced convection is blood circulation, which, as has been mentioned, can transfer heat from one part of the body to another. The circulatory system is highly adjustable; blood flow can be increased or decreased to specific areas depending on need. If a person becomes overheated, blood vessels to the surface dilate, carrying more blood to the surface for cooling. Air convection around the body also speeds the rate of heat transfer. Windchill factors quoted in weather reports give the temperature of still air that would conduct away the same amount of heat from a person as does the wind actually present (see Table 5.7). It is difficult to calculate the amount of heat transferred by convection

Table 5.6 Thermal Conductivities of Common Substances

Substance	Thermal Conductivity (J/s m °C)
Silver	420
Copper	380
Aluminum	200
Steel	40
Ice	2.0
Glass (average)	0.84
Concrete, brick	0.84
Water	0.56
Human tissue without blood	0.2
Asbestos	0.16
Wood	0.08–0.16
Snow (dry)	0.10
Cork	0.042
Glass wool	0.042
Down	0.025
Air	0.023

Figure 5.8 Convection currents transfer heat throughout a pan of water. The water at the bottom of the pan is heated by conduction and expands. Cooler, denser water is pulled down by gravity forcing the warmer air to rise, creating a convection heating loop.

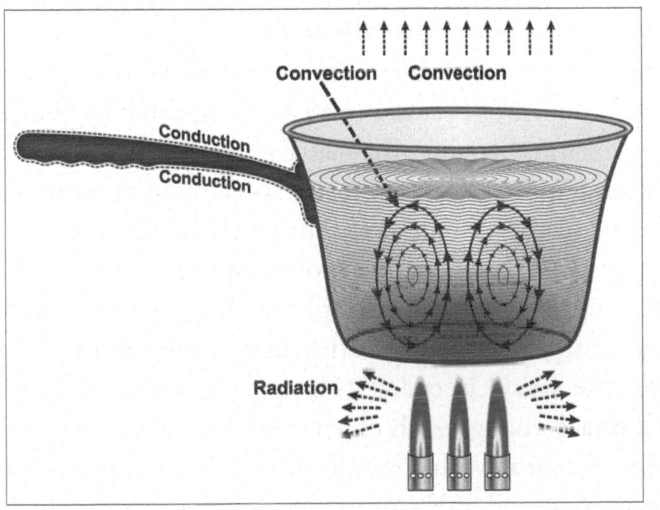

Figure 5.9 Natural convection currents can be used to heat a house. A fan is not always necessary.

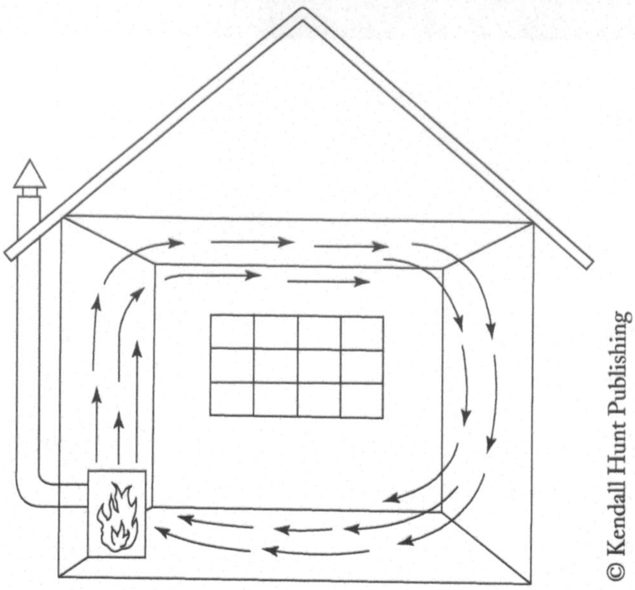

because it depends on so many factors. Data such as those in Table 5.7 usually come from measurements rather than from theoretical calculations.

Nature uses air as an insulator by cleverly controlling both conduction and convection. Hair and feathers are designed to trap air in small pockets and greatly slow the rate of heat transfer by convection. The density of uncompressed hair and feathers, which are mostly air, is quite small. Convection is going on in each little, as seen in Figure 5.10, but the distance that the air can move is very small and the rate of heat transfer is very small. Fiberglass household insulation works the same way. Because it is mostly air, it is cheaper to manufacture than many solid insulating materials.

An important extra twist to heat transfer by convection occurs when a phase change is involved. Evaporation of perspiration is a particularly good example. As was discussed in the previous section, evaporation cools the water and skin that it leaves behind. Without convection, evaporation would not continue for very long. The surrounding air would soon become saturated with vapor, halting the evaporation. Evaporation of perspiration must be accompanied by convection to be effective. In addition to avoiding saturation of the surrounding air, the vapor is removed so that, if it does condense, the energy it releases will not reenter the body.

It might appear that energy transfer by evaporation violates the rule that heat is always transferred from hot to cold. It does not because work is being done by the expanding water molecules, decreasing its internal energy and cooling the gas. Every process that moves energy from cold to hot involves work. Convection requires work because a force must be exerted to move the material. The work moves a material out of the system, and that material has a large amount of energy stored in it due to the phase change. Similarly, a refrigerator moves energy from cold to hot using a multistep process involving work. Refrigerators use both phase changes and convection in their cooling process. In the nineteenth century, before refrigeration units were invented, ice was cut from frozen lakes and transported all over the world in order to keep things cold. Refrigeration is another example of the effective use of a phase change with convection. Phase changes require so much energy that when they are used in combination with convection, very large amounts of energy can be transported quickly. This combination was used in Example 5.6, which considered cooling with an alcohol rub.

Table 5.7 Windchill Factor[a]

Moving Air Temperature (°C)	Wind Speed (m/s)				
	2	5	10	15	20
2	1.0	-6.6	-12	-16	-18
0	-1.3	-8.4	-15	-18	-20
-5	-7.0	-15	-22	-26	-29
-10	-12	-21	-29	-34	-36
-20	-23	-34	-44	-50	-52

[a]Temperature of still air that gives the same rate of cooling as winds at the speeds and temperatures listed.

Figure 5.10 Fur traps many small pockets of air, greatly slowing heat transfer by convection and conduction.

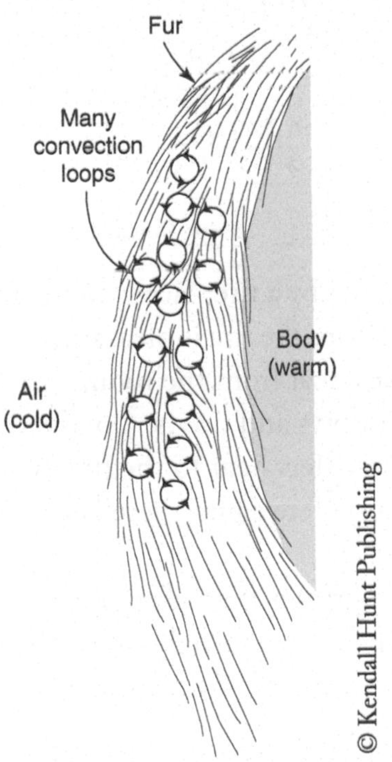

5.4.3 Heat Transfer by Radiation

Heat transfer by radiation is fundamentally different from either conduction or convection. Both of those processes transfer energy by molecular collision, and neither can take place in a vacuum. *Heat transfer by radiation* is the transfer of energy by electromagnetic radiation. By electromagnetic radiation, we mean light, both invisible and visible. Such radiation can travel through a vacuum, as from the sun to the earth, as well as through matter, such as a windowpane. Some examples of electromagnetic radiation are visible light, microwaves, IR, ultraviolet (UV), and x-rays. (Electromagnetic radiation will be discussed in detail in Chapters 16–18.)

Figure 5.11 A fireplace transfers heat by all three methods: conduction, convection, and radiation. Most of the heat put into the room is transferred by infrared radiation. Conduction also transfers heat into the room but at a lower rate. Convection removes heat from the room. The net effect of the fire may actually be a loss of heat from the room, depending on the source of air for the fire.

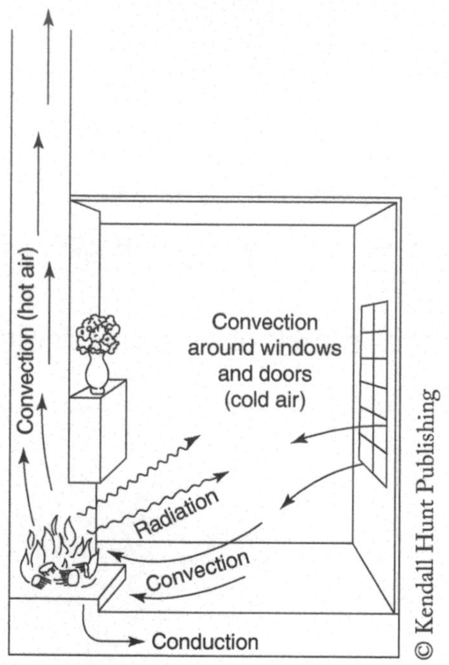

Most of the heat put into a room by a fireplace is in the form of infrared radiation, as Figure 5.11 illustrates. Although visible light from the fire is pleasant, it carries relatively little energy. Convection caused by the fire actually removes heat from the room. Hot air goes up the chimney and is replaced by cold air coming in around windows and doors. Conduction is too slow to be important. What is surprising is that the room also transfers heat to the fire by radiation. The fire, being hotter, transfers more radiation out than it receives. Here again there is a dependence on temperature; the net transfer of heat is from hot to cold.

Any body with a temperature greater than absolute zero emits heat by radiation. The amount of heat radiated per unit time depends very strongly on temperature.

The equation for the *rate of heat transfer by radiation* is

$$\frac{Q}{t} = \sigma e A\left[T_{\text{Hot}}^4 - T_{\text{Cold}}^4\right] \qquad \textit{Know}$$

The constant σ, sigma, has a value of

$$\sigma = 5.67 \times 10^{-8} \frac{\text{watts}}{\text{m}^2\text{K}^4}$$

Looking at the units of the constant, you should realize that when calculating the rate of heat loss by radiation, temperature must be measured in kelvin and that the area "*A*" must be in m^2. The values for *emissivity* "*e*" of the object ranges from zero to one. Objects that are perfect absorbers and perfect emitters of light will have an emissivity of 1. In the *IR* part of the light spectrum, the emissivity of a person's skin is 0.97.

EXAMPLE 5.11

What is the rate of heat transfer by radiation from a naked person with skin temperature of 33°C, standing in a darkened room with a temperature of 20.0°C? Note that the room is specified as being dark. This means that no visible light is present, so $e = 0.97$ can be assumed. The person has a surface area of 1.50 m^2.

Solution

Step by step 1: Examine the equation for rate of heat transfer by radiation.

$$\frac{Q}{t} = \sigma e A\left[T_{\text{Hot}}^4 - T_{\text{Cold}}^4\right]$$

Step 2: Convert the temperatures to kelvins by adding 273 to each Celsius temperature.

$$T_{\text{Hot}} = 33 + 273 = 306\ \text{K}\,;\ T_{\text{Cold}} = 20 + 273 = 293\ \text{K}$$

Step 3: Substitute the values into the rate of heat transfer by radiation equation.

$$\frac{Q}{t} = \sigma e A\left[T_{\text{Hot}}^4 - T_{\text{Cold}}^4\right]$$

$$\frac{Q}{t} = 5.67 \times 10^{-8}\,\frac{\text{watts}}{\text{m}^2\text{K}^4} \cdot 0.970 \cdot 1.5\,\text{m}^2 \cdot \left[(306\text{K})^4 - (293\text{K})^4\right] = 115\ \text{W}$$

Step 4: Use your calculator to verify the answer. You may need to practice on your calculator for that one. If you do not know how to enter these values correctly, seek help.

This is a significant rate of heat loss compared to the power produced by the body in leisurely activities—120 W when sitting at rest, for example.

Note that clothing plays an important role in slowing heat transfer by radiation, just as it slows heat transfer by conduction.

Things to know for Section 5.4

Rate of heat transfer by conduction: $\frac{Q}{t} = \frac{kA\Delta T}{L}$

Rate of heat transfer by radiation: $\frac{Q}{t} = \sigma e A\left[T_{Hot}^4 - T_{Cold}^4\right]$

Temperature in kelvins = Temperature in Celsius + 273

5.5 HEAT AND THE HUMAN BODY

This section considers the major aspects of heat and the body and describes certain medical uses of heat.

5.5.1 Effects of Heat on the Body

Heat transferred into or out of the body and thermal energy generated by the body itself can cause temperature changes. Normal body temperatures fall into a narrow range, near 97.7°F, 36.5°C, as Figure 5.12 shows. If body temperature becomes too high or too low, significant irreversible damage,

Figure 5.12 Ranges of body temperature under various conditions.

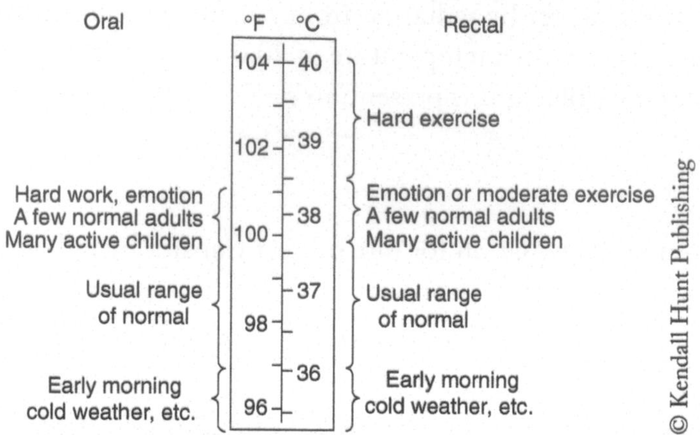

even death, can occur. One serious problem with the body's temperature-regulating mechanism is that cell metabolism increases with increasing temperature. Increased metabolism generates more than normal amounts of body heat, and this can cause temperature to increase further. A vicious circle results because the increased body temperature further speeds metabolism and the production of body heat. If body temperature rises above about 42°C (108°F), the body's cooling mechanisms cannot keep up, and external intervention, such as an alcohol rub, is necessary. An analogous problem exists when body temperature becomes too low; cell metabolism decreases, and insufficient body heat is produced to prevent body temperature from dropping further.

What are the body's mechanisms for getting rid of excess heat?

Does it have ways of producing increased body heat when necessary?

The first question can be answered by considering the three methods of heat transfer discussed in the previous section—conduction, convection, and radiation. Both conduction (as in Example 5.10) and convection are important in bringing heat to the body's surface. Blood circulation is forced convection, and when body temperature increases, there is increased blood flow to the skin to help remove heat. Similarly, if body temperature drops, blood flow to the surface is reduced to slow heat loss. Once heat reaches the surface, all three methods of heat transfer remove heat, provided that surrounding temperatures are lower than body temperature. (See Example 5.11 for a calculation of the rate of heat loss due to radiation from a person.)

As we have seen heat moves from hot to cold. Thus heat is transferred out of the body when surrounding temperatures are low and into the body when surrounding temperatures are high. Only the evaporation of perspiration keeps body temperature from rising uncontrollably when surrounding temperatures are high. The evaporation of perspiration relies on convection to carry away the energy used to make the perspiration change phase. In addition to the evaporation of perspiration from the skin there is a significant evaporation of water from the lungs.

EXAMPLE 5.12

Above and beyond—Although very instructive, this is a hard problem to do without help. It is ok to skip to the result.

On a day when air temperature is 34°C (the same as normal skin temperature) a cyclist is maintaining a speed of 15 km/hr. How many grams of water must this cyclist evaporate each minute to get rid of the body heat produced by his activity? The power consumption while cycling at 15 km/hr is 400 watts and cycling efficiency is 20%. Because the water begins at 34°C, it will require approximately 580 cal to vaporize 1 g of water.

Solution. Concept: Since air temperature is the same as skin temperature, conduction and radiation will not transfer any heat. It is correct, then, to conclude that all of the body heat must be transferred out of the person by the evaporation of perspiration from the skin and water from the lungs. First, we must determine the amount of energy that is given off as heat. Next, we calculate the amount of water that must be converted from liquid to gas in order to use up the extra energy.

Step by step 1: In table 4.2, we were told that the power consumption while cycling at this speed to be 400 W. Since cycling is 20% efficient, 20% of 400 W (or 80 W) is going to be useful for the work, and the remaining 320 W is going to be wasted through body heat.

Step 2: Recall the definition of power and solve it for energy.

$$\textit{Power} \equiv \frac{\textit{Energy}}{\textit{time}}\text{; therefore, } \textit{Energy} = \textit{power} \times \textit{time}$$

Step 3: Calculate the energy that has to be dissipated in 1 minute.

$$\textit{Energy} = \textit{power} \cdot \textit{time} = 320\frac{\text{j}}{\text{s}} \cdot 60.0\,\text{s} = 19{,}200\ \text{J}$$

Step 4: Convert the energy from joules to calories.

$$19{,}200\ \text{J} \cdot \frac{1.00\ \text{cal}}{4.19\ \text{J}} = 4580\ \text{cal}$$

Step 5: Use the heat of vaporization equation to determine the mass of water that must turn into a gas.

$$Q = H_V \cdot m\text{; solving for the mass, } m = \frac{Q}{H_V}.$$

$$m = \frac{4570\ \text{cal}}{580\frac{\text{cal}}{\text{g}}} = 7.90\ \text{g} \text{ of water is evaporated every minute.}$$

Note that this is the amount of water that evaporates every minute. This is a reasonable rate of evaporation, and the cyclist should be able to maintain a proper body temperature unless humidity is very high. The fact that most people would not cycle in hot humid weather is some indication that they realize the consequences. If the body cannot rid itself of the body heat it produces, then its temperature will rise, as seen in the following example.

EXAMPLE 5.13

How much will the temperature of the cyclist in the previous example increase in one hour if none of the body heat generated is lost to the surroundings? Assume the cyclist's mass is 76 kg and specific heat of a person is $0.83\frac{\text{kcal}}{\text{kg}\cdot{}^\circ\text{C}}$.

Solution: The energy required to cause a temperature change is given by

$$Q = mc\,\Delta T \qquad \textit{Know} \qquad \text{eq. 5.5}$$

Step by step 1: First, we will determine the amount of energy, *Q*, released in 1 hour. We saw in the last example that 4570 calories of energy are released every minute. In 1 hour, the energy "*Q*" will be 60 times that.

$$Q = 4570\frac{\text{cal}}{\text{min}} \cdot 60 \text{ minutes} = 274{,}000 \text{ cal}$$

Remember that diet calories are actually kilocalories.

Converting to kcal,

$$Q = 274{,}000 \text{ calories} \cdot \frac{1.00 \text{ kcal}}{1000 \text{ calories}} = 274 \text{ kcal kcal}$$

Step 2: Solve Equation 5.5 for ΔT and substitute the information.

$$Q = mc\,\Delta T\text{; therefore, } \Delta T = \frac{Q}{mc} = \frac{274 \text{ kcal}}{76\text{kg} \cdot 0.86\frac{\text{kcal}}{\text{kg} \cdot {}^\circ\text{C}}} = 4.35\ {}^\circ\text{C}.$$

Because we only entered the specific heat to two significant digits, we will round our answer to 4.4°C.

In Example 5.12, we saw that 80% of the cyclist's energy was given off as heat and that his temperature would rise 4.4°C if the heat was not dissipated. In Example 5.13, we calculated that the cyclist would have to have 7.9 g of water evaporated from his body every minute in order to prevent overheating. Note that 4.4°C is a very serious temperature increase. If the cyclist had been dehydrated, he would have stopped sweating and could not have dissipated the excess energy by evaporation. This can lead to a condition know as heatstroke. Please remember to drink plenty of fluids before, during, and after exercise. It would be even worse if the surrounding temperatures were higher than body temperature, in which case heat would be transferred into the body, adding to the body heat generated.

When air temperatures are moderate or low, heat will be transferred out of the body by all three methods of heat transfer. These methods can be manipulated to suit the situation, for example, with clothing and fans. The body has no difficulty in replacing heat transferred out of it when temperatures are moderate. However, this heat can have a noticeable effect on the environment; for example, a crowded room may become stuffy because of the heat and water vapor people give off. Consider the following example.

EXAMPLE 5.14

(a) Using data from Table 4.2, calculate the power in waste heat put into a classroom by 30 students.
(b) How many calories of energy do the students put into the room during a 50-minute lecture?

Solution: (a) Using the value of 210 W per person from Table 4.5, the power in waste heat is simply

$$\text{Total Power} = \frac{210\text{ W}}{\text{person}} \cdot 30 \text{ persons} = 6300 \text{ W}$$

$$\text{Total power} = 6300\frac{\text{J}}{\text{s}}$$

which is somewhat greater than the heat put out by an average electric clothes dryer on its highest setting.

(b) In order to determine the number of calories of energy that the students put into the room during a 50-minute lecture, please recall that the definition of power is

$$Power \equiv \frac{Energy}{time}$$

Solving for Energy, we get

$$Energy = Power \cdot time$$

$$Energy = 6300\frac{\text{J}}{\text{s}} \times 50 \text{ minutes} \times \frac{60\,\text{s}}{\text{min}} = 1.89 \times 10^7\,\text{J}$$

Note that this answer is in joules and that we were asked to determine the energy in calories.

The energy, $Q = 1.89 \times 10^7\,\text{J}\left(\frac{1\,\text{cal}}{4.19\,\text{J}}\right) = 4.51 \times 10^6\,\text{cal}$

This is enough heat to raise the temperature of 1000 kg of water to 4.5°C! Heating and air-conditioning systems in heavily populated buildings are specifically designed to handle heat put out by people.

In solving this example, it was correctly assumed that all the energy consumed by people sitting in a lecture goes into waste thermal energy. Most muscle action while sitting is that of one muscle against another, resulting in very little work. What little work is done, as in pushing a pencil across paper, is converted to thermal energy by friction.

When surrounding temperatures are low, it is necessary to limit the rate of heat transfer out of the body. The body does not sweat as much in cold weather, although there is still evaporation of water from the lungs. Thick layers of clothing reduce the rate of heat transfer by conduction, convection, and radiation. If clothing is insufficient, then the body loses more heat than it produces, its temperature drops, and uncontrollable shivering sets in. Shivering is a clever use of opposing muscles to do work against one another. Most skeletal muscles are involved in severe shivering, resulting in little work and a lot of thermal energy. Table 4.5 lists the energy consumption during shivering as 425 W (about the same as for playing tennis or cycling), all of which goes into body heat. Conscious physical activities are also used, such as walking about and rubbing hands together. All these physical activities produce waste thermal energy.

Any physical activity produces thermal energy since the body is not 100% efficient in converting food energy to work. Table 4.3 lists the efficiency of the body for various activities. The efficiency ranges from 20% down to 2%, meaning that from 80% to 98% of food energy is converted to thermal energy. Even when sleeping, the body pumps blood, digests food to nourish cells, etc., and produces thermal energy at the rate of about 80 W. By increasing the metabolic rate, as during a fever, the body can produce extra thermal energy without resorting to physical activity.

5.5.2 Diagnostic and Therapeutic Uses of Heat and Cold

Diagnostics. A person's overall temperature can indicate the presence and seriousness of an infection. One of the body's defense mechanisms against disease is to raise its temperature. Of course, if the temperature becomes too high, it can be of danger to the person, too. Low body temperature (hypothermia) also occurs in some medical situations and requires attention. Body temperature has an effect on the measurement of the gas content of blood (e.g., oxygen and carbon dioxide), so such measurements must be corrected for variations from normal temperature.

Skin temperature is lower than core temperature but higher than normal room temperature. It is therefore possible to measure the infrared radiation (IR) from a person. Since this radiation is proportional to absolute temperature to the fourth power, the amount of infrared radiation is a sensitive indicator of surface temperature. The technique of measuring IR radiation and thereby mapping temperature is called *thermography*. The picture obtained is called a *thermograph*, an example of which is shown in Figure 5.13.

Thermography gives an indication of blood supply, since one of the main methods of heat transfer in the body is blood flow. A depressed skin temperature indicates a deficiency in blood flow to a given region. This could be caused by clotting, stroke, and so on. A locally elevated temperature can indicate the presence of a malignant (cancerous) tumor. Such tumors grow very rapidly compared to other tissues and thus require an increased blood supply. There are difficulties in thermography, such as the effect of air currents and the fact that many healthy people have a nonuniform surface blood supply, which complicate its use as a diagnostic tool. Thermography has the advantage of being noninvasive, but it is not widely used at present. A structured evidence review of thermography for breast cancer (Kerr, 2004) reached the following conclusions: "The evidence that is currently available does not provide enough support for the role of infrared thermography for either population screening or adjuvant diagnostic testing of breast cancer. The major gaps in knowledge at this time can only be addressed by large-scale, prospective randomized trials. More robust research on the effectiveness and costs of technologically advanced infrared thermography devices for population screening and diagnostic testing of breast cancer is needed, and the conclusions of

Figure 5.13 Infrared of different frequencies correspond to different temperatures.

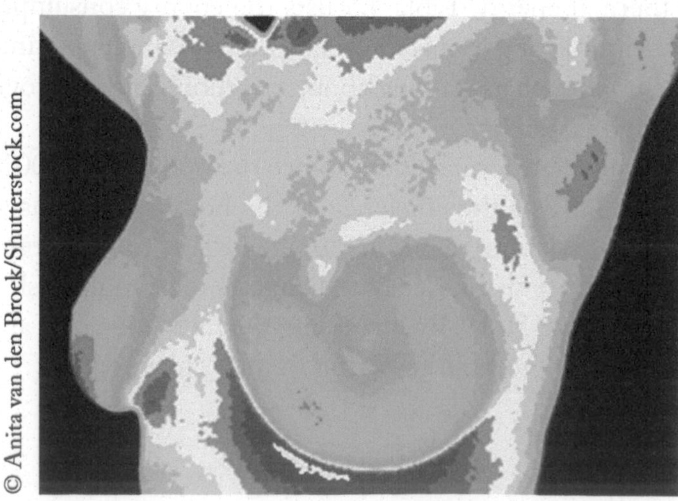

this review should be revisited in the face of additional reliable evidence." The American Medical Association has concluded that thermography has no proven medical value and some insurance companies no longer cover its use.

Therapeutic Uses of Heat. Local and general applications of heat have long been recognized for their therapeutic value. For example, heat applied to a sore muscle can significantly relieve pain. The mechanism of relief by elevated temperature seems to be twofold: relaxation of muscles and increased blood flow. When the temperature of a region is elevated, the body responds by increasing blood circulation to that region to cool it down—that is, to carry away heat by forced convection. The precise reasons that relaxation and increased circulation help are not well understood, but the effect is very real.

There are many methods of treating various ills with heat. The simplest is by conduction, using hot towels, heating pads, and the like. Heat transfer by radiation is also feasible; heat lamps emit most of their energy in the form of IR radiation. Newborn infants are sometimes placed under an IR heater to replace the heat they would have received from their mothers. Within a few hours of birth an infant's temperature-control mechanism works well enough that the heater is no longer necessary (as in Figure 5.13).

Forms of electromagnetic radiation other than IR are also used. Deep heating with microwaves and other forms of radio waves is called microwave (915 MHz and 2450 MHz) or radio diathermy (27.12 MHz). Microwave ovens carry this form of heat transfer a bit further. This application of heat transfer must be carefully controlled to affect only the intended area.

Ultrasound diathermy is another form of "heat" treatment. It is completely different from those using conduction of radiation. Ultrasound is sound having a frequency above the human audible range. Ultrasound can carry energy into the body, depositing it as thermal energy. If sufficiently intense, it can cause a significant local temperature increase. Because sound is a coordinated vibration of matter, it is not really heat transfer, but the energy carried in by the sound does end up as thermal energy when absorbed. (Recall from Section 5.1 that the motion associated with thermal energy is random and disordered. As will be seen in Chapter 8, sound energy is an ordered vibration.)

The applications of heat treatments vary since each method of treatment has its own advantages and disadvantages. For example:

Microwaves are thought to cause cataracts; ultrasound penetrates deeper than microwaves, but it can cause tissue burns near bones; and prolonged application of IR from a heat lamp can cause swelling.

Therapeutic Uses of Cold. The removal of heat from the body can also be of therapeutic value. The method of removing heat is most often conduction, sometimes convection. Lowered temperature acts as a local anesthetic. Children who are teething are fond of sucking on ice cubes to relieve pain. Swelling can sometimes be reduced by the application of ice packs.

Surgery using the application of cold is called cryosurgery. If the temperature of the entire body is lowered, the metabolic rate drops and most bodily functions are slowed. This can be advantageous in certain types of surgery. More often, however, cold is used to freeze small regions of the body. Warts and tumors can be treated in this way. Small parts of the brain can be frozen to treat Parkinson's disease, although this technique has given way to the use of drugs. A detached retina can be reattached by spot freezing it to the back of the eye. The frozen tissue forms scars, which serve as tiny welds. Freezing usually kills or injures tissue, but it can have therapeutic value.

Finally, lowered temperature can serve as a preservative, as in food refrigerators and freezers. Blood, bone marrow, and sperm are among the substances preserved by freezing. These can be thawed and revived, suggesting the possibility of placing people in suspended animation. It is not clear that this will ever be possible because the survival rate of various tissues depends on the rapidity of the freeze and thaw, and no single process works well for all types of tissues. Techniques are being developed for the freeze preservation of more complicated tissues and organs. Some success has been achieved in preserving corneas, for example.

Things to know for Chapter 5

$T_F = \frac{9}{5}T_C + 32 \quad T_{kelvins} = T_C + 273$

$\Delta L = \alpha L \Delta T \quad \%\text{Relative humidity} = \left(\frac{\text{vapor density}}{\text{saturation vapor density}}\right)100$

$$density \equiv \frac{mass}{volume}$$

Using symbols, we write the definition of density as follows: $\rho \equiv \frac{m}{V}$

Heat, $Q = mc\Delta T$; this equation is used to calculate the heat transferred in order to warm or cool a substance.

Specific heat capacity ≡ *The amount of heat required to change the temperature of 1 g of a substance, by 1°C is the property of a substance known as its specific heat capacity "c."*

H_f is called the *heat of fusion*. The *heat of fusion* is the amount of energy required to melt or freeze 1 g of a solid.

$$Q = H_f m$$

The *heat of vaporization* is the amount of energy required to vaporize or condense 1 g of a substance.

$$Q = H_V m$$

Rate of heat transfer by conduction: $\frac{Q}{t} = \frac{kA\Delta T}{L}$

Rate of heat transfer by radiation: $\frac{Q}{t} = \sigma e A[T_{Hot}^4 - T_{Cold}^4]$

QUESTIONS

5.1 What is the difference between atoms and molecules? How do compounds and elements differ?

5.2 Based on the submicroscopic structure of solids, liquids, and gases, explain why gases are easy to compress but liquids and solids are very difficult to compress.

5.3 If a bottle is filled with liquid with no air space at the top and is tightly capped, it will break when warmed up. If a few cubic centimeters of air are left above the liquid, the bottle will not break. Explain both situations.

5.4 Describe three devices that use thermal expansion to indicate temperature.

5.5 What is the difference between density and mass? Give an example of how the density of an object can be used to identify the substance of which it is composed.

5.6 What is the relationship between temperature and heat?

5.7 Heat is a form of energy but is not stored as heat. Explain what happens to heat energy that enters an object.

5.8 Does the presence of large amounts of water in the body help keep body temperature stable? Explain why or why not.

5.9 What would be more effective in keeping a bed warm in an unheated house, a hot water bottle or an equal mass of brick with the same temperature?

5.10 Temperatures in coastal areas are generally more moderate than temperatures inland. For example, temperatures in San Francisco are neither as hot in the summer nor as cold in the winter as in Sacramento, 90 miles inland. Explain why this is so.

5.11 Why is an ice bag at 0°C a more effective coolant than an equal amount of water at 0°C?

5.12 In cooking, more heat is required to keep an uncovered pot of water boiling than one that is covered. Explain why.

5.13 Relative humidity increases at night because the capacity of air to hold water vapor decreases with temperature. The dew point is the temperature at which relative humidity would be 100%. Evening temperatures usually do not fall much below the dew point. Why?

5.14 Some dehydrated foods are said to be freeze-dried. In this process, frozen food is rapidly dehydrated in a vacuum. Explain how the vacuum speeds dehydration and why the frozen food must be heated during the process.

5.15 Exercising in very cold weather can freeze lung tissue, but exposed skin may not be affected. Why does the temperature of the lungs drop more than that of the skin?

5.16 A cold room with high humidity feels less chilly than a dry room with the same temperature. Why?

5.17 The relative humidity of air in homes is usually very low in the winter. Why is this?

5.18 Which would be more comfortable in the summer, a house cooled by air-conditioning or one cooled by a "swamp cooler"? Standard air conditioners cool by passing it over cold coils. Swamp coolers cool the air by passing it through a fiber mesh and evaporating large amounts of water from the mesh. Hint: Consider the effects of each device on relative humidity.

5.19 How do the heats of fusion and vaporization of water tend to lessen climatic temperature extremes?

5.20 When camping out and sleeping on the ground, can you be warm with little or no insulation beneath you? Explain.

5.21 Closing window curtains on a cold night reduces heat loss. What methods of heat transfer are slowed by the curtains?

5.22 Why are clear nights generally colder than cloudy ones?

5.23 Why do desert dwellers wear loose-fitting white clothing that covers most of the body, rather than little or no clothing? (Note that this clothing is an advantage during both hot days and cold desert evenings.)

5.24 Will a swimming pool with dark-colored plaster be warmer than one with white plaster? Explain.

5.25 Describe the methods by which a solar blanket (a double layer of plastic enclosing many air bubbles) increases heat trapped by a swimming pool and reduces heat loss by the pool.

5.26 People who exercise vigorously outdoors on a cold day risk a reduced body temperature when they stop. Why is this?

5.27 Double-paned windows are designed to reduce heat transfer. However, the layer of air between the glass panes is never thicker than about 1 cm. A thicker layer of air would reduce conduction. Why isn't it used?

5.28 Thermos bottles are designed to reduce all forms of heat transfer. Figure 5.14 is a cutaway drawing of a thermos bottle. How do the various parts of the thermos bottle reduce the rate of heat transfer? For example, what is the function of the silvering? The long thin-walled neck? The use of glass?

5.29 Why is wet fur such a poor insulator?

5.30 What are the major methods of heat transfer within the body, between the body, and its surroundings?

5.31 When the body becomes warm, it responds by perspiring and by increasing circulation to the surface to carry away heat. What effect will this have on a person sitting in a 40°C (104°F) hot tub?

5.32 A person who is overheated will have a red, flushed appearance. A person who is chilled will be pale with blue lips and fingertips. Explain these appearances in of the body's control of one of the methods of heat transfer.

Figure 5.14 Cutaway drawing of a thermos bottle (see Question 5.28).

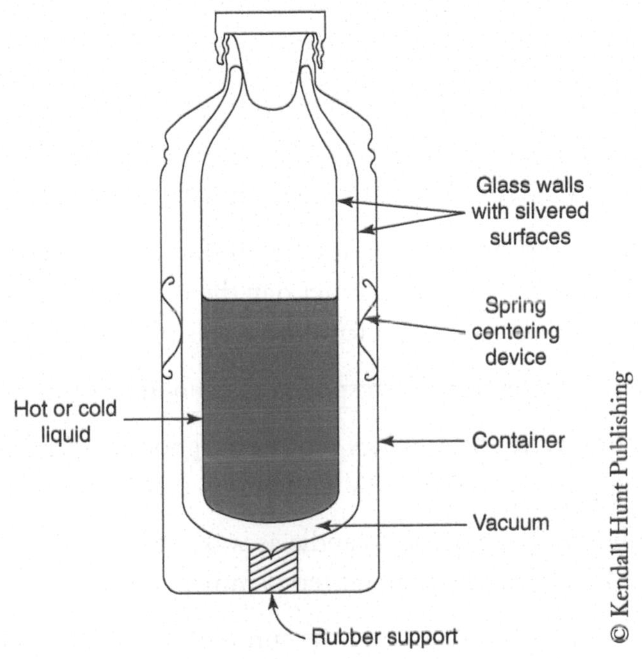

5.33 Smoking reduces the temperature of the extremities. What is the apparent effect of smoking on blood circulation?

5.34 Keeping in mind that fever is a defense mechanism against disease, explain why a person developing a fever will shiver as if cold? Is it advisable to cover this person with thick blankets?

Problems

SECTION 5.1

√5.1 (I) (a) Normal body temperature is 37°C. Convert this to Fahrenheit. (b) What is the temperature in degrees Celsius of a person with a fever of 104°F?

5.2 (I) (a) Room temperatures are kept at 68°F in the winter to conserve energy. What is this in degrees Celsius? (b) On a summer day the air conditioner is set to come on when the temperature rises above 27°C. How hot is this in degrees Fahrenheit?

5.3 (I) What Celsius temperature corresponds to absolute zero?

√5.4 (I) One of the properties that makes tungsten a good material for light bulb filaments is its high melting point of 3410°C. What temperature is this on the Fahrenheit scale?

5.5 (I) Frozen alcohol makes as good a candle as wax, with one significant disadvantage:·Alcohol melts at –114°C. What Fahrenheit temperature is this?

5.6 (I) The temperature of the surface of the sun is about 8000°F. What is this temperature on the Celsius and Kelvin scales?

5.7 (II) At what temperature do the Fahrenheit and Celsius scales have the same numerical value?

5.8 (I) A gap must be left between steel railroad rails to allow for thermal expansion. How large a gap is needed if the maximum temperature reached is 50°C more than the temperature at which the rails were laid? The length of a rail is 10 m. Refer to Table 5.1 for the coefficient of linear expansion of steal.

5.9 (I) The Golden Gate Bridge has an overall length of approximately 2800 m. If the bridge experiences temperature extremes from –20° to +40°C, what will its change in length be? The bridge is made primarily of steel. Refer to Table 5.1 for the coefficient of linear expansion of steal.

5.10 (I) The Great Pyramid of Cheops is 145 m high on a cold winter day when its temperature is 5°C. How high is it in the summer when its temperature is 20°C?

(Because of its size the pyramid warms and cools slowly and does not reach the temperature extremes of the area.) You may assume that the coefficient of expansion is the same as that for concrete.

5.11 (I) A man's height is measured by his standing next to a device made of aluminum, and he is found to be 1.80 m tall. If the original measurement was made at a temperature of 18°C, what height would be measured for the same man on a day when the temperature is 35°C? Is this difference enough to be of any concern? *Note that the man's temperature does not change, but that of the measuring device does.*

5.12 (III) If you have a 1 liter glass bottle full of water at 20°C, how much water will overflow the bottle if the temperature increases to 30°C? You may assume that the bottle is a cube 10 cm on a side.

SECTIONS 5.2 AND 5.3

√5.13 (I) How many calories of heat would be required to raise the temperature of a 45-kg person by 2.0°C? How many food calories is this? We know that the specific heat capacity of a human is $0.83\ \frac{\text{cal}}{\text{g}\cdot°\text{C}}$.

√5.14 (I) When a scalpel is sterilized, its temperature may rise to 150°C. Calculate the amount of heat in calories that must be removed from a 30-g steel scalpel to reduce its temperature from 150° to 20°C.

√5.15 (I) How many calories of heat are needed to thaw out a 0.30-kg package of frozen vegetables originally at 0°C, assuming the heat of fusion is the same as that of water?

√5.16 (I) A large power plant converts 1000 kg of water to steam every second to run its generators. (a) Calculate the amount of heat needed each second just to vaporize the water, assuming this is done at 100°C. (b) What is the power input in megawatts for this purpose?

√5.17 (I) A burn produced by live steam at 100°C is more severe than one produced by the same amount of water at 100°C. To verify this, (a) calculate the heat that must be removed from 5.0 g of water at 100°C to lower its temperature to 34°C (skin temperature); (b) calculate the heat that must be removed from 5.0 g of steam at 100°C to condense it and then lower its temperature to 34°C, and compare this with the answer to part (a).

5.18 (II) Suppose that 4.0 kg of shaved ice is needed to keep medication cold in a room that has no refrigerator. (a) What amount of heat must be removed from 4.0 kg of water to make ice once the water has reached 0°C? (b) How long will it be before all 4.0 kg of ice melts if heat enters it at the rate of 25 cal/s?

5.19 (II) On a summer day, sunlight may put 100 million calories into an 80,000-liter swimming pool. (a) What temperature rise occurs if no other heat enters or leaves the pool? (b) How many liters of water would have to evaporate from the pool to keep its temperature constant?

5.20 (II) A 0.50-kg block of material is heated from 20° to 35°C by the addition of 420 cal of heat. Calculate the specific heat of the block and identify the substance of which it is composed, assuming that it is made of a pure substance.

5.21 (II) (a) How much heat is needed to raise the temperature of a 1.0-kg steel pot containing 2.0 kg of water from 25°C to the boiling point and then to boil away 0.50 kg of the water? (b) If heat is supplied to the pot of water at the rate of 125 cal/sec, how long will this take?

5.22 (II) Following strenuous exercise a person has a temperature of 40°C and is giving off heat at the rate of 50 cal/s. (a) What is the rate of heat loss in watts? (b) How long will it take for this person's temperature to return to 37°C if his mass is 90 kg?

5.23 (II) If a 30-g lead bullet with a speed of 600 m/s strikes a practice target and half of the thermal energy generated is absorbed by the bullet, what is the temperature increase of the bullet?

5.24 (II) A 1000-kg car rolls down a hill starting from rest and loses 100 meters in altitude before braking to a stop. If half the thermal energy generated in the brakes is absorbed by 20 kg of material having a specific heat of 0.30 $\frac{\text{cal}}{\text{g}\cdot{}^\circ\text{C}}$, what is the temperature increase of the affected material? Assume that friction other than that of the brakes is negligible.

5.25 (II) A 0.20-kg aluminum bowl containing 0.75 kg of soup at 20°C is put into a freezer. If the freezer removes 80,000 cal from the bowl of soup, what is the final temperature? Assume that the soup has the same thermal properties as water.

5.26 (II) Warming cold hands by rubbing them together is a time-honored art. If a woman rubs her hands back and forth for a total of 20 rubs, calculate the temperature increase of her hands given the following information: The force exerted in each of the 20 rubs is 40 N, the distance moved in each rub is 0.075 m, and the total mass of the hands is 1.5 kg.

5.27 (II) An ice cube having a mass of 50 g and an initial temperature of −20°C is placed in 400 g of 30°C water. What is the final temperature of the mixture if the effects of the container can be neglected?

5.28 (II) Five grams of 20°C water is poured onto a 600-g block of ice that has an initial temperature of −15°C. What is the final temperature if the effect of the surroundings can be neglected?

5.29 (III) The formation of condensation on a cold glass of water will cause it to warm up faster than it would have otherwise. If 8.0 g of water condenses on a 100-g glass containing 300 g of water at 5.0°C, what will be the final temperature? Ignore the effect of the surroundings.

5.30 (I) What is the relative humidity on a day when the temperature is 25°C and the air contains 20.0 g/m^3 of water vapor? See Table 5.6 concerning the saturation density of water.

5.31 (II) What is the density of water vapor in grams per cubic meter in the desert when relative humidity is 10% and air temperature is 40°C?

5.32 (II) If the relative humidity is 75% on a summer morning when the air temperature is 20°C, what will it be later in the day when the temperature reaches 30°C? Assume that the water-vapor content of the air is constant.

5.33 (II) The relative humidity is 40% late in the day when the air temperature is 20°C. What will it be later that night when the temperature drops to 10°C? You may assume that the vapor content of the air is constant.

5.34 (III) What is the dew point on a day when the relative humidity is 39% and the temperature is 20°C?

5.35 (III) One day the relative humidity is 90% and the temperature is 25°C. (a) How many grams of water will condense out of each cubic meter of air if the temperature drops to 15°C? (b) How much energy does the condensation from each cubic meter release?

SECTION 5.4

√5.36 (I) Calculate the rate of heat conduction in watts out of an animal with 3.0-cm thick fur. The surface area of the animal is 1.5 m^2, its skin temperature is 35°C, and the temperature of the surrounding air is 0°C. Heat losses due to convection and radiation can be neglected, and the thermal conductivity of the fur can be assumed to be the same as that for air. Note that this is not much different than for a human in a warm room, as seen in Example 5.10.

√5.37 (I) Calculate the rate of heat conduction through the walls of a house if they are 8.0 cm thick and have twice the thermal conductivity as glass wool. The total area of the walls is 120 m^2 and the inside temperature is 20°C, whereas the outside temperature is 5°C. Note that this is only conduction through the walls and does not include conduction through the windows or ceiling.

5.38 (I) If a small iceberg with a mass of 20 million kg moves south from the Arctic, how much heat is required to melt the iceberg? Note that this amount of energy was released earlier at the origin of the iceberg.

√5.39 (I) What is the rate of heat loss by radiation from a man completely clothed in white (head to foot) if his skin temperature is 34°C and the surrounding temperature is 10°C? His surface area is 1.4 m^2 and the emissivity of the clothing is 0.2.

√5.40 (I) What is the rate of heat loss by radiation from a black roof of area 250 m^2 if its temperature is 20°C and that of the surroundings is 10°C? The emissivity of the roof is 0.95.

5.41 (I) Glowing embers in a fireplace have a temperature of 850°C, an emissivity of 0.98, and a surface area of 0.15 m^2. If the temperature of the surroundings is 23°C and 50% of the radiation from the fire enters the room, what is the rate of heat transfer into the room in kilowatts?

5.42 (II) Compare the rate of heat conduction through an 8.0-cm-thick wall of area 10 m^2 with that through a 0.75-cm-thick window of area 2.0 m^2, for the same temperature difference across each. The thermal conductivity of the wall can be assumed to be twice that of glass wool.

5.43 (II) It is usually cold next to a window on a winter day because heat conduction through the window is rapid enough to cool the air next to it. To see just how large the rate of heat conduction through a glass window is, calculate it for a window of area 3.0 m^2 and thickness 0.80 cm if the temperatures at the outer and inner surfaces are 5.0° and 10°C, respectively.

5.44 (II) If 5.0 g of water evaporate from 150 g of 90°C coffee, what will the final temperature of the coffee be? Assume the styrofoam cup is such a good insulator that all other forms of heat transfer can be ignored. Note that the "steam" above the coffee is really condensed water vapor droplets.

5.45 (III) What is the temperature difference across a double-paned window of area 3.0 m^2 if each pane of glass is 1.0 cm thick, as is the air gap between them, and the rate of heat conduction is 200 W. Hint: Calculate the temperature difference across each layer separately and add them together. You may assume that convection in the air gap is negligible.

5.46 (III) (a) A glass coffee pot has a bottom with an area of 400 cm^2, and the coffee in the pot has a temperature of 100°C. If the bottom is 0.75 cm thick, how hot must the underside be to conduct heat into the pot at the rate of 500 W? (b) How many grams of water boil away each second if convection plus boiling is the only method of energy transfer out of the pot?

5.47 (III) Considerable temperature decreases can occur when cold air blows through an open door. If 20 m^3 of 0°C air enters a room, how much heat is required to warm it to 20°C?

SECTION 5.5

√5.48 (I) How many calories of heat will a person lose by evaporating 1250 g of water and perspiration in a day? Use the value of 580 cal/g.

√5.49 (I) Calculate how many grams of perspiration a person must evaporate during physical exercise to get rid of 100,000 cal of heat. Use the value of 580 cal/g.

√5.50 (I) How many calories of heat would have to be put into the shoulder of a person receiving a heat treatment to raise the shoulder's temperature by 5°C? The mass of the shoulder region is 5.0 kg, and the effect of circulation and other factors can be neglected. The specific heat capacity of a human is $0.83\ \dfrac{\text{calories}}{\text{g}\cdot{}^\circ\text{C}}$.

5.51 (II) On a day when a man shows no visible perspiration, he will still evaporate about 600 g of water from his lungs. (a) How many calories of heat are removed by this evaporation? (b) What is the rate of heat loss in watts due to this process?

5.52 (II) How many grams of perspiration must a 50-kg woman evaporate to reduce her temperature by 1. 5°C?

5.53 (II) An average man may consume 3000 kcal per day. If he uses all these calories to perform work and produce heat (no storage of fat), how many grams of water and perspiration must be evaporate in order to get rid of half of the waste thermal energy, assuming his average efficiency is 5.0%?

5.54 (II) (a) Calculate the rate of heat loss by radiation from 10 cm^2 of skin if the skin temperature is 33°C and its emissivity is 0.97. (b) Compare this with the rate of heat loss if the skin temperature were 34°C, and comment on the sensitivity needed in a thermograph to observe variations in temperature of 1.0°C.

5.55 (II) One problem for astronauts is getting rid of waste body heat. If an astronaut is in a spacesuit in the vacuum of space and is awake but relaxed, calculate the temperature rise of her body in one hour assuming no heat escapes the suit. The total mass of the astronaut and suit is 100 kg, and the suit's specific heat is the same as her body's.

5.56 (III) An 80-kg patient is to be cooled to 29°C for surgery by being placed in ice water. The power output of this patient is 60 W. It takes 20 minutes to bring his temperature down and the surgery lasts 2 hours 40 minutes. How many kilograms of ice must melt to do this, assuming all other forms of heat transfer are negligible and that the ice water stays at 0°C?

5.57 (III) During heavy exercise, 2.0 L of blood are pumped to the surface of a person per minute to carry away core heat. If the blood is cooled by 2.0°C at the surface, what is the rate of heat transfer in watts due to blood flow (forced convection)? You may assume that the specific heat of blood is the same as that for water and that the density of blood is 1.05 g/cm^3. Note that although the core and skin normally differ in temperature by 3°C–4°C, the blood will not be cooled by that much. Skin temperature will rise during exercise, and the blood must have a greater temperature than the skin in order to transfer heat to it through vessel walls (conduction).

5.58 (III) Suppose a man is losing heat to the environment at the rate of 300 W. His body temperature is 2.0°C below normal, and he begins to shiver. If his mass is 76 kg, how long will it take for his temperature to rise to normal, given that the energy consumption rate during shivering is 425 joules per second.

6 Fluids and Pressure

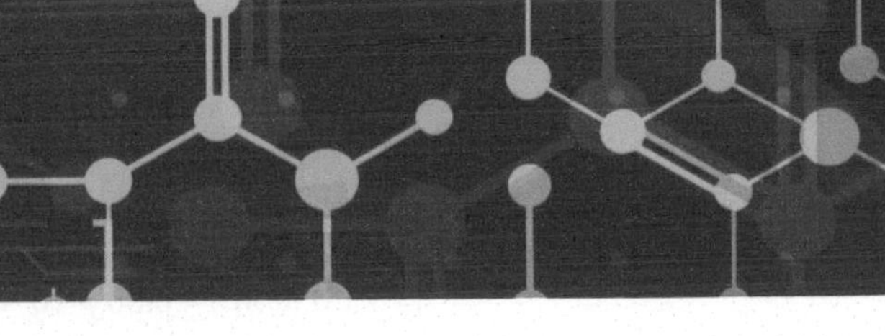

"If you thought that science was certain—well, that is just an error on your part."

—Richard P. Feynman

Liquids and gases exist in abundance on earth. The existence of life is intimately related to the characteristics of matter in these phases. (Phases of matter were discussed in Section 5.1.) Consequently, the physics of liquids and gases is not just another topic but is basic to life itself.

A *fluid* is either a gas or a liquid. This chapter explores the physics of both stationary and moving fluids, including many medical and biological examples. The next chapter treats in even greater detail certain biological and medical applications of fluids.

Although liquids and gases have many characteristics in common, they are distinguishable in a number of ways. For example, liquids are nearly incompressible, while gases are easily compressed. In addition, liquids tend to have much greater densities than gases. The gaseous phase of a substance usually exists at higher temperatures than the liquid phase. Because the gas has higher temperature, the molecules in it are able to break free from one another. Gases are thus able to escape from an open container, whereas liquids cannot. Whenever a distinction between liquids and gases is important, it will be mentioned specifically. Otherwise both *liquids and gases* will simply be referred to collectively as fluids. If it flows, its a fluid.

6.1 DEFINITION OF PRESSURE

Pressure is one of the most important concepts in fluids. Pressure is defined as the force applied per unit area, or $\text{Pressure} \equiv \dfrac{\text{Force}}{\text{Area}}$ Know eq. 6.1

Typical pressure units in the metric system are, $\dfrac{\text{newtons}}{\text{meter}^2}$ abbreviated as $\dfrac{\text{N}}{\text{m}^2}$; they are named pascals (Pa).

Pressure pushes. Vacuums don't suck!

Table 6.1 Conversion Factors for Various Units of Pressure

Conversion to N/m^2	Conversion to atm
1.0 atm = 1.013×10^5 N/m^2	1.0 atm = 1.013×10^5 N/m^2
1.0 dyn/cm^2 = 0.1 N/m^2	1.0 atm = 1.013×10^6 dyn/cm^2
1.0 kg/cm^2 = 9.8×10^4 N/m^2	1.0 atm = 1.03 kg/cm^2
1.0 $lb/in.^2$ = 6.90×10^3 N/m^2	1.0 atm = 14.7 $lb/in.^2$
1.0 mm Hg = 133 N/m^2	1.0 atm = 760 mm Hg
1.0 cm Hg = 1.33×10^3 N/m^2	1.0 atm = 76.0 cm Hg
1.0 cm water = 98.1 N/m^2	1.0 atm = 10.3 m water
1.0 bar = 1.000×10^5 N/m^2	1.0 atm = 1.013 bar

The definition is valid for all phases of matter: solid, liquid, or gas. There are many examples of pressure: A record player needle exerts pressure on a record, water near the bottom of a swimming pool exerts noticeable pressure on a swimmer's eardrums, and atmospheric pressure changes with weather conditions. Measurements of pressure are common: Tires must be inflated to the correct pressure, blood pressure should stay within a normal range, and too much pressure in the eye (glaucoma) can cause blindness. Unfortunately, there are many different units for pressure-an inconvenience that cannot be escaped. Table 6.1 lists some common pressure units, including atmospheres, where 1 atm is the atmospheric pressure or the average pressure due to the weight of the atmosphere at sea level.

Note that in 1954, 1 atm was defined as exactly 101,325 Pa and 1 torr $\equiv \frac{1}{760}$ atm.
1 mm of Hg is approximately equal to 1 Torr,
One standard atmospheric air pressure can push a column of mercury 760 mm (29.9 inches) up.
If you haven't already done so, you may want to review the pressure lesson before you proceed.
Let's look at an example that involves pressure. We hope that you find it instructive.

EXAMPLE 6.1

Calculate the pressure exerted by an old phonograph needle on a record if the needle supports 2.00 g on a circular area 0.250 mm in radius. Express this pressure in newtons per square meter and in atmospheres.

Solution:

Step by step 1: From the definition of pressure,

$$\text{Pressure} \equiv \frac{Force}{Area} = \frac{\text{weight of needle}}{\text{area of tip}}$$

Step 2: In order to proceed, we are going to have to *determine the weight of the needle* and the area of contact. If we want the weight of the needle in newtons, we will need to convert its mass to kilograms.

$$2.00 \text{ g} \cdot \frac{1 \text{ kg}}{1000 \text{ g}} = 2.00 \times 10^{-3} \text{ kg}$$

Weight of needle is,

$$Wt = \text{mg} = 2.00 \times 10^{-3} \text{ kg} \cdot 9.80 \frac{\text{m}}{\text{s}^2} = 0.0196 \text{ N}$$

Step 3: The *area of contact* is the area of the tip of the needle. Since the tip of the needle is circular, we will use the equation for the *area of a circle*.

$$Area_{\text{circle}} = \pi r^2 \quad \text{Know}$$

The radius of the needle is given as 0.250 mm. Converting to meters,

$$\text{radius} = 0.250 \text{ mm} \cdot \frac{1.00 \text{ m}}{1000 \text{ mm}} = 0.250 \times 10^{-3} \text{ m}$$

Area of needle tip is,

$$Area_{\text{circle}} = \pi r^2 = \pi \cdot \left(0.250 \times 10^{-3} \text{ m}\right)^2 = 1.963 \times 10^{-7} \text{ m}^2$$

Step 4: Substitute the values.

$$\text{Pressure} \equiv \frac{Force}{Area} = \frac{0.0196 \text{ N}}{1.963 \times 10^{-7} \text{m}^2} = 9.98 \times 10^4 \frac{\text{N}}{\text{m}^2}$$

Ex. 6.1(b) Convert the pressure from pascals to atmospheres. Use the fact that 1 atm = 1.01×10^5 pascals.

$$9.98 \times 10^4 \frac{\text{N}}{\text{m}^2} \cdot \frac{1 \text{ atm}}{1.01 \times 10^5 \frac{\text{N}}{\text{m}^2}} = 0.988 \text{ atm}$$

The answer is surprisingly large, and in fact records are apparently able to withstand very large pressures. The tone arm on most old record players can be adjusted to minimize pressure and reduce wear.

Pressure is often as important as the force creating it. *Consider* Figure 6.1. If someone pokes you in the arm with a finger, you will certainly feel it. If, however, a nurse pokes you with a hypodermic needle using the same force, you don't just feel it—the needle penetrates your skin. Pressure is inversely proportional to the contact area. The same force applied to a smaller area creates a larger pressure and has a much different effect. Fluids, as well as solids, can exert pressures. Consider the water in the container with straight sides shown in Figure 6.2(a). The water has a mass of 100 kg,

Figure 6.1 Two different results of applying the same force on different areas. In (a) the area is large and the pressure is small. In (b) the area is very small and the pressure is consequently large enough to break the skin.

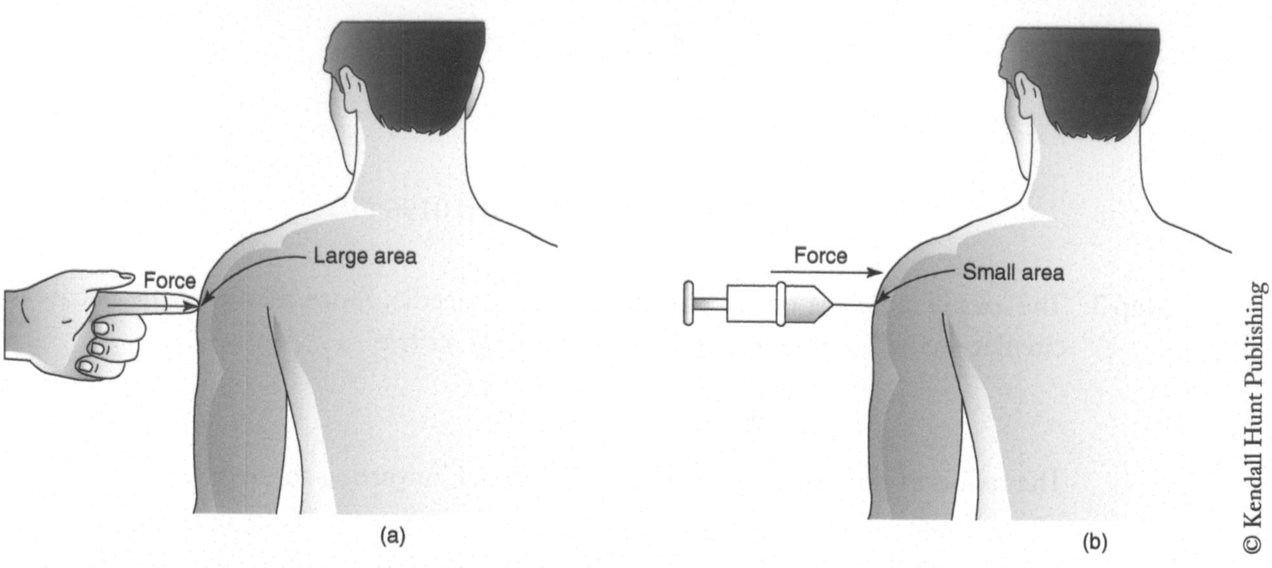

Figure 6.2 (a) The pressure due to the fluid at the bottom of a straight-sided container is its weight divided by the area supporting it. (b) Atmospheric pressure is due to the weight of the air over an area—in this case 1.0 m^2.

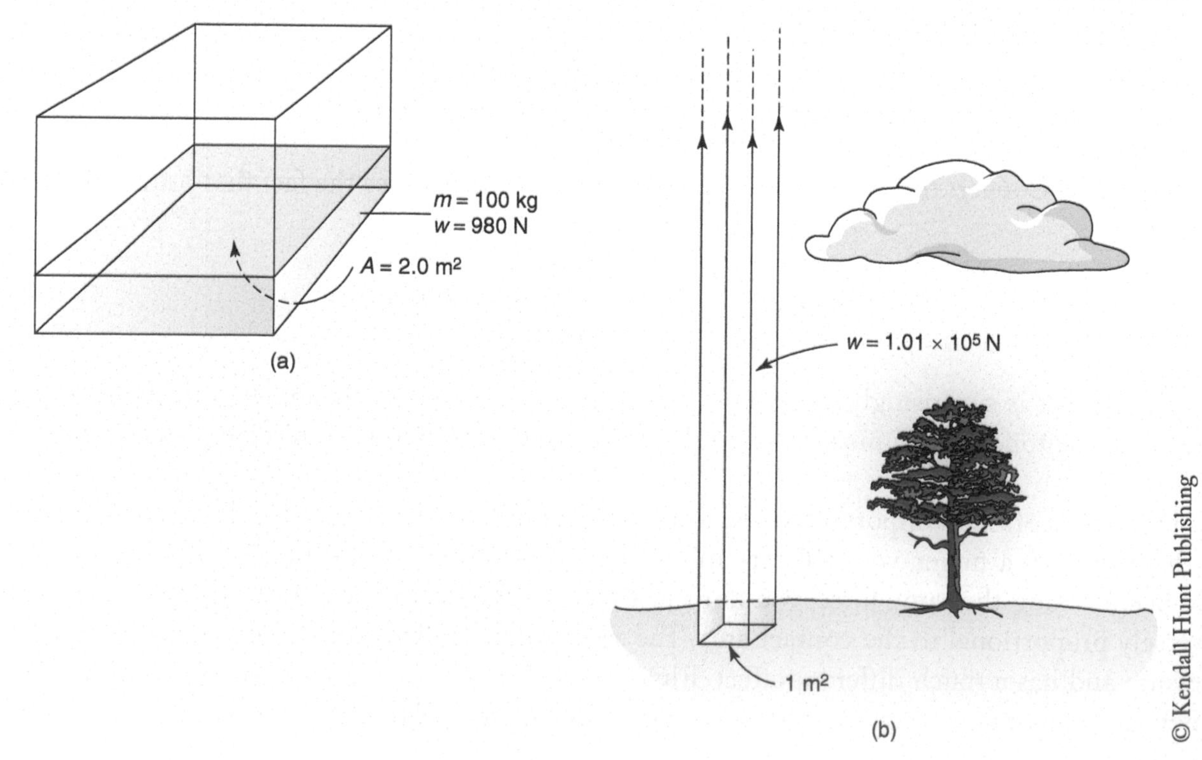

so its weight of 980 N (w = mg) must be supported by the bottom of the container. If the bottom has an area of 2.0 m^2, then the pressure due to the weight of the water on the bottom is 490 N/m^2 (Note: This calculation is valid only for containers with straight sides. More general situations are considered in Section 6.2.) Another example is atmospheric pressure, which we noted is caused by the weight of air, as illustrated in Figure 6.2(b).

Atmospheric pressure averages 1.01×10^5 N/m^2 at sea level, meaning that a column of air 1 m on a side extending to the top of the atmosphere weighs 1.01×10^5 N.

EXAMPLE 6.2

Calculate the force in newtons, exerted by the atmosphere on one side of a wall of area 10 m^2 (an average room wall in a house).

Solution: First write down the definition of pressure $\text{Pressure} \equiv \frac{\textit{Force}}{\textit{Area}}$. Next we solve it for force.

$F = P \cdot A$ Now we enter the value for atmospheric pressure from Table 6.1:

$$F = P \cdot A = 1.01 \times 10^5 \frac{\text{N}}{\text{m}^2} \cdot 10 \text{ m}^2 = 1.01 \times 10^6 \text{N}$$

A force this large would easily break down any wall in a house. How is it that the wall can withstand the force? The answer is that the atmosphere usually exerts an equal and opposite force on the other side of the wall, as illustrated in Figure 6.3. The total force on the wall is thus zero. *Except during hurricanes and tornados!*

Stationary fluids always exert forces perpendicular to surfaces whether that direction is up or down, left or right (e.g., see Figure 6.3). The reason the force is always perpendicular to the surface is that fluids cannot withstand shearing or sideways forces and therefore cannot exert sideways forces.

Figure 6.3 In most situations the atmosphere exerts a total force of zero on an object because pressure is exerted on all sides of the object. (See Example 6.1 and the discussion following it.)

6.1.1 The Dependence of Pressure in a Gas on Volume

6.1.1.1 A brief treatment

Gas in a container exerts pressure on the container walls in addition to that caused by gravity. This additional gas pressure is caused by the collisions of molecules with the walls, as illustrated in Figure 6.4(a). (As discussed in Chapter 5, molecules are in continuous motion unless the temperature is absolute zero.) Many millions of molecules strike the wall every second and thereby exert a nearly constant average force on it. The pressure exerted by the gas is this average force divided by the area of the wall. How does this pressure depend on the volume of the container?

Consider what happens if the same gas is put into a larger container, as in Figure 6.4(b), and kept at the same temperature. The pressure exerted by the gas will be less for two reasons. First, the number of collisions of molecules with the walls per unit time will be less because on average they have to go farther to find a wall. Second, the walls have larger area $\text{Pressure} \equiv \frac{Force}{Area}$, so pressure is smaller. The converse is also true; if the volume occupied by a gas is decreased, its pressure will increase, provided all other characteristics, such as the temperature and number of molecules, are unchanged. Pressure and volume are in fact exactly inversely proportional: $P \propto \frac{1}{V}$ where the symbol $\propto$ means proportional to. The pressure produced by gas molecules in a closed container is *inversely proportional to the volume* of the container. For example, if the volume of a gas is doubled, its pressure is cut in half. On the other hand, if the volume of the container is cut in half, the pressure inside the container is doubled.

This simple dependence of pressure in a gas on the volume it occupies has many applications, a number of which will be encountered in this and the following chapter. One example is the air pressure in a tire, which results from putting a large volume of ordinary air into the smaller volume of the tire, thus making its pressure greater than atmospheric pressure. When someone drinks a liquid through a straw, they expand the volume of the air in their mouth. This reduces the air pressure in the straw. The larger atmospheric pressure in the room *pushes* the fluid up the straw. *Vacuums don't suck! Pressure pushes!*

Figure 6.4 (a) Pressure is exerted by a gas on the walls of a container via collisions of the gas molecules with the walls. If the same amount of gas at the same temperature occupies a larger volume, then its pressure decreases.

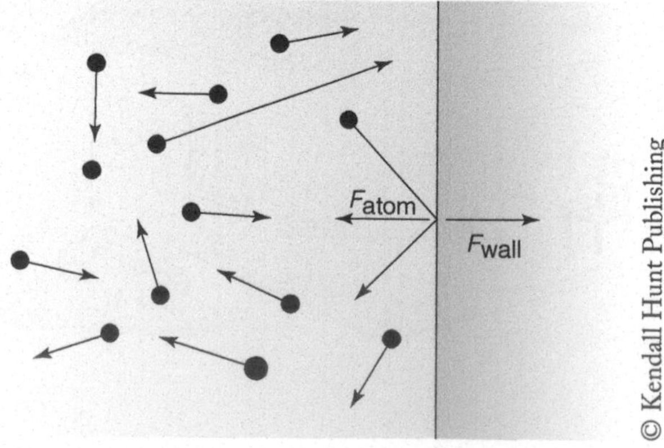

The relationship between pressure, temperature, volume, and the number of gas molecules in a container is described in the general gas law. $PV = nRT$

You will learn more about the relationship between pressure and volume in Lab 5.

Things to know for 6.1

Pressure pushes

$$\text{Pressure} \equiv \frac{Force}{Area}$$

Metric units of pressure are $\frac{\text{N}}{\text{m}^2}$**; they are called pascals.**

$Area_{circle} = \pi r^2$ **where** ***r*** **is the radius of the circle.**

6.2 PASCAL'S PRINCIPLE

One pioneer in the physics of fluids was the French philosopher and scientist Blaise Pascal (1623–1662). Pascal discovered an important property of stationary fluids: They can be used to transmit pressures to a place other than where the pressure is created. Furthermore, the pressure is not diminished in transit. This important experimental fact is expressed in Pascal's principle:

Any pressure applied to a confined fluid will be transmitted undiminished to all parts of the fluid.

This means that the applied pressure will be added to whatever pressures already exist in the fluid. One manifestation of Pascal's principle is that the total pressure at the bottom of a lake is the sum of the pressure due to the weight of the water plus atmospheric pressure. According to Pascal's principle, atmospheric pressure is transmitted undiminished to the bottom of the lake. Because water is not rigid, it cannot support the weight of the atmosphere without transmitting it to all parts of the lake.

Most of the phenomena considered in this and the following chapter occurs on earth and involves atmospheric pressure in addition to any other pressures involved. The effect of atmospheric pressure often cancels or is negligible, and it is tiresome always to add atmospheric pressure to get total pressure. *Gauge pressure* is therefore defined as the pressure above or below atmospheric pressure. *Total pressure, or absolute pressure, is gauge pressure plus atmospheric pressure:*

$$P_{\text{Total}} = P_{\text{gauge}} + P_{\text{atm}}$$

An easy way to remember this equation is to recall that a tire gauge reads zero when a tire is flat, even though a flat tire with a large hole obviously contains air at atmospheric pressure. The flat tire has a gauge pressure of zero and a total or absolute pressure of 1 atmosphere.

We shall use the convention that all pressures are gauge pressures unless otherwise specified. Most pressure-measuring devices yield gauge pressures (hence the name), and in most circumstances gauge pressure is of greater concern than total pressure.

Before we go on to apply Pascal's principle, it is important to note that applied pressure, not force, is transmitted undiminished to all parts of a fluid. This experimental fact makes pressure very important in fluids. Forces are transmitted by fluids, but they may be made larger or smaller by the fluid depending on circumstances. Nonetheless, Pascal's principle can be used to predict accurately the force exerted by a fluid, as seen in the following discussion of hydraulic systems.

6.2.1 Hydraulic Systems

Consider two cylinders connected to one another by a tube and filled with a fluid. The fluid is confined in the system by two pistons, as shown in Figure 6.5. For simplicity the fluid is assumed to be incompressible. When a force F_1 is exerted on the small piston, fluid flows from the small cylinder to the large one, moving the large piston up and exerting an upward force on that piston. *Pascal's principle* states that the pressure created by exerting the force F_1 on the small piston, is transmitted undiminished to all parts of the fluid. Therefore,

Because $P_1 = P_2$, it follows that $\dfrac{F_1}{A_1} = \dfrac{F_2}{A_2}$ eq. 6.3

Equation 6.3 holds true for any hydraulic system where the two pistons are at the same vertical height. Example 6.3 will demonstrate the usefulness of a hydraulic system in increasing input force.

EXAMPLE 6.3

The large piston in Figure 6.5 supports a dentist's chair, and the dentist wants to lift the patient by stepping on a pedal directly on top of the small piston. Calculate the force the dentist must exert if the patient plus chair have a mass of 120 kg and the small piston has a diameter of 1. 00 cm, while the large piston has a diameter of 5.0 0 cm.

Figure 6.5 Typical hydraulic system with two fluid-filled cylinders capped with pistons and connected by a hydraulic line. A downward force F_1 on the left piston creates an upward force F_2 on the right piston. The forces usually are not equal.

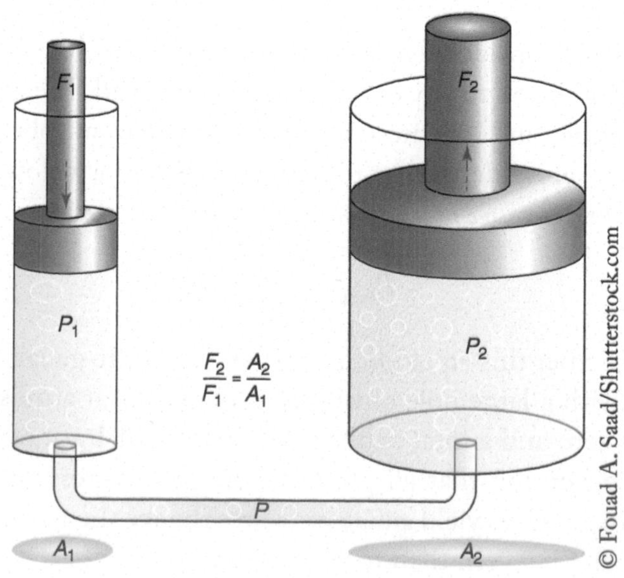

Concept. Because of Pascal's Principle, we know that the pressure in a sealed container is the same everywhere. $P_1 = P_2$. Also remember the definition of pressure. $P \equiv \frac{F}{A}$

Step by step 1: Since $P_1 = P_2$ therefore, $\frac{F_1}{A_1} = \frac{F_2}{A_2}$

Looking at figure 6.5, we are being asked to determine F_1, the force that the dentist must apply.

Step 2: Solving for F_1, yields our working equation.

$F_1 = \frac{F_2 A_1}{A_2}$. Where F_2 is the *weight* of the patient and chair.

Step 3: It is important to note that 120 kg is the *mass* of the patient and chair. We will have to calculate the weight, F_2, of the patient and chair.

$$F_2 = weight = \text{mg} = 120\ \text{kg} \cdot 9.8\frac{\text{m}}{\text{s}^2} = 1176\frac{\text{kg} \cdot \text{m}}{\text{s}^2}$$

The combined weight of the patient and chair equals 1176 newtons.

Step 4: The piston has a circular end, so we have to remember that the **area of a circle** = πr^2
The area, A_1 of the small piston end is,

$$A_1 = \pi r^2 = \pi(0.500\,\text{cm})^2 = 0.785\,\text{cm}^2$$

The area, A_2 of the large piston is,

$$A_2 = \pi r^2 = \pi(2.50\,\text{cm})^2 = 19.6\,\text{cm}^2$$

Step 5: Substitute these values into our working equation,

$$F_1 = \frac{F_2 A_1}{A_2} = \frac{1156\,\text{N} \cdot 0.785\,\text{cm}^2}{19.6\,\text{cm}^2} = 47.1\ \text{N}$$

We see that only 47.1 N of force were applied to the small piston resulted in a lifting force of 1180 N in the larger piston.

In most hydraulic systems a small force is put into the system and causes a large force to emerge from the other end of the system. The small cylinder is called a master cylinder and the large cylinder a slave cylinder. Car brakes, for example, have a master cylinder connected to four slaves, one at each wheel. A small force exerted by the driver creates larger forces in the wheels, pushing the brakes against a surface to slow the car with friction. Frictional force is given by $f = \mu_k F_N$, so the larger the force F_N created by the slave cylinder, the larger is the stopping force f of the brakes (see Figure 6.6).

Although a hydraulic system can increase the force put into it, it cannot increase the energy put into it without violating the conservation-of-energy principle. Remember that work = $F\Delta X$. In order to conserve energy, if the force increases, then the displacement, ΔX of the slave cylinder is decreased. For example, each time the dentist steps on the pedal in Example 6.3, the chair moves only $\frac{1}{25}$ as far as the dentist's foot. Similarly, the pistons in the slave cylinders in a car's brake system move a smaller distance than does the piston in the master cylinder. If there is no friction in the system (a good approximation in most hydraulic systems), then motion is decreased by the same factor that force is increased. To see this, recall that work and energy are equivalent and that work equals force

Figure 6.6 A car's hydraulic brake system increases the force F_1 exerted by the driver to create a large frictional force in the brakes. The brake pedal is connected to a master cylinder that is smaller in diameter than the slave cylinders.

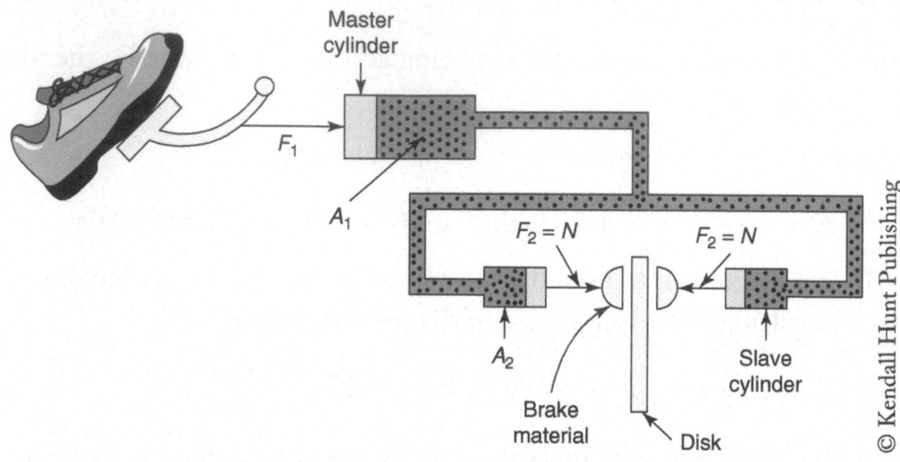

times displacement. If there is no friction, conservation of energy tells us that the work done *to* the system must equal the work output of the system.

$$Work_{in} = Work_{out}$$

That means that,

$$F_{in}\Delta X_{in} = F_{out}\Delta X_{out}$$

Solving for displacement out,

$$\Delta X_{out} = \frac{F_{In}\Delta X_{In}}{F_{out}} = \frac{47.1\text{N}\cdot\Delta X_{In}}{1176\text{N}} = \frac{1}{25}\Delta X_{In}$$

Thus if F_{out} is larger than F_{in}, then ΔX_{out} must be smaller than ΔX_{in}.

6.2.2 Pressure Due to the Weight of a Fluid-Pressure at Depth

As mentioned in Section 6.1, the force of gravity causes objects to have weight. Weight per unit area is called pressure due to the weight of a fluid. To find a general expression for this pressure, one that is valid for any shape container and for any depth in the fluid (not just on the bottom), consider Figure 6.7. The area A is at a depth h. Pressure at that depth is due to the weight of the column of fluid above the area A. Brace yourself, we are going to do a little derivation.

Step by step 1: $P \equiv \frac{Force}{Area} = \frac{\text{weight of fluid}}{Area} = \frac{\text{mg}}{A}$

That's a good start. Now we want to use the definition of density and make a substitution for the mass of the fluid "m_f."

$$\rho \equiv \frac{mass}{volume} \text{ therefore, } m = \rho V$$

Figure 6.7 The pressure due to a fluid depends on the depth *h* and the density *p* of the fluid.

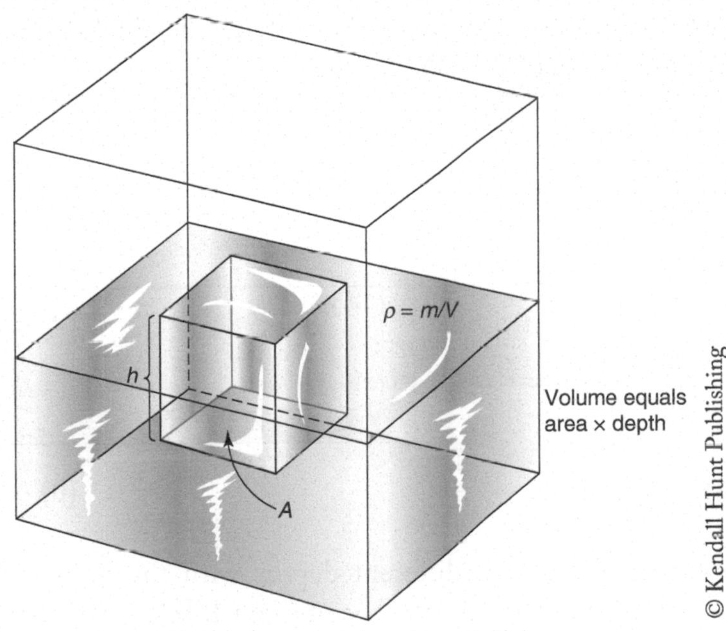

Step 2: Substituting for the mass of fluid, m, yields,

$$P = \frac{\rho_{\text{fluid}} \cdot V \cdot g}{A}$$

The volume of the column of fluid can be determined by multiplying the area A of one end of the column, by the depth.

$$Volume = A \cdot depth$$

Step 3: Substituting for Volume in the last expression yields,

$$P = \frac{\rho_{\text{fluid}} \cdot (A \cdot depth) \cdot g}{A}$$

Step 4: Dividing out the area *A*, results in an equation for the pressure at any depth—as long as the fluid is incompressible.

$$P_{\text{depth}} = \rho_{\text{fluid}} \cdot g \cdot depth \qquad \text{Know eq. 6.4}$$

Well there you have it. A thing of beauty. Take a moment to look at equation 6.4.

It is noteworthy that pressure due to the weight of a fluid depends only on the depth beneath the surface of the fluid and the density of the fluid. Any depth can be considered. For example, pressure at the surface is zero since *h* is zero at the surface. The surface area and shape of the container is unimportant. Consider the containers in Figure 6.8(a), which are filled to the same depth with the same fluid. Because all columns are the same depth, the pressure due to the fluid is the same at the bottom of each container and is given by $P_{\text{depth}} = \rho_{\text{fluid}} \cdot g \cdot depth$, and thus there is no flow.

Figure 6.8 (a) The pressure due to a fluid at a depth *h* is independent of the shape of the container. Although the containers are connected, no fluid flows and the depths remain the same. (b) If two containers originally filled to different depths (shaded) are connected, fluid will flow until the depths become equal.

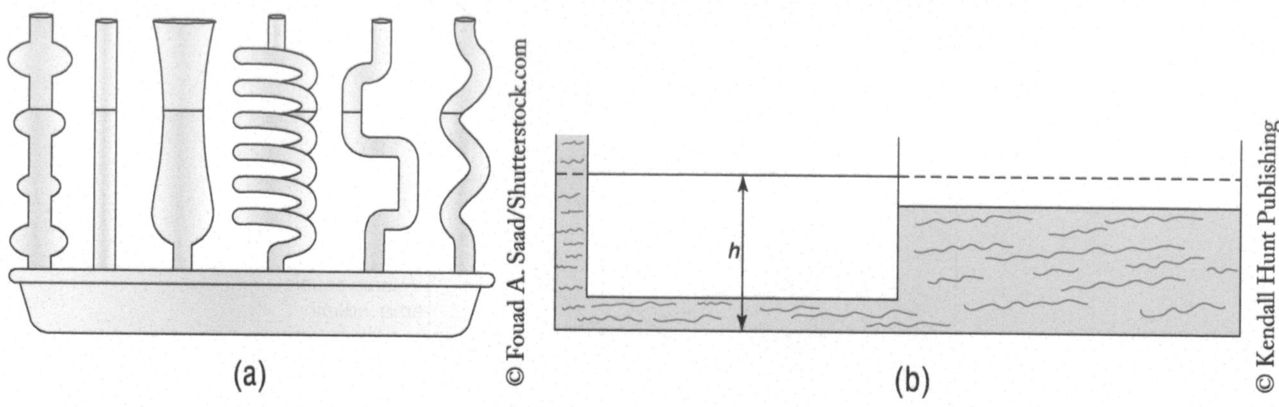

Conversely, if containers filled to different depths with the same fluid are connected as in Figure 6.8(b), the fluid flows until the depths become equal. This occurs because if the fluid depth is greater on the left, pressure at the left end of the tube will be greater than at the right and fluid will be pushed to the right. Pressure pushes the fluid until all pressures, that is, all heights, are equal. The pressure will be equal when the fluid is the same height on either side as shown by the dashed lines.

EXAMPLE 6.4

(a) Calculate the pressure in newtons per square meter at a depth of 2.50 m due to water in a swimming pool. (b) What is the total pressure at that depth?

Solution: (a) The pressure due to the water alone, as given by equation 6.4, is $P_{\text{depth}} = \rho_{\text{fluid}} \cdot g \cdot depth$ The density of fresh water is $1000\dfrac{\text{kg}}{\text{m}^3}$

$$P_{\text{depth}} = \rho_{\text{fluid}} \cdot g \cdot depth = 1000\frac{\text{kg}}{\text{m}^3} \cdot 9.80\frac{\text{m}}{\text{s}^2} \cdot 2.50\,\text{m} = 2.45 \times 10^4\,\frac{\text{N}}{\text{m}^2}$$

Ex. 6.4(b) The total pressure is the pressure due to the weight of the fluid plus atmospheric pressure:

$$P_{\text{total}} = P_{\text{depth}} + P_{\text{atm}} = 2.45 \times 10^4\,\frac{\text{N}}{\text{m}^2} + 1.01 \times 10^5\,\frac{\text{N}}{\text{m}^2} = 1.26 \times 10^5\,\frac{\text{N}}{\text{m}^2}$$

Another manifestation of how pressure depends only on depth and density is found in the intravenous (IV) administration of fluids, as shown in Figure 6.9. The pressure due to the IV fluid at the entrance of the needle is proportional to *h*, the height of the surface above the needle. The path taken is irrelevant; paths A and B both produce the same pressure. Along path B the pressure increases to a maximum at the lowest point and decreases at the needle, having a final value, $\text{P}_{\text{depth}} = \rho_{\text{fluid}} \cdot \text{g} \cdot \text{depth}$ exactly the same as that produced by path A. Applied pressure can be adjusted by raising or lowering the IV bottle relative to the patient.

Figure 6.9 The pressure due to the IV solution is proportional to the height *h* of the liquid surface above the needle. It is the same for either path A or B.

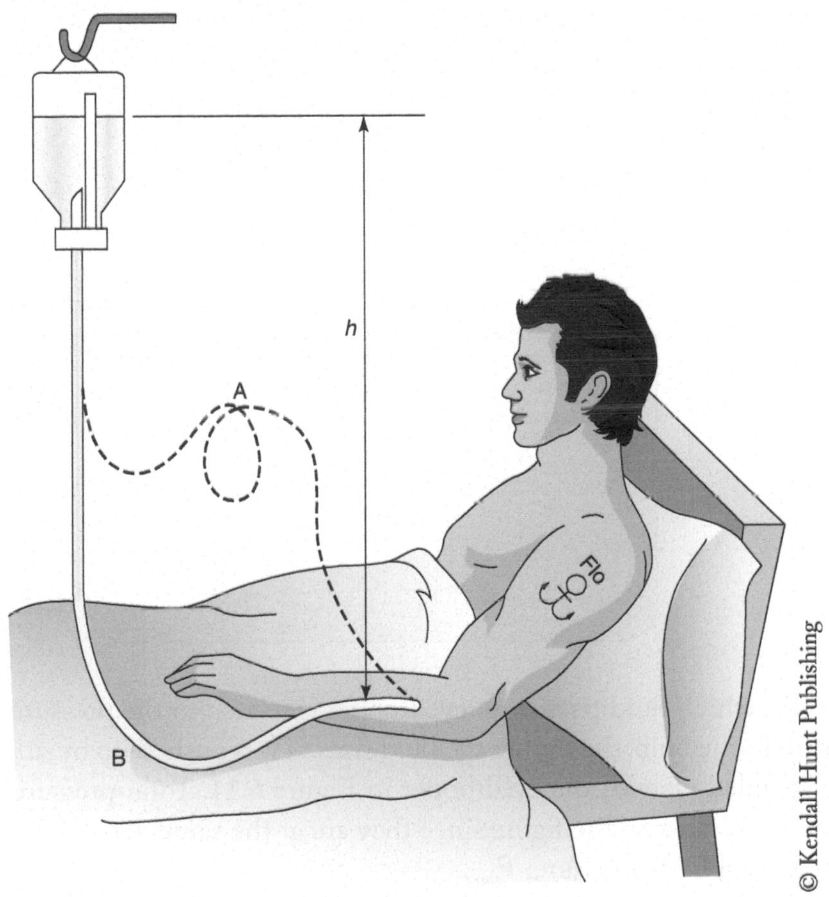

Consider the situation shown in Figure 6.10, where an eyedropper is filled with fluid. First air is squeezed out of the bulb with the tip beneath the fluid surface. The bulb is then released and, being elastic, returns to its original larger size. The volume of the gas in the bulb is thereby increased, and the pressure decreases to less than atmospheric pressure. Atmospheric pressure is then able to *push* fluid up the eyedropper to a height *h*. This is analogous to how hypodermic syringes are filled and how we drink through straws.

Figure 6.10 Sequence of events in filling an eyedropper as described in the text.

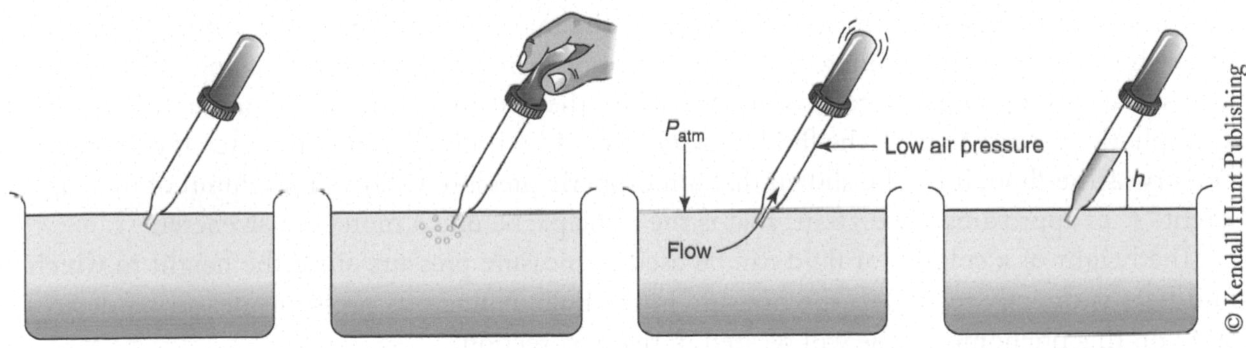

Figure 6.11 The total pressure at both points marked × must be the same hence $P_{atm} = h\rho g + P_{bulb}$.

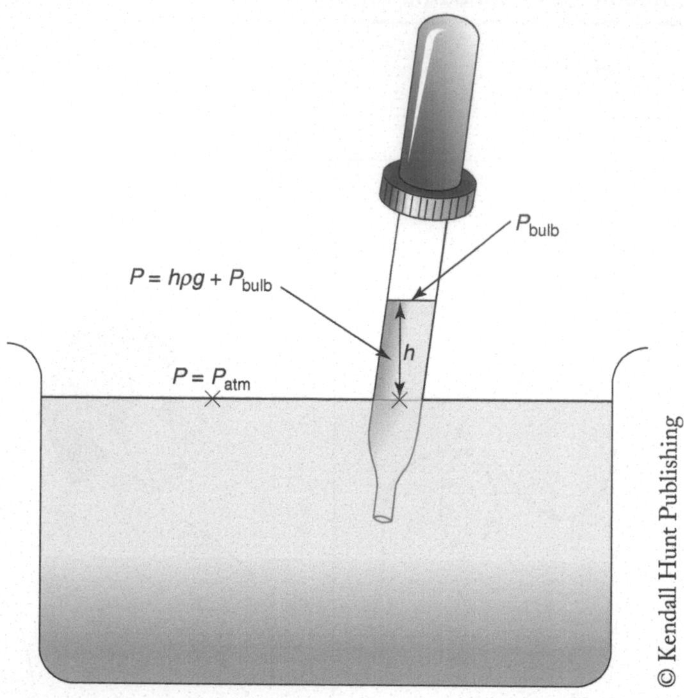

The height h to which fluid rises in an eyedropper is related to the pressure in the bulb, P_{bulb}; the lower the pressure in the bulb, the higher the fluid rises as it is pushed up by surrounding air pressure. Consider the expanded view of the eyedropper in Figure 6.11. Total pressure must be the same at both points marked by an x in the figure since they are at the same vertical height. Total pressure at the left x is just atmospheric pressure, P_{atm}.

The total pressure at the right, x, is the pressure due to the weight of the fluid plus the pressure of the air in the bulb, $P_{fluid} + P_{bulb}$. These two pressures must be equal. $P_{atm} = P_{fluid} + P_{bulb}$; $P_{atm} = \rho_{fluid}\, gh + P_{bulb}$

Solving for h, $h = \dfrac{P_{atm} - P_{bulb}}{\rho g}$

The pressure in the bulb is absolute pressure. The lower the air pressure in the bulb is, the larger the column of fluid "h" is. That column of water is pushed up by atmospheric pressure. Everything in the expression for h is constant except the pressure in the bulb; the smaller that pressure in the bulb, the higher air pressure will push the fluid up. Since the pressure in the bulb cannot be less than zero, there is a limit to how large h can be. For water, the maximum height h that atmospheric pressure can push water up is approximately 10.3 m. The same principle applies to "sucking" juice up a straw or to surface well pumps. Note also that the larger the density of a fluid, the more difficult it is for atmospheric pressure to push the fluid up. Mercury is 13.6 times denser than water. If one substitutes mercury as the fluid, it can be shown that atmospheric pressure will push a column of mercury to a height "h" of approximately 0.76 m. That is the principle behind a mercury barometer.

The height of a column of fluid can be used to measure pressure since the height to which the fluid rises is directly related to the pressure in the bulb. Numerous pressure-measuring devices are based on this phenomenon, as will be seen in the next section.

Things to know for section 6.2

Area of a circle = πr^2

$P_{\text{depth}} = \rho_{\text{fluid}} \cdot g \cdot depth$

Density $\rho \equiv \dfrac{mass}{volume}$

6.3 MEASUREMENT OF PRESSURE

6.3.1 Based on Pascal's Principle and $P_{\text{depth}} = \rho_{\text{fluid}} \cdot g \cdot depth$

Pascal's principle states that any pressure applied to a confined fluid is transmitted undiminished to all parts of the fluid. Fluids thus can be used to transmit pressure to a gauge at a convenient location. *Blood pressure*, for example, can be measured without putting a gauge into the body. Furthermore, pressure is transmitted undiminished, so the measurement can be very accurate.

Common blood pressure measurements illustrate this application of Pascal's principle in several respects. An inflatable cuff is placed on the upper arm, as shown in Figure 6.12, and inflated until blood flow is cut off in the brachial artery. Pressure is created by squeezing the bulb and is transmitted by the air in the tubes (a confined fluid) to the cuff and to the gauge. The wall of the cuff transmits the pressure to the arm (approximately a fluid) and through it to the artery. When the applied pressure exceeds the heart's output pressure, the artery collapses. The person making the measurement slowly releases air from the cuff, lowering its pressure, and listens for flow to resume when pressure in the cuff becomes lower than the maximum heart output.

Maximum blood pressure, called systolic pressure, is recorded together with a second, lower pressure called diastolic pressure. Diastolic pressure is the minimum pressure the circulatory system experiences and can be detected as a change in the sound of blood flow through the partially restricted artery. (These and other characteristics of the circulatory system are discussed in more detail in Chapter 7.) The main point here, based on Pascal's principle, is that all these pressures are transmitted undiminished, and the pressure read by the gauge is truly representative of the pressure in the heart.

It is important for the cuff to be at the same level as the heart in a blood pressure measurement (see Figure 6.12). If necessary, blood pressure can be measured in the leg, but a larger pressure is obtained for a standing patient because of gravity. The leg is at a greater depth in the circulatory system than the heart and experiences a greater pressure.

Any effect due to the weight of the air in the cuff and connecting tubes is negligible because the density of air is very small. If the gauge is placed a distance h below the cuff, any extra air pressure caused by this increase in height of the air column is added. This extra pressure is negligibly small because the density of air is so small. The gauge can be placed in any convenient location, such as on the wall, a table, the floor, or anywhere in between, but the cuff must be at the same level as the heart.

When a liquid rather than a gas is used to transmit pressure to a gauge, the situation is different. In spinal column pressure measurements a long needle is inserted between vertebrae into the

Figure 6.12

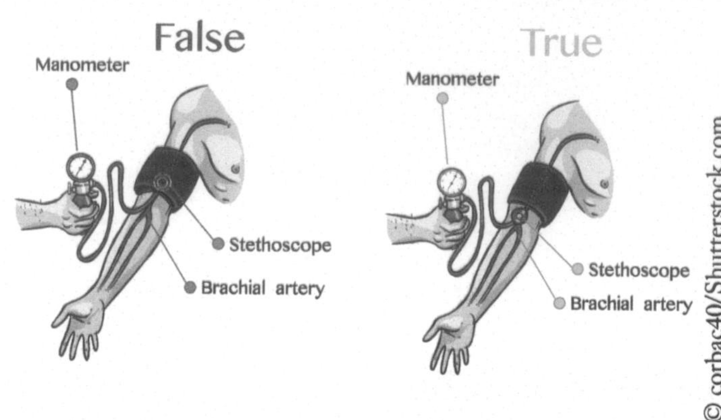

Figure 6.13 In a spinal column pressure measurement the fluid transmitting the pressure to the gauge is a liquid. The vertical position of the gauge must be the same as the needle to obtain an accurate reading.

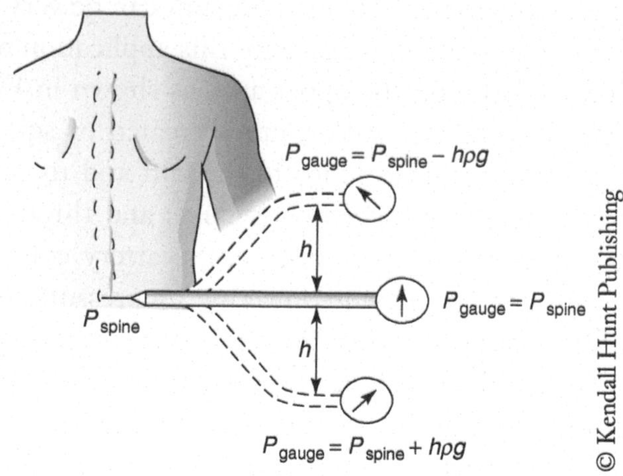

spinal column fluid, as shown in Figure 6.13 and pressure is transmitted to a gauge by a sterile saline solution. To obtain an accurate reading, the gauge must be placed at the same vertical height as the needle.

In summary, if a gas is used to transmit pressure to a gauge, then the vertical position of the gauge is unimportant because the densities of gases are small. If, however, a liquid is used to transmit pressure to a gauge, then the vertical position is critical since the pressure due to the weight of the liquid is not negligible.

6.3.2 Pressure Measuring Devices Including Those Based on $P_{depth} = \rho_{fluid} \cdot g \cdot depth$

A variety of pressure-measuring devices are in common use. Two mechanical pressure gauges are shown in Figure 6.14. In each gauge the force generated by a pressure is used mechanically to move an indicator needle. Other devices may use a transducer to convert movement into an electrical signal that can be displayed or recorded automatically. The variety and ingenuity of pressure measuring devices are impressive.

Figure 6.14 Two mechanical pressure.

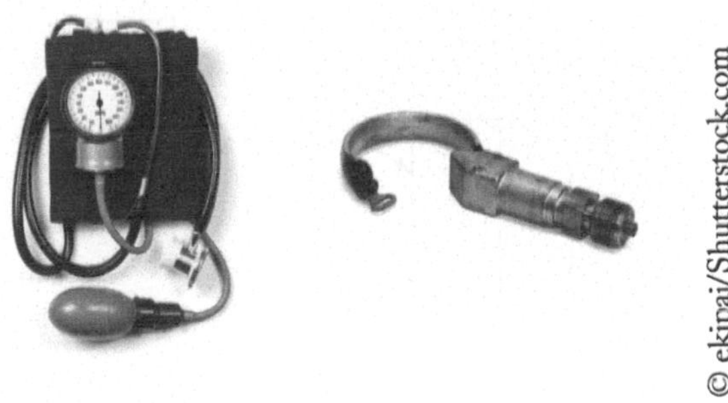

Pressure gauges based on fluids are also in common use and are usually simpler than mechanical devices. Consider the open-tube manometer (Figure 6.15). Manometers are often U-shaped tubes filled with a liquid, usually water or mercury. As usual, pressure is transmitted to a gauge through a fluid. The fact that the fluid level is not the same on both sides of the tube means that the pressure *P* is different from atmospheric pressure.

In Figure 6.15(a) the pressure is greater than atmospheric pressure, and in Figure 6.15(b) it is less. The pressure *P* differs from atmospheric pressure by an amount $P_{\text{depth}} = \rho_{\text{fluid}} \cdot g \cdot depth$, where "*h*" is the difference in height of the liquid on either side of the manometer and "ρ" is the density of the liquid used. In Figure 6.15(a), the pressure "*P*" must be greater than atmospheric pressure by an amount $P_{\text{depth}} = \rho_{\text{fluid}} \cdot g \cdot depth$ because it is able to raise the liquid a height "*h*" on the left side in spite of atmospheric pressure pushing down on the open surface. In Figure 6.15(b), the pressure "*P*" is apparently less than atmospheric pressure by an amount $P_{\text{depth}} = \rho_{\text{fluid}} \cdot g \cdot depth$ since the atmosphere pushing down on the left side is able to raise the liquid a height "*h*" on the right. A gauge pressure less than atmospheric is called a negative pressure. Total or absolute pressure is rarely negative, but gauge pressure can be as negative as −1.0 atm (for zero total pressure).

Manometers and analogous devices are in such common use that pressure units related to the use of these devices have been developed. A single-tube manometer filled with mercury, as shown in Figures 6.16 and 6.17 is commonly used in blood pressure measurements. The pressure in the cuff is $P_{depth} = \rho_{fluid} \cdot g \cdot depth$, where *h* is the height of the mercury and ρ is the density of mercury. Typical blood pressure measurements produce values of *h* in the range from 8 to 300 mm, so "*h*" is commonly recorded in millimeters. These units called mm Hg (read "millimeters of mercury"), are used almost universally in blood pressure measurements rather than perform the multiplication of $P_{\text{depth}} = \rho_{\text{fluid}} \cdot g \cdot depth$ for each measurement. A healthy blood pressure of 120 over 80 means systolic pressure is 120 mm Hg and diastolic pressure is 80 mm Hg. Units of mm Hg (also called Torr) are obviously not the same as force per unit area. If it is necessary to get units of mm Hg into units of force per unit area, then all that need be done is to multiply "*h*" times ρ times "*g*", making certain that all three quantities are in the same unit system (to get N/m^2, all units must be SI). Alternatively, conversion factors, such as those in Table 6.1, can be used.

Whenever using a manometer or similar device it is customary to quote pressure in terms of "*h*" alone, specifying the liquid used. For example, if the manometers in Figure 6.15 contain water, then the pressure would be recorded as *h* centimeters of water (cm H_2O) in case (a) and—*h* centimeters of water in case (b).

Figure 6.15 Open-tube manometers filled with liquid of density ρ.

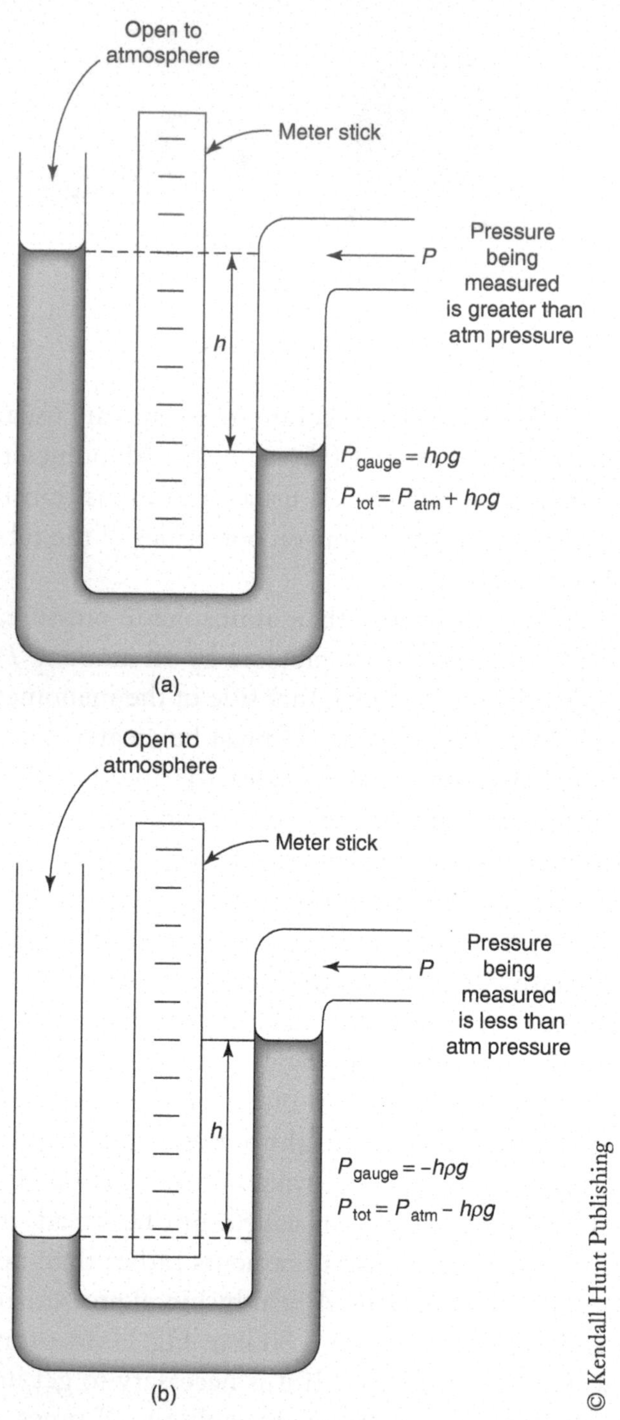

Mercury barometers measure atmospheric pressure in a manner analogous to manometers. A tube that is closed at one end is filled with mercury and inverted with its open end in a dish of mercury, care being taken not to allow any air into the tube. If the tube is long enough, part of the mercury will flow out of the tube leaving a vacuum at the top, as shown in Figure 6.18. The mercury barometer is an exaggerated version of the eyedropper discussed in the previous section and illustrated in Figures

Figure 6.16 A single-tube mercury manometer is frequently used to measure blood pressure. P is measured in units of mm Hg.

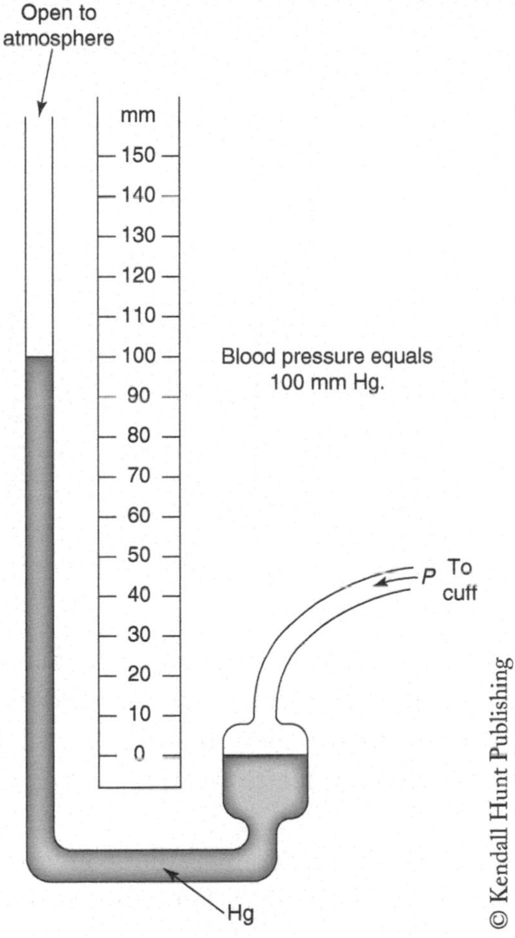

Figure 6.17

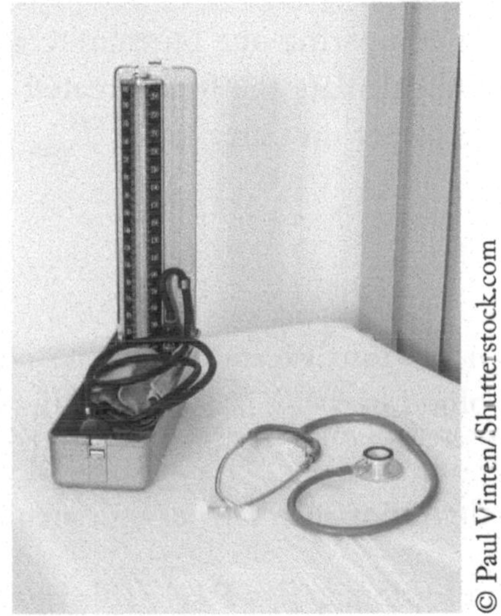

Figure 6.18 In a mercury barometer atmospheric pressure P_{atm} is able to force the mercury to a height such that $P = h\rho g = P_{atm}$. The height h can then be used as a measure of atmospheric pressure.

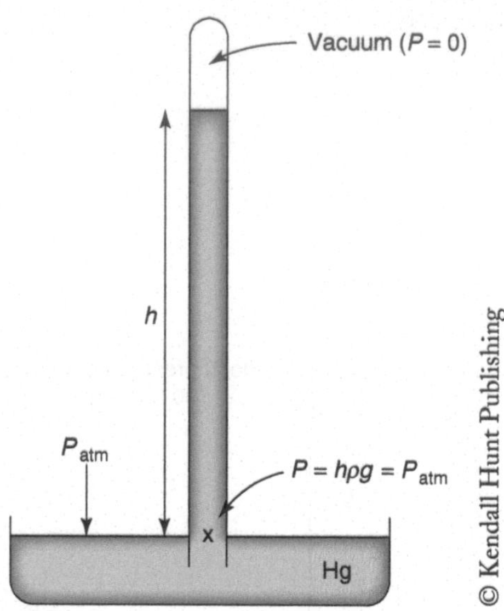

6.10 and 6.11, where P_{bulb} equals zero. (An absolute pressure of zero is a vacuum.) The pressure due to the weight of the mercury must then be equal to atmospheric pressure. Another way of looking at this is to say that since there is no pressure above the mercury the atmospheric is able to push the mercury to a height *h*, where the pressure due to its weight, $P_{depth} = \rho_{fluid} \cdot g \cdot depth$, equals atmospheric pressure.

Atmospheric pressure varies with altitude and weather conditions and is often quoted in units related to the construction of a mercury barometer. For example, normal atmospheric pressure produces an *h* of 760 mm and is quoted as 760 mm Hg or, equivalently, 29.9 inches of mercury.

Pressure measuring devices based on the weight of a column of liquid, such as those above and many others have been in use for many decades, and their various units of pressure, such as mm Hg, are much more commonly used then newtons per square meter or pascals. Units of mm Hg, cm water, and so on, are valid for comparing one pressure to another and can be used without any qualms. If, however, one wishes to calculate the force created by a pressure, it is necessary to convert the pressure into units of force per unit area using $P_{depth} = \rho_{fluid} \cdot g \cdot depth$, as in the following example.

EXAMPLE 6.5

Calculate the maximum *force* in newtons exerted by the blood on an aneurysm, or ballooning, in the aorta, given the maximum blood pressure is 140 mm Hg and the area of the aneurysm is 25.0 cm^2.

Solution: ***Concept.*** Pressure, by definition, is force per unit area. Start with the definition of pressure $P \equiv \frac{F}{A}$, then solve that equation for force, $F = P \cdot A$. This is our working equation.

Step by step 1: In order to determine the force in units of newtons, we need the pressure in newtons per square meter and we need the area in square meters. The specified pressure of 140 mm Hg is not in units of $\frac{\mathrm{N}}{\mathrm{m}^2}$, so we must convert them. We can accomplish this by using $P_{\text{depth}} = \rho_{\text{fluid}} \cdot g \cdot depth$

$$P_{\text{depth}} = \rho_{\text{fluid}} \cdot g \cdot depth = 13{,}600\ \frac{\mathrm{kg}}{\mathrm{m}^3} \cdot 9.8\ \frac{\mathrm{m}}{\mathrm{s}^2} \cdot 0.140\ \mathrm{m} = 1.87 \times 10^4\ \frac{\mathrm{N}}{\mathrm{m}^2}$$

Pressure at depth $= 1.87 \times 10^4\ \frac{\mathrm{N}}{\mathrm{m}^2}$

Step 2: Next we convert the area of the aneurism from square centimeters to square meters. We will use the fact that one square meter equals the same area as 10,000 square centimeter.

$$Area = 25.0\,\mathrm{cm}^2 \cdot \frac{1.00\,\mathrm{m}^2}{10{,}000\,\mathrm{cm}^2} = 2.50 \times 10^{-3}\,\mathrm{m}^2$$

Step 3: Now that we have pressure and area in their "correct" units, we can go back and complete our working equation.

$$F = P \cdot A = 1.87 \times 10^4\ \frac{\mathrm{N}}{\mathrm{m}^2} \cdot 2.50 \times 10^{-3}\,\mathrm{m}^2 = 46.8\ \mathrm{N}$$

This is a large force for the blood vessel to withstand, and there is considerable risk that the aneurysm will burst. Do your best to keep your blood pressure within acceptable limits.

6.4 BUOYANT FORCE AND ARCHIMEDES' PRINCIPLE

Have you ever wondered why your arms and legs feel lighter when they are under water? Or why some objects float and others don't? Why does a piece of steel sink in water while a steel ship floats? If you did, you are in good company. Archimedes (287 BC–212 BC) attempted to answer these and related questions, many centuries before the concepts of force and pressure were developed:

Archimedes discovered a rule of nature that is now named after him. *Archimedes' Principle states:*

Any object placed in a fluid will experience an upward or buoyant force equal to the weight of the fluid it displaces.

This principle applies both to liquids and gases and to objects which are completely or partially submerged. It can be used to determine the buoyant force on an object, but it does not explain the cause of the buoyant force. The buoyant force on an object submerged in a fluid is a result of the fact that pressure increases with depth.

$$P_{\text{depth}} = \rho_{\text{fluid}} \cdot g \cdot depth$$

When an object is submerged in a fluid, the bottom of the object is deeper than the top of the object. Nothing profound yet, but stay with me for a minute. Because the bottom of the object is deeper, the pressure from the surrounding fluid is greater on the bottom of the object than it is on the top.

Since Pressure 2 is larger than Pressure 1, there is an upward force on the object due to this pressure difference. The upward force caused by the pressure difference is called the *buoyant force*. If the buoyant force is larger than the objects weight, the object will accelerate toward the surface. If the upward force is less than the object's weight, the object will sink, but will weigh less while submerged in the fluid.

6.4.1 Deriving the Buoyant Force Equation

Above and beyond. *For those of you interested, we will derive the Buoyant force equation. If you don't care about the derivation, you may want to skip to the end of this discussion.*

As stated above, an object submerged in a fluid has an upward force applied to it caused by the pressure difference between the top and the bottom of the object.

The pressure difference is simply $P_2 - P_1$; Substituting the pressure at depth equation for the two pressures,

Step by step 1: $P_2 - P_1 = \rho_{\text{fluid}} \cdot g \cdot depth_2 - \rho_{\text{fluid}} \cdot g \cdot depth_1$

Step 2: Next we will do a little collecting of terms-if you get scared, you may want to call you mama.

$$P_2 - P_1 = \rho_{\text{fluid}} \cdot g \cdot \left(depth_2 - depth_1\right)$$

Step 3: In Figure 6.19b, what will we know about the submerged object if we actually subtract $depth_1$ from $depth_2$? I couldn't wait any longer. I will just tell you. We will know its height. Substitute height of object for $depth_2$ - $depth_1$,

$$P_2 - P_1 = \rho_{\text{fluid}} \cdot g \cdot h_{\text{object}}$$

We are getting close to the end. Hang in there.

Figure 6.19 Object submerged in a fluid.

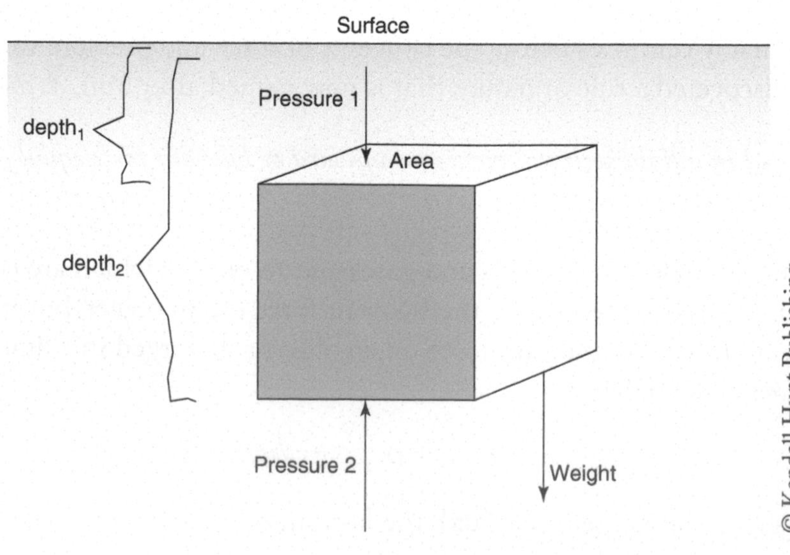

Step 4: The left side of our equation, $P_2 - P_1$, represents the difference in force per unit area on the bottom and top of the object. That is.

$$P_2 - P_1 = \frac{\Delta F}{A_{object}}$$

Step 5: Substituting for $P_2 - P_1$, in our previous equation yields,

$$\frac{\Delta F}{A_{object}} = \rho_{fluid} \cdot g \cdot h_{object} \text{ Therefore}$$

$$\Delta F = \rho_{fluid} \cdot g \cdot \left(h_{object} \cdot A_{object}\right)$$

Step 6: The height of the object, h_o multiplied by the area of the object, A_o, equals the volume of the submerged portion of the object.

$$\Delta F = \rho_{fluid} \cdot g \cdot V_{object\ submerged}$$

Well, there we have it.

The buoyant force on an object submerged in a fluid is the product of the density of the fluid, the value of 'g', and the volume of the submerged portion of the object. *Buoyant Force* $= \rho_{fluid} \cdot g \cdot V_{object\ submerged}$

We are using $9.80 \frac{m}{s^2}$ as the value for g near the earth's surface. The two variables that affect the buoyant force are the density of the fluid and the volume of the object, submerged in the fluid.

$$Buoyant\ Force = \rho_{fluid} \cdot g \cdot V_{object\ submerged} \qquad \text{eq. 6.5}$$

This is one of the coolest of all equations. It explains why objects are lighter underwater and why hot air balloons float. It also explains why hot air balloons that lift people, have to be so large. Hint: the density of air is much less than the density of water.

Floating objects often are not made of a single material. Most boats, for example, are made of numerous materials, many of which have densities greater than water. Boats float because their shape gives them enough volume underwater, so that the buoyant force equals the boat's weight, without submerging the entire boat. Its average density is less than that of water, as illustrated in Figure 6.20. The volume of the boat encompasses large amounts of air, giving it an average density less than that of water. If cargo is added, the volume of the boat beneath the surface increases and therefore the buoyant force increases. Ships are often rated by the maximum amount of water they can displace, typically thousands of tons, giving an indication of their cargo capacity.

$$Buoyant\ Force = \rho_{fluid} \cdot g \cdot V_{object\ submerged}$$

6.4.2 Floating Objects: Percentage Submerged

Above and beyond. *You may want to skip to the end of this derivation.*

You may have heard the expression, "just the tip of an iceberg." It means that what you are looking at, is actually a small percentage of the entire story. It comes from the fact that when ice is floating in water, approximately 90% of its volume is submerged (See Figure 6.20b).

Figure 6.20 The buoyant force equals the weight of a floating object.

(a)

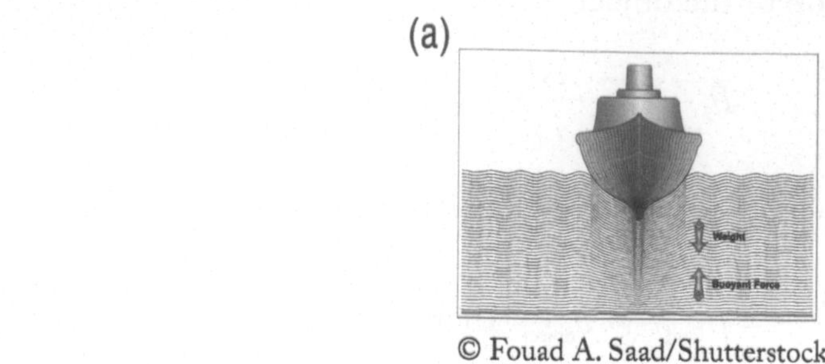

(b)

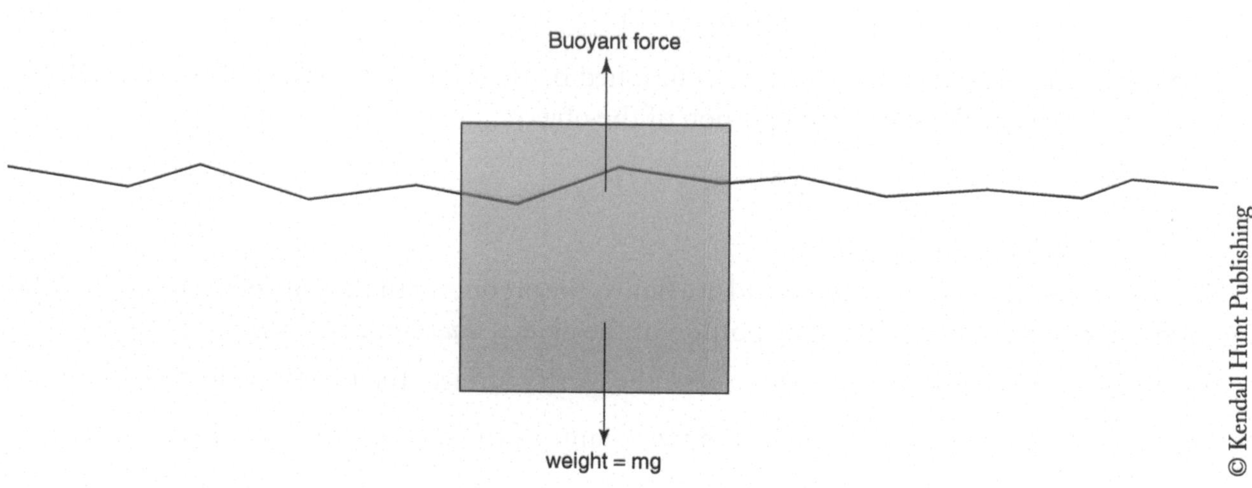

Step by Step 1: An object that is just floating is not accelerating. Because it is not accelerating, according to Newton's first law of motion, the net force on the object must be zero. Therefore, the buoyant force on a floating object is equal to the weight of the object.

$$Buoyant\ force = weight$$

Step 2: Substituting the equations for Buoyant force and weight yields;

$$Buoyant\ Force = \rho_{\text{fluid}} \cdot g \cdot V_{\text{object submerged}} = m_{\text{object}} \cdot g$$

Step 3: Solve the definition of density for the mass of the object.

$$\rho \equiv \frac{mass}{volume};\ \text{therefore m} = \rho \cdot \text{V}$$

Step 4: Substitute $\rho \cdot V$ for the mass of object.

$$\rho_{fluid} \cdot g \cdot V_{object\ \text{submerged}} = \rho \cdot V_{entire\ object} \cdot g$$

Step 5: Divide both sides of our equation by 'g'.

$$\rho_{\text{fluid}} \cdot V_{\text{object submerged}} = \rho \cdot V_{\text{entire object}}$$

$$\frac{V_{\text{object submerged}}}{V_{\text{entire object}}} = \frac{\rho_{\text{object}}}{\rho_{\text{fluid}}}$$

Therefore, the percentage of the object submerged is,

$$\%Submerged = \left(\frac{\rho_{\text{object}}}{\rho_{\text{fluid}}}\right) \cdot 100 \qquad \text{eq. 6.6}$$

EXAMPLE 6.6

Cork has a density of 0.24 g/cm^3. Calculate the fraction of a cork's volume that is submerged when it floats in water. Note that the density of fresh water is 1 g/cm^3.

Solution: Just use equation 6.6

$$\%Submerged = \left(\frac{\rho_{\text{object}}}{\rho_{\text{fluid}}}\right) \cdot 100$$

$$\%Submerged = \left(\frac{0.240\frac{\text{g}}{\text{cm}^3}}{1.00\frac{\text{g}}{\text{cm}^3}}\right) \cdot 100 = 24.0\%$$

Floating objects named hydrometers are used to measure the density of the fluid in which it floats. A hydrometer is illustrated in Figure 6.21. The hydrometer is composed of several materials and has an average density less than the fluid. Since the fraction of the hydrometer submerged is equal to the ratio of its density to the density of the fluid, the greater the density of the fluid, the smaller the fraction submerged and the higher the hydrometer floats in the fluid. The markings on the neck of the hydrometer are calibrated to give the density of the fluid. The numbers get larger toward the bottom of the neck because the buoyant force on a floating object is directly proportional to the density of the fluid that it is in.

So far, only examples involving liquids have been mentioned, but things such as helium-filled balloons can also float in gases. These balloons don't sit on top of the atmosphere as a boat sits on water, but they float in the sense that the buoyant force on them equals their weight. Hydrogen and helium are sufficiently less dense than air to be used in balloons (see Table 5.2). Hydrogen is explosively combustible, whereas helium is expensive and its supply is restricted by the US government. A viable alternative is to heat air, causing it to expand and its density to decrease. When hot air inflates a balloon, it increases the balloon's volume and therefore the buoyant force acting on it from the surrounding air pressure. Because the hot air is less dense than the surrounding air, the increase in the weight of the balloon is less than the increase in the buoyant force applied to it. Physical therapy for damaged muscles often takes place in water because the water supports most of the weight of the patient. Weakened muscles can move limbs in water and be exercised, whereas out of water they might be nearly immobile (see Figure 6.22).

Figure 6.21 A hydrometer floating in a fluid of density 0.87 g/cm³. The greater the density of the fluid, the higher the hydrometer will float.

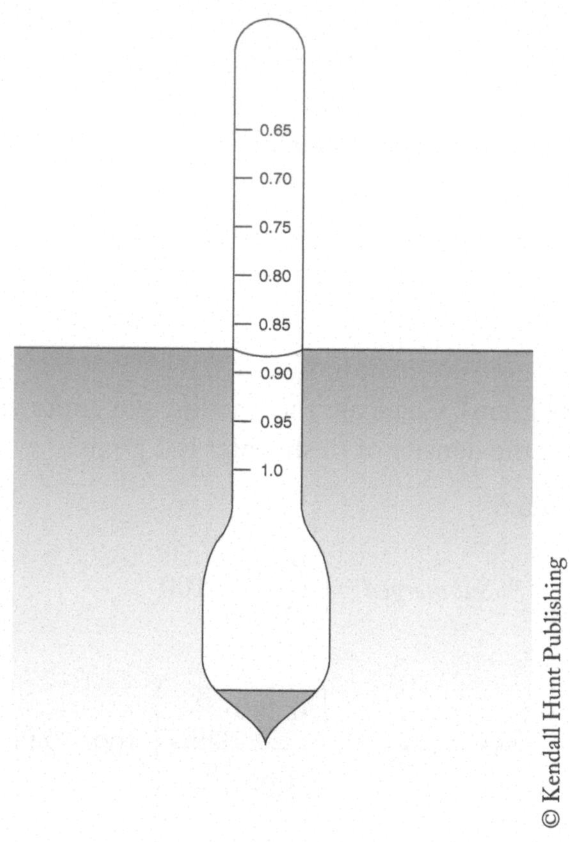

Figure 6.22 Cerebral palsy patient in water at therapy.

Objects That Sink; More Density Measurements

Legend has it that king Hiero II of Syracus, asked Archimedes to determine whether his new crown was pure gold or was mixed with silver, a cheaper metal. Archimedes decided to compare the density of the crown to that of pure gold. Density is mass per unit volume of an object. Archimedes could easily determine the mass of an object. The problem was that no one knew how to determine

Figure 6.23a Technique for measuring volume using Archimedes' principle.

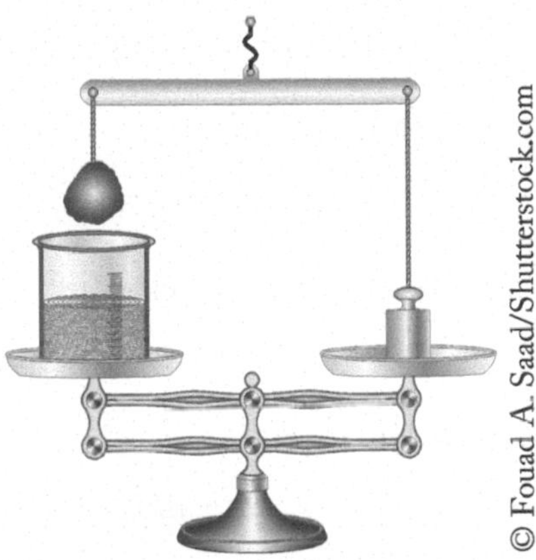

Figure 6.23b Technique for measuring volume using Archimedes' principle.

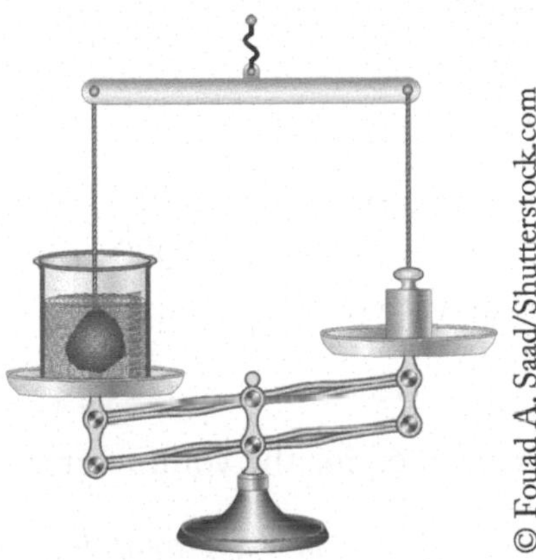

the volume of an irregularly shaped object. Legend asserts that Archimedes noticed the water level rise while he lowered himself into a public bath. Suddenly, Archimedes realized that a submerged object would displace a volume of water equal to the object's volume. He reportedly was so relieved and excited, that he jumped from the tub, and ran naked through the streets, exclaiming Eureka, meaning "I found it." Sixteen hundred years later, Galileo realized that a more accurate method to determine the volume of an object was to compare the mass of an object with its apparent mass while it is submerged in water as illustrated in Figure 6.23. A method known as *Archimedes Principle* could then used to determine the volume of the object, as described below.

If a solid object is weighed in air, the buoyant force due to the air is negligible, so the weight obtained (F) is the true weight of the object, as illustrated in Figure 6.23(a). However, if the object is weighed under water, as shown in Figure 6.23(b), fewer weights will be needed to make the scale balance because there is a buoyant force on the object, and the object appears to weigh less. This fact can be used to accurately calculate the volume of the object as follows.

Using Archimedes Principle to determine the volume of an irregularly shaped object

Above and beyond. *You may want to skip to the end of this derivation.*

Step by step 1: *Archimedes Principle* states that the difference between the object's actual weight and its apparent weight is equal to the buoyant force.

Actual weight—apparent weight = Buoyant force
Substitute mg for the actual weight of the object.

$$\text{mg} - \text{apparent mg} = \text{buoyant force}$$

Step 2: Substitute the buoyant force equation.

$$\text{mg} - \text{apparent mg} = \rho_{fluid} \cdot g \cdot V_{object\ submerged}$$

Step 3: Divide out the acceleration due to gravity, "*g*."

$$\text{m} - \text{apparent m} = \rho_{fluid} \cdot V_{object\ submerged}$$

Step 4: Solve for the volume of the object.

$$V_{\text{object}} = \frac{m - apparent\ m}{\rho_{\text{fluid}}} \quad \text{eq. 6.6b}$$

An example of this cleaver method for determining the volume of an irregularly shaped object follows.

EXAMPLE 6.7

A physiologist measures the mass of a person in air to be 80.0 kg and his apparent mass when submerged in water to be 2.0 kg. The density of water is 1000 kg/m^3.

Calculate the a) volume and b) density of the person.

Solution:

(a) Step by step 1: Write out *equation 6.6b* for the volume of the object.

$$V_{\text{object}} = \frac{m - apparent\ m}{\rho_{\text{fluid}}}$$

Step 2: Next, substitute the values.

$$V_{\text{object}} = \frac{80.0\,\text{kg} - 2.00\,\text{kg}}{1000\,\dfrac{\text{kg}}{\text{m}^3}} = 0.0780\,\text{m}^3$$

(b) Density of person.

$$\rho \equiv \frac{mass}{volume} = \frac{80.0\,\text{kg}}{0.0780\,\text{m}^3} = 1.02 \times 10^3\,\frac{\text{kg}}{\text{m}^3}$$

In practice, the subject whose density is determined must hold a mass in his lap and exhale in order to sink. The density obtained is the average density of the person, including the mass in his lap, residual air in his lungs, muscle, fat, bone, and so on. If enough is known about the person, the density measurement yields additional information. The percentage of body fat, for example, can be estimated. The same technique is used by geologists to measure the density of rocks and archaeologists to measure the density of coins.

General principles such as Archimedes' can be utilized to provide a variety of information. If, for example, an object of known volume is submerged in an unknown liquid, the density of the liquid can be determined. First the mass of the object is measured in and out of the fluid.

For example, if an object with a volume of 10.0 cm^3 has a mass of 27.0 g in air and an apparent mass of 20.0 g in the fluid, we can use Equation 6.6b to determine the density of the fluid?

$$V_{\text{object}} = \frac{m - apparent\ m}{\rho_{\text{fluid}}}$$

Solve for the density of the fluid.

$$\rho_{\text{fluid}} = \frac{m - apparent\ m}{V_{\text{object submerged}}} = \frac{27.0\,\text{g} - 20.0\,\text{g}}{10.0\,\text{cm}^3} = 0.700\frac{\text{g}}{\text{cm}^3}$$

Things to know for section 6.4

$$Buoyant\ Force = \rho_{\text{fluid}} \cdot g \cdot V_{\text{object submerged}}$$

$$\%Submerged = \left(\frac{\rho_{\text{object}}}{density\ of\ fluid}\right) \cdot 100$$

6.5 FLOW: POISEUILLE'S LAW, LAMINAR FLOW, AND TURBULENT FLOW

The distinguishing characteristic of a fluid is that it flows. Gases and liquids are both fluids. In many cases, flow is confined to tubes as, for example, in the circulatory system, lungs, garden hoses, and IV tubes. Although unconfined flow, such as ocean currents, will be considered here, only flow confined to tubes will be treated quantitatively.

The flow rate is defined as volume flowing per unit time,

$$Flow\ rate \equiv \frac{Volume}{time} \qquad Know \qquad \text{eq. 6.7}$$

Examples of units for flow rate are liters per minute and cubic centimeters per second.

Pressure plays a major role in determining the flow rate. Consider the tube in Figure 6.24. If the pressures are the same at either end of the tube, there is no flow. If P_1 is greater than P_2, flow occurs from left to right. Pressure always pushes fluids from higher pressure regions to lower pressure regions. Flow rate is directly proportional to the pressure difference between two regions and is inversely proportional to the resistance to flow. This statement can be summarized by the following equation.

$$Flow\ rate = \frac{\Delta P}{R} \qquad \text{eq. 6.8}$$

Resistance to flow is caused by friction between the fluid and the tube and friction within the fluid itself. There is a relatively simple expression for R if the fluid is incompressible and undergoing laminar flow. *Laminar flow* is smooth and quiet while turbulent flow has eddies, swirls, and ripples. (Another word for laminar is nonturbulent.) Sharp corners, constrictions, and partial obstructions in a tube all cause turbulence and it is turbulence that makes the resistance to flow much larger. A French scientist, J. L. Poiseuille (1799–1869), who was interested in the physics of blood circulation, found that for laminar flow of an incompressible fluid resistance depends on three things: the length of the tube, its radius, and the viscosity of the fluid. The expression for resistance to laminar flow of an incompressible fluid is called *Poiseuille's law*:

$$R = \frac{8\eta L}{\pi r^4} \qquad \text{eq. 6.9}$$

where η (the Greek letter eta) stands for viscosity, L is the length of the tube, and r is its radius. The viscosity η of a fluid is a measure of the friction within that fluid: the larger the viscosity, the slower it will flow. Honey has a large viscosity and flows slowly, whereas water has a relatively small viscosity and flows much more easily.

Poiseuille's law is intuitively reasonable. For example, viscosity appears in the numerator in equation 6.9, so that a viscous fluid offers more resistance to flow than a thin fluid. Similarly, the longer the tube, the higher will be its resistance. Finally, notice that tube radius appears in the denominator of the expression for resistance. We see that the resistance is inversely proportional to the radius raised to the forth power. This means that the resistance and therefore the flow rate, is extremely sensitive to the radius of the tube. In fact, the easiest way to adjust flow is to change the radius of a tube. The body's circulatory system does this by constricting and dilating blood vessels, often in specific organs. Water faucets regulate flow by changing the radius of the tube at the faucet. Flow rate in an IV system is regulated by a clamp on the IV tubing to adjust its radius.

Equations 6.8 and 6.9 taken together accurately describe laminar flow. (Equation 6.8 is always valid, but equation 6.9 holds only for laminar flow.)

It is often useful to combine equations 6.8 and 6.9 in the following manner.

$\textit{Flow rate} = \frac{\Delta P}{R}$ eq. 6.8; $R = \frac{8\eta L}{\pi r^4}$ eq. 6.9

Substitute for R in eq. 6.8,

$$\textit{Flow rate} = \frac{\Delta P}{R} = \frac{\Delta P}{\frac{8\eta L}{\pi r^4}} = \frac{\Delta P \pi r^4}{8\eta L}$$

$$\textit{Flow rate} = \frac{\Delta P \pi r^4}{8\eta L} \qquad \textit{Know} \qquad \text{eq. 6.10}$$

Looking at eq. 6.10 we see that flow rate is directly proportional to the pressure difference, ΔP, and to the radius raised to the fourth power. Flow rate is inversely proportional to the viscosity η and to the length L.

The following example examines the effect of each variable on laminar flow.

EXAMPLE 6.8

If the flow rate through a tube, as in Figure 6.24, is originally 50.0 cm^3/sec, *calculate the new flow rate* (a) if the pressure difference, (ΔP), doubles; (b) if the viscosity of the fluid, η, doubles; (c) if the length of the tube, L, doubles; and (d) if the radius of the tube, r, doubles. Assume that in each case only the factor mentioned differs from the original conditions. Use equation 6.10

Solution: We will use as our working equation for all parts of this solution. Because we are asked to calculate the new flow rate, we will use the equation in its present form. It is also important to realize that all other factors that might affect flow rate, are held constant.

$$Flow\ rate = \frac{\Delta P \pi r^4}{8\eta L}$$

a. What is the new flow rate if *ΔP doubles*. Looking at our working equation, we see that flow rate is directly proportional to ΔP, the pressure difference. Since ΔP is doubled, then the flow rate will be doubled. The new flow rate FR', will equal 2 times the old flow rate.

$$FR' = 2 \cdot FR = 2 \cdot 50 \frac{\text{cm}^3}{\text{s}} = 100 \frac{\text{cm}^3}{\text{s}}$$

b. What is the new flow rate if the viscosity of the fluid η, doubles. Looking at our working equation, $Flow\ rate = \frac{\Delta P \pi r^4}{8\eta L}$ we see that the flow rate is inversely proportional to the viscosity η, of the fluid. Doubles means 2. The inverse of 2 is 1/2. Therefore, the new flow rate is one half of the original flow rate:

$$\text{FR}' = \frac{1}{2}\text{FR} = \frac{1}{2}\left(50.0 \frac{\text{cm}^3}{\text{s}}\right) = 25.0 \frac{\text{cm}^3}{\text{s}}$$

c. What is the new flow rate if the length of the tube, *L*, *doubles*. Looking at our working equation, $Flow\ rate = \frac{\Delta P \pi r^4}{8\eta L}$ we see that the flow rate is inversely proportional to the length "*L*," of the tube. The inverse of 2 is 1/2. Therefore, $\text{FR}' = \frac{1}{2}\text{FR} = \frac{1}{2}\left(50.0 \frac{\text{cm}^3}{\text{s}}\right) = 25.0 \frac{\text{cm}^3}{\text{s}}$

d. If the radius of the tube, *r*, *doubles*. Looking at our working equation, $Flow\ rate = \frac{\Delta P \pi r^4}{8\eta L}$, we see that flow rate is directly proportional to the radius raised to the fourth power. Therefore,

$$FR' = 2^4 \cdot FR = 16 \cdot 50.0 \frac{\text{cm}^3}{\text{s}} = 800 \frac{\text{cm}^3}{\text{s}}$$

This example may have been new for some of you. Please study it until you feel that you can solve a similar problem on your own.

When two or more factors are changed simultaneously, each has its own effect and the overall result is the combined effect of all changes. For example, *if both the length and radius* of the tube are *doubled* in the above example, then the new flow rate is determined as follows:

As always, we look at our working equation in order to determine the relationship between the flow rate and the variables that are being changed. $Flow\ rate = \frac{\Delta P \pi r^4}{8\eta L}$. We see that the flow rate is directly proportional to the radius raised to the fourth power and inversely proportional to the length of the tube. Therefore,

$$FR' = FR \cdot \left(\frac{2^4}{2}\right) = 50.0\frac{\text{cm}^3}{\text{s}} \cdot \frac{16}{2} = 400\frac{\text{cm}^3}{\text{s}}$$

Changes other than factors of 2 are handled similarly. For example if the pressure difference (ΔP) is reduced to one-third its original value, then the flow rate is also reduced to one-third its original value.

Quantities other than flow rate can also be found. Consider the following example.

EXAMPLE 6.9

When a person eats a large meal, blood flow in the digestive system is increased by dilating blood vessels supplying that system. *By what factor must the flow regulating vessels dilate* to increase flow rate from 1.0 liter/min to 5.0 liter/min?

The first thing that we have to do is make sure that we understand the question. We are being asked to compare the new radii to the old radii. Our previous questions asked us about flow rate, so our working equation started off with "flow rate = ." Because we are now being asked about the new radius, we need a new working equation. We need one that begins with "radius =."

We will solve the flow rate equation 6.10 for "*r*."

$$Flow\ rate = \frac{\Delta P \pi r^4}{8\eta L}$$

$$r = \left(\frac{FR \cdot 8\eta L}{\Delta P \pi}\right)^{0.25}$$

This is our working equation

In the problem statement we were told that the new flow rate is five times the old flow rate.

$$\frac{FR'}{FR} = \frac{5\frac{\text{L}}{\text{min}}}{1\frac{\text{L}}{\text{min}}} = 5$$

From our working equation, we also see that the radius is directly proportional to the flow rate raised to the 0.25 power.

Therefore, the new radius is:

$$r' = r \cdot 5^{0.25} = 1.50r$$

We see that if the radius is increased to one and one-half of its original size, the flow rate will be five times its original value! So a 500% increase in flow rate is accomplished by a 50.0% increase in the radii of the blood vessels. It is not possible to get a value for the radius from the given information—only the factor by which it changes.

EXAMPLE 6.10

The flow rate in an IV setup, such as the one illustrated in Figure 6.9, is observed to be 2.0 cm^3/min for a glucose solution of density 1050 kg/m^3. The surface of the solution is a height h = 1.0 m above the entrance of the needle. If the person's blood pressure is 8.0 mm Hg, (a) what is the pressure

Figure 6.9 The pressure due to the IV solution is proportional to the height *h* of the liquid surface above the needle. It is the same for either path A or B.

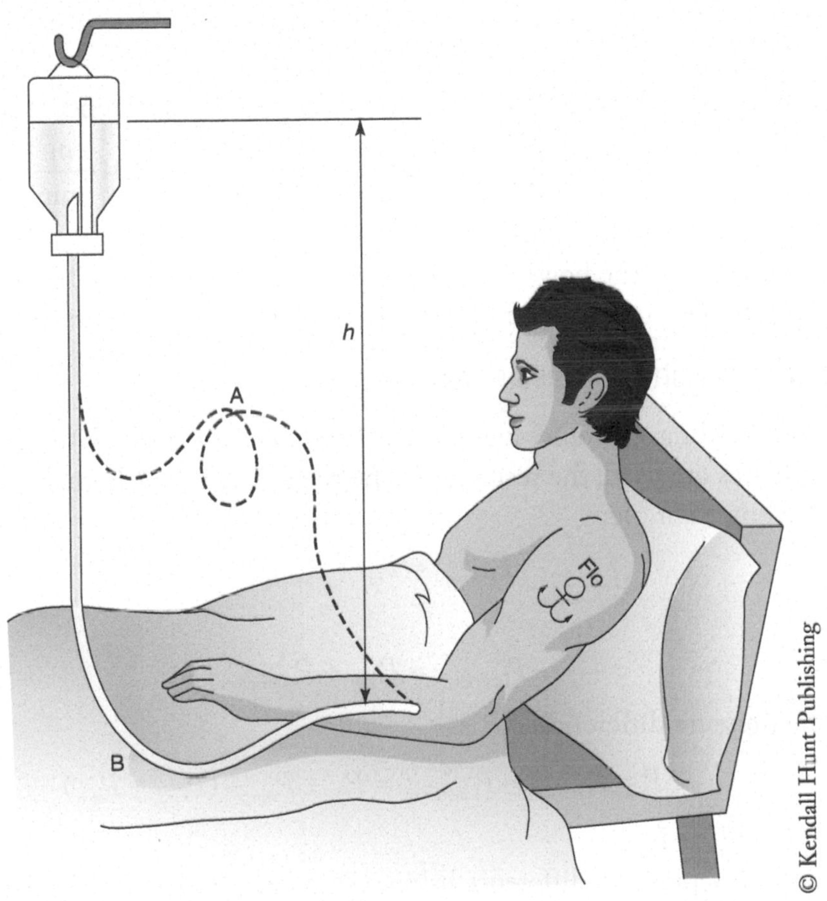

difference in Pascals? (b) If the height is increased to 1.5 m, what is the new pressure difference in Pascals? (c) What is the new flow rate at a height of 1.5 m? Note that the person's blood pressure (P_2) remains at 8.0 mm Hg.

a. What is the pressure difference in

Solution: The pressure from the weight of the 1.00 m of glucose solution is,

$$P_{depth} = \rho_{fluid} \cdot g \cdot depth = 1050\frac{kg}{m^3} \cdot 9.80\frac{m}{s^2} 1.00\,m = 10{,}300\,Pa$$

The person's blood pressure is, $P_{depth} = \rho_{fluid} \cdot g \cdot depth = 13{,}600\frac{kg}{m^3} \cdot 9.8\frac{m}{s^2} \cdot 8 \times 10^{-3}\,m = 1070\,Pa$

The *pressure difference* $\Delta P'$ is, $10{,}300Pa - 1070\,Pa = 9230\,Pa$

b. If his blood pressure is still 1070 Pa and the height is increased to 1.50 m, what is the new pressure difference?

Solution: The pressure from the weight of 1.50 m the glucose solution is,

$$P_{depth} = \rho_{fluid} \cdot g \cdot depth = 1050\frac{kg}{m^3} \cdot 9.8\frac{m}{s^2} \cdot 1.50\,m = 15{,}400\,Pa$$

The *new pressure difference* $\Delta P'$ is, $15{,}400\,Pa - 1070\,Pa = 14{,}300\,Pa$

c. What is the new flow rate at a height of 1.5 m?

Solution: Our working equation for flow rate will be our old stand by, eq. 6.10

$$Flow\ rate = \frac{\Delta P \pi r^4}{8\eta L}$$

We see that flow rate is directly proportional to the pressure difference. Therefore,

$$FR' = FR\left(\frac{\Delta P'}{\Delta P}\right) = 2.00\frac{\text{cm}^3}{\text{min}} \cdot \frac{14{,}300\ \text{Pa}}{9230\ \text{Pa}} = 3.10\frac{\text{cm}^3}{\text{min}}$$

We see that raising the glucose solution from 1.0 m to 1.5 m increased the flow rate from 2.0 cm^3/min to 3.1 cm^3/min. If the height of the solution is less than about 10 cm, blood will flow out of the patient into the IV bottle. Some IV setups are equipped with antibackflow valves.

6.5.1 Effect of Atmospheric Pressure on Flow

Pressures in the preceding example, like most others we have dealt with, are gauge pressures. If, instead, total pressures are used, the result has to be the same. To verify this, consider the total pressures in the preceding example:

$$P_{\text{Hi(total)}} = P_{\text{Hi}} + P_{\text{atmosphere}}$$

and

$$P_{\text{Low(total)}} = P_{\text{Low}} + P_{\text{atmosphere}}$$

Therefore the pressure difference, ΔP is

$$\Delta P = P_{\text{Hi(total)}} - P_{\text{Low(total)}} = P_{\text{Hi}} + P_{\text{atm}} - (P_{\text{Low}} + P_{\text{atm}})$$

Therefore, $\Delta P = P_{\text{Hi}} - P_{\text{Low}}$

This proves that the pressure difference using just gauge pressure are the same as if we used total pressure. The vent tube in the IV bottle allows atmospheric pressure to push on the surface of the IV solution, aiding flow. Atmospheric pressure also pushes on the patient and is transmitted through the body to the circulatory system, inhibiting flow. The effect of the atmosphere thereby cancels. When a flexible bag without a vent tube is used to supply IV solution, atmospheric pressure is transmitted through its walls to the IV solution just as it would be through a vent tube.

More generally, because the total pressure is defined (equation 6.2) as

$$P_{\text{Total}} = P_{\text{gauge}} + P_{\text{atm}},$$

then any time that two pressures are subtracted, atmospheric pressure will subtract out just as it did above. Therefore it does not matter whether gauge or total pressures are used as long as all pressures are either gauge or total. The only caution that need be exercised is to be certain that the gauge pressures are correct. For example, if there is no vent in a rigid IV bottle, then the gauge pressure is not simply the *h* of the fluid; it is *h* plus the negative pressure of the air above the fluid.

6.5.2 Pressure Difference Causes Fluids to Flow

Consider Figure 6.24 once more. Solving equation 6.8 for the pressure drop

$$Flow\ rate = \frac{\Delta P}{R} \qquad \text{eq. 6.8}$$

$$\Delta P = FR \cdot R \qquad \text{eq. 6.11}$$

Figure 6.24 A fluid of viscosity η flowing through a tube of length L and radius r. Flow will be in the direction shown only if P_1 is greater than P_2.

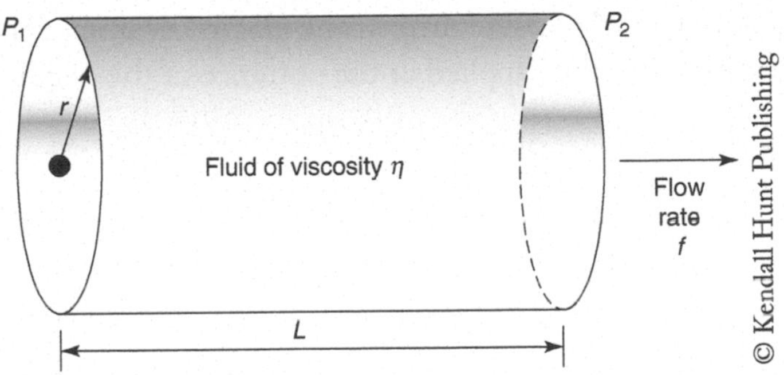

With the equation in this form, the pressure drop from P_{Hi} to P_{Low} can be interpreted as being caused by flow and resistance: the greater the resistance, the greater the pressure drop. If flow is laminar, then the resistance R is given by equation 6.9 (Poiseuille's law) and depends only on the dimensions of the tube and the viscosity of the fluid.

Equation 6.11 can be useful in explaining certain phenomena, such as drops in household water pressure when many people are watering their lawns.

Figure 6.26 is a schematic of a water main supplying several houses. The pressure at the entrance of the pipes supplying the houses is P_2. If no water flows, P_2 is the same as P_1, the pressure at the entrance of the water main. This is confirmed by equation 6.8, since Flow rate must equal 0 and $\Delta P = 0$. If all the houses use water simultaneously, then resistance is decrease, causing an increased flow rate. Since the flow rate is large, the pressure difference ΔP is also large, implying that P_2 is much smaller than P_1. There are two ways to keep P_2 large during heavy use. One is to increase P_1. The other is to increase the size of the water main and reduce its resistance. In practice, neither is done for water mains serving homes. However, the human body does both in the circulatory system during times of heavy use. During strenuous exercise, the blood pressure increases (analogous to an increase in P_1) and the major blood supplying arteries dilate (analogous to an increase in the size of the water main).

Equation 6.11, ($\Delta P = FR \cdot R$), can also be used to analyze where pressure drops occur in tubing systems. In an IV system, for example, the IV tubing is much larger in radius than the needle. The

Figure 6.26 A water main supplying a number of users. P_2 drops if flow rate increases.

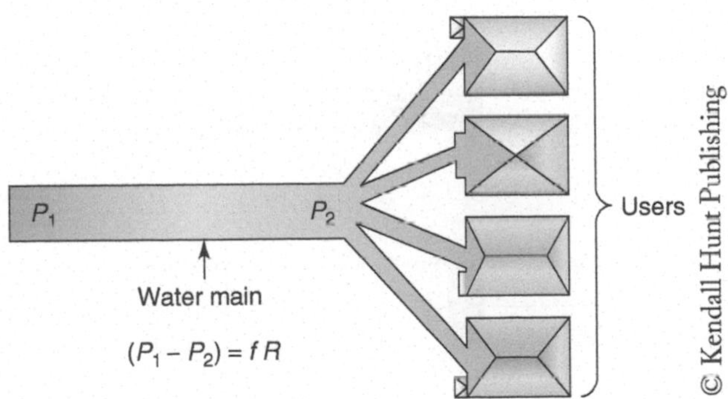

resistance of the needle is far greater than the resistance of the IV tubing because R depends so strongly on radius. The pressure drop in the IV tubing is therefore negligible. Almost all the pressure drop occurs in the needle, because its resistance is large. If a clamp is used on the IV tubing to control the flow rate, there is a large pressure drop at the clamp because of the reduced radius and large resistance. A smaller pressure is thus applied at the entrance of the needle and a smaller flow results.

6.5.3 Turbulent Flow

Turbulent or nonlaminar flow is characterized by eddies and swirls in the fluid stream. Resistance for turbulent flow is significantly larger than it is for laminar flow. Equation 6.8 still describes the relationship between flow rate, pressure, and resistance, but resistance is no longer given by Poiseuille's law (equation 6.9).

Turbulence has a number of causes. Sharp bends and partial obstructions, particularly those with irregular shapes, can cause eddies to form as seen in Figure 6.26. It is often possible to hear turbulent flow because of the sounds made by the eddies. When you hear water running somewhere in your home, the sound is turbulence, usually in the valve. Heart murmurs are sounds produced by turbulent blood flow through a defective heart valve or hole between heart chambers. An aneurysm (ballooning of an artery) can also cause audible turbulence because of its irregular shape.

What about the effects of turbulence? Turbulent flow always exhibits more resistance than laminar flow. The onset of turbulence reduces flow rate if the pressure difference in the tube is constant. In the case of an aneurysm the body may respond by increasing blood pressure to maintain normal flow rate. Unfortunately, increased blood pressure only worsens the aneurysm.

Fluid flow that is not confined to tubes also exhibits laminar and turbulent flow. Modern cars are designed to cause as little turbulence as possible to avoid the extra resistance and increased fuel consumption. Aircraft, ships, and submarines are similarly designed, and the shapes of the fastest birds and sea life are such as to create very little turbulence.

Turbulence is also caused by large fluid speeds. (Note that speed and flow rate are not the same thing. The speed v of a fluid in two different tubes may be the same, but the tube with the larger diameter will have a larger flow rate.) Above a certain critical speed, friction between the fluid and the walls of the tube starts eddies. Commercial aircraft are flown at less than top speed for this

Figure 6.26 Turbulent flow exhibits eddies and swirls and offers more resistance than laminar flow. An obstruction is one possible cause of turbulent flow.

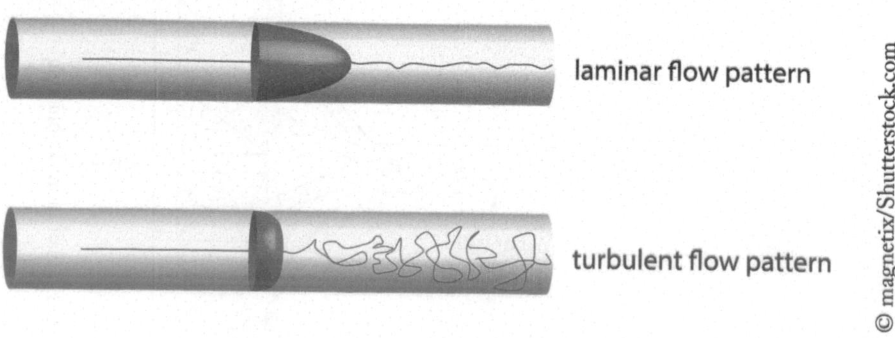

reason. One reason supersonic aircraft consume inordinate amounts of fuel is that at those speeds it is extremely difficult to avoid creating turbulence.

Other causes of turbulence have to do with the properties of the fluid itself. Blood cells are not entirely fluid in their characteristics. In small blood vessels, especially capillaries, interaction of the cells with vessel walls causes turbulence. In larger vessels blood flow is approximately laminar and Poiseuille's law applies fairly well.

It is very difficult to calculate resistance for turbulent flow. While resistance during laminar flow depends on the radius and length of the tube, and the viscosity of the fluid, resistance during turbulent flow also depends on other factors, such as the speed v of the object moving through the fluid. Fuel consumption in cars driving at steady speed on level ground is due entirely to work done against friction. Fuel consumption does not greatly depend on the speed of a car up to about 90 km/hr (54 mi/hr). Above that speed, the air flow becomes noticeably turbulent and resistance increases. In addition, the greater the speed above the point where flow becomes turbulent, the greater is the resistance. The drag force, F_d on any object is directly proportional to the density of the fluid and proportional to the square of the relative speed between the object and the fluid.

Drag force $F_d = \frac{1}{2}\rho v^2 c_d A$ where c_d is the drag coefficient.

(At very low Reynolds numbers, without flow separation, the drag force is proportional to v instead of v^2).

6.5.4 Turbulent Flow and the Reynold's Number

Turbulent flow is identified as random and chaotic. When flow becomes turbulent, resistance increases and flow rate decreases. As the speed of a fluid increases, turbulent flow will eventually result. The onset of turbulent flow is predicted by a dimensionless number, called the Reynold's number. Although the number was introduces by George Stokes, it was named after Osborne Reynolds (1842–1912), who popularized its use. The Reynold's number depends on the density of the fluid, the fluids viscosity, the average speed of the fluid, and the diameter of the vessel. The factors affecting the Reynold's number are summarized in the following equation.

$R_e = \frac{\rho d \bar{V}}{\eta}$ the phrase "rho d vessel nancy" might help you to remember this rule.

Reynold's number
Where,

ρ represents the density of the fluid.
d is the diameter of the vessel
$\bar{V}$ is the average speed of the fluid in the section of vessel being studied
η is the viscosity of the fluid

Depending on tube geometry, the onset of turbulent flow occurs when the Reynold's number is greater than 2000 to 2500. The boundary layer will transition from laminar to turbulent providing the Reynolds number of the flow around the body is high enough. Larger velocities, larger objects, and lower viscosities contribute to larger Reynolds numbers

Example 6.11 Calculate the flow speed above which turbulent flow should occur in the aorta. Assume a diameter of 2.0 cm and a Reynold's number of 2000. For the density of blood, use 1.05 $\frac{\text{g}}{\text{cm}^3}$

The viscosity of blood is 0.035 $\frac{\text{g}}{\text{cm}\cdot\text{s}}$

Step by step 1: Write down the Reynold's number equation.

$$R_e = \frac{\rho d\bar{V}}{\eta}$$

Step 2: Determine our "working equation" by solving for $\bar{V}$.

$$\bar{V} = \frac{R_e\eta}{\rho d}$$

Step 3: Enter the values and then do the math.

$$\bar{V} = \frac{2000 \cdot 0.035\frac{\text{g}}{\text{cm}\cdot\text{s}}}{1.05\frac{\text{g}}{\text{cm}^3}\cdot 2.00\,\text{cm}} = 33.3\frac{\text{cm}}{\text{s}}$$

We see that under these conditions, a blood flow speed greater than $33.3\frac{\text{cm}}{\text{s}}$ results in a Reynold's number above 2000 and may result in turbulent flow.

Things to know for 6.5

$\textit{Flow rate} \equiv \frac{\textit{Volume}}{\textit{time}}$ typical units are $\frac{\text{liter}}{\text{minute}};\frac{\text{cm}^3}{\text{s}}$

$$1\,\text{L} \equiv 1000\ \text{cm}^3 \quad \textit{Flow rate} = \frac{\Delta P\pi r^4}{8\eta L}$$

Reynold's number: $R_e = \frac{\rho d\bar{V}}{\eta}$

Here is a little memory aide: "rho d vessel nancy"

6.6 THE BERNOULLI EFFECT AND ENTRAINMENT

What happens to the velocity of a fluid when the diameter of a tube or vessel is decreased? (see Figure 6.27) The velocity of the fluid will increase and surprisingly, its internal pressure will decrease.

The situation described above is one example of the Bernoulli effect, first explained by Swiss scientist Daniel Bernoulli (1700–1782). Bernoulli's principle may be stated this way: *Where the speed of a fluid is high the internal pressure is low, and where the speed of a fluid is low the pressure is high.*

Figure 6.27 When the diameter of a tube or vessel is decreased, the velocity of the fluid is increased and its internal pressure decreases.

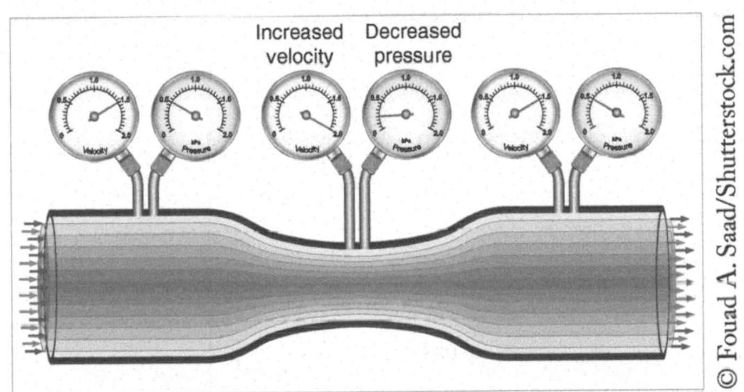

As in the previous section, it should be noted that the speed v of a fluid is not the same as the flow rate. Flow rate is the *volume* of fluid moving per unit time. The difference between speed and flow rate can be seen by comparing a slow moving river current and a garden hose. The "lazy river" may have a large flow rate and a small fluid speed, while a garden hose will have a small flow rate with a much higher fluid speed than the "lazy river."

Whenever one fluid is pushed into the flow of another, it is said to have been entrained, and the process is called entrainment. One fluid is pushed into the other because the pressure in the rapidly moving fluid is smaller than outside pressure. The Bernoulli effect and entrainment have many applications; a few are illustrated in Figure 6.28. The fluids involved in entrainment may be liquids or gases of any type. Numerous other applications exist, each using the Bernoulli effect to create a low pressure.

Many entrainment devices have a jet or constriction designed to increase the speed of the fluid and enhance its entrainment of another fluid. A venturi is a tube with a constriction designed to increase the Bernoulli effect. The schematic in Figure 6.29 illustrates pressures in a venturi. Pressure is lowest in the narrowest part of the tube, where the speed of the fluid is greatest. Pressure increases downstream in the larger part of the tube, where speed decreases. Automobile carburetors, for example, entrain gasoline into air in a venturi.

How much does the fluid speed change in going from a large radius to a small one or vice versa? That question can be answered for an incompressible fluid by referring to Figure 6.30, which shows a section of tube of length L, where L is the distance the fluid moves in a time t. That is, the average speed of the fluid is

$$\bar{v} = \frac{L}{t}$$

The entire volume of the section of tube shown will flow by in time t.

Flow rate can be determined by multiplying the cross-sectional area by the average velocity.

Flow rate $= A \cdot \bar{v}$ Derivation to follow.

Above and beyond. *You may want to skip to final result of this derivation.*

Step by step 1: Please recall the definition of flow rate.

$$\textit{Flow rate} \equiv \frac{\textit{Volume}}{\textit{time}}$$

Figure 6.28 Examples of entrainment due to the Bernoulli effect: (a) A Bunsen burner; (b) a perfume atomizer; (c) an aspirator suction pump; (d) liquid entrainment for respiration therapy.

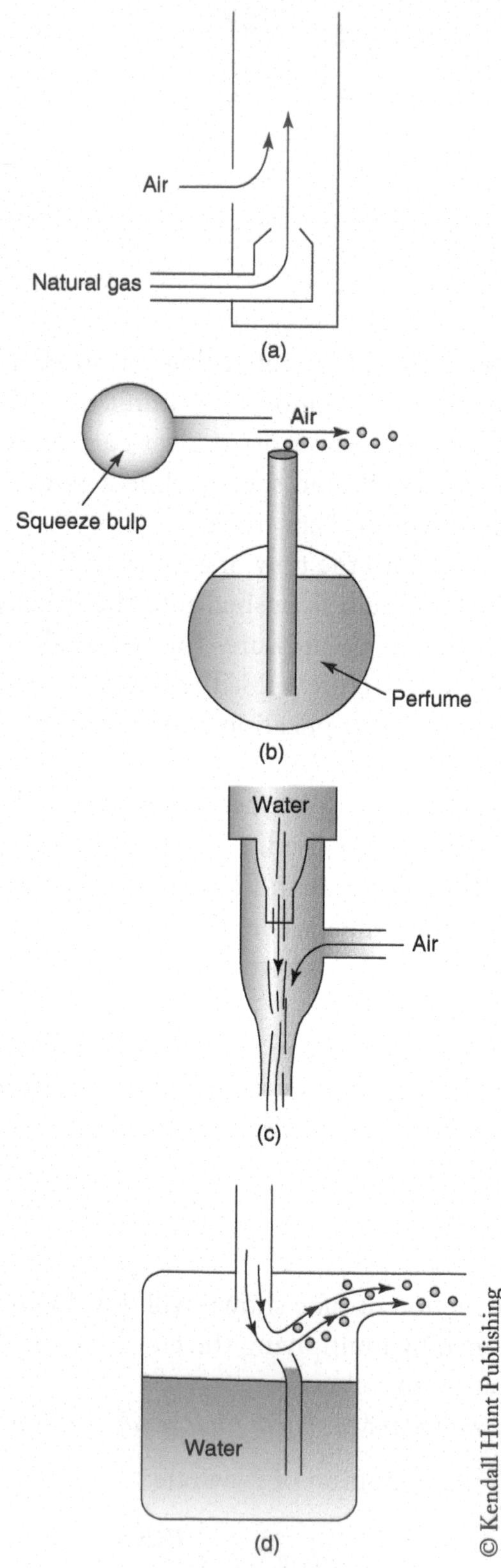

Step 2: The volume of fluid will equal the cross-sectional area of the tube A, multiplied by the length of the tube L.

$$Volume = A \cdot L$$

Step 3: Substitute for volume.

$$Flow\ rate \equiv \frac{Volume}{time} = \frac{A \cdot L}{t} = A\bar{v} \text{ know eq. (6.12)}$$

Where A is equal to the cross-sectional area and $\bar{v}$ is the average velocity of the fluid. So we see that the flow rate can be determined from the speed of the fluid when the cross-sectional area is known.

Final result: *Flow* rate = $A\bar{v}$ know eq. (6.12)

This relationship is valid for any fluid flowing through a tube with cross-sectional area A at an average speed v.

If we refer back to Figure 6.29 and assume the fluid is incompressible, then the flow rate must be the same in all parts of the tube: What flows through the large part of the tube also flows through the narrow part of the tube. Therefore, the fluid must move faster through the narrow parts than it does when the cross-sectional area is larger. We can prove this mathematically as follows:

Flow rate through area one (A_1) = Flow rate through area two (A_2)

$$A_1 v_1 = A_2 v_2$$

Solving for the velocity in the narrow part of the tube yields,

$$v_2 = \frac{A_1 v_1}{A_2}$$

Since A_1 is larger than A_2, then speed v_2, must be larger than v_1. According to Bernoulli's principle, the pressure is thus smaller in the narrow part of the tube. This last result accurately gives the relationships between speeds when the fluid is incompressible—that is, a liquid. The sense of the

Figure 6.29 The pressure drops in the narrowest part of a venture because the speed is greatest there. The pressure rises again in the larger part of the tube downstream, but P_3 is less than P_1 owing to resistance to flow in the tube.

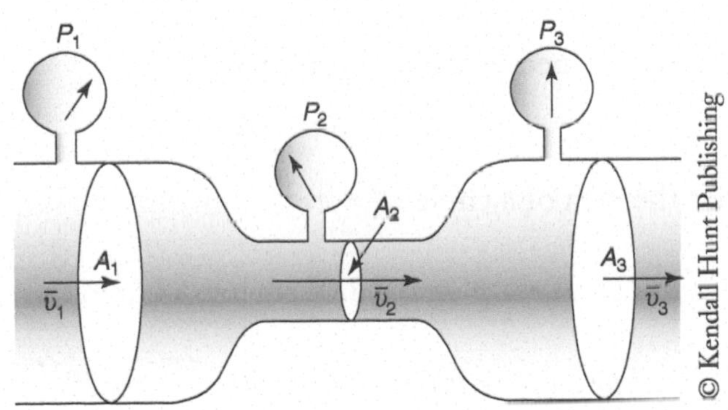

Figure 6.30 The relationship between flow rate, cross-sectional area, and speed.

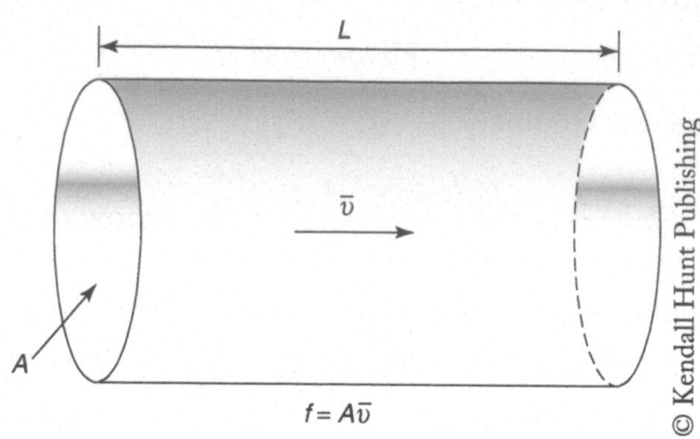

result also is correct for compressible fluids (gases), in which case the pressure decreases significantly in the narrow part of the tube.

What causes pressure to decrease when fluid speed increases? One way to answer this question is to consider energy. When a fluid goes from a large tube to a smaller one as in Figure 6.29, its speed increases. Its kinetic energy therefore increases. Where does that energy come from? The tube is horizontal, so it is not coming from gravitational potential energy. But there is potential energy associated with pressure, and this energy is used to increase the kinetic energy of the fluid. *Conservation of energy tells us that as the velocity increases, the internal pressure of the fluid must decrease.* Finally, when the fluid enters another larger section of tube, as it does on the right of Figure 6.29, the process is reversed and pressure increases. Note that the pressure on the right in Figure 6.29 (P_3) is slightly smaller than P_1. It has dropped because of resistance in the tubing. Part of the energy of the fluid has been converted to thermal energy by friction, and less is left to raise the pressure back to its original value of P_1.

In addition to helping explain the Bernoulli effect, equation 6.12 can be used analytically to describe the relationship between flow rate, cross-sectional area, and fluid speed.

Please review Tremblay's 'Flow Rate' lecture **before proceeding.**

EXAMPLE 6.11

The average speed of a liquid in a tube of radius 2.00 cm is 50.0 cm/sec. (a) Calculate the flow rate. (b) If the tube narrows to a radius of 1.00 cm, as in a venturi, what is the new average speed of the liquid?

Solution:

a. The flow rate can be determined by using equation 6.12.

$$\textit{Flow}\ \text{rate} = A\bar{v}$$

Please recall that the area of a circle $A = \pi r^2$

$$A = \pi r^2 = \pi \cdot (2.00\,\text{cm})^2 = 12.6\,\text{cm}^2$$

$$\textit{Flow}\ \text{rate} = A\bar{v} = 12.6cm^2\, 50.0\frac{cm}{s} = 628\frac{cm^3}{s}$$

b. If the tube narrows to a radius of 1.00 cm, as in a venturi, what is the new average speed of the liquid? First recall equation 6.12.

$$Flow\ \text{rate} = A\overline{v}$$

In an in-compressional fluid the flow rate will always equal the cross-sectional area multiplied by the average velocity of the fluid.

We can determine the average fluid velocity by solving equation 6.12 for average velocity and substitute the values.

$$\overline{v} = \frac{Flow\ rate}{Area} = \frac{628\frac{\text{cm}^3}{\text{s}}}{\pi \cdot (1.00\,\text{cm})^2} = 200\frac{\text{cm}}{\text{s}}$$

Some devices used for measuring flow rates actually measure the speed of the fluid and then calculate the flow rate based on the size of the tube. The calculation may be done automatically by the device so that the user is unaware of the process.

6.6.1 Bernoulli Effect in Flow Not Confined to Tubes

The examples of the Bernoulli effect and entrainment given so far involve fluids flowing in tubes, but the effect is general and need not be limited to such situations. Pressure in a flowing fluid is always lowest where the fluid speed is highest. Consider the wing shown in Figure 6.31. As the wing cuts through the air, that part of the air going over the top must go farther in the same amount of time it takes the wing to pass by than the air going under the wing. The air going over the wing therefore has a higher speed relative to the wing, and the pressure above the wing is less than the pressure below the wing. Because a faster moving fluid has a lower internal pressure than a slower moving fluid, the pressure on the bottom of the wing pushing up is higher than the pressure on the top of the wing pushing down. Therefore, the net force on the wing is in the upward direction. Similar explanations describe in part the functioning of sails on boats.

Burrowing animals always have at least two entrances to their homes. The purpose is to cause air circulation by using the wind currents passing over the holes. Wind speed over one of the holes will occasionally be larger than over another, creating a smaller pressure and causing air to flow through the burrow. Chimneys and flues also benefit from air currents passing over them, creating a lower pressure and aiding the upward flow of fumes. Some flues are designed with a cover piece that not only keeps out rain but also channels wind across the top of the flue. The list of applications of the Bernoulli effect is long and varied.

Figure 6.31 The Bernoulli effect explains part of the lift created by a wing. The shape of the wing increases the velocity of the air moving across its upper surface. That decreases the air pressure on the top. Higher air pressure beneath the wing applies a force, pushing it upward.

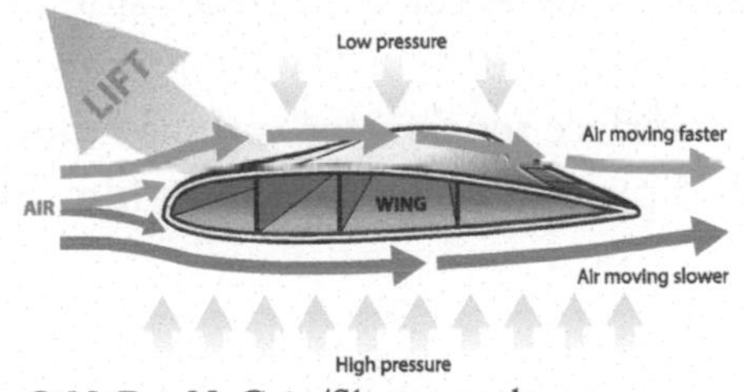

Things to know for 6.6

$$Flow\ \text{rate} = A\bar{v}$$

6.7 COHESION AND ADHESION

The forces acting between molecules in gases, liquids, and solids were described briefly in Chapter 5. This section examines the major effects of those forces in fluids.

Forces between molecules can be either attractive or repulsive; the forces are repulsive when molecules get very close together and are otherwise attractive. The repulsive nature of the forces at close distances accounts for the near-incompressibility of liquids and solids. The attractive nature of the forces is responsible for a number of phenomena, among which are surface tension, capillary action, and fluid viscosity. Attractive forces between molecules of the same type are called cohesive forces; those between dissimilar molecules are called adhesive forces. The strengths of cohesive and adhesive forces depend on the substances involved.

6.7.1 Surface Tension

Cohesive forces in liquids cause them to bead up, as in raindrops. The liquid surface acts like a stretched rubber sheet that tries to contract and make the surface area as small as possible, hence forming beads. The surface is under tension and the overall effect is called surface tension.

Figure 6.32 illustrates the effect of surface tension created by cohesive forces in two different circumstances. Consider a nonspherical drop. Molecules in the interior of the drop experience a net force of zero since they have neighbors on all sides. Molecules on the surface, particularly those at the ends, experience a net inward force, making the drop spherical. Once the drop becomes spherical, inward forces on all surface molecules are equal and are balanced by repulsive forces that prevent the molecules from getting too close together.

Figures 6.33a and 6.33b depict an insect and paper clip supported by water tension.

In both circumstances shown in Figure 6.33 the surface behaves like a stretched rubber sheet. This analogy works so well that it can be used to describe a variety of phenomena involving liquids. Some insects, for example, can walk on water even though their density is greater than water. The weight of the insect on the surface causes a dent, as shown in Figure 6.33a. Surface tension opposes the stretching of the surface, yielding an upward force that can support the weight of the insect. The same force exerted on a smaller area, such as the weight of a needle placed point down, would break the surface. Yet a needle carefully placed on its side on water can be supported by surface tension and will appear to float, even though the density of iron is significantly greater than water. Note that the surface tension forces are parallel to the surface on either side of the object, and the sum of those forces is vertical in these cases.

One device used to measure surface tension for various liquids is shown in Figure 6.34. A wire frame with one moveable side is dipped in liquid forming a film. The force needed to stretch the film

Figure 6.32 Surface tension is created by cohesive forces and always tends to make the surface area as small as possible. Under the influence of surface tension (a) a nonspherical drop becomes spherical and (b) a bump on the surface of a liquid is flattened.

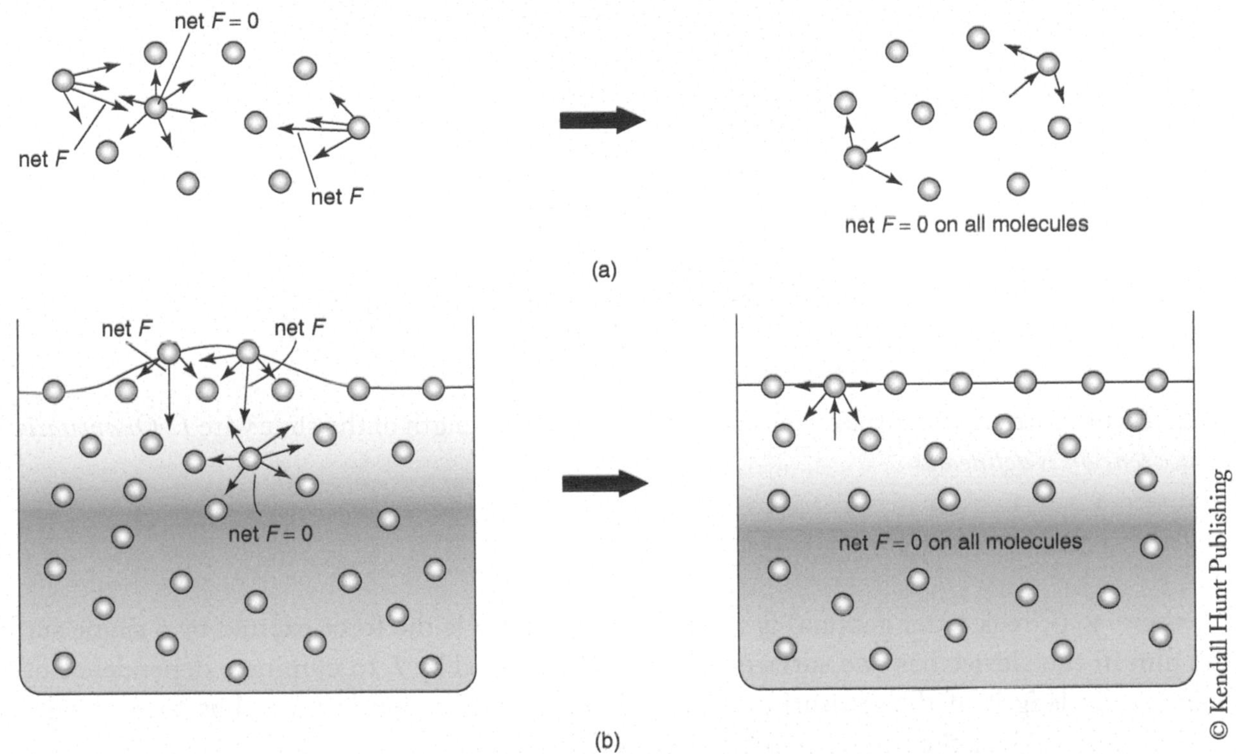

Figure 6.33 Surface tension supporting the weight of (a) and insect and (b) a paper clip.

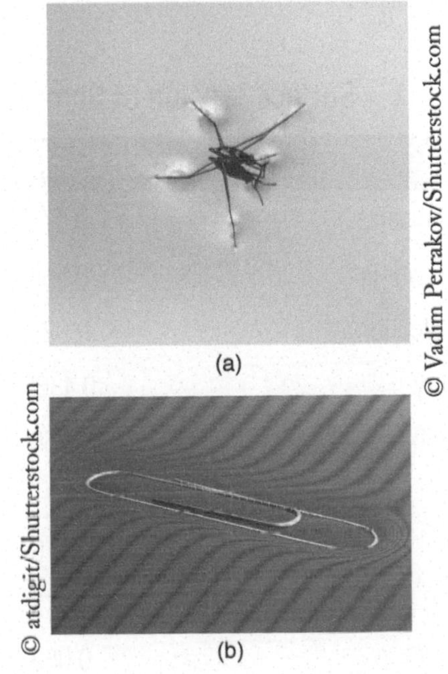

Figure 6.34 Device for measuring surface tension. The force necessary to stretch the film is measured under controlled conditions.

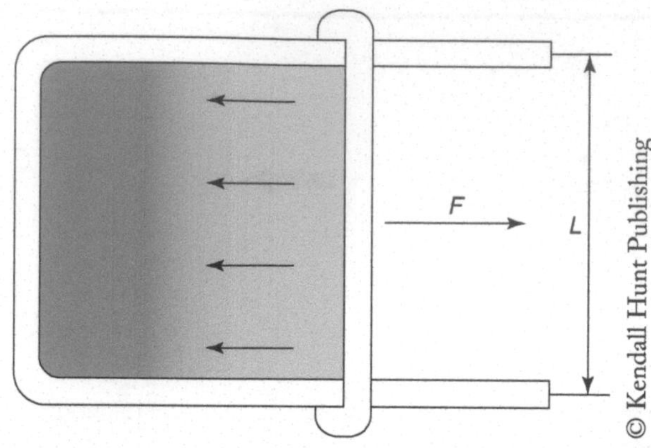

is directly proportional to surface tension in the film and the length of the slide wire L. *Quantitatively, surface tension is defined as*

$$\gamma \equiv \frac{F}{L} \qquad \text{eq. 6.13}$$

where γ (Greek letter gamma) is surface tension and F is the force exerted by a single surface (the film in this device has two surfaces). The force is divided by L to eliminate dependence of the force on the length of the measuring device. Units for surface tension are thus force divided by length. Table 6.2 lists the surface tension γ for a few liquids.

Surface tension causes bubbles as well as droplets. The inward force of surface tension makes bubbles spherical, if other forces are balanced, and raises the pressure of the gas inside. The pressure

Table 6.2 Surface Tension of Some Liquids

Liquid	Surface Tension γ (N/m)
Blood plasma	0.073
Blood, whole	0.058
Ethyl alcohol	0.023
Mercury	0.4355
Tissue fluids	0.050
Water	
at 0°C	0.076
at 20°C	0.072
at 100°C	0.059
Soapy water (surfactant)	0.037

inside a bubble is therefore larger than atmospheric, and that extra pressure is due to surface tension. (This increase in pressure inside bubbles has important effects in the respiratory system, as we shall see.) It can be shown that the extra pressure (i.e., gauge pressure) inside a bubble is given by

$$P = \frac{4\gamma}{r} \qquad \text{eq. (6.14)}$$

where r is the radius of the bubble. It is reasonable that pressure inside a bubble is directly proportional to surface tension, but the dependence of pressure on the radius of the bubble is less obvious. One common experience that supports the fact that a small-radius bubble creates a large internal pressure is the effort needed to blow up a balloon. Much more effort is needed when the balloon is small than when it is larger. The rubber sheet in a balloon behaves in a manner analogous to a liquid film.

When a bubble is pierced, pressure inside it causes the gas to flow out. This process is actually crucial to normal exhalation in breathing. Adult lungs contain about 300 million tiny mucus-lined sacs called alveoli. The exchange of gases such as carbon dioxide and oxygen with blood occurs in these sacs, and the large number of sacs increases the surface area available for the interchange.

When a person exhales, the alveoli contract under the effect of surface tension in the mucous lining. Pressure created in the alveoli by surface tension is largely responsible for exhalation; it helps to create a larger than atmospheric pressure in the lungs.

The importance of surface tension in the lungs is seen in victims of emphysema. Many of the alveoli in a person suffering from emphysema join together to form fewer and larger alveoli, as shown in Figure 6.35. One effect of the larger sacs is a reduction in pressure because of their larger radii. When a person with emphysema attempts to exhale, his alveoli create a small pressure and air flow is less than normal. The lungs have effectively lost some of their elasticity.

Surface tension in the alveoli also can be too large, making them very difficult to inflate. Such a problem, called respiratory distress syndrome or hyaline membrane disease, occurs in newborn

Figure 6.35 Alveoli and their air and blood supply. Normal, asthma, and emphysema.

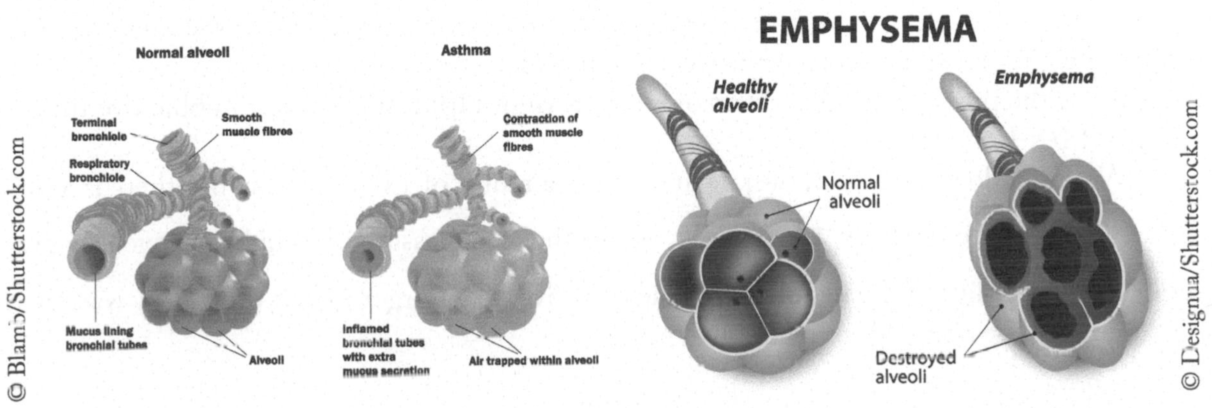

infants, particularly premature infants, and can be fatal. Normally, a substance called a *surfactant* reduces the surface tension of the mucous lining of the alveoli so that they can be inflated properly. Some infants lack surfactant and can only breathe with great effort-hence the name respiratory distress syndrome. A similar problem occurs in drowning victims: Expelling water from the lungs only partially restores the ability to breathe since the water increases surface tension in the alveoli (see Table 6.2), making them difficult to reinflate. The breathing mechanism will be discussed in more detail in Chapter 7.

EXAMPLE 6.12

Calculate the pressure in millimeters of mercury (mm Hg) inside (a) a water bubble 2.00×10^{-2} cm in radius; (b) a water bubble five times larger, 1.0×10^{-1} cm; and (c) a bubble of water 2.00×10^{-2} cm in radius with a surfactant in it that reduces the surface tension to 0.037 N/m.

Solution: Equation 6.14 and the value of γ for water in Table 6.2 are used to find the solutions.

a. In order to calculate the pressure inside a water bubble 2.00×10^{-2} cm in radius, we will use equation 6.14. *Note that 2.0×10^{-2} cm equals 2.00×10^{-4} m.*

$$P = \frac{4\gamma}{r} = \frac{4 \cdot 0.072\frac{\text{N}}{\text{m}}}{2.00 \times 10^{-4}\,\text{m}} = 1.44 \times 10^{3}\,\frac{\text{N}}{\text{m}^2}$$

To convert from $\frac{\text{N}}{\text{m}^2}$ to mm Hg, we will use the pressure at depth equation.

$$P_{\text{depth}} = \rho_{\text{fluid}} \cdot g \cdot depth$$

Note that the fluid is mercury and the density of mercury is $13{,}600\frac{\text{kg}}{\text{m}^3}$
Solve for depth and substitute the known values.

$$depth = \frac{\text{Pressure}}{\rho_{\text{fluid}} \cdot g} = \frac{1.44 \times 10^{3}\,\frac{\text{N}}{\text{m}^2}}{13{,}600\frac{\text{kg}}{\text{m}^3} \cdot 9.80\frac{\text{m}}{\text{s}^2}} = 0.0108\,\text{m}$$

Converting meters to millimeters tells us that the depth and therefore the pressure in terms of the height of a column of mercury is 10.8 mm Hg.

b. Calculate the pressure in millimeters of mercury (mm Hg) inside a water bubble five times larger, 1.0×10^{-1} cm.

We could do the problem over again and use a radius of 1.0×10^{-1} cm, but there is a quicker way. Looking at equation 6.14, $P = \frac{4\gamma}{r}$, we see that the pressure in a bubble is inversely proportional to the radius of the bubble. If the radius is five times larger than it previously was, then the pressure will be one-fifth of its original value.

$$P' = \frac{P}{5} = \frac{10.8\,\text{mm}}{5} = 2.16\,\text{mm Hg}$$

c. Calculate the pressure in millimeters of mercury (mm Hg) inside a bubble of water 2.00×10^{-2} cm in radius with a surfactant in it that reduces the surface tension to 0.037 N/m.

$$P = \frac{4\gamma}{r} = \frac{4 \cdot 0.037\frac{\text{N}}{\text{m}}}{2.00 \times 10^{-4}\,\text{m}} = 740\frac{\text{N}}{\text{m}}$$

To convert from $\frac{\text{N}}{\text{m}^2}$ to mm Hg, we will use the pressure at depth equation.

$$P_{\text{depth}} = \rho_{\text{fluid}} \cdot g \cdot depth$$

Note that the fluid is mercury and the density of mercury is $13{,}600\frac{\text{kg}}{\text{m}^3}$

Solve for depth and substitute the known values.

$$depth = \frac{\text{Pressure}}{\rho_{\text{fluid}} \cdot g} = \frac{740\frac{\text{N}}{\text{m}^2}}{13{,}600\frac{\text{kg}}{\text{m}^3} \cdot 9.80\frac{\text{m}}{\text{s}^2}} = 0.0555\,\text{m}$$

Converting meters to millimeters tells us that the depth and therefore the pressure in terms of the height of a column of mercury is 5.55 mm Hg.

The bubbles in this example are similar in size to the alveoli. In the first case the pressure is a bit high for easy inflation of an alveolus, whereas in the second it is a bit low to cause normal exhalation; in the third it is about right.

6.7.2 Adhesion and Capillary Action

Adhesive forces vary in strength depending on the combination of materials. Water adheres fairly well to glass and rather poorly to wax. The strength of adhesive forces has a visible effect on how well a liquid wets a surface. If adhesion between the liquid and the surface is strong, the liquid will wet the surface easily, sometimes flowing to all parts of the surface under the attraction of the adhesive force. High surface tension in the liquid tends to inhibit the wetting process since there is a strong tendency to form drops instead (see Figure 6.36). Water has a relatively large surface tension and does not easily wet many substances, such as polyester fabrics. Soap (a surfactant) reduces the surface tension of water, allowing it to penetrate and wash away particles of dirt much more effectively than plain water.

Capillary action is the rise or fall of a liquid in a tube owing to the cohesion of the liquid to itself and its adhesion to the walls of the tube. If open-ended glass tubes of various diameters are placed into a dish of water, as in Figure 6.37, the water forms a curved surface called a meniscus at the wall of each tube. The water rises in each tube to a height that depends on the diameter of the tube; the smaller the diameter, the higher the water rises. If the tube is larger than about 0.5 cm in diameter, the rise will be insignificant and the water inside will be the same height as outside (as expected from the extensive discussions of pressure due to the weight of a fluid).

Figure 6.38 shows expanded views of glass tubes inserted into water and mercury. Water is raised and mercury depressed by capillary action. Furthermore, the meniscus of the water is curved

Figure 6.36 (a) Water on a waxed surface forms a bead because the cohesive forces in the drop are much stronger than the adhesive forces between water and wax. (b) Water spreads out on many surfaces because the adhesive forces between water and the surface are larger than the cohesive forces.

© Tharnapoom Voranavin/Shutterstock.com

© ananaline/Shutterstock.com

Figure 6.37 Water rises up narrow tubes because of capillary action. The narrower the tube, the higher the water rises.

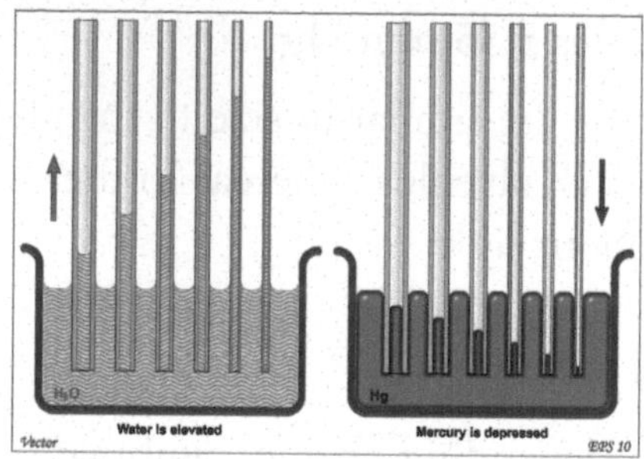

© Fouad A. Saad/Shutterstock.com

upward, while that of mercury is curved downward. The behavior of both liquids is due to the relative strengths of the cohesive and adhesive forces. Mercury has a very high surface tension (i.e., very large cohesive forces) and forms a surface curved downward like a drop. Adhesive forces between mercury and glass are weak compared to the cohesive forces in mercury, so adhesion between the tube walls and mercury is unable to spread out the surface. The shape is reminiscent of water on a waxed surface-an analogous situation. Water has both smaller cohesive forces and much larger adhesion to glass. In the water-filled tube the adhesive forces are stronger than the cohesive forces and pull the water up the walls of the tube, inverting the curvature of the surface.

Now, given the shapes of the surfaces, consider what further effect surface tension has. As always, surface tension makes the area of the surface as small as possible. This can be done by flattening the surface in the tube. In the case of mercury this means a downward force on the surface. The surface of the water curves in the opposite sense; consequently there is an upward force on its surface. Mercury is thus forced downward and water upward.

The height to which capillary action can raise or depress a liquid in a tube depends on a number of factors. The stronger the cohesive force in the liquid the more effective surface tension is in flattening the surface and forcing the liquid up or down the tube, so the height is proportional to y. The

Figure 6.38 (a) Water rises in small glass tubes (sometimes called capillary tubes) because adhesive forces are large compared to cohesive forces. (b) Mercury is depressed in a capillary because the adhesive forces are small compared to the cohesive forces. The dashed lines show what the shape of the surface would be without the flattening effect of surface tension.

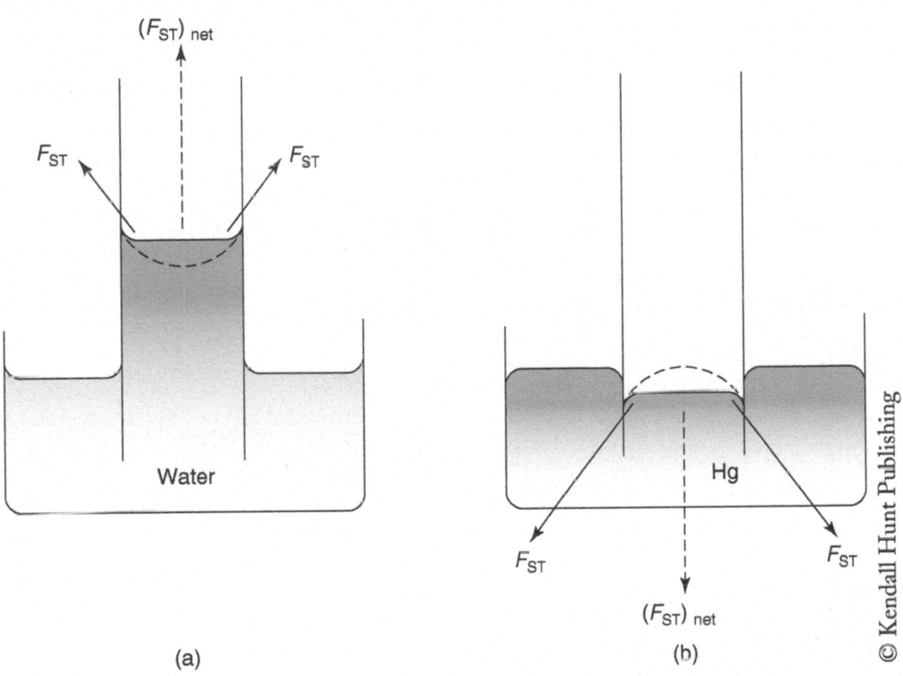

smaller the radius of the tube, the more molecules are in contact with the sides and the less mass there is to be raised, so height is inversely proportional to radius. The more dense a fluid is, the more mass there is to raise, so height is inversely proportional to density. In summary,

$$h \propto \frac{\gamma}{\rho r} \quad \text{eq. 6.15}$$

where h is the height to which the liquid rises, γ is the surface tension, ρ is the density of the liquid, and r is the radius of the tube.

6.7.3 Can Cohesion Produce a Negative Absolute Pressure?

The question of how tall trees raise sap to their uppermost branches is a controversial one. Veins in trees are not small enough to raise sap more than about 10 m by capillary action. Osmosis (to be discussed in Chapter 7) produces root pressure which can also be shown to be inadequate to force sap to the tops of tall trees. The actual mechanism is not well established, but one possibility is that sap is raised in stages.

It also has been suggested by some researchers that cohesive forces pull sap up in long chains. As one molecule evaporates during transpiration, another is pulled up by cohesive forces in the chain. Pulling is the opposite of pushing, which is what positive pressure does. Pulling thus corresponds to negative absolute pressure. Some researchers have reported negative pressures as large as −25 atm in the sap at the tops of trees, but it is difficult to reconcile this with the ease with which water molecules normally can be pulled apart, especially when impurities are present that can break the cohesive forces. It has not been possible to recreate sap behavior under laboratory conditions or to understand the phenomenon theoretically. The question remains open and under active discussion.

Things to know for section 6.7

Surface tension $\gamma \equiv \frac{F}{L}$

Things to know for chapter 6

$$Pressure \equiv \frac{Force}{Area}$$

Metric units of pressure are $\frac{\text{N}}{\text{m}^2}$; they are called pascals.

$$Area_{\text{circle}} = \pi r^2$$

$$1\,\text{L} \equiv 1000\,\text{cm}^3$$

$$P_{\text{depth}} = \rho_{\text{fluid}} \cdot g \cdot depth$$

Density $\rho \equiv \frac{mass}{volume}$

$$Buoyant\ Force = \rho_{\text{fluid}} \cdot g \cdot V_{\text{object submerged}}$$

$$\%Submerged = \left(\frac{\rho_{\text{object}}}{fluid}\right) \cdot 100$$

$$Flow\ rate \equiv \frac{Volume}{time} = A\bar{v}$$

units are $\frac{\text{liter}}{\text{minute}}; \frac{\text{cm}^3}{\text{s}}$

$$Flow\ rate = \frac{\Delta P \pi r^4}{8\eta L}$$

Reynold's number: $R_e = \frac{\rho d\bar{V}}{\eta}$

$$\text{Surface tension } \gamma \equiv \frac{F}{L}$$

Questions

6.1 What is a fluid?

6.2 Why are liquids and solids more difficult to compress than gases?

6.3 What factors other than the volume a gas occupies affect its pressure? Explain.

6.4 Define gauge pressure, total pressure, and pressure due to the weight of a fluid.

6.5 Why is it that you are able to get up from sunbathing if the weight of the atmosphere is resting on your body?

6.6 Name and state the principle upon which hydraulic systems are based. How does a hydraulic system achieve an increase in force? Can one be used to decrease force?

6.7 The two pistons in Figure 6.5 are at the same height. What would be the effect of having the piston on the left at a larger height than the one on the right? A smaller height?

6.8 Give an example in which the density of an object could be used to identify the substance of which it is composed. Could the density of an object composed of many substances also be used to identify which substances are in it? Explain.

6.9 Why can't an absolute pressure in a gas be less than zero?

6.10 Explain in terms of ***equation 6.4*** why you can notice a change in atmospheric pressure when taking an elevator in a tall building. About how large is the total "*h*" (height), creating atmospheric pressure at sea level?

6.11 Why does the curved tube in the Bourdon gauge in Figure 6.14 move to the right if pressure is increased? Explain in terms of the forces the pressure increase when exerted on the various surfaces in the gauge.

6.12 What is a negative pressure? Can a single-tube manometer, such as the one in Figure 6.16, be used to measure a negative pressure? Explain how or why not.

6.13 Why is mercury rather than water used in barometers and in blood pressure measurements? If a manometer is used to measure spinal column pressure, why is it preferable to use a saline solution rather than mercury in the manometer?

6.14 How are units of pressure such as millimeters of mercury, centimeters of water, and inches of mercury converted into units of force per unit area, such as newtons per square meter?

6.15 Why can some people float in a swimming pool if they take a deep breath and sink if they empty their lungs? Why is it difficult to swim under water in the Great Salt Lake?

6.16 Does the fact that more force is needed to pull the plug in a bathtub when it is full than when it is empty contradict Archimedes' principle? Explain.

6.17 Is Archimedes' principle of any use in a weightless environment such as the space shuttle?

6.18 What are the differences between laminar and turbulent flow? What can cause turbulence to occur?

6.19 Write down Poiseuille's law for laminar flow. What are the factors upon which the flow rate depends and to which factor is the flow rate most sensitive?

6.20 If resistance to flow is a result of friction, then energy is being expended to keep the flow going. What are the final forms the expended energy may take?

6.21 In the situation shown in Figure 6.39, what are the pressures P_1 and P_2? Express those pressures first as gauge pressure in centimeters of water and then as total pressures.

6.22 The pressure at a given depth below the surface of a static fluid is the same everywhere, but in a fluid moving horizontally the pressure decreases along the direction of motion. What causes this decrease in pressure?

6.23 When a clamp is used to decrease the flow rate of a fluid in a flexible tube, what are the two distinct physical causes of the decrease?

6.24 What is the dependence of fluid pressure on fluid speed? Name the principle that gives this dependence.

6.25 How does a venturi or other constriction in a tube enhance entrainment?

6.26 Give an example in which a large fluid speed accompanies a small flow rate. How are flow rate and fluid speed related?

6.27 Why is it that a shower curtain is pulled into the spray of the shower?

6.28 Use the Bernoulli effect to explain how the sinuses are drained when a person blows her nose.

6.29 If the air exerts an upward force on a wing, such as in Figure 6.31, then by Newton's third law the wing must exert a downward force on the air. Examine the direction of flow of the two air streams, one above and one below the wing, as the air comes off the trailing edge. From this can you argue that some.

Figure 6.40 See problem 6.50.

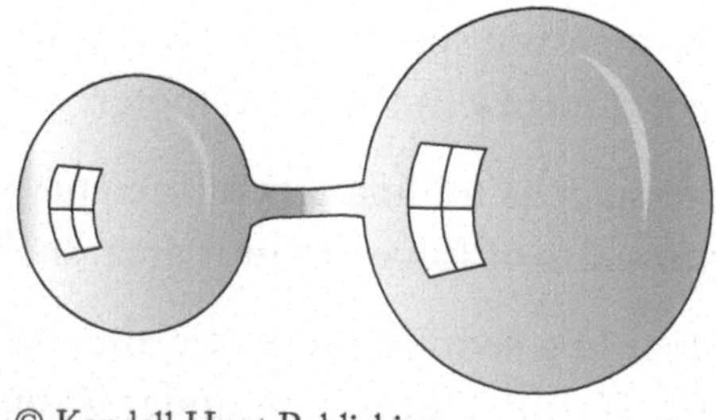

Problems

SECTION 6.1

√6.1 (I) As a woman walks, her entire weight is momentarily supported on the heel of one shoe. Calculate the pressure exerted on the floor by a high-heel shoe worn by a 50-kg woman if the heel has an area of 2.00 cm^2. Express the pressure in newtons per square meter and atmospheres.

√6.2 (I) What pressure is exerted by the tip of a nail struck with a force of 20,000 N? Assume the tip is a 1.5-mm-radius circle.

6.3 (I) What pressure is created when 2.0 liters of air are forced into a bicycle tire originally at atmospheric pressure and containing 0.45 liter of air? The inside volume of the tire remains 0.45 liter.

6.4 (II) A pressure cooker is shaped like a can with a lid 25 cm in diameter. If the pressure in the cooker can reach 3.0 atm, how much force must the latches holding the lid onto the pot be designed to withstand?

6.5 (III) Assuming car tires are perfectly flexible and support the weight of a car by air pressure alone, calculate the total area of the tires in contact with the ground for a 1000-kg car having a tire pressure of $2.5 \times 10^5 \frac{N}{m^2}$.

SECTIONS 6.2 AND 6.3

6.6 (I) Calculate the force that must be exerted on the master cylinder of a hydraulic system to lift a 10,000-kg truck with the slave cylinder. The master cylinder has a diameter of 1.50 cm and the slave has a diameter of 30.0 cm.

6.7 (I) The deepest place in any ocean on earth is in the Marianas Trench near the Philippines. Calculate the pressure at the bottom of this trench due to the ocean water, given that its depth is 11.0 km and assuming that the density of ocean water is $1025 \frac{kg}{m^3}$.

6.8 (II) (a) Calculate the distance to the top of the atmosphere, assuming that the density of air is constant from sea level up. (b) Although there is no exact place where the atmosphere ends, if we assume an altitude of about 130 km calculate its average density and compare this with $1.29 \frac{kg}{m^3}$ (approximate density at sea level).

6.9 (II) (a) Calculate the pressure in newtons per square meter due to whole blood in an IV system, such as the one shown in Figure 6.9, if h = 1.5 m. (b) Noting that there is an open tube, so that atmospheric pressure is exerted on the blood in the bottle, calculate the total pressure exerted at the needle by the blood.

6.10 (II) Water towers are used to store water above the level of homes. If a user observes that the static water pressure at home is $3.00 \times 10^5 \frac{N}{m^2}$, how high above the home is the surface of the water in the tower?

6.11 (II) How tall would a water-filled manometer have to be in order to measure blood pressures up to 300 mm Hg?

6.12 (II) Pressures in natural gas supplied to homes and businesses are sometimes measured using a water filled manometer. If a gas pressure is typically 10.0 cm of water, calculate the pressure in mm Hg. Is it practical to measure this pressure accurately with a mercury-filled manometer?

6.13 (II) The heart exerts a maximum force of 25 N over an effective area of 17.5 cm^2. Calculate the pressure produced (in N/m^2 and mm Hg).

6.14 (III) (a) Using the pressure at depth equation, show that the maximum height to which water can be raised in a syringe or similar device is 10.3 m. (b) Calculate the maximum height to which gasoline can be raised in such devices.

6.15 (III) You wish to design a hydraulic system so that a force put into it would be increased by a factor of 100. (a) What must the ratio of the area of the slave cylinder to the area of the master cylinder be? (b) What must the ratio of the diameter of the slave cylinder to the diameter of the master cylinder be? (c) By what factor is motion reduced in the system?

6.16 (III) The plunger on a hypodermic syringe has a diameter of 1.0 cm. What force must be exerted on the plunger to inject medication into a vein that has a blood pressure of 10.0 mm Hg?

6.17 (III) A submarine is trapped at the bottom of the ocean with its hatch at a depth of 20 m below the surface. The hatch is circular and 0.40 m in diameter. Calculate the force that would have to be exerted against it to open it. The air inside the submarine has a pressure of 1 atm.

SECTION 6.4

√6.18 (I) A certain ideal hydrometer has a uniform density of 0.90 g/cm^3. What percentage of the hydrometer will be submerged if it is placed in a fluid of density 0.95 g/cm^3? In a fluid of density 1.05 g/cm^3? *See* Figure 6.21.

√6.19 (I) What is the density of a piece of wood that floats in water with 70% of its volume submerged?

6.20 (I) Calculate the density of a fluid in which a hydrometer of uniform density 0.80 g/cm^3 floats with 90% of its volume submerged.

6.21 (I) Calculate the density of a person who floats in fresh water with 5.0% of her volume above the surface.

6.22 (II) An ornithologist weighs a bird bone in air and in water to find the bone's density. The mass of the bone in air is 25.0 g, and its apparent mass in water is 2.00 g. (a) What mass of water does the bone displace? (b) Calculate the volume of water displaced. (c) Assuming the bone's volume to be equal to that of the water displaced, calculate the density of the bone. Note that the answer is significantly less than that given for bone in Table 5.2. The bones of birds frequently have air pockets that make them lighter than solid bones, so their average density is also less than solid bone.

6.23 (II) In a measurement of a man's density he is found to have a mass of 65.0 kg in air and an apparent mass of 1.50 kg when completely submerged in water with his lungs empty. (a) What mass of water does he displace? (b) What volume of water does he displace? (c) Calculate his density. (d) If his lung capacity is 2.00 L, is he able to float with his lungs filled with air without treading water?

6.24 (II) A piece of iron with a mass of 1560 g in air is found to have an apparent mass of 1402 g when submerged in a fluid of unknown density. (a) Find the volume of iron, using the value for its density given in Table 5.2. (b) What mass of fluid does it displace? (c) Calculate the fluid's density and identify the most likely fluid, using Table 5.2.

6.25 (II) Suppose a man has a mass of 85.0 kg and a density of 0.980 g/cm^3. (a) Calculate his volume. (b) Calculate the mass of air that he displaces, using the density of air from Table 5.2. (c) Calculate the buoyant force on him due to the air he displaces and compare this to his weight in newtons.

6.26 (III) Referring to Figure 6.19, prove Archimedes' principle that the buoyant force on the cylinder is equal to the weight of the fluid displaced. To do this you may assume that the buoyant force is equal to $F_2 - F_1$ and that the ends of the cylinder have equal areas. Note that the volume of the cylinder (and that of the fluid displaced) is equal to $(h_2 - h_1)A$.

6.27 (III) A woman of mass 50 kg floats in fresh water with 3.0% of her volume above water when her lungs are empty and with 5.0% of her volume above water when her lungs are full. Calculate her lung capacity. Hint: Calculate her density and volume in each case.

SECTION 6.5

√6.28 (I) Normal blood flow rate in a resting adult is approximately 5.00 L/min. Convert this to cubic centimeters per second.

√6.29 (I) If the flow rate in an IV setup is 2.0 cm^3/min, how long does it take to empty a 1.00-L bottle of IV solution?

√6.30 (I) A glucose solution is being administered intravenously (in an IV as in Figure 6.9) with a flow rate of 3.00 cm^3/min. If the glucose solution is replaced by blood plasma having a viscosity of 1.5 times that of glucose, what is the new flow rate?

√6.31 (II) A fluid is flowing through a tube with a flow rate of 100 cm^3/sec. Calculate the new flow rate if each of the following changes are made with all other factors remaining the same as the original conditions: (a) The pressure difference ΔP is one-third its original value. (b) The length of the tube is 1.5 times its original length. (c) The radius of the tube is 20% of its original value. (d) A fluid with a viscosity one-half the original is substituted. (e) The radius is 90% of its original value and the pressure difference is 1.5 times its original value.

√6.32 (II) A flow rate of 20.0 cm^3/sec is obtained through a tube of length 100 cm and radius 0.500 cm when there is a pressure difference of 25 cm of water along the tube. Calculate the new flow rate when the following changes are made, assuming all other factors remain the same as the original conditions. (a) A 350-cm long tube replaces the original. (b) The tube radius is decreased to 0.400 cm. (c) The pressure difference ΔP is increased to 40.0 cm of water. (d) The pressure difference is increased to 50.0 cm of water and the radius is decreased to 0.200 cm.

√6.33 (II) Early on a summer day water pressure is high and produces a flow rate of 20 L/min through a garden hose. Later in the day, when it is hot and water use is heavy throughout town, pressure drops and the flow rate through the same hose is only 8.0 L/min. If the original pressure at the entrance of the hose was 30 m of water, how large is the pressure later in the day?

√6.34 (III) If the radius of a blood vessel is reduced by cholesterol deposits to 90% of its original value, flow rate will decrease. By what factor would the pressure difference along the vessel have to be increased to get the flow rate up to its original value?

√6.35 (III) By what factor would the radii of the arteries have to decrease to reduce blood flow rate by a factor of 2? (Reducing by a factor of 2 means to cut the flow rate in half).

√6.36 (II) Calculate the flow speed above which turbulent flow should occur in the aorta. Assume a diameter of 2.00 cm and a Reynold's number of 1900. For the density of blood, use $1.05\frac{g}{cm^3}$.
The viscosity of blood is $0.035\ \frac{g}{cm \cdot s}$

6.37 (III) A given pressure difference produces a flow rate of 2.00 L/min through one tube and 3.0 L/min through another tube. What would the flow rate be if these tubes were hooked together so that the fluid has to flow through them sequentially? The pressure difference remains the same.

SECTION 6.6

√6.38 (I) The aorta is 1.00 cm in radius and the average speed of blood in the aorta is 30 cm/sec in a normal resting adult. Calculate the blood flow rate in the aorta in cubic centimeters per second and liters per minute.

6.39 (I) If turbulent flow begins in the aorta when the average blood speed reaches 80 cm/sec, what is the maximum flow rate that can be achieved without turbulence in an aorta 0.90 cm in radius? Give the flow rate in cubic centimeters per second and liters per minute.

6.40 (I) What is the average speed of water in a water main 0.25 m in diameter that is carrying 1000 L/min?

6.41 (I) Calculate the flow rate of blood through a capillary in cubic centimeters per second, given that the capillary has a diameter of 4.00×10^{-4} cm and the average speed of the blood in the capillary is 3.00×10^{-2} cm/sec.

6.42 (II) Fire hoses have nozzles that increase the speed of the water to reach higher in a fire. (a) If the flow rate in the fire hose is 5000 L/min and the opening of the fire hose nozzle is 2.0 cm in diameter, what is the speed of the water as it leaves the nozzle? (b) If the diameter of the hose itself is 10 cm, what is the speed of the water in the hose?

6.43 (II) In a certain Venturi the diameter of the tube decreases from 2.00 to 0.500 cm. Calculate the speed of a liquid in the constriction if it has a speed of 50.0 cm/sec in the larger part of the Venturi. The liquid is assumed to be incompressible.

6.44 (II) The speed of an incompressible fluid increases from 0.250 to 10.0 m/sec in a Venturi. Calculate the diameter of the constricted part of the tube, given that the diameter of the larger part of the tube is 0.300 m.

6.45 (III) Show that the speed of an incompressible fluid flowing through a constriction in a tube increases from 1.00 to 100 cm/sec if the diameter of tube decreases from 3.00 to 0.300 cm.

6.46 (III) Prove that the speed of an incompressible fluid through a venturi constriction increases by a factor equal to the square of the factor by which the radius decreases.

SECTION 6.7

6.47 (I) Calculate the total surface area of the alveoli in the lungs of an adult, given that there are 300 million alveoli and their average radius is 1.00×10^{-4} m. How does this compare with the area of a sphere the size of the lung (about 0.100 m in radius)?

6.48 (I) What is the pressure in mm Hg inside a soap bubble of diameter 0.100 m?

6.49 (I) The smallest capillaries in trees are 0.0100 mm in radius. Given that capillary action can raise sap to a height of only 0.400 m in trees, calculate the radius of a capillary that would raise sap to the top of a 100-m tall redwood.

6.50 (I) If two water bubbles are connected, as in Figure 6.40, the small one deflates, making the large one larger. Explain by calculating the pressures inside the bubbles, given their radii to be 2.00 and 3.00 cm, respectively. This also explains why two bubbles join to form a single large bubble when they collide, rather than share their air to make two bubbles of equal size.

6.51 (II) Calculate the pressures inside 1.00-cm-diameter bubbles of water, alcohol, and soapy water. Which do you think would be the most stable, and why does soapy water make larger bubbles easier than plain water?

6.52 (III) Calculate the force exerted on the slide wire in Figure 6.34 if the length of the wire is 5.00 cm and the fluid is water.

Figure 6.39 See question 6.21.

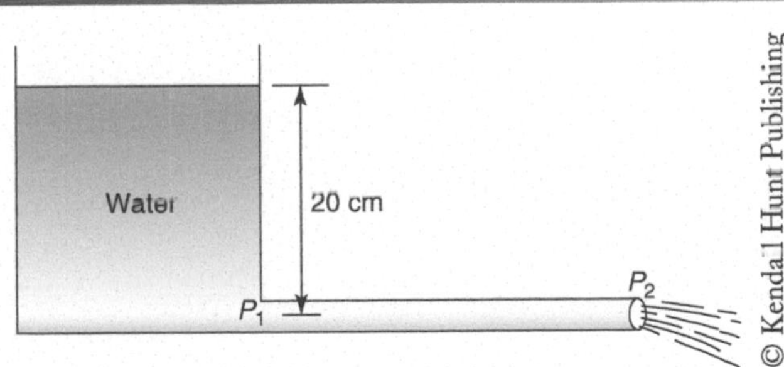

7

Biological and Medical Applications of Pressures and Fluids

"The noblest pleasure is the joy of understanding."

Leonardo da Vinci

7.1 EXAMPLES OF PRESSURE IN HUMANS

This chapter continues the treatment of fluids but with particular emphasis on biological and medical applications. Many of the basic physical principles related to fluids were developed in Chapter 6, and most of the material in this chapter is based on applications of those principles. The major exception to this is the topic of diffusion, covered in Section 7.2, which is crucial to understanding the functioning of biological organisms on a cellular level.

A few examples of pressures in the human body were included in Chapter 6, the most notable being blood pressure in the major arteries. There are many more examples of pressures in humans, including blood pressures other than in arteries. Table 7.1 lists typical values for fluid pressures in the body. As before, all pressures quoted are gauge pressures. Whenever pressures differ significantly from these typical values, a problem is indicated. For that reason, pressure measurement is an important diagnostic tool in medicine.

Blood pressures will be considered in Section 7.3, on the cardiovascular system. Thoracic cavity pressure and other pressures related to respiration will be considered in Section 7.4, on the physics of respiration. The remainder of this section is devoted to the discussion of biological pressures other than cardiovascular and respiratory.

7.1.1 Bladder Pressure

One of the most noticeable of bodily pressures, bladder pressure varies over, quite a large range. It is zero when the bladder is empty and climbs steadily to about 25 mm Hg when the bladder reaches its normal capacity of some 500 cm^3. The micturition reflex is triggered by a bladder pressure of about 25 mm Hg. That reflex stimulates the feeling of needing to urinate, and it further triggers muscle contractions around

Table 7.1 Typical Fluid Pressures in Humans

Arterial blood pressures	mm Hg
Maximum (systolic):	
adult	100–140
infant	60–70
Minimum (diastolic):	
adult	60–90
infant	30–40
Venous blood pressures	
Venules	8–15
Veins	4–8
Major veins (CVP)	4
Capillary blood pressure	
Arteriole end	35
Venule end	15
Bladder	
Average	0–25
During micturition	110
Brain, lying down (CSF)	5–12
Eye, aqueous humor	12–24
Gastrointestinal	10–20
Intrathoracic	−4 to −8
Middle ear	<1

the bladder that can raise bladder pressure to 110 mm Hg, accentuating the sensation. Coughing, straining, sitting up, tight clothes, and simple nervous stress also can increase bladder pressure and trigger the micturition reflex long before the bladder is full. Students studying for exams and authors striving to meet deadlines make many trips to the toilet. Pregnant women experience increased bladder pressure due to the weight of the fetus resting on the bladder and find it necessary to urinate frequently. The capacity of their bladders is also less than the normal 500 cm^3 because of the space that the fetus occupies.

Bladder pressure while urinating is normally 15–30 mm Hg, but an obstruction of the urinary tract, such as from a swollen prostate gland, can necessitate pressures as large as 70 mm Hg. The larger the resistance of a tube, the larger is the pressure difference needed to cause the same flow rate.

Bladder pressure can be measured by catheterization through the urinary tract or by insertion of a needle through the abdominal wall into the bladder (called *direct cystometry*). Both techniques transmit bladder pressure through a liquid to a measuring device, commonly a water manometer. Because it is most convenient to use water to transmit the pressure and to fill the manometer, bladder pressures are normally given in centimeters of water. The normal range is from 0 to 30 cm of water, rising to 150 cm of water during the micturition reflex.

7.1.2 Cerebrospinal Pressure

The skull and spinal column contain cerebrospinal fluid (CSF), as shown in Figure 7.1. The CSF supports the weight of the brain with buoyant force, acts as a protective cushion, and supplies nutrients filtered from the blood. CSF is generated in the skull and circulates around the brain, through cavities in the brain called ventricles, and down the central canal of the spinal cord. Normally, CSF is absorbed in the spinal column as fast as it is generated in the skull. However, the narrow ventricle called the cerebral aqueduct (see *Figure 7.1*) can become blocked, causing pressure to build inside the skull. This is a moderately common problem in infants called hydrocephalus (literally "water head") and can cause an enlarged head, mental retardation, or death. When detected early, the pressure and its effects can be minimized surgically.

Pressure in the CSF can be measured as shown in *Figure 7.2* and earlier in Figure 6.13. Unfortunately, this method does not detect hydrocephalus, because the blockage in the aqueduct prevents the excess pressure in the brain from being transmitted to the spinal column. It is not convenient to measure pressure in the brain directly because of its bony structure, so other methods of detecting hydrocephalus are generally used. These include shining a light through the soft skull of an infant, ultrasonic scans, and x rays.

Figure 7.1 Brain, spinal cord, and cerebrospinal fluid (CSF). Black arrows indicate the circulation of CSF.

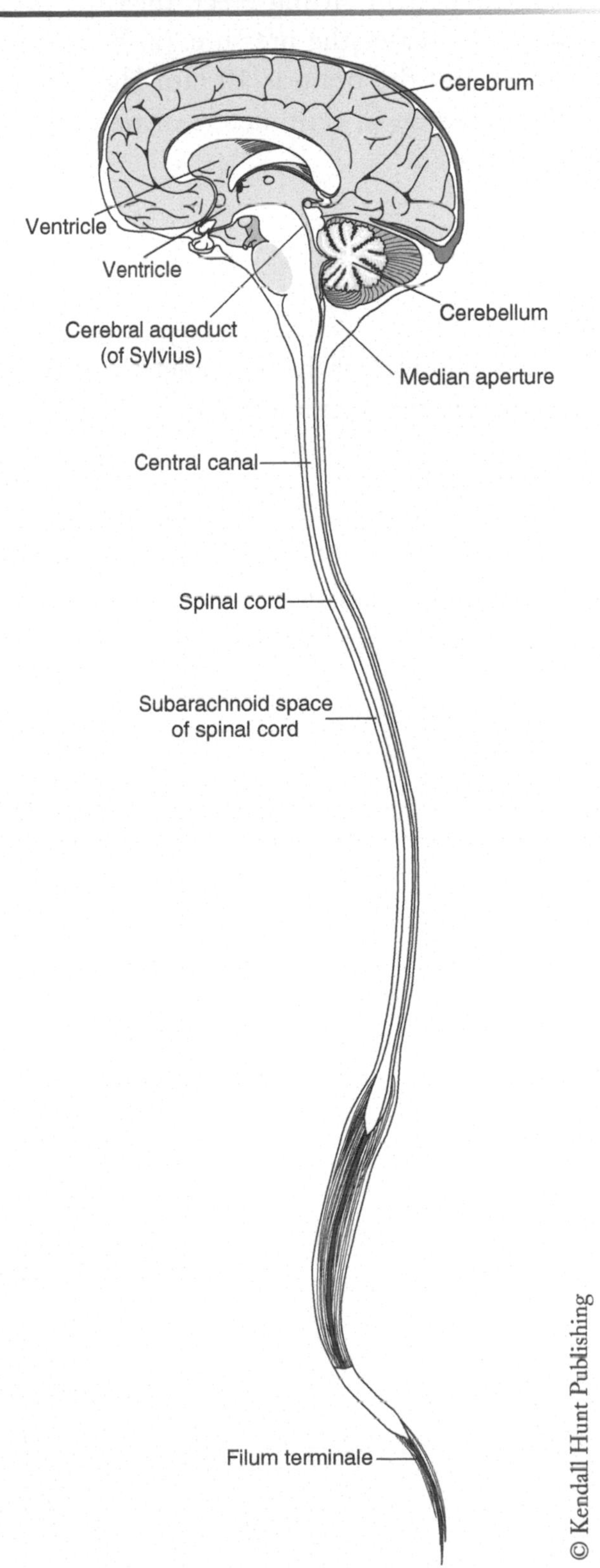

EXAMPLE 7.1

If pressure in the cerebrospinal fluid (CSF) is measured as shown in Figure 7.2, using a spinal tap with the patient sitting erect, then the pressure due to the weight of the CSF in the spinal column increases the pressure. (a) What pressure is measured in centimeters of water if the pressure around the brain is 10 mm Hg, and the tap is at a point 60 cm lower than the brain? (b) What pressure in centimeters of water is measured if the patient lies down? The density of CSF is 1.05 g/cm^3.

$$1.05 \times 10^3 \frac{\text{kg}}{\text{m}^3}$$

Solution.

a. Since the bottom of the manometer is at the same vertical height as the needle, the pressure measured in the manometer is

$$P_{\text{total}} = P_{\text{brain}} + P_{\text{CSF}}$$

where P_{brain} is the pressure in the brain and P_{CSF} is the pressure due to the 60-cm column of cerebrospinal fluid. Putting everything into SI units yields a pressure in N/m^2. We will use the *pressure at depth* equation to determine the pressure caused by both the pressure around the brain and pressure caused by the weight of the cerebrospinal fluid:

$$P_{\text{depth}} = \rho_{\text{fluid}} \cdot g \cdot depth$$

Figure 7.2 A water manometer used to measure CSF pressure, as in Example 7.1.

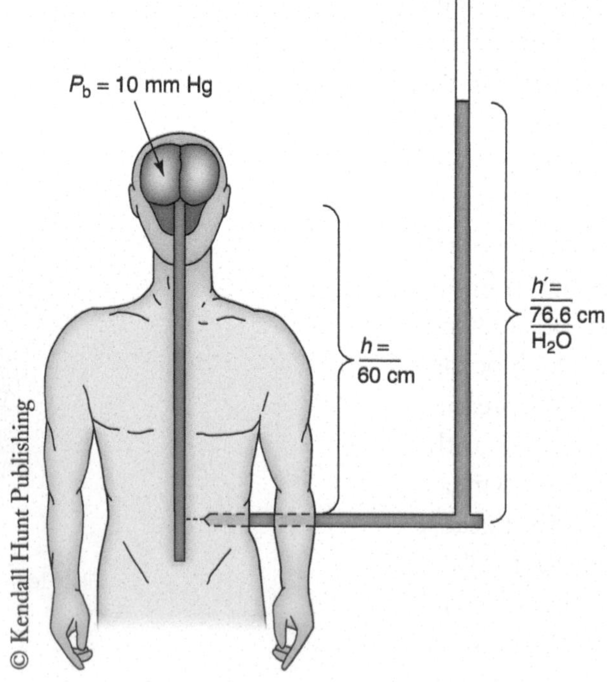

Step by step 1: the pressure around the brain, P_{brain}. Note that 10 mm of mercury equals 0.0100 m

$$P_{depth} = 13.6 \times 10^3 \frac{kg}{m^3} \cdot 9.80 \frac{m}{s^2} \cdot 0.0100m = 1.33 \times 10^3 \frac{N}{m^2}$$

Step 2: the pressure due to the weight of the cerebrospinal fluid.

$$P_{CSF} = 1.05 \times 10^3 \frac{kg}{m^3} \cdot 9.80 \frac{m}{s^2} \cdot 0.600\,m = 6.17 \times 10^3 \frac{N}{m^2}$$

The total pressure = $P_{total} = P_{brain} + P_{CSF}$

$$1.33 \times 10^3 \frac{N}{m^2} + 6.17 \times 10^3 \frac{N}{m^2} = 7.50 \times 10^3 \frac{N}{m^2}$$

To determine the equivalent height of a column of water, we use the pressure at depth equation as follows:

$P_{depth} = \rho_{fluid} \cdot g \cdot depth$ where "*depth*" is the height of a column of water. Solving for depth,

$$depth = \frac{P_{depth}}{\rho_{fluid} \cdot g} = \frac{7.50 \times 10^3 \frac{N}{m^2}}{1000 \frac{kg}{m^3} \cdot 9.80 \frac{m}{s^2}} = 0.765\,m$$

This is equal to 76.5 cm of water.

When the patient is sitting up, the total pressure measured is equivalent to the pressure at the bottom of a *76.5 cm column of water.*

Solution.

b. What pressure in centimeters of water is measured if the patient lies down? The density of water is 1000 kg/m^3.

If the patient lies down, then there is no excess pressure due to gravity and the pressure measured will be the pressure surrounding the patient's brain. We have already calculated the pressure surrounding the brain in part (a):

$$P_{depth} = 13.6 \times 10^3 \frac{kg}{m^3} \cdot 9.80 \frac{m}{s^2} \cdot 0.0100\,m = 1.33 \times 10^3 \frac{N}{m^2}$$

Solving the pressure at depth equation for the height (depth) of the column of water:

$$P_{depth} = \rho_{fluid} \cdot g \cdot depth$$

$$depth = \frac{P_{depth}}{\rho_{fluid} \cdot g} = \frac{1.33 \times 10^3 \frac{N}{m^2}}{1000 \frac{kg}{m^3} \cdot 9.80 \frac{m}{s^2}} = 0.136\,m$$

When the patient is lying down, the pressure is equivalent to a *water column that is 13.6 cm tall.* Notice that there is an extra pressure equivalent to a 63 cm column of water when the person sits erect, due to the weight of the fluid in the spinal column.

Spinal taps are performed for many reasons: to administer anesthetics; to introduce dye for contrast on an x-ray, or to withdraw CSF for density measurements, for example. One side effect of spinal taps is severe headache. Patients are instructed to lie motionless on their backs for several hours following spinal taps to prevent loss of CSF at the entry point of the needle. As seen in the example above, pressure at the entry point is significantly less if the person lies down, reducing the chance of leakage. The skull and spinal column contain only about 125 cm^3 of CSF, so even a small loss is significant. Because the brain is deprived of some of its buoyant force support and of its protection from physical blows, headaches ensue. The CSF is slowly replenished over a period of four or five days, and in the meantime lying down distributes the remaining cerebrospinal fluid evenly.

7.1.3 Pressure in the Eye

The shape of the eye is maintained by fluid pressure. This pressure is called the *intraocular pressure*; its normal range is 12–24 mm Hg. The eye has two fluid-filled chambers as shown in Figure 7.3. The front chamber contains aqueous humor and the rear chamber contains vitreous humor. About 5 cm^3 of aqueous humor is produced daily in the eye. Excess fluid flows out through a canal and is absorbed into the bloodstream. *Aqueous humor* is similar in character to cerebrospinal fluid and carries nutrients to the lens and cornea of the eye, neither of which has blood vessels. Pressure created by the aqueous humor is transmitted throughout the vitreous humor. *Vitreous humor* is a jellylike substance that does not circulate and is not replenished. Pressure in the vitreous humor holds the retina flush against the interior parts of the eyeball and helps to maintain the shape of the eye.

Partial blockage of the drainage canal for the aqueous humor results in pressure buildup in the entire eye. (When partially blocked, the canal has a higher resistance to flow, so fluid pressure rises.)

Figure 7.3 The shape of the eye is maintained by pressure in the aqueous and vitreous humors. Pressure in the vitreous humor also helps to hold the retina in place. Aqueous humor is continuously replenished, exiting through the canal of Schlemm.

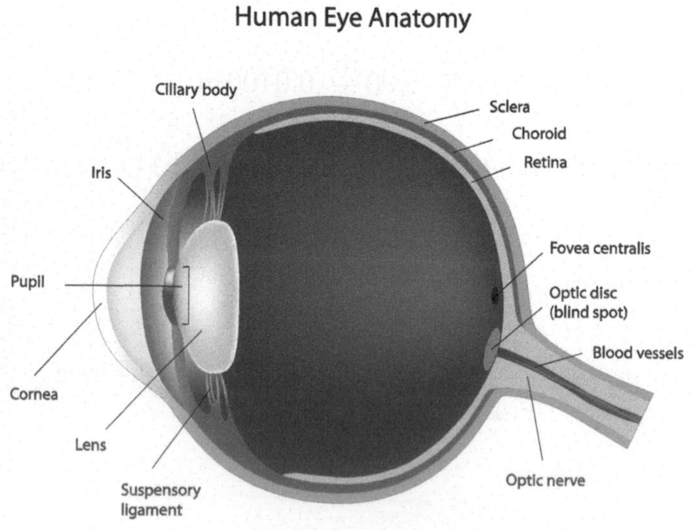

Human Eye Anatomy

Intraocular pressure can rise to as much as 85 mm Hg, the average arterial blood pressure. Above that pressure, aqueous humor penetrates arterial walls and is absorbed into the blood stream, so the pressure rises no higher. Excessive intraocular pressure is a symptom of *glaucoma* and can result in blindness due to deterioration of the retina.

Glaucoma is most common in people over 40, who consequently are routinely tested for high intraocular pressure. Intraocular pressure is measured by a number of techniques, most of which involve exerting a force on the eye over a certain area (a pressure) and observing its resulting deformation and rebound. Measurements of intraocular pressure are complicated by the nonfluid character of the wall of the eye, which has a resilience of its own. A normal range of measurements is obtained by testing a large population, and excess intraocular pressure can be detected reliably. Glaucoma may be treated with prescription eye drops that improve the draining of fluid, drugs that suppress the production of aqueous humor, insertion of drain tubes or by surgery to supply a drainage path. Although the damage caused by glaucoma can't be reversed, early treatment can slow or prevent vision loss.

7.1.4 Pressure in the Gastrointestinal System

Food, drink, and waste products moving through the 6-m-long digestive tract or gastrointestinal (GI) system are fluid or fluid like in character. Their flow is regulated by pressure and especially by valves and sphincter muscles in the system. Table 7.1 indicates that pressures in the GI system are usually positive. The esophagus is an exception; its pressure is directly related to thoracic (chest) cavity pressure and is negative. Thoracic cavity pressure is sometimes monitored by measuring pressure in the esophagus. A sphincter is needed at the junction of the esophagus and stomach to prevent back flow of stomach fluids, the most common cause of "heartburn." During swallowing, muscle action in the esophagus forces fluids into the stomach.

Pressures in the GI system are increased by swallowed air or by flatus produced by bacterial action, causing cramps. This is very noticeable in infants, who often swallow air while eating. Blockages in the GI system also cause pressures to increase, even to the point of rupture, due to the buildup of fluids.

The stomach is elastic, so pressure in it increases gradually, becoming large only when the stomach is overfilled. The sensation of hunger occurs when stomach pressure is low. The pressure is dependent on the capacity of the stomach, which can change with eating habits. The stomach stretches considerably when a person consistently overeats, and a large stomach's relative emptiness triggers the sensation of hunger before the person really needs more food.

A method of feeding moderately ill patients is shown in Figure 7.4. A tube is inserted through the patient's nose and down the esophagus into the stomach (called a nasogaastric tube). Liquid can be fed down the tube by gravity since pressure in the stomach is not very large. This feeding method is useful for patients who have difficulty swallowing but are not liable to vomit.

7.1.5 Pressure in the Skeleton

Pressures in the skeletal system are far larger than any of the fluid pressures listed in Table 7.1, being as large as 7600 mm Hg under ordinary circumstances. Skeletal pressures depend on physical activity. In several examples in Chapter 3, we calculated forces in the musculoskeletal system, occasionally noting that these forces could be very large.

Pressure in the skeletal system is the force carried by a bone or joint divided by the area on which it is exerted. Figure 7.5(a) shows the knee joint surface at the upper end of the lower leg bone (the

Figure 7.4 Feeding and administration of medicine can be accomplished through a nasogastric tube. The pressure due to the fluid, *hpg*, exceeds the pressure in the stomach.

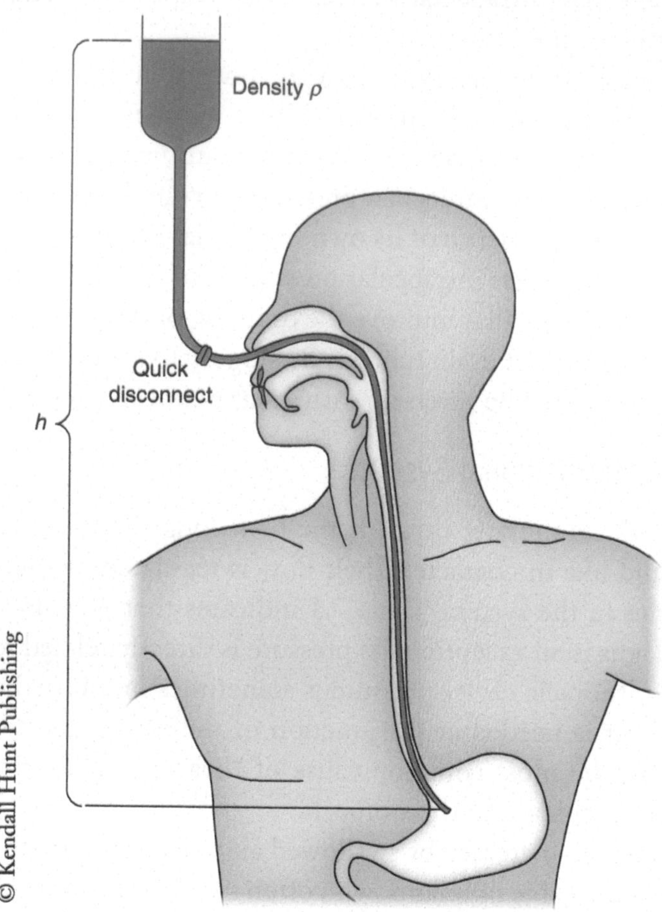

tibia). Note that the surface of the joint is larger than the cross-sectional area of the bone below. This large area reduces the pressure at the joint and prevents its deterioration. In Figure 7.5(b), a cross section of a finger bone is shown. The bone is flat on the gripping side. The large flat area reduces pressure on the tissues covering the bone when forces are exerted by the hand.

Extensive discussions of pressures in the cardiovascular and respiratory systems follow, in Sections 7.3 and 7.4, respectively. Before we proceed to those topics, Section 7.2 covers molecular phenomena basic to cellular processes and intimately related to pressures in biological systems.

7.2 MOLECULAR PHENOMENA AND BIOLOGICAL PROCESSES

A number of biological processes, such as the movement of wastes and nutrients across cell membranes, can be explained in terms of the behavior of molecules. Questions as diverse as how soaking a sprained ankle in Epsom salts reduces swelling and why patients on kidney dialysis get headaches can be answered on the basis of a few facts about molecular motion. The central theme of this section is the movement of substances by molecular motion at a cellular level rather than by large-scale flow.

Figure 7.5 Reduction of skeletal pressures by exerting forces on large areas. In the case of the tibia and typical of many joints in the body, the knee joint has a contact area larger than the bone to which it is attached.

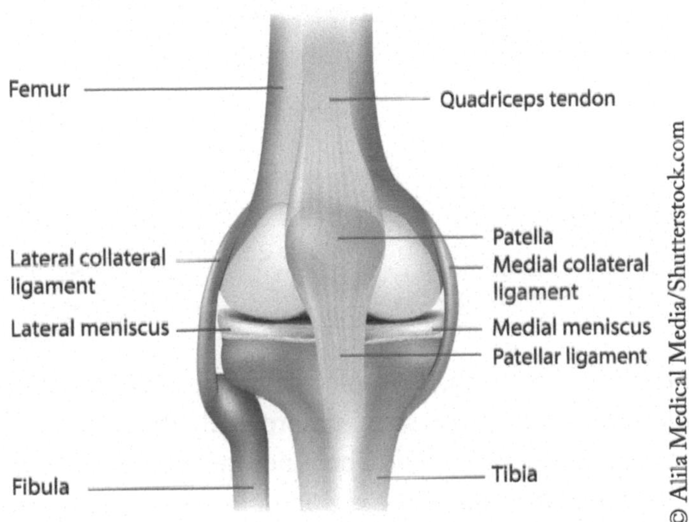

7.2.1 Diffusion

If a drop of food coloring is placed carefully into a glass of still water, it very slowly spreads out until the entire glass of water is a uniform color. This is an example of *diffusion*, the movement of substances due to random thermal molecular motion. Even in the absence of large scale flow, molecules move about, bounce off one another, and mix continuously. The average speed of molecules depends on their mass and temperature, being higher for lighter molecules and for higher temperatures. (Recall from Chapter 5 that temperature is associated with the average kinetic energy, $\frac{1}{2}mv^2$, of molecules in an object.) Oxygen molecules, for example, have an average speed of 475 m/s at room temperature, and each has about 5×10^9 collisions with other molecules every second. These numerous collisions slow down diffusion since molecules cannot travel very far before having a collision that may scatter them in any direction, including back in the direction from, which they came. Materials moved by diffusion thus move *much* more slowly than the average speed of a molecule.

The direction and rate of diffusion are described by *Fick's first law*, named after Adolf Fick (1829–1901), a German physiologist:

a. The direction of diffusion is always from a region of higher concentration to one of lower concentration.
b. The rate of diffusion is directly proportional to the difference in concentration between the two regions.

Both aspects of the law can be explained in terms of the random thermal motion of molecules. Figure 7.6 shows a box having more molecules on the right than on the left.

Figure 7.6 A box with more molecules on the right of an imaginary dashed line than on the left. Pure chance dictates that more molecules will move from right to left than vice versa.

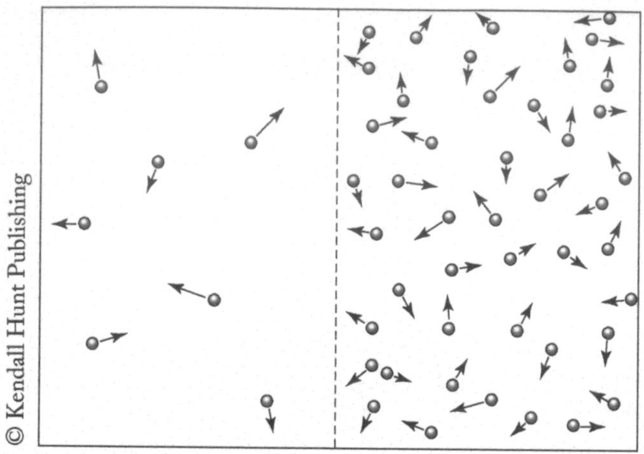

Because molecular motion is random, the molecules move in all directions with equal probability. But there are more molecules on the right, so it is more likely that one of these will move to the left than vice versa. On the average then there will be a net movement of molecules from right to left, from the region of higher concentration to the region of lower concentration. This is the first part of Fick's law. Furthermore, the greater the imbalance, the faster is the migration. If there are 10 times as many molecules on the right, then it is ten times as likely for one to move from the right to the left than vice versa. The greater the difference in concentrations, the greater is the rate of migration. This is the second part of Fick's law. Migration continues until there are equal numbers on the right and left, at which time there will be equal probability of a molecule moving from right to left and vice versa—hence no net movement.

The rate of diffusion depends on other factors in addition to concentration differences. Cohesive and adhesive forces also affect the rate of diffusion. If oil, for example, is placed in water, random thermal motion is not sufficient to mix the oil and water since cohesive forces among the oil molecules are larger than adhesive forces between oil and water molecules. Sugar, however, easily dissolves and diffuses in water since its cohesive forces are less than the adhesive forces. The limit to how much sugar can be dissolved in a given volume of water is tied to the relative strengths of the cohesive and adhesive forces. This raises another point: The rate of diffusion increases with increasing temperature. This is because the average molecular speed increases with temperature and the energy available to break cohesive bonds increases. Thus, sugar diffuses more rapidly in hot water than in cold, and more sugar can be dissolved in hot water than in cold.

The rate of diffusion is always so slow that it is usually unimportant, except at the cellular level where distances are small. (The same is true for osmosis and related processes.) Because of this, transport processes in biological systems become increasingly more complicated in larger organisms. Unicellular organisms use diffusion and related processes exclusively to transfer food and waste products in and out of their systems and to move substances within their systems. However, an organism as large as an earthworm requires a tube through it to move food and waste products. Larger organisms have correspondingly more complex systems to transport materials in addition to diffusion at the cellular level. Humans, for example, have many billions of capillaries, veins, arteries, lymph ducts, and intestines and a host of other tubes in several complicated systems.

Plants and insects use diffusion to exchange carbon dioxide, water, and oxygen with their surroundings. Plants consume carbon dioxide and excrete oxygen. Carbon dioxide diffuses into plants whenever the concentration inside is less than it is outside. Similarly, excess oxygen and water diffuse out of plants whenever their concentrations are greater than those in the surrounding air. Conversely, insects consume oxygen and excrete carbon dioxide. Both are diffused from high to low concentration along breathing tubes called tracheae.

7.2.2 Diffusion Through Membranes

Permeability—Most diffusion in biological organisms takes place through membranes. All cells and some structures within cells, such as the nucleus, are surrounded by membranes. These membranes are very thin, from 65×10^{-10} to 100×10^{-10} m across (less than 100 atoms thick). Most membranes are *selectively permeable*; that is, they allow only certain substances to cross them. (A shorter equivalent expression for "selectively permeable" is *semipermeable.*) There are several explanations for why most membranes are semipermeable. One is that there are pores in some membranes through which substances diffuse. These are so small (from 7×10^{-10} to 10×10^{-10} m) that only small molecules get through (see Figure 7.7). Other explanations have to do with such factors as the chemistry of the

Figure 7.7 Schematic of a semipermeable membrane. Smaller molecules are able to pass through the pores while larger molecules cannot.

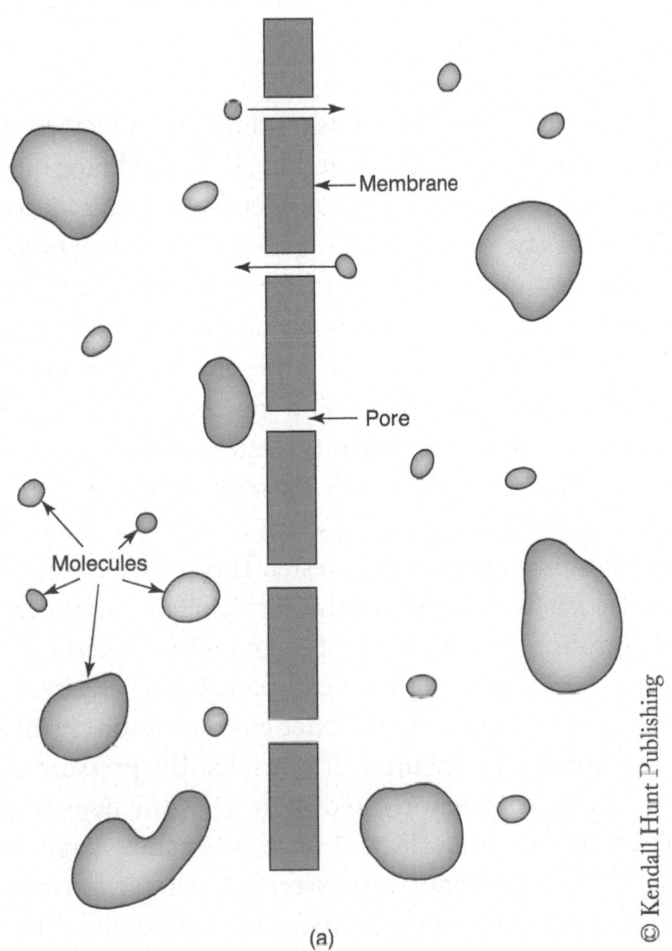

Figure 7.7b Simple diffusion of liquid soluble molecules through plasma membrane.

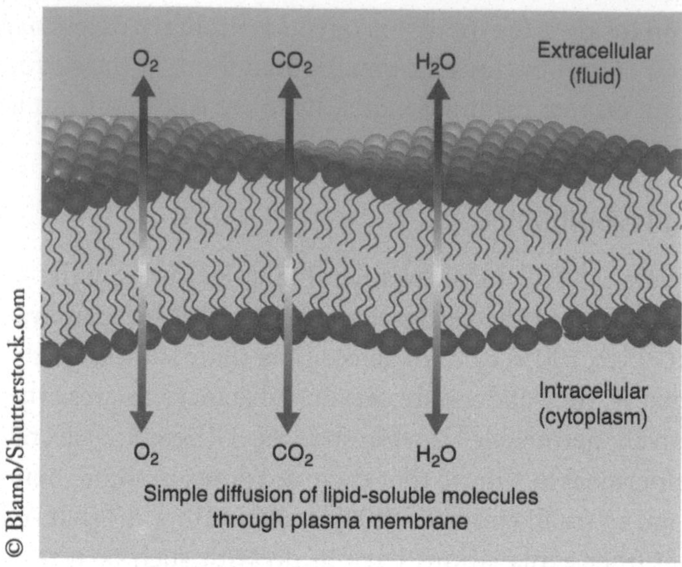

membrane, cohesive and adhesive forces, charges on the ions involved, and the existence of carrier molecules. The subject is not completely understood and will not, be of further concern here except for the fact that most membranes are indeed semipermeable, which has important consequences.

7.2.3 Osmosis

Both aspects of Fick's first law also apply to diffusion through a membrane, although the membrane may slow the process. In most examples of diffusion through a membrane many different types of·molecules are present. It is therefore very important to consider the concentration of the molecule of interest. For example, Figure 7.8(a) shows two solutions of sugar water separated by a semipermeable membrane that allows only water to diffuse through it. There is more sugar on the right than on the left; in which direction will the water diffuse? The correct answer is that the water will diffuse from the region where it is high in concentration to the region where it is lower in concentration. Water will therefore move from the left to the right in this situation. (Confusion sometimes arises since the solution on the right is commonly referred to as more concentrated, but this refers to the sugar, which is not the molecule of interest.) The reason for the direction of diffusion is the same as before. There are more water molecules on the left side of the membrane than on the right, so the chance of a water molecule crossing the membrane from the left to the right is proportionally greater. This situation is the first example of osmosis.

Osmosis is usually defined as the transport of water through a semipermeable membrane due to an imbalance in its concentration on either side of the membrane. Osmosis may be by diffusion, but it may also be a bulk flow through pores in the membrane. In either case, water moves from a region of high water concentration to a region of low water concentration. If the osmosis in Figure 7.8(a) is allowed to continue, the situation in Figure 7.8(b) results. The pressure on the right is then greater than the pressure on the left by an amount ρgh, where ρ is the density of the liquid on the right. This extra pressure was created by osmosis, and once h becomes large enough, the back pressure forces water back through the membrane at the same rate that it moves through by osmosis. That pressure is called the *relative osmotic pressure* for two water solutions, as in *Figure 7.8*, where neither

Figure 7.8 Two solutions of sugar water are separated by a semipermeable membrane that water molecules can penetrate but sugar cannot. If osmosis continues, the fluid rises on the right until the pressure due to the extra weight equals the osmotic pressure.

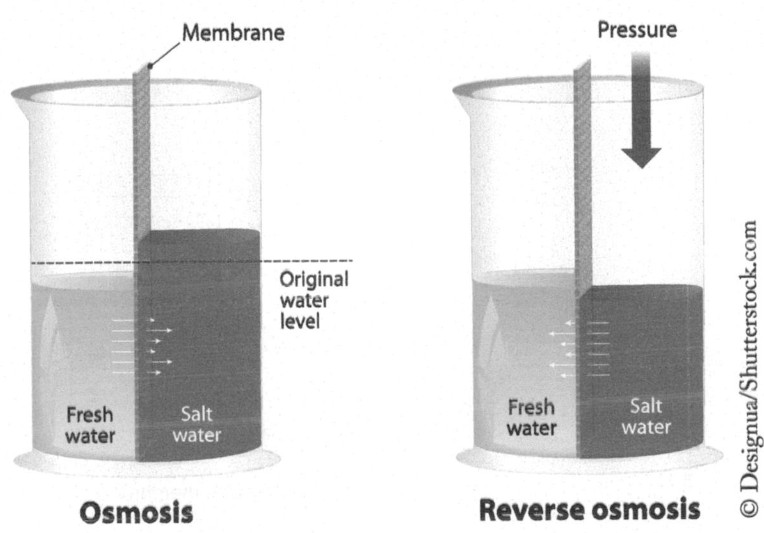

is pure water. If one of the solutions is pure water, then the pressure necessary to stop the flow of water from the region of high water concentration to the region of lower water concentration is called simply the *osmotic pressure* (the back pressure necessary to stop osmosis). Osmotic pressures can be rather large. For example, if pure water and seawater are separated by a perfectly efficient semipermeable membrane that will not pass salt, the pressure necessary to prevent osmosis is 25.9 atm. This means that water will diffuse through the membrane until the salt water surface is 261 m above the fresh water surface!

Reverse osmosis takes place when the pressure opposing osmosis exceeds the osmotic pressure. As shown in Figure 7.9, the net flow of water is in the direction opposite to osmosis when a pressure greater than the osmotic pressure (or relative osmotic pressure) is exerted. Reverse osmosis can be used to desalinate seawater but is not economical due to the lack of efficient and durable membranes.

7.2.4 Examples of Osmosis in Biological Organisms

Water is one of the most universal solvents. It is a major component of most fluids in biological organisms and is crucial to life as we know it. The following are a few examples of osmosis in biological organisms.

Water, Intake by Roots and Turgor in Plants—Osmosis between roots and ground water is thought to be responsible for the transfer of water into many plants. Groundwater is purer and hence has a higher water concentration than sap, so osmosis moves water into roots. Water in sap in then transferred by osmosis into cells, causing them to swell with increased pressure. This pressure is called *turgor pressure* and is partly responsible for the ability of many plants to stand up.

Relative osmotic pressure is not large enough to cause sap to rise to the top of a tall tree, however. To do this, the sap would have to have more dissolved materials in it than is found to be the case. Other processes in addition to osmosis are probably involved in the uptake of water, since some plants grow in salt water that would remove water from the plant if only osmosis were present.

Figure 7.9 Reverse osmosis occurs when a pressure greater than the osmotic pressure is applied. Osmosis still occurs in the expected direction, but backflow due to the applied pressure exceeds the flow from osmosis.

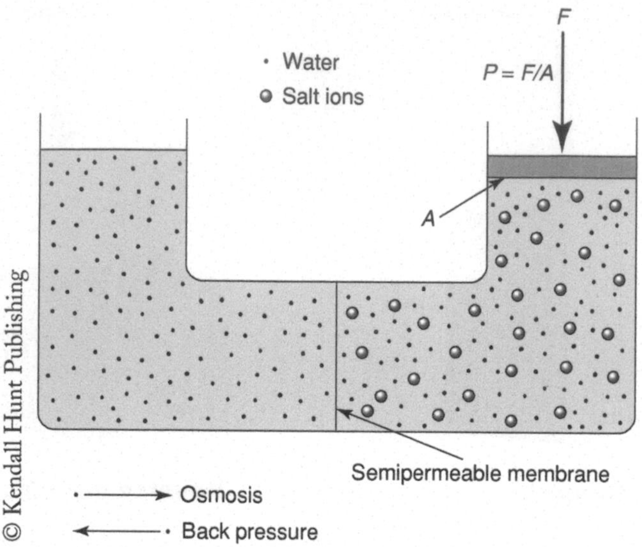

Epsom Salts Soaking a sprained ankle in Epsom salts is a common method of reducing swelling. The concentration of water in the swollen part of the ankle is greater than in the Epsom salts solution. Osmosis therefore transports water out of the ankle into the soaking solution.

Regulation of the Fluid Between Cells (Interstitial Fluid) In humans and other animals the interstitial fluid is regulated by exchange of substances with blood in capillaries. Many substances are moved across capillary walls, but the transport of water (i.e., osmosis) is of immediate interest. *Figure 7.10* shows the process under average conditions. Blood is a complicated fluid. It contains water, glucose, electrolytes (dissolved salts), gases, proteins, red and white blood cells, waste materials such as urea, and so on. Blood enters from the top right in *Figure 7.10*, and because the capillary is

Figure 7.10 Water transfer between a capillary and the interstitial region.

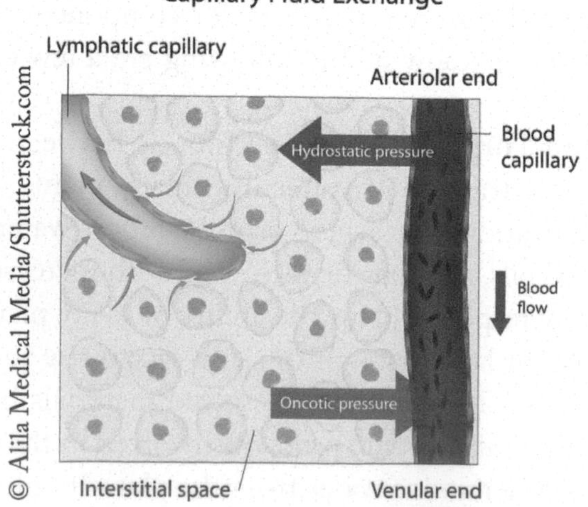

small in diameter, there is a significant pressure drop along it. Pressures are typically 35 mm Hg at the entrance and 15 mm Hg at the exit.

Water is more concentrated in the interstitial fluid than in the blood, but the typical osmotic pressure difference between blood and interstitial fluid is 22 mm Hg. Reverse osmosis therefore occurs near the entrance of the capillary. Blood pressure near the capillary's exit is less than the relative osmotic pressure, and osmosis carries water from the interstitial region into the capillary.

The overall result is an exchange of water while the total amount of water in the interstitial region remains constant. Departures from this normal behavior can occur for a number of reasons and result in either too much water in the interstitial regions or too little. Both conditions are medically unfavorable. Edema, or the buildup of interstitial fluid, creates swelling in those regions.

Swelling (Edema) Due to Heart Failure The right side of the heart receives blood from the veins. If a patient suffers right heart failure, then the heart is less effective in taking up the blood sent to it, and. pressure at the exit of capillaries rises. This causes reverse osmosis all along the capillary and the buildup of interstitial fluid, called edema. The fluid usually collects in the legs, back, and buttocks when there is right heart failure.

The left side of the heart receives blood from the lungs. Left, heart failure increases pressure in the lungs, causing reverse osmosis there.

This water buildup is called pulmonary edema and is an extremely dangerous condition, for the patient can literally drown in his own fluids.

Effect of Electrolyte Balance Too much salt consumption causes water retention because osmosis carries water from the salty blood into the interstitial region. Similarly, electrolyte imbalances (electrolytes are essentially dissolved salts) of any type have effects on osmosis and are particularly worrisome in heart patients.

7.2.5 Dialysis

Most people have heard of dialysis because of the advent of artificial kidney dialysis machines in the 1960s. Dialysis is the diffusion of substances other than water through semipermeable membranes and occurs in kidneys and many other places in biological organisms. Small molecules are usually involved in dialysis since membranes tend to be impermeable to large molecules. Fick's law applies to the direction and rate of dialysis. Reverse dialysis, also called filtration; occurs when pressure on the high concentration side is large enough to reverse the normal direction of dialysis. Kidney function and the effect of diuretics are two examples of dialysis in humans. Let us consider these in greater depth.

Kidneys use dialysis, osmosis, and active transport (not yet discussed) to remove waste products from the blood and maintain its proper balance of fluids and electrolytes. Artificial kidney dialysis machines use mainly dialysis to remove waste products from blood. Artificial kidneys are a poor substitute for healthy kidneys, but they do prolong life.

The basic kidney unit, a nephron, is illustrated in Figure 7.11. Blood circulates through a bundle of capillaries that is enclosed in a small membrane sac. Substances removed from the blood must cross two membranes: the capillary wall and Bowman's capsule wall. The transport process is reverse dialysis caused by arterial blood pressure. Only blood cells and large molecules such as proteins are left behind.

The substances filtered out of the blood then move into a drainage tube, where many are reabsorbed into the bloodstream. None of the processes discussed so far completely describes the reabsorption of substances in the drainage tube. Active transport (to be discussed) is responsible for much

Figure 7.11 Schematic illustration of a kidney nephron. A two-step process of filtration followed by partial reabsorbtion is used by the two million nephrons in a normal adult to remove wastes and maintain fluid balance.

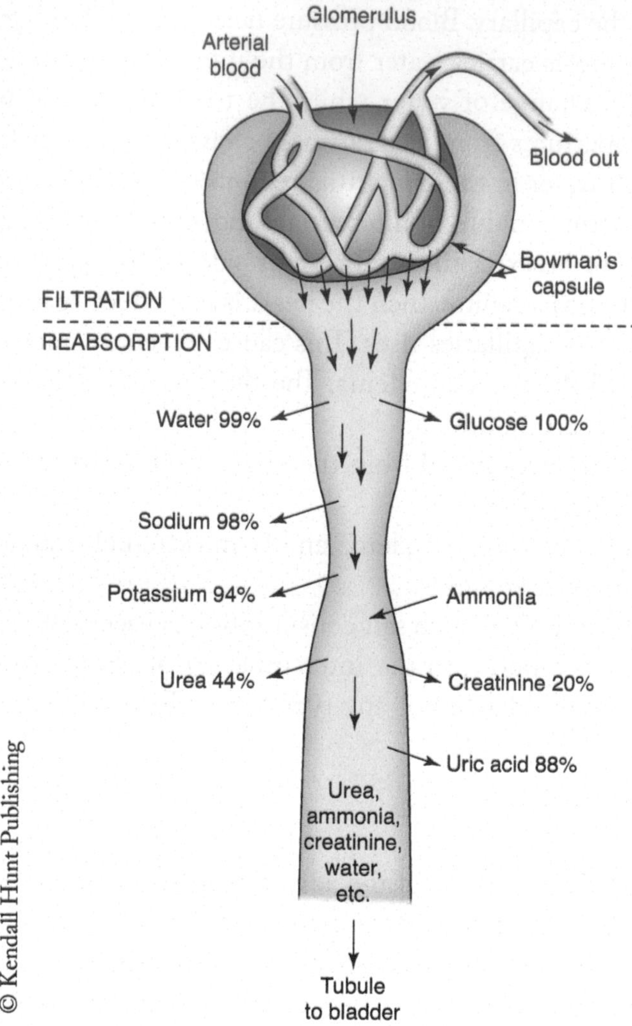

of the reabsorption. The net effect of the two-step process of filtration followed by reabsorption is to rid the body of waste products while maintaining a proper balance of electrolytes, water, and so on.

Kidney failure can cause high blood pressure. If filtration slows or stops in the glomerulus, the kidney emits a hormone called renin that causes blood pressure to rise so that filtration recommences. Unfortunately, high blood pressure has many ill effects on the body and can even cause the capillaries in the glomerulus to burst.

Any substance that causes increased urine output is called a diuretic. Some of these work by altering the amount of water reabsorbed in the drainage tube. They penetrate the glomerulus and Bowman's capsule but cannot be reabsorbed in the drainage tube. The presence of such a molecule decreases the concentration of water in the tube and its reabsorption, resulting in a greater production of water in the urine.

One symptom of diabetes is unusual thirst and frequent urination. Diabetes victims ineffectively utilize glucose and its concentration rises in the blood. This in turn causes an excessive amount of glucose to pass through the glomerulus and Bowman's capsule into the proximal convoluted tubule, where it acts as a diuretic and appears in the urine.

7.2.6 Active Transport

The discussion of molecular phenomena and biological processes so far has included the general categories of diffusion, osmosis, reverse osmosis, dialysis, and reverse dialysis (filtration). All these processes for transporting substances on the cellular level are passive in nature. The driving energy for passive transport comes from molecular kinetic energy or pressure. There is another class of transport phenomena, called active transport, in which the living membrane itself supplies energy to cause the transport of substances.

Biological organisms sometimes need to transport substances from regions of low concentration to high concentration-the direction opposite to that in osmosis or dialysis. Of course, sufficiently large back pressure causes reverse osmosis or reverse dialysis, but there are known instances in which substances move in the direction that reverse osmosis or reverse dialysis would take them even though existing pressures are insufficient to cause reverse osmosis or reverse dialysis. In these instances, active transport must be taking place, which means that living membranes expend their 'own energy to transport substances. Active transport can also aid ordinary osmosis or dialysis and explains why some transport proceeds faster than expected from osmosis or dialysis alone.

Active transport is a very complicated process·that is only now beginning to be understood. A great deal can be said of what happens in active transport, but not as much can be said about why it happens. The reabsorptive phase of a kidney nephron is one example of active transport. Another example is the red blood cell. The concentration of potassium in red blood cells is about 20 times larger than that of sodium, whereas the blood plasma surrounding them has about 20 times as much sodium as potassium. Active transport maintains this against the tendency of dialysis to transport sodium into the cell and potassium out of the cell. Active transport is known also to occur in the intestines.

Active transport is extremely important in nerve cells. Changes in the concentration of electrolytes across nerve cell walls are responsible for nerve impulses. After repeated nerve impulses, significant migration has occurred and active transport "pumps" the electrolytes back to their original positions. (See Chapter 13 for a detailed discussion of these processes in nerves.)

7.3 THE CARDIOVASCULAR SYSTEM

Many of the characteristics of the cardiovascular system can be explained in terms of the laws of physics. The cardiovascular system consists of two pumps (the right and left sides of the heart) and a complex arrangement of vessels that transports blood through nearly every part of the body.

7.3.1 The Heart as a Double Pump

The purpose of any pump is to create pressure in a fluid, usually by exerting a force directly on the fluid. The heart in higher animals, including humans, consists of two pumps and two reservoirs preceding those pumps. The right side of the heart pumps blood through the lungs to the left side of the heart. The left side of the heart pumps blood through the rest of the circulatory system, returning it to the right side of the heart, where the process begins again. The operation of the human heart is shown in Figure 7.12. The atria are reservoirs for the ventricles, which are the main pumps. (The atria

Figure 7.12a Blood flow of the human heart.

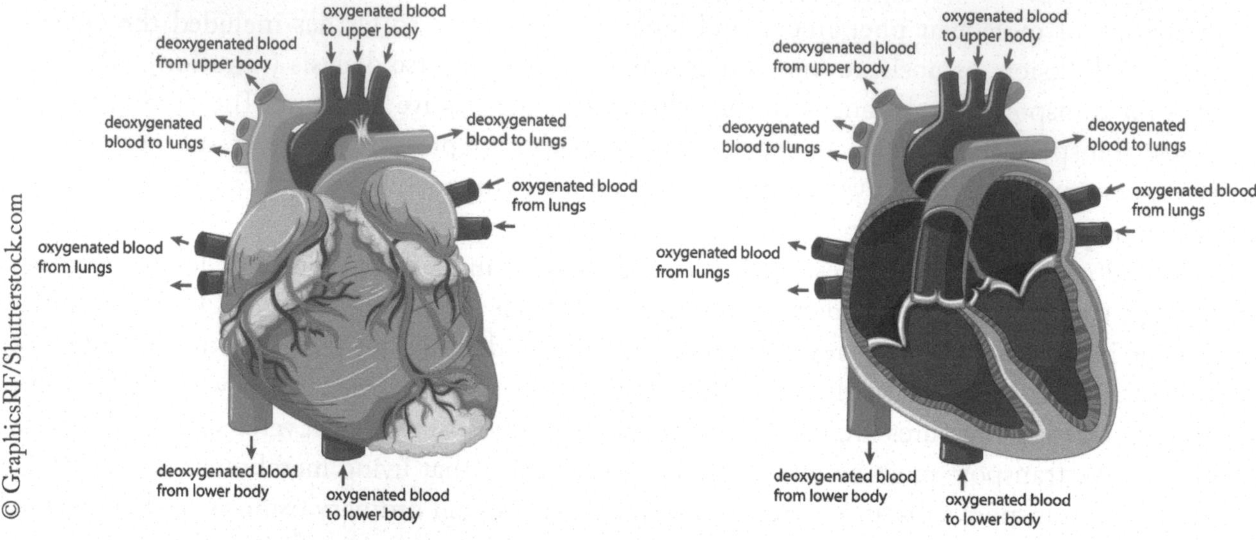

Figure 7.12b Anatomy of a human heart.

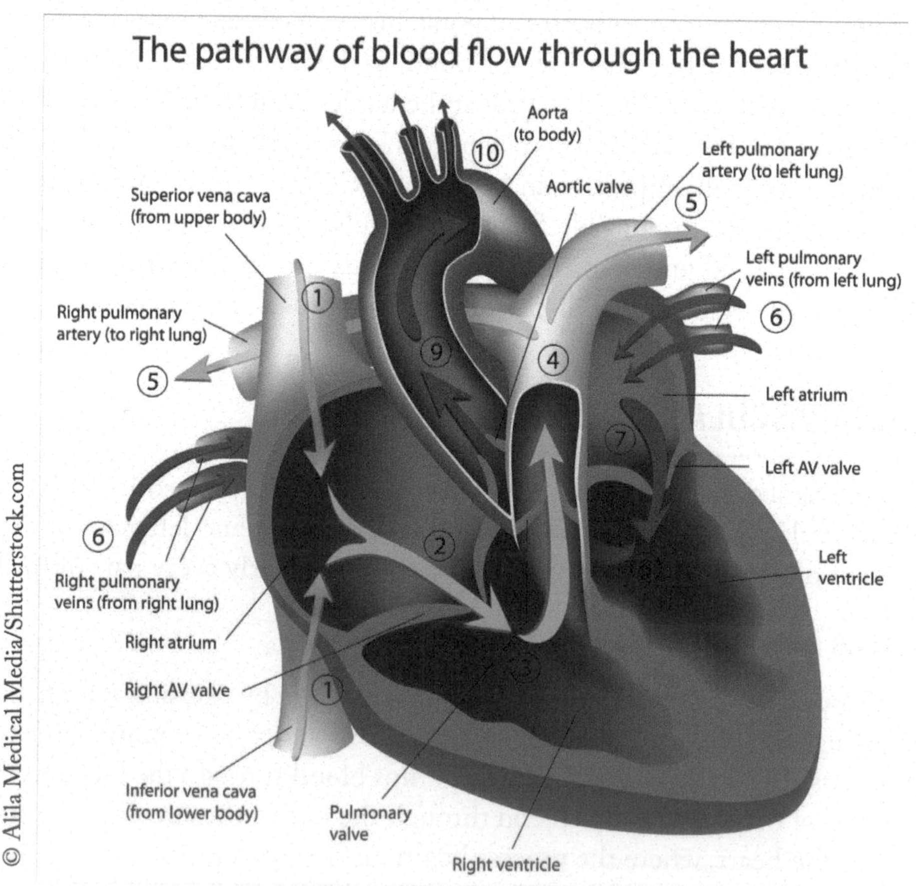

pump blood into the ventricles.) Heart valves are opened or closed by pressure differences across them. In addition, the valves between the atria and ventricles are aided in opening by cords attached to the interior of the ventricles. Blood is nearly incompressible, so when the ventricles pump blood out of the heart, the aorta balloons to take up the extra blood.

7.3.2 Overall Circulatory System

Major Features—The circulatory system can be represented approximately by the schematic in Figure 7.13. The system is vastly complex in detail, and certain of its aspects, such as its interconnection with the lymph system, will not be considered. The major features of the system are the two sides of the heart, the lungs, and the arteries, arterioles, capillary beds, venules, and veins. Blood is pumped into the aorta, the largest artery in the body, by the left ventricle. The aorta branches in to the major arteries, which branch into arterioles and then into capillaries in various organs. The capillaries rejoin to form venules, which join to form the major veins, returning blood to the right atrium of the heart. The right atrium pumps blood into the right ventricle, which then pumps it through the lungs into the left atrium. From the left atrium, blood enters the left ventricle and starts another round trip.

Figure 7.13 Schematic of the major features of circulatory system.

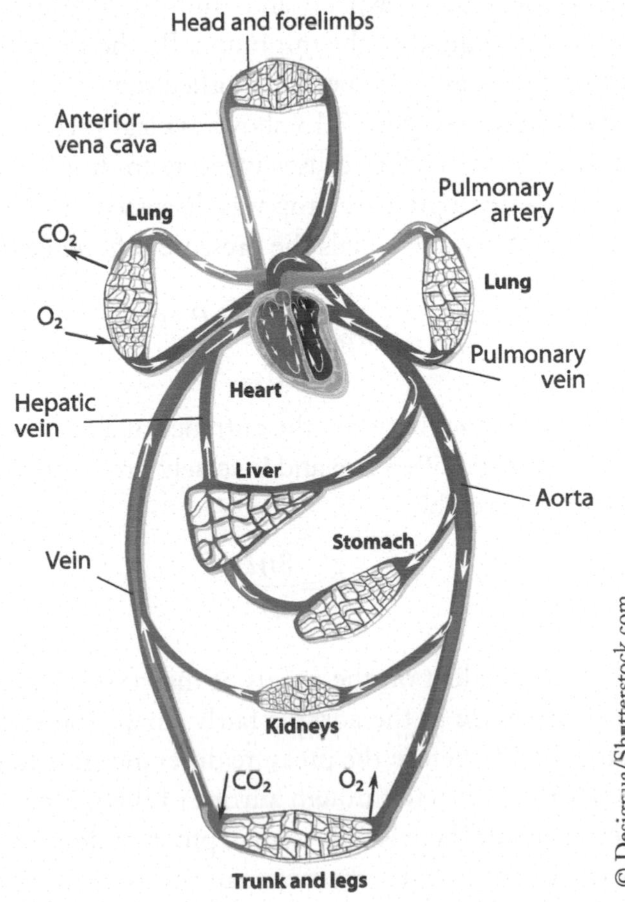

Figure 7.14 Graph of blood pressure versus time in a major artery.

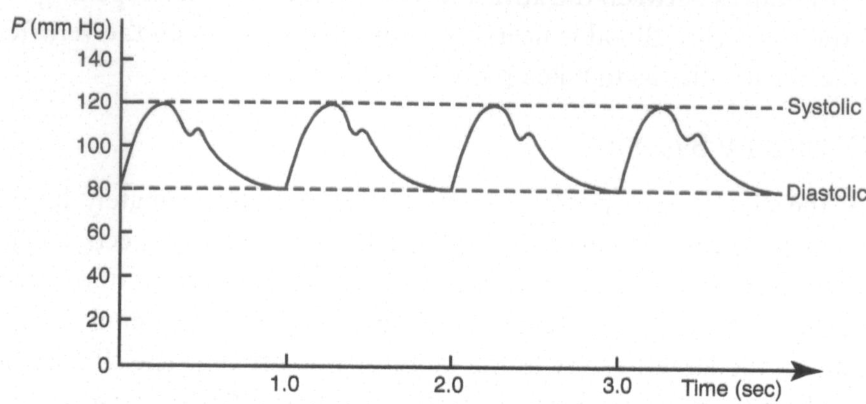

Pressure Variations in Heart Output—The output pressures of the left and right sides of the heart given in Figure 7.13 are the maximum values of each heartbeat. Figure 7.14 is a graph of blood pressure versus time in one of the major arteries (not, in the lungs).

The maximum and minimum pressures are the systolic and diastolic pressures defined in Section 6.3. Diastolic pressure is due to the elasticity of the arteries. They rebound between contractions, typically maintaining a minimum pressure of 80 mm Hg. The bump on the curve in Figure 7.14 is caused by blood rebounding from the left ventricle valve when it closes. A pulse cannot be detected in the venous system, since the arteries expand and contract with each heartbeat and alternately reduce the maximum pressure and increase the minimum. By the time blood emerges from the capillaries the variations in blood pressure have been smoothed out.

Pressures Around the System—Figure 7.13 shows typical pressures in the major parts of the circulatory system. Resistance in the system causes pressure to drop as blood flows around the system. The idea that resistance causes a pressure drop was discussed in Chapter 6. Recall that for both laminar and turbulent flow, the flow rate equals the pressure difference divided by the resistance.

$$Flow\ rate = \frac{\Delta P}{R}$$

where ΔP is the pressure difference between the entrance of a tube and at its exit. The flow rate is directly proportional to the pressure difference and inversely proportional to the resistance, R.

Recall that the resistance is given by,

$$R = \frac{8\eta L}{\pi r^4}$$

and is therefore strongly dependent on the radius of the vessel. The large radius aorta offers little resistance; the pressure difference in the aorta is fairly small. The small radii capillaries produce significant resistance to flow and therefore the pressure difference it is fairly large. However, because there are so many capillaries, the flow rate through anyone of them is small, and so the pressure drop is not as large as might be expected. There is only a small pressure drop in the venous system from the capillaries back to the heart. This is because the two major veins returning blood to the right side of

Figure 7.15 Capillary bed. Arterioles branch into capillaries that rejoin to form venules. Some capillaries have small sphincters to control flow. Larger capillaries act as shunts.

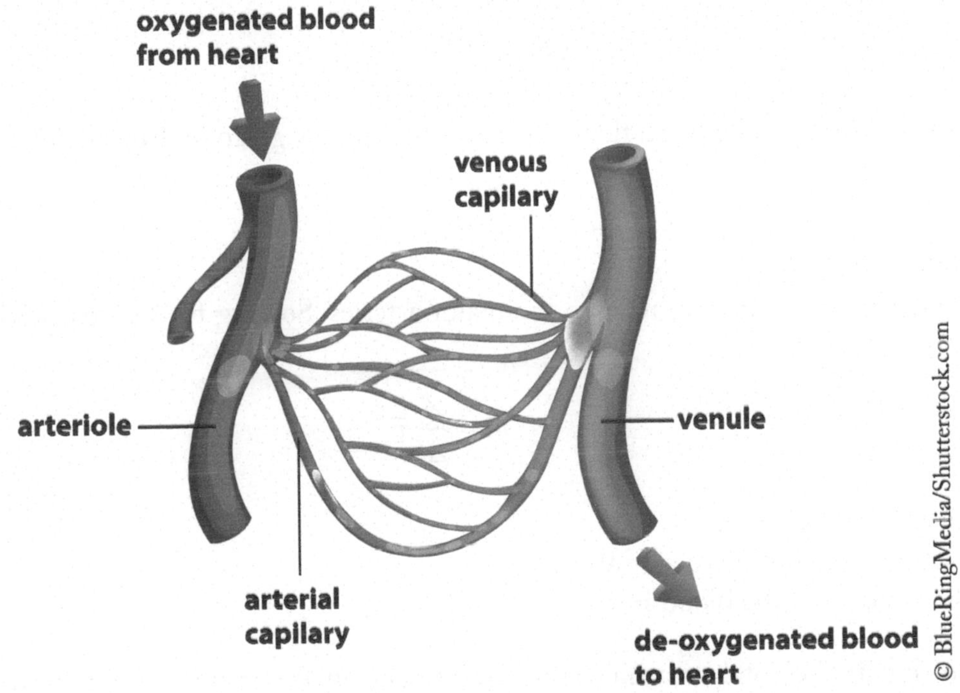

the heart are each larger than the aorta. The right side of the heart increases blood pressure to pump blood through the lungs. Lung resistance reduces the pressure again before the blood enters the left side of the heart, where the process continues. (See Figure 7.13 for additional information.)

Capillary Beds—Many of the blood's duties are performed in capillaries, including the removal of waste products and the transport of nutrients and oxygen to cells. Capillaries are so small (5–20 μm) in diameter that blood cells actually deform when passing through smaller ones. The processes of diffusion, osmosis, dialysis, and active transport are effective through capillary walls, which are thin and in close contact with the blood cells. A schematic of a typical capillary bed is shown in Figure 7.15.

An incoming arteriole branches into capillaries, some of which have sphincter muscles to control flow. Another important method of flow control is constriction and dilation of arterioles. Pathways through capillary beds are often tortuous and interwoven, giving the body considerable latitude in determining how much blood flows to which organs. Although Figure 7.15 doesn't show it, most capillaries are 50–100 times longer than their diameter, typically 1000 μm long (1.0 mm), compared to 5–20 μm in diameter.

7.3.3 Blood Velocities and Branching

The velocity of blood is much greater in the major arteries (about 30 cm/s) than in the capillaries (about 0.030 cm/s). Surprisingly, the blood velocity increases again when the capillaries rejoin to form veins. Most changes in average blood velocity other than those produced by the heart are due to changes in the total cross-sectional area of the system during *branching*. For example, the aorta branches into the major arteries, each of which has a smaller cross-sectional area, but whose combined area is larger than the area of the aorta. The total flow rate in the major arteries is the same as in the aorta since all the blood which passes through the aorta must also pass through the major arteries. Therefore,

$$FR_{\text{aorta}} = FR_{\text{major arteries}}$$

From the relationship between flow rate and fluid velocity given by Equation 6.12,

$$FR = A\overline{v}$$

we see that $A\overline{v} = A'\overline{v}'$

where the primed quantities refer to the major arteries. Solving for "v," *the velocity in the major arteries*, yields

$$\overline{v}' = \frac{A}{A'}\overline{v}$$

A is the area of the aorta.
A' is the total area of the major arteries.
$\overline{v}$ is the velocity of the blood in the aorta.

We see that if the total area increases, the velocity in the major arteries must decrease (see Example 7.2). The circulatory system continues to branch out in a similar manner; blood vessels become smaller in size and much more numerous at each branching. When the vessels begin to recombine on the venous side of the system, the process reverses and the total area of the system decreases. Consequently, the average blood velocity increases when the blood passes from the capillary system into the venules and from there into larger veins, and so on. Figure 7.13 illustrates this branching schematically.

EXAMPLE 7.2

(a) Calculate the average blood velocity in the major arteries, given that the aorta has a radius of 1.00 cm, the blood velocity is 30.0 cm/s in the aorta, and the total cross-sectional area of the major arteries is 20.0 cm^2. (b) What is the total flow rate in cm^3/s and liters/minute? (c) On the assumption that all the blood in the circulatory system goes through the capillaries, and given that the average velocity of the blood in the· capillaries is 3.00×10^{-2} cm/s, what is the total cross-sectional area of the capillaries?

Solution.

(a) Step by step 1. *Calculate the average blood velocity in the major arteries.* The velocity of the blood in the major arteries is given by, $\overline{v}' = \frac{A}{A'}\overline{v}$ Where, A is the area of the aorta.

A' is the total area of the major arteries.

$\overline{v}$ is the velocity of the blood in the aorta.

Step 2. Because the vessels have a circular cross section, we can determine the cross-sectional area with,

$$A_{\text{circle}} = \pi r^2$$

Area of aorta = $A_{\text{circle}} = \pi(1.00\text{cm})^2 = 3.14\text{cm}^2$
Total Area of major blood vessels = 20 cm^2 (*Note that if each artery has a cross-sectional area of 0.500 cm^2, then there are approximately 40 of them.*)
v **Velocity of blood in the aorta = 30 cm/s**

Step 3. Substitute the values and do the math. T*he average blood velocity in the major arteries is,*

$$\bar{v}' = \frac{A}{A'}\bar{v} = \frac{3.14\,\text{cm}^2}{20.0\,\text{cm}^2}\cdot 30.0\frac{\text{cm}}{\text{s}} = 4.71\frac{\text{cm}}{\text{s}}$$

We see that the average velocity of the blood was reduced as it moved from the aorta to the major arteries.

Solution (b): What is the total flow rate in cm^3/s and liters/minute? The total flow rate must be the same in each system. Calculating it in the aorta gives

$$FR = A\bar{v} = 3.14\,\text{cm}^2 \cdot 30.0\frac{\text{cm}}{\text{s}} = 94.2\frac{\text{cm}^3}{\text{s}}$$

Converting to liters per minute,

$$FR = 94.2\frac{\text{cm}^3}{\text{s}}\cdot\frac{\text{liter}}{\text{minute}}\cdot\frac{1\,\text{L}}{1000\,\text{cm}^3}\cdot\frac{60\,\text{s}}{1\,\text{min}} = 5.65\frac{\text{L}}{\text{min}}$$

Solution (c): On the assumption that all the blood in the circulatory system goes through the capillaries, and given that the average velocity of the blood in the· capillaries is 3×10^{-2} cm/s, what is the total cross-sectional area of the capillaries in cm^2?

$FR = A\bar{v}$ Therefore the total cross-sectional area of the capillaries is,

$$A = \frac{FR}{\bar{v}} = \frac{94.2\frac{\text{cm}^3}{\text{s}}}{3.00\times 10^{-2}\frac{\text{cm}}{\text{s}}} = 3.14\times 10^3\,\text{cm}^2$$

One point needs to be emphasized. The decrease in velocity going from the aorta to the capillaries is, not caused by resistance to flow. Resistance causes pressure to drop but does not affect velocity. This can be seen from the fact that blood velocity increases again in the veins while pressure continues to drop in the veins leading back to the heart. A similar branching and rejoining of blood vessels takes place in the lungs.

It is necessary to have a low blood velocity in the capillaries and to have many, many capillaries in order to have an effective transport of substances· between blood and the interstitial regions by diffusion and related slow processes. An estimate of the number of capillaries in the body (see Problem 7.12) gives about 10^{10}, implying that there is a capillary every few cells on the average.

Figure 7.16 Muscle pumps in the venous system help return blood to the heart from the extremities.

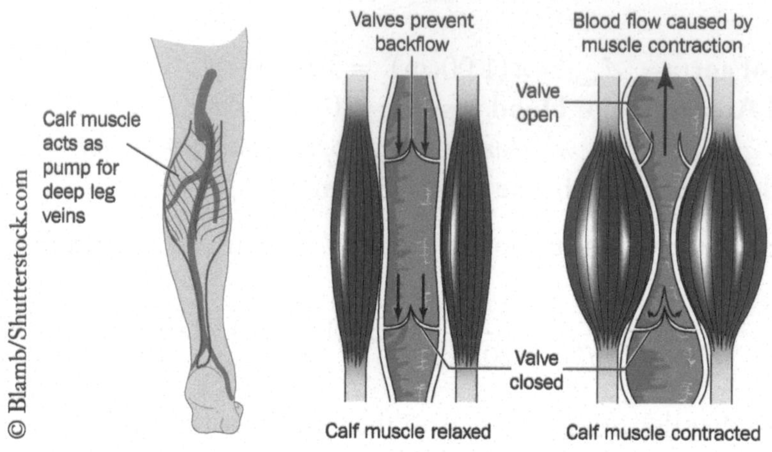

The return of blood to the heart through the veins is aided by the contraction of skeletal muscles, especially in the extremities. When muscles contract, they bulge and force blood out of the larger veins, as shown in Figure 7.16. The one-way valves only permit flow toward the heart. People are rarely completely relaxed, even when sitting quietly. Small motions, such as shifting in a seat or tensing, aid circulation with this muscle pump. In people that are completely motionless, such as soldiers at attention, the return of blood to the heart is slow enough that they may faint.

7.3.4 Effect of Gravity on the Circulatory System

Gravity affects pressure but not flow rate in the circulatory system. In any closed system, pressure due to gravity has no net effect on flow rate, much as the atmosphere has no effect on flow in an intravenous (IV). Consider flow from the heart to the legs and back to the heart in a standing person. Gravity is in the same direction as downward flow, and in the opposite direction to upward flow, so it has no net effect; it neither aids nor hinders flow in a closed loop such as the circulatory system.

Gravity can have an effect on fluid balance. Recall from Chapter 6 that the pressure due to the weight of a fluid is

$$P_{\text{depth}} = \rho_{\text{fluid}} \cdot g \cdot depth$$

It is convenient in the circulatory system to measure depth or height of the fluid relative to the heart. For a standing person, pressure in the major arteries increases below the heart and decreases above the heart by an amount $\rho g h$, where h is positive for any point below the heart and negative for any point above the heart. (The pressure drop due to resistance is negligibly small in the major arteries.) The blood pressure in the major arteries in the head is then

$$P_{\text{head}} = P_E - \rho g h_{\text{head}}$$

This can cause fainting if the blood pressure in the person's head, P_{head}, is too low. Pressure in the major arteries in the legs is

$$P_{legs} = P_{heart} + \rho g h_{legs}$$

(see Figure 7.17). The larger pressure in the legs can cause fluid buildup (edema) by reverse osmosis through capillary walls, especially in people who are on their feet many hours a day. One remedy is to sit down and elevate the legs, while having your feet and legs massaged.

Figure 7.17 Blood pressure in the major arteries increases below the heart and decreases above the heart because of the effect of gravity. The pressure drop due to resistance to flow in the major arteries is only about 5 mm Hg and is neglected in the figure.

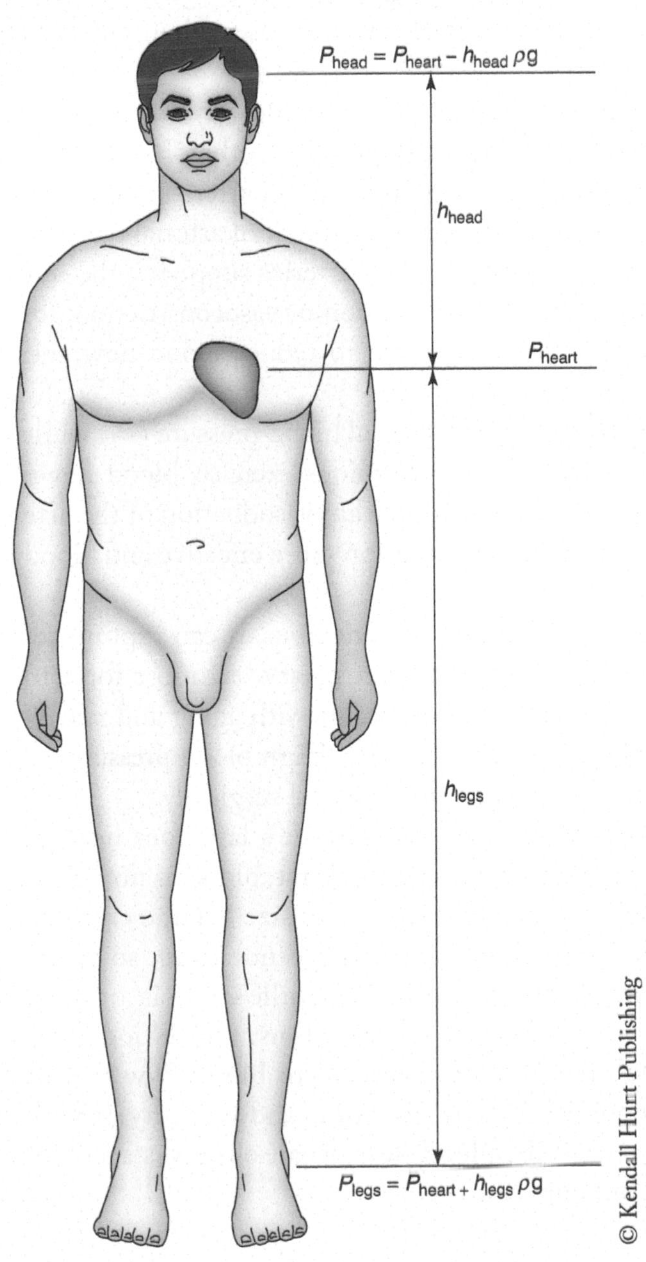

7.3.5 Flow Regulation in the Circulatory System

The body has considerable latitude in blood flow regulation. Blood flow to individual organs and total blood flow through the body can be varied. Blood flow is adjusted both by changes in vessel radii and by changes in blood pressure. In a tube of uniform diameter, fluid flow obeys the relationship

$$FR = \frac{\Delta P}{R}$$

As discussed in Section 6.5, this relationship is always true, but the expression for the resistance R may be simple or complex, depending on the type of flow. If the flow is laminar (nonturbulent), then the resistance R is given by $R = \frac{8\eta L}{\pi r^4}$ This is Poiseuille's law. Its applicability to blood flow is discussed later in this section, but it does give the correct sense of the dependence of resistance on radius and viscosity. That is, resistance always decreases when radius increases and increases when viscosity increases.

It is not surprising that the body adjusts blood flow by changing vessel radii, since resistance to flow is more sensitive to this parameter than to any other. Most vessel dilation (vasodilation) and constriction (vasoconstriction) take place in the small arteries and arterioles, with some occurring in those capillaries that have sphincters. The adjustment of arterial radii may be general, as when a person goes into shock, or very specific, as when arteries supplying the intestines are dilated following a large meal. Drugs also can induce vasodilation or vasoconstriction; for example, a local anesthetic may be administered with a vasoconstrictor to reduce blood flow and prolong the time that the anesthetic remains in place.

Flow is also adjusted by changes in arterial blood pressure (P_1) while venous blood pressure (P_2) remains constant. For example, during strenuous exercise, blood flow may quadruple with a 50% increase in arterial blood pressure and significant vasodilation of the arterial systems throughout the body. Athletic conditioning makes vasodilation more effective and blood pressure increases need not be as great.

High blood pressure can, of course, be dangerous. It can rupture vessel walls, perhaps causing a stroke or heart attack. Furthermore, the heart must exert more force to create high blood pressure and is strained. High blood pressure often occurs with aging and may be designed to maintain flow in vessels narrowed by plaque (arteriosclerosis). High blood pressure is controlled with medication, and narrowed blood vessels are sometimes replaced surgically.

Applicability of Poiseuille's Law—Poiseuille's law does not quantitatively describe blood flow very accurately for a number of reasons. First, blood is not an ideal fluid. It contains blood cells, which are not fluid in character and whose size is large enough to affect flow in arterioles, capillaries, and venules. Second, vessel walls are not rigid, so flow is affected as they expand and contract with each heartbeat. Third, Poiseuille's law is only valid for nonturbulent flow. High blood velocity, sharp bends or constrictions, and blood cells can all cause turbulence. Nevertheless, Poiseuille's law is widely applied to blood flow and does give a good qualitative description of the dependence of flow on radius and viscosity. It will continue to be used quantitatively in this book to describe blood flow because it gives acceptable approximations in view of the lack of a simple but superior theory.

Turbulence: A Diagnostic Indicator—Turbulence can sometimes be detected by the sound it makes. For example, the noise that a water faucet makes is the sound of turbulent flow around its valve. The sound of turbulence is a valuable diagnostic indicator in the circulatory system.

Turbulence in the heart, major arteries, and veins is easily detected by a stethoscope. Normal heart sounds are caused mostly by valves closing. Flow through leaky valves is turbulent since a partially closed valve forms an irregular obstruction. Sounds made by leaking heart valves are called heart murmurs. A hole between heart chambers is another cause of turbulent flow. Such a hole exists in newborn· infants and normally closes within a few hours after birth, sometimes before birth. Holes between chambers of the heart and malfunctioning valves are corrected surgically when they are severe enough to impair proper circulation.

An aneurysm is a ballooning of a vessel due to a weakened wall. The irregular shape of an aneurysm can cause turbulence, and it is possible to detect aneurysms by the sound of turbulent flow. Aneurysms are quite dangerous. If one bursts, it is often fatal owing to stroke, heart failure, or severe hemorrhaging. Arteriosclerosis can also cause audible turbulence.

Turbulence is used to make common blood pressure measurements (see Figure 6.12 and the discussion in Section 6.3). Two pressures are recorded; the maximum heart pressure (systolic) and the minimum arterial pressure (diastolic). Systolic pressure is measured by noting the cuff pressure when flow first resumes as the pressure in the cuff is lowered. The first flow is turbulent. The cuff pressure is then reduced until there is flow during all parts of the cycle. Flow remains turbulent because the artery is partially constricted. The turbulent sounds made by blood flow in this measurement are called the *Korotkoff sounds.* The pressure at which flow occurs during all parts of the cycle but remains turbulent is called the diastolic pressure and is normally recorded. The pressure at which turbulence ceases is sometimes recorded as a third, lower, pressure, but it seems to be a less important indicator than either diastolic or systolic pressure.

7.3.6 Laplace's Law and Wall Tension in the Circulatory System

How thick must the wall of a tube be to withstand fluid pressure? Obviously, the higher the pressure, the thicker the wall must be, but small-diameter tubes, such as capillaries, have thinner walls than might be expected. Blood pressure averages 25 mm Hg in the capillaries and 100 mm Hg in the aorta. Why is it then that the aorta has rather thick walls (about 0.2 cm) while capillary walls are only about 0.5×10^{-5} cm thick? The answer is that capillaries have small diameters, and therefore there is less area for pressure to act upon, and hence less force on the capillary wall.

It can be shown that the tension the walls of a tube must withstand is given by

$\gamma = P \cdot r$ Laplace law

where γ is the tension in the wall (given the symbol γ in analogy to surface tension discussed in Section 6.7), P is the pressure in the tube, and r is the radius of the tube. Typical units are N/m. This is Laplace's law, first proven by the French mathematician and astronomer Pierre Simon Laplace (1749–1827) (*see Figure 7.18*).*

Laplace's law can be used to explain how a tiny capillary with very thin walls tolerates pressures as large as a fourth of those in the aorta. The capillary radius r is so small that the tension γ in its wall is also small.

Figure 7.18 Laplace's law. A tension $\gamma = Pr$ is created in the walls of a cylindrical tube of radius r by the fluid pressure P.

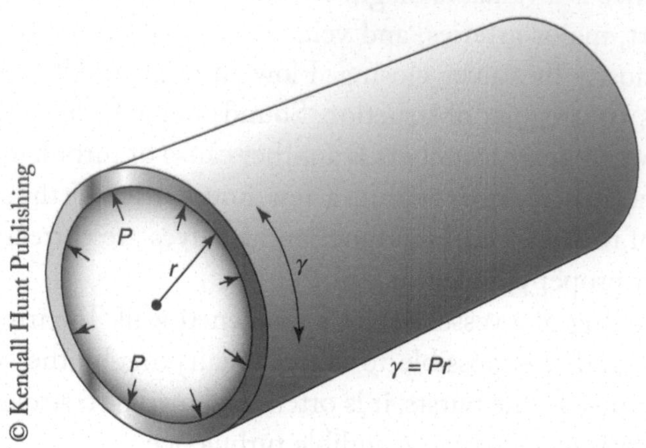

*Laplace's law is of the same form as Equation 6.14, differing only by a factor of 4. Equation 6.14 is Laplace's law applied to a spherical bubble. In Section 6.7, we were interested in the pressure created in the bubble by wall tension. In the present context we are interested in the wall tension created by the pressure in a tube.

EXAMPLE 7.3

Calculate the wall tension γ in N/m created by blood pressure in the aorta and a capillary under the following circumstances. (a) Average blood pressure in the aorta is 100 mm Hg, and the radius of the aorta is 1.0 cm. (b) Average blood pressure in the capillary is 25 mm Hg, and the radius of the capillary is 5.00×10^{-4} cm (5 μm).

Solution.

a. Calculate the wall tension γ in N/m created by blood pressure in the aorta and a capillary under the following circumstances given an average blood pressure in the aorta of 100 mm Hg, and the radius of the aorta of 1.00 cm.

Step by step 1. Use the pressure at depth equation to convert the pressure to newtons per square meter,

$$P_{\text{depth}} = \rho_{\text{fluid}} \cdot g \cdot depth = 13{,}600\frac{\text{kg}}{\text{m}^3} \cdot 9.8\frac{\text{m}}{\text{s}^2} \cdot 0.100 = 1.33 \times 10^4 \frac{\text{N}}{\text{m}^2}$$

Step 2. Use Laplace's law to determine the tension in N/m.

$$\gamma = P \cdot r = 1.33 \times 10^4 \frac{\text{N}}{\text{m}^2} \cdot 0.0100\,\text{m} = 133\frac{\text{N}}{\text{m}}$$

b. **Solution (b)** Calculate the wall tension γ in N/m created by blood pressure in the aorta and a capillary given an average blood pressure in the capillary of 25 mm Hg, and the radius of the capillary is 5.00×10^{-4} cm (5 μm).

Step by step 1. Use the pressure at depth equation to convert the pressure to newtons per square meter,

$$P_{depth} = \rho_{\text{fluid}} \cdot g \cdot depth = 13{,}600\,\frac{\text{kg}}{\text{m}^3} \cdot 9.80\,\frac{\text{m}}{\text{s}^2} \cdot 0.0250\,\text{m} = 3.33 \times 10^3\,\frac{\text{N}}{\text{m}^2}$$

Step 2. Use Laplace's law to determine the tension in N/m.

$$\gamma = P \cdot r = 1.33 \times 10^3\,\frac{\text{N}}{\text{m}^2} \cdot 5.00 \times 10^{-6}\,\text{m} = 6.65 \times 10^{-3}\,\frac{\text{N}}{\text{m}}$$

which is a factor of 20,000 less than the wall tension in the aorta.

The dependence of wall tension in a tube on pressure and radius as expressed by Laplace's law has important ramifications in the circulatory system. For example, aneurysms tend to grow dramatically once they begin, because the ballooning of the vessel makes its radius larger and consequently increases tension in the already weakened wall. Conversely, an enlarged heart (congestive heart failure) must exert a larger force in its walls to create sufficient blood pressure, damaging it further. In a normal heart the walls of the left ventricle are thicker and the ventricle is smaller in radius than the right ventricle, as shown in Figure 7.19. This is consistent with the fact that the left ventricle creates pressures several times greater than the right ventricle.

7.3.7 Energy and Power

Energy and Power Supplied by the Heart—The heart obviously performs work and expends energy. Ultimately, all the energy that the heart expends ends up as thermal energy. Initially, however, the heart gives energy to the blood in three forms. Blood is given kinetic energy as it is accelerated out of the heart into the circulatory system. The heart also supplies gravitational potential energy to blood

Figure 7.19 Cutaway drawing of the heart, showing the left and right ventricles. The left ventricle has thicker walls and a smaller radius, allowing it to produce larger pressures than the right ventricle.

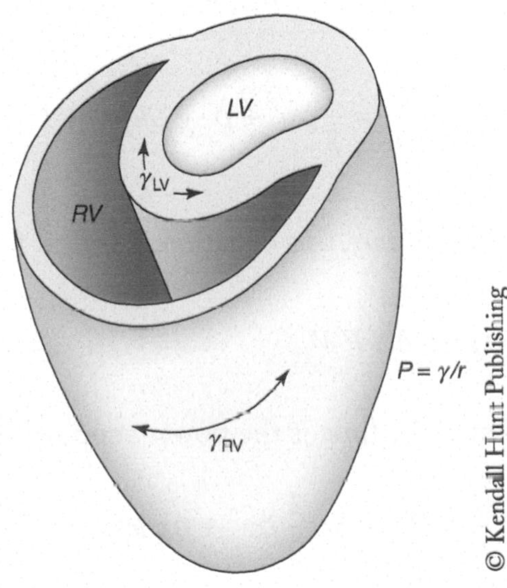

that is pumped to a higher point. Another form of potential energy is supplied to blood when the heart increases blood pressure. As the blood circulates through the body, its energy is converted to thermal energy by resistance to flow—literally by fluid friction.

For fluids, it is convenient to consider the kinetic and potential energies per unit volume. Potential energy per unit volume in a fluid is given by

$$\frac{P.E.}{\text{Volume}} = \rho_{\text{fluid}} \cdot g \cdot depth + P$$

where, as usual, P.E. is potential energy, ρ_{fluid} is the density of the fluid, g is the acceleration of gravity. Consider what happens to the potential energy of blood during a round trip through the circulatory system. Blood pressure is reduced by resistance from an average of 100 mm Hg in the aorta to about 4 mm Hg when blood reenters the heart. The term $P_{\text{depth}} = \rho_{\text{fluid}} \cdot g \cdot depth$ is unchanged in the round trip, but pressure is reduced, so potential energy is lost; it is, in fact, converted to thermal energy as pressure is reduced by resistance. No energy is converted to thermal energy if there is no motion; therefore, $\frac{P.E.}{Volume}$ is constant in a static fluid and pressure depends only on depth in a static fluid, as previously discussed.

The kinetic energy per unit volume in a fluid is

$$\frac{K.E.}{Volume} = \frac{\frac{1}{2}mv^2}{Volume}$$

Because $\frac{\text{m}}{\text{V}}$ is just the density ρ_{fluid}, the kinetic energy per unit volume in a fluid is $\frac{K.E.}{Volume} = \frac{1}{2}\rho_{\text{fluid}}v^2$ eq. 7.3

Adding the kinetic and potential energies per unit volume in a fluid gives the *total energy per unit volume* in a fluid:

$$\frac{Energy}{Volume} = \frac{K.E.}{Volume} + \frac{P.E.}{Volume}$$

thus

$$\frac{Energy}{Volume} = \frac{1}{2}\rho v^2 + \rho \cdot g \cdot depth + P \qquad \text{eq. 7.4}$$

where E is total energy.

To find the *power supplied by the heart*, this equation is solved for the total energy E supplied by the heart:

Step by step 1. $E = (\frac{1}{2}\rho v^2 + \rho g depth + P)V$

Step 2. This energy divided by time is the power supplied by the heart:

$$\frac{E}{t} = Power = (\frac{1}{2}\rho v^2 + \rho g depth + P)\frac{V}{t}$$

Step 3. Volume divided by time is flow rate; therefore,

$$Power = (\frac{1}{2}\rho v^2 + \rho g depth + P) \cdot F.R.$$

where F.R. is the flow rate of the fluid.

Each of the three terms inside the parentheses represents power supplied for a specific purpose. The first and third terms represent power supplied to increase the speed of the blood, v, and its pressure, *P*, respectively. The power used for these two purposes is referred to as the *kinetic power* (associated with increases in speed) and the *pressure power* (associated with increases in pressure), respectively. The second term, $\rho g depth$, is negligible for the heart because depth, the height of the column of blood, is not increased directly by the heart.

Kinetic Power output of the heart is,

$$Kinetic\ Power = (\frac{1}{2}\rho v^2) \cdot F.R.$$

Pressure Power output of the heart is,

$$\text{Pressure Power} = \text{P} \cdot \text{F.R.}$$

EXAMPLE 7.4

Calculate the (a) kinetic power, (b) pressure power, and (c) generated by the left ventricle of the heart for a typical resting adult. Give your answer in watts. The speed of the blood emerging from the left ventricle is 30 cm/s, the flow rate is 83 cm^3/s, the density of blood is 1.05 g/cm^3, and the pressure is 120 mm Hg (1.60×10^4 N/m^2). Note that these values are typical for the left ventricle, and the results include neither the power generated by the right ventricle in pumping blood through the lungs nor that of the atria pumping into the ventricles.

Solution (a). The kinetic power output is

$$Kinetic\ Power = (\frac{1}{2}\rho v^2) \cdot F.R.$$

Step by step 1. Convert density, velocity, and flow rate to standard SI units.

$$\rho = 1.05 \times 10^3 \frac{\text{kg}}{\text{m}^3}; v = 0.300 \frac{\text{m}}{\text{s}}; F.R. = 83.0 \times 10^{-6} \frac{\text{m}^3}{\text{s}}$$

Step 2. Substitute the values and do the math.

$$Kinetic\ Power = \left(\frac{1}{2}\right) 1.05 \times 10^3 \frac{\text{kg}}{\text{m}^3} \left(0.300 \frac{\text{m}}{\text{s}}\right)^2 \left(83.0 \times 10^{-6} \frac{\text{m}^3}{\text{s}}\right) = 3.92 \times 10^{-3}\,\text{W}$$

Solution (b): The Pressure output is

$$\text{Pressure Power} = P \cdot F.R.$$

Step by step 1. Recognize that we must us standard SI units for pressure in order to get power in Watts.

Step 2. Substitute the values and do the math.

$$\text{Pressure Power} = 1.60 \times 10^4 \frac{\text{N}}{\text{m}^2}(83.0 \times 10^{-6} \frac{\text{m}^3}{\text{s}}) = 1.33\ \text{W}$$

Solution (c): The total useful power output is the sum of these two results.

The total power is $3.92 \times 10^{-3}\ \text{W} + 1.33\ \text{w} = 1.33\ \text{W}$ which, rounded to three digits, is the same as the pressure power since the kinetic power is relatively small.

The power output of the left ventricle is not large, but it must supply this power continuously for the entire life of the person. Because the heart rests about two-thirds of the time, the actual power output of the ventricle during contraction is about three times greater.

The total power output of the heart is somewhat larger than that of the left ventricle alone. Table 7.2 gives typical power outputs of the heart for resting and active adults. The values given for an active person are obtained by assuming that the person's blood pressure has increased by 50% and that the total flow rate is four times greater than when the person rests.

These are not the extreme possibilities. During very stressful physical activity the values may exceed those in the table by a factor of 5, and during sleep the values may be less than those shown for a resting adult. For a person at rest, kinetic power is only 10% of pressure power. During activity kinetic power exceeds pressure power, and the total power output of the heart may be as large as 100 W. Muscles are at best 25% efficient, which implies a minimum of 400 W of power produced by the heart alone during extreme conditions.

Kinetic and Potential Energy in a General Fluid—Equation 7.4, $\frac{Energy}{Volume} = \frac{1}{2}\rho v^2 + \rho \cdot g \cdot depth + P$ is more important than the preceding discussion of energy and power of the heart might imply. It can be used, for example, to explain the Bernoulli effect, the decrease in pressure when fluid speed v increases. Consider what happens if resistance is negligible (and thus $\frac{Energy}{Volume}$ is constant) and the tube is horizontal, so that h. is also constant. Then when v increases, P must decrease so that all the terms on the right-hand side of Equation 7.4 add up to

Table 7.2 Typical Power Outputs of the Heart

	Resting (W)	Active (W)
Kinetic power	0.15	9.6
Pressure power	1.50	9.0
Totals	1.65	18.6

Underline indicates addition of values.

the same value of $\frac{Energy}{Volume}$. In order to conserve energy, the energy needed to increase the speed of the fluid comes from a decrease in the pressure. Thus, when the velocity of a fluid increases, its internal pressure must decrease.

Equation 7.4 also applies to 1:1 fluid in which resistance is not negligible, as in the circulatory system. When there is significant resistance, then the energy per unit volume, $\frac{Energy}{Volume}$, decreases as the fluid flows; the energy is converted to thermal energy. If p and h are constant in a flowing fluid, then the speed v is determined solely by the tube diameter via $F.R. = A\bar{v}$ where A is the cross-sectional area and $\bar{v}$ is the average velocity. The reduction in energy therefore comes from a reduction in pressure. In other words, Equation 7.4 is the basis for the by now familiar statement that resistance decreases pressure.

The flow of gases is more complicated than that of liquids because gas densities vary as a result of their being compressible. However, Equation 7.4 is still valid. Finally, in a static fluid $v = 0$, and Equation 7.4 can be used to calculate the relationship between pressure and depth in a fluid, be it a liquid or a compressible gas.

7.4 THE PHYSICS OF RESPIRATION

The respiratory system's most important function is to oxygenate blood and to remove carbon dioxide from it. Oxygen is needed to combine chemically with food, and carbon dioxide is the primary gaseous waste product of cells. The respiratory system also exchanges heat with the environment by the evaporation of water and by forced convection. Heat is usually transferred out of the body by the respiratory system, but it may bring heat in if the surrounding temperature and humidity are high. Of course, another function of the respiratory system is to make sounds, as in talking and singing.

7.4.1 The Breathing Mechanism

Air flows into the lungs when the pressure in them is less than atmospheric and out of the lungs when the pressure is greater than atmospheric (provided that the breathing passages are unobstructed). The flow of air obeys the familiar equation

$$Flow \text{ rate} = \frac{\Delta P}{R}$$

where ΔP is the pressure difference between the lungs atmospheric pressure. R is the resistance of the breathing passages to air flow. Air flows out of the lungs when the pressure is greater than atmospheric and into the lungs when pressure is less than atmospheric. Remember, pressure pushes. Air is pushed from higher pressure regions to lower pressure regions.

The body changes pressure in the lungs by increasing or decreasing the volume surrounding the lungs. During inhalation, the lungs are expanded by muscle action in the diaphragm and rib cage (see Figure 7.20). The diaphragm is a sheet of muscle lying below the lungs. When relaxed it has an upward curvature, and when contracted it moves downward, expanding the lungs. At the same time, certain muscles in the upper body also contract and increase the size of the rib cage, further increasing the volume of the lungs. Pressure in the lungs falls and is pushed into them.

Figure 7.20 There stages in the breathing cycle. (a) At rest, the pressure in the lungs in zero. A negative thoracic pressure keeps lungs open. (b) During inhalation the chest expands, the diaphragm moves downward, and pressures drop. (c) During exhalation the chest contracts, the diaphragm rises, and pressures rise.

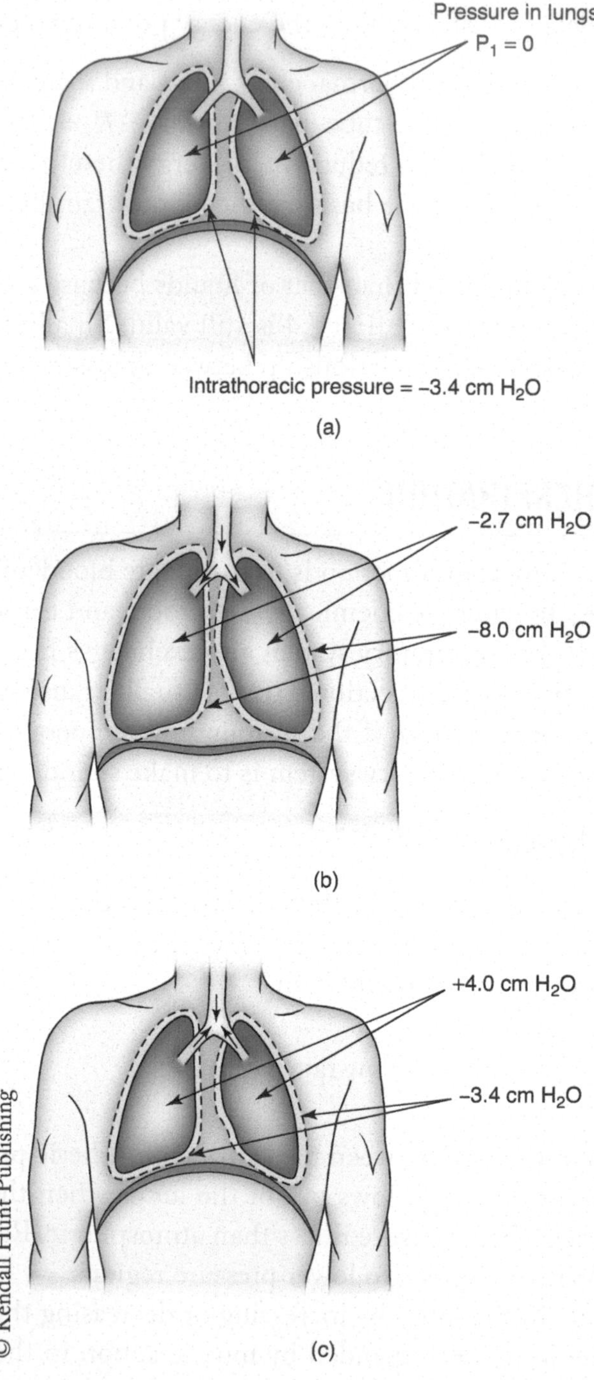

During exhalation the muscles relax, and the diaphragm and rib cage return to their original shapes and sizes. The volume of the lungs decreases and the pressure in them consequently rises. Air is thus pushed out of the lungs. No muscle action is required to exhale, but it may be used to speed exhalation by forcibly contracting the rib cage.

Cohesive and adhesive forces in the liquids in and around the lungs play important roles in the breathing mechanism. The lungs are attached to the chest walls by liquid contact—that is, by adhesive forces. The lungs are pulled open by these adhesive forces when the rib cage expands and the diaphragm moves downward. As seen in Figure 7.20, the pressure in the region between the lungs and the chest wall rises and falls but remains negative. The pressure is called intrathoracic pressure. Intrathoracic pressure may become positive during very forceful exhalation, such as in blowing up a balloon.

If intrathoracic pressure outside the lungs is negative during normal breathing, how does pressure inside the lungs become positive to cause exhalation? The answer is that surface tension in the alveoli (see the related discussion in Section 6. 7) makes the lungs elastic like an inflated balloon. Intrathoracic pressure must in fact be negative to keep the lungs expanded. Any injury that allows air into the chest cavity increases pressure to atmospheric pressure and forces the lungs to collapse due to their surface tension. This condition is called pneumothorax; it can be caused by an accident or by surgery or may occur spontaneously (see Figure 7.21(a) and (b)). Air is suctioned out of the intrathoracic region to remedy severe pneumothorax, although the body can slowly reabsorb the air in less severe cases.

7.4.2 Interaction between the Respiratory and Circulatory Systems

Gas transfer into and out of the blood is a primary function of the respiratory system. Figure 7.21 illustrates the branching of breathing passages in the lungs into smaller and smaller tubes ending in alveoli, where the exchange of gases takes place. The hundreds of millions of alveoli are each about 200 μm in diameter and have walls about 0.4 μm thick. Blood passing through capillaries adjacent

Figure 7.21a Normal and collapsed lung.

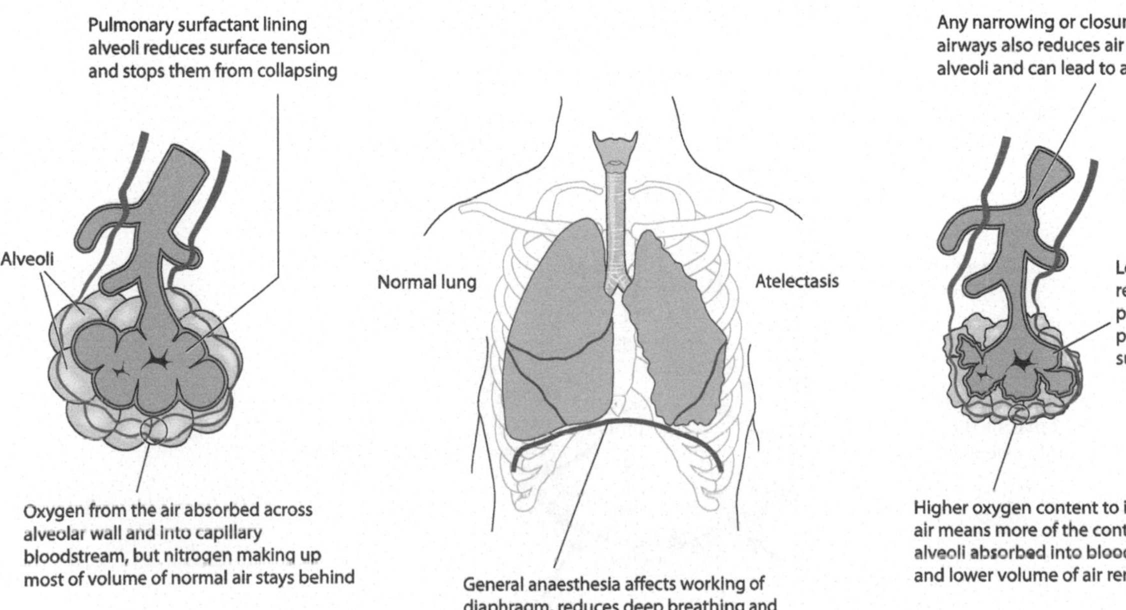
Pulmonary surfactant lining alveoli reduces surface tension and stops them from collapsing
Alveoli
Oxygen from the air absorbed across alveolar wall and into capillary bloodstream, but nitrogen making up most of volume of normal air stays behind
Normal lung
Atelectasis
General anaesthesia affects working of diaphragm, reduces deep breathing and coughing (affecting expansion of alveoli)
Any narrowing or closure of airways also reduces air reaching alveoli and can lead to atelectasis
Less expansion reduces production of pulmonary surfactant
Higher oxygen content to inspired air means more of the contents of alveoli absorbed into bloodstream and lower volume of air remains

Figure 7.21b Pneumothoreax: a collapsed lung caused by air entering the thoracic cavity. Surface tension in the lung causes it to contract like a punctured balloon.

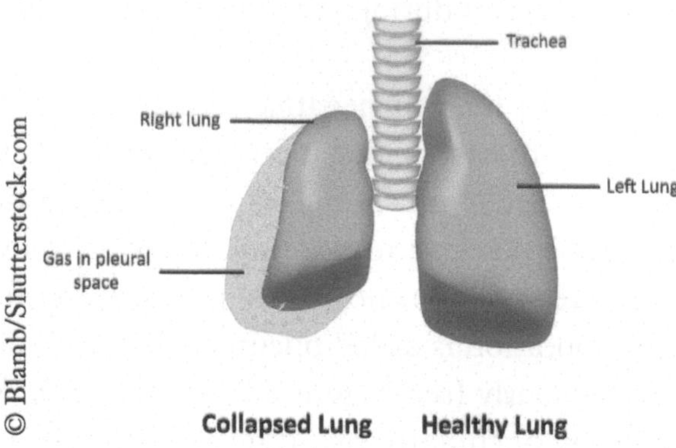

to alveoli exchanges gases with air in the alveoli by diffusion through capillary and alveolar walls (see Figure 7.22).

Diffusion of gases between air in the lungs and blood proceeds in the direction from high to low concentration, and the rate of diffusion is greatest when the difference in concentration is greatest. Diffusion obeys Fick's law here, but the actual rate of exchange is greatly affected by the presence of hemoglobin in the blood. Hemoglobin has a chemical affinity for oxygen, and its presence allows the blood to carry far more oxygen than it otherwise could. Because of hemoglobin, the rate of exchange also depends on blood acidity, the presence of carbon dioxide, and temperature. All these

Figure 7.22 Detail of the transfer of gases between an alveolus and capillary.

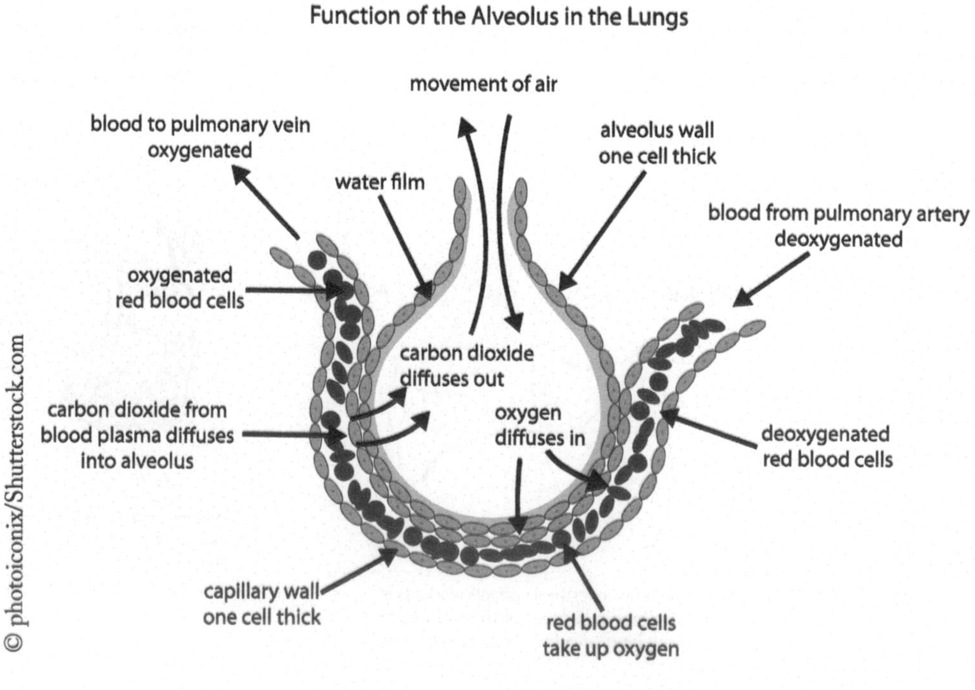

dependencies are cleverly designed to enhance oxygen transfer when it is most needed. For our purposes it is sufficient to think of the gas transfer as if it were simple diffusion.

Since the relative concentration determines the direction of gas transfer, it is useful to have a way of expressing concentrations numerically. One way is to use the concept of partial pressure. If a mixture of gases occupies a given volume, the partial pressure of a gas is defined as the pressure that it would exert if it alone occupied the entire volume. An experimental law formulated by John Dalton (1766–1844), known as Dalton's law of partial pressures, states that the total pressure due to a mixture of gases is equal to the sum of the partial pressures due to each gas. Another equivalent statement of Dalton's law is that each gas in a mixture exerts a partial pressure proportional to its molecular concentration. This last statement of Dalton's law makes it clear that partial pressure is directly related to concentration and can be used as a measure of concentration.

In practice, partial pressures in millimeters of mercury are often quoted to indicate the concentration of a gas. For example, about 20% of atmospheric molecules are oxygen molecules. The partial pressure of oxygen is thus 20% of the total atmospheric pressure. If atmospheric pressure is 760 mm Hg, then the partial pressure of oxygen would be about 150 mm Hg. This is often expressed as P_{O_2}. $P_{O_2} = 150 \text{ mm Hg}$. If air in the lungs has a $P_{O_2} = 105 \text{ mm Hg}$ and the blood has $P_{O_2} = 40 \text{ mm Hg}$, then the concentration of oxygen in the lungs is much greater than in the blood, and oxygen will be transferred into the blood.

Figure 7.23 schematically represents gas exchange with the circulatory system. The partial pressure of oxygen in the lungs is only 105 mm Hg, less than that in the atmosphere, because air in the

Figure 7.23 Schematic of the interchange of gases between the lungs and blood and between the blood and tissue.

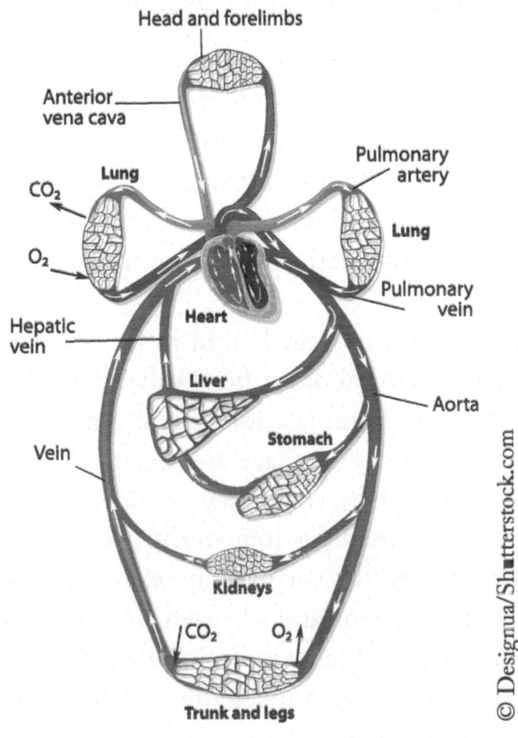

lungs is only partially replaced with each breath. Similarly, the partial pressure of carbon dioxide in the lungs is greater than in the atmosphere. Gases move in the directions of the arrows, in each case from high to low concentration (partial pressure). Oxygenated blood travels through the arteries to the capillaries, where it is transferred to cells having lower oxygen concentration. Oxygen-poor blood travels back through the veins to the lungs, where the process begins again. A similar cycle for the removal of carbon dioxide is also shown.

Diffusion in liquids and through membranes is a slow process. Hemoglobin greatly increases the oxygen capacity of blood, and furthermore excess carbon dioxide causes hemoglobin to release oxygen. Diseases that affect membrane thicknesses (such as fibrosis of the lungs) or the affinity of blood for oxygen (such as carbon monoxide poisoning) reduce the effectiveness of the respiratory system. Measurements of P_{O_2} and P_{CO_2} can indicate the presence of such diseases and other problems. Corrective measures are sometimes possible. For example, hyaline membrane disease in premature infants is due to a lack of surfactant in the lining of the lungs. A surfactant reduces surface tension in the alveoli so that they can inflate properly. Infants suffering this disease show low arterial P_{O_2} and can be treated by enriching the oxygen content of their air. In the early years of this treatment the oxygen level was sometimes raised too high and brain damage resulted. The blood P_{O_2} of such infants is today closely monitored to ensure that it is neither too low nor too high.

7.5 MEDICAL INSTRUMENTATION AND DEVICES RELATED TO FLUIDS

A number of devices associated with fluids have already been discussed in this and Chapter 6. Progress in medical instrumentation is occurring so rapidly that any list of specific instruments quickly becomes obsolete. Rather than emphasize details of specific instruments, this section will look at general features and basic principles of operation.

7.5.1 Devices Associated with the Circulatory System

Central venous pressure (CVP) is the pressure in the major veins that return blood to the heart. As seen in Figure 7.13, CVP is normally about 4 mm Hg. High CVP may indicate that too much fluid has been given in a transfusion, and low CVP is an indication of low fluid balance, as in a patient with internal hemorrhaging. Because the major veins lie deep within the body, CVP must be measured by inserting a catheter. Very small pressure-sensing devices that create an electrical output are sometimes built into the catheter. Alternatively, CVP is transmitted through liquid in the catheter to an external measuring device.

The measuring device must be at the same height as the right atrium, as in Figure 7.24a.

Heart valve replacements using natural or artificial valves have been practical for many years. (see Figure 7.25) Both types function well and last for several years. Natural valves from pigs or human donors can be used since they are mostly cartilage and need not be tissue-matched. Muscle action is not needed to operate replacement valves, since they are opened and closed by pressure differences.

Artificial hearts are being developed for long-term use. The Jarvik artificial hearts implanted in Barney Clark and William Schroeder are famous examples of efforts in this direction. Most artificial hearts are external to the patient and are used to pump and oxygenate blood during open heart surgery. There are engineering problems in designing artificial hearts to replace a natural heart for long periods of time. One is designing a mechanical pump that could do the work of the heart

Figure 7.24 Measurement of central venous pressure (CVP). The IV bottle supplies sterile saline solution to the manometer to a height larger than CVP. Valve A is closed and valve B is opened; fluid flows into the patient until fluid height in the manometer is indicative of CVP. This procedure is followed to prevent flow of blood into the manometer.

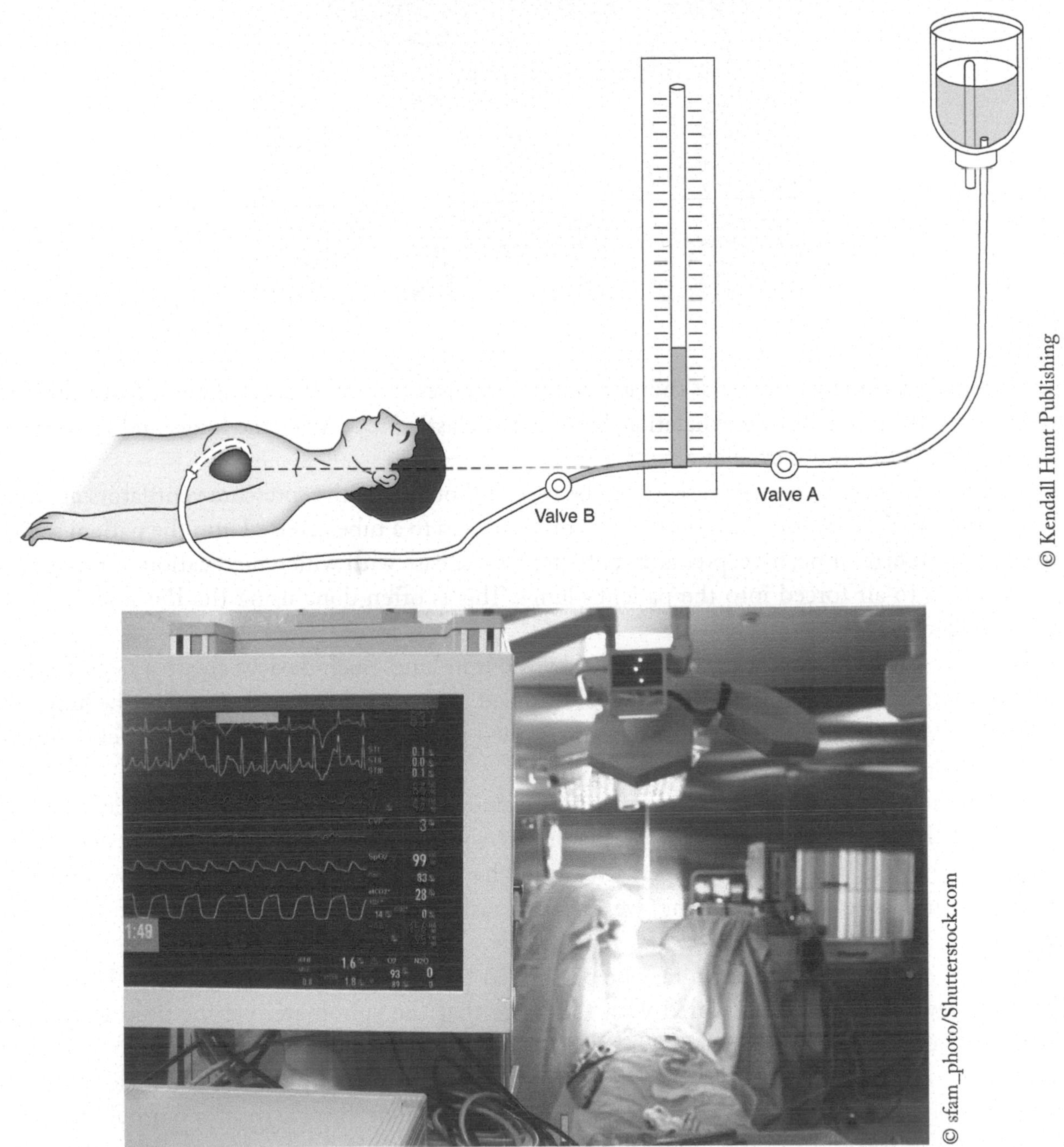

continuously for several years. Another problem is that of supplying power to the heart at the necessary rate and keeping the power supply small enough to be portable.

Devices Associated with the Respiratory System Ventilators and respirators are used to move air in and out of the lungs. Any circumstance that seriously impairs the ability of a person to breathe,

Figure 7.25 Implanting a mechanical heart valve.

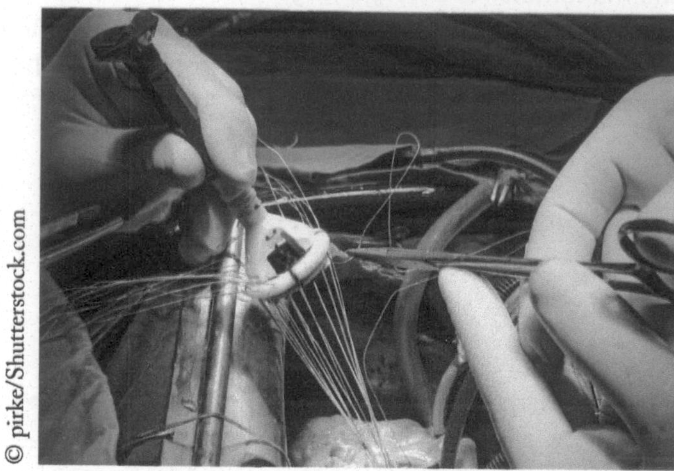

© pirke/Shutterstock.com

such as injury, a comatose state, or thoracic surgery, requires the use of a ventilator. Most ventilators do not assist the lungs during exhalation, because the elastic property of the lungs and chest wall is usually sufficient to cause exhalation.

Either positive or negative pressure can be used, although positive-pressure ventilators are much more common. Positive-pressure ventilators are connected to a tube inserted into the patient's windpipe. An advantage of positive-pressure ventilators is the ease with which medication or oxygen, say, can be added to air forced into the patient's lungs. This is often done using the Bernoulli effect or with nebulizers.

Negative-pressure respirators are typified by the iron lung. Such devices create a negative pressure over the entire body of the patient except the head. Air pressure is thus lowered in the lungs, and air flows in. As with positive-pressure devices, applied pressure is returned to atmospheric pressure to allow the patient to exhale.

Respiration therapy has become very sophisticated, as have so many areas of medical care. An example of this sophistication is that ventilators are sometimes programed to cause a deep breath every few minutes. This is done to open alveoli that have closed during ordinary shallow breathing. During normal unassisted breathing, we unconsciously take periodic deep breaths for the same reason.

Measurement of lung volumes may be performed using a spirometer, as shown in Figure 7.26. There are several volumes associated with the lungs that the spirometer can be used to measure. Tidal volume is defined as the volume inhaled in a normal breath. This is about 500 cm^3 at rest but significantly more during exercise. After taking a normal breath, people can still inhale quite a bit more air. This additional volume is the inspiratory reserve volume. Similarly, after normal exhalation, more air can be forced from the lungs. This additional volume is the expiratory reserve volume. The maximum amount of air that can be exhaled is called the vital capacity. These lung volumes can indicate the health of the lungs, their ability to tolerate strenuous exercise, and so forth.

Figure 7.27 is a picture of a *rotameter*, a device for measuring gas flow rates. Pressure on the bottom of the float is greater than on the top, so the net force due to the gas is upward. The tube is slightly tapered so that the higher the float gets, the easier it is for gas to get around it. The actual

Figure 7.26 Spirometer testing pulmonary capacity.

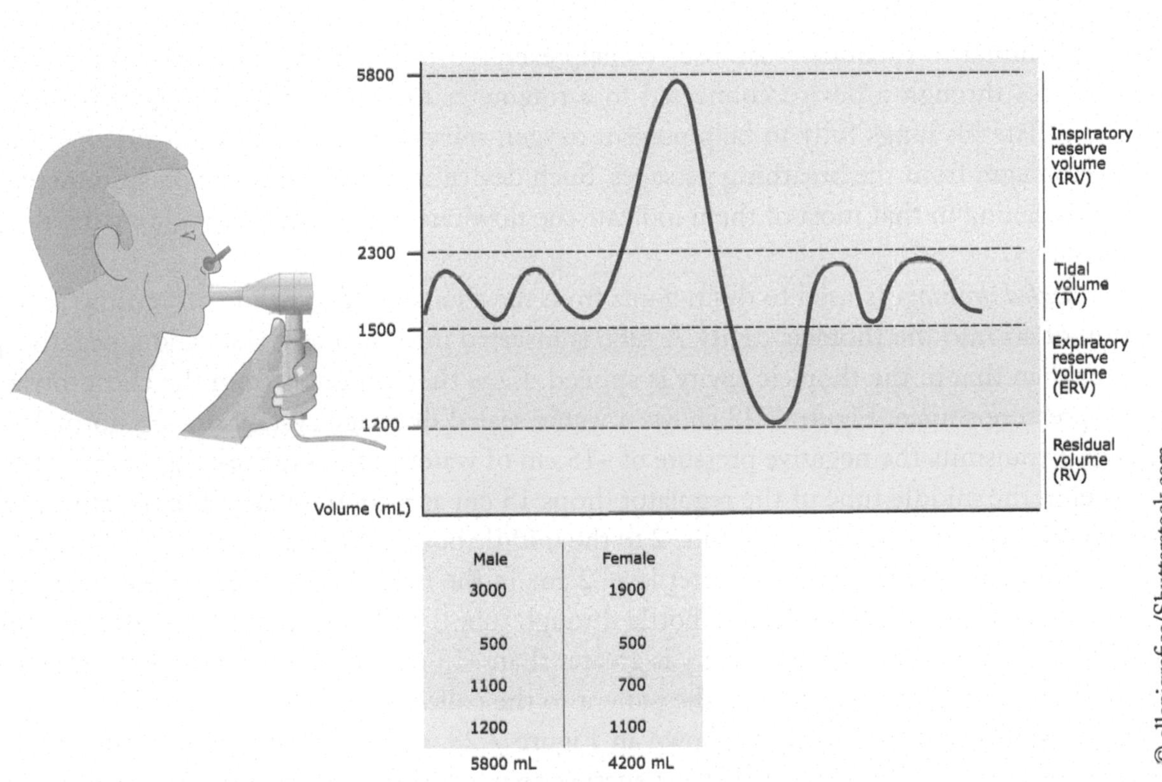

Male	Female
3000	1900
500	500
1100	700
1200	1100
5800 mL	4200 mL

Figure 7.27 Rotameter used to test gas flow rates.

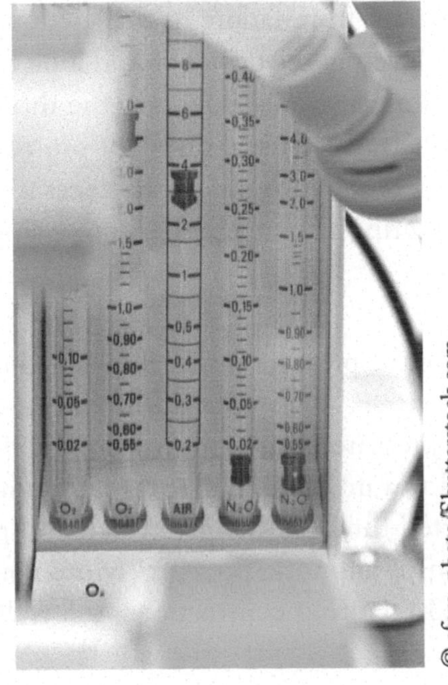

height to which the float rises depends on the flow rate and many other factors. Among them are the density of the gas, the size of the tube, and the size, shape, and mass of the float. Because of their sensitivity to gas density, rotameters designed for one gas must not be used to measure the flow rate of another gas.

Devices similar to rotameters are used to help exercise the lungs of bed-confined patients. The patient inhales through a device connected to a rotameter to raise a float. The goal is to get the patient to inflate his lungs fully to help prevent oxygen starvation and carbon dioxide buildup and to remove phlegm from the breathing passages. Such devices are called inspiration spirometers. The name is misleading in that most of them indicate the flow rate, rather than the volume inhaled as in a spirometer.

Water-sealed drainage is used to drain fluids from the thoracic cavity while preventing accidental admission of air into the thoracic cavity. A tube is inserted into the chest wall, and a negative pressure lower than that in the thoracic cavity is applied. Flow then proceeds from the chest toward the negative-pressure source. Figure 7.28 shows a water-sealed drainage system during normal operation. Tube 1 transmits the negative pressure of -15 cm of water to the pressure-regulator bottle. The water level in the middle tube of the regulator drops 15 cm as a consequence. The negative pressure is transmitted undiminished through tube 2 to the middle bottle, the water seal. Part of the pressure is used in the water seal to lower the water level 2 cm, in the right-hand tube. A pressure of -13 cm of water is transmitted to the collection bottle through tube 3 and from there to the patient through tube 4. The pressure in the thoracic cavity is greater than -13 cm of water, varying from about -3 to -8 cm, so flow is in the direction from the patient to the collection bottle.

The pressure-regulator bottle, as shown in Figure 7.28, prevents the pressure from becoming more negative than -20 cm of water. If the suction source becomes more negative than -20 cm of water, then the atmosphere pushes air into the regulator bottle through the center tube. Air flows in so quickly that the pressure can become no lower. This only limits the pressure, but if the center tube is raised or lowered, it too can regulate the pressure. For example, if the center tube is raised so that 12 cm of it is below water, then the most negative pressure that can be exerted is -12 cm of water.

The water seal protects the patient from failure of the suction device or a break in tubes 1 or 2. Figure 7.29 shows what the water levels in the device would look like if tube 2 broke or if the suction source failed. Without the water seal, air would flow into the thoracic cavity, raising the pressure to atmospheric (zero gauge pressure). The water seal does not protect the patient from breaks in tubes 3 or 4 or from the loss of a seal where the tube enters the chest wall. Modern water-sealed drainage devices are built as a unit, incorporating all three bottles into a single structure.

7.5.2 Labor Monitors

Labor monitors are used to indicate the pressure in the fluid surrounding a fetus (the amniotic fluid) during labor. The uterine walls contract repeatedly during labor, creating pressure in the amniotic fluid. The timing and amount of pressure created during contractions are valuable indicators of the progress of the mother and child toward birth. Labor monitors may be applied internally or externally. Internal monitors are inserted through the birth canal and placed on the amniotic sac. These provide a nearly direct indication of amniotic pressure. More commonly, labor monitors are used externally, as shown in Figure 7.30. During contractions, the wall muscles exert forces along their

Figure 7.28 Water-sealed drainage operating normally. A pressure of –15 cm of water is transmitted through the regulator bottle to the water seal. Two centimeters of pressure is used to lower the water level 2 cm in the water seal; thus –13 cm of water pressure is transmitted to the patient.

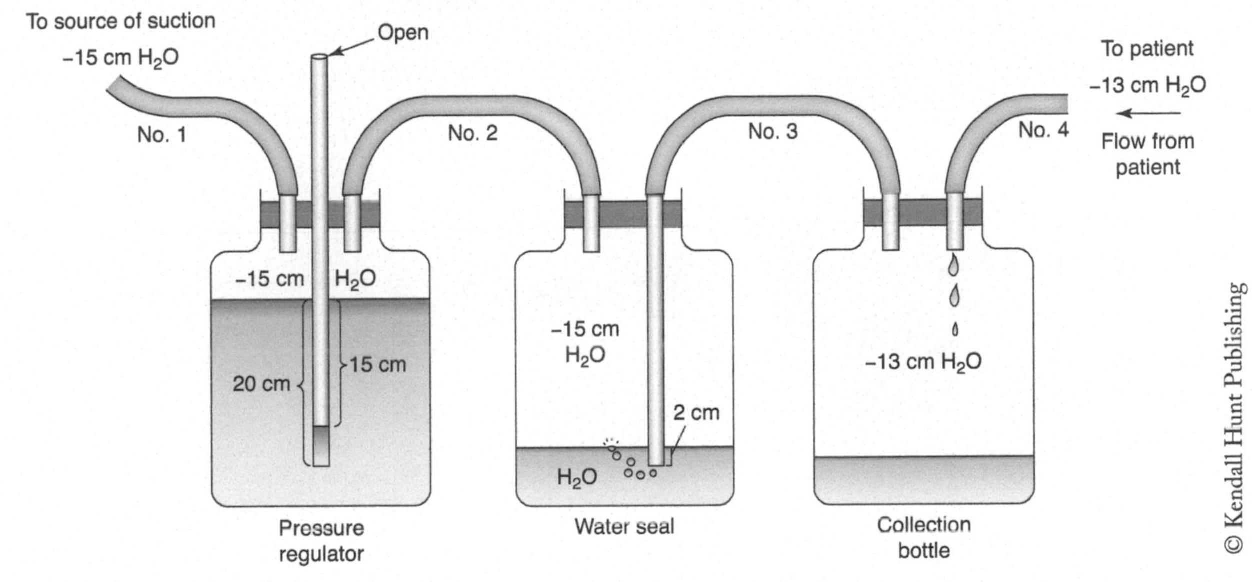

Figure 7.29 The water seal protects the patient from a loss of suction ahead of the seal—in this case a break in tube 2. Intrathoracic pressure will be transmitted back to the water seal, causing the water level to vary in height from 3 to 8 cm during exhalation and inhalation, respectively. Atmospheric pressure cannot force the water any higher, and the flow of air into the thoracic cavity is prevented.

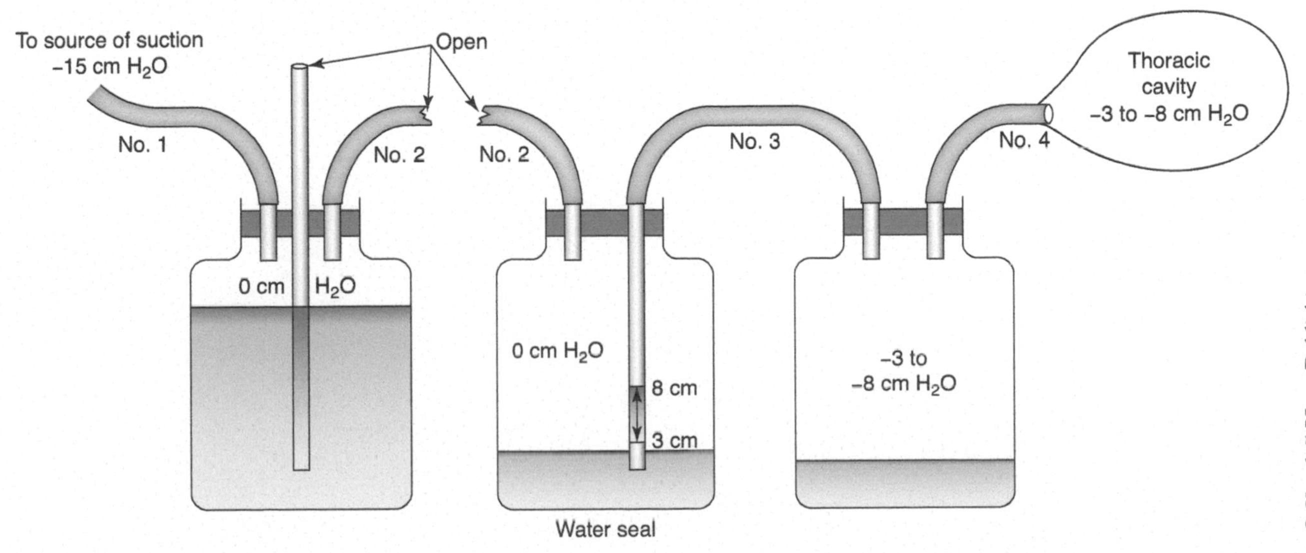

Figure 7.30 An external labor monitor is strapped to the abdomen of a pregnant woman.

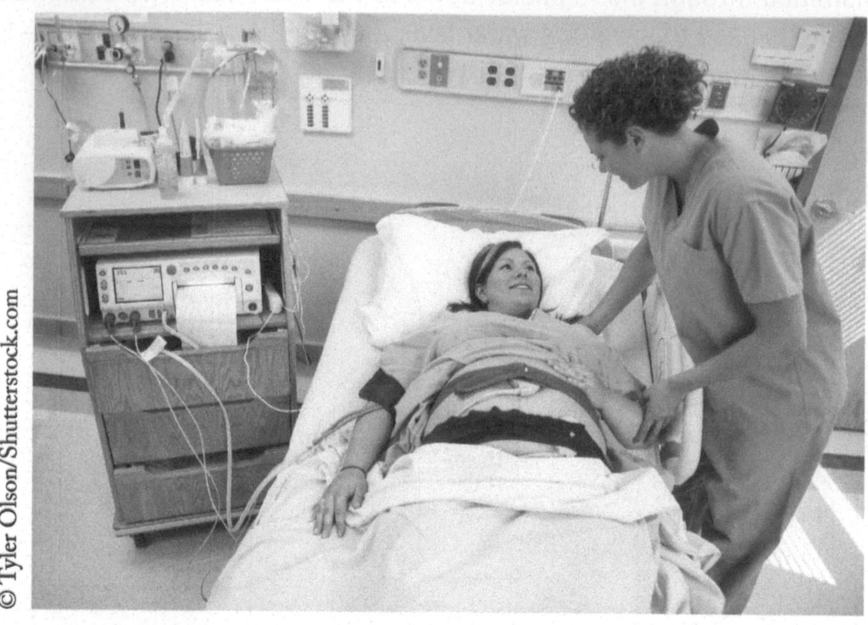

lengths. This is considered a tension that creates pressure in the fluid. The external monitor is pushed upward, and a transducer converts the movement to an electrical signal that is recorded. Several other instruments and devices related to fluids have been discussed in this and Chapter 6. Many others exist as well but are not included because of space limitations. The basic principles of fluids can be used to understand the functioning of all such devices.

Questions

7.1 When a blockage of fluid flow occurs in the body, such as in the eye, brain, or gastrointestinal (GI) system, why does the pressure upstream build to higher than normal levels?

7.2 Explain why the pressure in the cerebrospinal fluid surrounding the brain cannot be measured by placing a pressure sensor on the skull. Is this an exception to Pascal's principle? Conversely, why is it possible to measure arterial blood pressure with an external sensor?

7.3 New parents spend a great deal of time burping their babies. How does belching help relieve the babies' cramps and allow them to eat more at one feeding?

7.4 Toe dancing is much harder on toes than normal walking. Explain in terms of the pressure on the toes.

7.5 One danger in surgically removing an intestinal blockage is that the intestine may burst explosively when cut, risking abdominal infection. To avoid this danger, such surgery is sometimes performed in a pressurized chamber. Explain how this helps.

7.6 The Eustachian tube runs from the middle ear to the nasal passage. The purpose of the Eustachian tube is to equalize pressure in the middle ear with atmospheric pressure. (a) Explain in terms of the physics of fluid flow how the equalization would occur if a person is suddenly subjected to a higher or lower air pressure. (b) Table 7.1 lists the gauge pressure in the middle ear to be less than 1 mm Hg. Why is it important that the pressure in the middle ear be neither too positive nor too negative?

7.7 Define diffusion, osmosis, dialysis, filtration, and active transport.

7.8 There are no blood vessels in the cornea. Suggest a method for oxygen to reach the cornea from the atmosphere through, the tear layer. Is there a potential problem for people who wear nonporous contact lenses? If so, explain why it is possible to wear contacts during the day but not while sleeping.

7.9 Water molecules are small compared to most·other molecules. Is this consistent with the fact that most membranes are permeable to water?

7.10 What is the difference between osmotic pressure and relative osmotic pressure?

7.11 The salinity of ocean water is greater than that of blood, and the salinity of fresh water is less than that of blood. If the red blood cells in the lungs of a drowning victim have burst because of osmotic transfer of water into them, was the victim drowned in the ocean or in fresh water?

7.12 Compare the processes used by an artificial kidney with·those of a real one, and comment on the reasons why the real one is far superior.

7.13 Under·what circumstances can it be said with certainty that active transport is taking place?

7.14 Why is it inadvisable to drink ocean water? What is the net effect on the water content of the body if a, large amount of salt water is consumed?

7.15 People whose legs are paralyzed (paraplegics) often suffer from swelling in the lower extremities. Why is this, and what is a possible remedy?

7.16 Why don't people breathe by absorbing oxygen through the skin and. diffusing carbon dioxide out through the skin? (Actually, a small percentage of normal oxygen intake is accomplished by this method.)

7.17 Explain how the periodic variations in blood pressure due to heartbeats are smoothed out by the time blood reaches the venous system (the veins).

7.18 If a person's diastolic pressure rises during exercise, it is considered a sign that the arterioles are not dilating. How would this cause the diastolic pressure to rise during exercise?

7.19 What causes blood velocity to increase when blood goes from the capillaries to the veins?

7.20 The veins on the back of a person's hands bulge noticeably when the hand is lower than the heart but not when the hand is held over the head. Explain why.

7.21 It seems natural that newborn infants should have lower blood pressure than full-grown adults. Why don't infants need blood pressure as high as an adult's?

7.22 Some people suffer from a disease in which they have an abnormally high blood cell count, meaning that the number of blood cells per unit volume of liquid is very high. What is the likely effect of this on the viscosity of the blood and its flow rate?

7.23 What is the effect of turbulence on flow rate? What are some circumstances in which turbulence is easily detected by the sound it makes?

7.24 What are two possible reasons that plaque buildup in blood vessels (arteriosclerosis) causes turbulence?

7.25 If a fluid is flowing through a tube at a given flow rate, will the tube heat up more when the flow is turbulent or when it is laminar? Explain.

7.26 What role does streamlining play in the design of cars and airplanes?

7.27 If you wished to pump air into a patient's lungs by squeezing on a balloon (connected to their ventilator), would you choose a balloon of large or small diameter to get the maximum pressure? To get the maximum volume?

7.28 Large balloons, such as weather balloons, do not need large pressures to be inflated. Why is this?

7.29 Major veins passing through leg muscles are not as large in radius as some other major vessels in the body. Does this hinder or aid the action of the "muscle pump" in returning blood to the heart (as illustrated in Figure 7.16)?

7.30 An increase in pressure requires energy. What is the source of the energy that increases pressure with depth in the oceans? In the body?

7.31 If a container springs a leak, as shown in Figure 7.31, to what maximum height h can the fluid squirt? Explain in terms of energy.

7.32 During inhalation do the lungs pull air in or does the atmosphere push air in?

7.33 The Heimlich maneuver is a process by which an obstruction in the windpipe (trachea) can be dislodged by applying a sudden upward force to the victim's abdomen just below the diaphragm. Explain how the force is transmitted from the person who applies it to the obstruction in the windpipe.

7.34 Intrathoracic pressure normally remains negative. Explain how pressure in the lungs can become positive (to cause exhalation) without muscle action.

7.35 In asthma, the large airways are restricted and reduced in diameter, particularly during an attack. What are the likely effects of asthma on the ability of the victim to breathe?

7.36 Very large doses of radiation, such as those used to treat lung cancer, can cause fibrosis of the lungs. Fibrosis is a thickening of alveolar walls. What is the likely effect of this condition on the ability to inhale and exhale? On gas exchange?

7.37 In emphysema, the alveoli join to form fewer and larger sacs. Victims of emphysema are frequently barrel-chested because their lungs are more than 50% inflated at all times. They are also unable to blowout a match. What is the explanation for these symptoms?

7.38 Mouth-to-mouth resuscitation is possible because there is more oxygen in exhaled air than in blood in the lungs. That is, the P_{O_2} of exhaled air is greater than the P_{O_2} of blood in the lungs. Why is this so? (Note that the victim has not been breathing for some time.)

7.39 When an external manometer is used to measure CVP, as in Figure 7.24, sterile saline solution is used to transmit pressure to the manometer. Why not use air or mercury?

7.40 Intrathoracic pressure can become positive when a person makes a large effort to blow up a balloon. If the effort is prolonged, the veins on the neck will bulge and blood flow in the venae cavae will be diminished. Explain in terms of the magnitude of central venous pressure and the pressure in the venae cavae (the major veins returning blood to the heart).

Figure 7.31

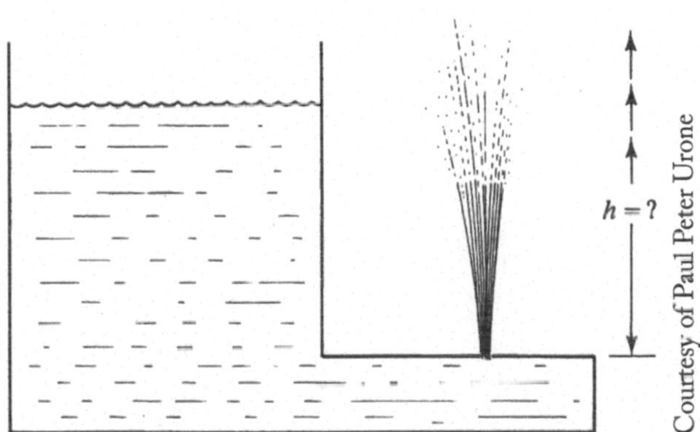

7.41 Why is the pressure on the bottom of a float in a rotameter, as in Figure 7.27, larger than the pressure on the top of the float?

7.42 Explain why the pressure transmitted to the patient in the water-sealed drainage system shown in Figure 7.28 is changed from −15 to −13 cm of water during passage through the water seal. Explain in terms of the pressure due to the weight of a fluid.

7.43 By comparing pressures inside and outside of the water-seal bottle in Figure 7.29, explain why the water level in the tube fluctuates between 3 and 8 cm.

7.44 Why can't atmospheric pressure force water back into the patient in the failed water-sealed drainage system shown in Figure 7.29?

PROBLEMS

SECTION 7.1

7.1 (I) The pressure in the esophagus is 10 mm Hg. Convert that pressure to centimeters of water, as it would be measured by a water manometer.

7.2 (I) Calculate the average pressure exerted on the palm of a shot-putter's hand by a shot if the area of contact is 50 cm^2 and if he exerts a net force of 800 N on the shot. Calculate the pressure in N/m^2 and convert it to mm Hg, and compare it with the 7600-mm Hg pressures sometimes encountered in the skeletal system.

7.3 (II) As discussed in, Section 3.4 and illustrated in Figure 3.20, proper lifting should be done with the torso erect. (a) Calculate the pressure exerted on a disk in the lower back having an area of 20 cm^2 if a woman stands erect and supports the weight of her upper torso on that disk. The mass of the upper torso is 25 kg. Express this pressure in mm Hg. (b) If this same person lifts a load with her back, the force exerted on the disk may be as large as 5000 N! Calculate the pressure on the disk for the 5000-N force and compare this pressure with that calculated in part (a). Is it surprising that such a pressure can cause a slipped or crushed disk?

7.4 (II) If the pressure in the cerebrospinal fluid (CSF) surrounding the brain of an infant rises to 85 mm Hg (5–12 mm Hg is normal), the resulting outward force will cause the skull to grow large. Calculate the outward force in newtons on each side of such an infant's skull if the effective area of each side is 75 cm^2.

7.5 (II) If the maximum force that an eardrum can withstand without breaking is 3.0 N and the area of the eardrum is 1.0 cm^2: (a) Calculate the maximum tolerable pressure in the middle ear in N/m2 and convert this to mm Hg. When might the pressure in the middle ear become greater than the air pressure outside the ear? (b) To what maximum depth could a person dive in fresh water without bursting an eardrum?

7.6 (II) (a) Calculate the force, in newtons, on the, back wall of a normal eye of area 10 cm^2 subjected to an intraocular pressure of 18 mm Hg. (b) How large is this force if the pressure rises to 85 mm Hg?

7.7 (II) Suppose the pressure in a patient's stomach is 15 mm Hg. To what height h must the food in a nasogastric feeding arrangement be raised in order for the pressure due to the weight of the food to be twice the pressure in the stomach? The density of the food is 1.2 g/cm^2 (see Figure 7.4).

7.8 (II) A full-term fetus may have a mass of 3.5 kg. If the entire weight of the fetus rests momentarily on, the mother's bladder, supported on an area of 90 cm^2, calculate the pressure exerted in mm Hg. Is this pressure great enough to trigger the micturition reflex?

7.9 (III) A patient is being fed through a nasogastric tube as shown in *Figure 7.4*. A flow rate of 100 cm^3/min is observed when the food is a height $h = 45$ cm above the level of the stomach and the pressure in the stomach is 10 mm Hg. Calculate the new flow rate if the food is raised to a new height of 75 cm. The food density is 1.2 g/cm^3.

SECTION 7.3

7.10 What is the flow rate in cubic centimeters per second and liters per minute in an aorta of radius 0.90 cm if the average blood velocity in it is 75 cm/s?

7.11 Calculate the flow rate through a capillary if its radius is 5.0 μm and the average blood velocity is 3.0×10^{-2} cm/s.

7.12 (I) Referring to Example 7.2, part (c), in which the total cross-sectional area of the capillary system was found to be 3.1×10^3 cm^2, calculate the total number of capillaries in the body given their average radius is 5.0 μm. Note that the result of this calculation overestimates the number of capillaries by about a factor of 2 since there is an assumption that all of the blood in the circulatory system flows through a capillary.

7.13 (II) What is the average velocity of blood in the two major veins that return blood to the heart (the venae cavae) if their radii are each 1.4 cm and the combined flow through them is 80 cm^3/s?

7.14 (II) (a) Calculate the average blood velocity in an artery supplying the brain if its radius is 0.40 cm and the flow rate through it is 5.0 cm^3/s. (b) What is the average velocity at a constriction in the artery if that constriction reduces the radius to 0.15 cm? Assume the same flow rate as in part (a). Is there a possibility that the constriction could be detected using a stethoscope? Explain.

7.15 (II) (a) Neglecting the pressure drop due to resistance, calculate the blood pressure in mm Hg in an artery in the brain 30 cm above the heart. The pressure at the heart is 120 mm Hg and the density of blood is 1.05 g/cm^3 (b) Using the same assumptions, calculate the blood pressure in an artery in the foot 1.6 m below the heart.

7.16 (II) What minimum blood pressure in mm Hg at the heart is required to produce an arterial blood pressure of 60 mm Hg in the brain if the brain is 32 cm above the heart and blood density is 1.05 g/cm^3? You may neglect pressure drops due to resistance.

7.17 (I) Blood flow to the digestive system increases after a large meal. If the flow to the digestive system is normally 0.75 L/min and the arteries that supply the system dilate to 1.67 times their normal radii, what is the new flow rate? Assume that all other factors remain constant and that the flow is completely regulated by these arteries. Note that the result is consistent with the fact that your heart pounds with indigestion after a very large meal.

7.18 (I) An artery supplying heart muscle with blood is occluded to 25% of its original radius by a clot. By what factor is blood flow through this artery reduced?

7.19 (I) Blood viscosity in the fingers and toes is usually slightly higher than in the rest of the circulatory system because temperature is lower and blood, cell count is higher in those places. (Blood cells tend to move straight ahead at branching points, so slightly more plasma branches off than cells, making the cell count slightly higher in the extremities.) If the viscosity of blood in the fingers and toes is 10% greater than normal, by what factor is the flow rate changed?

7.20 (II) Blood flow is increased during exercise by the dilation of vessel radii together with an increase in pressure. Suppose that the flow rate increases by a factor of 5 and blood pressure increases by 50%. Calculate the factor by which the average blood vessel radius must have increased to produce this flow.

7.21 (II) Smoking causes blood vessels to constrict. If the average vessel radius of a smoker decreases by 8% and the body increases blood pressure to keep the flow constant, calculate the needed percent increase in pressure.

7.22 (II) Blood flow to the skeletal muscles varies greatly depending on activity. Calculate the flow rate to these muscles during heavy activity if blood pressure increases by 50% and their average vessel radii increase by a factor of 2. The resting flow rate is 0.65 L/min.

7.23 (II) (a) Calculate the wall tension y in an artery of radius 0.50 cm for a blood pressure of 110 mm Hg. (b) What would this wall tension become if an aneurysm formed with a radius of 3.0 cm?

7.24 (II) If the heart muscle can produce a maximum wall tension in the left ventricle of 1.2×10^3 N/m, calculate the pressure (in N/m^2 and mm Hg) that the 3.5-cm-radius ventricle can produce.

7.25 (II) What maximum blood pressure in mm Hg can be tolerated by a blood vessel of radius 0.25 cm if that vessel can withstand a wall tension of no more than 75 N/m?

7.26 (III) (a) Calculate the pressure (in N/m^2) that must be supplied by the "muscle pump" to create a pressure equivalent to 150 cm of blood. The density of blood is 1.05 g/cm^3. (b) Calculate the wall tension γ that must be created in the vessel by muscle action to produce this pressure. The radius of the vessel is 0.20 cm.

7.27 (III) Under extreme circumstances total flow rate in the circulatory system may reach 38 L/min, (633 cm^3/s). (a) Calculate the kinetic power output of the left ventricle if the average blood velocity is consequently 230 cm/s and the density of blood is 1.05 g/cm^3 (b) Calculate the pressure power output of the left ventricle if the average pressure is 150 mm Hg.

7.28 (III) Confirm the values for the power output of the heart given in Table 7.2 for an active person by starting with the values for a resting person and assuming a 50% increase in blood pressure and a factor of 4 increase in flow. Note that if flow increases by a factor of 4 the average velocity does, too.

SECTION 7.4

7.29 (I) Calculate the total surface area of the alveoli in the lungs assuming that there are 600 million alveoli, each being a 200-μm-diameter sphere.

7.30 (I) If nitrogen molecules comprise about 80% of the atmosphere, what is the partial pressure of nitrogen in mm Hg on a day when atmospheric pressure is 760 mm Hg?

7.31 (II) In deep-sea dives, the pressure of the air that a diver breathes may be quite large. The mixture of gases breathed is adjusted so that the partial pressure of oxygen is the same as on the surface of the earth. If a person is breathing air with a pressure of 10 atm, what fraction of the, molecules in that air should be oxygen in order for the partial pressure (P_{O_2}) to be 150 mm Hg?

SECTION 7.5

7.32 (I) To what height, does saline solution rise in a manometer used in a central venous pressure (CVP) measurement if the CVP is 4.0 mm Hg? The saline solution density is 1.01 g/cm^3 and the bottom of the manometer is at the same vertical height as the right atrium.

7.33 (I) If saline solution in a central venous pressure (CVP) measurement, such as that illustrated in Figure 7.24, rises to a height of 10 cm above the level of the right atrium, what is the CVP in mm Hg? The saline solution density is 1.01 g/cm^3.

7.34 (II) A patient is connected to a positive-pressure ventilator that pushes air into the lungs. If this is done by exerting a force on a piston of radius 10 cm, how much force in newtons must be exerted?

7.35 (I) Determine values for the following lung volumes from the chart in Figure 7.26(b): (a) tidal volume; (b) inspiratory reserve volume; (c) expiratory reserve volume; (d) vital capacity.

7.36 (III) A spirometer is to be used to determine the total volume of a subject's lungs. The volume of gas in the spirometer is 20 L, 10% of which is helium. She takes a deep breath of pure air and then pressure of 3.0 cm of water breathes in and out of the spirometer, diluting the helium to a final concentration of 7.7%. (a) What is the total volume of this subject's lungs? (b) If the vital capacity of her lungs is 5.0 L, what is her residual volume?

7.37 (III) If the effective areas of the top and bottom of a 0.5-g float in a rotameter are 1.0 cm^2, how much larger must the pressure on the bottom of the float be than on the top to support its weight? (Give your answer in N/m^2 and mm Hg.)

8 Elasticity and Waves; Sound

"For a successful technology, reality must take precedence over public relations, for nature cannot be fooled."

—Richard P. Feynman

If a tree falls in the forest and no one is there to hear it, does it make a sound? The answer to this old question depends on what is meant by sound. If sound is defined as a physical vibration carried by air or other substances, then the falling tree makes a sound whether anyone hears it or not. If sound is defined as the human perception of the physical phenomenon, then there is no sound. We shall take sound to be the physical phenomenon, independent of observers. In this chapter, the characteristics of sound are discussed as one example of a broader class of phenomena known as waves. Hearing, the human perception of sound, will be studied in Chapter 9.

Sound is a type of wave. So are light and earthquake tremors.

These phenomena differ vastly from one another, yet they have certain basic characteristics in common. All exhibit behavior analogous to water waves and hence are classified as waves. We shall be concerned mainly with continuous waves—that is, waves that are periodic and go through several cycles before dying out. For example, the sound from a tuning fork is a continuous wave; the sound from an explosion is not. The cause of continuous waves is *periodic motion.*

8.1 HOOKE'S LAW AND PERIODIC MOTION

When a guitar string is plucked, it vibrates back and forth, creating a sound wave—the organized vibration of air molecules. All sound waves have a source of vibration, which could be a guitar string, a voice, or a radio speaker. These vibrations are due to the elastic properties of substances, which cause them to return to their original shapes when deformed. The vibrations are an example of periodic motion—that is, the same motion is repeated over and over before the vibration dies out. If the deformations are small, the restoring forces are simple. We shall begin by examining the nature of the restoring forces.

Forces applied to an object deform it until restoring forces in the object grow large enough to balance the applied forces. For example, when you lie on your bed, you deform it until restoring forces in the bed equal your weight. All materials exert restoring forces that increase with deformation up to their elastic limit, beyond which the material is permanently deformed. For small deformations most materials exert a restoring force that is directly proportional to the amount of deformation. Quantitatively this is expressed in Hooke's law:

$$F = -k\Delta x$$

where F is the restoring force, k is a constant depending on the object, and Δx is the amount of deformation or change in length of the object. The minus sign indicates that the force is in the direction opposite to the deformation and tends to restore the object to its original shape.

Equation 8.1 is named after Robert Hooke (1635–1703), who first stated the principle.

Consider the diving board in Figure 8.1. After the swimmer jumps, the board vibrates up and down because of its elastic properties.

The vibrating motion of the board is similar to that of so many vibrating objects that it is useful to consider it in some detail. In Figure 8.1(a) the board is deformed from its normal shape, and there is an upward restoring force on the board. In Figure 8.1(b) the board has returned to its normal undeformed shape but now has an upward velocity. The restoring force, following Hooke's law, is zero, so the board continues to move upward. In Figure 8.1(c) the board has moved above its normal position. The restoring force is now downward, and the velocity of the board has been reduced to zero by the restoring force. In Figure 8.1(d) the board has again returned to its normal position but now has a downward velocity. In Figure 8.1(e), the board has moved below its normal position and the restoring force is accelerating it upward again. The cycle of events repeats itself over and over again until friction in the board and air resistance damp out the motion.

$$\text{Frequency} \equiv \frac{\text{number of vibrations}}{\textit{time}} \qquad \textbf{\textit{Know}}$$

The vibrating motion of the diving board in Figure 8.1 is typical of objects deformed from their normal shapes and then released. Under the influence of restoring forces, usually obeying Hooke's law, the objects vibrate back and forth about their normal shape in periodic motion. The time it takes for one complete vibration depends on the size, shape, and rigidity of the object. For example, different guitar strings vibrate with different frequencies when plucked.

The frequency f of a periodic vibration is defined as the number of vibrations per unit time:

Figure 8.1 The vibrating motion of a diving board after a swimmer leaves it. The restoring force F_{rest} is largest when the deformation is largest and is in such a direction as to return the board to its normal position. Note that the conditions in **(e)** are similar to the starting conditions in **(a)**, so the motion repeats itself.

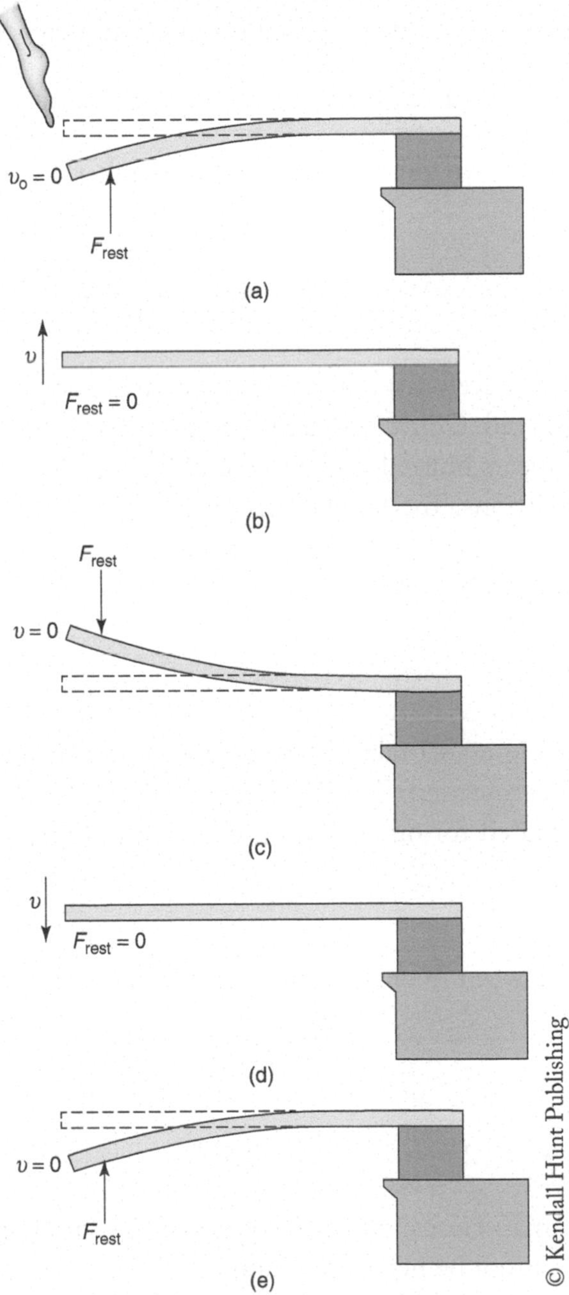

It is very common to use the SI unit for frequency, the hertz (Hz), named after *Heinrich Hertz* (1857–1894). When the time is measured in seconds the units of frequency are called hertz.

$$1\ \text{Hz} \equiv \frac{1\ \text{vibration}}{\text{second}} \qquad \textbf{\textit{Know}}$$

The words vibration, oscillation, waves, and cycle are considered synonymous for periodic motion, and are often used interchangeably.

The period T is defined as the time required for one complete vibration:

$$T \equiv \text{time for one complete vibration}$$

The higher the frequency, the less time each vibration takes. Period and frequency are the inverse of one another:

$$f = \frac{1}{T} \qquad \textit{Know}\ \text{eq. 8.2}$$

A diving board has a relatively large period and low frequency, so the individual vibrations of the board usually can be observed. Conversely, a guitar string has a small period and high frequency. A vibrating guitar string appears blurred, and the sound it produces is perceived as a tone rather than a series of individual vibrations. The higher the frequency, the shorter the period. High frequency sounds are said to have a high pitch. If the frequency is less than 20,000 Hz and above 20 Hz, a young, healthy human ear can hear it.

EXAMPLE 8.1

(a) If the diving board in Figure 8.1 takes 0.080 seconds to make one complete vibration, how many vibrations does it make each second? That is, if its period T is 0.080 seconds, what is its frequency? (b) What is the time required for one vibration of a guitar string that has a frequency of 264 Hz (middle C)?

Solution.

(a) The relationship between frequency and period given in equation 8.2 is,

$$f = \frac{1}{T}$$

Therefore, $f = \dfrac{1}{0.080\ \text{s}} = 12.5 \dfrac{\text{vibrations}}{\text{seconds}} = 12.5\text{Hz}$

Because a healthy human ear can only detect sounds in the frequency range of 20–20,000 Hz, this frequency, 12.5 Hz, cannot be heard by a human.

Ex. 8.1(b) Here the frequency is given as $f = 264$ Hz

Solve $f = \dfrac{1}{T}$ for the period, T.

$$T = \frac{1}{f} = \frac{1}{264\,/\,\text{s}} = 0.00379\ \text{seconds}$$

Things to know for section 8.1

$$f \equiv \frac{\text{number of vibrations}}{time}$$

$$1\ \text{Hz} \equiv \frac{1\ \text{vibration}}{\text{second}}$$

Period $T \equiv$ time for one complete vibration

$$f = \frac{1}{T}$$

8.2 TRANSVERSE AND LONGITUDINAL WAVES

The periodic motion and vibrations discussed in the previous section are sources of waves. Waves travel; for example, ocean waves created in a storm may travel thousands of kilometers before coming to shore. There is also motion within the wave itself; an ocean buoy moves up and down, back and forth, as a wave passes.

Figure 8.2 illustrates two types of motion within waves.

A *transverse wave* is one in which motion within the wave is perpendicular to the travel of the wave, as in Figure 8.2(a). The wave moves to the right, but the individual coils move up and down, perpendicular to the travel of the wave. A *longitudinal wave* is one in which motion within the wave is parallel to the travel of the wave, as in Figure 8.2(b), where the same spring is vibrated along its

Figure 8.2 **(a)** A transverse wave; individual coils move up and down, perpendicular to the travel of the wave (to the right). **(b)** A longitudinal wave; individual coils move right and left parallel to the travel of the wave.

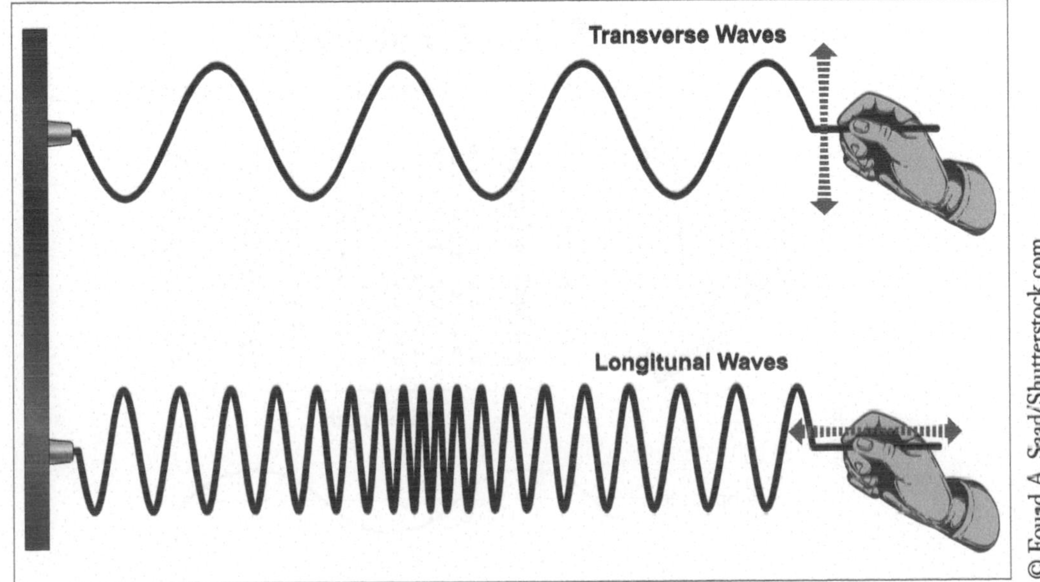

length. The restoring force is in the opposite direction of the deformation, so the compressed coils are forced apart, moving the compression to the right.

There are many examples of transverse and longitudinal motions in waves. Ocean waves and earthquake tremors have both transverse and longitudinal components; sound is a pure longitudinal wave in fluids; waves on a string can be transverse or longitudinal but are usually transverse. If you could see a sound wave, you would notice that it resembles an expanding sphere with alternating regions of compression and expansion. Figure 8.3 is a two dimensional illustration of the periodic vibrations in a sound wave.

A graph of pressure versus distance from a source at one instant in time.

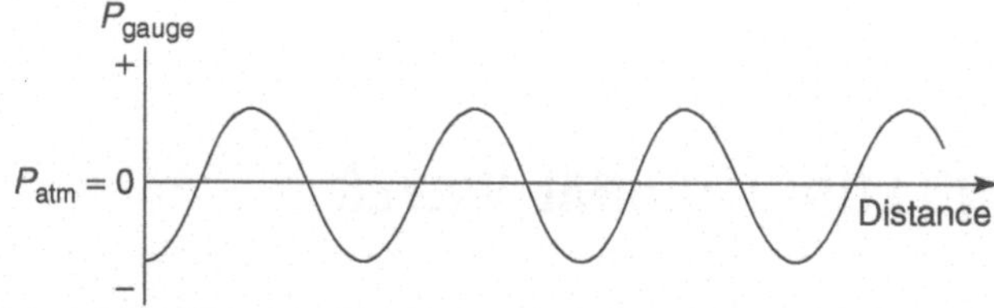

When air is compressed, its pressure rises. This forces some molecules to move away from the region of increased pressure, thus creating a longitudinal wave analogous to the one in Figure 8.2(b). The repeated compressions and rarefactions move away from the string and have the same frequency as the vibrating string. The organized vibrations of air molecules are the sound wave.

Sound waves can force objects to vibrate. As a sound wave impinges on an object, such as the eardrum in Figure 8.4, air pressure rises and falls. Pressure behind the eardrum remains constant (nearly zero gauge pressure), so there is a net force on the eardrum with each compression and rarefaction. Pressure in each compression is slightly positive, so the net force is to the right when a compression arrives. Pressure in each rarefaction is slightly negative, so the net force is to the left

Figure 8.3 Sound waves created by a vibrating speaker. **(a)** A compression being made by the vibrating cone moving to the right.

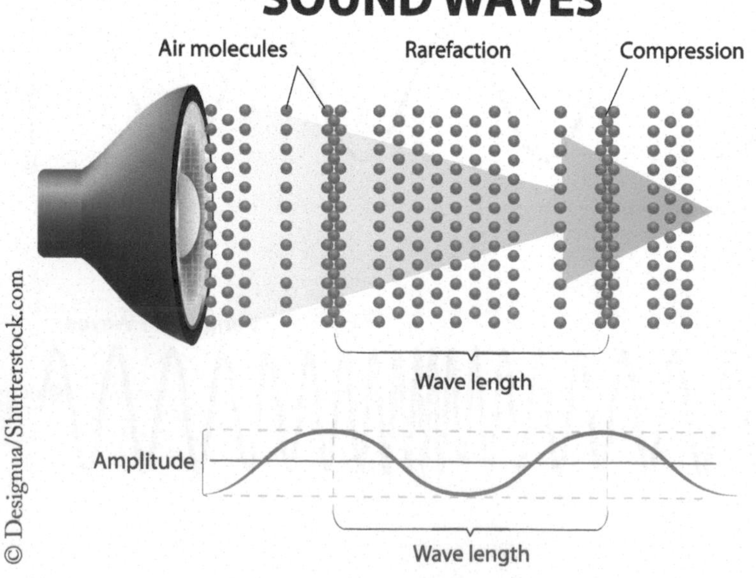

Figure 8.3bc **(b)** A rarefaction being made by the vibrating cone moving to the left. The earlier compression has moved to the right. (c) Repeated motion sends out periodic sound waves that expand spherically away from its source.

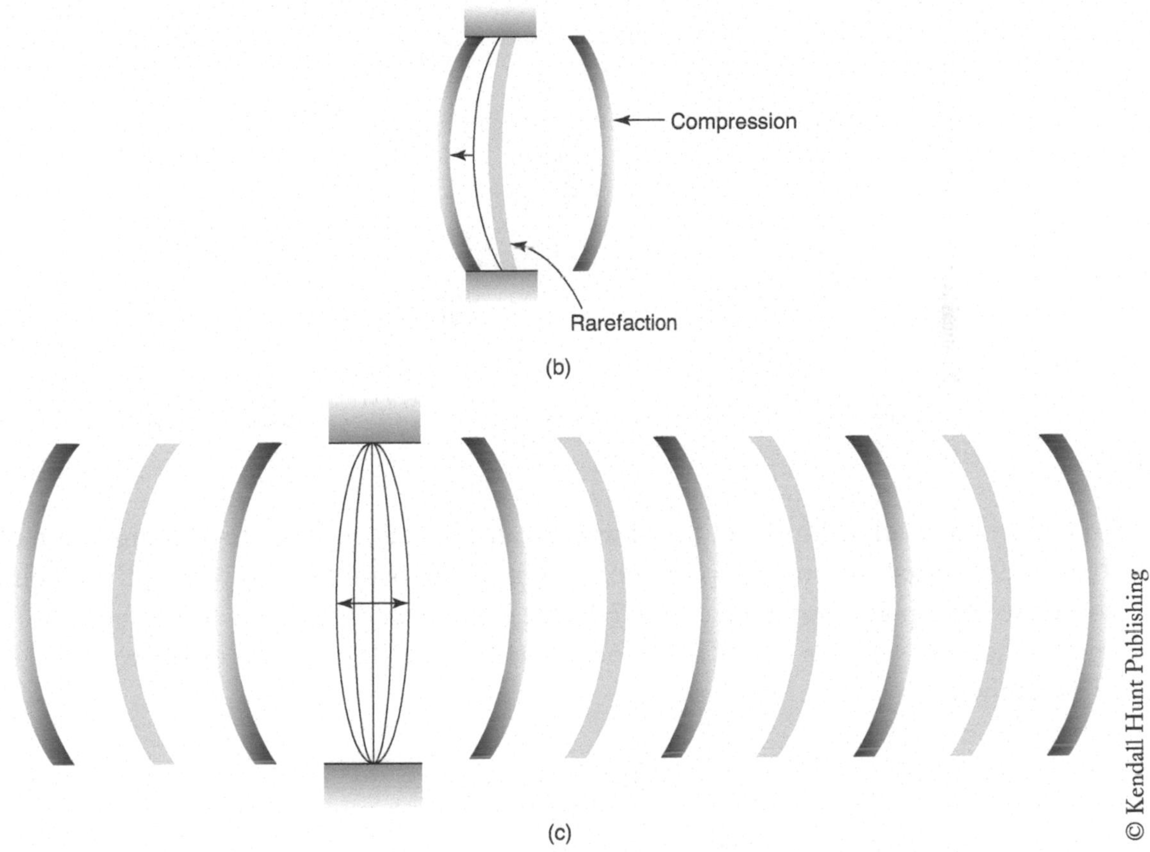

when a rarefaction arrives. The eardrum is thus forced to vibrate at the same frequency as the sound wave. If the sound is intense enough and at a frequency in the normal range of hearing, nerve cells in the ear are stimulated by eardrum motions and the sound is perceived.

$$P = \frac{F}{A}$$

8.2.1 Wavelength and Other Wave Characteristics

Frequency and period are two of the characteristics that all waves possess. What other features do all periodic waves have? One is *wavelength* λ, defined as the distance between consecutive similar parts of a wave. For the ocean wave shown in Figure 8.5 the wavelength is the distance between adjacent crests. The wavelength of sound waves is the distance between consecutive compressions or rarefactions. Another common characteristic of all waves is that they travel with some speed of propagation, labeled V_w for the ocean wave in Figure 8.5.

The seabird in the figure bobs as the waves pass with an amplitude X, the maximum displacement of the medium from equilibrium. For sound waves, the amplitude is proportional to the maximum gauge pressure in the compressions.

Figure 8.4 Anatomy of the human ear.

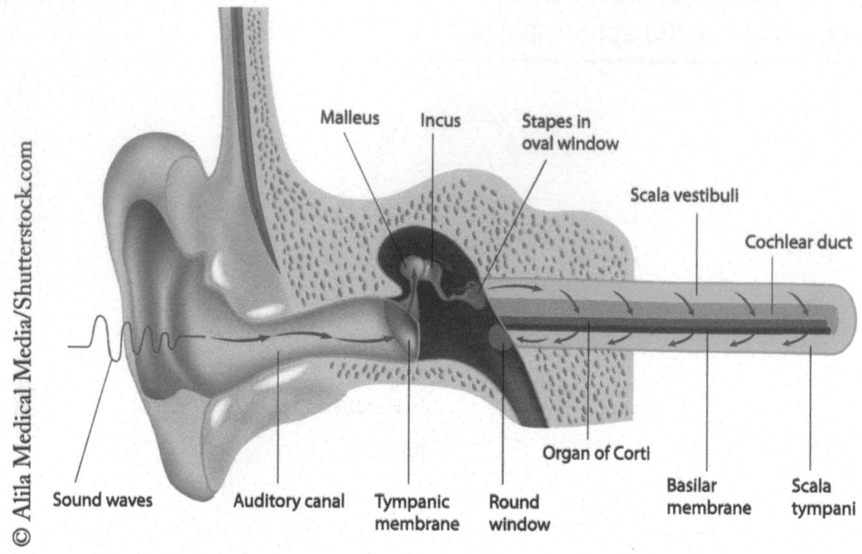

An important relationship, valid for all waves, can be obtained by further examination of Figure 8.5. The time required for one complete vibration is T, the period of the wave. (The seabird makes one complete trip up and down in a time T.) One full wavelength passes to the right in this time. This means that the wave has moved a distance λ in a time T. The velocity of any wave can be determined as follows:

$$v = \frac{\Delta X}{\Delta t} = \frac{\lambda}{t} = f\lambda$$

Therefore $v = f\lambda$ **Know eq. 8.3**

f is the *frequency* of vibration
λ is the Greek letter lambda and represents the wave's *wavelength*.

EXAMPLE 8.2

Calculate the speed of propagation of the ocean wave in Figure 8.5 if its crests are 20 m apart and the bird takes 8.0 seconds to complete one trip up and down.

Solution: From the given information the wavelength is 20 m and the frequency of the vibration is 1/8.0 s, so that:

$$v = f\lambda = \frac{1}{8} \cdot 20 \text{ m} = \frac{20}{8}\frac{\text{m}}{\text{s}} = 2.5\frac{\text{m}}{\text{s}}$$

EXAMPLE 8.3

What is the wavelength of sound of frequency 528 Hz (high C) given the speed of sound to be 341 m/s?

Figure 8.5 Water waves are combinations of transverse and longitudinal waves. As the ocean wave moves to the right with a propagation velocity, the seabird bobs up and down, back and forth, always returning to its original position. The wavelength is the distance between similar parts of the wave. The amplitude *X* of the transverse part is the maximum vertical displacement from equilibrium, shown by the dotted line.

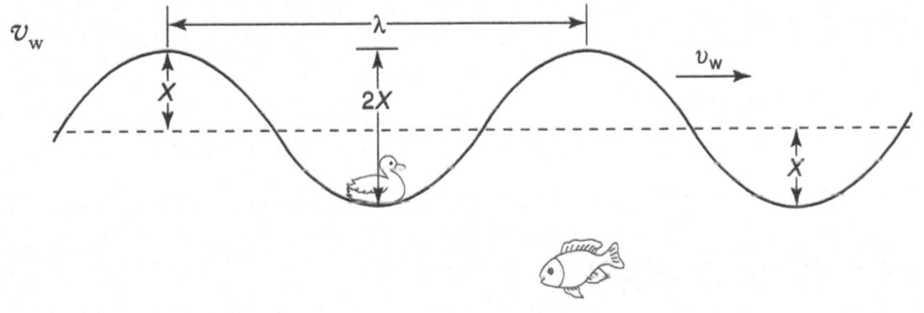

Solution: We solve Equation 8.3 for the wavelength, λ.

$$v = f\lambda\text{; dividing both sides by } f \text{ yields } \lambda = \frac{v}{f}$$

$$\lambda = \frac{v}{f} = \frac{341\frac{\text{m}}{\text{s}}}{\frac{528}{\text{s}}} = 0.646 \text{ m}$$

The wavelength of an ocean wave is easily observed, but that of a sound wave must be observed indirectly or calculated. If the compressions and rarefactions in a sound wave were visible with the naked eye, then we would see something like the situations depicted in Figures 8.3 and 8.4. If the frequency of the sound were 528 Hz, as in Example 8.3, then the compressions would be 0.646 m apart and would stay that distance apart as they sped through the air at 341 m/s.

The speed of propagation of a wave, v_w, is usually independent of its frequency and wavelength. Different sound frequencies, for example, all travel at the same speed in a given medium. Music from an orchestra sounds the same whether observed at large or small distances, indicating that the various frequencies arrive together. The speed of sound does, however, vary with temperature. The speed of sound in air at sea level is 331 m/s at 0°C and increases by 0.607°C for every degree above 0°C. Air pressure has very little effect on the speed of sound.

EXAMPLE 8.4

What is the speed of sound in air when the air temperature is 23.0°C?

Solution: The temperature dependence of the speed of sound in air is given by the following rule:

$$V_{\text{sound}} \text{ in air} = 331\frac{\text{m}}{\text{s}} + \frac{0.607}{°\text{C}}\frac{\text{m}}{\text{s}} \cdot T_c$$

Therefore,

$$V_{sound} \text{ in air at } 23.0°C = 331\frac{m}{s} + \frac{0.607 m}{°C}\frac{m}{s} \cdot 23.0°C = 345\frac{m}{s}$$

You will determine the speed of sound in air in Lab 8.

Things to know for section 8.2

The velocity of a wave $v = f\lambda$

f is the *frequency* of vibration
λ is the Greek letter lambda and represents the wave's *wavelength*

8.3 ENERGY IN WAVES; INTENSITY

All waves carry energy. Ocean waves erode shores, sound can break a glass, and an earthquake may produce the destruction of many nuclear bombs. Some people even define waves as disturbances in which energy is transferred without the transfer of mass. The energy associated with a wave is proportional to its amplitude squared. For example, if the amplitude of a wave is tripled, its energy is $3^2 = 9$ times larger than it used to be. Because waves usually spread out from a source, their energy also spreads out. The energy per unit area of a wave at a given location is therefore as important as the total energy carried by the wave. The intensity of a wave is defined as its power per unit area:

$$Intensity \equiv \frac{Power}{Area} \quad \text{eq. 8.5 } Know$$

Its units are: $\frac{\text{watts}}{\text{m}^2}$ *Know*

This definition of intensity is used for all types of waves and makes good sense intuitively. A very intense sound, for example, damages the ear because it carries energy in faster than the ear can dissipate it harmlessly. The eardrum is forced to vibrate with a large amplitude that can damage both the eardrum and the hearing mechanisms inside the ear.

One unit for intensity is the watt per meter squared (W/m^2). The human ear can detect sounds of intensity as low as 10^{-12} W/m^2, and can be briefly subjected to sounds as intense as 1 W/m^2 without being damaged. This is a remarkable range of sensitivity; the ear can detect sounds that vary in intensity by a factor of 10^{12}. Equally remarkable is how small an intensity the ear can detect. One-trillionth of a watt per meter squared is a very small amount of power per unit area. Molecules in a sound wave of this intensity vibrate over a distance of less than one molecular diameter and the gauge pressures involved are less than 10^{-9} atmospheres.

Although people can distinguish between sounds of different intensities, they are unaware of the factor of 10^{12} in sound intensity from the threshold of hearing to the sounds that damage the ear. The ear's response to sounds is approximately logarithmic rather than linear. Partly for that reason another intensity scale is used for sound: the decibel (dB) scale. The following equation defines the decibel scale and allows sound intensities to be converted from watts per square meter to sound level in decibels:

Table 8.1 Sound Levels and Intensities of Various Sounds

Sound Level (dB) B	Intensity (W/m²) *I*	
0	1×10^{-12}	Threshold of hearing at 1000 Hz
10	1×10^{-11}	Rustle of leaves
20	1×10^{-10}	Whisper 1 m distant
30	1×10^{-9}	Quiet home
40	1×10^{-8}	Soft music, average home
50	1×10^{-7}	Average office
60	1×10^{-6}	Normal conversation
70	1×10^{-5}	Noisy office, busy traffic
80	1×10^{-4}	Loud radio
90	1×10^{-3}	Inside subway train: Damage after prolonged exposure
100	1×10^{-2}	Average factory, siren at 30 m: Damage from 8 hour exposure per day
110	1×10^{-1}	Damage from 30 minutes exposure per day
120	1	Pneumatic chipper at 2 m, loud rock concert indoors: Threshold of pain, damage in minutes
140	1×10^{2}	Jet airplanc at 30 m: Severe pain
160	1×10^{4}	Bursting of eardrums

Sound level: $\beta = 10 \cdot \text{Log}\left(\frac{I}{I_0}\right)$ decibels eq. 8.6 *Know*

where $I_o = 1 \times 10^{-12}$ W/m² is the threshold of hearing at a frequency of 1000 Hz. I_o is referred to as the reference intensity. Log is base 10.

β is the Greek letter beta and is called the sound level. It is important to note that β in decibels gives the level of the sound relative to some standard, rather than in terms of power per unit area. (A discussion of logarithms is given in Appendix A.) Note that each 10 dB corresponds to a change in intensity by a factor of 10.

Table 8.1 gives the intensity of various sounds in watts per meter squared and decibels.

Please make sure that you have listened to the *sound level lecture* before proceeding.

Note that the threshold of hearing has a sound level of zero decibels. Sounds less intense than threshold have negative sound levels on the decibel scale. One advantage of the decibel scale is that the numbers are easy for us to relate to different sound levels. A difference in sound level of 10 dB is easily sensed. For example, the 90-dB sound level inside a subway train is clearly louder than the 80-dB sound level of busy traffic. Although it doesn't seem 10 times louder, it is.

EXAMPLE 8.5

(a) Calculate the sound level in decibels of a sound of intensity 10^{-9} W/m^2. (b) Calculate the intensity in watts per meter squared of a sound level of 90 dB.

Solution (a) Use Equation 8.6.

$$\beta = 10 \cdot \text{Log}\left(\frac{I}{I_0}\right)$$

$$\beta = 10 \cdot \text{Log}\left(\frac{1\times 10^{-9}}{1\times 10^{-12}}\right) = 10 \cdot \text{Log}(1000)$$

Since the log of 1000 is 3, $\beta = 10 \cdot 3 = 30$ decibles

Solution (b) is a little trickier. We are given that $\beta = 90\,\text{dB}$ and we know that the reference intensity, I_0 always equals 1×10^{-12} W/m^2.

Step by step 1.

First enter the known values into Equation 8.5: $\beta = 10 \cdot \text{Log}\left(\frac{I}{I_0}\right)$

$$90 = 10\text{Log}\left(\frac{I}{1\times 10^{-12}\frac{\text{W}}{\text{m}^2}}\right)$$

Step 2.

Divide both sides of the equation by 10.

$$9 = \text{Log}\left(\frac{I}{1\times 10^{-12}\frac{\text{W}}{\text{m}^2}}\right)$$

Step 3.

Take the antilog of both sides of the equation.

$$1\times 10^{9} = \frac{I}{1\times 10^{-12}\frac{\text{W}}{\text{m}^2}}$$

Step 4.

Multiply both sides of the equation by 1×10^{-12} W/m^2

$$I = 1\times 10^{9} \cdot 1\times 10^{-12}\,\frac{\text{watts}}{\text{m}^2} = 1\times 10^{-3}\,\frac{\text{watts}}{\text{m}^2}$$

We have found that 90 dB is equivalent to an intensity of

$$1\times 10^{-3}\,\frac{\text{W}}{\text{m}^2}$$

EXAMPLE 8.6

(a) How many times more intense than threshold is a sound level of 60 dB?

Solution: (a) We are asked to compare the *intensities* of a 60 dB sound to the threshold intensity of 0 dB sound. Using Table 8.1, we see that a 60 dB sound has an intensity of 1×10^{-6} W/m^2. A 0 dB sound has an intensity of 1×10^{-12} W/m^2.

Comparing the two intensities:

$$\frac{I}{I_0} = \frac{1\times 10^{-6}\,\frac{\mathrm{W}}{\mathrm{m}^2}}{1\times 10^{-12}\,\frac{\mathrm{W}}{\mathrm{m}^2}} = 1\times 10^{6}$$

We see that a 60 dB sound has 1 million times the intensity of the threshold sound.

EXAMPLE 8.6 (B)

How many decibels higher is a sound intensity of 10^{-1} W/m^2 than a sound intensity of 10^{-10} W/m^2?

Solution. (b): From Table 8.1, we see that a sound with the intensity of 10^{-1} W/m^2, has a sound level of 110 dB. A sound with an intensity of 1×10^{-10} W/m^2, has an intensity of 20 dB.

In order to calculate the difference in intensities, we subtract.

110 dB - 20 dB= 90 dB

The 110 dB sound is *90 dB higher* than the 20 dB sound.

Sound level and intensity are physical, not human, characteristics. They are also valid, for example, for ultrasound, or sound that is too high in frequency to be heard. You should try the "Sound level and Intensity practice quiz." Don't worry, it comes with a solution.

8.3.1 Reflection and Attenuation of Waves

The intensity of a wave is related to its amplitude squared, but ultimately the intensity is dependent on the energy put into the wave at its source. As the wave spreads out from its source its energy is spread over an ever increasing area. If the source of the wave is isolated (no walls, ground, or obstructions nearby) and broadcasts uniformly in all directions, then the intensity of a wave will be inversely proportional to the square of the distance from the source. *Reflection* of the wave, *attenuation* by the atmosphere or objects, and nonuniformity of the source, however, all change the dependency of intensity and distance from the simple inverse-square relationship.

Reflections of waves, such as sound echoes, are very common. Reflections can be caused by a solid object or just a change in character of the medium in which the wave is traveling. Ultrasonic scanners are used to look inside the body. They work by sending ultrasound pulses into the patient and listening for echoes (see Figure 8.7).

Reflections of the sound occur at the interfaces between different tissues. The depth at which the reflection occurred is calculated from the time it took the echo to return. The probe is passed over the region of interest and a picture is electronically constructed from the echo and position information. The process is very similar to that used in sonar and by dolphins and bats to navigate and locate prey. Submarines, dolphins, and bats use ultrasound in the frequency range of 30–100 kHz, while medical diagnostics employ frequencies in the range of 1–20 MHz.

Figure 8.7a An ultrasound transducer (speaker-microphone) is passed across the abdomen of a pregnant woman. A computer analyzes the echoes as a function of position and constructs a picture.

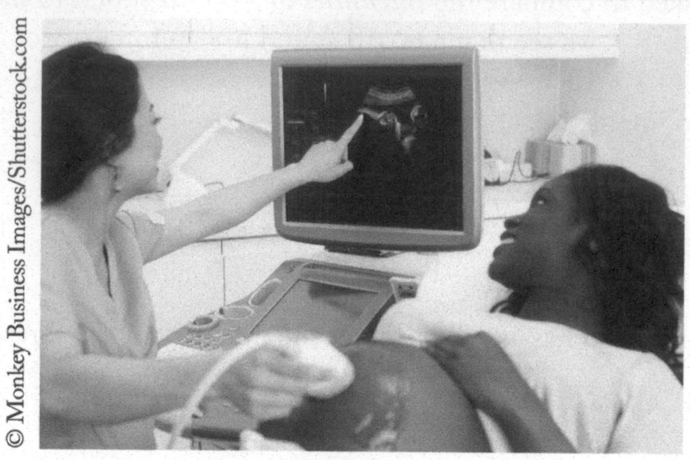

Figure 8.7b Some of the ultrasound is reflected from each interface, and the return time of an echo depends on the depth of the interface.

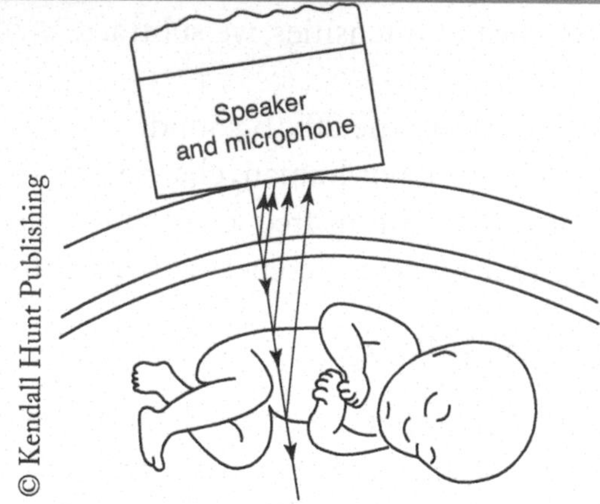

The intensity of ultrasound used for medical diagnostics is kept low to avoid tissue damage. Intensities of about 10^{-2} W/m^2 are used and seem to cause no ill effects. Most of the energy carried in by the ultrasonic wave is converted to thermal energy, but 10^{-2} W/m^2 causes negligible heating of tissues, unlike x-rays, which always do some tissue damage. Ultrasound is therefore often used in obstetrical applications.

Ultrasound of considerably higher intensity is used for therapeutic purposes. Ultrasonic diathermy is deep heating using ultrasound of intensities of that range from 1 to 10 W/m^2. Ultrasound diathermy is especially useful in the delivery of heat to selected muscles and structures because there is a difference in the sensitivity of various fibers to the acoustic vibrations; some are more absorptive and some are more reflective. For example, in subcutaneous fat, relatively little energy is converted into heat, but in muscle tissues there is a much higher rate of conversion to heat. Care must be taken in ultrasonic diathermy to avoid hot spots, which destroy tissue rather than just warm it a few degrees above body temperature, although ultrasound of intensity 10^3 W/m^2 is used in some medical procedures to destroy cancerous tissues or gallstones.

Figure 8.7c Ultrasonic scan of fetus.

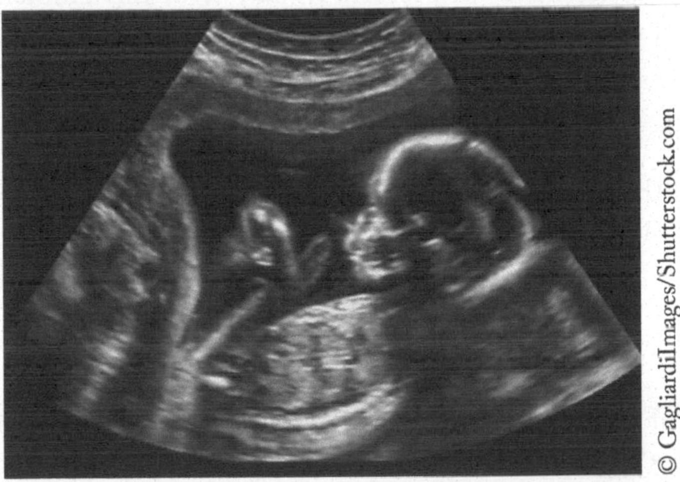

Finally, why is ultrasound rather than audible sound used in diagnostics? The major reason is that *the smallest detail observable when using a wave as a probe is about one wavelength*. Ultrasound wavelengths can be small enough to see needed detail. One drawback, however, is that ultrasound is absorbed and is useful only to depths as great as 400 wavelengths. A balance between high frequency, small wavelength for detail, and low frequency, large wavelength for penetration is possible. Frequencies of 1–5 MHz are most common, giving sufficient detail and penetration, as the next example shows.

EXAMPLE 8.7

Ultrasound has a speed of 1540 m/s in tissue. (a) Calculate the smallest detail visible with 2.0-MHz ultrasound. (b) The effective maximum depth of ultrasound waves is approximately 400 wavelengths. To what depth can the sound probe effectively? (c) How much time is required for the sound to travel to the depth determined in part b and return to the transducer?

Solution. (a): The smallest detail visible has a size approximately equal to one wavelength of the sound. In Equation 8.5, we see that the relationship between propagation speed, frequency, and wavelength is:

$$v = f\lambda$$

Solving for wavelength, and substituting the information given yields:

$$\lambda = \frac{v}{f} = \frac{1540\,\frac{\text{m}}{\text{s}}}{\frac{2\times 10^{6}}{\text{s}}} = 0.77\times 10^{-3}\ \text{m} = 0.77\ \text{mm}$$

Detail larger than 0.77 mm should be visible.

Solution. (b): Since the effective depth is approximately 400 wavelengths,

$$400\cdot(0.77\text{mm}) = 308\text{mm} = 0.308\ \text{m}$$

This is sufficient to probe a normal size person.

Solution. (c): We are dealing with an echo. We are asked to determine the time for the sound to travel to a point of reflection and then return. In part "a" of this problem, we determined that the reflector could be at a maximum depth of 0.380 m. When there is no acceleration, the relationship between displacement, velocity, and time is:

$$\Delta x = vt$$

Because we want the time to travel down to the reflecting organ and back to the surface, the total distance traveled is twice the depth. Solving for time, our working equation becomes;

$$t = \frac{\Delta x}{v} = \frac{2 \cdot 0.308\text{m}}{1540\frac{\text{m}}{\text{s}}} = 4.00 \times 10^{-4} \text{ seconds}$$

Ultrasound is sent out in short bursts with a pause to listen for echoes. In this case, the pause must be at least 4.00×10^{-4} seconds before sending another pulse of sound into the patient.

Distance measurements made by timing the round trip of a wave are used in a wide variety of situations, from finding the distance to the moon by bouncing laser light off it, to measuring the depth of the ocean by bouncing sound off the bottom. Remember that you must use the round trip distance to determine the time required for the echo to go out and back.

Things to know for section 8.3

$Intensity \equiv \frac{Power}{Area}$ It units are $\frac{\text{watts}}{\text{m}^2}$

Sound level: $\beta = 10 \cdot \text{Log}\left(\frac{I}{I_0}\right)$ Units are decibels.

The reference intensity $I_o = 1 \times 10^{-12}$ W/m^2

8.4 SUPERPOSITION AND RESONANCE

What happens if two or more waves pass through the same region of space at the same time? Will they combine to form larger waves, or will they cancel one another? This section is devoted to the description of what happens when waves come together.

There are many examples of multiple sound waves occurring simultaneously. Vocal sounds are a combination of waves of different frequencies, usually making the voice individually recognizable. Musical instruments often are designed to combine sound waves in such a way as to make them more intense. Echoes and reverberations are very common, so normal sounds are made up of many waves added together even if the source emits only a single wave.

Waves add in a simple way. The disturbance of the resultant wave is the sum of the disturbances of the combined waves. This is the principle of *superposition* and is illustrated in Figure 8.8. Figure 8.8(a) shows two waves having equal wavelengths and amplitudes superimposed crest to crest and trough to trough. The waves produce another of equal wavelength and greater amplitude. This is an example of *constructive interference.*

Figure 8.8(b) shows what happens if the same two waves are superimposed crest to trough; they cancel one another. Crests are positive disturbances, troughs negative disturbances, so adding a crest

Figure 8.8 Superposition. **(a)** Pure constructive interference. Crests add to crests and troughs add to troughs. **(b)** Pure destructive interference. Crests and troughs add to zero.

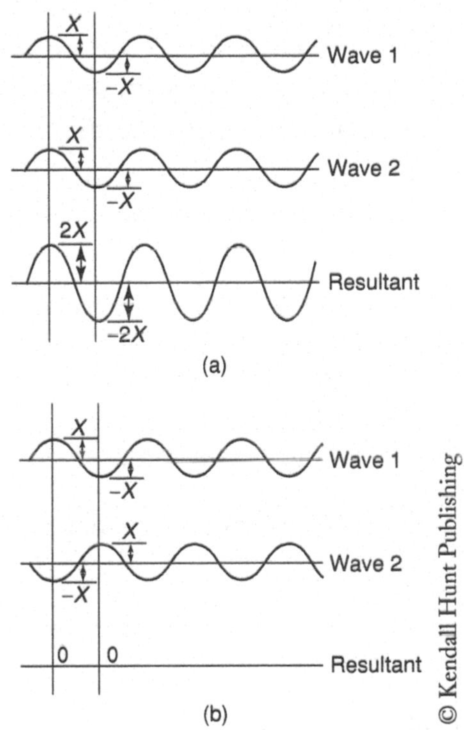

and a trough tends to decrease the magnitude of the disturbance. When added waves produce a larger disturbance, the process is called constructive interference; when they produce a smaller disturbance, the process is called *destructive interference*. If two waves completely cancel each other, the result is no wave. This is an example of complete, destructive interference.

8.4.1 Beats

When two waves of nearly equal frequency are combined, a phenomenon called beats results. Figure 8.9 shows what happens. The waves get in and out of step as time goes by and alternate between constructive and destructive interference. The resulting wave has an amplitude that increases and decreases periodically. Beats can occur for any types of waves; if they are sound waves, the varying amplitude is easily perceived as a warbling sound. One place where beats are very noticeable is in jet aircraft with two engines. The sounds put out by the engines have similar frequencies, and the engines make an irritating warble.

Destructive interference Constructive interference

Beats can be used to determine a frequency very accurately. Piano and guitar tuners listen for beats between the sound from a tuning fork and the string they are adjusting (or between two strings). When the beats disappear, the two frequencies are the same. The number of times per second that two sounds get in and out of step is called the beat frequency. The *beat frequency* is equal to the difference between the frequencies of the combined waves.

The beat frequency, $f_B = |f_2 - f_1|$ eq. 8.7

The beat frequency, f_B, is equal to the absolute value of the difference between the two frequencies. Listening for beats is much more accurate than comparing individual frequencies. If the piano

Figure 8.9 Beats: When waves of slightly different frequencies are superimposed the resultant has a periodically varying amplitude.

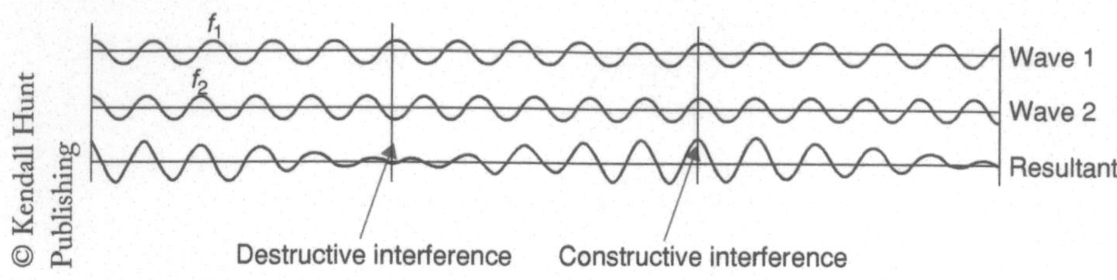

Figure 8.10 The superposition of waves of different frequencies and amplitudes exhibits partially constructive and partially destructive interference.

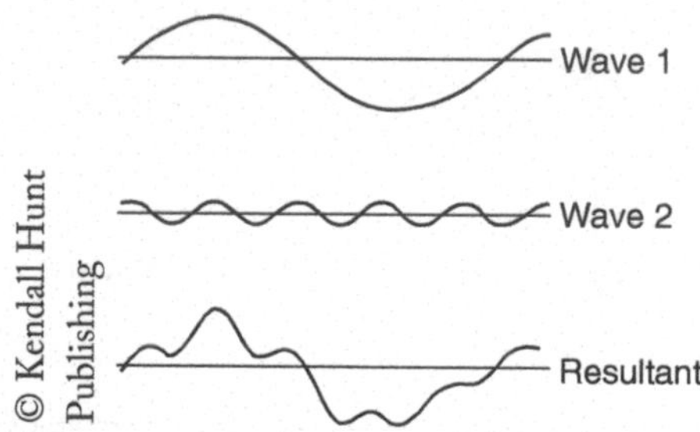

tuner uses a 264-Hz tuning fork and the string vibrates at 266 Hz, the beat frequency is 2 Hz and is easily heard. However, the average person would not be able to distinguish between the 264- and 266-Hz tones if played separately.

8.4.2 Multiple Frequencies

Figure 8.10 shows the superposition of two waves having different wavelengths, amplitudes, and frequencies. Again the disturbances add, this time both constructively and destructively.

The resultant wave can appear quite complex when only a few waves have been added together, as in Figure 8.11.

If the waves are sound waves, the resultant tone may be recognizable as a particular musical instrument or a person's voice because of the various frequencies present. (see Figure 8.12). People are often able to unfold a complicated sound and tell what composed it. For example, at a party the sounds of music and several conversations are all individually distinguishable even though the sound arriving at the ear is a complicated sum of these. The human brain is an amazing thing.

8.4.3 Resonance

Most objects prefer to vibrate at one or more natural frequencies if stimulated. The natural frequencies are determined by the size, shape, and composition of the object. If a sufficiently intense wave having the natural frequency of an object comes along, then the object will vibrate with a relatively large amplitude. Each vibration of the incoming wave arrives at just the right time to increase the amplitude of the object's vibrations. One example of this is a person pushing a child on a swing; pushing with the same frequency as the swing increases or maintains the amplitude of the motion. The forces are applied to an object by a wave having the same frequency as the natural frequency of

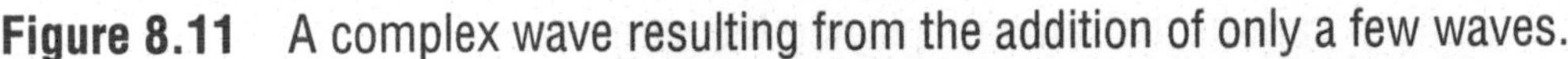

Figure 8.11 A complex wave resulting from the addition of only a few waves.

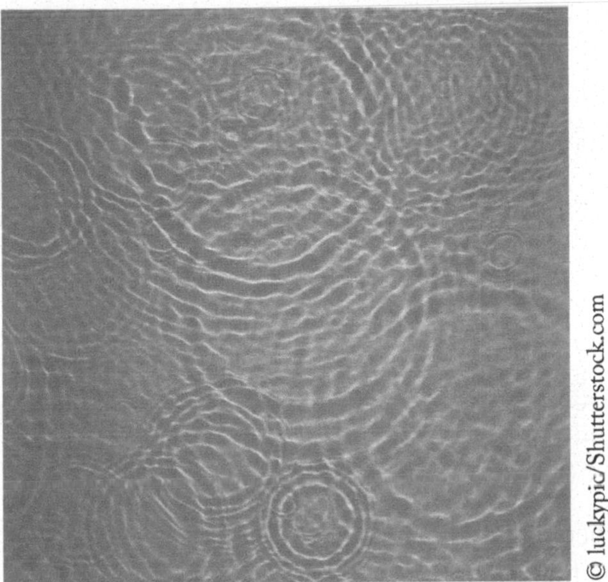

© luckypic/Shutterstock.com

the object, a large increase in the amplitude of the wave can result. This is called *resonance*. Any type of periodic motion or wave can be involved.

Resonance plays an important role in sound production. Partially enclosed volumes of air, for example, have natural vibration frequencies. Children love to blow across the mouth of a soda bottle to make it toot. The turbulent flow of breath across the bottle top causes sounds of many frequencies, and the column of air in the bottle resonates vigorously at its natural frequency. Similarly, stringed instruments have sounding boxes with complicated shapes so that the air inside has many natural frequencies. The vibrating string causes the air column in the sounding box to vibrate at a natural frequency. Resonance creates a sound wave of much larger amplitude than the string alone.

The resonance of an air column is illustrated in Figure 8.13. A compression leaves the tuning fork, travels down the tube, and is reflected from the water. If the compression returns to the opening of the tube at the same time as the next vibration is entering it, they add constructively and the sound is more intense than it would be without the column of air. The distance to the water can be adjusted until the timing is just right. A column of air of the right length has the same natural frequency as the tuning fork and is therefore is caused to resonate by the tuning fork. In Lab 8, we will use a resonance tube to determine the speed of sound in air.

Vocal sounds are also produced using resonance. The vocal cords are analogous to the tuning fork in Figure 8.13 but produce many frequencies simultaneously. The upper breathing passages form a resonant cavity whose natural frequencies depend on the shape of the mouth, the position of the tongue, and so on (see Figure 8.14). Vocal sounds depend greatly on resonance in addition to vibrations created by the vocal cords.

The lowest frequency at which an object vibrates is called its fundamental frequency. If it has other natural frequencies, they are called overtones. Voices and many musical instruments make air resonate to produce sounds and are distinguished by their overtones even when producing the same fundamental tone (see Figure 8.12). The size of a musical instrument is an indication of the frequencies it can produce—larger instruments producing lower frequencies. This is true to an extent for people; a high-frequency voice is expected in small people, for example.

Figure 8.12 Sound waves from various instruments playing the same note.

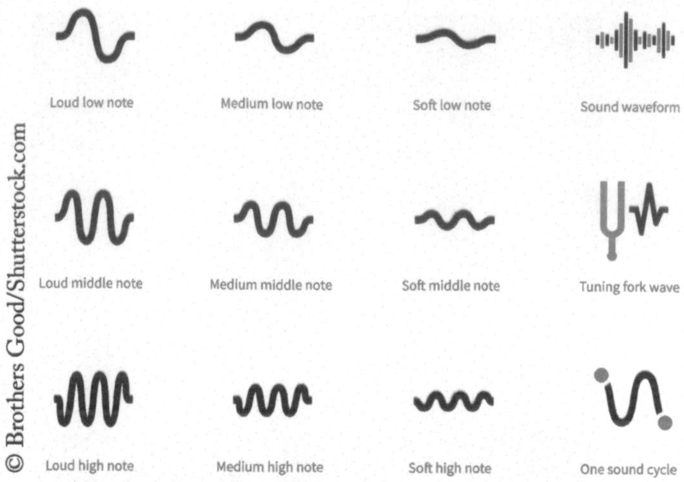

Figure 8.13 A resonance tube. The air column resonates if its natural frequency is the same as that of the tuning fork.

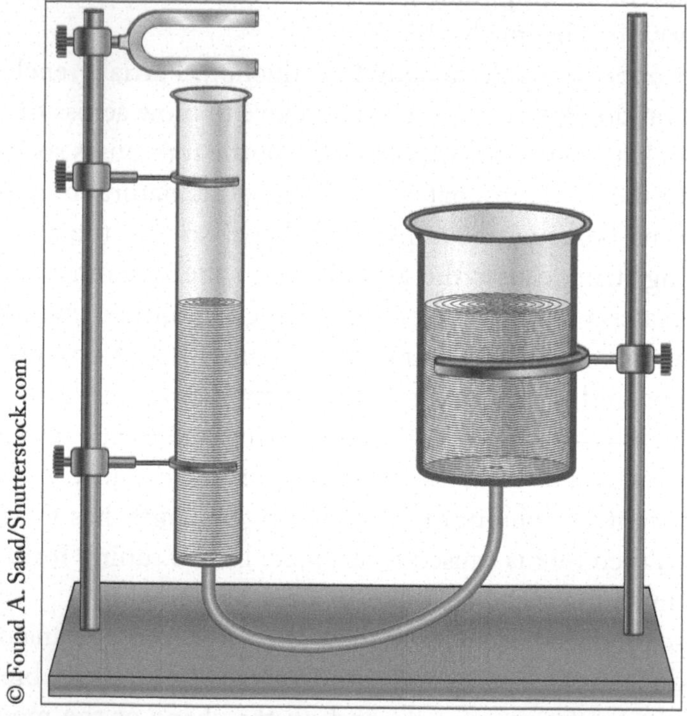

8.5 THE DOPPLER EFFECT: MOVING SOURCES AND OBSERVERS

A race car emits a distinctive roar when it passes. The observed frequency of the sound it makes decreases noticeably as the car passes by and moves away from the observer—sort of an eeeeaar-rhoooom. This is one example of the *Doppler effect*, the change in observed frequency of a wave depending on the relative motion of the source and observer. The observed frequency is higher than the source frequency if the source and observer are moving toward each other; the observed

Figure 8.14 The vocal cords moderate exhaled air to produce a complicated sound wave causing the upper breathing passages to resonate much like the tuning fork and air column in Figure 8.13.

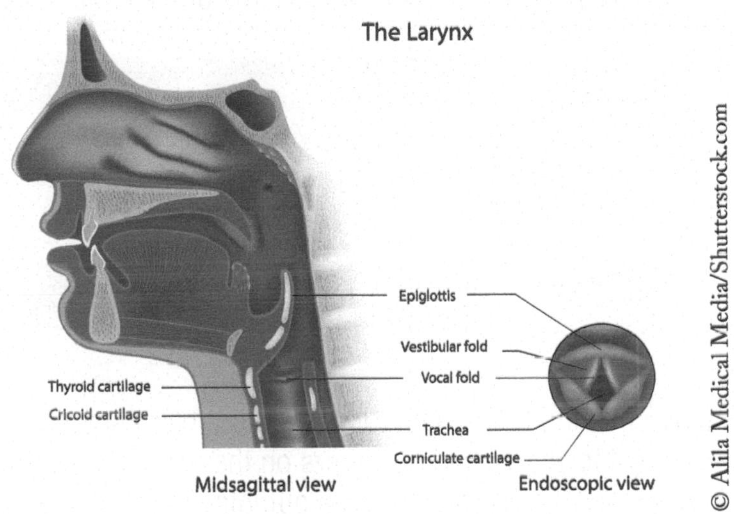

frequency is less than the source frequency if the source and observer are moving apart. The Doppler effect occurs with all types of waves but is most commonly observed with sound waves. It is familiar in the observed sound of passing trains, motorcycles, low-flying planes, vehicles with sirens, and so on. The effect is named for the Austrian physicist and mathematician Christian Johann Doppler (1803–1853), who did experiments with sound waves.

Figure 8.15 shows why the frequency of an observed wave depends on the relative motion of source and observer. (The figure shows the effect for a moving sound source, but the overall effect occurs whether the source, observer, or both are moving; the crucial factor is the relative motion of the source and observer.) In Figure 8.15(a), a stationary source emits a periodic sound wave in still air. The maxima of the wave are represented by circles.

Figure 8.15 (b) The Doppler effect: the source is moving to the right, approaching person 1 and receding from person 2. Each sound wave moves out in a sphere from the point where the source emitted it.

The maxima are closer together on the right than on the left because the source moves to the right before making each successive vibration. The wavelength is thus smaller on the right than on the left. The smaller wavelength λ means a higher frequency since the frequency and wavelength are inversely proportional to each other.

$$V_w = f\lambda; \text{ therefore } f = \frac{V_w}{\lambda}$$

(We are assuming that the speed of the wave, V_w, is constant and the air is still.) *Observer 1* therefore receives a higher frequency sound than observer 2. As the source passes by observer 1 and then moves away, the frequency smoothly shifts from being higher than the source to being lower than the source, making the familiar eeeeaarrhoooom.

The Doppler effect also occurs for a moving observer. If an observer moves toward a sound source, then each maximum is received earlier than if the observer is stationary because each successive maximum has less distance to travel than its predecessor. Since the maxima arrive sooner, the

Figure 8.15a No Doppler effect: a stationary source. The larger circles were emitted earlier than the smaller ones and have had time to travel farther. The distance between maxima is the same in all directions and is the wavelength of the sound. All listeners observe the same frequency since the wavelengths of the sounds reaching them are identical.

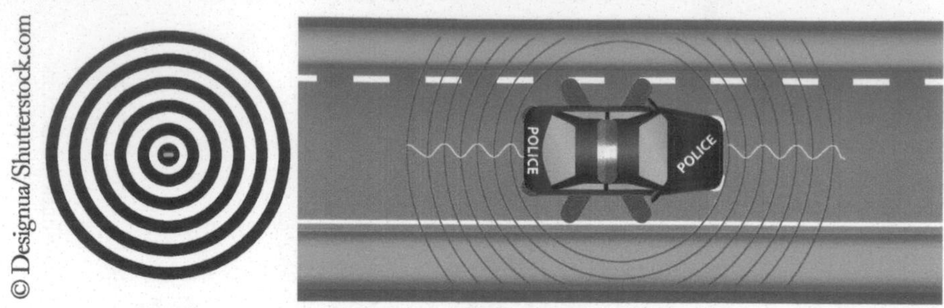

Figure 8.15b A source moving to the right; observers on the right receive a higher frequency than put out by the source and observers on the left receive a lower frequency than put out by the source. Each circle is an outgoing maximum centered about the point where the moving source emitted it.

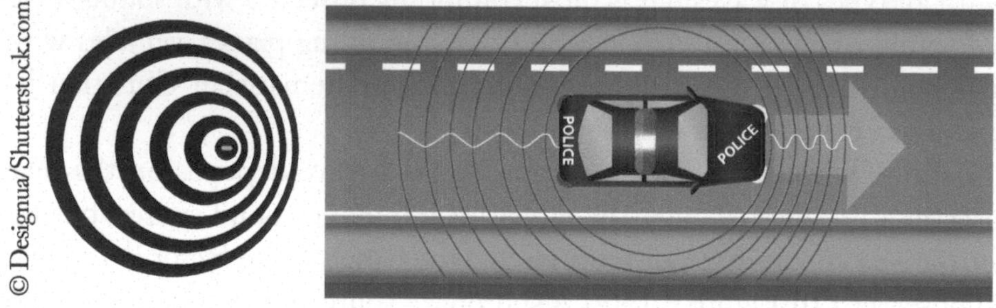

observed frequency is higher. Similarly, if the observer moves away from the source, each successive maximum has farther to travel and is received later than if the observer is stationary. The observed frequency is therefore lower when the observer moves away from the source. A train passenger can observe the Doppler effect in the sound of a railroad-crossing bell. The observed frequency of the bell shifts from high to low as the observer on the train approaches and then moves away from the bell.

The mathematical expression for the observed frequency of a moving source is given by:

$$f_{\text{observed}} = \left(\frac{V_{\text{w}}}{V_{\text{w}} \pm V_{\text{source}}}\right) f_{\text{emitted}} \quad \text{know Eq. 8.8}$$

V_{w}= the speed of the wave. Notice the $\pm$ sign in front of the speed of the source, V_{source}. The plus sign is used when the source moves away from the observer, and the *minus sign is used when the source moves toward the observer*. This equation is valid only for a moving source and stationary observer. No echo is involved. However, it is accurate within 5% for a moving observer and stationary source if the velocity of the observer is less than 20% of the velocity of the wave. Reminder, Eq. 8.8 is not valid when an echo is involved., no echo.

EXAMPLE 8.8A

In some of Doppler's experiments, musicians riding on an open train car played notes that were observed by a stationary observer as the train approached and passed by. Calculate the observed frequency (a) as the train approaches and (b) as the train moves away, given that the frequency the musicians are playing is 220 Hz, the velocity of the train is 15 m/s, and the velocity of sound in air is 341 m/s.

Solution.(a): Since the source is moving toward the observer, the minus sign is used in Equation 8.8:

$$f_{\text{observed}} = \left(\frac{V_{\text{w}}}{V_{\text{w}} \pm V_{\text{source}}} \right) f_{\text{emitted}}$$

$$f_{\text{observed}} = \left(\frac{341 \frac{\text{m}}{\text{s}}}{341 \frac{\text{m}}{\text{s}} - 15 \frac{\text{m}}{\text{s}}} \right) 220 \text{ Hz} = 230 \text{ Hz}$$

Solution.(b): Since the source is moving away from the observer, the plus sign is used in Equation 8.8, resulting in a lower frequency than the original.

$$f_{\text{observed}} = \left(\frac{341 \frac{\text{m}}{\text{s}}}{341 \frac{\text{m}}{\text{s}} + 15 \frac{\text{m}}{\text{s}}} \right) 220 \text{ Hz} = 211 \text{ Hz}$$

8.5.1 Use of the Doppler Effect to Measure Velocities

The Doppler effect can be used to calculate the velocity of a moving source. (In Example 8.7, the source velocity could have been calculated if the observed frequency had been given.) Doppler shifts in the light from stars and other astronomical objects are measured to calculate the motion of the object relative to earth.

Ultrasonic sound waves sent into the body are Doppler shifted by any motion in the objects that reflect them. It is possible, for example, to measure blood velocity by observing the Doppler shift of ultrasound reflected from blood cells. The Doppler shift of ultrasound is also used to monitor fetal heart motion. It is possible to detect fetal heart motion at least four weeks earlier in pregnancy by using ultrasound than by listening with a stethoscope because there is very little ultrasound background noise. Furthermore, the velocity of the fetal heart is higher than any other velocities in the abdominal region, giving a distinctly greater Doppler shift than other movements. These examples involve reflected sound waves. An echo.

So how do we solve problems *when an echo is involved*, as is the case for ultrasound imaging of a beating heart? When an echo is involved, the following equation is used:

$$f_{\text{D}} = \frac{2 f_{\text{o}} \cdot V_{\text{scatterer}} \cdot \text{Cos}(\theta_{\text{D}})}{V_{\text{w}} - V_{\text{scatterer}}}$$

f_D, The Doppler shift, which is the difference between the source frequency and the observed frequency. This is also equal to the beat frequency.
θ_D is the Doppler angle, and is measure relative to the direction that the reflecting object is moving and the direction that the wave is traveling.
f_o is the operating frequency of the transducer, often given in MHz.
V_w is the speed that the sound travels through the medium. The speed of sound in soft tissue is 1540 m/s.
$V_{scatterer}$ is the speed at which the reflecting body is approaching the transducer.

EXAMPLE 8.8B

If 2-MHz ultrasound is reflected from a soft tissue boundary moving at 10 m/s toward the transducer, calculate the Doppler shift. Note that the Doppler angle is 0°.

$$f_D = \frac{2 \cdot f_0 \cdot V_{scatter} \cdot \mathrm{Cos}(\theta_D)}{V_w - V_{scatter}} = \frac{2 \cdot 2\mathrm{MHz} \cdot 10\frac{m}{s}.\mathrm{Cos}(0)}{1.54 \times 10^3 \frac{m}{s} - 10\frac{m}{s}} = 0.026\,\mathrm{MHz}$$

It was noted in the previous section that beats can be used to determine accurately the frequency of a wave. Beats are used both in radar and in ultrasonic Doppler-shift stethoscopes to measure the frequency of the reflected wave. The reflected wave is combined with the original frequency and the beat frequency, $f_B = |f_2 - f_1|$ is measured. The beat frequency will increase as the difference in frequencies increases. (This fact can be used when tuning a guitar.) One great advantage of using beats is that the frequency of the source and reflected waves need not be measured accurately. The shift may be quite small and difficult to measure except by comparing the reflected frequency with the original and then observing the beats. The beat frequencies observed using ultrasonic Doppler-shift stethoscopes are in the audible range and are sometimes amplified for the benefit of the person doing the diagnostics. The sounds are not the actual sounds of the fetal heart, for instance, but rather the sound of the beat frequency of the ultrasound reflected from the moving heart.

8.5.2 Sonic Booms and Bow Wakes: Sources Moving Faster Than the Velocity of the Wave

Sonic booms still seem a little exotic, but they are very similar to an ordinary bow wake. Both occur when the source of a wave moves faster than the wave does. Figure 8.16 shows a wave source moving through a medium at a greater speed than the waves it is creating. The waves move out in circles from their points of origin, just as in Figure 8.15. There is constructive interference of the maxima along the lines drawn tangent to the circles, and the wave intensity is very large along those lines. The waves interfere mostly destructively between the lines, and the wave intensity is much smaller in that region. The angle θ between the lines of constructive interference depends on the velocity of the source: The faster the source, the smaller the angle. If the source of waves is a boat or even a duck, then a bow wake is formed, as seen in Figure 8.17.

Bow wakes from boats can rock other boats and are a serious source of riverbank and levee erosion. If the wave source is a supersonic airplane, then the constructive interference causes a sonic

Figure 8.16 A wave source moving faster than the propagation speed of the wave creates a shock wave, or bow wake. There is constructive interference along the lines drawn tangent to the circles.

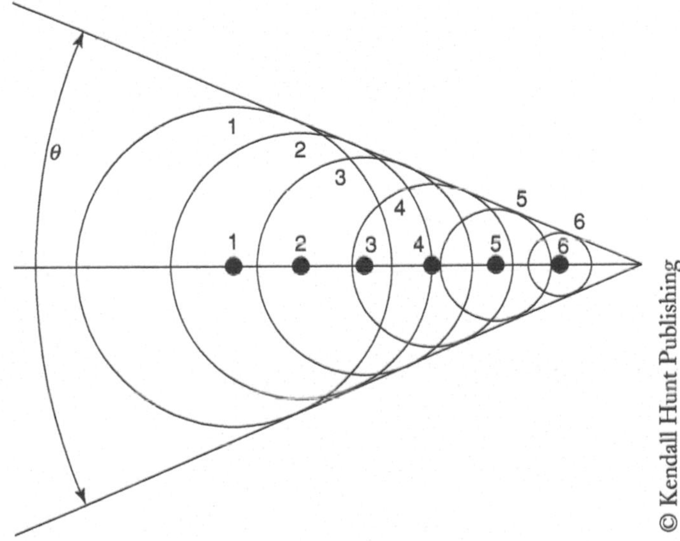

Figure 8.17 Bow wake created by a ducks moving faster than the water waves.

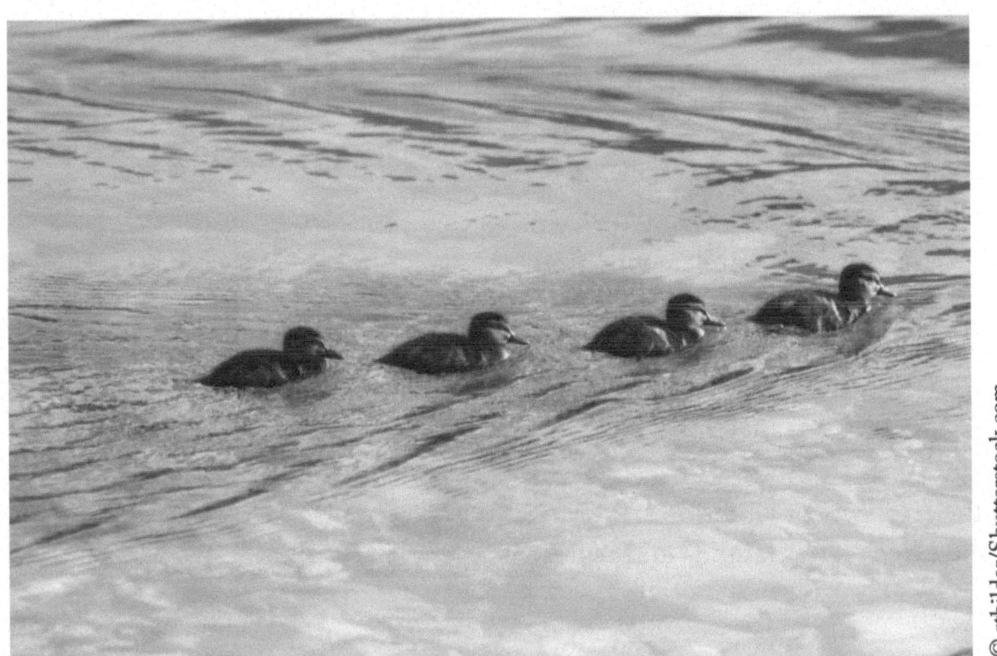

boom as seen in Figure 8.18. No sound is heard by an observer on the ground until the plane has passed by, after which the observer hears a loud boom. Sonic booms are not the sound of a plane going through the "sound barrier" but rather are created continuously by the plane and trail along the earth like a woman's wedding dress.

Figure 8.18 A sonic boom generated by an airplane moving faster than the speed of sound. There are actually two booms: one caused by the nose of the plane, the other by its tail.

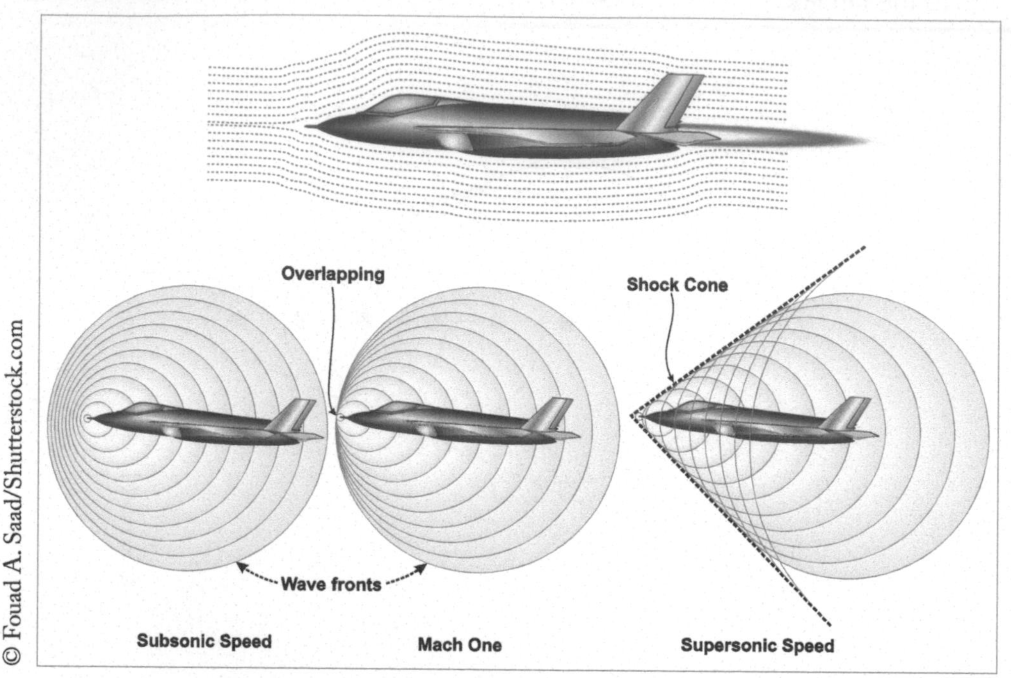

Things to know for section 8.5

$$f \equiv \frac{\text{number of vibrations}}{\textit{time}}$$

$$1\ \text{Hz} \equiv \frac{1\ \text{vibration}}{\text{second}}$$

Period $T \equiv$ time for one complete vibration

$$f = \frac{1}{T}$$

The velocity of a wave $V = f\lambda$

f is the *frequency* of vibration
λ is the Greek letter lambda and represents the wave's *wavelength*

$\textit{Intensity} = \dfrac{\textit{Power}}{\textit{Area}}$ Its units are: $\dfrac{\text{watts}}{\text{m}^2}$

Sound level: $\beta = 10 \cdot \text{Log}\left(\dfrac{I}{I_0}\right)$ Units are decibels.

The reference intensity $I_0 = 1 \times 10^{-12}\ \text{W/m}^2$

Doppler equation with no echo: $f_{\text{observed}} = \left(\frac{V_w}{V_w \pm V_{\text{source}}}\right) f_{\text{emitted}}$

V_w; the velocity of the wave.

Doppler equation with an echo as used in ultrasound: $f_D = \frac{2 \cdot f_o \cdot V_{\text{scatter}} \cdot \text{Cos}(\theta_D)}{V_w - V_{\text{scatter}}}$

Where f_D is the Doppler shift or beat frequency. f_o the operating frequency.

Note that when V_{scatter} is slow compared to $V_{\text{propagattion}}$,

V_w - V_{scatter} is approximately equal to V_w

Things to know for Chapter 8

Frequency $f \equiv \frac{\text{number of vibratons}}{time}$ Units are Hertz $1\text{Hz} \equiv 1 \frac{\text{vibration}}{\text{s}}$

Period $T \equiv$ Time for one complete vibration

Frequency $f = \frac{1}{T}$

The velocity of a wave $V = f\lambda$

f is the *frequency* of vibration
λ is the Greek letter lambda and represents the wave's *wavelength*

$Intensity = \frac{Power}{Area}$ Its units are: $\frac{\text{watts}}{\text{m}^2}$

Sound level: $\beta = 10 \cdot \text{Log}\left(\frac{I}{I_o}\right)$ Units are decibels.

The reference intensity $I_o = 1 \times 10^{-12}$ W/m^2

Doppler equation with no echo: $f_{\text{observed}} = \left(\frac{V_w}{V_w \pm V_{\text{source}}}\right) f_{\text{emitted}}$

Doppler equation with an echo as used in ultrasound: $f_D = \frac{2 \cdot f_o \cdot V_{\text{scatter}} \cdot \text{Cos}(\theta_D)}{V_w - V_{\text{scatter}}}$

Where f_D is the Doppler shift or beat frequency.

Note that when V_{scatter} is slow compared to V_w,

V_w - V_{scatter} is approximately equal to V_w

f_o; the operating frequency.

Questions

8.1 Which do you expect to have the greater natural frequency, a rigid object or a spongy object (identical except for the material from which they are constructed)? Explain your reasoning.

8.2 Sound travels faster through rigid materials like steel than through spongy materials like rubber. Why?

8.3 If the coils of the spring in Figure 8.2(a) were not connected together, then a transverse wave could not be transmitted. How does this help explain why sound waves in fluids are longitudinal and cannot be transverse? (Note: Surface waves on water have a transverse component, but transverse waves occur only at the surface and are much slower than the longitudinal sound waves transmitted by water.)

8.4 Sound velocities in gases are of the same order of magnitude as the average random velocities of molecules in the gas. Explain this in terms of how sound waves are transmitted on a molecular scale. Is this consistent with the fact that the speed of sound in air increases with increasing temperature?

8.5 Pascal's principle states that any change in pressure exerted on an enclosed fluid is transmitted to all parts of the fluid. At what speed would you expect the pressure to be transmitted?

8.6 How can the distance to an object be measured by bouncing a wave off it?

8.7 How does the frequency of a wave affect its wavelength, assuming that wave velocity is independent of frequency? Illustrate with an example of waves on a spring, as in Figure 8.2.

8.8 Ultrasonic scans of the eye are made using higher frequencies than scans of the abdomen (about 5 versus 2 MHz). Why is this possible and what is its advantage?

8.9 What is the primary difference between ultrasound used for diagnostic purposes and that used for therapeutic purposes? What are the known hazards of each?

8.10 Which of the following characteristics of waves are most important in determining the damage done by earthquakes-wavelength, frequency, amplitude, intensity, or velocity of propagation?

8.11 In some circumstances one wave, such as the vibration of a guitar string, can cause another wave, such as a sound wave. The frequency of the second wave is the same as the first, but its wavelength is not. Explain why.

8.12 What are the advantages of the decibel scale? Does a sound level of 0 dB correspond to an intensity of zero? Explain.

8.13 Ultrasound used for deep heating (diathermy) can cause "bone burns" or tissue damage due to overheating near the bones. Explain how reflection and interference might play a role here.

8.14 Give several examples of resonance.

8.15 Loudspeaker boxes are designed so that they do not resonate at normal audible frequencies. What would the effect be on music produced in a loudspeaker box that resonated very well at 300 Hz?

8.16 What wave characteristic is more crucial when an operatic soprano breaks a wine glass by singing at it, the frequency or the intensity of the note she sings? Identify and describe the physical phenomenon involved.

8.17 When two waves interfere constructively as in Figure 8.8(a), the resultant amplitude is twice as great as that of each of the two, meaning its intensity is four times as great as either. Furthermore, when total destructive interference occurs, as in Figure 8.8(b), the resultant amplitude is zero, meaning the intensity is zero. Explain how energy is conserved in each case. (Hint: consider Figure 8.11 and think of other forms that the wave energy could take.)

8.18 What is the relationship between the length of an air column and its lowest natural frequency (the fundamental)? Give examples, such as musical instruments, that are consistent with your assertion.

8.19 When a wave is reflected from a moving object its frequency is Doppler shifted. Are two shifts involved or one? Explain.

8.20 Ultrasound can be used to determine if a blood vessel is occluded by measuring the blood velocity at and near the occlusion. Would you expect the Doppler shift of the reflected sound to be greatest at the occlusion or nearby?

8.21 A sonic boom can break windows. What is the mechanism by which the sound does its damage?

Problems

SECTION 8.1

√8.1 (I) What is the time required for one complete vibration of an ultrasonic stethoscope that operates at a frequency of 2.0 MHz?

√8.2 (I) What is the frequency of a sound wave made by

a tuning fork that requires 2.44×10^{-4} seconds for one complete vibration?

√8.3 (1) What are the periods of the lowest and highest frequencies normally audible to humans (20 and 20,000 Hz)?

8.4 (II) If a tire has a tread pattern with crevices every 2.0 cm, what is the frequency of the sound it makes while moving at 30 m/s? You may assume that each crevice makes a vibration as it comes in contact with the pavement.

8.5 (II) What is the frequency of sound put out by the engine of a car moving at 90 km/hr if its four cylinder engine makes 2000 revolutions per kilometer?

8.6 (III) How fast is a race car going if its eight-cylinder engine emits a sound of frequency 1400 Hz, given that the engine makes 2000 revolutions per kilometer?

SECTIONS 8.2 AND 8.3 (Unless otherwise noted, use 344 m/s for the speed of sound in air.)

√8.7 (I) Calculate the wavelengths of the lowest and highest frequency sounds normally audible to humans (20 and 20,000 Hz), given the speed of sound to be 344 m/s.

√8.8 (I) (a) If the speed of sound in seawater is 1560 m/s, what is the wavelength of a 100-kHz sound made by a dolphin? (b) What is the wavelength of a 100-kHz sound made by a bat in air?

√8.9 (I) A gun is fired on a day when the speed of sound is 335 m/s and an echo is heard 0.75 seconds later. How far away is the object that created the echo?

√8.10 (I) At a fireworks display, a person noticed that the sound of an exploding shell arrived 0.30 seconds after the flash from the explosion. How far away from the person was the explosion if the speed of sound is 340 m/s? Can the time needed for the light to arrive be neglected? (The speed of light is 3.0×10^8 m/s.)

√8.11 (I) A storm in the South Pacific creates ocean waves traveling at 10 m/s. How long does it take these waves to reach the California coast 12,000 km away?

√8.12 (I) How many times a minute does a boat bob up and down on ocean waves that have a wave length of 50 m and a speed of 6.0 m/s?

√8.13 (I) What is the frequency of ultrasound with a wavelength of 0.25 mm in tissue if the speed of sound in tissue is 1540 m/s?

√8.14 (I) If the person in Figure 8.2(a) shakes the spring up and down three times a second and the crests on the spring are observed to be 0.50 m apart, what is the speed of propagation of the wave on the spring?

√8.15 (I) What is the sound level β of a sound with an intensity of 1×10^{-7} W/m^2?

√8.16 (I) What is the sound level β of ultrasound with an intensity of 10 W/m^2 used for deep heat treatments (diathermy)?

√8.17 (II) Calculate intensity in watts per square meters of a 40-dB sound?

8.18 (II) Some people can hear sounds below the normal threshold, perhaps as low in sound level as −10 dB. Calculate the intensity in watts per meter squared of a −10 dB sound.

8.19 (II) What is the difference in sound level between a 1×10^{-5} W/m^2 sound and a 1×10^{-11} W/m^2 sound?

8.20 (II) In factors of 10, how much more intense is the 90-dB sound level in a factory than the 30-dB sound level in a quiet hospital room?

8.21 (II) A housefly at a distance of 5 m creates a noise level of 30 dB. Neglecting any effects due to interference, what is the sound level of 1000 flies at that distance?

8.22 (II) Ten cars at an intersection are blowing their horns simultaneously, creating a 100-dB sound in the middle of the intersection. What is the average sound level made by each horn? You may neglect any interference effects.

√8.23 (II) (a) What frequency ultrasound should be used to see details as small as 1.0 mm in tissue? The speed of sound in tissue is 1540 m/s. Is this a maximum or minimum frequency? (b) Given that the maximum effective depth is approximately 400 wavelengths, to what depth is this sound effective as a diagnostic probe?

8.24 (II) 5.0-MHz ultrasound is used to look for objects in the eye, such as fragments of glass. (a) What is the smallest detail detectable with this frequency sound, given the speed of ultrasound in tissue is 1540 m/s? (b) Is the effective depth of penetration great enough to probe the entire eye? (About 3.0 cm is needed.)

8.25 (II) Ultrasonic scanners determine distances to objects in a patient by measuring the times for echoes to return. What is the difference in time for echoes from tissue layers in a patient that are 2.00 and 2.10 cm beneath the surface? The speed of sound in tissue is 1540 m/s. (This is the minimum pulse repetition period needed for the scanner to see details as small as 1 mm and still avoid range ambiguity.)

8.26 (III) (a) How much power must be supplied to a speaker to create a 60-dB sound if the speaker is 10 cm in diameter and 1% efficient? Is your answer consistent with the fact that battery-operated radios can play for many hours before wearing out their batteries? (b) If the efficiency is increased to 10% (barely possible), what is the increase in sound level?

8.27 (III) A 90-dB sound is absorbed by an eardrum 0.75 cm in diameter for 2 hours. How much energy in joules does the eardrum absorb in that time?

8.28 (III) Earthquakes create both longitudinal and transverse waves in the crust of the earth. The longitudinal waves travel faster than the transverse waves, averaging 8.2 and 4.8 km/s, respectively. A distant observer of earthquake waves will receive the two different components at different times. If the longitudinal wave arrives 150 seconds before the transverse wave, how far away is the origin of the earthquake?

SECTIONS 8.4 AND 8.5

8.29 (I) Each key on a piano causes either two or three strings to vibrate. If striking one key on a piano vibrates three strings having frequencies of 263.8, 264.0, and 264.3 Hz, what beat frequencies result?

8.30 (I) If a piano tuner hears five beats each second when listening to a 440-Hz tuning fork and a single string in a piano, what is the frequency of the string? (There are two possibilities.)

8.31 (I) Many cars have two horns. Suppose a car has horns that emit frequencies of 150.0 and 150.4 Hz. (a) What beat frequency is heard when the horns blow? (b) What is the time between beats?

√8.32 (I) An ambulance with a siren emitting a steady frequency of 1200 Hz is moving at 120 km/hr. What frequency is observed by a stationary observer as the ambulance approaches? As it moves away? Use 344 m/s as the speed of sound in air.

√8.33 (I) A low-flying jet aircraft approaches a stationary observer on the ground at a speed of 1000 km/hr on a day when the speed of sound is 345 m/s. (a) If the engine emits a sound of frequency 4000 Hz, what frequency does the stationary observer receive as the plane approaches? (b) As the plane flies away?

8.34 (II) A train blows its horn as it approaches a crossing. (a) If the train engineer observes a frequency of 125 Hz and a stationary observer at the crossing observes a frequency of 132 Hz, what is the speed of the train? (The speed of sound is 330 m/s on this day.) (b) What frequency is observed by the person at the crossing as the train moves away?

8.35 (II) The speed of sound in tissue is 1540 m/s. An ultrasonic wave sent into blood will be partly reflected back toward the source by blood cells. If the returning echo has a frequency 400 Hz higher than the original 2.0-MHz frequency, what is the velocity of the blood? (Note that because there is an echo involved, there are two Doppler shifts here.)

8.36 (II) An ultrasonic wave is sent into a patient and reflected back to the source by the moving wall of the heart. Calculate the beat frequency obtained if the reflected sound is mixed with sound of the original frequency given the following information: The wall of the heart is moving at 10 cm/s straight away from the source, the original frequency is 2.5 MHz, and the speed of ultrasound in tissue is 1540 m/s. (Note that because there is an echo involved, there are two Doppler shifts here)

8.37 (III) (a) An ultrasound stethoscope detects a beat frequency of 250 Hz when it mixes an echo with sound of the original frequency of 2.0 MHz. Calculate the speed of the object, in centimeter per second, reflecting the ultrasound given the speed of sound in soft tissue is 1540 m/s. Use a Doppler angle of 0°. (Note that because there is an echo involved, there are two Doppler shifts here). (b) Can this information determine whether the object is moving toward or away from the stethoscope?

9

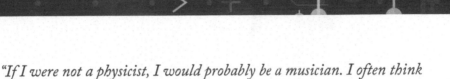

Sound and Hearing

"If I were not a physicist, I would probably be a musician. I often think in music. I live my daydreams in music. I see my life in terms of music."

—Albert Einstein

The physical phenomenon of sound was introduced in the previous chapter. Sound is one type of wave, and many wave properties were illustrated using sound as an example. This chapter is devoted to hearing, the human perception of sound, one of our most important senses. Normal hearing spans a remarkable range of sound frequencies and intensities. Microphones are just barely able to detect intensities as low as the ear can. The perception of sound is limited, however.

The hearing mechanism, its limitations, and the distinction between perception and actual physical quantities are discussed in Sections 9.1 and 9.2. A brief description of hearing loss and its correction is given in Section 9.3.

9.1 THE HEARING MECHANISM

What we normally call the ear is not an ear. Figure 9.1 is a diagram of the human ear. The external part is called the pinna or auricle. The pinna may slightly increase the sensitivity of the ear by reflecting sound waves into it. In humans this is almost negligible unless a hand is cupped behind the pinna. In some animals the pinna is much larger and can be rotated toward a sound source, both locating the source and increasing the ear's sensitivity.

The ear proper is divided into three parts: the outer ear, the middle ear, and the inner ear. The outer ear is just the ear canal, which terminates at the eardrum (tympanum). Pressure variations in sound waves exert forces on the eardrum and cause it to vibrate. The middle ear contains three small bones called the hammer, anvil, and stirrup (malleus, incus, and stapes). These bones transmit force exerted on the eardrum to the inner ear through the oval window. Because they form a lever system with a mechanical advantage of about 2, the force delivered to the oval window is multiplied by 2. Furthermore, the oval window has an area about loth that of the eardrum; thus the pressure created in the fluid-filled inner ear is about 40 times that exerted by sound on the eardrum (see Figure 9.2). This system enables the ear to detect very low-intensity sounds.

The middle ear also offers some protection against damage from very intense sounds. Muscles supporting and connecting the three small bones contract when stimulated by very intense sounds

Figure 9.1 Gross structure of the human ear.

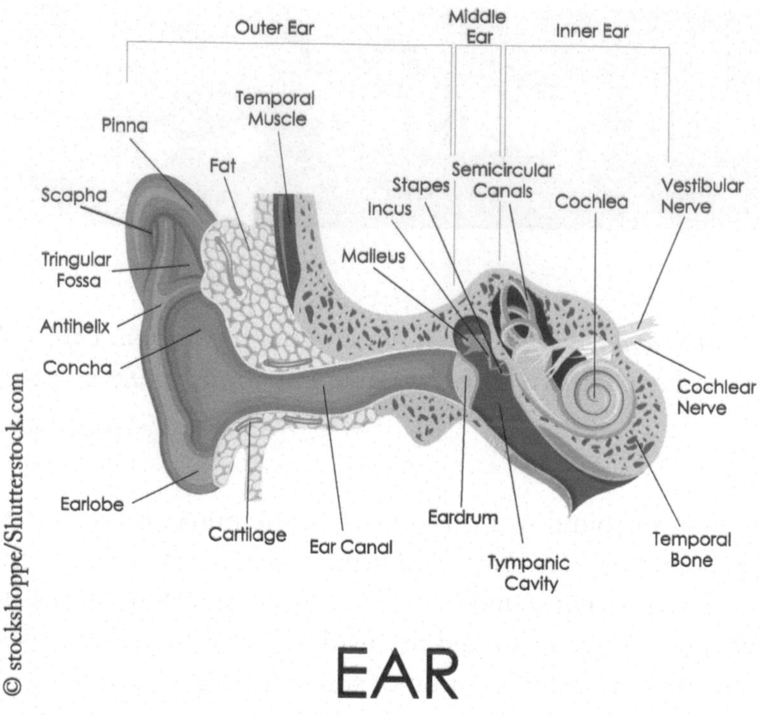

Figure 9.2 Diagram showing how the middle ear increases the sound pressure waves it conducts to the inner ear. The hammer, anvil, and stirrup form a lever system with a mechanical advantage of about two: F_2 is about twice F_1. The area of the oval window, A_2, is about $\frac{1}{20}$th that of the eardrum; therefore P_2 is about 40 times P_1. (Adapted from John R. Cameron and James G. Skofronick, *Medical Physics*, Wiley-Interscience, New York, 1978, by permission.)

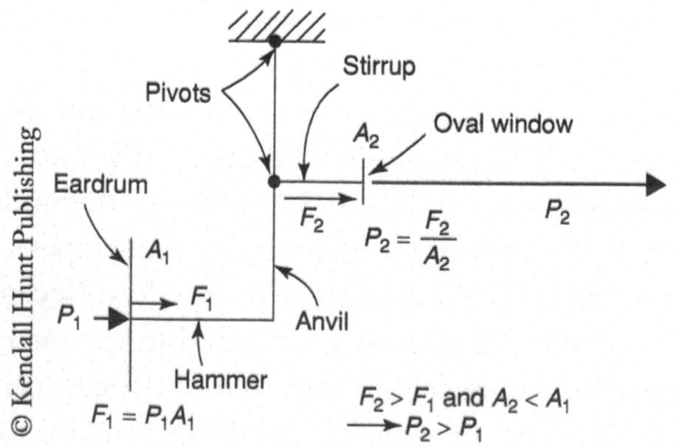

and reduce the force transmitted to the oval window by a factor of about 30. The reaction time for this defense mechanism is at least 15 milliseconds, so it cannot protect against sudden increases in sound intensity, such as from gunfire. The Eustachian tube is another protective structure in the middle ear. This tube allows air pressure in the middle ear to be equalized with atmospheric pressure to avoid large pressure differences across the eardrum, such as might be experienced with changes in altitude in plane flight. The Eustachian tube is normally closed, but the acts of chewing or yawning can open it, sometimes causing a distinct pop as air passes through it.

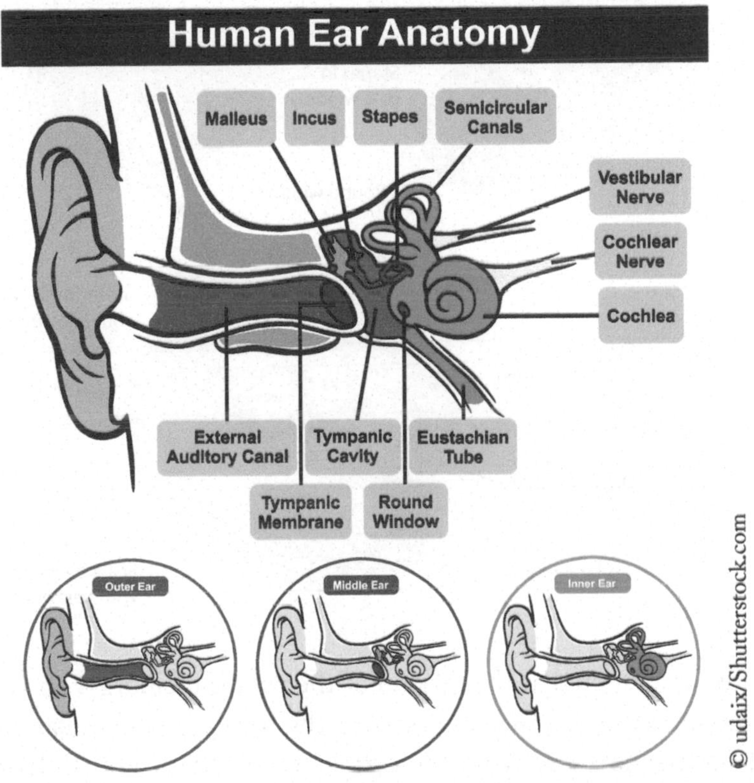

The *inner ear* contains the cochlea, the organ that converts sound waves into nerve signals to the brain. It is a coiled, tapered tube about 3 mm in diameter and 3 cm long (if uncoiled).

The cochlea has three fluid-filled chambers that run its entire length; sound enters one of them through the oval window, see Figure 9.3. Because liquids are almost incompressible, the inward motion of the oval window is transmitted through the cochlea to the round window, which then bulges out into the middle ear. The process is reversed when the oval window is pulled out by the stirrup. The central chamber is called the cochlear duct and contains the sound sensing structures of the ear. Sound vibrations cause the tectorial membrane to rub against hairs, stimulating nerves at their bases. Some 30,000 nerve endings participate in sending sound information to the brain from the cochlea.

The exact operation of the cochlea is not yet completely understood. However, certain aspects of how it sends signals to the brain can be stated with some confidence. One mechanism converts sounds having frequencies less than about 1000 Hz to nerve impulses of the same frequency. For example, a 264-Hz sound stimulates nerves to fire 264 times per second. Frequencies higher than 1000 Hz cannot be sent this way because individual nerves cannot fire more rapidly than about 1000 times per second. Furthermore, it is difficult to explain how the ear tells the difference between a frequency change and a change in intensity using this method. Normally, nerves indicate intensity by sending more rapid pulses, which here might be confused with higher frequency.

Another mechanism is also in operation, by which certain nerves are stimulated only by certain, frequency sounds. The brain then interprets a signal from these nerves as a sound of that frequency. High frequency sounds are detected at the near end of the cochlea (near the oval window), while

Figure 9.3 Anatomy of the cochlea.

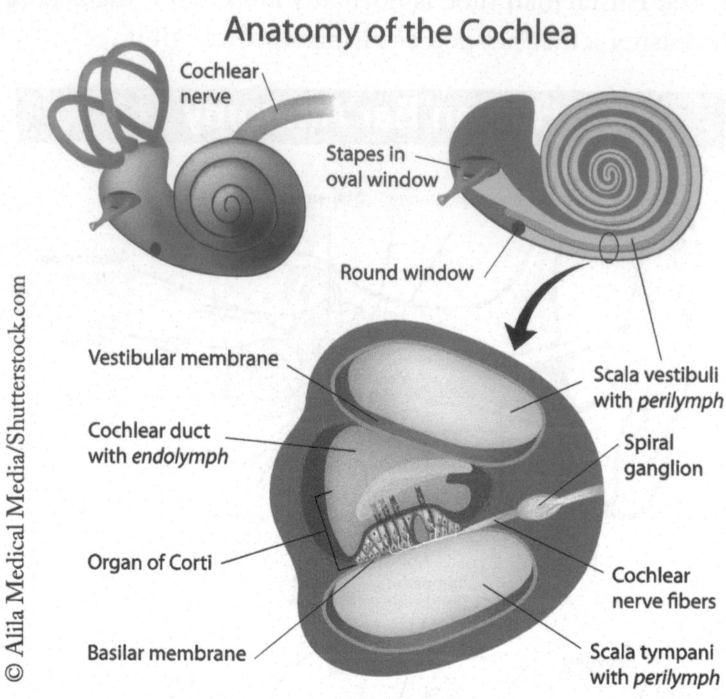

low-frequency sounds are detected at the far end of the cochlea. Georg Von Bekesey (1900–1970), a communications engineer who became interested in the hearing mechanism, did much to show that various parts of the cochlea are stimulated by various frequencies. Von Bekesey received the Nobel Prize in 1961 for his many contributions to the understanding of the ear. While this mechanism is accepted as a partial explanation of how sounds get converted into nerve impulses, it does not explain the entire frequency range of human hearing. Variations in rigidity along the length of the cochlear duct make it resonate at different frequencies in different locations, but this explains only a range of frequency detection about one-tenth the actual range. It is likely that when hearing is understood more completely, it will involve both mechanisms in modified form.

9.2 SOUND PERCEPTION

One meaning of the word *perception* is awareness through the senses. The perception of sound is particularly, interesting, because hearing is one of our basic senses. To what extent does hearing accurately detect the physical characteristics of sound? The human ear has a remarkable sensitivity and range. It can detect sounds varying in intensity by a factor of 10^{12} from threshold to those causing damage, and in frequency from 20 to 20,000 Hz. Yet the perception of sound differs in important ways from the actual physical properties of sound. The vocabulary used so far has been related to physical properties of sound, such as frequency and intensity. We shall now define words related to the perception of sound and then discuss the relationships of perception and physically measurable quantities.

9.2.1 The Perception of Single and Multiple Frequencies

Pitch is the perception of frequency. Most people have good relative pitch; that is, they can tell that one sound has a higher or lower frequency than another. Frequencies usually must differ by 0.3% or more to be told apart. For example, 1000 and 1003 Hz are noticeably different in pitch. A few people have what is called perfect pitch; that is, they can identify musical notes in addition to telling which note has the higher frequency. This ability is rare even among musicians. A person with a poor sense of pitch is called tone deaf. Skill in sensing pitch can be improved with training, but it is in part an innate ability. Pitch perception does not depend on sound intensity; high- and low-intensity sounds of the same frequency are sensed to have the same pitch. Of course, there is no pitch perception for sounds outside the audible range, those either too low in intensity or too high or low in frequency.

Multiple-frequency sounds are often perceived subjectively. A number of terms are used to describe multiple-frequency sounds, such as noise, music, rich, shrill, and mellow. These terms often imply value judgments that vary with the individual and his or her cultural background. There is not always a strong correspondence between the perception of multiple frequencies and any physically measurable quantities, for the same sound may be interpreted differently by different individuals. However, a few generalizations are possible. Sounds consisting of several frequencies are most often described as rich. A singer whose voice has many overtones may or may not be appreciated but will usually be characterized as having a rich voice. High frequency overtones are often called shrill or brassy. Singers whose voices have only a few overtones are said to have very pure voices. A voice with too few overtones is usually said to be thin.

What constitutes music is often hotly debated, but most people consider sounds with very large numbers of frequencies to be noise. An important factor in Western cultures is whether nearly equal frequencies are present simultaneously. Most music utilizes frequencies whose ratios are integers or simple fractions. Two tones played together usually sound pleasant when the ratio of their frequencies is not too close to unity. If the simultaneous frequencies are too close—for example, C and D on a piano then most people find them disharmonious. This taste for frequencies in simple ratios was first discussed by the ancient Greeks, who related it to the lengths of strings on musical instruments. (Frequencies could not be easily measured in those days). On the other hand, in the music of some cultures nearly equal frequencies are intentionally played together; the resulting beats may be perceived as richness of tone.

As noted in Chapter 8, humans are able to, recognize individual frequencies played simultaneously even though the combined sounds may be complicated in appearance (see Figure 8.12). In addition to recognizing a musical instrument or voice, most people find it easy to tell that several keys are being played in a piano chord, for example. There is perception of the parts as well as the whole. This ability to unravel a sound depends to some extent on the person, whether nearly equal frequencies are involved, and the relative intensity of each frequency. The ability is almost certainly due to the mechanism of sound conversion into nerve impulses in the cochlea and the fact that different parts of the cochlea are sensitive to specific frequencies.

9.2.2 The Perception of Intensity

Loudness is the perception of intensity, a well-defined physically measurable quantity. At a given frequency the more intense a sound is, the louder it seems (assuming the frequency is in the audible range). The ear does not respond linearly to intensity; a sound 10 times as intense as another does not sound 10 times as loud. (Most people would say it sounds about twice as loud.) The decibel scale defined in Section 8.3 corresponds fairly well to the human perception of loudness. Decibels are physically measurable and are fairly representative of the comparative numbers that people, would assign to the loudness of sounds. The smallest difference in intensity an average person can sense is about 1 dB, and an intensity difference of 3 dB is easily discernible.

Loudness depends strongly on frequency as well as intensity. Two sounds of different frequencies but equal intensities rarely sound equally loud. This is because the ear is more sensitive to some frequencies than others. The lowest curve in Figure 9.4 gives the intensity level needed for sound of different frequencies to be barely audible.

The curve is lowest between about 2000 and 5000 Hz, meaning that a low intensity is more audible' for those frequencies than others. Very large intensities are needed for sound to be audible near the extremes of the normal range of hearing—approximately 100 dB at 20 or 20,000 Hz, for example. The lowest curve in Figure 9.4 represents very good hearing; only 1% of the population can hear sounds at such low intensities. Also shown are curves representing the intensities that can be heard by 50% and 99% of the population. The threshold for normal hearing is often defined as 0 dB at 1000 Hz, corresponding to 10^{-12} W/m^2 (This serves as the reference intensity, I_o in the definition of decibels).

One cause of the sensitivity of the ear to frequencies in the 2000 to 5000-Hz range is resonance of air in the outer ear. The length of the ear canal is such that sounds of about 3000 Hz will cause the air in it to resonate, amplifying the sound and making the ear more sensitive to frequencies around 3000 Hz. Structures in the ear are forced to vibrate at the same frequency as the sound entering it. In any forced vibration some energy is converted to other forms, such as heat, depending on the characteristics of the object being forced to vibrate. The rigidity and mass of the structures in the

Figure 9.4 Hearing sensitivity profiles as a function of frequency tor the US population. The number to the right of the curves is the percentage of people who can hear sounds at that level. Also shown is the average threshold of feeling. Dashed lines indicate a relatively large uncertainty in the values graphed. (Adapted from Peter B. Denes and Elliot N. Pinson, The Speech Chain, Bell Telephone Laboratories, 1963, by permission.)

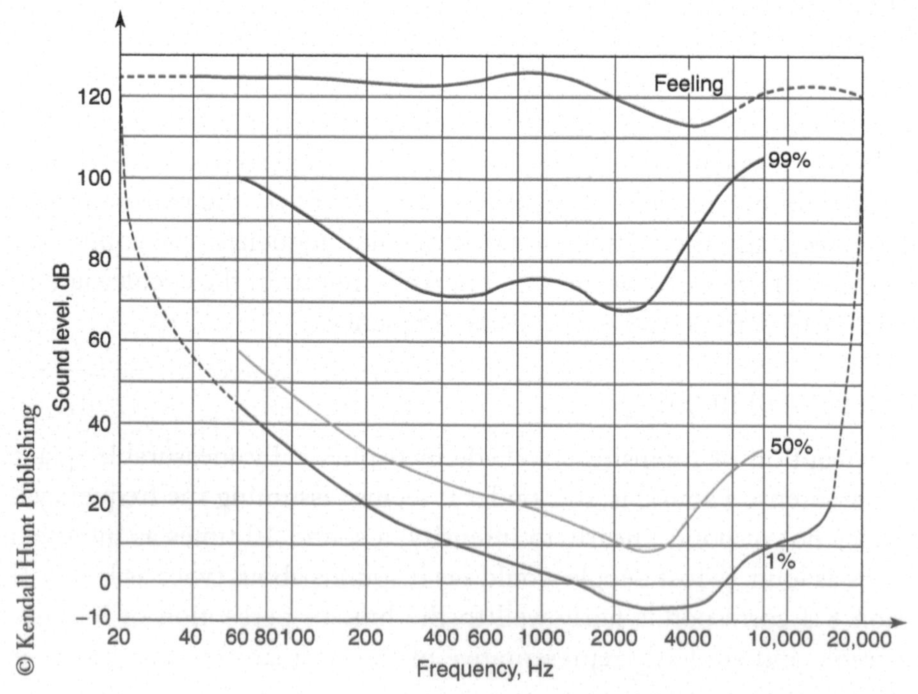

ear are such that they vibrate easily in the middle range of audible frequencies. At lower and higher frequencies the conversion of sound energy into vibrational energy in the ear is less efficient and hence less effective in stimulating nerves in the cochlea. It is not surprising that a single mechanism will not vibrate easily at all frequencies; it is surprising that the ear responds to frequencies over as large a range as it does.

Loudness, then, differs from intensity in some important respects, particularly because the perception of loudness depends on frequency as well as intensity. When considering how sounds affect humans, particularly psychologically, loudness is at least as important as intensity. Loud sounds can be very irritating, even to the point of causing an increase in blood pressure. It is important to keep the sound level low in hospitals, for example, but it is not important to reduce the intensity of all frequencies by the same amount. For example, according to the graph in Figure 9.4, a 40-dB 60-Hz sound is not audible and need not be further reduced in intensity to make a person comfortable. In contrast, a 40-dB 4000-Hz sound is very audible and has a much different loudness than the 60-Hz sound of the same intensity.

A unit, the *phon*, has been developed for loudness. The relationship between phons and decibels is shown in Figure 9.5. Phons and decibels are defined to be the same at 1000 Hz. A 60-dB 1000-Hz sound has loudness of 60 phons; for- example. The curved lines in the figure represent sounds of equal loudness. The lowest curve in Figure 9.5 is the 0-phon curve; it is very similar to the 1% curve in Figure 9.4. From the 0-phon curve one can determine the intensity in decibels needed for a sound to barely audible. For example, a 400-Hz tone is audible at an intensity of 10 dB, a 1000 Hz tone at 0 dB. Note the bottom curve dips below 0 dB in the region of 1500-6000 Hz; sounds of less than 0 dB are audible in that frequency range.

Figure 9.5 The relationship between loudness in phons and intensity in decibels. Each curve is an equal loudness curve; all frequencies along a curve are perceived as having equal loudness.

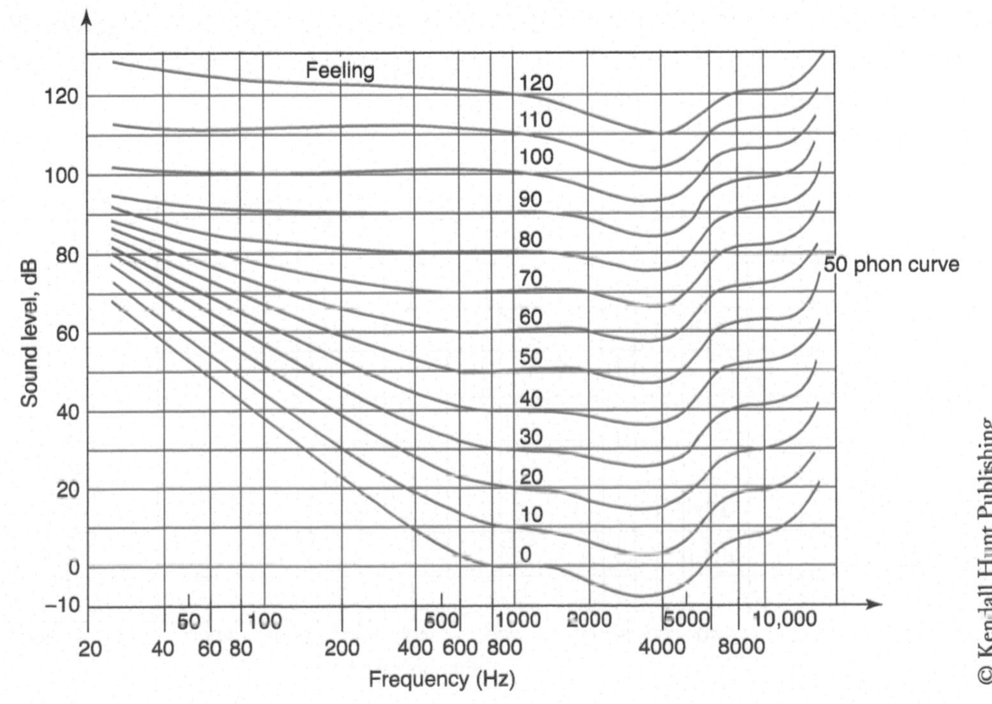

Now consider the 50-phon equal-loudness curve in Figure 9.5. A 1000-Hz tone with an intensity of 50 dB has a loudness of 50 phons, but a 100-Hz tone must have an intensity of about 67 dB to sound as loud to the average person. A 3000-Hz tone needs an intensity of only 47 dB to sound as loud. The ear is still more sensitive near 3000 Hz and less sensitive near the extremes of the' audible frequency range. Note that as intensity increases the equal-loudness curves become relatively flat. At high intensities the ear responds equally well to most frequencies, although it remains more sensitive in the region around 3000 Hz. For example a 100-dB 60-Hz sound and a 100-dB 800-Hz sound are both of equal loudness; both are on the 100-phon curve.

The curves in Figure 9.5 are averages obtained by asking large numbers of people to evaluate the loudness of sounds of different frequencies. Not everyone responds in the same way, but the general features of Figure 9.5 hold for anyone without a hearing impairment.

9.2.3 Sound-Level Measurements and Environmental Noise

There is a growing awareness that noise pollution and sound levels must be controlled in certain environments. Intense sound damages hearing, and laws have been and continue to be enacted to protect workers from excessive sound intensities. Psychologically negative effects occur at much lower levels than physical effects and are related more to perceived loudness than intensity. Sound-level meters, often called decibel meters, are designed to measure sound levels as humans would respond to them; their output is representative of loudness rather than intensity.

The three internationally accepted weightings given to sounds in sound level measurements are shown in Figure 9.6. The A *weighting* makes the meter respond most like the ear at low intensities since it suppresses the response to low frequencies. For example, a sound of frequency 40 Hz must have an intensity of at least 35 dB to affect the meter reading. The A weighting is most useful for

Figure 9.6 Frequency-dependent weightings used in sound level meters. Negative decibel values mean that the meter reduces its response to a given frequency to be more representative of the human ear's response.

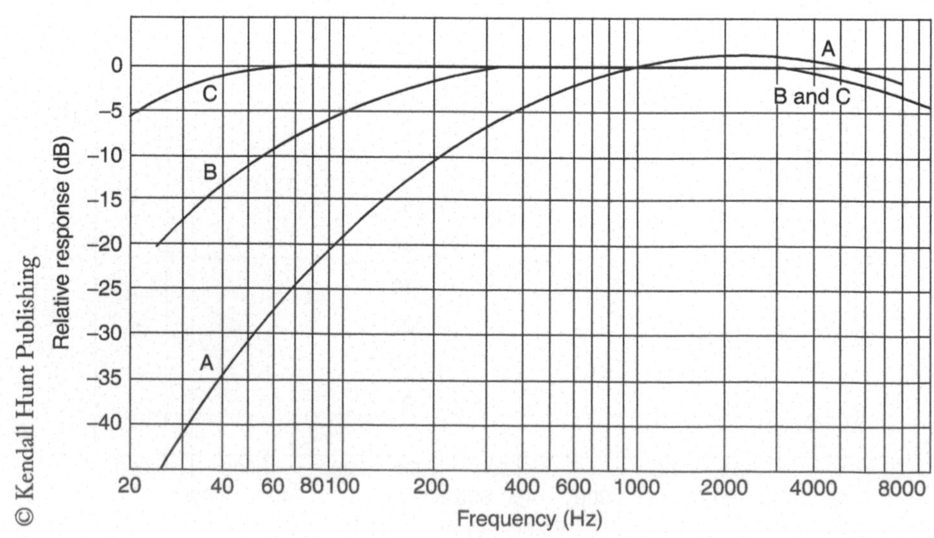

measuring sound levels in hospitals, offices, and classrooms, where low frequency sounds are nearly inaudible and needn't be of concern. The *B* and *C weightings* are more representative of the response of the ear to moderate and high intensities, with the C weighting being nearly equal at all frequencies. These weightings are of most value in measuring high sound levels along highways, at airports, and in certain industries. At high sound levels low frequencies are not only audible but may damage hearing.

It is much more difficult to reduce sound levels than it is to measure or complain about them. There are four aspects to noise reduction: the distance from a source, attenuation by absorption, destructive interference of waves, and the reduction of sound output from the source. The last of these is perhaps the most effective. Remember that sounds come from a vibrating object. The smaller the area of the object, the less air it can interact with and the lower the sound level it will create. The more rigid the materials of which the vibrating object is made, the smaller amplitude its surface will have, producing smaller pressure waves in the air. The vibrating object should be cushioned from contact with other objects it might cause to resonate. Absorption is used in earmuffs and acoustical ceiling tile. Absorptive materials usually are soft, with holes or crevices, so that the sound must make multiple reflections to bounce off them or get through them. Distance usually helps, unless the source broadcasts mostly in one direction. Destructive interference of waves is used in noise cancelling headphones.

Perception is traditionally the domain of disciplines other than physics. Nevertheless, it is interesting to see the role of physics in certain perceptions and the correspondence of perception to physically measurable 'quantities. Knowledge of sound perception aids greatly in treating hearing loss, designing musical instruments, and reducing noise.

9.3 HEARING LOSS AND CORRECTION

Hearing loss has many causes, the most common being age. Trauma (sudden injury), prolonged exposure to high sound levels, disease, and congenital birth defects are among other causes. Knowledge of the hearing mechanism and the nature of sound waves gained in this and the previous chapter is useful in understanding the general features of hearing loss and its possible correction.

There are two basic types of hearing loss. The first is called conductive hearing loss. It is caused by defects in the structures that conduct sound to the inner ear. The second is called neural hearing loss (sometimes called sensorineural, for sensory and neural). Neural hearing loss results from damage to the cochlea or neurons that send sound information to the brain. Neural hearing loss, like any nerve damage, is generally difficult to correct.

9.3.1 Hearing Tests

One step in evaluating hearing loss is to administer hearing tests. These tests not only determine the severity of a hearing loss but also aid in determining its type and methods of correction. The most common testing procedure is to place the patient in a sound proof room and ask him or her to signal when a sound becomes audible. The intensity of sound is raised and lowered to determine the threshold of hearing for that person. Each ear is tested individually, usually using a headset. A range of discrete frequencies is tested-typically 250, 500, 1000, 2000, 4000, and 8000 Hz, but occasionally frequencies intermediate to these also are tested. If a hearing loss is detected, a second test is often administered using bone conduction of sound rather than normal air conduction. In bone conduction tests a probe is placed against the skull behind the ear and sound vibrations of various frequencies and intensities are sent to the inner ear. Bone conduction tests bypass the outer and middle ear structures; if hearing is significantly better by bone conduction, then the hearing loss is conductive rather than neural.

Figure 9.7 is an audiogram—a graph of the results of a hearing test. The hearing threshold levels graphed on the vertical axis are the number of decibels above the normal threshold needed to be barely audible to the person tested. (A person with normal hearing will have a test result of 0 dB at every frequency.) The open brackets are test results using bone conduction, and the circles are results using air conduction. This person has a hearing loss of 45 dB at 250 Hz, 50 dB at 500 Hz, and so on.

The audiogram shown in Figure 9.7 is indicative of a conductive hearing loss since the results using bone conduction are nearly normal. One possible treatment is the use of a hearing aid that sends sound to the inner ear by bone rather than air conduction. Other treatments may be used, depending on what part of the conduction mechanism is affected and how it is affected. Surgery and ordinary amplification through air are other possibilities.

Before examining other hearing test results, it is instructive to note the difficulty of making precise hearing measurements and also to comment on what constitutes a severe hearing loss. Bone conduction tests suffer from a number of difficulties. Attenuation of sound in bone varies with frequency and differs from attenuation in air. This makes it difficult to obtain precise intensities at all frequencies and to compare bone conduction results with those of air conduction tests. At high intensities, bone conduction carries the sound to both ears, and the testing device may make significant air noise. It can be difficult to tell which ear is responding and whether the sound is carried by bone or air. Most of these difficulties are overcome by careful and consistent technique and by putting noise into the ear that is not being tested (called masking). The result is an uncertainty of 10 dB or more in the values obtained in bone conduction tests. Air conduction tests are more accurate because air attenuation is negligible for all frequencies and the equipment can be calibrated more

Figure 9.7 Audiogram of a person with a conductive hearing loss. Rectangles are bone conduction results; circles are air conduction results.

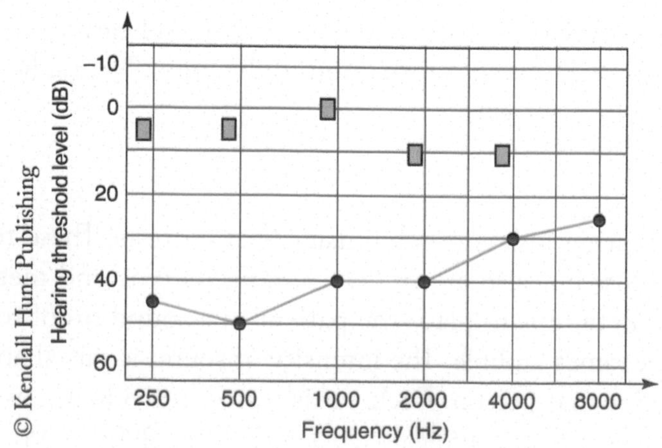

easily. Sound conducted by air into one ear is attenuated by about 50 dB before it gets to the other ear, so there is little confusion as to which ear is responding.

A hearing loss usually is considered severe when it interferes with a person's ability to understand conversational speech. The shaded area in Figure 9.8 gives the frequency and intensity distribution of normal speech. For example, sounds of 2000 Hz at 45 dB and 500 Hz at 60 dB fall in the shaded region. Also shown are the equal-loudness curves for 0, 40, and 60 phons.

A person with an overall hearing loss of 40 dB can understand conversational speech without a hearing aid. One with a 60-dB overall loss cannot, although some of the lower frequencies may be audible. More quantitatively, hearing losses classified by the hearing in the best ear are sometimes given a numerical rating. An overall 30-dB hearing loss is about a 5% hearing disability, and an overall 90-dB hearing loss is considered a 100% hearing disability.

Figure 9.9(a) is an audiogram of a typical hearing loss caused by aging, called presbycusis (literally "elder hearing"). Hearing is normal at low frequencies but falls off rapidly at higher frequencies. The loss is neural, since hearing by bone conduction is no better than by air. This person should not have much difficulty understanding conversational speech but is likely to have difficulty if the losses worsen. Hearing loss with age is very common. A 45-year-old person usually has a 10-dB loss and cannot hear frequencies over 12,000 Hz at all. A person 65 years old typically has a 30-dB loss for frequencies above 3000 Hz.

Correction by amplification may be useful for moderate losses only. Severe losses cannot be corrected by amplification because sound intensities of 80 dB or more may be needed. These may accelerate the deterioration of the ear and may be intense enough to cause discomfort. This is true for all severe hearing losses, not just those caused by aging.

Figure 9.8 Frequencies and intensities found in conversational speech. Hearing losses of greater than 40 dB in the indicated frequency range will impair the ability to understand conversation.

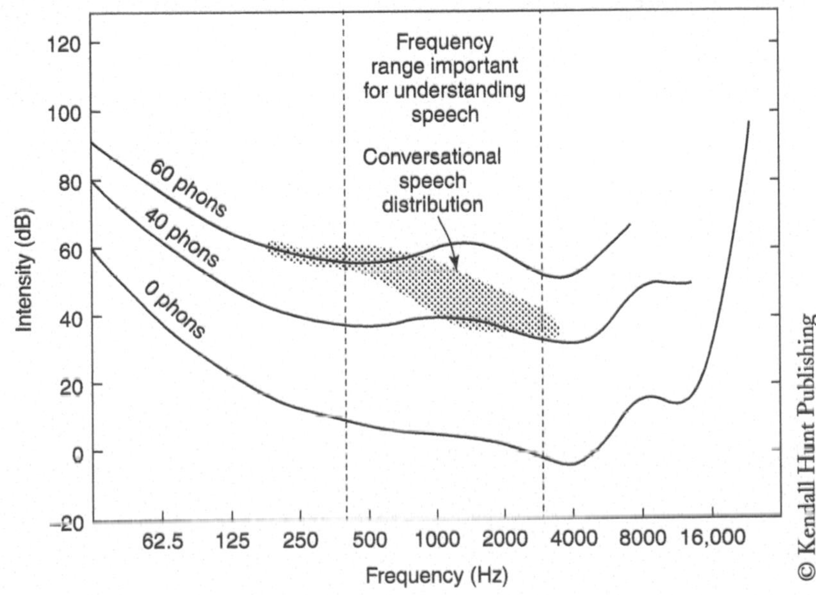

Figure 9.9 **(a)** Audiogram to neural hearing loss caused by aging (presbycusis). Open brackets are bone conduction results; the circles and Xs are air conduction results. The increased loss at higher frequencies is typical to presbycusis. **(b)** Audiogram of neural hearing loss in a 9-year-old due to loud noise (cap pistol). Circles and Xs are air conduction results for the right and the left ear, respectively. Bone conduction tests were not performed. Isolated dips in audiograms are typical to noise-induced nerve damage.

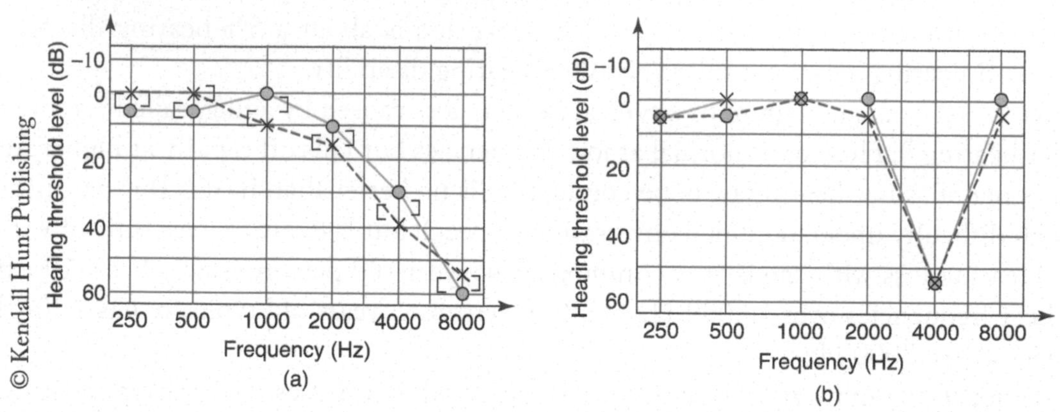

Noise-induced hearing loss often damages hearing only over a narrow range of frequencies, as is the case in the audiogram shown in Figure 9.9(b).

Isolated dips in audiograms, such as the one shown, are almost certainly due to neural damage. The reason for this certainty is the sensitivity of the cochlea to specific frequencies at specific locations. The location sensitive to about 4000 Hz was damaged in this case. Furthermore, conduction failures affect more than just a narrow range of frequencies. Bone conduction tests are considered unnecessary for such conditions. The loss is not easily treatable, because sound amplification at one frequency is not easy in a device as small as a hearing aid, and neural damage cannot be repaired surgically.

Questions

9.1 What is the function of the pinna (other than to serve as a support for eyeglasses)?

9.2 What effect does the ear canal have on hearing sensitivity?

9.3 How does the middle ear increase hearing sensitivity? Does the middle ear increase the energy of sound waves? Explain?

9.4 How does the middle ear protect the inner ear from loud sounds? Explain why this protection is not perfect, especially for sudden noises and prolonged very high-intensity sounds.

9.5 What are the two mechanisms used by the cochlea for coding sound information in nerve impulses? Why is it that neither mechanism alone describes the total range of hearing? (In fact, both mechanisms together do not.)

9.6 What is the difference between a perception and a physically measurable quantity?

9.7 Give an example of a sound that is intense but not loud.

9.8 How are pitch and frequency related? How are they different?

9.9 How is loudness related to intensity? How is it different? What are the physical factors upon which loudness depends?

9.10 Suppose you are deciding between two stereo systems. One reproduces frequencies with an accuracy of 0.3% and intensities with an accuracy of 1 dB. The other reproduces frequencies with an accuracy of 0.1% and intensities with an accuracy of 0.2 dB. Would it be wise to spend a great deal more money on the more-accurate system if all other features are the same? Explain.

9.11 The "cocktail party effect" is the ability to listen to only one conversation when several are audible simultaneously. Is it likely that this is a function of the ear or the brain? (Consider the hearing mechanism.)

9.12 The ear is not very sensitive to low-frequency sounds, such as those made by the heart. How does a stethoscope, such as the one shown in Figure 9.10, amplify heart sounds?

9.13 Environmental noise usually is not dangerous to hearing if it is inaudible, as is the case with fairly intense low-frequency sounds. Why is this, and' can you think of an exception?

9.14 How do drapes, stuffed furniture, and carpeting help make a room quiet?

9.15 How can a hearing test detect whether a hearing loss is conductive or neural?

9.16 Considering the ability of the ear to discriminate between sounds of different frequencies, explain why it makes sense to test hearing in steps of increasing size (250, 500, 1000, 2000, . . .) rather than uniform steps (250, 500, 750, . . .).

9.17 Describe the nature of a hearing loss in which a person could perceive a conversation as being reasonably loud but not be able to understand what is being said.

9.18 Why is hearing loss due to excessive noise most likely to be in the middle of the audible frequency range?

9.19 When your neighbor plays the stereo too loud, the frequencies you hear are predominantly the lowest ones. Why do the low frequencies penetrate better than high frequencies? Consider the wavelength of sound. (Understanding this doesn't make it any more pleasant.)

9.20 Why can't hearing aids be used for every type of hearing loss?

PROBLEMS

9.1 (I) What beat frequencies will be present (a) if the musical notes C and D are played simultaneously at frequencies of 264 and 297 Hz, respectively? (b). If C and F are played at 264 and 352 Hz? (c) If C, F, and A are played at 264, 352, and 440 Hz?

9.2 (I) Suppose a person is trying to distinguish between two sounds of different frequencies. If one has a frequency of 2000 Hz, what are the closest frequencies that can be distinguished as having a different pitch? (The sounds are not present at the same time.)

9.3 (I) What is the smallest frequency difference between sounds near 100 Hz that can be sensed using pitch perception? Between sounds near 4000 Hz? Between sounds near 15,000 Hz?

9.4 (I) Using the graph in Figure 9.5 find the threshold of hearing in decibels for frequencies of 60, 440, 1000, 4000, and 15,000 Hz. Note that a 60-Hz sound is made by many electrical appliances, 440 Hz is a very common frequency in music, 4000 Hz is close to the region of maximum sensitivity of the ear, and a 15,000-Hz tone is produced by most televisions.

9.5 (I) Suppose you have a 30-dB sound of frequency 1000 Hz. What sound levels in decibels are needed for frequencies 60, 3000, and 8000 Hz respectively, to seem as loud as the 1000-Hz tone?

9.6 (I) What is the sound level in decibels of a 600-Hz tone if it has a loudness of 10 phons? If it has a loudness of 80 phons?

9.7 (I) What are the loudnesses in phons of sounds of frequencies 200, 1000, 4000, and 10,000 Hz, all having intensity levels of 50 dB? What if they all have levels of 100 dB? 0 dB?

9.8 (I) A person has an overall hearing loss of 40 dB.

By how many factors of 10 must sound be amplified for it to seem normal in loudness to this person?

9.9 (II) What is the intensity in watts per square meter of a just barely audible sound of frequency 400 Hz? Of 6000 Hz?

9.10 (II) Calculate the intensity in watts per meter squared of an 80-Hz sound having a loudness of 30 phons. Do the same for a 10,000-Hz sound of loudness 10 phons.

9.11(II) Suppose a woman has a 10-dB hearing loss at 250 Hz and a 50-dB hearing loss at 4000 Hz. How many factors of 10 more intense must a 4000 Hz tone be to sound as loud to her as a 250-Hz tone?

9.12 (II) If a man needs an amplification of 106 at all frequencies, what is his overall hearing loss?

9.13 (II) Will a 60-Hz sound of intensity 20 dB register on a sound-level meter if the A-weighting is used? If the B-weighting is used? If so, what will its effective sound level be?

9.14 (III) (a) Calculate the maximum net force on an eardrum due to a sound wave having a maximum gauge pressure of 10^{-8} atmosphere (1×10^{-3} N/m^2) if the diameter of the eardrum is 0.75 cm. (b) Assuming the mechanical advantage of the hammer, anvil, and stirrup system is 2, calculate the pressure created in the cochlea. The stirrup exerts its force on the oval window, which has an area loth that of the eardrum.

9.15 (III) The old-fashioned ear trumpet increases sound intensity because it has a large sound-gathering area compared to the eardrum. What decibel increase would an ear trumpet produce if its area is 884 cm^2, the area of the eardrum- is 0.442 cm^2, and the trumpet is 5% efficient in transmitting all sound to the eardrum?

9.16 (III) A stethoscope is placed directly against the skin of a patient. It is reasonable to assume that sound is transmitted 100 times as efficiently into the stethoscope by contact than by air. The area of the stethoscope in contact with the patient is 12.5 cm^2, and the stethoscope is 40% efficient in transmitting sound to an eardrum of area 0.50 cm^2. What is the gain in decibels?

10 Introduction to Electricity and Magnetism

"You can know the name of a bird in all the languages of the world, but when you're finished, you'll know absolutely nothing whatever about the bird . . . So let's look at the bird and see what it's doing—that's what counts."

—Richard P. Feynman

Modern living relies heavily on electricity and magnetism—from the electricity that lights your home to the magnet that helps remove bits of iron from an injured eye. The nature of living organisms is dependent on electrical and magnetic phenomena. The human central nervous system, for example, is an electrical system far more complex than is found in the most advanced computers.

All electrical and magnetic phenomena are manifestations of the electromagnetic force, one of only four basic forces found in nature. Such diverse topics as pressure, sound waves, and heat transfer all involve electromagnetic forces since molecular forces are electromagnetic. Most types of force, including friction, cohesion, adhesion, and tension, are manifestations of the electromagnetic force. (See the discussion at the end of Section 3.2.)

Chemistry is based on the electromagnetic interactions of atoms and molecules. This means that something as simple as carbon combining with oxygen in a fire—or as complex as DNA replication and heredity—depends on the nature of the electromagnetic force. The very structure (size, shape, etc.) of atoms is dependent on the nature of the electromagnetic force since it is the primary force acting inside them. Furthermore, visible light is but one type of electromagnetic wave; radio, infrared, ultraviolet, and x-rays are others. Vision, heat lamps, sunburn, and cancer therapy thus all directly involve electromagnetic effects. The list could be extended indefinitely.

This chapter, which begins the direct study of electromagnetism, examines electric and magnetic forces, as well as such associated topics as electric current and voltage. Chapters 11–13 consider simple electric circuits, electrical safety, and bioelectricity. Electromagnetic waves are covered in some detail in Chapter 16.

10.1 ELECTRIC CHARGES AND FORCES

One of the first electrical phenomena studied was static electricity. Most of us have encountered static electricity when removing clothing from clothes dryer or upon being shocked by a metal doorknob after walking across a nylon carpet. The ancient Greeks noticed that static electricity can be generated by rubbing cloth on amber (petrified resin). After being rubbed, the amber will attract light objects, such as dust and feathers. Some force other than gravity is exerted by the amber. That force is electric. In fact, the word electricity comes from the Greek elektron, meaning amber.

How does rubbing a piece of amber with cloth cause a force to appear? That question remained unanswered until the substructure of atoms was determined thousands of years later. Figure 10.1 shows a simplified view of an atom with electrons arranged in energy levels, surrounding a tiny nucleus. The nucleus is positively charged. It contains at least one proton. When objects are rubbed together, surface atoms come into contact with one another and some electrons may be transferred between objects. If this happens, both objects will exhibit the ability to attract dust, feathers, or bits of paper and are said to be charged.

Charge is a basic physical property carried by certain subatomic particles, such as electrons. It is the basis of the electromagnetic force. Experiments have shown that charge has the following properties.

1. There are two types of charge. These are arbitrarily designated positive (+) and negative (-). All ordinary objects contain large and usually equal numbers of positive and negative charges. Electrons carry negative charge, and particles called protons carry positive charge. (Protons

Figure 10.1 A simplified view of an atom showing electrons in regions around the nucleus. The electrons are arranged in energy configurations that are different for every element. **This diagram is not to scale.**

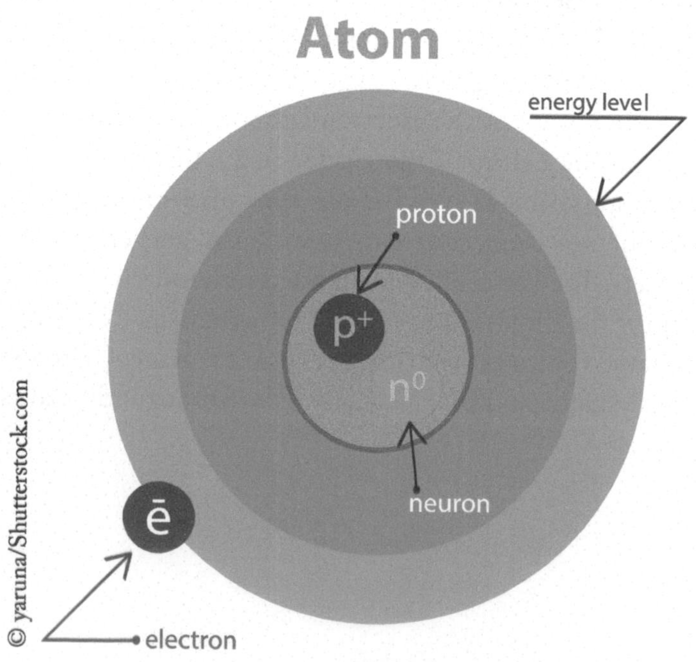

are found in atomic nuclei and are discussed in Chapter 18.) Electrostatic forces result from the separation of positive and negative charges. For example, amber has a greater affinity for electrons than does cloth and becomes negatively charged when rubbed. The cloth is left with a deficiency of electrons and hence a net positive charge (see Figure 10.2).

2. Charge is conserved. That is, charges can be separated, but they can neither be created nor destroyed. Neutral objects contain equal numbers of positive and negative charges. Both objects in Figure 10.2(a) are neutral.
3. Like charged objects repel each other and objects with opposite charges attract each other.

The quantitative expression for the force between stationary charges (the *electrostatic force*) is named Coulomb's law after the French physicist Charles Coulomb (1736–1806), although he was probably not its original discoverer. The Coulomb force is proportional to the magnitude of the charges involved. Although the force gets weaker with the square of the distance, it can reach out to

Figure 10.2 **(a)** Amber and cloth are original neutral, containing equal numbers of positive and negative charges (only a few are shown for illustration), **(b)** When rubbed together some electrons (–) are transferred to the amber, **(c)** The amber is left with a net negative charge and the cloth with a net positive charge. Electrostatic effects may now be observable.

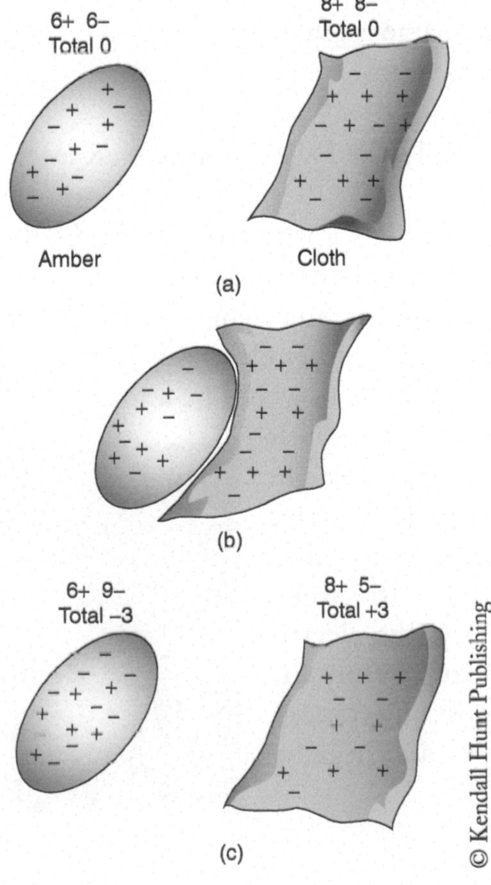

any distance. The direction of the Coulomb force is along a line joining the two charges. Coulomb's law for the static electric force is written as:

$$F = k\frac{q_1 q_2}{d^2} \quad \textbf{\textit{Know}} \text{ eq. 10.1}$$

where F is the electric force (in Newtons) between charges q_1 and q_2, d is the distance between the charges, and k is a constant determined by experiment.

$$k = 9 \times 10^9 \frac{N \cdot m^2}{\text{coulomb}^2}$$

The metric unit of charge, the coulomb (C), is a large amount of charge. The charges of the electron and proton are the smallest amounts of charge that are observed in nature. They are exactly equal in magnitude and opposite in sign, each carrying a charge of magnitude q = 1.6×10^{-19} C.

The *charge on an electron* is,

$$q_c = -1.6 \times 10^{-19} \, coulombs \quad \textbf{\textit{Know}}$$

The *charge on a proton* is,

$$q_p = +1.6 \times 10^{-19} \, coulombs$$

It takes 6.25×10^{18} electrons or protons to make a charge of 1.0 Coulomb. Because a coulomb is such a large amount of charge, charge is often given in microcoulombs. The Greek symbol μ (mu) is often used as a prefix that means micro.

$$1 \, \mu C \equiv 1 \times 10^{-6} \, coulombs \quad \textbf{\textit{Know}}$$

The electric force is fundamentally a very large force. As Richard Feynman pointed out, if two people were separated by 1 m and one of the people transferred just 1% of her electrons to the other person, what would happen? Since she lost 1% of her electrons, she would be positively charged. Because he gained 1% of her electrons, he would be negatively charged. Because they are oppositely charged objects, there would be a force of attraction. Do you want to guess how large the force would be? Give up? It would be approximately equal to the weight of the entire earth! They would not survive it. The *electric force is extremely large compared to gravity*. It is approximately 1×10^{38} times larger than the force of gravity. The electric force would completely dominate in every situation if it were it not that most objects contain equal amounts of positive and negative charge.

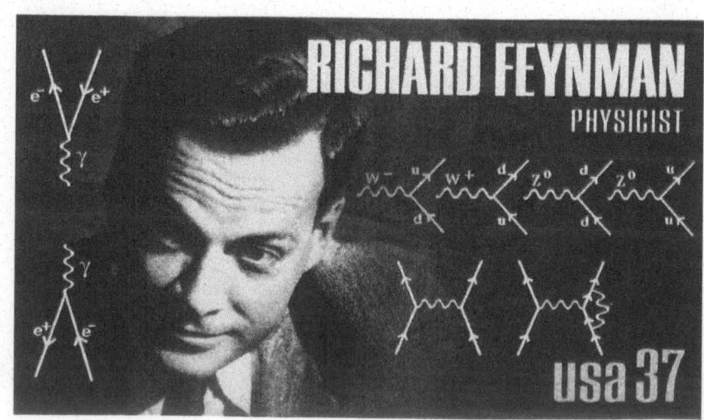

Consider the following example.

EXAMPLE 10.1

If all the positive and negative charges in 1 g of water could be separated, they would total 5.35×10^4 C and -5.35×10^4 C, respectively. Calculate the electric force between such charges when separated by a distance of 1.0 m.

Using Coulombs law,

$$F = k\frac{q_1 q_2}{d^2}$$

The *magnitude* of the force would be,

$$F = \frac{9\times10^9 \dfrac{N\cdot m^2}{C^2}(5.35\times10^4 C)(5.35\times10^4 C)}{(1m)^2} = 2.58\times10^{19}\text{ N}$$

The *direction* of the force can be determined from the signs of the charges. Because the charges were of the opposite type, one positive and one negative, the objects would be *pulled toward each other*.

Because 2.58×10^{19} Newtons is such a large force, it is in fact impossible to separate all the charges in a single gram of water to a distance of a meter. When static electricity is generated in situations such as those discussed at the beginning of this section, only about 1×10^{-9} C of charge is typically separated.

10.1.1 The Electric Field; Fields of Force

The electric force can act over extremely large distances. It is natural to ask whether the force acts instantaneously or takes time to be transmitted to some other place. For example, suppose two charges are originally separated by some distance and one of them is moved closer to the other. The Coulomb force will increase because the distance has decreased, but how soon will the other charge feel the increased force? Furthermore, what transmits the force between the charges? There is no "physical contact" as with a string or stick.

To answer these and other questions, physicists have theorized that every charge creates an electric field around itself. It is the force field that carries the force to other charges. If a charge changes position, then the field follows it and changes too. If the magnitude of the charge changes, then the field changes proportionally. Changes in the electric field travel at a finite speed: the speed of light (3.0×10^8 m/sec). The idea of an electric field is very useful to physicists.

The idea of an electric field can be generalized to all other forces, such as magnetic and gravitational forces: All magnets create a magnetic field and all masses create a gravitational field. Changes in these fields also travel at the speed of light. In fact, theoretical and experimental physicists are currently striving to show that all force fields are different manifestations of the same thing.

Things to know for section 10.1

Coulomb's law:	$F = k\frac{q_1 q_2}{d^2}$
Charge of an electron:	$q_e = -1.6\times10^{-19}$ coulombs
1 micro Coulomb:	$1\mu C \equiv 1\times10^{-6}$ coulombs

10.2 VOLTAGE: ELECTRIC POTENTIAL

Most cars use 12 Volt batteries, flashlights use one or more 1.5-V batteries, and household appliances run on 120- or 220- V electricity. Volts and voltage are common terms, but their exact definitions are not so familiar. Voltage, by definition is,

$$Voltage = \frac{Energy}{charge} \quad \textbf{Know} \qquad \text{eq. 10.2}$$

The SI unit for voltage is the volt (V), named for Alessandro Volta (1745–1827), the Italian scientist who invented the battery. Its units are directly related to its definition:

The units of voltage are,

$$\frac{Joules}{\text{coulomb}} \equiv volts \quad \textbf{Know}$$

The energy given to a charge by a voltage can be calculated by solving equation 10.2 (the definition of voltage) for the change in electrical potential energy:

Figure 10.3 (a) A CRT: Electrons "boil" off a hot filament and are accelerated to a large velocity by a high voltage. Those electrons that pass through the hole are focused into a narrow beam and are steered by applying voltage to the vertical and horizontal plates. When the electrons strike the phosphor, part of their energy is converted to light. The brightness of the spot can be varied by adjusting the temperature of the filament to obtain greater or fewer numbers of electrons. The tube is evacuated so that the electrons will not be stopped by collisions with air molecules.

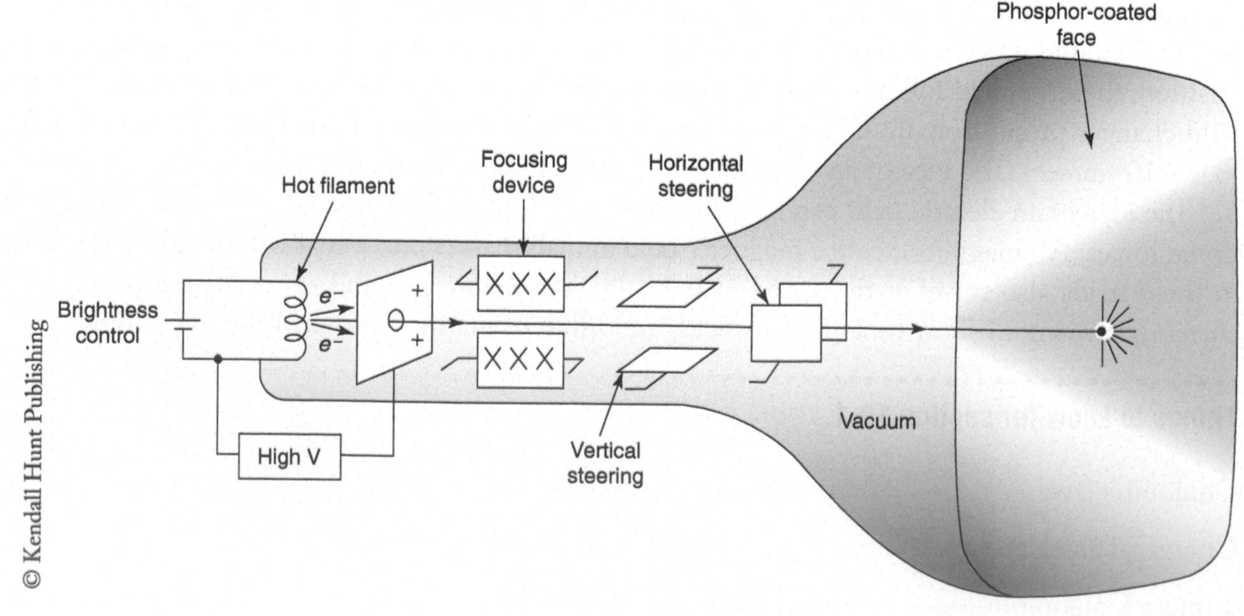

Figure 10.3 (b) An electron accelerating across a 20,000-V electrical potential difference.

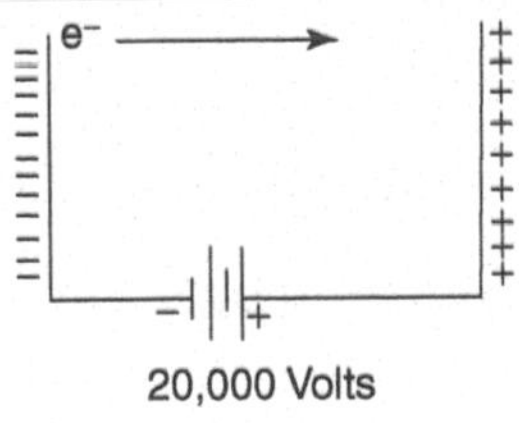

$$v \equiv \frac{E}{q}; \text{ therefore } E = vq$$

$$\textit{Electrical Potential Energy} = V_q \qquad \text{eq. 10.3}$$

where q is the amount of charge that moves and V is the electrical potential difference, or voltage, between its starting point and final position. The change in the electrical potential energy ΔPE goes into other forms, such as kinetic energy of the moving charge. Equation 10.3 is the exact relationship between voltage and energy. Energy is directly proportional to the voltage and to the amount of charge moved. Thus, for example, a 12-V truck battery contains more energy than a 12-V motorcycle battery because the truck battery is larger and can store and move much more charge.

One place where charges are moved by high voltages is inside of cathode-ray tubes (CRT). Inside of a CRT, the charged objects are electrons that are freed from atoms and accelerated toward a phosphor-coated screen. When the electrons strike the phosphor, part of their energy is converted to visible light. Cathode ray tubes are inside of x-ray machines, oscilloscopes, video games, older TVs, ECG monitors, and so on. Figure 10.3 is a schematic of the internal construction of a CRT.

Please view Tremblay's 'Electrical Potential Energy' lecture before proceeding.

EXAMPLE 10.2

Calculate the speed of an electron accelerated by the 20,000-V potential difference found in the CRT in Figure 10.3. The mass of an electron is 9.11×10^{-31} kg.

Solution: In this example, we are asked to determine the velocity of the electron when it strikes the positive plate. The electrical potential energy given to the electron is converted almost entirely to kinetic energy as it moves through the vacuum. Therefore, the electron's electrical potential energy, qV, at the negative plate, will equal the kinetic energy of the electron as it hits the positively charged screen.

Using conservation of energy, we write the following equation:

$$\textit{Electrical P.E.} = v_q = K.E. = \frac{1}{2}mv^2$$

$$Vq = \frac{1}{2}mv^2$$

In case your algebra is a little rusty, we will solve this equation for velocity v, using step by steps.

Step 1. Is v in the numerator? Yes.

Step 2. Is v alone? No. It is being multiplied by 1/2 and by the mass, m.

Step 3. Divide both sides of the equation by m and multiply both sides of the equation by 2.

We now have: $\frac{2Vq}{m} = v^2$

Step 4. Is v alone? Yes.

Step 5. Is v to the first power? No. It is being squared. Take the square root of both sides. Therefore the velocity of the electrons as they move through the hole in the positive plate is,

$$v = \sqrt{\frac{2Vq}{m}} = \sqrt{\frac{2(20{,}000)1.6\times10^{-19}\,\text{C}}{9.11\times10^{-31}\,\text{kg}}} = 8.4\times10^{7}\,\frac{\text{m}}{\text{s}}$$

This is a surprisingly large speed, but it is correct to within 3%. (At these speeds, relativity theory must be used to obtain better accuracy.) The electron's mass is so small that it is easy to accelerate it to high speeds.

In the preceding example a voltage caused charges to move from higher electrical potential energy to lower electrical potential energy. As the electron lost its electrical potential energy it gained kinetic energy. This is similar to a falling object losing gravitational potential energy and gaining kinetic energy. Energy is always conserved. When one type of energy appears to be lost, it has only transferred to another type of energy or to another object.

The most common uses of electricity involve the movement of charges. Light bulbs, computers, nerves, and CRTs, all utilize moving charges to perform their respective functions. Voltage can cause charge to move and that motion constitutes a flow of electricity, or current, the subject of the next section.

Things to know for section 10.2

Definition of voltage: $\quad Voltage \equiv \frac{Energy}{charge}$

Metric units of Volts: $\quad \frac{\text{joules}}{\text{coulomb}} \equiv \text{volts}$

10.3 CURRENT: THE FLOW OF CHARGE

Anyone who has reset a circuit breaker or changed a burned-out fuse has noticed that they are rated in amperes. An ampere is a unit of electric current, defined as the amount of charge flowing per unit time:

$$Current \equiv \frac{Charge}{time}; \quad I \equiv \frac{q}{t} \quad \textit{Know}$$

where I is the symbol for current and q is the amount of charge flowing past a point in a time t. The SI unit of current is the ampere (A). One ampere of current is defined as the flow of one coulomb of charge per second:

$$1 \text{ amp} \equiv \frac{1 \text{ coulomb}}{\text{second}} \quad \textit{Know}$$

André Marie Ampère.

© Nicku/Shutterstock.com

The ampere is named after Andre Ampere (1775–1836), a French physicist and mathematician who studied the relationship between current and magnetism.

How large is an ampere? Table 10.1 gives representative currents and voltages in certain situations.

A 100-W incandescent light bulb has a current of about 1 A, a moderate amount of current in electrical devices, but even the smallest current listed in the table involves very large numbers of charges moving per unit time. In most situations electric current is a flow of

TABLE 10.1 Representative Currents and Voltages

Device	$I(A)$	$V(V)$
Pocket calculator	0.0003	3
Cauterizer	0.002	15,000
Pacemaker	0.010	5
Clock	0.025	120
Flashlight	0.1	3
Table radio	0.3	120
100-W bulb	0.83	120
TV set	1.5	120
ECG monitor	2.0	120
Toaster	10	120
Clothes dryer	20	220
Defibrillator	10–20	10,000
Lightning bolt	20,000	100,000,000

electrons or ions. (An ion is a charged atom or molecule with an excess or deficiency of electrons.) The amount of charge in coulombs on an electron or ion is extremely small, so very large numbers of them are required in currents like those listed. Consider the following example.

EXAMPLE 10.3

The current in most electronic devices consists of a flow of electrons. How many electrons move through a pocket calculator during one hour of use?

Solution: In order to determine the number of electrons, it is first necessary to find the amount of charge; the number of electrons can then be calculated by dividing by the charge per electron. From the definition of current, equation 10.4:

$$Current \equiv \frac{Charge}{time}; \quad I \equiv \frac{q}{t} \quad \textbf{Know}$$

Solve the above equation for charge and substitute the given values. Note that 0.0003 amps can be written as 0.300×10^{-3} amps.

$$q = I \cdot t = 0.300 \times 10^{-3} \text{ amp} \cdot 1 \text{ hour}$$

Next, replace amp with coulombs per second and replace 1 hour with 3600 seconds.

$$q = 0.300 \frac{\text{C}}{\text{sec}} (3600 \text{ sec}) = 1.08 \text{ coulombs}$$

Now we convert the charge in coulombs to the number of electrons that are required to give us that charge.

And our *final answer* is:

$$1.08 \text{ C}\left(\frac{1 \text{ electron}}{1.6 \times 10^{-19} \text{C}}\right) = 6.75 \times 10^{18} \text{electrons}$$

Wow! That's a lot of electrons, especially considering that the calculator uses a relatively small current.

Commonly encountered currents involve so many electrons or ions that individual charges are not very noticeable. In fact, it was about 100 years from the time electric currents were first studied until it was proven that currents consist of the gross movement of large numbers of individual charges. Electric current is caused by an electrical potential difference between two locations, voltage.

Electric current moves much more easily through some materials than others. These materials are selected for wiring houses for example, and are called good conductors. Copper is one of the most commonly used conductors because it is relatively cheap and is the second best conductor of electricity at normal temperatures. (Silver is the best.) Wires are often insulated with plastic or rubber to prevent shock and short circuits. Insulators are poor conductors of electric current. The primary difference between conductors and insulators is the number of free charges in them. Metals, for example, are usually good conductors because they contain large numbers of electrons that are not attached to individual atoms. Other materials, such as amber, have very few free charges and the electrons in them are strongly attached to atoms and molecules. Static electricity can be built up on such poor conductors more easily than on a conductor, where the charges might be able to move away rapidly.

The surface of the earth itself is a good conductor because of the presence of water and dissolved salts. When salts dissolve, they break up into positive and negative ions that are relatively free to move, making the water a relatively good conductor of electric current. Even in dry regions there is enough water for the ground to be a good conductor. Any contact with the earth through a good conductor is called a ground and offers a path for electric current to flow (see Figure 10.4). This is why it is so dangerous to work on electrical devices when standing on wet ground. The water may lower your electrical resistance, resulting in a larger current might flow through you (and hazard) than if you were insulated by dry rubber-soled shoes.

These and other topics related to electric current will be discussed at length in later sections and chapters. The next section considers one effect of major importance: *Every time a current flows, a magnetic field is created.* This fact and other relationships between current and magnetism underlie all devices using magnetism.

Figure 10.4 Grounding: An object with an excess of negative charge is **(a)** brought near to and **(b)** in contact with a metal water pipe leading into the earth. Both the pipe and the earth are good conductors so Coulomb repulsion can move charges through them. The charges separate to large distances in the ground, leaving a negligible excess charge on the object.

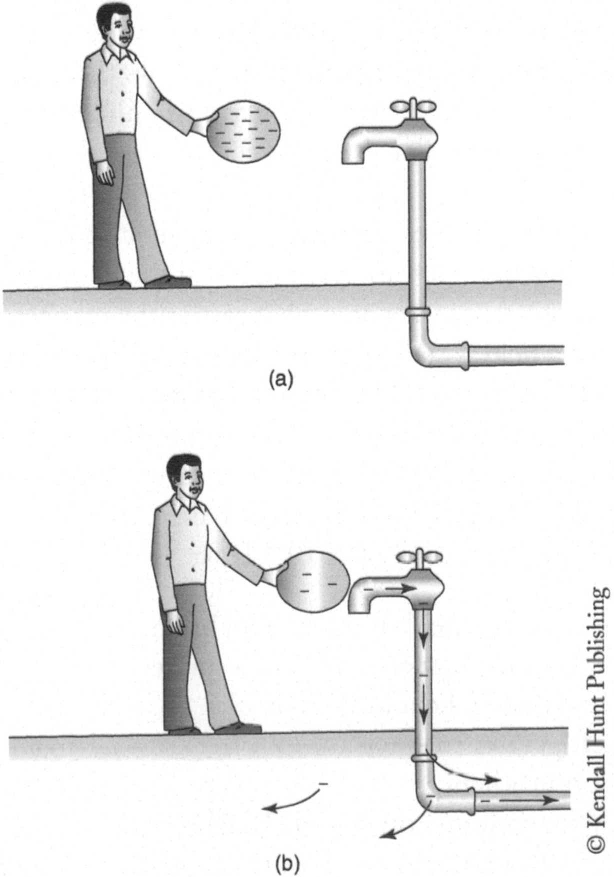

Things to know for section 10.3

$$Current \equiv \frac{Charge}{time}; \quad I \equiv \frac{q}{t}$$

$$1 \text{ amp} \equiv \frac{1 \text{ coulomb}}{\text{second}}$$

10.4 MAGNETISM

Perhaps the most familiar magnets are the small ones used to stick notes on refrigerators. Of course, compass needles also are magnets; mounted on low-friction pivots, they align themselves along the earth's magnetic field. The earth itself is in fact a gigantic magnet.

Magnets exert forces on one another and on some materials, such as iron. This force is the magnetic force, which is another manifestation of the electromagnetic force. Magnets and magnetism are named after Magnesia, a region in Asia Minor where ancient peoples found rocks that would attract or repel one another. Those rocks (called lodestones) contain iron, one of only a few elements that can be permanently magnetized. Magnets attract even unmagnetized iron. Iron and other materials that can be permanently magnetized, such as nickel, cobalt, and gadolinium, are called ferromagnetic materials after the Latin word for iron.

All magnets have two poles. These are named north and south poles because one end of a magnet is attracted almost toward the magnetic north pole of the earth and the other end toward the magnetic south pole of the earth. A little experimenting shows that like poles repel and unlike poles attract, analogous to the Coulomb force between like and unlike charges. Magnetic force is quite distinct from electric force, however. For example, compass needles usually have a total charge of zero but still feel magnetic forces.

A small compass can be used to test or map out a magnetic field, as shown in Figure 10.5. Lines are often drawn to indicate the direction and magnitude of the magnetic field, as in Figure 10.5. The magnetic field is not the force itself; rather, it is what carries the magnetic force to and exerts it upon another object. (The idea of a field was discussed at the end of Section 10.1.) Presently, the strength of the earth's magnetic field varies from 31 microtesla in most of South America and South Africa to more than 58 microtesla near the magnetic poles in northern Canada and part of Siberia, and south of Australia. The earth's magnetic field protects us from energetic, charged particles contained in cosmic rays. Its strength, location, and orientation change over time.

Typical values-Wikipedia. 1 gauss = 1.00×10^{-4} Tesla.

- **10^{-9}–10^{-8} gauss**—the magnetic field of the human brain
- **0.31–0.58 gauss**—the Earth's magnetic field at its surface
- **25 gauss**—the Earth's magnetic field in its core
- **50 gauss**—a typical refrigerator magnet
- **100 gauss**—a small iron magnet
- **2000 gauss**—a small neodymium-iron-boron (NIB) magnet
- **600–70,000 gauss**—a medical magnetic resonance imaging machine

- **10^{12}–10^{13} gauss**—the surface of a neutron star
- **4×10^{13} gauss**—the quantum electrodynamic threshold
- **10^{15} gauss**—the magnetic field of some newly created magnetars
- **10^{17} gauss**—the upper limit to neutron star magnetism; no known object in the universe can generate a stronger magnetic field

For centuries no one understood how magnetic fields are created in magnetic materials. It was only in the 1800s that some of the most basic properties of magnetic fields began to be understood. It was then discovered that all electric currents create magnetic fields. Figure 10.6 shows how small test compasses react in the vicinity of a current carrying wire. (Note that conventional current is in the direction that positive charge would flow, which is opposite to the direction of electron flow.) An even more important subsequent discovery was that electric currents are the only cause of magnetic fields. Those occurring in permanent magnets are created on the submicroscopic scale by a special form of electron motion—the "spin" of electrons.

In most materials the orientation of atoms is random, so their magnetic fields cancel, and most macroscopic objects have a net magnetic field of zero.

In ferromagnetic materials a complicated cooperative mechanism aligns the spins of a fraction of the electrons in a local region, creating a local magnetic field. Such regions are called domains. There usually are many randomly oriented domains, and consequently the net magnetic field of

Figure 10.5 Lines are often drawn to represent the direction and magnitude of the magnetic field. Arrows give the direction of the field, which is the same as the direction of the force on the north pole of a compass. The density or closeness of the lines is proportional to the strength of the field.

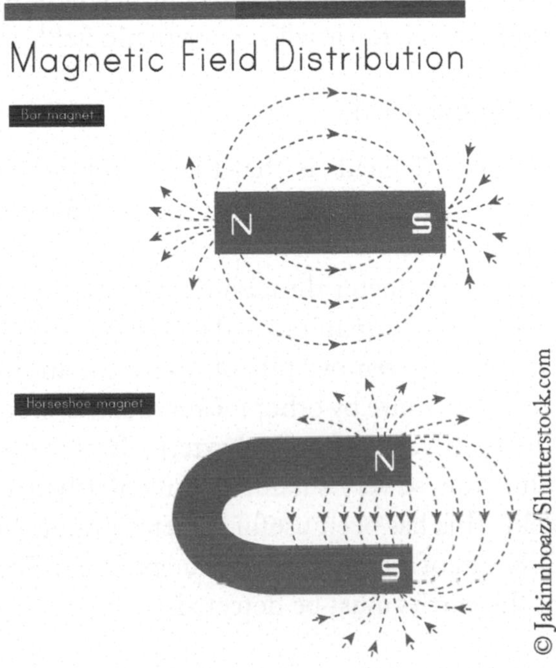

Figure 10.6 Compasses close to a long, straight wire carrying a current will point in the directions shown, Note that the direction of the current is the direction in which a positive charges would flow if it could flow. Circular lines representing the direction of the magnetic field due to a current in a straight wire.

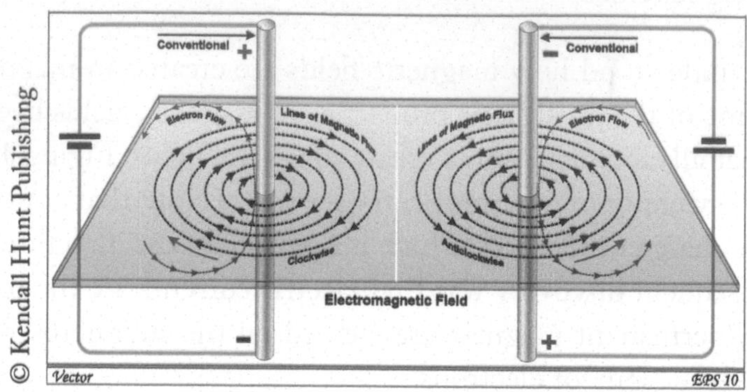

a macroscopic ferromagnetic object is often very small, as shown in Figure 10.7(a). However, an externally applied magnetic field can align these domains in the same direction as the field, giving the object a large net magnetic field, as illustrated in Figure 10.7(b). (The domains tend to become unaligned once the external field is removed because of thermal agitation. A permanent magnet can be produced by a very strong external field or by heating the object and then cooling it in an externally applied field.) Note that the alignment of the domains is in a direction such that the object is attracted by the external magnetic field, explaining why even unmagnetized ferromagnetic materials are attracted by magnets. Nonferromagnetic materials are also affected by external magnetic fields, but the effects are extremely weak compared with those in ferromagnetic materials.

To summarize, on the macroscopic scale there are two basic sources of magnetic fields: permanent magnets and electric currents. Those from permanent magnets are also due to electric currents at the atomic level. Lab 6 will help you to visualize a magnetic field.

10.4.1 Medical Uses of Magnetic Fields

One convenient property of magnetic fields is that they penetrate tissue and have relatively little effect on stationary nonferromagnetic materials. This means that magnetic fields can be used to probe the body with few adverse effects.

Magnets can be used to direct a catheter through the circulatory system. One magnet is located in the tip of the catheter, and the other is manipulated externally to direct the catheter to the desired location. Another use of magnets is to remove bits of iron from an injured eye. These bits of metal may be too small and numerous to remove by other means. Furthermore, the small magnetic fields of ferromagnetic materials in the body can be detected externally, thereby helping to locate iron slivers and the like. Asbestos contains iron, so very small amounts of asbestos can be detected in the body by observing its magnetic field. This has been useful in determining whether asbestos workers have as little as one-tenth of a milligram of asbestos in their lungs. The technique is difficult because fields only one-thousandth that of the earth's must be detected.

Figure 10.7 **(a)** The domains in an unmagnetized iron bar are randomly oriented. The total magnetic field of the bar is consequently small. **(b)** A strong external magnetic field aligns most of the domains, giving the iron bar a net magnetic field in the same direction as the external field.

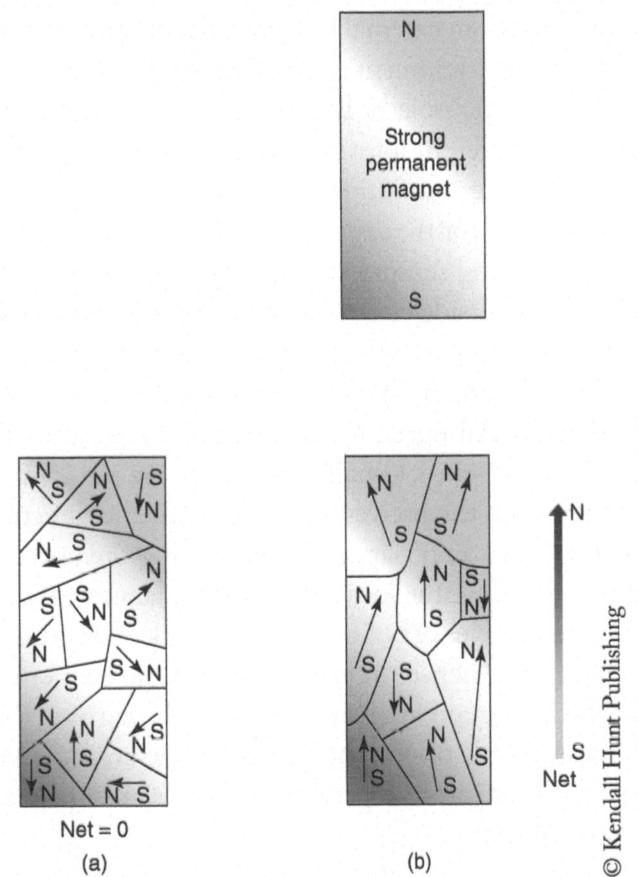

There are electric currents in the nerves and muscles, and like all currents these create magnetic fields. The body's magnetic fields can be measured noninvasively, yielding medically useful information. The body's largest magnetic field is created by the heart and it is only about one-millionth the strength of the earth's magnetic field. Nevertheless, it is possible to measure the magnetic field of the heart using sophisticated equipment. Such a recording is called a magnetocardiogram (MCG). MCGs give information not obtained in electrical measurements of the heart. They use extremely sensitive devices such as superconducting quantum interference devices known as SQUIDs to map the magnetic field over a patient's chest. The SQUID is a very low noise detector of magnetic fields, which converts the magnetic flux threading a pickup coil into voltage allowing detection of weak neuro magnetic signals. Once the field is mapped, it is then possible to locate the source of abnormal rhythms known as arrhythmia. Since the SQUID relies on physical phenomena found in super conductors it requires cryogenic temperatures for operation. Magnetocardiography is used in laboratories and clinics worldwide, in order to do research on normal heart function and for clinical diagnosis.

The magnetic field of the brain also can be measured, but this field is even smaller than the heart's, averaging less than one hundred millionth of the earth's field. Recordings of magnetic brain waves, called magnetoencephalograms (MEGs), are conducted externally, using an array of more than 300 SQUIDS, contained in a helmet shaped, liquid helium containing vessel, called a dewar. This allows for simultaneous measurements at many points over the head. The MEG system is operated in a shielded room that minimizes interference from external magnetic disturbances, including the Earth's magnetic field, noise generated by electrical equipment, radio frequency signals, and low frequency magnetic fields produced by moving magnetic objects like elevators, cars, and trains. As with the heart, the magnetic field of the brain gives different information than electric signals from the brain. The MEG system has high precision, millimeter resolution, providing detailed information. Additionally it has millisecond resolution, which allows for the capture of epileptic spikes within the brain. MEG is useful in studying many functions, including sensory, motor, language, and memory cortex. It is noninvasive and when properly shielded from outside electromagnetic interference, easy to use.

A diagnostic technique that uses an effect called nuclear magnetic resonance (NMR) allows detailed imaging of the body's soft tissues. The nuclei of some atoms have small magnetic fields because of their spins, just as electrons do. All nuclei that contain odd numbers of nucleons have a magnetic moment and angular momentum ("spin"). Usually these spins are randomly oriented, but when placed in a strong magnetic field they align themselves with that field. The directions that these tiny magnets point can be altered by sending in a radio signal. By measuring the amount of radio waves absorbed and reemitted it is possible to measure the location and abundance of certain elements. It is a type of resonance since only certain frequencies of radio waves will work, depending on the type of nucleus and the strength of the magnetic field—hence the name NMR. One element that can easily be detected using NMR is hydrogen, which is found in great abundance throughout the body.

To make an NMR image, one places the patient in a strong magnetic field as shown in Figure 10.8. Radio signals are sent into the patient, and their absorption and reemission are measured and then analyzed by a computer.

An image is constructed by the computer; one such image is shown in Figure 10.9. NMR has two distinct advantages over x-rays. First, it has none of the hazards associated with x-rays.

Figure 10.8 An NMR image is taken by placing a patient in a strong magnetic field.

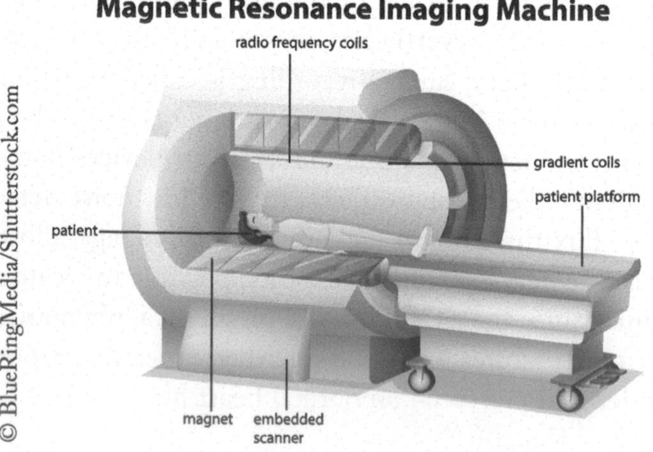

Figure 10.9 A computer-constructed MRI image of the head.

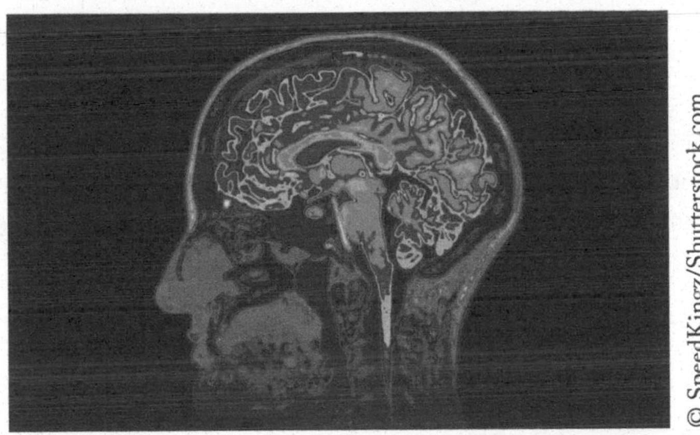

© SpeedKingz/Shutterstock.com

Second, NMR gives information related to the functioning of organs, whereas x-rays are sensitive to their densities, which is not always a good indication of organ function. NMR imaging is very useful in detecting cancer and other degenerative diseases and has become an important diagnostic tool.

10.4.2 Other Uses of Magnetic Fields

Magnetism is a common phenomenon. Its use ranges from mundane door latches to sophisticated medical imaging. Magnetic fields rather than electric fields are often used in the CRTs of x-ray machines to focus and direct the beam of electrons. Because every electric current creates a magnetic field, a force can be exerted on that current by another magnetic field (employing, in simple terms, the fact that like poles repel and unlike poles attract). The beam of electrons in a CRT essentially becomes a magnet and can be moved about by other magnetic fields. Similarly, if an electric current is confined to a wire, force can be exerted on the wire by a magnetic field. This very useful fact is employed routinely in all electric motors. Electric motors are designed so that current passed through them creates magnetic fields that exert forces on other magnets in them. These forces usually cause a shaft to turn, and that shaft can be used to do any variety of jobs, from running an elevator to powering a wheelchair.

Forces on charged objects moving through a magnetic field

When a charged object moves across magnetic field lines, there will be a force applied to the object. The *magnitude* of the force on a charged object, moving across a magnetic field can be determined with the following equation:

$$|\mathrm{F}| = B \cdot q \cdot V_{\perp} \quad \textit{\textbf{Know}}$$

B The magnetic field.

q The amount of charge that the object has.

$V_{\perp}$ The right angle component of the object's velocity

Magnetic field is often measured in a unit called Tesla, named after the Serbian inventor and engineer, Nikola Tesla (1856–1943).

$$1 \text{ tesla} = \frac{1 \text{ newton}}{\text{coulomb} \cdot \frac{\text{m}}{\text{s}}} \quad \textit{\textbf{Know}}$$

The following example will demonstrate how to calculate the magnitude of the force applied to a charged object, moving through a magnetic field.

EXAMPLE 10.4

Suppose that an object with a charge of $q = +1.6 \times 10^{-19}$ coulombs is moving with a velocity of 1.00×10^5 m/s at right angles through a magnetic field as shown below.

The magnetic field is $B = 55\,\mu\text{tesla}$. What is the magnitude of the force applied to the moving, charged object?

Please note that: $1\text{ tesla} = \dfrac{1\text{ newton}}{\text{coulomb}\cdot\dfrac{\text{m}}{\text{s}}}$ and that $1\text{ micro} = 1\mu = 1\times10^{-6}$.

Therefore $55\mu\text{ tesla} = 55\times10^{-6}\dfrac{\text{N}}{\text{C}\cdot\dfrac{\text{m}}{\text{s}}}$

Solution: The equation for the *magnitude* of the force is,

$$|\text{F}| = B\cdot q\cdot V_{\perp}$$

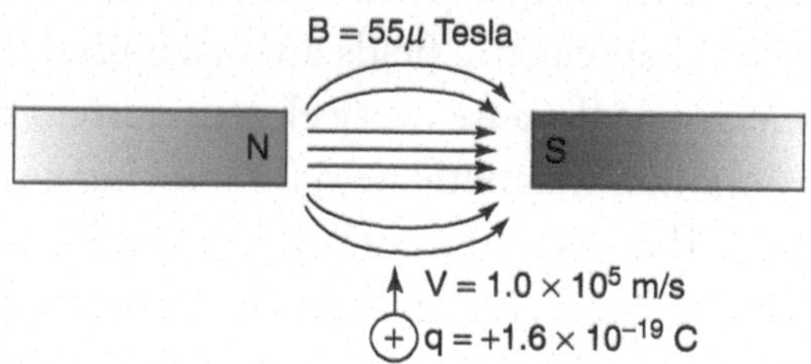

$$|\text{F}| = BqV_{\perp} = 55\times10^{-6}\frac{\text{N}}{\text{C}\cdot\dfrac{\text{m}}{\text{s}}}\cdot1.6\times10^{-19}\text{C}\left(1.00\times10^{5}\frac{\text{m}}{\text{s}}\right) = 8.8\times10^{-19}\text{N}$$

The direction of the force applied to the moving charge is into the page. This is determined by using the "right hand rule."

Things to know for section 10.4

Magnitude of the force on a charged object moving through a magnetic field:

$$|\text{F}| = BqV_{\perp}$$

$$1\text{ tesla} = \frac{1\text{ newton}}{\text{coulomb}\cdot\dfrac{\text{m}}{\text{s}}}$$

$$1\text{ micro} = 1\mu = 1\times10^{-6}$$

10.5 METERS, MOTORS, GENERATORS, AND TRANSFORMERS; FARADAY'S LAW OF INDUCTION

10.5.1 Meters

A loop of wire carrying a current creates a magnetic field perpendicular to the plane of the loop at its center (see Figure 10.10a). The field can be made stronger by having a coil of many loops and by placing a ferromagnetic core inside the loops. The induced magnetic field of the core adds to that created by the current in the wire (see Figure 10.10b). If such a coil is mounted on a pivot and placed near permanent magnets as shown in Figure 10.11, magnetic forces will cause it to rotate. This device, called a *galvanometer*, is the basic measuring device of many meters. Magnetic force in the galvanometer depends on the strength of the magnetic field created by the coil, which in turn depends on the amount of current flowing through the coil. Thus the deflection of the needle is a measure of the current flowing through the coil.

A display that uses a needle is called an analog display. The great majority of electric meters with analog readouts employ galvanometers. Galvanometers are basically current-measuring devices, but they can readily be adapted to measure other quantities. It is necessary to have another designed to

Figure 10.10 (a) The large arrow represents the magnetic field at the center of the loop of wire carrying a current in the direction shown.

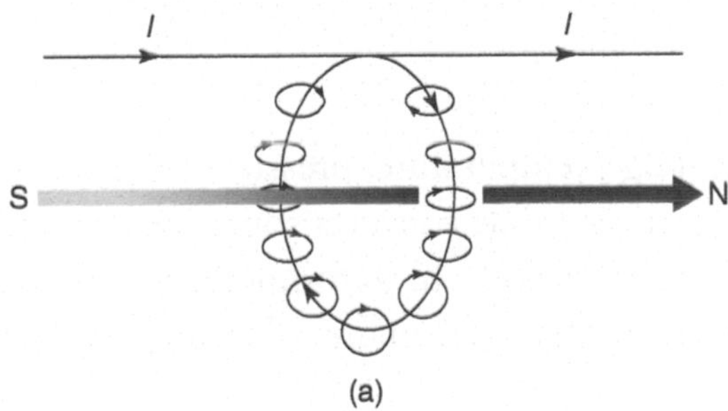

Figure 10.10 (b) The magnetic field can be made stronger by having a coil of many loops and adding an iron rod in the center.

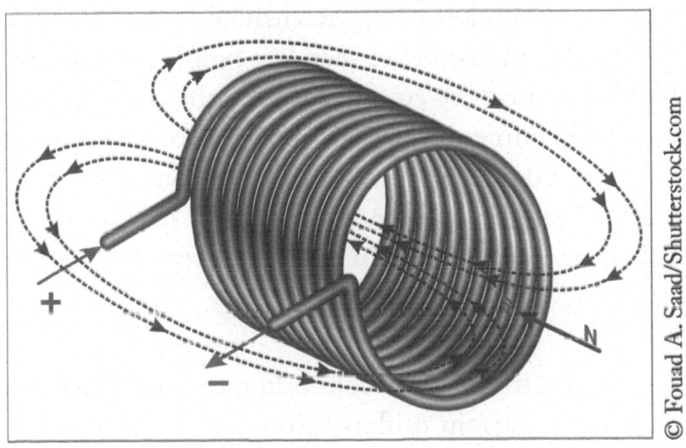

Figure 10.11 A galvanometer used as a current meter. A coil of wire with an iron core carrying a current creates a magnetic field that causes the coil to rotate clockwise against the restoring force of a spring. The greater the current, the greater the deflection of the pointer.

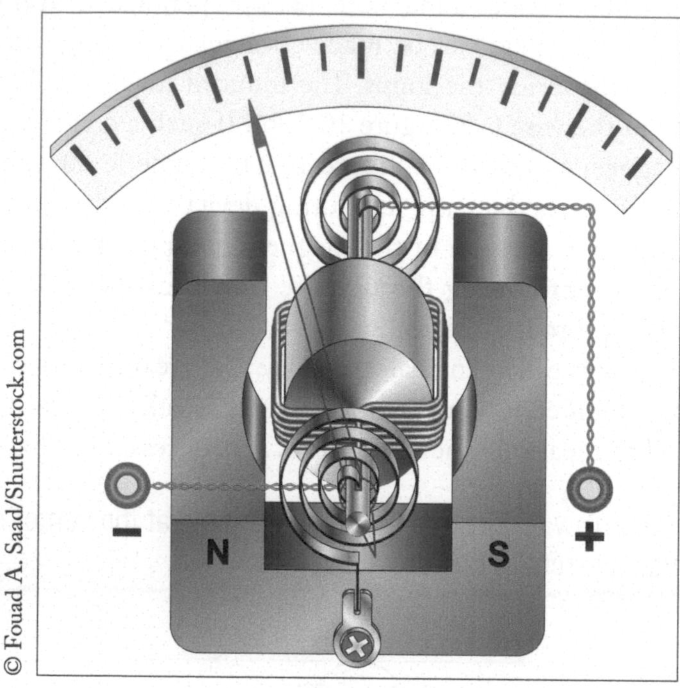

convert the measured quantity (e.g., temperature, pressure, or position) to a current which is then sent through a galvanometer. A device that converts some measurement into an electric current is called a *transducer*. Galvanometers are used in automobile fuel gauges, heart-defibrillator energy meters, stereo amplifiers, Geiger counters, and many other instruments.

10.5.2 Motors

An electric *motor* is similar to a galvanometer in that forces among magnets are used to cause a coil to rotate. Figure 10.12 is a schematic of a motor that is similar to the galvanometer but lacks a spring, thus allowing repeated rotations. In the position shown, magnetic forces cause the coil to rotate clockwise. To keep the coil rotating clockwise, the current in the coil is reversed as the coil passes the permanent magnets; then the repulsion between like poles continues to rotate the coil clockwise. The commutators and brushes reverse the current through the coil twice per rotation to maintain its clockwise motion. Motors have numerous applications and can be made to operate on nearly any type of electric power. (The work done by a motor comes, of course, from electric energy put into the motor by the current passing through it.) To be efficient, most motors are more complicated than the one shown in Figure 10.12, but all motors use magnetic forces.

10.5.3 Generators

Nature is often beautifully symmetric, and the common motor provides an example of such symmetry. A voltage is required to cause current to flow through the coil of a motor, which in turn creates

Figure 10.12 A motor. Magnetic forces cause the coil to rotate clockwise. The commutator and brushes reverse the current through the coil as it passes by the permanent magnets, so that its field will reverse and it will continue to rotate clockwise.

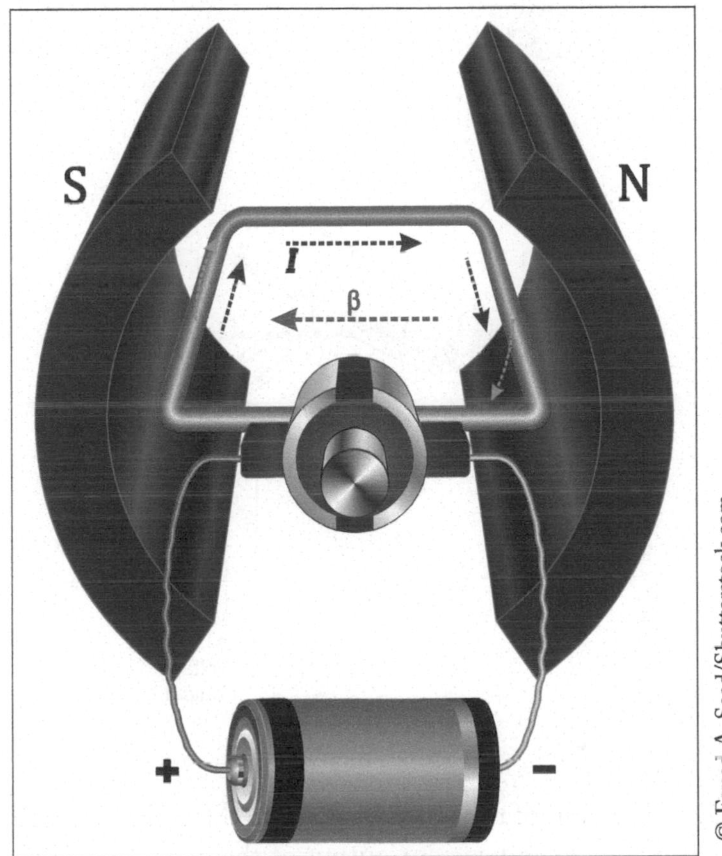

a magnetic field and causes the coil to rotate. If the situation is reversed and the coil is rotated by something external to the motor, as shown in Figure 10.13, then something very important happens—a voltage is created in the coil of the motor. The motor becomes a generator of electric voltage. Whenever wire is moved across a magnetic field, forces on its electrons create an electric current. Thus a motor run backward generates voltage and electric energy rather than using them. The work done by whatever rotates the coil is the source of the electric energy put out by the generator. The same device can be used as both a motor and a generator. Some early automobiles used an electric motor to start their gasoline engines and then used the same device to generate electricity to recharge their batteries. It usually is more efficient to have separate motors and generators specifically designed for a particular application.

10.5.4 Faraday's Law of Induction

The following symmetry exists in nature: A current causes a magnetic field and a magnetic field can cause a current. However, a magnetic field does not always create a current. Only when the magnetic field in a certain region changes is a voltage induced that may cause a current to flow. For example,

Figure 10.13 A generator is a motor run backward. When a coil is rotated in a magnetic field, a voltage is induced in the coil. The source of the energy put out by the generator in this case is a person turning a crank.

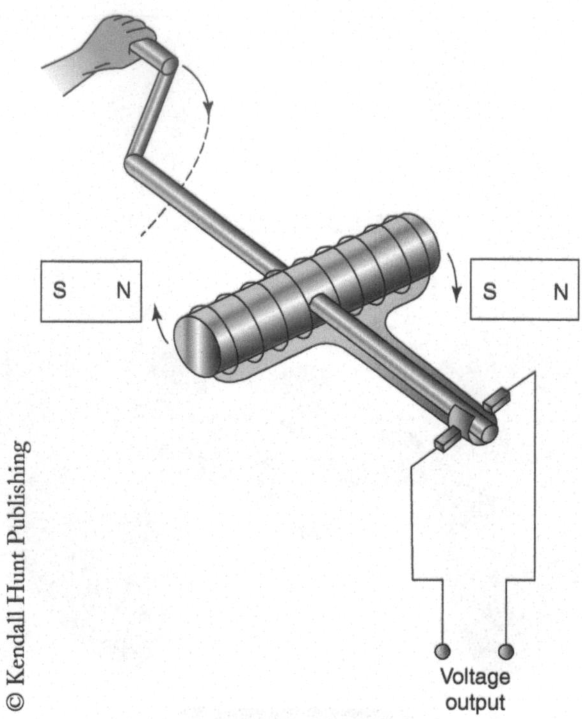

in a generator the relative direction of the magnetic field to the coil of wire changes because the wire is rotated. This creates the voltage and current output of the generator. The same effect is obtained by holding the coil still and rotating the permanent magnets around it. Another way to change the magnetic field is to change its strength by moving the magnets toward or away from the coil. Yet another way to induce a voltage is to change the shape of the coil—for example, by stretching it into a long thin loop instead of a circular one. All methods for inducing voltage with a magnetic field involve motion of the wire relative to the field or vice versa. The law describing this is called *Faraday's law of induction*, after the English physicist Michael Faraday (1791–1867), who discovered the effect slightly after and independently of the American physicist Joseph Henry (1797–1878). The law states that whenever magnetic fields change direction or strength or move across a wire, a voltage is induced. Furthermore, the sign of the voltage is such that the current it produces creates a magnetic field that reduces the change in the parent field. That is, nature opposes changing magnetic fields; it induces a current that creates a field to lessen the change. Faraday's law is the underlying principle behind electric power generators. It describes how they can convert any type of mechanical work into electric energy. This law also describes the functioning of many other electrical devices, among them the transformer.

Joseph Henry

Michael Faraday

10.5.5 Transformers

Electrical *transformers* convert an input voltage to an output voltage that may be either higher or lower than the input. For example, an x-ray machine may need 75,000 V to operate, yet it is plugged into a 120-V outlet. The needed high voltage is produced by a transformer inside the x-ray machine. Another example is found on many power poles; a transformer reduces high line voltage to the 120 or 240 V most commonly used domestically and commercially. It is much more efficient to produce and transmit electric power at high voltages (sometimes as high as 500,000 V) and then use transformers to reduce the voltage to desired levels at the location where the electricity is actually used.

The basic construction of a transformer is shown in Figure 10.14. Two coils of wire, called the primary and the secondary, are placed near one another. If the current in the primary coil is changed, then its magnetic field changes everywhere, including at the location of the secondary coil. Faraday's law then requires that a voltage be created in the secondary coil. Iron is used to concentrate and add to the magnetic field and to carry the field to the secondary. Ferromagnetic materials tend to trap magnetic fields. In a step-up transformer, the voltage in the secondary coil will be greater than in the primary. If the secondary has 10 times as many loops than the primary, it will have 10 times the voltage of the primary. However, the secondary will have one-tenth the amount of current that the primary has. Thus energy is conserved. Conversely, in a step-down transformer, the secondary has few loops than the primary. The voltage in the secondary will be less than in the primary but the secondary will have more current than the primary, again conserving energy.

Figure 10.14 A transformer utilizes Faraday's law of induction. If the voltage input varies, then the magnetic field at the secondary coil varies inducing an output voltage.

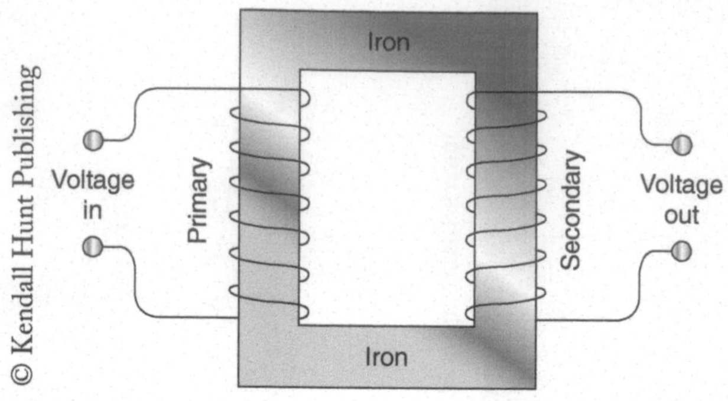

Transformer theoretical output voltage:

$$V_{out} = \frac{\text{no. of coils in the secondary}}{\text{no. of coils in the primary}} \cdot V_{in}$$

If the current through the primary coil is constant, as it would be with a battery or other direct current (dc) source, then no voltage is created in the secondary. The current in the primary must vary, as it would with any alternating current (ac) source. One of the reasons that ac power is so widely used is that its voltage is easily increased or decreased by transformers.

It is important to remember that the *energy* carried from one side of a transformer to the other does not increase if the voltage is increased. Transformers do not violate the conservation-of-energy principle. If the transformer increases voltage, then the secondary coil puts out less current than is put into the primary. Conversely, if the transformer decreases voltage, then the secondary puts out more current than is put into the primary. The reasons behind this will be explained in Chapter 11. Energy is always conserved.

Things to know for Chapter 10

Coulomb's law $F = k\frac{q_1 q_2}{d^2}$

Charge of 1 electron: $q_e = -1.6 \times 10^{-19}$ c

Voltage: $Voltage \equiv \frac{Energy}{charge}$; $\frac{\text{Joules}}{\text{coulomb}} \equiv \text{volts}$

$Current \equiv \frac{Charge}{time}$; $I \equiv \frac{q}{t}$; $1 \text{ amp} \equiv \frac{1 \text{ coulomb}}{\text{second}}$

Force on a charged object, moving across a magnetic field: $|F| = qV_{\perp}B$

Units of magnetic field strength: $1\text{ tesla} = \dfrac{1\text{ newton}}{\text{coulomb}\cdot\dfrac{\text{m}}{\text{s}}}$

1 micro: $1\mu = 1\times10^{-6}$

Transformer: $V_{out} = \dfrac{\text{no. of coils in the secondary}}{\text{no. of coils in the primary}}\cdot V_{in}$

Questions

10.1 What is the planetary model of the atom?

10.2 Considering the nature of the Coulomb force, why would you expect an atom to have the same number of electrons in orbit as it has protons in its nucleus?

10.3 What is an ion?

10.4 The Coulomb force can be extremely strong compared to gravity and can act over very large distances. Why, then, is the orbit of the moon around the earth determined by gravity and not by electric forces?

10.5 Describe a situation in which static electricity is generated. What physical process separates the charges in the situation you describe?

10.6 How are electrical potential energy and gravitational potential energy similar? How are they different?

10.7 What is the relationship of voltage (potential difference) to energy?

10.8 In what ways is voltage (potential difference) analogous to pressure?

10.9 Explain how atoms in a substance at high temperature might lose electrons, as happens in the filament of a CRT. Further, why would more electrons be freed from their atoms at higher temperatures than at lower temperatures?

10.10 Consider the vertical steering plates in the CRT in Figure 10.3. If a voltage (potential difference) is applied to those plates to steer the beam of electrons upward, should the upper plate be positive or negative relative to the lower plate?

10.11 What is a force field and how does it differ from a force?

10.12 In what ways is electric current analogous to fluid flow? In what ways is it different?

10.13 Explain how grounding removes all excess charge from a person. If you have a charged object that is a very poor conductor of electricity, will grounding it remove the charge? Explain.

10.14 What is the cause of all magnetism?

10.15 Describe the property that all ferromagnetic substances have in common.

10.16 How would heating and cooling a ferromagnetic object in the presence of a magnetic field help magnetize it permanently?

10.17 Give two examples of the use of magnetism not mentioned in the text.

10.18 The core of the earth is thought to be mostly iron and nickel. How might this explain the magnetic field of the earth? Furthermore, the core is fluid or semifluid throughout. Might this explain why the earth's magnetic field disappears and itself over periods of thousands of years?

10.19 The magnetic fields of the body are small. What does this imply about electric currents in the body?

10.20 What is NMR imaging?

10.21 Under what circumstances can a magnetic field create a voltage?

10.22 How are galvanometers and motors similar? How do they differ?

10.23 Describe how a generator creates a voltage.

10.24 What is a transformer? On what physical principle is it based? Describe its operation.

10.25 What is the output of a transformer when a dc voltage is used as an input? Explain.

Problems

SECTION 10.1

√10.1 (I) How many excess electrons does an object with a total charge of -1.0×10^{-9} coulombs have on it? This is approximately the amount of charge involved in commonly encountered static electricity.

√10.2 (I) How many electrons must be removed from an object to leave it with a charge of 2.0×10^{-10} C? This is a fairly small, though not negligible, static charge.

√10.3 (I) (a) Calculate the force in Newtons between two identical charges of 1.0×10^{-9} C separated by 1.0 cm. These charges are typical of those involved in static electricity. (b) What is the force if the separation is doubled? tripled? (c) What is the force if the charges are increased by a factor of 10 and the separation is kept at 1.0 cm?

10.4 (II) Clothes often cling together when removed from clothes dryer due to static charges built up during drying. If a force of 5.0 N must be exerted to pull apart two articles of clothing, calculate the charge on each. Assume that the charges are equal in magnitude but opposite in sign and that the average distance between charged parts of the two articles is 0.80 mm.

10.5 (II) Calculate how far apart two identical charges of 1.0 C must be in order for the Coulomb force between them to be less than 500 N (about 112 lb). (This gives some indication of just how large the coulomb is.)

10.6 (II) A hydrogen atom consists of a single proton and a single electron separated on the average by 0.53×10^{-10} m. (a) Calculate the Coulomb force between them. (b) Calculate the acceleration of the electron due to this force and compare the acceleration with that of gravity (g).

SECTION 10.2

√10.7 (1) The energy expended by a battery is related to both battery voltage and the amount of charge it moves. (a) Calculate the energy in joules put out by a 9.0 V calculator battery that moves 3.0 C of charge through a handheld calculator after 1 hour of operation. (b) Calculate the energy put out by a 12.0 V car battery that moves 1000 C of charge (as it may when starting the car's engine).

√10.8 (I) About 20,000 V is required to create a spark 1.00 cm long through dry air. If you walk across a rug and a spark 1.20 cm long jump between your finger and a doorknob, how much energy is released in the spark given that 5.00×10^{-9} C of charge flows?

√10.9 (III) (a) Calculate the speed of an electron accelerated by a voltage of 12,000 V in the CRT of an old black and white TV. The mass of an electron is 9.11×10^{-31} kg. (b) Compare this with the speed obtained in the CRT of an old color TV that uses 25,000 V to accelerate electrons.

10.10 (I) Coulomb forces can be used to attract charged paint droplets to an object. This method is sometimes used to apply paint or felt to odd-shaped objects. Calculate the speed of a 1.0×10^{-3} g droplet of paint just before it strikes an object, if its charge is 1.0×10^{-5} C, neglecting air resistance. There is a 500-V potential difference between the object and the droplet and the droplet starts from rest.

10.11 (II) What voltage is needed to accelerate electrons to 1.0% of the speed of light, in a situation similar to that of the CRT in Example 10.2? The speed of light is

$$3.00 \times 10^8 \frac{\text{m}}{\text{s}}.$$

SECTION 10.3 AND GENERAL PROBLEMS

√**10.12 (I)** Sparks between charged objects, such as between hair and a comb on a dry day, can move a charge of about 2.0×10^{-9} C. The spark may last only about $1\,\mu$ second. What current in amperes flows under these circumstances?

√**10.13 (I)** A heart defibrillator can sometimes restore a normal heartbeat to a heart attack victim. Suppose the defibrillator passes 6 A of current through the heart for 0.01 second. How much charge passes through the heart in this event? (The total current and charge passing through the torso are larger. See Table 10.1.)

10.14(II) Suppose a battery used to power a clock wears out after moving 10,000 C of charge through the clock. How long did the clock run off this battery if it used 0.50×10^{-3} A of current?

10.15 (II) A 100-W light bulb has an average lifetime of 800 hr. How many electrons move through the bulb in its life? You may use the current given in Table 10.1.

10.16 (III) Nonnuclear submarines use batteries when submerged. The current supplied by these batteries is 1000 A at full speed ahead. How long will it take for the batteries to move 1×10^{25} electrons at that current?

10.17 (III) When starting a truck engine a 12-V battery supplies 250 A of current for 12.0 seconds. (a) Calculate the charge in coulombs moved in that time. (b) Calculate the energy in joules expended by the battery.

10.18 (II) Suppose that an alpha particle with a charge of 3.2×10^{-19} coulombs is moving with a velocity of 3.6×10^5 m/s, directly across a magnetic field. If the mass of the object is 6.68×10^{-27} kg and the magnetic field strength β equals $46\,\mu$tesla, determine (a) the magnitude of the force on the alpha particle and (b) the amount of acceleration that the particle undergoes.

10.19 (I) If the magnetic field in problem 10.18 was directed from the left side of this page to the right side of this page, and if the positively charge alpha particle was moving from the top of this page toward the bottom, what is the direction of the force that is applied to the alpha particle?

11

Simple Electric Circuits

"I'm smart enough to know that I'm dumb."

—Richard P. Feynman

Simple electric circuits are the subject of this entire chapter. Their study enables one to understand the operation of simple devices such as light bulbs as well as more complex entities, such as nerve cells. Chapters 12 and 13 then consider applications of both simple electric circuits and electromagnetism.

By definition, moving charges constitute an electric current (see Section 10.3). If there is a complete path for charges to follow, then a circuit exists. In a complete circuit, there is an unbroken path for charges to follow.

In a very few situations, the flow of charges without a complete circuit is of interest, as in grounding an object charged with static electricity. If there is not a complete circuit, the current lasts only a very short time. The battery moved excess charges into the bulb and wires in a very short time when the wires were first attached. After that time, perhaps 10^{-9} seconds, no charge flowed because the break in the wire is such a poor conductor and because excess charge accumulated in the wire and repelled any further charge from entering.

Most applications of electricity, however, utilize complete circuits so that current may flow continuously for long periods of time. These will be the subjects of this chapter.

Current is caused by an electrical potential difference between two locations in a conductor. The amount of *current through a given circuit is directly proportional to the electrical potential difference and is inversely proportional to the resistance.* Written mathematically as follows:

$$I = \frac{V}{R}$$

This experimental fact was established by George Simon Ohm (1787–1854), a German physicist. While voltage is the cause of electric current, anything that impedes current is called resistance; the greater the resistance, the lower the current. For example, the resistance of the break in the wire is so large that essentially no current flows and the bulb does not glow. Resistance is defined so that current is inversely proportional to it.

The dependence of current "I" on voltage "V" and resistance is referred to as Ohm's Law:

Ohm's law

$$I = \frac{V}{R} \qquad \textit{Know} \qquad \text{eq. 11.1a}$$

where "R" is the symbol for resistance. This important equation will be used repeatedly throughout the discussion of electricity. Voltage is defined as energy per unit charge.

$$V \equiv \frac{\text{Energy}}{\text{Charge}} \qquad \textit{Know} \qquad \text{eq. 11.1b}$$

Voltage has units of joules per coulomb.

$$1\,\text{V} \equiv 1\frac{\text{J}}{\text{C}} \qquad \textit{Know} \qquad \text{eq. 11.1c}$$

The unit of resistance is the ohm, symbolized by Ω, the Greek letter omega.

$$\text{units of an ohm}: \Omega \equiv \frac{\text{V}}{\text{A}}$$

Lab 7, Ohm's law, should give you further insight into the relationship between current, voltage, and resistance.

EXAMPLE 11.1

What current flows through a flashlight if its resistance is 15 Ω and its batteries have a total electrical potential difference of 3.0 V?

Solution: By Ohm's law,

$$I = \frac{V}{R}, I = \frac{3.0\text{V}}{15\Omega} = 0.20\,\text{A}$$

EXAMPLE 11.2

What is the resistance of a calculator that uses a 3.00-V battery and through which 0.300 mA flows? Please note that 1.00 mA (1 mA), is the same thing as 1×10^{-3} A. There are 1000 mA in 1 A.

Solution:

Step by step 1: The first thing that we have to do is convert the current from milliampere to ampere.

$$0.3\text{mA}\left(\frac{1\,\text{A}}{1000\text{mA}}\right) = 0.0003\,\text{A}$$

The current, I, is 3.0×10^{-4} A.

Step 2: Next, we solve Ohm's law for the resistance R and substitute the given values.

$I = \frac{V}{R}$ therefore $R = \frac{V}{I}$

$$R = \frac{V}{I} = \frac{3.00\,\text{V}}{3.00 \times 10^{-4}\,\text{A}} = 1.00 \times 10^{4}\,\Omega$$

11.1 SCHEMATIC REPRESENTATION OF A CIRCUIT

Electric circuits are only rarely drawn pictorially. They usually are represented schematically, as in Figure 11.1.

The straight lines that connect the battery and resistors represent wires, which are normally assumed to have negligible resistance. This means that all the resistance in a simple circuit is represented by the resistance symbol.

The direction of so-called conventional current dates back to 1750, with Ben Franklin. He incorrectly guessed that the charge carrier was positive. Later, when electrical current was studied, the direction of the current was assumed to be from positive to negative, as shown in Figure 11.1. Current direction is defined as the direction that a positive charge would move if it was free to move. In 1879, Edwin Hall devised an experiment that determined that negative charges, not positive charges, moved through a metal conductor. In 1897, J. J. Thompson identified these negatively charged particles as electrons. Since they are negatively charged, the electrons therefore flow in a direction opposite to that of conventional current (which was defined before the electron was discovered). In some situations, as in nerve cells and inside fluorescent lights, both positive and negative ions move. In those situations, positive ions flow in the direction of conventional current and negative ions in the opposite direction.

In simple electric circuits, the direction of flow is usually not of major importance. For example, a flashlight bulb works the same no matter which direction electrons move through it. However, the direction of flow may be crucial in more complex situations, such as the movement of ions across biological membranes. In this text, you may assume that the direction of current flow is in the direction that a positive charge would go if it could go.

Figure 11.1 Schematic representation of a simple electric circuit having a battery as a voltage source. Also shown are the directions of conventional current and electron flow.

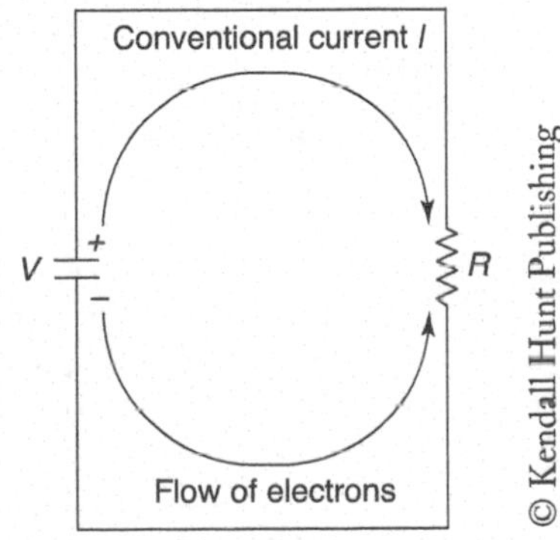

11.1.1 A Simple Electrical Circuit with Resistors in Series

Figure 11.2(a) is a diagram of a simple electrical circuit. It consists of two resistors in series with each other, a switch, and a 12-V battery. The resistors are said to be in series because the current must go through one to get to the other.

There is no alternate path for the current to take. The total resistance in the circuit is the sum of the two resistors, R_1 and R_2. Total resistance equals 6 Ω.

$$R_{\Delta saf} = R_1 + R_2 = 2\Omega + 4\Omega = 6\Omega$$

We can determine the current, I, in the circuit, by using Ohm's law.

The total voltage drop across resistances in an electrical resistance will add up to the total voltage supplied. This is another example of conservation of energy. To see this we can solve Ohm's law for voltage: $V = IR$. In Figure 11.2(b), we determine the voltage across each resistor. There is an 8-V drop across R_1 and a 4-V drop across R_2, indicating a total voltage drop of 12 V.

$$I = \frac{V}{R} = \frac{12V}{6\Omega} = 2\text{ A}$$

Figure 11.2a Schematic of a simple electrical circuit with resistors in series with each other.

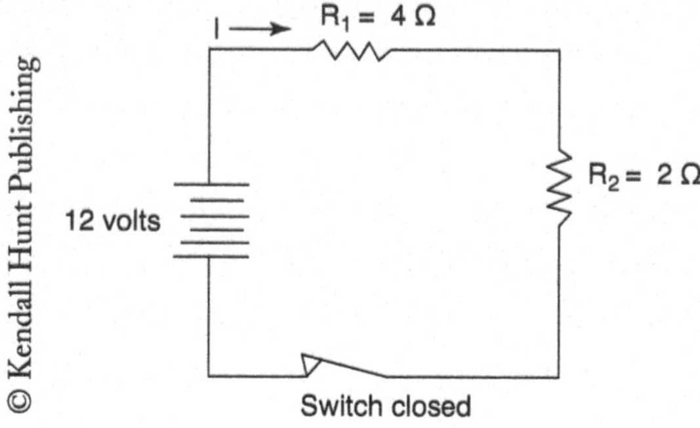

Figure 11.2b Schematic of a volt meter used to determine the voltage drop across R_1 and R_2.

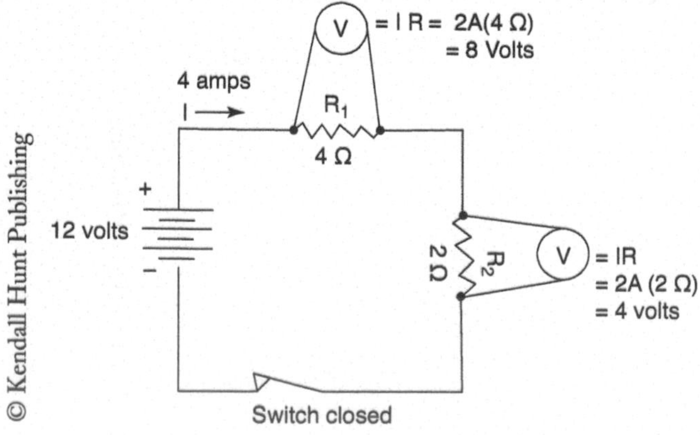

It is important to remember that the definition of voltage is energy per unit charge. We can see that the voltage drop across the two resistors adds up to 12 V and is equal to the total voltage supplied by the battery. Energy is conserved.

Things to know for Section 11.1

Ohm's law:

$$I = \frac{V}{R} \quad \text{Units are } \frac{\text{Coulombs}}{\text{second}} = \text{Ampere}$$

Definition of voltage:

$$\text{Voltage} \equiv \frac{\text{Energy}}{\text{charge}} \quad \text{Units are } \frac{\text{joules}}{\text{Coulomb}} = \text{Volts}$$

11.2 POWER IN ELECTRIC CIRCUITS

Power is the rate of doing work or expending energy. Its unit is joules per second or watts. Light bulbs and many other electrical devices are rated in terms of their rate of energy consumption (in watts). The most useful thing about electricity is that its energy can be converted to so many other forms. For example, motors convert this electrical energy to work, light bulbs to visible and invisible light. The more powerful the device, the faster it converts electric energy into another form.

What determines how much power an electrical device produces or requires? To answer this question, recall the definition of power:

$$\text{Power} \equiv \frac{\text{Energy}}{\text{time}} \qquad \textit{Know}$$

In order to calculate the power of an electrical system in terms of current and voltage, we will do a short derivation. Please take out a pencil and follow along.

Step by step 1: From the definition of voltage, $V \equiv \frac{E}{q}$, we solve it for energy,

$$E = Vq$$

Step 2: Next, substitute this equation for electrical energy back into the power equation.

$$\text{Power} \equiv \frac{\text{Energy}}{\text{time}} = \frac{Vq}{t}$$

Step 3: Recall that the definition of current is $I = \frac{q}{t}$. That means that we can substitute "I" for $\frac{q}{t}$ in our power equation.
Therefore,

$$\text{Power} \equiv \frac{\text{Energy}}{\text{time}} = \frac{Vq}{t} = IV$$

So, if we want to *calculate the power* required by an electrical device, we can use the following equation:

$$\text{Power} = IV \qquad \textit{Know}$$

This simple equation gives the power consumed by any electrical device that has a current "*I*" flowing through it and that has a potential difference "*V*" applied to it. Equation 11.2a also gives the power generated by any source of electricity producing a current, *I*, and a potential difference (voltage), *V*.

If you are asked to determine the power of an electrical device but you do not know the voltage, you can use Ohm's law to determine it.

$$I = \frac{V}{R};\ \text{therefore } V = IR$$

You can then determine an actual numerical value for V and substitute it into Equation 11.2a, or you can use Ohm's law to derive another equation as follows:

When you do not know the voltage, substitute IR for V

$$\text{Power} = IV = I(IR) = I^2R$$

$$\text{Power} = I^2R \qquad \text{eq. 11.2b}$$

Ok smarty, what if you know the voltage, but you do not know the current? Then what do you do? Ohm's law to the rescue again. You can use Ohm's law to determine a numerical value for the voltage, or you can do another derivation as follows:

Since $I = \frac{V}{R}$,

$$\text{Power} = IV = \left(\frac{V}{R}\right)V = \frac{V^2}{R}$$

So, when you do not know the current,

$$\text{Power} = \frac{V^2}{R} \qquad \text{eq.11.2c}$$

EXAMPLE 11.3

How much power does a calculator require if it uses a 9.0-V battery and 0.50 mA flows through it? Do not forget that 1 mA is the same thing as 1×10^{-3} A.

Solution: Since current and voltage are given, Equation 11.2a is most convenient:

$$\text{Power} = IV = 0.5\times10^{-3}\ \text{A}(9.0\ \text{V}) = 4.5\times10^{-3}\ \text{W}$$

Let us double check our units. In the equation above, we have amperes multiplied by volts. You should already have memorized that one ampere is one coulomb/sec. Those of you who are memorizing the "Things to know," also realize that a volt is a joule/coulomb. Therefore, 4.5×10^{-3} W, is a very small amount of power. This low rate of energy consumption enables the calculator to run for many hours before depleting the energy in a battery.

$$\text{Ampere} \times \text{Volts} = \frac{\text{C}}{\text{s}} \cdot \frac{\text{J}}{\text{C}} = \frac{\text{J}}{\text{s}} \equiv \text{watts}$$

For those of you who believe in the conservation of energy, you should not be surprised by the fact that the power consumed by the calculator is identical to the power supplied by the battery. It is always the case that whatever total power is consumed by resistances is exactly equal to the power supplied by the voltage source. The voltage source is often called a source of power and energy, and a resistance is called a sink (or consumer) of power and energy.

EXAMPLE 11.4

Calculate the resistance of a 100-W light bulb that uses 120 V electricity.

Solution: Because voltage and power are given and resistance is wanted, Equation 11.2c is most convenient: $\text{Power} = \frac{V^2}{R}$. Solving for resistance gives

$$R = \frac{V^2}{P} = \frac{(120\,\text{V})^2}{100\,\text{W}} = 144\,\Omega$$

Typical household appliances have resistances that range from a few ohms to a few hundred ohms.

11.2.1 Energy Consumption and the Cost of Using Electricity

One thing that is often on our minds these days is the cost of energy; it is energy that we pay for rather than power. The cost of using an electrical device depends on two things: its power rating and the time that it is on. The easiest way to see this is to take the definition of power and solve it for energy, yielding

$$\text{Power} \equiv \frac{E}{t}; \text{therefore Energy} = \text{Power} \cdot \text{time}$$

Thus, a large electrical device (meaning one with a large power rating) uses more energy in a given amount of time than a small one. In addition, small electrical devices, such as light bulbs, that are on for many hours also consume large amounts of energy.

Electric bills usually list the energy consumed in units of kilowatt-hours, abbreviated kWh, rather than using more direct units; like joules. The reason is partly tradition, but it is also because other energy units, such as joules, are small compared to the typical amounts of energy consumed (1 kWh = 3.6×10^6 J). Consider the following example.

EXAMPLE 11.5

The average cost of electric energy is about 11.0 cent/kWh. *Calculate the cost* of operating (a) a 100-W light bulb for 16 hour, and (b) a 6.0-kW clothes dryer for 2.5 hour.

Solution: (a) The problem states that it costs 11 cents for 1 kWh. of energy. We are given the electrical power required to operate the devices and the amount of time that they are on. Because power involves energy and time, we will solve the definition of power, for energy.

$$\text{Power} \equiv \frac{\text{Energy}}{\text{time}}\text{; therefore } E = \text{Power} \cdot \text{ time}$$

a. The energy consumed by the light bulb is

$$E = P \cdot t = 100\,\text{W} \times 16 \text{ hour} = 1600\,\text{W} \times \text{hour} = 1.6\,\text{kW} \times \text{hour}$$

The cost of the electrical energy is 11.0 cent/kWh.
The cost of leaving the light on for you is

$$1.6 \text{ kW} \cdot \text{hour}\left(\frac{11\,\text{cent}}{\text{kW} \cdot \text{hr}}\right) = 17.6\,\text{cents}$$

b. The energy consumed by the clothes dryer is

$$E = P \times t = 6.0\,\text{kW} \times 2.5 \text{ hour} = 15 \text{ kW} \times \text{hour}$$

The cost of the electrical energy is 11.0 cent/kWh.
So, the cost of cleaning your dirty clothes is

$$15 \text{ kW} \cdot \text{hour}\left(\frac{11\,\text{cent}}{\text{kW} \cdot \text{hr}}\right) = 165\,\text{cents} = \$1.65$$

With energy becoming progressively more expensive, various measures are being taken to cut down its consumption. All such measures either reduce the power rating of a device, the time it is used, or both. Devices with reduced power ratings, such as efficient air conditioners, may be more expensive to construct, but that extra expense is usually quickly recovered in reduced energy costs. Power ratings of electrical devices are commonly written somewhere on the device or are otherwise available. The cost per kilowatt-hour of electricity in a given region can be determined by examining your latest bill. Then, by using the method of the last example, it is easy to calculate how much an appliance costs to operate and how long would it take to save the extra cost of a more efficient model. Manufacturers of some appliances supply this cost information for you.

Things to know for Section 11.2

Ohm's law $I = \dfrac{V}{R}$

Power in electric circuits: Power $= IV$

Useful derivations:

Using Ohm's law and substituting I × R for V in Power = IV, we get

$$\text{Power} = I^2R$$

Using Ohm's law and substituting $\frac{V}{R}$ for I in Power = IV, we get

$$\text{Power} = \frac{V^2}{R}$$

11.3 ALTERNATING CURRENT

Only one type of voltage source, the battery, has been mentioned specifically so far. Batteries put out a steady or direct voltage. Current flowing through a system powered by a battery is also steady and is called direct current (dc). However, nearly all electricity that is transmitted from power plants to homes and businesses fluctuates periodically in time: it is called alternating current (ac). Figure 11.3 compares voltage as a function of time for dc and ac power.

ac voltage oscillates periodically in time with a period T. In the United States and Canada, the period is $\frac{1}{60}$ second, so the frequency is 60 cycles/s, or 60 Hz. Some European countries use a frequency of 50 Hz. The voltage alternates smoothly between positive and negative. This causes the current through a device to alternate in the direction it flows—hence the name ac. Fluorescent light bulbs using ac actually blink on and off 120 times a second, too fast for the eye to follow. If you look at one and wave your hand back and forth between your eyes and the light, you might notice a stroboscopic effect. Try it.

Figure 11.4 is a schematic of an ac voltage source connected to a resistance. The symbol ⏚ stands for a connection to the earth, called a ground. The ground connections force the voltage of the wire to be zero between points A and B. (To have a voltage other than zero, the wire must have an excess charge. That charge would quickly escape into the earth through the ground connections.) The voltage at point C alternates between positive and negative, causing the current to reverse direction continually. The lines in the schematic are not to scale; for example, the ac voltage source is usually a power plant many miles distant. The ground connection on the left is at the power plant, and the ground connection on the right is at the user's location.

Figure 11.3 (a) dc voltage is constant in time; (b) ac voltage oscillates in time with a period T. V_{eff} is the effective voltage; V_0 is the peak voltage.

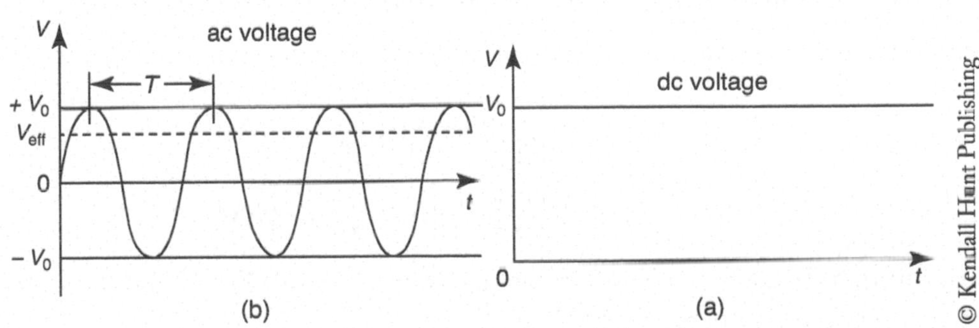

Figure 11.4 Schematic of a simple electric circuit using an ac voltage source and grounded on one side.

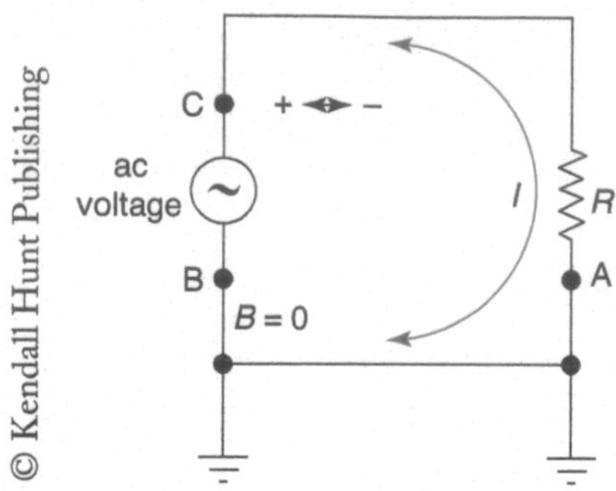

The *average* voltage and current in an ac circuit are zero since they are positive and negative equal amounts of time. Yet light bulbs, electric heaters, refrigerators, and TV s, all work on ac. That is because current is almost always flowing, be it positive or negative. The direction in which current moves through a light bulb does not matter; the tungsten filament in the bulb heats up and produces visible and invisible light. Meaningful effective voltage and effective current for *ac* are given by

$$V_{\text{eff}} = \frac{V_0}{\sqrt{2}} \qquad \text{eq. 11.3a}$$

$$I_{\text{eff}} = \frac{I_0}{\sqrt{2}} \qquad \text{eq. 11.3b}$$

where V_0 and I_0 are the peak voltage and peak current, respectively (see Figure 11.3).

It is almost always the *effective voltage* and current that are quoted for ac circuits or devices. For example, residential electricity in the United States is 120 V ac; that 120 V is the effective voltage. Most meters that measure ac voltages and currents are designed to indicate effective voltage or current rather than peak voltage or current. Furthermore, the power rating of ac devices is the effective power consumption of that device. Ohm's law and the equations for power (Equation 11.2) hold for both ac and dc (keep in mind that the voltage, current, and power for ac are the effective values). It can be assumed that the values given for ac voltage, current, and power are effective values unless otherwise stated.

EXAMPLE 11.6

Calculate the peak voltage V_0 for 120-V ac electricity.

Solution: The effective voltage is 120 V. To find the peak voltage, solve Equation 11.3a, $V_{\text{eff}} = \frac{V_0}{\sqrt{2}}$.

$$V_{\text{eff}} = \frac{V_0}{\sqrt{2}}; \text{ therefore } V_0 = V_{\text{eff}}\sqrt{2}$$

$$V_0 = V_{\text{eff}}\sqrt{2} = 120\,\text{V} \cdot \sqrt{2} = 170\,\text{V}$$

Thus, common household electricity swings from +170 to −170 V and back 60 times per second, producing an effective voltage of 120 V.

EXAMPLE 11.7

(a) What is the effective current through a 120-V light bulb that has a resistance of 144 Ω when hot? (b) What is the power rating of the bulb?

Solution: (a) Using Ohm's law,

$$I = \frac{V}{R} = \frac{120\,\text{V}}{144\,\Omega} = 0.833\ \text{A}$$

Subscripts have been omitted from *I* and *V*, implying that these are effective values.

Solution: (b) Any of the equations for power may be used. Choosing Equation 11.2a gives

$$\text{Power} = IV = 0.833\,\text{A} \cdot 120\,\text{V} = 100\ \text{W}$$

This is then a 100-W light bulb, and Power is the effective power consumption of the bulb. Recall that in Example 11.4 the resistance of a 100-W bulb was found to be 144 Ω. In that example, the effective voltage of 120 V was used without explicit mention. We are always using the effective values.

11.3.1 Why Use ac Electricity?

In the early days of electricity, there were bitter arguments over whether ac or dc power should be used in large-scale applications, such as lighting cities and supplying household power. Thomas Edison argued for dc whereas George Westinghouse argued for ac. The "war on currents" was settled in favor of ac primarily because it can be transmitted over large distances with greater efficiency. When electricity is transmitted long distances through wires, the power loss is proportional to the current squared. With ac the voltage can be "stepped up" at the electrical company. This is done to reduce the current and thus reduce power loss during transmission. When electrical current is low, much less energy is lost to heating the transmission wires. Therefore, the most efficient way to transmit electricity is to send it as low current at high voltages. However, it is dangerous to use high voltages in homes or businesses. A "step down" transformer is then used to reduce the voltage again at the user's site. This is most easily done with ac.

To see why high-voltage power transmission is more efficient than low-voltage power transmission, we need to consider the actual resistance of the lines carrying power cross-country. (Up to now all wires have been assumed to have negligible resistance, but cross-country wires are very long.) Current flowing through those wires heats them and thereby loses energy. The higher the voltage used, the less heat is generated in the wires carrying a given amount of power. This is not particularly obvious at first glance. To show that it is true, consider the following two examples.

EXAMPLE 11.8

(a) Calculate the amount of current needed to transmit 1.0 MW of power at a voltage of 120 V. (b) Calculate the amount of current needed to transmit 1.0 MW of power at a voltage of 12,000 V.

Solution: (a) From Equation 11.2a, $P = IV$. Note that 1.0 MW = 1.0×10^6 W. Solving for the current, $I = P/V$, gives

$$I = \frac{\text{power}}{\text{voltage}} = \frac{1.0\times10^6\,\text{W}}{120\,\text{V}} = 8.33\times10^3\ \text{A}$$

Example 11.8 (b) Similarly,

$$I = \frac{\text{power}}{\text{voltage}} = \frac{1.0\times10^6\ \text{W}}{12{,}000\ \text{V}} = 83.3\ \text{A}$$

The message here is that for constant power, when the voltage is increased, the current is decreased. This is a consequence of the conservation of energy. Now, the power lost as heat in the transmission line is given by $P = I^2R$, where R is the resistance of the line. We see that the power lost is directly proportional to the square of the current in the line. The lower the current, the less power is lost in the line. With ac current, the power company is able to step up the voltage, which then steps down the current, and saves the company a lot of lost energy. The next problem will illustrate this fact.

EXAMPLE 11.9

Using the currents found in the previous example, find the power loss in a 0.01-Ω transmission line (a) for the 120-V transmission of 1 MW of power, and (b) for the 12,000-V transmission of 1 MW of power.

Solution: (a) From Equation 11.2b, $P = I^2R$. The power lost at low voltage and high current is

$$\text{Power} = I^2R = (8.33\times10^3\ \text{A})^2 0.01\Omega = 6.94\times10^3\ \text{W} = 0.694\ \text{MW}$$

Solution: (b) From Equation 11.2b, $P = I^2R$. The power lost at high voltage and low current is

$$\text{Power} = I^2R = (83.3\ \text{A})^2 0.01\Omega = 69.4\ \text{W}$$

You can see that there is a significant power savings when the current is low. In summary, it is more efficient to transmit electrical power long distances at high voltages because the same amount of power can be transmitted with less current. Using a smaller current drastically reduces the amount of power lost to heat in the transmission wires.

11.4 MULTIPLE-RESISTANCE CIRCUITS

The vast majority of electric circuits contain more than one resistance. Just a glance inside a TV or a look at all of the appliances running simultaneously in a hospital room is convincing. In this section, the ideas developed in the previous sections are applied to multiple resistance circuits. Time and space do not allow the treatment of electrical components other than resistances, such as transistors and capacitors. Nevertheless, multiple resistance circuits are very common and an understanding of them is quite useful.

Two of the ways resistances may be connected together are called *series and parallel* connections and are illustrated in Figures 11.5 and 11.6. Note that it is not necessary for resistances to be connected to a voltage source to be in series or parallel. Furthermore, any number of resistances may be connected in series or parallel, not just three as illustrated.

Figure 11.5 Series connection of three resistances. The same current flows through each resistance.

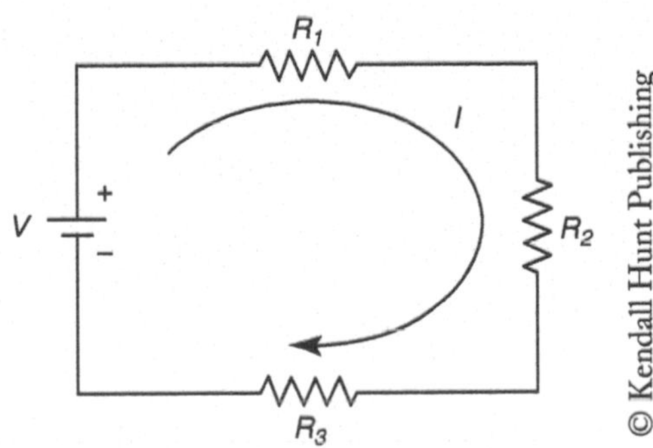

11.4.1 Series Connection of Resistances

The series connection is so named because the resistances are connected end to end and all the charge flowing in the circuit must go through each resistance in sequence (serially). There is no alternate path for the current to take. Some Christmas tree lights are connected in series. When just one light burns out, all of the lights go dark.

The same current thus exists in each resistance in series because there are no alternative paths and charge cannot be stored in the wires. (Remember that charge cannot be created or destroyed, so each charge must move on or be Figure 11.5 stored.) Since the current goes through each resistor in the circuit, the total resistance is the sum of the individual resistances. That is, adding resistors that are in series with one another,

$$R_{series} = R_1 + R_2 + R_3 + \ldots \qquad \textit{Know} \qquad \text{eq. 11.4}$$

where R_{series} is the total resistance of a series connection of resistors.

EXAMPLE 11.10

Consider a *series* circuit like the one shown in Figure 11.5. Suppose the battery has a voltage of 6.0 V and the resistances are 6.0, 4.0, and 2.0 Ω. (a) Calculate the total resistance of these resistances in series. (b) Calculate the current flowing in the circuit. (c) Find the voltage drop across each resistance. (d) Using Ohm's law, find the total voltage drop of the three resistances together.

a. Calculate the total resistance of these resistances *in series.*

Solution: (a) Looking at Figure 11.5(a), we can see that the resistors are in series with each other. In order to add resistors in series, use Equation 11.4.

$$R_{total} = R_{series} = R_1 + R_2 + R_3 = 6\Omega + 4\Omega + 2\Omega = 12\Omega.$$

b. Calculate the current flowing in the circuit.

Solution: (b) Let us use Ohm's law for this part.

$$I = \frac{V}{R} = \frac{6.0V}{12\Omega} = 0.50\,\text{amps}$$

c. Find the voltage drop across each resistance.

Solution: (c) Again, we can use Ohm's law. We will solve it for voltage.

$$I = \frac{V}{R}; \text{ therefore } V = IR$$

$$V = IR$$

$$V_{6\Omega} = 0.50\text{A}(6\Omega) = 3 \text{ V}$$

$$V_{4\Omega} = 0.50\text{ A}(4\Omega) = 2 \text{ V}$$

$$V_{2\Omega} = 0.50\text{ A}(2\Omega) = 1 \text{ V}$$

Notice that the total voltage drop across all resistors in the circuit equals the battery voltage, 6 V.

d. Find the total voltage drop of the three resistances together.

Solution: (d) Because we believe in the conservation of energy, we already know that the answer must be 6 V. We can check this by adding the voltages from part c: 3 V + 2 V + 1 V = 6 V. Just for fun, we can solve Ohm's law for voltage and substitute the current and total resistance as follows:

$$I = \frac{V}{R}.\text{ Therefore, } V = IR = 0.50\text{ A}(12\Omega) = 6\text{V}$$

Notice that we still get 6 V for total voltage drop.

Two important points of physics are made in the preceding example. The first is the interpretation of $V = IR$ as the voltage drop in a resistor (first mentioned in Section 11.1). As current passes through a resistance, voltage is reduced by the resistance. The current is the same at the entrance and exit of the resistance, but the voltage drops across the resistance. The voltage drop represents a decrease in electrical potential energy as the current passes through a resistor.

Second, the total voltage drop across all resistances in a circuit, is equal to the voltage supplied by the source. This is always true in a closed loop and is related to the *conservation of energy* principle. Whatever energy is supplied by the battery is used by the resistances in the circuit. The energy used by the resistances is the product of the voltage drop and the charge moved: $E = Vq$. The total voltage drop across all resistors in the circuit must equal the voltage supplied by the battery. Finally, note that in series each resistance does not receive the full voltage of the source.

In some cases, the small resistance of wires supplying voltage to an appliance has a noticeable effect on the operation of the appliance. This resistance can be thought of as being in series with the appliance, as shown in Figure 11.7, where R_w is the resistance of the wires and R is the resistance of an appliance. Consider the following example.

EXAMPLE 11.11

Let the appliance represented by R in Figure 11.7 be a refrigerator, and let the resistance of the wires leading to it be $R_w = 0.25\ \Omega$. The voltage source supplies 120-V ac. It is typical of appliances

Figure 11.6 Schematic of a simple electric circuit using an alternating current (ac) voltage source and grounded on one side. The small, usually negligible resistance of the wires supplying an appliance with voltage is in series with the resistance of the appliance.

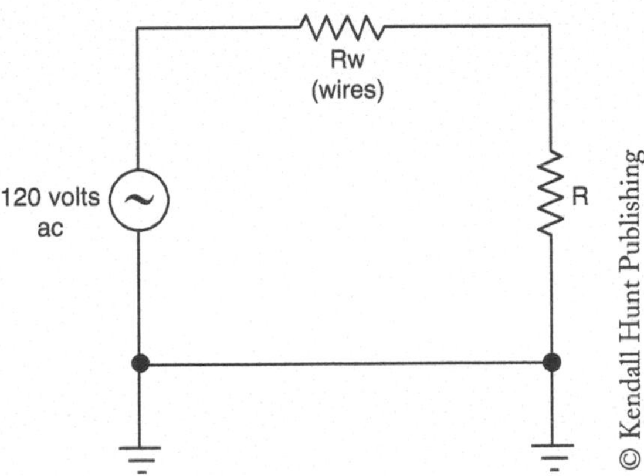

with motors in them to draw significantly more current when starting than when operating steadily. (a) When starting, 20 A of current flows in this circuit. Calculate the voltage that the refrigerator receives. (b) When the appliance is running steadily, 3.0 A flows. Calculate the voltage the refrigerator receives.

Solution: (a) We are told that when the refrigerator is first starting, 20 A of current flows in this circuit. Remember that the total voltage drop must be 120 V. Use Ohm's law to determine the current in the circuit.

$$I = \frac{V}{R}; \text{ therefore } V = IR$$

The voltage drop across R_w is

$$V = IR = 20\,\text{A}(0.25\,\Omega) = 5\text{ V}$$

If you believe in the conservation of energy, you must conclude that the voltage drop at the refrigerator is 120 V - 5 V = 115 V.

(b) When the appliance is running steadily, 3.0 A flows. Calculate the voltage the *refrigerator* receives.

Solution: (b) The voltage drop across R_w is

$$V = IR = 3\,\text{A}(0.25\ \Omega) = 0.75\text{ V}.$$

Because the circuit has a total voltage of 120 V and you believe in the conservation of energy, we must conclude that the voltage drop across the refrigerator is 120 V - 0.75 V = 119.25 V ≈ 119 V.

The voltage to the refrigerator when it starts is decreased enough to have an effect, while the voltage when the refrigerator is running steadily is not significantly less than the full voltage of the source. If you happen to have the refrigerator door open when it comes on, you will notice that the light inside dims temporarily. The light can be considered part of the total resistance R of the refrigerator.

Figure 11.7 Parallel connection of three resistances. Each resistance receives the same voltage but may have different currents. The total current supplied by the battery is the sum of the currents flowing through each resistance.

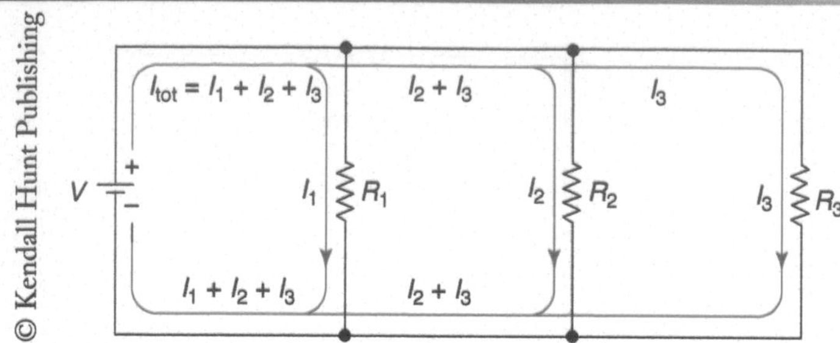

The following is a brief summary of major points for a *series connection of resistances:*

1. $R_{series} = R_1 + R_2 + R_3 + ... R_j$
2. There is the same current through each resistor.
3. Voltage drop in each resistor, R_j is $V = IR_j$.
4. Each resistor does not receive the full voltage of the source. The sum of the voltage drops across all of the resistors will equal the full voltage of the source.

11.4.2 Parallel Connection of Resistances

The three resistances in Figure 11.6 are connected in parallel. The first thing to notice about a parallel connection is the low resistance path from the voltage source to each resistance. (The straight lines represent wires of negligible resistance.)

Resistors in parallel

Each resistor in parallel has the same voltage drop. *When resistors are connected in parallel, their total resistance is always less the smallest value.* The current through each resistance is then given by Ohm's law:

$$I_j = \frac{V_j}{R_j}$$

where V is the voltage across the resisters and the subscript j = 1, 2, 3, and so on. The total current supplied by the voltage source is the sum of the individual currents flowing through the resistors. That is,

$$I_{total} = I_1 + I_2 + I_3 \ldots + I_j$$

The total current is greater than it would be if only one of the resistances was connected, implying that the total resistance of a parallel connection is less than the resistance of any of the individual resistors. Algebraic manipulation of the last equation and Ohm's law yields the following expression for *the total resistance of a parallel connection:*

$$\frac{1}{R} = \frac{1}{R_1} + \frac{1}{R_2} + \frac{1}{R_3} \ldots + \frac{1}{R_j} \qquad \textit{Know} \qquad \text{eq. 11.5}$$

This equation will always give a smaller value for $R_{parallel}$ than any of the resistances R_1, R_2, R_3, and so on.

The reason that the resistance is smaller in parallel is that current has a choice of paths to follow. This is analogous to cars on a highway crossing a bridge over troubled waters. The number of lanes on the bridge are similar to resistors in parallel. Increasing the number of lanes, reduces the resistance to traffic flow.

Homes, businesses, hospitals, and so on are wired so that all their electrical appliances and devices are connected in parallel. A parallel connection enables all of the devices to get the full voltage of the source and to operate independently. Figure 11.8 is a schematic of a typical household circuit. In it, four devices are plugged in and one receptacle is open. Three of the devices are switched on and operating, but any combination could be operated. This is true even if a new device is plugged into the available receptacle.

EXAMPLE 11.12

Suppose a 1.0-kW microwave oven and a 500-W coffee maker are in parallel with each other are operated simultaneously on 120-V electricity. (a) What is the current through each appliance and what is the total current? (b) Calculate the resistance of each appliance and (c) the total resistance of the circuit.

Solution: (a) What is the current through each appliance and what is the total current?

The current can be calculated using Equation 11.2a:

$$\text{Power} = IV;\ \text{therefore the current } I = \frac{\text{power}}{\text{voltage}}$$

The current through the 1000 W **microwave** is

$$I = \frac{P}{V} = \frac{1000\text{ W}}{120\text{ V}} = 8.33\text{ A}$$

The current through the 500 W **coffee maker** is

$$I = \frac{P}{V} = \frac{500\text{ W}}{120\text{ V}} = 4.17\text{ A}$$

So the total current is

$$I_{total} = 8.33\text{ A} + 4.17\text{ A} = 12.5\text{ A}$$

Solution: (b) Calculate the resistance of each appliance and the total resistance of the circuit.

The resistance can be determined from Ohm's law:

$$I = \frac{V}{R};\ \text{therefore } R = \frac{V}{I}$$

The resistance of the microwave is

$$R = \frac{V}{I} = \frac{120\text{ V}}{8.33\text{ A}} = 14.4\Omega$$

The resistance of the coffee maker is

$$R = \frac{V}{I} = \frac{120\ \text{V}}{4.17\ \text{A}} = 28.8\Omega$$

Note that the full voltage is used since the resistances are connected in parallel. This cannot be done for resistances in series.

Solution: (c) Calculate the total resistance of the circuit.

Step by step 1: Because these resistors are in parallel, the total resistance can be found from Equation 11.5:

$$\frac{1}{R} = \frac{1}{R_1} + \frac{1}{R_2} + \frac{1}{R_3} \ldots + \frac{1}{R_j}$$

$$\frac{1}{R} = \frac{1}{14.4\Omega} + \frac{1}{28.8\Omega}$$

Step 2: We can add these fractions using the common denominator of 28.8 Ω as follows:

$$\frac{1}{R} = \frac{2}{28.8\Omega} + \frac{1}{28.8\Omega} = \frac{3}{28.8\Omega}$$

Step 3: Notice that so far, we have solved for $\frac{1}{R}$. Next, we need to solve for R. Because $\frac{1}{R} = \frac{3}{28.8\Omega}$,

$$R = \frac{28.8\Omega}{3} = 9.6\Omega$$

As mentioned, the total resistance is less than either of the resistances in the circuit. An alternative method of finding the total resistance is to solve Ohm's Law for R and use the total current found in part (a) for *I*. You might want to try this to see if the same answer is obtained.

The following is a brief summary of major points for parallel connections of resistances:

1. Adding resistors in parallel, $\frac{1}{R} = \frac{1}{R_1} + \frac{1}{R_2} + \frac{1}{R_3} \ldots + \frac{1}{R_j}$ *Know*
2. Each resistor in parallel has the same voltage across it.
3. More current will go through the smaller resistors in parallel.
4. The sum of the currents going through the parallel resistors will equal the total current in the main part of the circuit.

11.4.3 Combinations of Series and Parallel Connections

A circuit with a combination of series and parallel connections can be analyzed using the ideas already developed in this section. In Figure 11.9, R_1, R_2, and R_3 are in parallel with each other. They are said to be in parallel, because the current has more than one path that it can take. Some of the current goes through each resistance and recombines as the current exits. The group of parallel resistors are in series with the R_4 resistor. All of the current will go through R_4.

Figure 11.8 Household and business wiring connects devices in parallel so that each receives the full voltage of the source. Any device can be switched on or off without affecting the others.

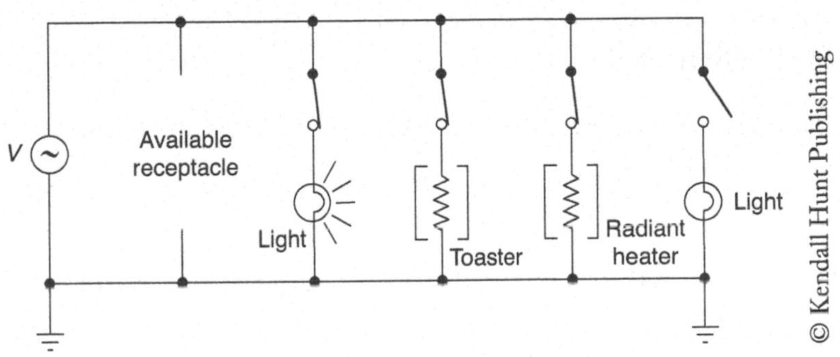

Figure 11.9

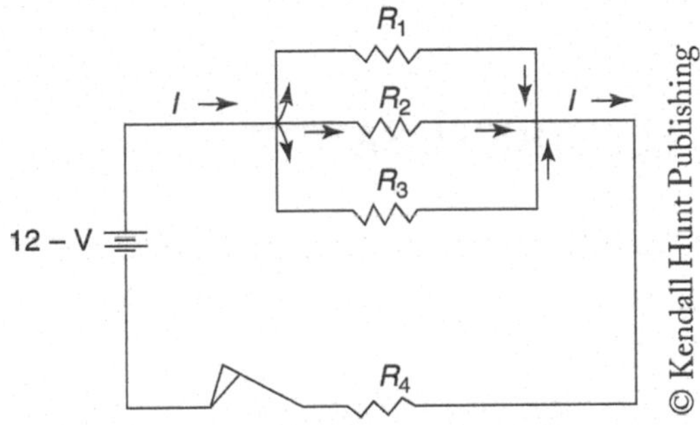

EXAMPLE 11.13

In Figure 11.9, let $R_1 = 2\ \Omega$, $R_2 = 3\ \Omega$, $R_3 = 2\ \Omega$, and $R_4 = 1.25\ \Omega$.

a. Determine the series equivalent of the parallel resistors.
b. Determine the total resistance in the circuit.
c. Determine the current that goes through R_4.
d. Determine the power required by the circuit.

Solution: (a) $\frac{1}{R} = \frac{1}{R_1} + \frac{1}{R_2} + \frac{1}{R_3}$, $\frac{1}{R} = \frac{1}{2} + \frac{1}{3} + \frac{1}{2} = \frac{3}{6} + \frac{2}{6} + \frac{3}{6} = \frac{8}{6}$

Therefore, $R = \frac{6}{8}\Omega = 0.75\,\Omega$

Solution: (b) Since the parallel group of resistors is in series with R_4, the total resistance is $0.75\,\Omega + 1.25\,\Omega = 2\,\Omega$.

Solution: (c) Use Ohm's law to determine the current through R_4. Note that it will be the current that flows through the main body of the circuit. Therefore, we can use the total voltage and the total resistance as follows:

$$I = \frac{V}{R} = \frac{12\,\text{V}}{2\,\Omega} = 6\,\text{A}$$

Solution: (d) The power required by the circuit is

$$P = IV = 6\,\text{A} \cdot 12\,\text{V} = 72\,\text{W}$$

Things to know for Section 11.4

In order to determine the total resistance of resistors in **series**, just add them up:

$$R_{\text{series}} = R_1 + R_2 + R_3 + \ldots$$

In order to determine the total resistance of resistors in *parallel*, use the following rule:

$$\frac{1}{R_{\text{parallel}}} = \frac{1}{R_1} + \frac{1}{R_2} + \frac{1}{R_3} \ldots + \frac{1}{R_j}$$

11.5 THE BASICS OF ELECTRICAL SAFETY

Circuit breakers, fuses, and the three-wire system are all designed to minimize damage if something goes wrong. This section presents the basics of electrical safety. Chapter 12 goes into more detail, particularly as regards medical situations.

Electricity has two basic hazards: *thermal hazards* from overheating wires or appliances and *shock hazards* from electric current passing through a person.

11.5.1 Thermal Hazards: Circuit Breakers and Fuses

Thermal hazards result from heat being produced faster than it can be carried away. It is the rate of heating, or power, transformed to thermal energy that is important. Power in electric resistances is given by Equations 11.2a–11.2c, repeated here for convenience:

$$\text{Power} = IV \quad \text{(a)}$$

$$\text{Power} = I^2R \quad \text{(b)}$$

$$\text{Power} = \frac{V^2}{R} \quad \text{(c)} \qquad (11.2)$$

Thermal hazards occur when there is a drastic increase in the power consumed by a circuit. One example of this is the heating of wires that supply current to parallel connections of resistances, as shown in Figure 11.10. If more resistances are added, then the current flowing in the circuit increases. This is because each resistance connected in parallel reduces the total resistance in the circuit. From Ohm's law, we know that the current in the circuit is inversely proportional to the current.

$$I = \frac{V}{R}$$

As the resistance decreases, the current increases.

From Equation 11.2a, Power $= IV$, we see that as the current increases, the power increases.

Normally the thermal energy produced in wires is negligible, but if resistance is drastically reduced, then the current becomes large enough to produce a large amount of thermal energy. Under these circumstances, the heat produced can become large enough to ignite materials close to the wires.

In order to prevent large currents from starting fires, a circuit breaker or fuse can be included in the circuit, as shown in Figure 11.10. Both devices are designed to interrupt the flow of current if it exceeds a specified level.

Figure 11.11 shows the construction of a circuit breaker and a fuse.

Another problem that creates a thermal hazard is the short circuit.

A short circuit is any very small resistance receiving full voltage. For example, as the cord of an appliance wears, the insulation may break and allow the wires to touch, as shown in Figure 11.12.

The contact between wires, has a very small resistance, allowing a very large current to flow. The large current heats the wires to the outlet as described in the preceding paragraphs, but it also creates a great deal of heat at the short.

Consider Equation 11.2c:

$$\text{Power} = \frac{V^2}{R}$$

where R is the resistance of the short. *Power* will be extremely large since R is very small. Note that R receives the full voltage of the source because it is in parallel with the appliance. Of course, the circuit breaker or fuse will interrupt the large current of the short, but there is often some thermal damage to the cord in the brief time the current does flow.

11.5.2 Shock Hazards: the Three-Wire System

A shock hazard exists whenever there is a possibility for current to pass through a person. The severity of a shock depends primarily on how much current passes through the person. For 60-Hz electricity,

Figure 11.10 Typical household circuit with two electrical devices in use, with extra room to plug others in. If there is too much current, the circuit breaker, or fuse, will open, since that could overheat the wires (whose resistance is represented by R_W).

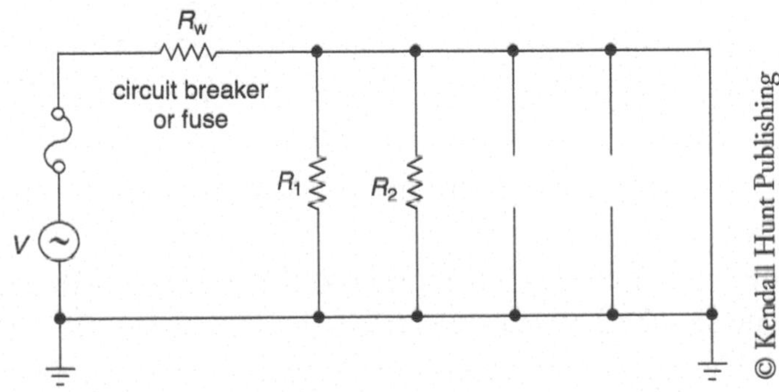

Figure 11.11(a) A circuit breaker: The bimetallic strip moves to the right when heated. If it moves into the notch, the compressed spring forces the metal strip downward, breaking the connection at the fixed contact point.

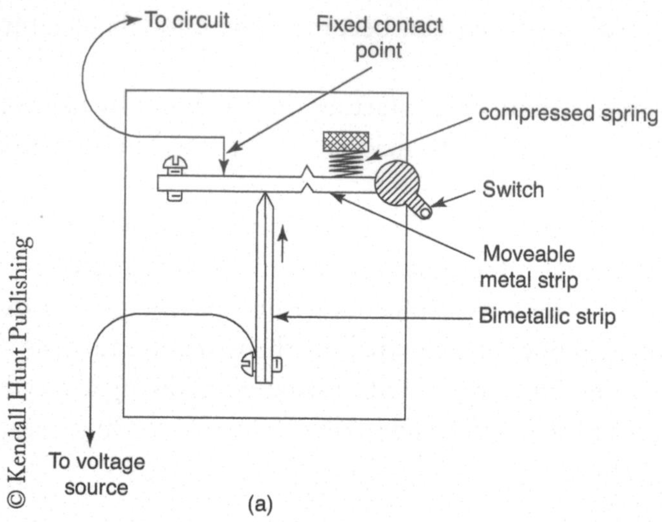

Figure 11.11(b) A fuse: The metal strip has a low melting point and breaks the connection if it is overheated by a large current and then melts.

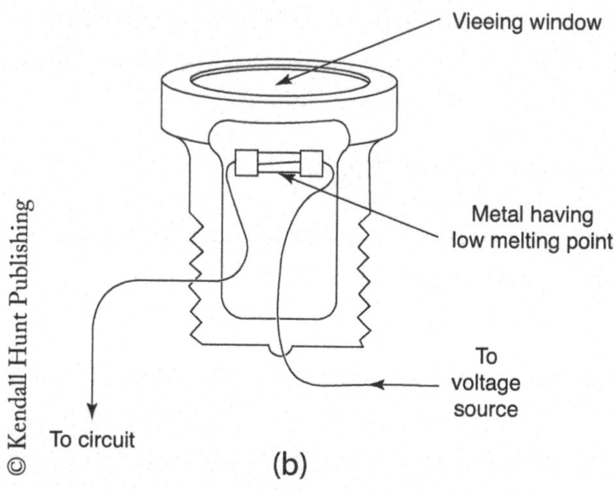

the maximum harmless current through a normal person is 5 mA (5 milliamperes). Currents of less than 1 mA usually cannot even be felt, so a good rule of thumb is that if a shock can be felt, it is likely to be harmful. (Some medical patients are rendered microshock sensitive by attachment to medical equipment and must be protected from currents 1000 times smaller than these. The interested reader should refer to Chapter 12.)

The most common type of electric power is ac with the return wire grounded as shown in Figure 11.10. A person can be shocked by such electricity by touching only one wire because the feet can complete the circuit by being in contact with the ground. The lower the resistance of the person, the greater is the hazard.

Most wires are insulated, and most appliances have protective cases to prevent people from coming into contact with a source of voltage.

The large resistance of the insulation makes the current through the person very small [see Figure 11.13(a)].

The current passing through the person in the figure is

$$I_{\text{person}} = \frac{V}{R_{\text{Low person}} + R_{\text{person}}}$$

since the resistance of the insulation and the person are in series. Note that this assumes that the case of the appliance is a good conductor, such as steel.

Of course, insulation can wear out, especially where wires enter an appliance. If the hot wire comes into contact with the metal case of an appliance, then the situation shown in Figure 11.13(b) may occur. The person is in parallel with the appliance and receives the full voltage of the source.

$$I_{\text{person}} = \frac{V}{R_{\text{person}}}$$

The amount of current passing through the person is the resistance of the person, R_p, has a great effect on the severity of the shock. That is why it is very dangerous to touch electrical appliances when standing barefoot on wet ground since this greatly reduces R_p. On the other hand, the person might not even notice the shock if he or she is standing on a rubber mat and wearing shoes, thus greatly increasing R_p.

The *three-wire* system was developed to help prevent shock hazards. In the three-wire system the case of the appliance is grounded, as shown in Figure 11.14. The ground connection to the case forces its voltage to be zero, making the case safe to touch. Insulation on the three wires is usually color coded, as indicated in the figure. The black wire is "hot" because it supplies voltage to the appliance; the white wire is "neutral" since it is grounded at the home or business and at the source. The white wire supplies a return path for the current flowing in the circuit (remember a circuit must have a complete path). The three-wire system adds a green "ground" wire to ground the case of appliances.

Figure 11.12 A short circuit in the cord to an appliance. The circuit breaker or fuse will interrupt the resulting high current to minimize thermal damage.

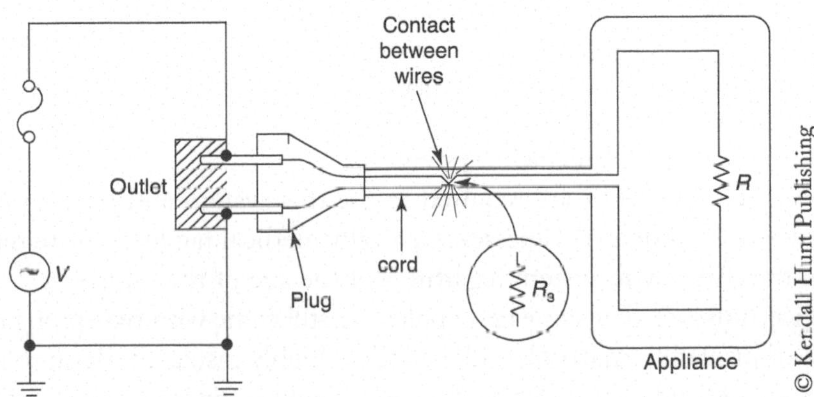

Figure 11.13a (a) Normally, the case of an appliance prevents the person from coming into direct contact with a voltage source.

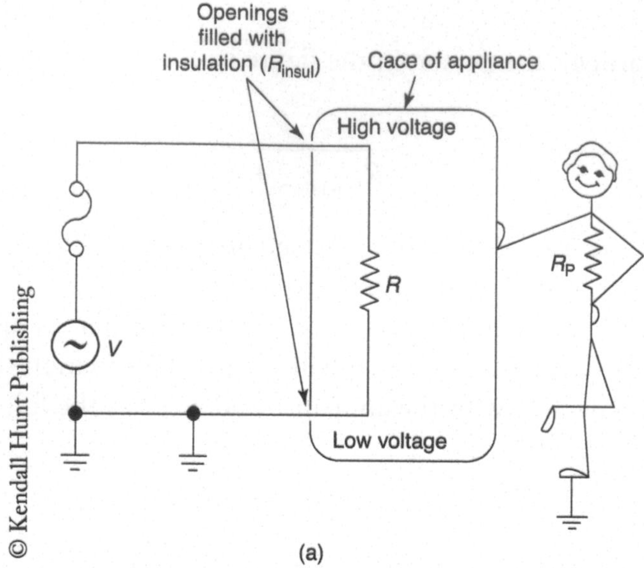

Figure 11.13(b) A shock hazard exists if the insulation fails and brings the high-voltage wire into contact with the metal case of an appliance, The victim receives the full voltage of the source.

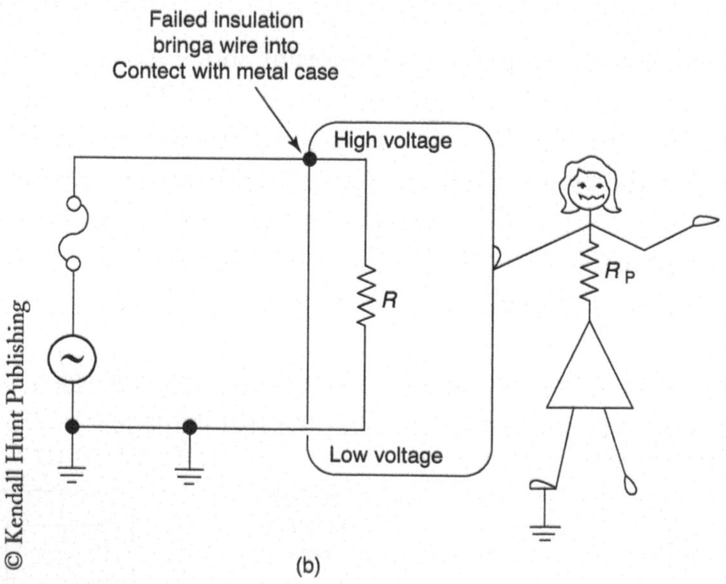

Figure 11.15 represents a three-wire system.

Homes built since the 1960s are required to use this system and have three-hole outlets rather than the older two-hole outlets. Businesses, especially medical facilities, were required to rewire their buildings to use the three-wire system regardless of the age of the building.

You may notice, however, that some new appliances still come with two-prong plugs. Such appliances have cases made of insulating materials and are called doubly insulated. Because the case is a very poor conductor, it acts as a second insulator if the hot wire should come into contact with the case.

A circuit breaker or fuse is still necessary in the three-wire system. If a hot wire does come into contact with a grounded case, a short circuit occurs because the green wire is a very low-resistance path to ground. The short will trip a circuit breaker or blow a fuse, and the appliance will consequently not operate. This forces the user to repair the short.

The basics of electrical safety have now been covered. Chapter 12 will go into more detail, particularly as regards medical situations.

Figure 11.14 The three-wire system adds a wire to ground the case of the appliance, thereby fixing its voltage at zero.

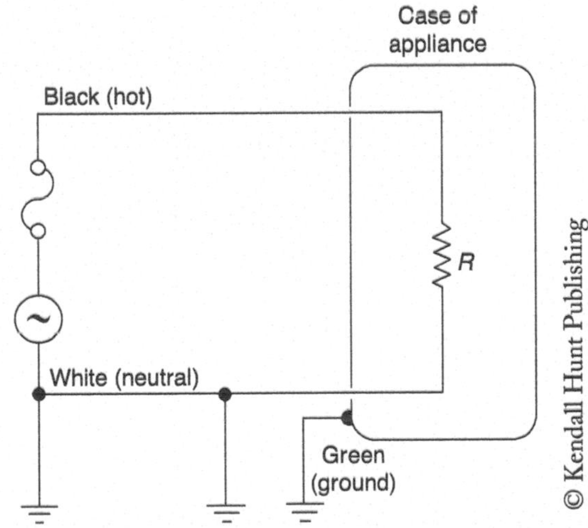

Figure 11.15 Household and business wiring using the three-wire system distributes electric power from a circuit breaker box (only one breaker is shown here) to three-hole outlets with the wires attached as shown. The holes are of different sizes and shapes so that three prong and modern two-prong plugs can be inserted in only one way.

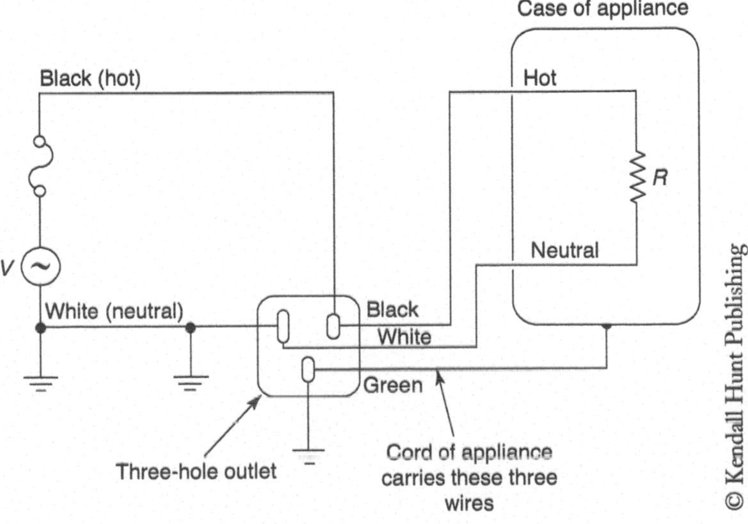

Things to know for Chapter 11

$$V \equiv \frac{E}{q} \quad 1\,\mathrm{V} \equiv 1\frac{\mathrm{J}}{\mathrm{C}}$$

$$I \equiv \frac{q}{t} \quad 1\,\mathrm{A} \equiv 1\frac{\mathrm{C}}{\mathrm{s}}$$

Ohm's law $I = \frac{V}{R}$

$$\text{Power} \equiv \frac{\text{Energy}}{\text{time}} \quad 1\,\mathrm{W} \equiv 1\frac{\mathrm{J}}{\mathrm{s}}$$

Power in electric circuits:

$$\text{Power} = IV$$

$$\text{Power} = I^2 R$$

$$\text{Power} = \frac{V^2}{R}$$

Adding resistors in series:

$R_{\text{series}} = R_1 + R_2 + R_3 + \ldots$

where R_{series} is the total resistance of a series connection of resistors.

Adding resistors in parallel:

$$\frac{1}{R} = \frac{1}{R_1} + \frac{1}{R_2} + \frac{1}{R_3} \ldots + \frac{1}{R_j}$$

Questions

11.1 Why are two conducting paths needed from any voltage source? Why cannot an electrical device operate by just being connected to a single hot wire?

11.2 Explain why the direction of electron flow in a circuit is from negative to positive as shown in Figure 11.1.

11.3 Electric current passing through a resistance produces thermal energy, increasing the resistance's temperature (as in an incandescent light bulb). Explain how the moving charges might transfer some of their energy to the atoms in a resistor. Is your explanation consistent with the fact that the resistance of most substances increases with temperature?

11.4 In what ways are simple electric circuits analogous to fluid-circulation systems? In what ways do they differ?

11.5 Two of the equations for electric power (Equations 11.2a and 11.2c) involve voltage. When is the full voltage of the source used to calculate power? When is some voltage other than the full voltage of the source used?

11.6 What are kilowatt-hours and what is their most common use?

11.7 What are ac and dc power? Give an example of the use of each.

11.8 What are effective voltage, current, and power?

11.9 Why is one side of a circuit frequently grounded as shown in several of the figures in the chapter? (For example, see Figure 11.7.)

11.10 In Figure 11.4, why is the voltage zero between points A and B? What is the advantage of having zero voltage there?

11.11 What is the advantage of using high voltages when transmitting power long distances?

11.12 Why do your house lights dim momentarily when you start a vacuum cleaner?

11.13 Why doesn't a bird sitting on a high-voltage transmission line get electrocuted?

11.14 Are appliances in homes and businesses connected to the voltage source in series or parallel? Explain.

11.15 If one light in a string of Christmas lights burns out and all the others go dark, is the string wired in series or parallel? Explain.

11.16 Explain why the total resistance of a series connection of resistances is greater than any of the individual resistances. Explain why the total resistance of a parallel connection of resistances is less than any of the individual resistances.

11.17 Why do all the resistances in a parallel connection receive the same voltage?

11.18 Make an analogy between a series circuit and a fluid circulating system. Include an explanation of why the voltage drop in each resistance is not the full voltage of the source and why the pressure drop in each resistance in the fluid system is not the full pressure of the pump.

11.19 Identify, the two major hazards of electricity. Describe situations in which each exists separately and one in which they exist together.

11.20 Name and describe the function of each wire in the, three-wire system.

11.21 What electrical hazard is minimized by a circuit breaker or fuse? By the ground wire to the case of an appliance? Why are both a ground wire and a circuit breaker or fuse needed?

11.22 Explain why a short circuit is a thermal hazard rather than a shock hazard.

11.23 Why do some appliances have a fuse in them even though the circuit they are plugged into is protected by a circuit breaker?

11.24 Why is it inadvisable to replace a circuit breaker or fuse with one having a larger current rating (one that will not interrupt the flow of current until it reaches a higher value)?

11.25 What is the relationship between the resistance of wires supplying electricity to a typical business and the current rating of the circuit breaker used? That is, if the wires used have a smaller than normal resistance, should the circuit breaker have a smaller or larger trip current? Explain.

11.26 The severity of a shock depends on the current passing through a person. Why, then, is it not possible to say how large a voltage is dangerous?

11.27 What hazard, if any, is associated with a white wire coming into contact with the metal case of an appliance? Does your answer depend on whether the three-wire system is being used?

PROBLEMS

SECTION 11.1

√11.1 (I) How much current flows through an air conditioner that uses 220-V electricity and has a resistance of 7.50 Ω?

√11.2 (I) Power transmission lines are hung from metal towers on glass insulators. What current flows through one insulator if it has a resistance of 1.00×10^9 Ω and the line has a voltage of 2.00×10^5 V?

√11.3 (I) How much current flows through a 2.4-Ω headlight in a car with a 12.0-V electrical system?

√11.4 (I) The maximum safe ac current through a normal person is 5.00 mA. A repair technician touches the hot wire in an appliance opened for servicing and thereby comes into contact with 120 V. Will he receive a harmful shock (a) if his hands are wet and the resistance of the path through him to ground is 8000 Ω? (b) if his skin is dry and he is wearing rubber-soled shoes giving a resistance of 500,000 Ω to ground?

√11.5 (II) Calculate the resistance of a vacuum cleaner that uses 14.0 A of current and operates on 120-V electricity.

11.6 (II) What is the resistance of a 120-V electric clock that uses 5.00 mA of current?

11.7 (II) Find the resistance of indicator lights on a stereo amplifier if each operates on 14.0 V (produced inside the amplifier) and draws 0.0500 A of current.

11.8 (II) (a) What is the voltage drop in an extension cord having a resistance of 0.0600 Ω and through which 5.00 A of current is flowing? (b) A smaller diameter wire in a cheaper extension cord gives it a larger resistance, say, 0.300 Ω. What is the voltage drop in this extension cord if 5.00 A flows through it?

11.9 (II) What is the voltage of a battery that produces a 5.00-A current when attached to a 2.40 Ω resistance?

11.10 (II) A given electric motor requires a minimum of 10 A to run properly and has a resistance of 20 Ω when running. What minimum voltage is required to operate the motor?

SECTIONS 11.2 AND 11.3

11.11 (I) Calculate the electric power of a lightning bolt having a current of 20,000 amps and a voltage of 1.00×10^8 V.

√11.12 (I) When a truck with a 12.0-V electrical system starts its engine, a current of 500 A flows through the starter motor. How much power is supplied by the battery to the starter alone?

√11.13 (I) Cords to appliances sometimes warm up because of the thermal energy created by their resistance. How much power is used by the 0.300-Ω cord if an appliance through which 12.0 A flow? (Such a cord is faulty because its resistance should be lower.)

√11.14 (I) (a) Calculate the current through a short-circuit resistance of 0.100 Ω that experiences a voltage of 120 V. (b) Calculate the power that is converted to heat by such a short circuit.

√11.15 (I) What is the power rating in watts of an indicator light on a heart monitor if the light is known to have a resistance of 10,000 Ω and operates on 120-V electricity?

√11.16 (I) *(I)* The resistances of most substances increase when they are heated. A bulb for a flashlight using 3.00 V has a resistance of 2.00 Ω when cold and 24.0 Ω when hot. Calculate the current through and the power consumed by the bulb (a) when it is first turned on and is still cold, and (b) when it is hot.

11.17 (I) An unsuspecting tourist takes an electric razor designed to operate with 120-V electricity to Europe and plugs it into a 220-V outlet. (A special adapter is needed to do this.) The razor is ruined by overheating. If the razor normally consumes 20.0 W of power, how much power does it consume using 220-V electricity?

11.18 (I) An oscilloscope can be used to make a graph of ac voltage similar to that in Figure 11.3(b). If the peak voltage is measured to be 311 V, what is the effective voltage?

11.19 (II) What resistance must the heating element of an electric clothes dryer have to produce heat at the rate of 5.00 kW when a current of 20.0 A flows through it?

11.20 (II) A small office building is cooled by an air conditioner that operates on 408-V electricity and uses 50.0 kW of power when running steadily. (a) What is the resistance of the air conditioner? (b) How much does it cost to run the air conditioner for one summer month if it is on 8.0 hour/day for 30 days and electricity costs 11.0 cent/kWh?

11.21 (II) (a) How much current flows through the 60.0-W headlight of an automobile that has a 12.0-V electrical system? (b) What is the resistance of the light?

11.22 (II) Newborn infants are often kept warm with an electric radiant heater. (a) If the heater produces a maximum of 500 W of power and operates on 120 V electricity, what maximum current passes through the heater? (b) What is the resistance of the heater?

11.23 (II) A power transmission line is limited to a maximum of 100 A of current. At what minimum voltage should it be operated to transmit 10.0 MW of power?

11.24 (II) (a) Calculate the warm resistance of a 25-W light bulb designed to operate with 120-V electricity. (b) This bulb is used in the refrigerator in Example 11.11. Calculate the power it consumes if the voltage it receives is 119 V and if it is 115 V. This difference is enough for your eye to detect. Assume the resistance of the bulb does not change appreciably.

11.25 (II) What is the cost of operating a hot-water heater for 1 year if it consumes 5.00 kW of power and is on an average of 2.0 hour/day? Assume that the cost of electricity is 11.0 cent/kWh.

11.26 (II) Modern solid-state electronics consume less energy than older vacuum tube devices. Compare the cost of operating a 500-W oscilloscope to monitor a heart patient's ECG for a week to the cost of doing the same thing with a solid state oscilloscope consuming 80 W. Assume the cost of electric energy is 11.0 cent/kWh.

11.27 (II) The average TV set is reputed to be on for 6.0 hour/day. How much does it cost to run the nation's TVs for a year if there are 100 million of them and the average cost of electric energy is 12.0 cent/kWh? What is the cost per household if there is one set in each? The average TV has a power rating of 150 W.

11.28 (II) Military aircraft use 400-Hz ac power because it is possible to design lighter-weight equipment using this frequency. (a) What time is required for one complete cycle of this ac power? (b) What is the effective voltage if the peak voltage is 300 V?

11.29 (III) Power companies reduce their output voltage to reduce power output in times of extremely high usage. What percentage reduction in power is achieved by reducing the output voltage of a power plant from 120 V to 110 V, assuming the resistance of the load is unaffected by the change?

11.30 (III) A microwave oven operates on 120-V electricity and draws 10.0 A of current. (a) Calculate the effective power (simply the power). (b) Calculate the peak power consumed by the oven.

11.31 (III) An x-ray tube is similar to a cathode ray tube in an oscilloscope, but it uses a higher voltage and its beam of electrons strikes a metal target (Tungsten), producing x-rays. (a) Calculate the power in the beam of electrons if 100,000 V is used and the beam current is 15.0 mA. (b) Assuming that a 100% efficient transformer can be used to increase an input voltage of 120 V to the 100,000 V needed, what current must be supplied to the transformer?

SECTIONS 11.4 AND 11.5

√11.32 (I) (a) Calculate the resistance of a series connection of resistances of 6.0 Ω and 12.0 Ω. (b) Do the same for a parallel connection of the same resistances.

√11.33 (I) (a) What is the resistance of ten 100-W light bulbs connected in parallel? (b) in series? The resistance of a 100-W, 120-V light bulb is 144 Ω. (Ignore the temperature dependence of resistance).

√11.34 (I) What total resistances can be obtained by connecting a 100 Ω and a 20.0 Ω resister together?

11.35 (II) (a) Calculate the currents used by and resistances of 25.0-W and 60.0-W light bulbs designed to operate on 120-V electricity. (b) If these two bulbs are connected in series to a 120-V source, what power is consumed by each, assuming that their resistances are the same as found in (a)?

11.36 (II) In figure 11.16, $R_1 = 3\ \Omega$, $R_2 = 4\ \Omega$, and $R_3 =$. The battery voltage is 12 V.

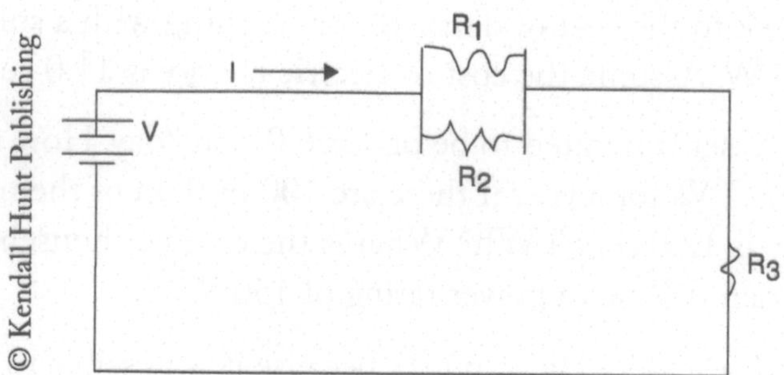

a. Determine the series equivalent of the parallel resistors. ___________
b. Determine the total resistance in the circuit. ___________
c. Determine the current through R_3. ___________
d. Determine the power required by the circuit. ___________

Bring your completed solutions to class.

11.37 (II) Will a 120-V hospital circuit protected by a 15.0-A circuit breaker be able to operate a 200-W ECG monitor, a 1200-W microwave oven, and eight 40.0-W lights simultaneously?

11.38 (II) It is common practice to have a separate circuit for a refrigerator alone so that no other appliance can trip the circuit breaker and shut off the refrigerator. What happens if a 1000-W toaster is operated on the same outlet as a refrigerator that draws 15.0 A when it comes on? The circuit breaker is rated at 20.0 A.

11.39 (II) How many 40.0-W light bulbs can be operated on a 120-V circuit protected by a 15.0-A fuse?

11.40 (III) A 120-V circuit in an intensive care unit is protected by a 20.0-A circuit breaker. It has four 100-W lights, two 200-W ECG monitors, a 500-W electric bed, and two 25.0-W IV pumps operating on it. What is the maximum power that can be supplied by this circuit to another appliance? What would the resistance of that appliance be? Is it prudent to use the circuit to its maximum?

11.41 (III) A 120-V radiant heater has two heating elements. (a) What resistance should the elements have so that one consumes 1.00 kW and the other 0.500 kW? (b) What are the five possible power consumptions of this heater ignoring any temperature dependence of the resistances?

11.42 (III) Batteries have an internal resistance that acts like another resistor in series with the load of the battery. As a rechargeable battery (such as that in an automobile) ages, its internal resistance increases. Suppose the internal resistance of a new car battery is 0.0100 Ω and it increases to 0.100 Ω with age. (a) Calculate the resistance of the starter motor of the car if it draws 200 A of current from the new battery. (b) What current flows through the starter when the battery is old? (c) To show that the effect of aging is less on smaller devices, calculate the current through the car radio (which has a resistance of 12.0 Ω) when the battery is new and old. Assume that no other devices are being used simultaneously.

11.43 (III) A 20 meter long extension cord used to power a 1000-W circular saw has a resistance of 1.00 Ω (higher than normal, so it is worn). (a) Assuming that 120-V electricity is used, calculate the voltage that actually gets to the saw. (First calculate the resistance of the saw assuming it gets the full voltage; then add the two resistances together and find the current in the extension cord.) (b) What is the actual power consumption of the saw if its resistance is constant? (c) Repeat both (a) and (b) assuming that the saw can be modified to use 220-V electricity. This is often done to save on power lost in extension cords and is a major advantage of using 220-V electricity.

GENERAL PROBLEMS

11.44 (III) A heart defibrillator passes 10.0 A of current through a patient's torso for 5.00×10^{-3} second to restore normal beating of the heart. (a) How much charge passed through the patient's torso? (b) What voltage was used if a total energy of 500 J was dissipated by the current? (c) What was the resistance of the path through the person?

11.45 (III) A 120-V electric immersion heater is able to increase the temperature of a 100-g aluminum cup containing 400 g of water from 20°C to 95.0°C in 2.00 minute. Calculate the resistance of the heater assuming that it is constant throughout this process. The specific heats of aluminum and water are $0.215\ \frac{\text{calorie}}{\text{g°C}}$ and $1.00\ \frac{\text{calorie}}{\text{g°C}}$, respectively.

11.46 (III) The resistance of a short circuit in a 120-V appliance cord is 0.500 Ω. (a) Calculate the power dissipated by the short circuit. (b) Calculate the temperature rise of the materials surrounding the short, given the following information. It takes 0.050 second for the circuit breaker to interrupt the current. Two grams of matter in the immediate vicinity of the short is heated, and that matter has a specific heat of $0.200\ \frac{\text{calories}}{\text{g} \cdot \text{°C}}$. What is the likely effect on the materials involved?

12 Electrical Safety

"Genius is one percent inspiration and 99 percent perspiration."

—Thomas Edison

Thermal hazards and shock hazards are the two major dangers of electricity. The proper use of circuit breakers and fuses (described in Section 11.5) limits thermal damage in electric circuits. Thermal hazards are caused by thermal energy generated by excessive amounts of current. Such currents are much larger than those that pose a shock hazard if passed through a person. Therefore, circuit breakers and fuses do not prevent many shock hazards; at best, they can switch off a defective piece of equipment.

This chapter is concerned primarily with shock hazards. Shock hazards are particularly severe in medical situations because patients are commonly connected to electrical equipment. Furthermore, some patients are made exceptionally sensitive to electrical shock by electrical connections to them that may allow current to flow directly to the heart. This chapter begins with a description of the physiological effects of electrical shocks and then considers various situations that present shock hazards. Also included are descriptions of safety procedures and devices designed to prevent shocks.*

12.1 PHYSIOLOGICAL EFFECTS OF ELECTRICAL SHOCK

The severity of an electrical shock depends on many factors, the most dominant being the amount of current passing through the person. The effect of a shock also depends on the path the current takes, the duration of the shock, and whether the voltage causing the shock is alternating current (ac) or direct current (dc). If the voltage is ac, there is also a dependence on the frequency of the ac voltage. Table 12.1 gives the effects of electrical shock as a function of current on the assumption that the shock has a duration of 1 second, passes through the trunk of the body, and is caused by 60-Hz ac electricity.

* The reader should be familiar with Section 11.5, an introduction to electrical safety, before proceeding with this chapter.

Table 12.1 Effects of Electrical Shock as a Function of Current[a]

Current (mA)	Effect
1	Threshold of sensation
5	Maximum harmless current
10–20	Onset of sustained muscular contraction; can't let go for duration of shock; contraction of chest muscles may stop breathing during shock (fatal if continued)
50	Onset of pain; heart still unaffected
100–300+	Ventricular fibrillation possible; very often fatal
300	Onset of burns (thermal hazard); depends on concentration of current
6000 (6 A)	Onset of sustained ventricular contraction and respiratory paralysis; both cease when shock is over; heartbeat often returns to normal

[a] For an average male shocked through intact skin for 1 second by 60 Hz ac. Values for females are 60%–80% of those listed.

At a current of 1 mA the sensation of being shocked becomes noticeable. Although not painful at this current, it is usually not considered a pleasant sensation. Somewhat higher currents can cause muscles to contract involuntarily.

The onset of this effect occurs at 10–20 mA. People sometimes say that they were "knocked across the room" by a shock. What really happened was that their muscles contracted, propelling them in a manner not of their own choosing, making it seems that the electricity exerted a force on them. The reason muscles contract in response to electric currents is that nerves controlling them utilize electric currents to do so. The effect of current-induced sustained muscular contraction is sometimes called the "can't let go" effect. A person being shocked by a wire grasped in the hand will be unable to release it. The contractor muscles that close the hand are stronger than the extensor muscles that open it (see *Figure 12.1*).

Figure 12.1 Electric current can cause muscular contractions with varying effects. (a) Victim is "thrown" backward by involuntary contraction of muscles, extending legs and torso. (b) Victim can't let go of the wire because the muscles that close the hand are stronger than those that open it.

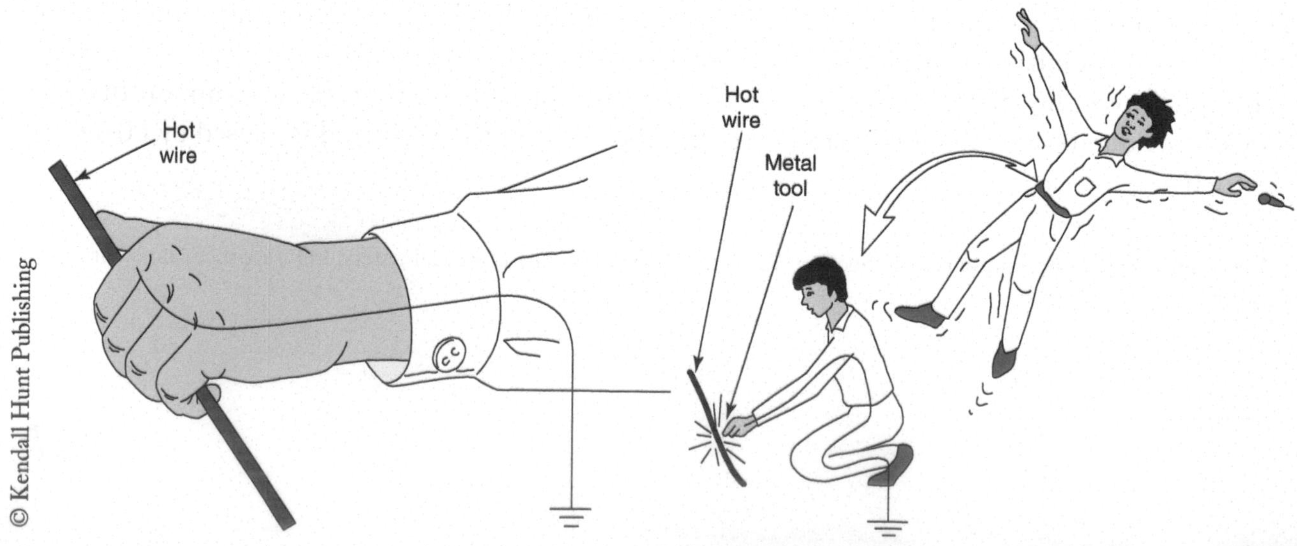

Of course, the precise effect of electrically induced muscular contraction depends on the path the current follows through the body.

The leading cause of death from electrical shock is ventricular fibrillation, an irregular and uncoordinated beating of the heart. The threshold for this effect is between 100 and 300 mA, depending on the path taken by the current. The heart is controlled by a pattern of electrical impulses that spread over it (see Chapter 13). When the electrical pattern on one part of the heart is disrupted by an externally created current, the disruption tends to spread over the heart, causing it to beat randomly and ineffectively. The victim dies from lack of blood circulation.

Interestingly, the remedy for ventricular fibrillation is the passage of a much larger current through the heart than the smallest that can cause ventricular fibrillation. Currents of about 6 A are passed through the chest by a device called a defibrillator. The defibrillator delivers a maximum energy of 360 joules. As can be seen in Table 12.1, currents of this magnitude cause sustained ventricular contraction and respiratory paralysis. Both effects cease after the shock. Such a large current through the heart affects its overall electrical pattern in such a way that normal beating often resumes following the shock. Very large currents are needed to affect the entire heart–lung region because it is large and deep within the body. Heart defibrillators have saved many lives and are common equipment in medical facilities.

You will note that there has been no mention of dangerous voltages. This is because the severity of a shock depends on current, which depends on both resistance and voltage.

According to Ohm's law, the current "*I*" depends on voltage and resistance as follows:

$$I = \frac{V}{R}$$

We see that because current depends on a combination of voltage and resistance, when resistance is low, voltages as small as 0.02 V may be lethal, whereas in other circumstances when resistance is high, 5000 V may not be the least damaging. The amount of energy involved is also an important consideration. Under normal circumstances, a maximum of 360 J of energy delivered by a defibrillator is considered safe.

EXAMPLE 12.1

Calculate the *current* through a person and identify the *likely effect* of touching with one hand a wire carrying 120 V ac (a) if the person is standing on a rubber mat and has a resistance of 150,000 Ω to ground. (b) Convert the current to milliamps and then refer to Table 12.1 for likely effects.

Ans. a) Using Ohm's law, we calculate the current:

$$I = \frac{V}{R} = \frac{120\ \text{V}}{150{,}000\ \Omega} = 0.0008\ \text{amps}$$

Ans. b) Convert the current to milliamps and then refer to Table 12.1 for likely effects.

$$0.0008\ \text{A}\ \frac{1000\ \text{mA}}{1\ \text{A}} = 0.80\ \text{mA}$$ This is less than the threshold of sensation.

12.1.1 Frequency Effects

In terms of the possibility of shock, the choice of 60 Hz as the frequency for ac power could not have been worse. The body is most sensitive to electrical shock at frequencies near 60 Hz, possibly because this is similar to the firing frequency of many nerves. Figure 12.2 is a graph of the threshold current for two shock effects plotted as a function of frequency. Both curves have their lowest points near 60 Hz, meaning that the smallest currents that can cause those effects have frequencies near 60 Hz. Nerves and muscles in the body are less sensitive at both higher and lower frequencies.

The effects of high-frequency shocks are mostly thermal; a large high-frequency current may cause burns but is less likely to cause ventricular fibrillation. Warts, for example, can be burned off by passing high-frequency current through them. Electrosurgery and electrocautery also use high-frequency currents.

12.1.1.1 Microshock Sensitivity

As noted earlier, the physiological effects in Table 12.1 are based on the assumption that the current passes through the trunk of the body, spreading out as it does so. If there is a direct path to the heart, such as through exposed pacemaker leads, then much smaller currents can cause ventricular fibrillation. Under this circumstance, a person is said to be microshock sensitive. Currents as small as 20 μA may be able to cause ventricular fibrillation in a microshock-sensitive person. These currents are less than a thousandth of those necessary to cause ventricular fibrillation in a normal person, and they are also far below

Figure 12.2 Average values for threshold of sensation and "can't let go" currents as a function of frequency.

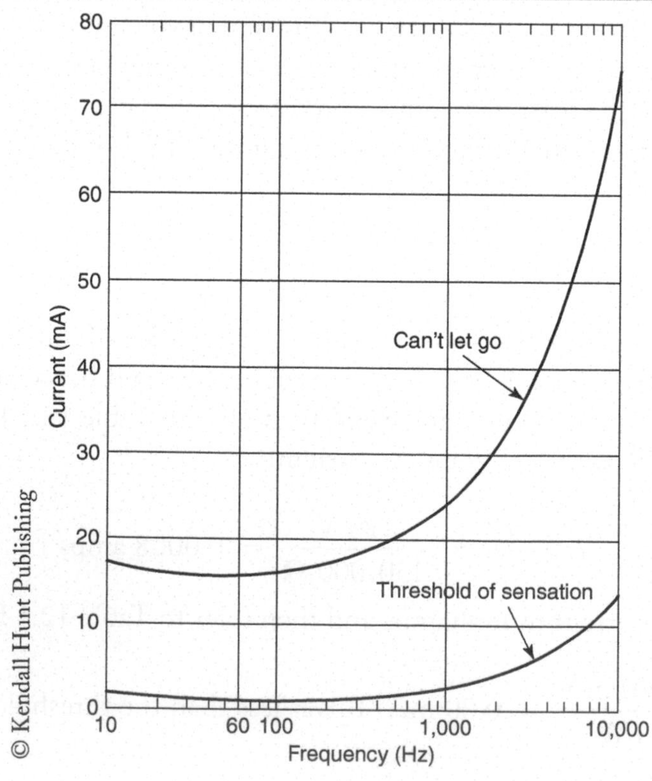

the threshold of sensation. It is therefore possible that a person touching exposed pacemaker leads of a patient could conduct a current too small for them to feel but which is lethal to the patient. Note that a patient with an implanted pacemaker is not microshock sensitive since the leads are under the skin.

Pacemaker leads are only one cause of microshock sensitivity.

Catheters are often inserted into the circulatory system, sometimes into the heart itself, for a variety of purposes. The resistance of blood and saline solutions is small and, of course, blood vessels give a direct path to the heart. Even urinary catheters and chest drainage tubes may make the patient microshock sensitive even though the path of current through them is not directly to the heart.

Any device that bypasses the resistance of the skin poses a severe electrical hazard to a patient. The resistance of the body is normally rather large, perhaps several hundred thousand ohms. Most of this resistance is in the top, dead layers of skin. (When wet, the body's resistance decreases to as little as a few thousand ohms.) The interior of the body has a small resistance because of the presence of dissolved salts. Using Ohm's law and referring to Table 12.1, you can see that low resistance means that even a low voltage can cause lethal currents.

A device that gives a direct electrical path to the heart is thus something of a double-edged sword. The current spreads out less and the resistance is smaller than if the current passes through intact skin. Smaller currents are required because of the direct path, and even smaller voltages will cause these currents because of the small resistance.

Microshock sensitivity is the reason that electrical safety standards are stringent in medical institutions. Great efforts are made to limit current that may accidentally pass through a patient. Current is limited by two means: prevention of contact with voltage sources and maximization of electric resistance of any path that current might take through a patient. These precautions are elaborated in Section 12.2.

The most commonly encountered electrical hazards were discussed in Section 11.5. In this section, other hazards are considered, including those threatening to patients who have been made microshock sensitive. References at the end of this chapter consider many other hazardous situations. Anyone who works with critically ill patients should be aware of the general precautions outlined at the end of this section. The growing number of health science professionals who participate in planning intensive care units or make decisions regarding the purchase of electrical equipment, for example, need a more detailed knowledge of electrical safety.

Two important precautions can be stated at the outset. First, metal cases of all appliances should be properly grounded to keep the voltage on them close to zero. Second, *patients should never be grounded.* Grounded persons have a lower resistance path for current to flow through them than ungrounded persons. Lower resistance means higher current if the person does contact a voltage source (see Figure 12.3). These and other precautions are summarized at the end of this section.

12.1.2 Leakage Currents

The Problem: A leakage current is one that flows to the case of an appliance. Leakage currents present shock hazards, particularly to people who have been made microshock sensitive. One cause of leakage currents is a partial breakdown of the insulation between the hot wire and case of an appliance.

12.1.3 Danger, Broken Ground Wire

Another cause, almost impossible to prevent, is the induction of currents by the oscillating magnetic and electric fields of ac electricity. The movement of charge through an appliance creates oscillating

Figure 12.3 (a) Relatively safe situation in which the skin is intact and the patient is not grounded. The current spreads out and experiences the resistance of the skin, body interior, and mattress. (b) Very dangerous situation. Patient is rendered microshock sensitive by pacemaker lead and is also grounded. The current spreads out less and experiences only the resistance of the interior of the body.

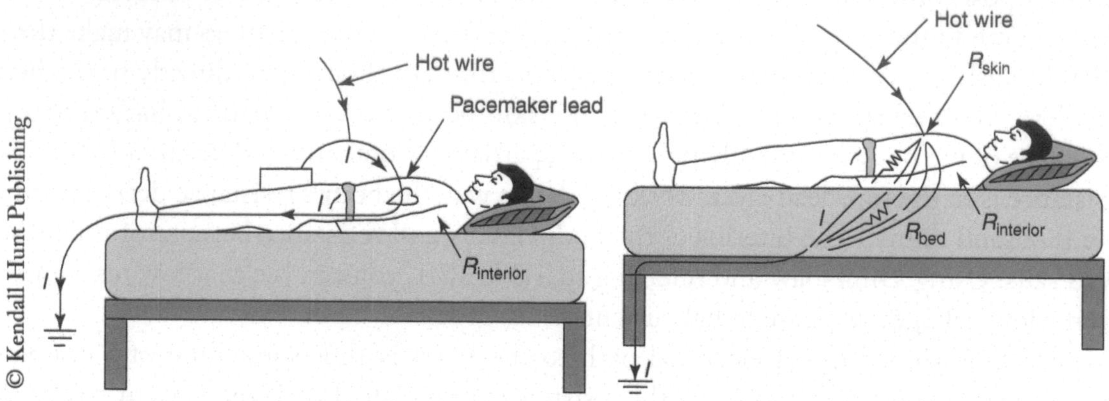

Figure 12.4 Leakage current hazards caused by broken ground wire.

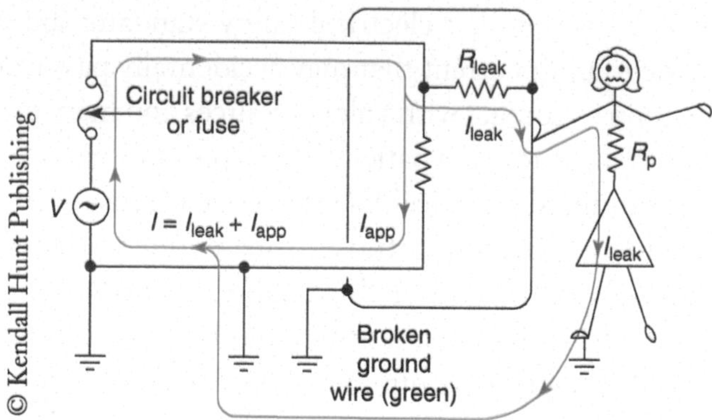

magnetic fields, which can induce currents in nearby conductors. (See *Section 10.5* for more discussion of magnetically induced currents.) Similarly, the charges that make up a current exert Coulomb forces on nearby charges, also contributing to currents in a metal appliance case. The ground wire (green) in the three-wire system will conduct leakage currents safely to ground. If the ground wire is absent or broken, a hazard from leakage currents exists. Figure 12.4 shows a hazardous situation involving a leakage current. The leak is represented by a finite resistance between the hot wire and the case of the appliance. The current could be lethal to a microshock-sensitive person.

Remedies and Prevention: The best solution is to use battery-operated appliances or appliances with plastic cases. With such doubly insulated appliances, the three-wire system is not required.

If it is not possible to use doubly insulated appliances, then the remedy is more difficult. One device that reduces but does not eliminate leakage current hazards is a GFI. A GFI is similar to a circuit breaker in that it cuts off voltage to an appliance when it detects a hazard; GFIs compare the current supplied to an appliance through the hot (black) wire and the current returning through the neutral (white) wire. These currents should be the same; if they are not, a leakage current may exist. Figure 12.5 shows an appliance plugged into a GFI. If the currents in the hot and neutral wires differ

Figure 12.5 The ground fault interrupter (GFI) compares the currents in the wires to an appliance and opens the circuit if they differ by as little as 5 mA. Faulty ground wire and leakage of current to the case combine to cause a current through the man.

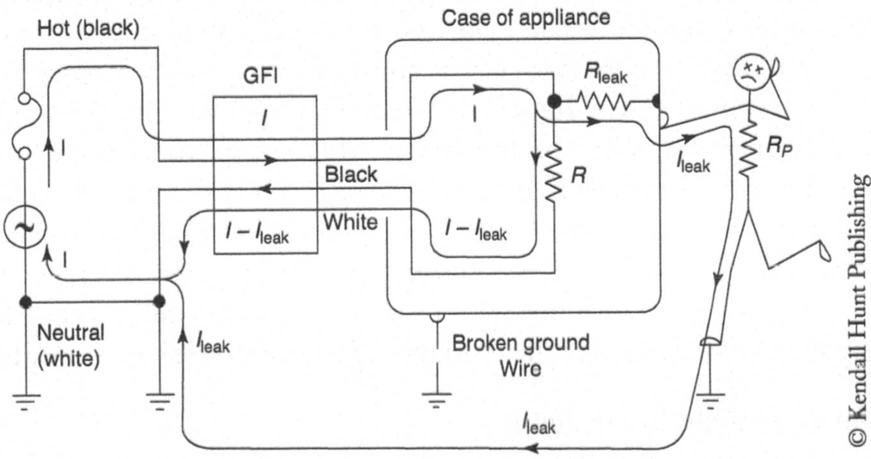

Figure 12.6 An isolation transformer: There is no complete circuit for current to flow through the person.

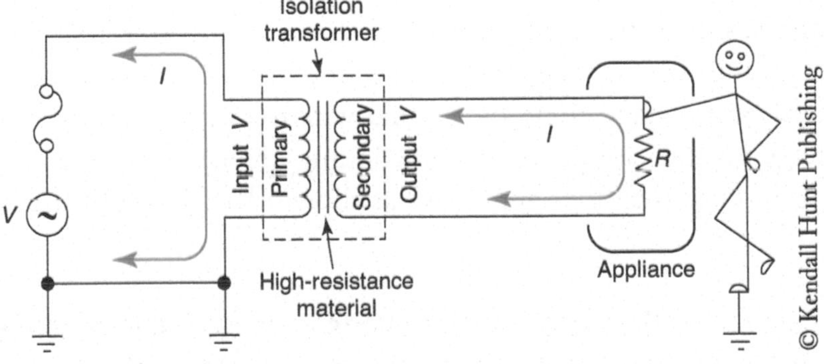

by as little as 5 mA, the GFI will interrupt the current. Five milliamps is the maximum harmless current but is far above the lethal level for microshock-sensitive people. As usual, the resistance of the person should be kept as high as possible. Those who are microshock sensitive should never be grounded.

The *isolation transformer* is a device that greatly reduces the hazard of leakage currents. Briefly, transformers are devices that take an input voltage and change it to some output voltage. There is no flow of charge between the input and output of a transformer; the output voltage is induced by electromagnetic fields. (For more discussion of transformers see *Section 10.5*.)

Figure 12.6 shows an appliance powered by an isolation transformer. There is a complete circuit for current to flow through the appliance, but there is no complete circuit for current to flow through the person in the figure, who is touching only one of the transformer's output wires, and neither of those wires is grounded. The appliance is isolated from the original voltage source by the high resistance of the material between the coils of the transformer.

For current to flow through the person it would have to pass through this high-resistance material, through the wire and person, and then back through the ground. The resistance of the material is so high that the current that flows is extremely small and harmless even to microshock-sensitive people.

A person who touches both wires coming out of a transformer receives its full output voltage, but this requires touching two wires rather than just one and is less likely. Leakage to the case of an appliance powered by an isolation transformer is negligible. Current cannot flow through a person touching the case because there is no complete path for it to follow. As in Figure 12.6, the case of an appliance powered by an isolation transformer need not and should not be grounded.

Yet another solution to the leakage current problem is to use only those appliances approved by Underwriters' Laboratories (UL). The UL requirements, upgraded in 1974, allow various leakage currents in new devices, depending on their intended use. Appliances made before 1974, even UL approved, were allowed 5 mA of leakage current, which is potentially lethal to microshock-sensitive patients. It is likely that leakage currents increase as appliances age. If an appliance is old or suspicious for any reason, the ground connection should be checked by a qualified technician. Again, if the appliance is properly grounded, there is no hazard because the resistance of the ground connection is very small when compared to the resistance of even a microshock-sensitive person.

12.1.4 Different Ground Connections

The Problem. A hazard arises when a person comes into contact with two or more appliances at a time. Even when all the appliances are properly grounded, there is still a hazard if they are plugged into separate circuits having different ground connections. To illustrate the hazard, let us consider the situation, shown in Figure 12.7, of a microshock-sensitive patient in contact with two properly grounded appliances. On of the appliances shorts to its case and a large current flows through its ground wire. The ground wire is many meters long and has a finite resistance, perhaps as much as an ohm, so there must be a voltage along the ground wire to cause the current to flow through it. That is, there is a voltage on the case of the malfunctioning appliance that causes a current through the patient and the ground of the other appliances as shown. The circuit breaker on the malfunctioning

Figure 12.7 When one of the two appliances shorts to its case, the person receives a potentially lethal shock because the two appliances have different ground connections. The voltage on the case of the faulty appliance rises while the voltage on the case of the other appliance remains close to zero.

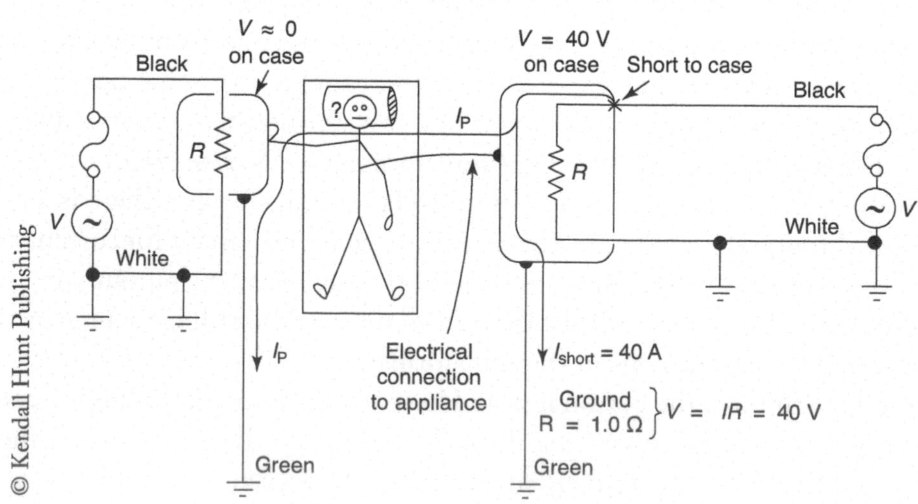

Figure 12.8 illustrates two open circuits, one safe and the other hazardous. The white wire is expected to have a voltage of zero since it is grounded. In Figure 12.8(a), however, the white wire inside the appliance has the full voltage of the source. There is no voltage drop in the resistor *R* because there is no current flowing through it. The person touching an exposed white wire is then in series with the resistance of the appliance, and a current flows through him, where R_{person} is the resistance of the person to ground (see Figure 12.9). If the break is as shown in Figure 12.8(b), there is no shock hazard in touching either the black or white wires in the appliance, because the break is between the voltage source and the appliance.

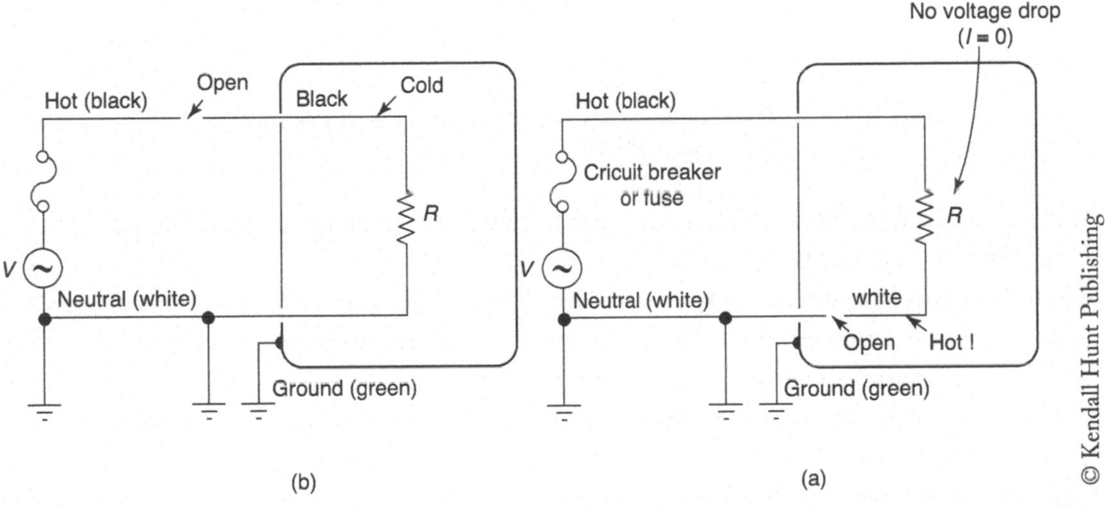

appliances will trip, but a finite amount of time passes before it does. It typically takes a second or more for the breaker to trip when the current is double its minimum trip rating. A shock of this duration is potentially lethal.

Remedies and Prevention: Contact with Two or More Appliances at the Same Time. The appliances connected to the same circuit in Figure 12.7 should have ground wires connected by a low resistance wire. Such a connection forces the voltages on the cases of the appliances to be the same. Then, if one appliance fails as described here, the voltage of both cases will rise momentarily so that there will be no potential differences across the patient. Hospital often goes to great trouble to make certain that all equipment around critically ill patients is connected to the same ground. For example, beds used by such patients are often equipped with an electrical panel from which all nearby equipment receives its power. It is highly inadvisable to string an extension cord to a distant outlet to supply power to an additional appliance. The distant outlet is almost certain to have a different ground.

12.1.5 Open Circuits

The Problem. An open circuit is one in which there is no closed path for current to follow, perhaps owing to a break in one of the wires leading to an appliance. Naturally, appliances do not function if there is an open circuit. An unexpected hazard exists to anyone trying to repair the appliance while it is still plugged in.

$$I = \frac{V}{\left(R + R_{\text{person}}\right)}$$

EXAMPLE 12.2

Suppose the situation pictured in Figure 12.9 occurs.

Calculate the current through the person, given the voltage of the source to be 120 V and the resistance of the appliance to be 100 Ω (144-W device). Let the resistance of the person be 10,000 Ω, which is a bit low but very possible with wet skin or sweaty hands.

Solution: Since the person is in series with the appliance, the total resistance is $R + R_{person}$.

$$I = \frac{V}{\left(R + R_{person}\right)}$$

$$I = \frac{120\ \text{V}}{10{,}000\ \Omega} = 0.0119\ \text{A} = 11.9\ \text{mA}$$

This current is large enough that the person may not be able to let go. If the person is microshock sensitive, the shock could be lethal.

Remedies and Prevention: Open circuits. Both GFIs and isolation transformers help prevent shock hazards from open circuits. These devices were described earlier in relation to leakage currents; their effectiveness and operation is the same in the present situation as with leakage currents.

It is better never to work on an appliance while it is plugged in. Some devices, however, such as TVs, have components that store charge and can have high voltages even when unplugged. Such appliances should have warning labels, but this cannot be counted on. Voltages can be detected with a voltmeter or a cheaper device called a voltage tester. When the two probes of a voltage tester are put into contact with voltages in the range of 5.0–600 V, a small light glows. This indicates the presence of a voltage but not its magnitude. If in doubt, the best thing to do is to call a qualified technician to fix the appliance.

Figure 12.9 Shock received by a person working on a plugged-in appliance with an open circuit. The person completes the circuit by touching the white wire. See Example 12.1.

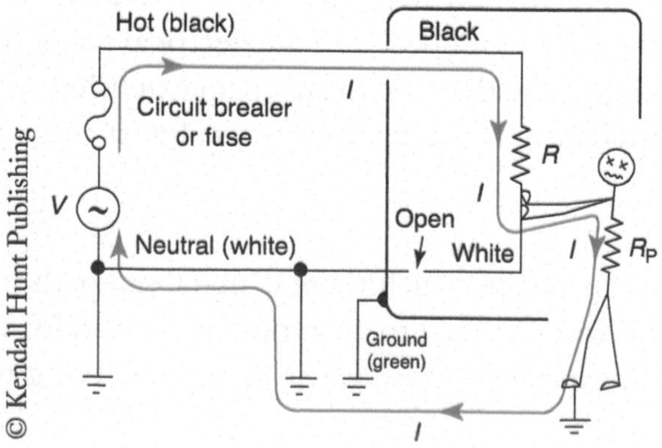

12.2 SUMMARY OF PRECAUTIONS

1. ***Use the three-wire system or doubly insulated appliances.*** This ensures that the case of an appliance is grounded and safe. Isolation transformers are an expensive way to eliminate the need for the three-wire system, but they do protect microshock-sensitive patients from failures in ground connections.
2. ***Use a common ground connection for all appliances.*** This is necessary for microshock-sensitive patients, but it is prudent in all medical situations.
3. ***Never ground a patient.*** The method for ensuring that the patient is not grounded is situation dependent. For example, if electrical measurements are being made, none of the wires connected to the patient should be grounded. (Older ECG equipment may intentionally ground one lead to the patient.) Metal bed frames should not be grounded, and appliances with grounded metal cases should be kept out of reach.

These three general precautions cover most situations. No list of specific precautions can be long enough to cover every specific situation. Another general suggestion is to treat electricity with the respect it deserves, but that respect should be informed and not a phobia. For example, a great deal of money can be wasted on expensive equipment to protect people from hazards dangerous only to the microshock sensitive. The use of the three-wire system is sufficient for most domestic situations. Knowledge is the greatest protection.

Things to know for Chapter 12

Prefix of "m" means milli and equals 10^{-3}

Prefix of "μ" means micro and equals 10^{-6}

Questions

12.1 What major factors determine the severity of an electrical shock? Why can't we say whether a particular voltage is hazardous or safe without additional information?

12.2 According to Table 12.1 the minimum "can't let go" current is smaller than that which can cause ventricular fibrillation. Explain why this is so.

12.3 Before working on a power transmission line, linemen will touch the line with the back of the hand as a final check that the voltage is zero. Why the back of the hand?

12.4 Warts are burned off by passing a current of high-frequency electricity through them. A complete circuit is made through patients by having them sit on a large metal plate (called a butt plate) electrically connected to the voltage source. Why is a large metal plate used?

12.5 Why is the electrical resistance of wet skin smaller than that of dry skin? Explain why blood and other bodily fluids are expected to have low resistance.

12.6 What conditions render a person microshock sensitive? What is it about those conditions that make the person sensitive to smaller currents than other people?

12.7 Might a person receiving an IV be microshock sensitive? If the person's IV tube touched the frame of her bed, might she be grounded?

12.8 What role, if any, do circuit breakers and fuses play in preventing shock hazards?

12.9 Why is it dangerous to touch a person who is being shocked and can't let go of the wire shocking him? How might you help him without endangering yourself?

12.10 Why should a medical patient never be grounded?

12.11 How do doubly insulated appliances avoid the need for using the three-wire system?

12.12 What is an open circuit and when is it a hazard?

12.13 What is a leakage current? What are its causes and what hazards does it present? How can the hazards of leakage currents be overcome?

12.14 Why should all appliances in the vicinity of a critically ill patient utilize the same ground connection?

12.15 Explain how a GFI helps eliminate the hazard of an open circuit. Will the GFI offer enough protection for microshock-sensitive people?

12.16 Does a properly attached ground (green) wire prevent the hazard of an open circuit? Explain.

12.17 How does the use of an isolation transformer prevent the hazard of an open circuit? Explain.

12.18 What hazard is presented by grounding the case of an appliance powered by an isolation transformer?

12.19 Could the hazard illustrated in Figure 12.7 be prevented by the use of an isolation transformer? Would it sufficiently protect microshock-sensitive people?

12.20 Would the use of a GFI prevent the hazard illustrated in Figure 12.7? Would it sufficiently protect microshock-sensitive people?

PROBLEMS

√12.1 (I) Calculate the current through a person and identify the likely effect of touching with one hand a wire carrying 120 V ac (a) if the person is standing on a rubber mat and has a resistance of 200,000 Ω to ground; (b) if the person is standing barefoot on a wet bathroom floor and has a resistance of 5000 Ω to the ground.

√12.2 (I) Suppose a person touches the case of an electrical appliance while taking a bath. The total resistance through the person's hand and body to the metal drainpipe and into the ground is 4000 Ω. What is the smallest voltage on the case of the appliance that could cause ventricular fibrillation?

√12.3 (I) A man trying to fish a burning slice of bread out of a toaster foolishly uses a metal butter knife and comes into contact with 120-V ac. Luckily, he is wearing rubber-soled shoes and doesn't even feel the current passing through him. (a) What is the minimum resistance of this person to ground? (b) What current would pass through him, and what effect is it likely to have if his hands are wet and he is barefoot with a resistance of only 6000 Ω to ground?

√12.4 (I) During experiments to determine the sensitivity of sheep hearts to electric current it is found that a current of 30.0 μA applied directly to the heartproduces ventricular fibrillation. If the resistance of a sheep heart is 500 Ω, what minimum voltage could produce this current?

√12.5 (I) A current of 20 μA applied directly to the heart (perhaps by accident during surgery) may cause ventricular fibrillation. If the resistance of a human heart is 300 Ω, what is the smallest voltage that poses a danger when the heart is exposed during surgery?

√12.6 (I) Example 12.1 mentions that an appliance using 120 V and having a resistance of 100 Ω consumes 144 W of power. (a) Show this. (b) Calculate the current through the person in Example 12.1 if his resistance is 100,000 Ω rather than 10,000 Ω. What is the likely effect of such a current upon him?

12.6 (II) An electronics technician might work on an appliance while it is plugged in, but she will take care that her resistance to ground is very large. Her body has a resistance of 200,000 Ω, her rubber-soled shoes have a resistance of 500,000 Ω, and the rubber mat on which she stands has a resistance of 1,000,000 Ω. What is the maximum harmless voltage she can contact?

12.21 (II) Suppose a physician who is well insulated from the ground touches the ungrounded metal case of an appliance that has shorted to its hot wire and thus has a voltage of 120-V ac. With her other hand she simultaneously touches the pacemaker lead of a patient who is grounded. Her resistance is 100,000 Ω and that of the patient is 1000 Ω. Calculate the current through the two people (they are in series), and identify the likely effect on each.

12.21 (II) Suppose an old appliance has a leakage current of 5 mA to its properly grounded case. Someone touching the case is in parallel with the resistance in the ground (green) wire. (a) Calculate the voltage on the case of the appliance if the resistance of the ground wire is 2.0 Ω (higher than normal). (b) What current flows through a microshock-sensitive person with a resistance of 1000 Ω who touches the case? Is this a dangerous current?

12.21 (II) (a) Calculate the current through a person standing barefoot on a wet floor who grasps a 220-V ac wire and has a resistance of 4400 Ω to ground. (The current is greater than the "can't let go" current but not enough to cause ventricular fibrillation.) (b) What current will flow through a would be rescuer who grasps the first person by the wrist and has a resistance of 8000 Ω to ground? You may assume the resistance of the part of the original victim between the source of voltage and the would-be rescuer is 500 Ω. What is the likely effect on the second person?

12.21 (III) If the hot wire in a properly grounded appliance shorts to its case, a large current flows through the ground wire, raising the voltage on the case. (a) Calculate the voltage on the case of an appliance if the current through it and the ground wire is 50 A and the resistance of the ground wire is 0.75 Ω (higher than normal). (b) Calculate the current passing through a normal person touching the case who has a resistance of 50,000 Ω to ground. What is the effect on the person? (c) Calculate the current through a microshock-sensitive person who touches the case and has a resistance of 2000 Ω to ground. (d) What maximum resistance can the ground wire have and still be safe for the microshock-sensitive person?

12.21 (III) The resistance of the body of a certain microshock-sensitive man is only 2500 Ω. Suppose he comes into contact with the case of a faulty appliance that has 5.0 V on it. He is not grounded but is lying in bed. What minimum resistance of the bed to ground is necessary to protect him from ventricular fibrillation?

12.21 (III) (a) The hot wire of a 120-V ac appliance shorts to its properly grounded case, producing a current of 40 A. What is the resistance of the short if the ground wire has a resistance of 1.0 Ω (higher than normal)? (b) What is the voltage on the case of the appliance? (c) What minimum resistance must a normal person have to be safe from shock by touching the case?

12.21 (III) A short circuit in an appliance creates a current of 50 A through its case to ground. (a) Calculate the voltage on the case of the appliance if the resistance of the ground (green) wire is 1.50 Ω (higher than normal). (b) Calculate the current through a woman who touches the case of the shorted appliance at the same time as she touches the case of another appliance on a separate ground. The resistance of the path through her is 50,000 Ω. What is the likely effect on this person? (c) If the cases of the two appliances are connected to one another by a wire of resistance 0.10 Ω and the ground wire of the second appliance has a resistance of 2.0 Ω (higher than normal), what will be the voltage on the case of the second appliance? (Assume the total current is still 50 A.) (d) Is it safe for a microshock-sensitive person with a resistance of 1000 Ω to be touching both appliances?

13

Bioelectricity

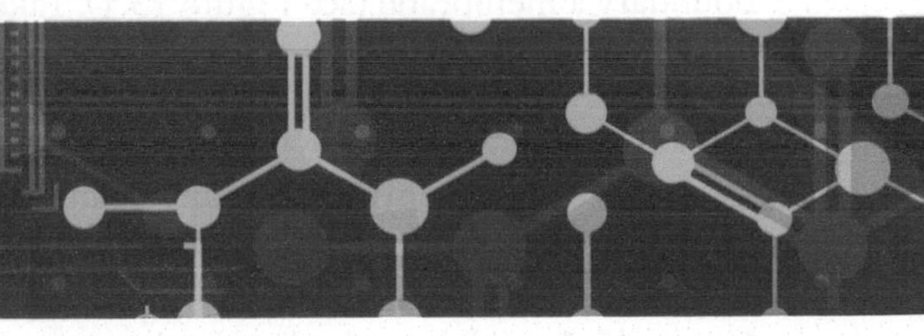

"Either write something worth reading or do something worth writing."

—Benjamin Franklin

This chapter is an introduction to the fascinating and complex subject of *bioelectricity*. Voltages are created by almost all types of animal cells and are largest in nerve and muscle cells. In fact, a surprisingly large fraction of the energy used by cells, and hence a large fraction of the energy requirement of the body, is needed to create and maintain cell potentials (voltages). At least 25% of the energy used by cells goes toward the creation and maintenance of bioelectricity.

Cell potentials may remain constant for long periods of time but can also be changed by various internal or external stimuli. Changes in cell potentials can amount to voltage pulses (e.g., nerve impulses), with effects that depend on the type of cell involved. Nerve cells use these voltage pulses as communication signals for a variety of purposes, including gland control, vision, and muscle control. Nerve potentials are sometimes measured in individual nerves, indicating a specific type of nerve activity. The combined activity of many cells is also measured, such as in an electroencephalogram (EEG), a recording of the entire brain (billions of nerve cells) that is used as a gross indication of its activity.

Changes in cell potentials can cause muscle cells to contract. Such potentials are sometimes measured to diagnose muscle activity, most often for the heart in recordings called electrocardiograms (ECGs).

There are many other examples of bioelectricity. The infamous electric eel stuns its prey with electrical shocks. Some fish use electric fields to detect objects around them. Another example is that the healing of broken bones may involve electric currents.

The brief treatment of bioelectricity in this chapter concentrates on two of its major aspects. The first is the generation and transmission of bioelectricity at the cellular level. The second is the detection of bioelectricity and its use as a diagnostic indicator.

13.1 GENERATION AND TRANSMISSION OF BIOELECTRICITY

13.1.1 Generation; the Resting Potential

Bioelectricity is a cellular phenomenon. Although the fluids inside and outside cells are essentially neutral, there are differences in ionic concentrations that result in an electric potential, across cell boundaries. Those cells that create electric potentials (most living cells in animals do) generate a thin layer of positive charge on their outside surface and a thin layer of negative charge on the inside surface of their boundary a membrane (see Figure 13.1). Taken together, these two layers of charge are called a dipole layer. The two factors to be considered in explaining the creation of cell potentials are

1. The diffusion of ions across a semipermeable* membrane (the cell boundary); *For other effects caused by the semipermeability of membranes, see Section 7.2.
2. The Coulomb force, in which like charges repel and unlike charges attract.

Figure 13.1 (a) Animal cell with a layer of positive charge on its exterior surface and a layer of negative charge on its interior surface. (b) Enlarged view of a section of cell membrane with its layers of positive and negative charges. The membrane itself consists of a lipid bilayer sandwiched between two layers of protein molecules. Not shown are holes through the membrane and large protein molecules that may penetrate the membrane.

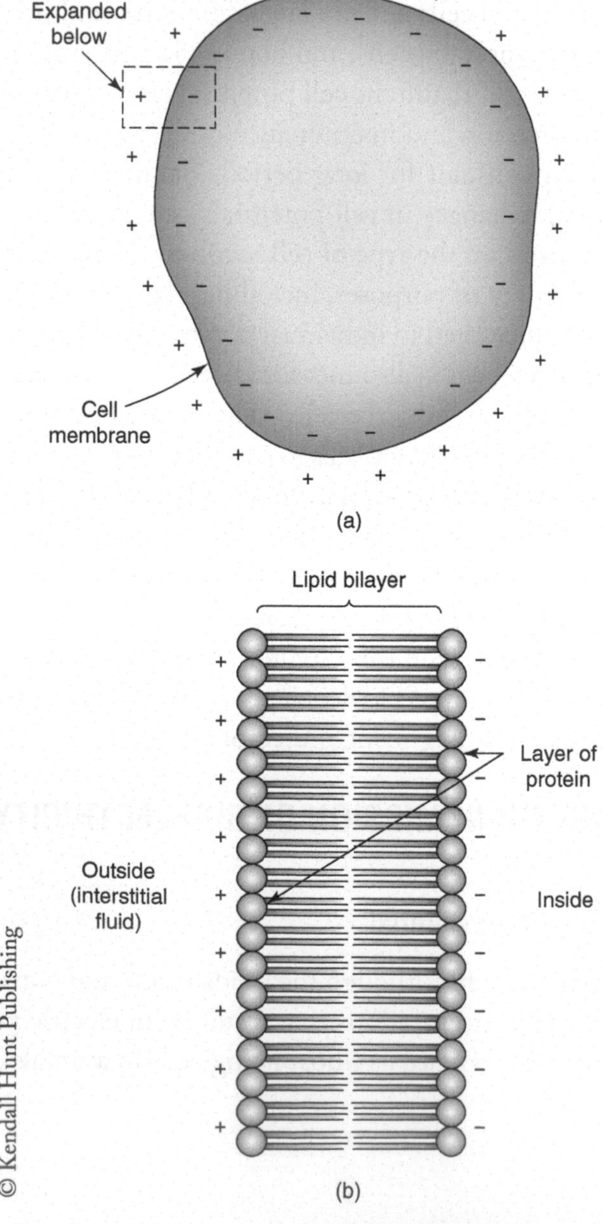

13.1.2 Differences in Ionic Concentrations and Semipermeability of the Cell Boundary

There are many types of ions in the fluids inside and outside cells. The most important of these in creating cell potentials are Na^+, K^+, and Cl^- ions. There are large differences in the concentrations of these ions inside and outside cells, as Figure 13.2(a) shows. Negative ions other than Cl^- are designated as A^-. Note that the total charge inside and out is zero; that is, the fluids are electrically neutral.

To see how the cell potential is established, consider what happens in the example of an initially neutral cell membrane having the ionic concentration differences shown in Figure 13.2(a). The cell membrane is normally permeable to K^+ and Cl^- ions; it is about 100 times less permeable to Na^+, and is highly impermeable to other ions. Only K^+ and Cl^- diffuse across the cell membrane in significant numbers. As usual, the direction of net diffusion is from regions of high concentration to regions of low concentration. Therefore, K^+ and Cl^- ions diffuse in opposite directions, as shown in Figure 13.2(b). The K^+ and Cl^- ions have opposite charges, so there is a very strong attractive Coulomb

Figure 13.2 (a) Ion concentrations outside and inside a cell. Positive ion concentrations are graphed above the line and negative ion concentrations below to help illustrate that the fluids are electrically neutral. (Adapted from John R. Cameron and James G. Skofronick, *Medical Physics,* Wiley-Iriterscience, 1978, by permission.) (b) The membrane is permeable to K^+ and Cl^- ions, which diffuse in opposite directions as shown. Small arrows indicate that the Coulomb force resists the continued diffusion of both K^+ and Cl^- ions. If the membrane becomes permeable to Na^+ ions, both the diffusion gradient and the Coulomb force act to move Na^+ into the cell.

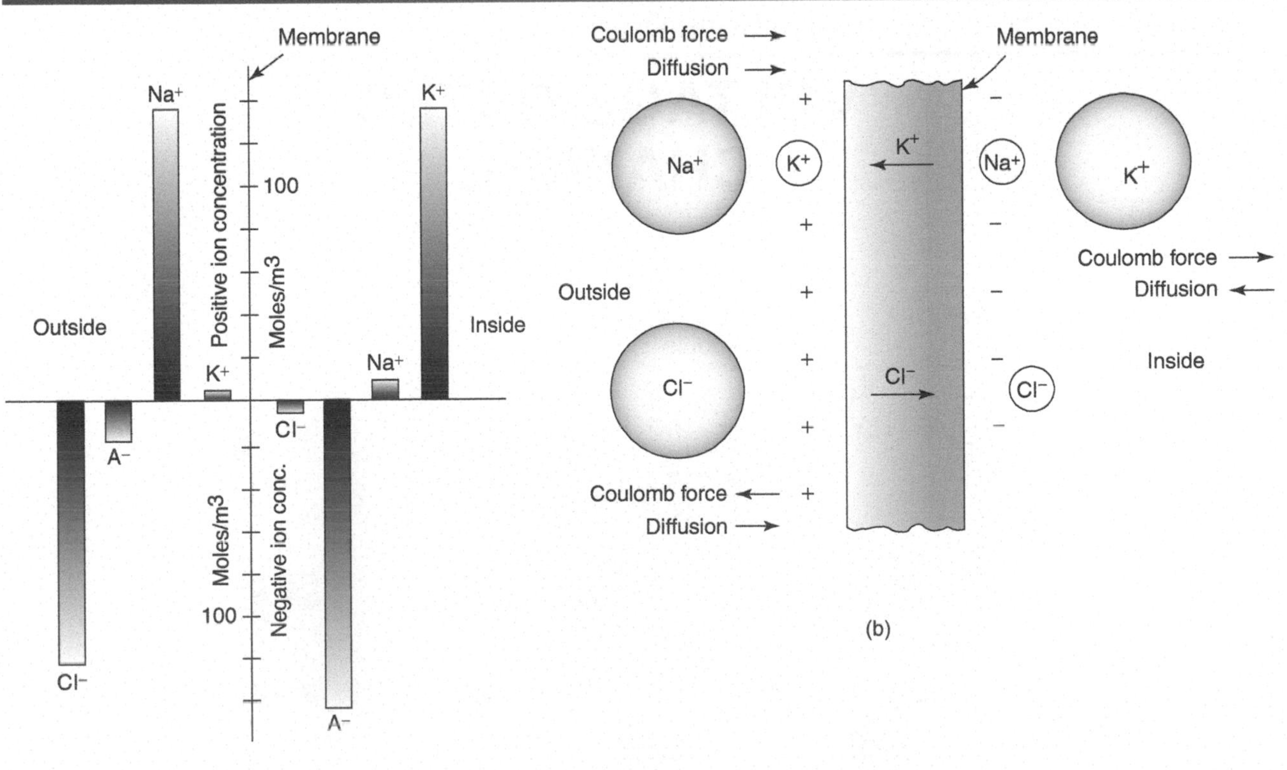

force between them that causes them to form two thin layers of charge right on either side of the cell membrane, as seen in Figure 13.2(b).

Diffusion of K^+ and Cl^- continues until both attractive and repulsive Coulomb forces stop it. As the charge layers build up on the cell membrane, the Coulomb forces grow. The attraction of unlike charges acts to pull K^+ and Cl^- ions back to their regions of high concentration. Furthermore, the repulsion of like charges acts to keep further ions from moving out of their regions of high concentration. A. balance is quickly reached between diffusion from high to low concentration and the opposing Coulomb forces. Once this balance is reached, the cell is in its ***resting state*** (Figure 13.2c).

If more K^+ and Cl^- ions diffuse across the cell membrane, the Coulomb forces grow and move some ions back. The net transfer is zero and the equilibrium is thus stable.

The Coulomb force is quite strong and only about 1 out of 100,000 of the K^+ and Cl^- ions move across the membrane. This is such a small fraction that the overall ion concentrations are not altered. In addition, the fluids inside and outside the cell remain essentially neutral although some charge has been separated. That small—separation of charge is very important, however, because it is the source of bioelectricity. The inside of cells is at a potential 70–90 mV less than the outside, being about 90 mV in nerve and muscle cells. The outside of the cell is usually taken to be at 0 V, so the interiors of nerve and muscle cells have a resting potential of about -90 mV.

Quantitative Description of Potential Due to Concentration Differences; The Nernst Equation The balance between the concentration gradient and the Coulomb force is also an energy balance. Electrical potential energy just balances the potential energy due to the concentration difference—that is, the ability of the concentration gradient to do work by moving ions against the

Figure 13.2 (c) The resting potential of a membrane.

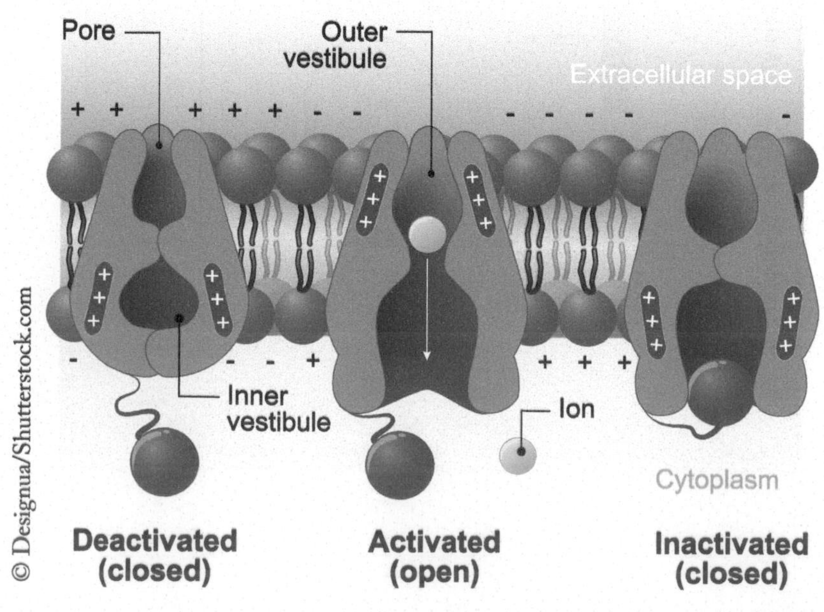

electric potential. The equation describing this energy balance is the Nernst equation. The Nernst equation gives the voltage that will be created by a concentration difference, but it does so only for a membrane that is completely permeable to one type of ion and completely impermeable to all others. Under those circumstances the Nernst equation is

$$V = V_{in} - V_{out} = -2.30\frac{kT}{Ze}(\log C_{in} - \log C_{out}) \quad \text{eq. 13.1}$$

where V is the potential difference (inside minus outside), C_{in} and C_{out} are the concentrations of the ion to which the membrane is permeable, k is the Boltzmann constant, T is the absolute temperature, and Ze is the charge on the ion in multiples of electron charge (Z is the valence of the ion). The minus sign indicates that an excess of diffusible positive ions in the neutral fluid inside a cell produces a negative voltage inside the cell. The most important feature of the Nernst equation is that the potential difference is proportional to the concentration difference. Now, using the concentrations given in Figure 13.2(a), we can calculate what potential each type of ion would create if the membrane were permeable to it alone. For Na^+, the result is an inside potential of +59 mV. Since· the actual potential in nerves is about −90 mV, this indicates that the membrane is not very permeable to Na^+. For K^+ and Cl^- the results are −88 and −70 mV, respectively. These are close to actual values in many types of cells, but that does not necessarily imply the membranes are permeable to K^+ and Cl^-. It only implies that if the membranes are permeable to them, then there will be little movement under resting conditions.

Effect of Resting Potential on Membrane Structure The structure and characteristics of membranes are topics of intense research efforts. Much is as yet not understood, but it can be stated with certainty that the resting potential must have a major influence on membrane structure and characteristics (Figure 13.2d). Although the resting potential is only 90 mV, it exists across membranes

Figure 13.2 (d) Ionic basis of the resting membrane potential.

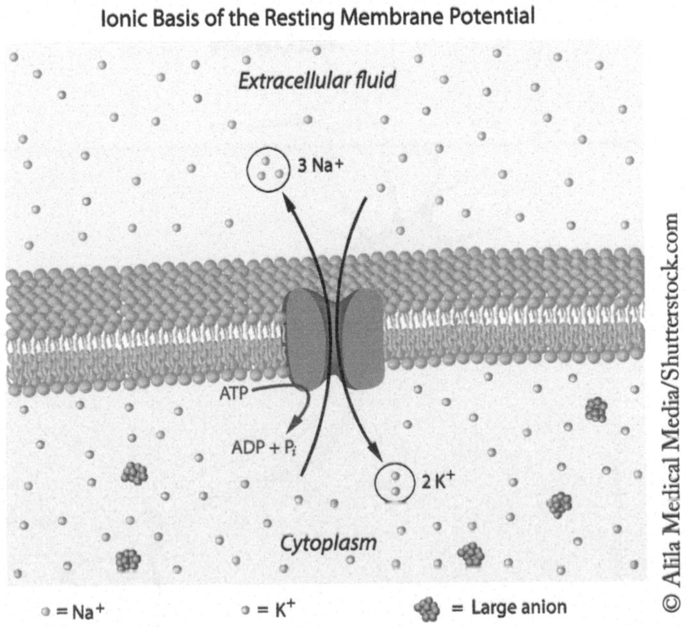

averaging only 8 nm thick. The voltage per meter is thus extremely large—on the order of 11 MV per meter. Such large voltages per meter can align molecules and have effects on pores and membrane permeability. For example, if the potential across a membrane is reversed, the permeability of the membrane changes drastically, suddenly becoming about 1000 times more permeable to Na^+ than it had been. No one understands exactly why the permeability changes, but one reason is the effect of the potential on the structure of the membrane. Nature has found a way to use such permeability changes to transmit bioelectrical signals.

13.1.3 Transmission of Bioelectrical Signals: The Action Potential Depolarization and Repolarization

The ability of nerve cells, also called neurons, to transmit bioelectrical signals is, of course, crucial to life in higher animals. When a neuron fires, it sends out a bioelectrical signal that consists of a temporary reversal in the potential across the neuron membrane. This reversal in potential starts in some localized region of the neuron and propagates along the cell membrane to other locations. The reversal in potential is called depolarization. The cell potential soon returns to its normal or resting polarity of negative inside and positive outside. The return to the resting state is thus called repolarization.

Figure 13.3 shows the structure of a neuron. It is notable for its very long tail, called an axon or nerve fiber. Axons, which may be as long as 1 m, carry bioelectrical signals or nerve impulses away from the cell body to muscles, glands, or other neurons. Dendrites can also be long and tend to transmit signals from sensors inward to the body of the neuron. Several types of stimuli can trigger a neuron to fire and transmit a nerve impulse to some other location. Among these are pressure, temperature changes, electrical signals from other neurons, externally applied electric currents, and chemicals sent across junctions between neurons (called synapses).

Figure 13.3 The neuron. Nerve impulses can be generated in the cell body by various stimuli. Once generated, the impulses travel down the axon.

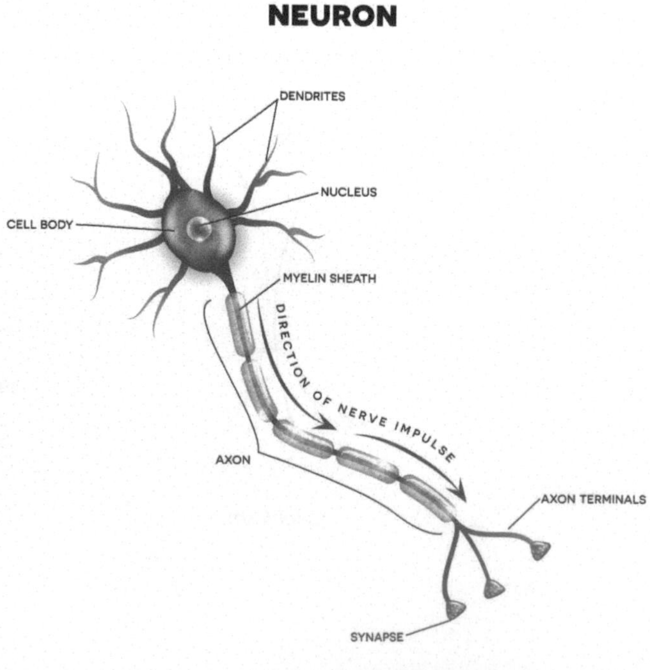

NEURON

DENDRITES

NUCLEUS

CELL BODY

MYELIN SHEATH

DIRECTION OF NERVE IMPULSE

AXON

AXON TERMINALS

SYNAPSE

The process by which a nerve fires and propagates a nerve signal can be understood in terms of the same two factors considered above to describe the resting potential of cells: namely, diffusion through a semipermeable membrane and the Coulomb force. In addition, active transport processes must be included to maintain the cell potential over long periods of time.

When a stimulus causes a nerve to fire, it does so by greatly increasing the permeability of the cell membrane to Na^+. The cell membrane becomes about 1000 times more permeable to Na^+ than normal, making it about 10 times more permeable to Na^+ than K^+. Because of the large concentration difference of Na^+, a rapid diffusion of Na^+ ions into the cell now occurs. The influx of Na^+ makes the interior of the cell positive instead of negative. The potential inside the cell goes from its resting potential of –90 mV to about +40 mV. This is depolarization, the first step in the entire process. It is illustrated in Figure 13.4(a). The dipole layer on the membrane has essentially reversed, but only a small fraction of the ions need move to reverse the potential.

The reversal of potential across the cell membrane (depolarization) apparently alters membrane structure in such a way that two further changes in permeability take place. First, the permeability to Na^+ returns to normal (very small), and second, permeability to K^+ increases temporarily by a factor of 30. This first change cuts off further influx of Na^+ and the second change allows a rapid outward diffusion of K^+. The membrane potential is thus returned to its normal resting value; that is, it is repolarized. The dipole layer has been reestablished with only a small net loss in the concentration gradients driving the system. Finally, permeability to K^+ returns to normal, ending the process. These events are illustrated in Figure 13.4(a), and a graph of the permeability of the membrane to Na^+ and K^+ as a function of time is shown in Figure 13.4(b). Although the changes in permeability are known to occur as described, their exact cause is not well understood.

The change in cell potential from negative to positive and back during depolarization-repolarization amounts to a voltage pulse. This voltage pulse is the nerve impulse and is called the *action potential.* It is graphed for a nerve in Figure 13.4(a). The action potential can occur in most animal cells, but it has some of its most interesting effects in nerve and muscle cells. Neurons can transmit the action potential to other locations, creating a variety of responses. In muscle cells the action potential causes a muscle contraction. The action potential in muscle cells, as in neurons, can originate in the cell itself or be induced by an outside source.

The Sodium-Potassium Pump Each time a cell fires there is a net loss of Na^+ and K^+ ions from their respective regions of high concentration. However, it is important to note that the number of these ions crossing the cell membrane during each firing is a very small fraction of those available. Remember that only one in 100,000 of the available Na^+ and K^+ ions are needed to create the resting potential, and a similarly small fraction moves across the cell membrane during each firing. Nerves can thus fire rapidly and repeatedly (many hundreds of times) before the concentrations of Na^+ and K^+ are depleted enough to make a difference.

Over the long term the cell must find a way to move Na^+ out of its interior and K^+ back into its interior to maintain the concentration differences that create the resting potential and drive the action potential. It must utilize active transport to do this (see Section 7.2) because it is moving both Na^+ and K^+ against their concentration gradients and Na^+ against the Coulomb force (see Figure 13.2). Tracer studies have shown that the cell moves one Na^+ ion out for every K^+ ion it moves in. The active transport doing this is thus called the *sodium-potassium pump.* Active transport requires energy and a sizable fraction of a cell's metabolism is devoted to maintaining the resting potential and driving the action potential.

Figure 13.4 (a) Action potential at a fixed location on a nerve. Voltage inside the nerve is plotted as a function of time. Rapid movement of ions takes place only during depolarization and repolarization. Active transport is used over the long term to maintain concentration gradients. (b) Associated changes in membrane permeability during the action potential. Note the logarithmic scale for permeability.

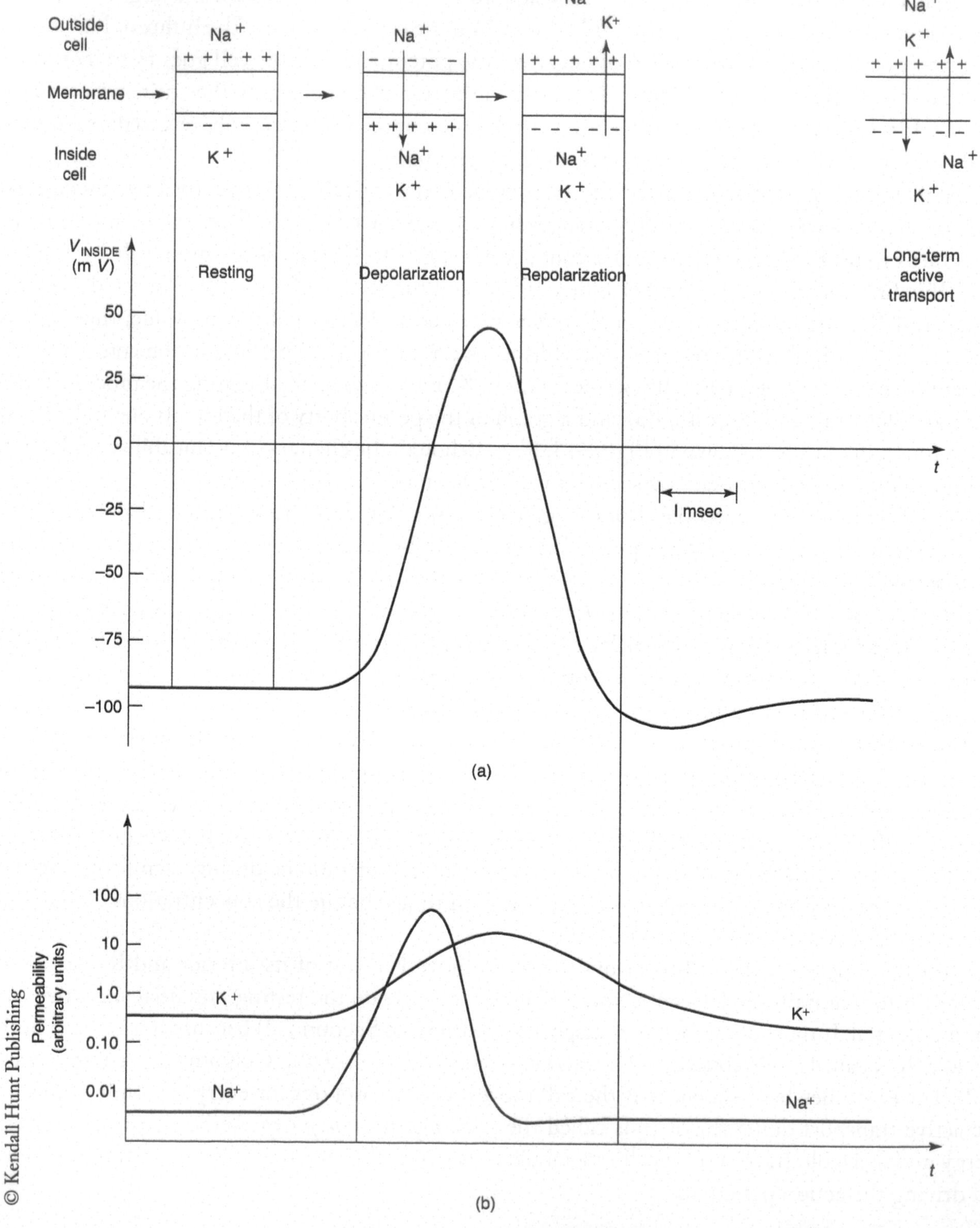

Active transport is not needed to maintain the concentration gradient for Cl^- in neurons. Neuron membranes are very permeable to Cl^- and the Coulomb force moves it back and forth across the membrane during depolarization-repolarization. There is thus a small influx of Cl^- when the interior becomes positive during depolarization and a similar out flux of Cl^- when the interior returns to negative.

Propagation of the Action Potential The description of the action potential and the graphs in Figure 13.4 tell what happens at a fixed location on a cell membrane. How is the action potential transmitted to other locations? The answer is that the voltage change during depolarization of one region is sufficient to change the permeability and thereby depolarize adjacent regions of cell membrane. The depolarization-repolarization process shown in Figure 13.4 is thus stimulated in adjacent cell membrane a short time after starting at some point. These adjacent regions in turn stimulate additional cell membrane farther away, and the action potential is propagated along the cell membrane as shown in Figure 13.5. Of course, the propagation of action potentials is not limited to the cell in which they originate. Neurons can transmit them to other neurons, glands, and muscles. The brain is an extreme example of complex interconnections of neurons.

Myelinated and Unmyelinated Axons Figure 13.3 shows the axon of a nerve cell with segments covered by myelin sheaths and small gaps between the myelin sheaths called nodes of Ranvier. The axon in Figure 13.3 is myelinated; a bare axon without the sheaths is called unmyelinated. Myelinated axons have certain advantages over unmyelinated axons. The myelin sheath electrically insulates the axon from being fired by other axons. This allows bundles of nerves to carry signals without "cross talk" between axons of different nerves. The speed of propagation in myelinated axons is much higher than in unmyelinated axons—as high as 130 m/s versus as low as 0.5 m/s in some unmyelinated axons. Furthermore, the energy required to send a signal down a myelinated axon is much less than for an unmyelinated axon. Both types of axons exist throughout the body, and it has been speculated that the development of the myelin sheath was an important evolutionary step.

The reason that myelinated axons conduct signals at larger speeds and using less energy has to do with the insulating properties of myelin. Consider Figure 13.6, which shows the propagation of a nerve impulse along a myelinated axon. When the nerve impulse or action potential reaches the myelin sheath, it does not cause many ions to cross the axon membrane because the myelin sheath separates the axon from the fluid outside. The action potential travels along the myelin sheath much as any voltage pulse would travel through an ordinary resistor, losing voltage as it goes along (recall that $V = IR$ is the voltage drop in a resistor) but moving very rapidly.

The voltage pulse would eventually become too small to stimulate anything at the end of the axon, so it is periodically regenerated in the nodes of Ranvier. When the voltage pulse reaches a node of Ranvier, it is still large enough to stimulate the depolarization-repolarization cycle, generating a full voltage action potential that is sent on down the axon to be regenerated at each successive gap. The nodes of Ranvier act as small amplifiers along the axon. The energy-consuming sodium-potassium pump does not have to work as hard to maintain the concentration gradients, because so few charges cross the axon membrane in the myelinated regions. The speed of propagation is larger than in unmyelinated axons because most of the axon is covered with myelin. In fact, the speed in the myelin sheaths is so large compared to the gaps that the action potential seems almost to leap from one gap to the next. Such propagation is called *saltatory* propagation.

Figure 13.5 Propagation of an action potential along a membrane. A stimulus changes the permeability of the membrane to Na^+ ions, triggering the action potential. This in turn changes the permeability of adjacent membrane, and the action potential propagates outward from the starting point as shown.

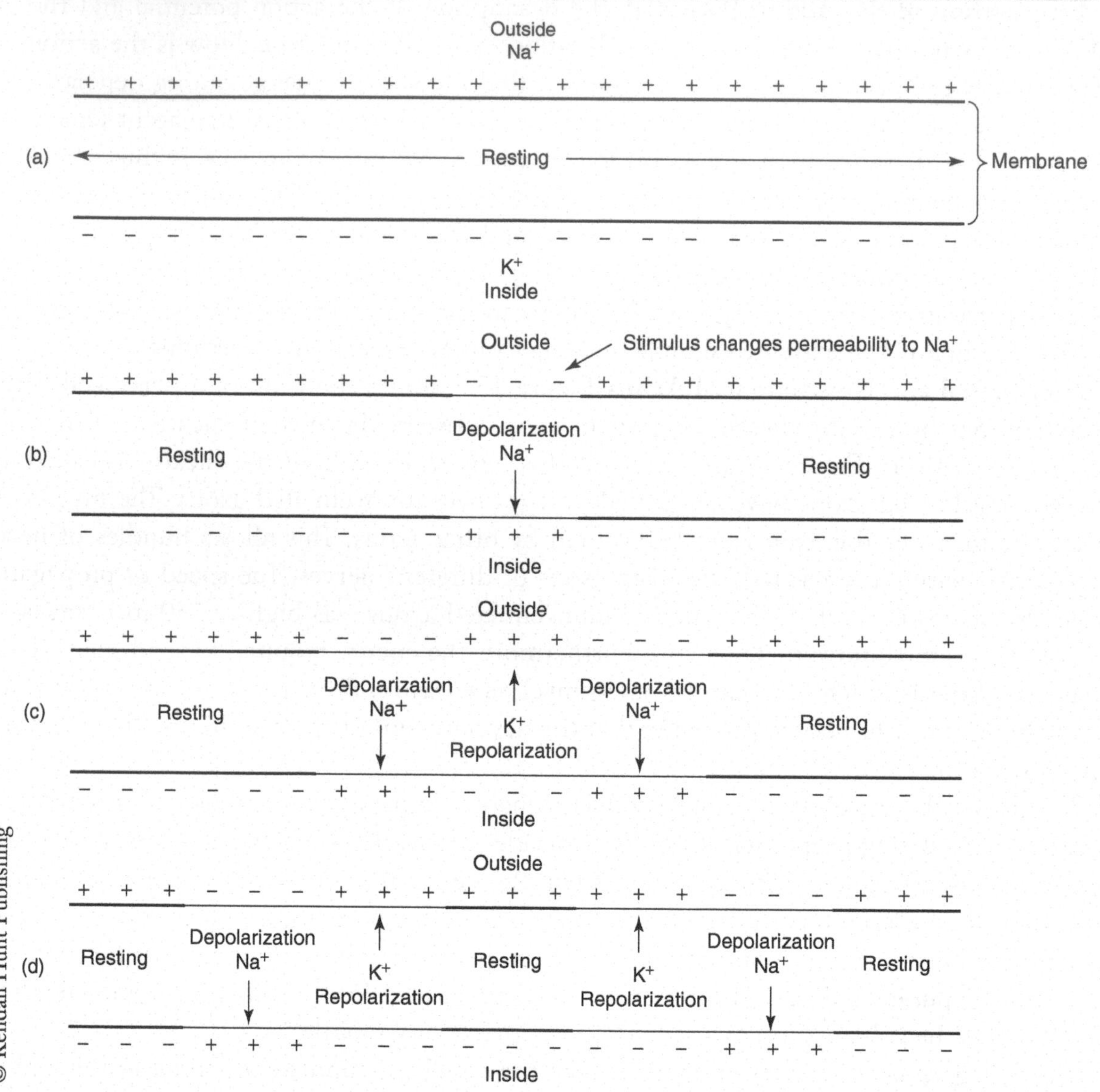

Both the generation and propagation of bioelectricity are complex processes involving several simultaneously occurring physical phenomena, such as diffusion, Coulomb forces, and active transport. Nevertheless, many of the basic properties of bioelectricity on the cellular level can be understood in terms of principles of physics. Of course, a detailed understanding requires a great deal of chemistry and biology and a dollop of more advanced physics, but some features are still mysterious.

Figure 13.6 Propagation of a nerve impulse down a myelinated axon. The nerve impulse travels very fast in the myelinated regions but loses voltage. It is regenerated in each gap.

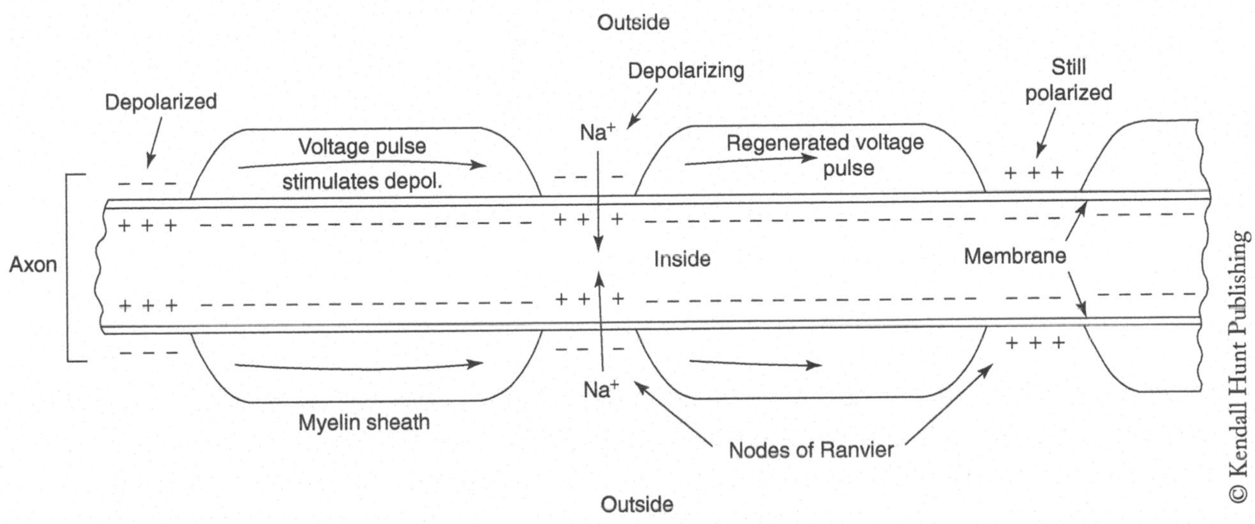

13.2 DETECTION OF BIOELECTRICITY AND ITS USE IN MEDICAL DIAGNOSTICS

Action potentials created by *individual* cells can be measured using very small probes. This usually is done for research purposes, such as investigating the function of a particular bundle of nerve fibers. It also can be done for therapeutic purposes. For example, such measurements can help determine which nerves an artificial eye should stimulate to be most effective. Coordinated electrical activity of entire systems of cells, as in the brain or heart, can also be measured. Large-scale electrical activity is much easier to observe than that of individual cells and is consequently more commonly measured.

13.2.1 Measurement of Bioelectricity in Individual Cells

Nerve cells action potentials in nerve cells appear as graphed in Figure 13.4(a). To obtain such data a tiny probe is inserted into the interior of the nerve cell and another is placed in the fluid between cells. The potential difference between these probes is graphed in Figure 13.4(a), with the exterior probe taken to be at zero volts. Care must be exercised that the probes themselves do not alter the behavior of the cell they monitor. For example, one variety of probe inserted into cells is filled with a conducting solution of potassium chloride (KCl), so that ion concentrations inside the cell are not disturbed. Many early measurements of nerve potentials were made on axons removed from squids because those axons are exceptionally large. Today it is possible to measure individual nerve potentials in other animals, including humans, and it is not necessary to remove the nerves from the organism.

A variety of information has been gained by monitoring individual nerves. For example, the response of nerves in the hand has been investigated by monitoring nerve bundles carrying signals from the hand to the brain. It was found that some nerves respond to touch, rub, or pinch, others to temperature changes. Some of this information was expected because of earlier; less direct measurements, but measurement of individual nerves provides invaluable direct evidence.

Figure 13.7 Action potential in one type of muscle cell. Note the much longer time required and the difference in shape from nerve action potentials (compare with Figure 13.4a).

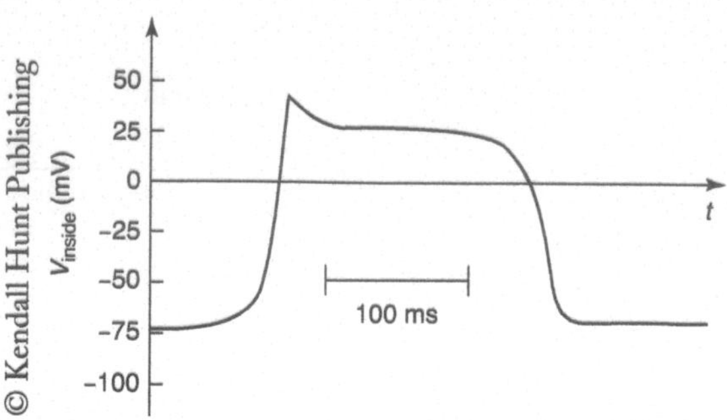

Firing characteristics of nerves are found to differ. Some nerves fire continuously while stimulated and others fire only when stimulation starts or stops; in some the firing rate is proportional to the strength of the stimulus. Alternatively, the probe can be used to stimulate specific nerves electrically and have the subject relate what he or she feels. And, as has been mentioned, the basic characteristics of nerve action potentials, graphed in Figure 13.4(a), were determined from observations of individual nerves.

Muscle Cells Action potentials in individual muscle cells exhibit some important differences from action potentials in nerve cells. Depending on the type of muscle cell, the action potential may last much longer than in nerve cells—from a few milliseconds to as long as several hundred milliseconds. A representative pulse for one type of heart cell is shown in Figure 13.7. Permeabilities of muscle cell membrane to various ions differ from those in nerves, and the details of action-potential generation in muscle cells are not as well known. However, one major characteristic is the same. An action potential consists of depolarization of the cell membrane followed by its repolarization.

An action potential in a muscle cell causes it to contract. Muscle action potentials may be generated inside a cell or triggered from the outside, such as by a nerve or another muscle cell. Large-scale muscle contraction sometimes occurs because there is good electrical contact among cells within a group of cells. The action potential of one muscle cell in a group can thus be transmitted from cell to cell within the group.

13.2.2 Measurement of Bioelectricity in Large Collections of Cells

Muscle Activity: The Electromyogram It is easy to measure electrical activity associated with muscle action by inserting probes into the muscle or by placing probes (often called electrodes) on the skin. This is because the coordinated action of many thousands of individual muscle cells is involved. Such a recording is called an electromyogram (EMG). One use of an EMG is to train a person to use diseased or injured muscles. Because the EMG is very sensitive to muscle movement, patients can determine whether they are successful in moving a muscle by watching the EMG output. Athletes are sometimes trained to develop specific muscles by using an EMG.

EMGs are used in a variety of research areas. Because the EMG is more sensitive than touch or vision in determining which muscles are used in a particular activity, they are useful in *kinesiology*, the study of human motion. Electrical stimulation of muscle movement is also used in conjunction with EMGs to diagnose damaged muscles, to determine which muscles are controlled by which nerves, or to explore the tiring rate of muscles.

The Electrocardiogram The most common and most important of all measurements of bioelectricity is the recording of electrical signals from the heart, called an ECG. A great deal can be determined about the heart from ECGs; however, the electrical nature of the heart is complex, and its electrical signals are not simple. (A detailed treatment of ECGs can easily occupy an entire book.) Only the basics of how the ECG is measured and some of its implications are discussed here.

First, it is important to note that electrical activity can start in one part of a muscle and spread out like a wave on a pond. This is particularly true of the heart. An array of electrodes on the skin of a person can determine the timing, magnitude, and direction of motion of these electrical patterns. Figure 13.8 shows how this is done and how the addition of more electrodes gives more complete information. In Figure 13.8(a), depolarization of the muscle has started at the top of the muscle and is moving downward. When cells depolarize (the first step in the action potential), their exteriors go from positive to negative. Electrode B is therefore positive compared to electrode A; that is, the potential difference between B and A is positive. Conversely, if the potential difference between B and A is positive, then the depolarization is moving toward B.

The exact direction of the electrical wave is not determined by two electrodes. Only the component of the arrow representing the magnitude and direction of the electrical wave (a vector) is determined. Adding a third electrode, as in Figure 13.8(b), helps because the components of the vector along three directions are observed. Vector addition is used to gain further information. However, three points determine a plane and the muscle is three-dimensional, so more electrodes are necessary to obtain complete information.

Figure 13.8 (a) Wave of depolarization spreading across the heart is represented by an arrow pointing in the direction of motion of the wave and having a length proportional to the strength of the wave (a vector). An electrode placed at B will be positive compared with an electrode at A. (b) The Einthoven triangle configuration of electrodes for measuring electrical signals from the heart in an ECG. Each pair of electrodes measures one component of the voltage wave.

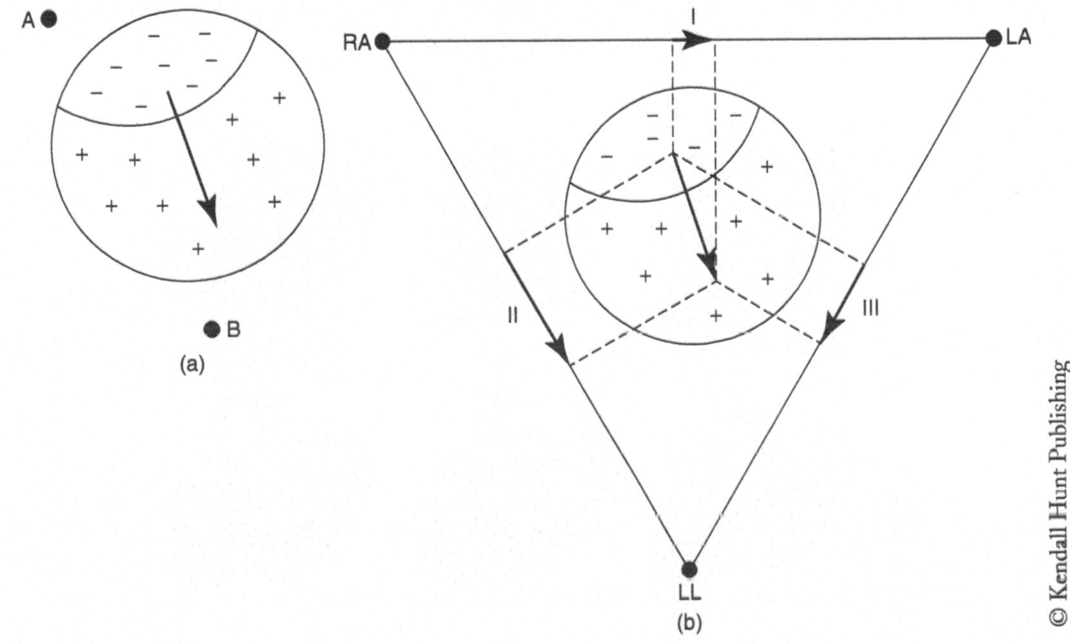

Standard ECGs are measured using a triangular configuration of electrodes like that in Figure 13.8(b). This configuration is called the *Einthoven triangle* after Willem Einthoven, a Dutch physiologist who pioneered its use at the beginning of this century. The electrodes sometimes are placed on the right and left arms and the left leg and are traditionally labeled RA, LA, and LL. The potential differences between the three pairs of electrodes are traditionally labeled I, II, and III, as shown in the figure. All three are usually recorded, and the measurements are often called leads I, II, and III. (Each lead is the potential difference between two electrodes.) It is now common in a detailed ECG to have six additional electrodes attached across the chest of the subject to give three-dimensional information. Typically, skin potentials on the order of 1 mV are observed.

Figure 13.9 shows how electrical patterns spread out on the heart.

Each heartbeat starts with the right and left atria pumping blood into the right and left ventricles. The ventricles then pump blood through the lungs and the rest of the body. The electrical signal that starts this sequence of events originates in an area of the upper left part of the heart called the sinoatrial or *SA node*. This wave travels across the atria, depolarizing muscle cells and causing the atria to contract (see Figure 13.9(1+2)).

When this depolarization wave reaches the bottom of the atria, it stimulates an area called the atrioventricular or AV node. The AV node then fires and begins the depolarization of the bundle of His, which carry the depolarization to the Purkinje fibers, a special conducting system of muscle fibers. These muscle fibers have lost the ability to contract and act as a rapid conduction system, bringing depolarization to all parts of the ventricles nearly simultaneously and causing their contraction (see Figure 13.9(3-6)).

Figure 13.9 ECG and electrical activity of the myocardium.

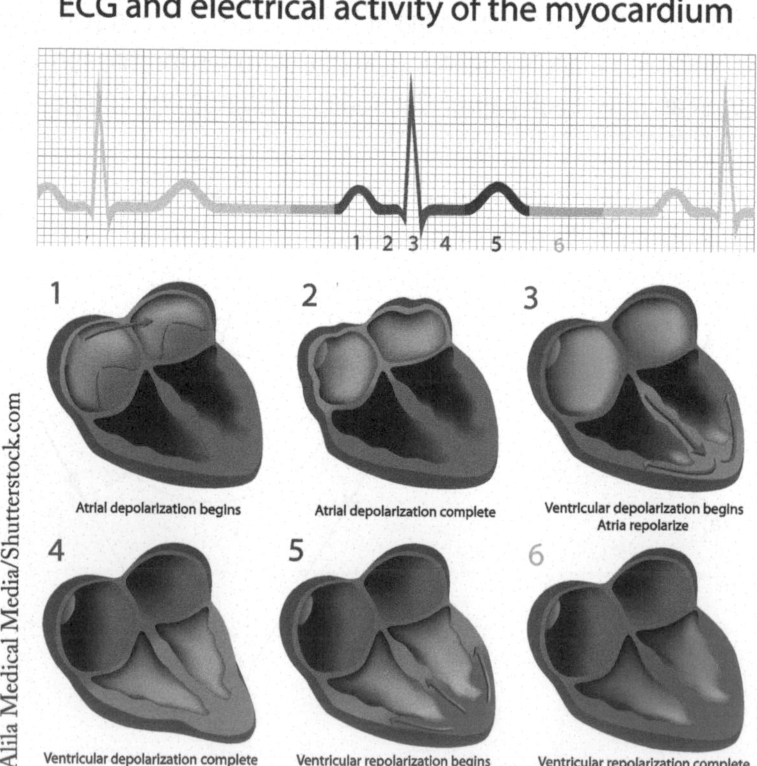

Figure 13.10 An ECG with corresponding blood pressure. The P wave is the depolarization of the atria. The QRS complex is the depolarization of the ventricles, followed shortly by ventricular contraction as indicated by the subsequent maximum in blood pressure. The T wave is the repolarization of the ventricles to ready them for the next beat. See also Figure 7.14.

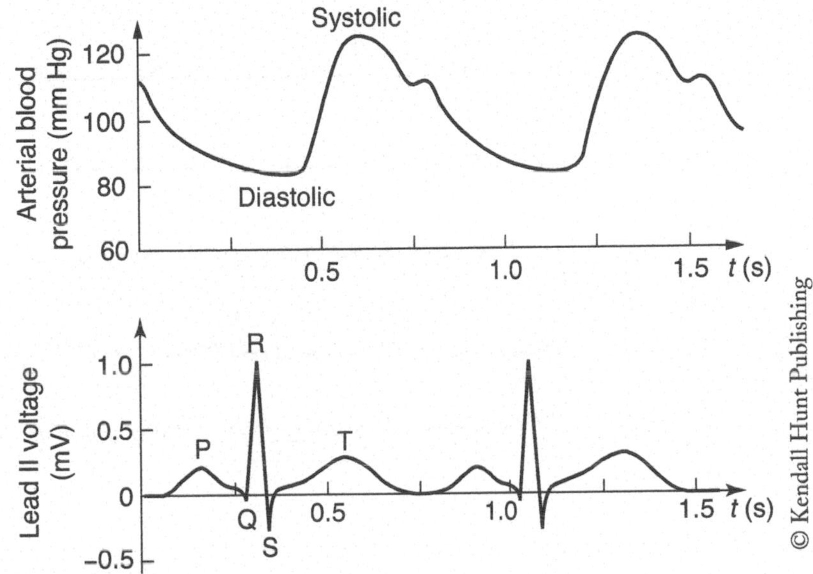

After the heart muscle has depolarized and contracted, it repolarizes and is then ready for the next beat. The SA node is the natural pacemaker for the heart and will cause it to beat about 72 times a minute without outside stimulation. However, there are nerves leading into the SA node that can adjust the heart rate in response to bodily needs such as during exercise, anxiety, or sleep.

Figure 13.10 shows an ECG trace for lead II (the potential difference between the left leg and right arm) and a graph of the corresponding arterial blood pressure. The major features of the ECG are labeled P, Q, R, S, and T. The *P wave* corresponds to the depolarization and contraction of the atria. The *QRS* complex accompanies the depolarization and contraction of the ventricles. The *T wave* is the repolarization of the ventricles to prepare them for the next contraction. Repolarization of the atria occurs earlier and is not observed because it is masked by the large QRS complex.

Traces from other pairs of electrodes differ from the lead II trace shown, because they get different components of the electrical waves moving across the heart. Taken together, the several traces (3–12) of an ECG can give very detailed information on the functioning of the heart. For example, dead regions of the heart, called infarcts, reflect the electrical waves. In an ECG such reflections can be observed, and the affected parts of the heart sometimes can be identified.

An AV *block* can also be observed in an ECG. In an AV block the depolarization wave from the atria is blocked by dead or scarred tissue from reaching the AV node. In terms of the ECG this means the P wave has trouble triggering the QRS complex. This can appear as an abnormally long time between the P wave and the QRS complex. A severe AV block prevents triggering of the ventricles by the electrical wave from the atria. The AV node then takes over, and acts as its own pacemaker, but the heart rate becomes unacceptably low. The remedy is the implantation of an artificial

Figure 13.11 Normal and pathological electrocardiograms.

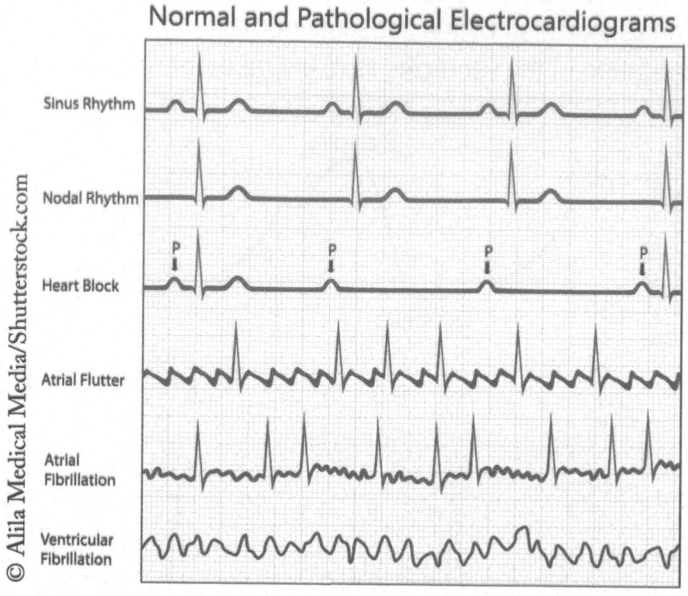

pacemaker. The artificial pacemaker has an electrical connection to the ventricles and stimulates them to contract at an acceptable rate. Artificial pacemakers are rather common and enable many people to lead nearly normal lives instead of being greatly incapacitated.

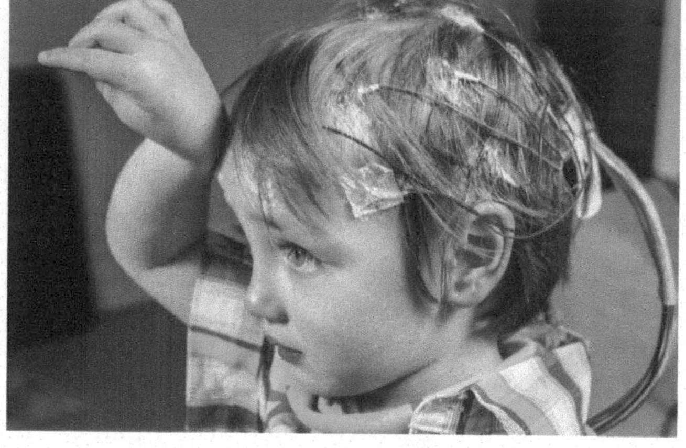
© Steve Buckley/Shutterstock.com

Another heart problem easily diagnosed with an ECG is ventricular fibrillation, a malfunctioning of the signals from the AV node to the ventricles that causes the ventricles to contract chaotically and ineffectively. It is fatal within minutes if uncorrected, so ECGs are useful in detecting ventricular fibrillation only when a patient is already being monitored, as during surgery or in intensive care. An ECG taken during ventricular fibrillation, such as that in Figure 13.11, shows normal and pathological electrocardiograms, including complete distortion of normal P, QRS, and T features during ventricular fibrillation. The remedy is electrical defibrillation by passing a large current through the heart. This apparently completely depolarizes the heart muscle, so that the heart is often able to restore its normal electrical patterns. Defibrillators are now constructed with microcomputers that initiate defibrillation at the time of a QRS, if any are observable. It is also interesting that some artificial pacemakers have built-in defibrillation capabilities. These pacemakers use a microcomputer to recognize when the heart is fibrillating and to administer the defibrillating shock.

Brain Activity: The EEG Electrodes placed on the scalp can detect complex electrical patterns associated with nerve activity in the cortex of the brain. The recording of these potentials is an *electroencephalogram* (EEG) and is a valuable tool both in research and in medical diagnostics.

Numerous electrodes are placed on the scalp to record a detailed EEG. Potential differences between various pairs of electrodes are recorded, and a reference electrode is attached to one ear. Signals from the left and right sides of the brain are often compared because asymmetry in brain activity is frequently an indication of brain disease. Voltages of about 50 μV are typically observed on the scalp. These are small compared to the millivolt potentials found on the skin associated with muscle movement. Electrical signals from eye movements, heart activity, and noise from electrical devices in the room are problematic but can be overcome. Figure 13.12 shows a typical EEG for a normal subject.

EEG traces are obviously more complex and appear to be less regular than ECG traces. This is not surprising, to say the least, since hundreds of millions of neurons are involved and their activity is neither as simple nor as coordinated as that of heart muscle cells. Analysis of EEG traces shows that there are certain basic frequencies present in the electrical signals. The presence or absence of certain frequencies is strongly correlated with certain types of brain activity. Consequently, frequency ranges are given the names listed in Table 13.1. A skilled observer can detect the presence of the various *brain waves*, as they are called, in an EEG. The advent of computers has made that task somewhat easier.

Mental alertness is one thing that can be detected with an EEG.

Beta waves become dominant during concentrated mental activity. *Alpha waves* decrease and disappear as a person falls asleep and sometimes are monitored to determine the depth of anesthesia

Figure 13.12 An EEG for a normal patient. The letters on the left identify the locations of electrodes.

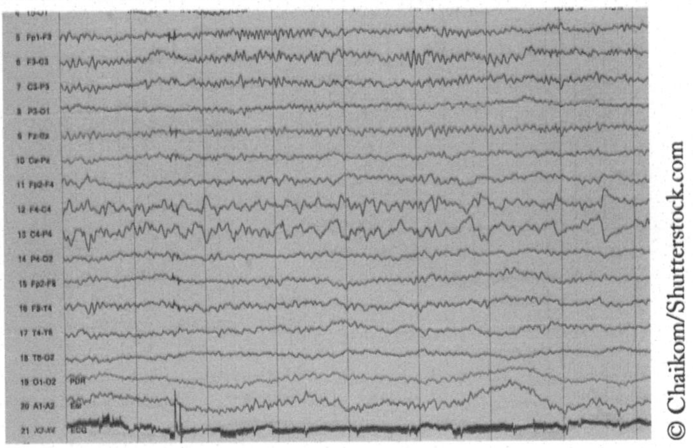

Table 13.1 Brain Wave Frequency Ranges

Band	Frequency Range (Hz)
Delta (δ)	0.5–4
Theta (θ)	4–8
Alpha (α)	8–13
Beta (β)	13–22
Gamma (γ)	22–30

during surgery. Delta waves are indicative of deep sleep and become dominant for a person in a coma. Lack of brain activity as indicated by an EEG is sometimes used as a criterion for determining when a person is dead.

Certain brain disorders, such as epilepsy, and diseases that affect brain function can be detected with an EEG. The presence of brain tumors can be determined because there is less brain activity in the region of the tumor. In fact, any disease that affects blood flow or glucose supply to the brain, as do sickle-cell anemia and hypoglycemia, may cause an abnormal EEG. The presence of certain psychoactive drugs also causes an abnormal EEG.

The correlation between EEG and specific brain activities or malfunctions is not perfect. For example, a significant percentage of epileptics, depending on the type of seizure they suffer, have normal EEGs between seizures.

QUESTIONS

13.1 Are the concentrations of the following ions greatest inside or outside of cells: Na^+, K^+, Cl^-? What process is used to maintain the concentration gradients of Na^+ and K^+ ions over long periods of time?

13.2 In what way is energy expended to create bioelectricity?

13.3 Note that in Figure 13.2(b) both the concentration gradient and Coulomb force tend to move Na^+ ions into the cell. What prevents significant amounts of Na^+ from moving into the cell?

13.4 If the cell membrane is permeable to both Na^+ and Cl^- ions, why don't they diffuse across the membrane until their concentrations are equal on either side?

13.5 How can a cell (and the fluid around it) be electrically neutral and still create a potential difference across the cell membrane?

13.6 Explain why a large concentration of K^+ ions inside a cell causes it to become negative inside.

13.7 Define depolarization, repolarization, and action potential.

13.8 Describe the sequence of events occurring in an action potential in terms of diffusion, the Coulomb force, and membrane permeability.

13.9 Why can a cell fire many times before needing active transport to maintain K^+ and Na^+ concentration gradients?

13.10 Explain why there is a minimum time interval between firings of a cell. Does that minimum time depend on the type of cell—such as nerve or muscle?

13.11 Estimate the maximum firing rates for neurons and for heart cells. Hint: Consider the graphs of their action potentials.

13.12 What is the relationship between the maximum firing rate of nerves and the hearing mechanism as discussed in Chapter 9? Similarly, what· is the relationship between the maximum firing rate of heart cells and the maximum heart rate?

13.13 Note in Figure 13.4(a) that the potential in a nerve cell overshoots a bit when it returns to its resting potential, briefly becoming −110 mV. Explain this overshoot in terms of the increased permeability of the membrane to K^+ ions graphed in Figure 13.4(b).

13.14 Draw a cell membrane as it would appear if the sequence shown in Figure 13.5 is carried one step further.

13.15 What is the function of the nodes of Ranvier, and why are these gaps between myelin sheaths necessary?

13.16 Explain the advantages of myelinated nerves in terms of the insulating properties of myelin.

13.17 Give an example of an action potential that originates inside a cell. Give another example in which an action potential is transmitted to another cell, causing it to fire.

13.18 How can the direction in which a depolarization wave moves across a muscle be measured?

13.19 Why are more than three electrodes needed in an ECG to obtain three-dimensional information about the heart?

13.20 Explain in terms of efficient use of a muscle why simultaneous depolarization of all parts of the ventricles is desirable.

13.21 Depolarization of a muscle cell causes it to contract a few tenths of a second later. Is this consistent with Figure 13.10? Explain.

13.22 What physical quantities (e.g., current, voltage, resistance, velocity, time, mass, etc.) are actually measured in EMGs, ECGs, and EEGs?

13.23 Which yield more specific types of information, ECGs or EEGs? Explain why.

13.24 Why do altered blood flow and blood glucose levels affect EEGs?

13.25 What is the physical distinction between the various types of brain waves (alpha, beta, etc.) measured in EEGs?

PROBLEMS

The Nernst equation (equation 13.1) gives the membrane potential that would result from the concentration gradient of a single type of ion in the presence of a membrane perfectly permeable to it. The purpose of the following three problems is to verify the potentials given for K^+, Na^+, and Cl^- ions in the text. The value of Boltzmann's constant is $k = 1.38 \times 10^{-23}$ J/K. A temperature of 310 K (37°C) is assumed.

$$V = V_{in} - V_{out} = 2.30\frac{kT}{Ze}(\log C_{in} - \log C_{out}) \qquad \text{eq. 13.1}$$

√13.1 (II) Show that the membrane potential for K^+ ions is −88 mV, given the concentrations of K^+ ions to be 140 mol/m^3 inside the cell and 5 mol/m^3 outside.

√13.2 (II) Show that the membrane potential for Na^+ ions is +59 mV, given the concentrations of Na^+ ions to be 15 mol/m^3 inside the cell and 140 mol/m^3 outside.

√13.3 (II) Show that the membrane potential for Cl^- ions is −70 mV, given the concentrations of Cl^- ions to be, 9 mol/m^3 inside the cell and 125 mol/m^3 outside.

14 Geometric Optics

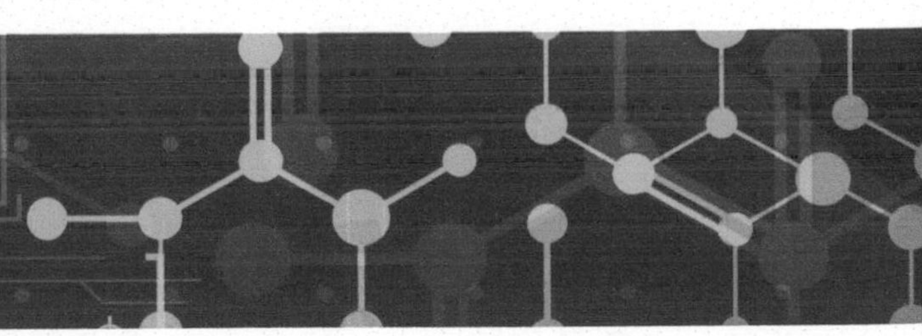

"I have not failed. I have found 10,000 ways that don't work."
—Thomas A. Edison

Optics is the branch of physics that deals with light. Geometric optics is a subfield of optics that deals with the interaction of light and macroscopic objects (roughly speaking, anything larger than the thickness of a human hair). The rules of geometric optics are relatively simple. Those rules and some of their applications in nonbiological systems are the subject of this chapter. The application of geometric optics to the human eye, including certain common vision defects, is treated briefly in Chapter 15. The behavior of light when it interacts with microscopic objects is considered in Chapter 16.

14.1 LIGHT AS A RAY: REFLECTION AND REFRACTION

Light comes to an observer's eyes either directly or indirectly from some source, as illustrated in Figure 14.1. Light may originate in the sun, light bulbs, cathode ray tubes (CRTs), fireflies, and so on. Light can travel through empty space, air, water, or the lens of the eye, for example. The substance that light, or any wave is traveling in, is called a *medium*.

When light bounces off of an object, such as the lake in Figure 14.1, it is said to have been reflected.

Figure 14.1 An example of light reaching an observer's eye by a direct path from the original source and also by reflection.

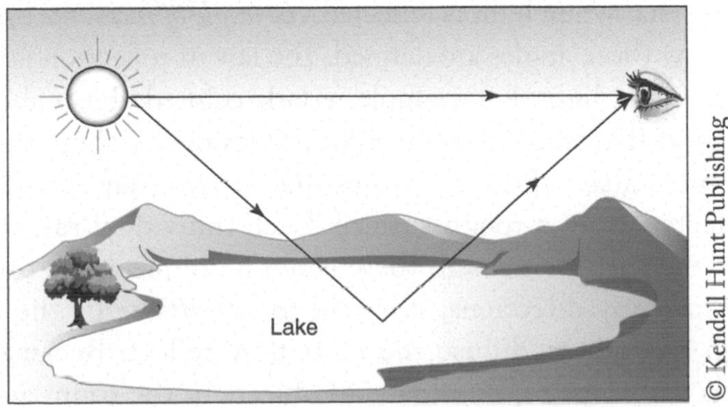

Figure 14.2 Light refracting as it moves from air to glass.

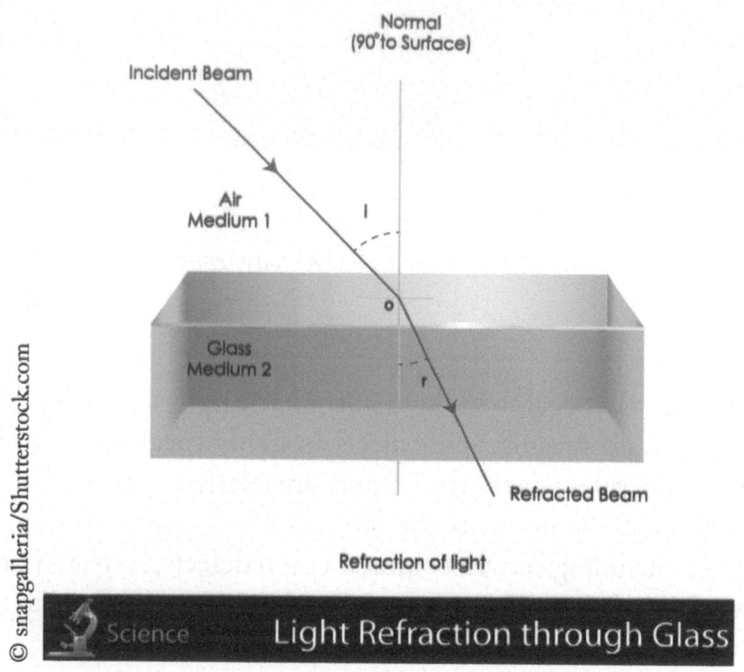

When light changes directions while passing from one medium to another, as in Figure 14.2, it is said to have been *refracted.* An important property of light is that it moves in straight lines, provided that it interacts only with macroscopic objects, and it changes direction only if it is reflected or refracted. The technical term for a straight line that starts at some point and radiates outward is ray, and whenever we deal with light and macroscopic objects we speak of light rays. The study of light rays is geometric optics, which is also called *ray optics.*

14.1.1 Reflection

The law of reflection says that when light is reflected, *the angle of incidence equals the angle of reflection.* Figure 14.3 illustrates how these angles are defined. The law of reflection is valid whether or not the surface reflects all the incident light. For example, a dark-colored object absorbs rather than reflects most light falling on it, but that fraction of the light it does reflect obeys the law of reflection.

The law of reflection is also valid for surfaces that are not flat. Consider Figure 14.4, which shows light being reflected from a rough surface. Each individual ray of light obeys the law of reflection, but because the surface is rough there is not a unique angle for all rays. The light thus comes off the surface in many directions; it is said to be *diffused* by the surface. The surfaces of most things are rough enough to diffuse the light they reflect. Because of this, a single light bulb in a room often is sufficient to illuminate all objects in the room and make them visible to observers in most parts of the room.

Figure 14.3 Law of reflection: $\theta_i = \theta_r$. The thin line *i* perpendicular to the surface at the point where the light strikes it. The angles of incidence θ_i and reflection θ_r are measured relative to the perpendicular.

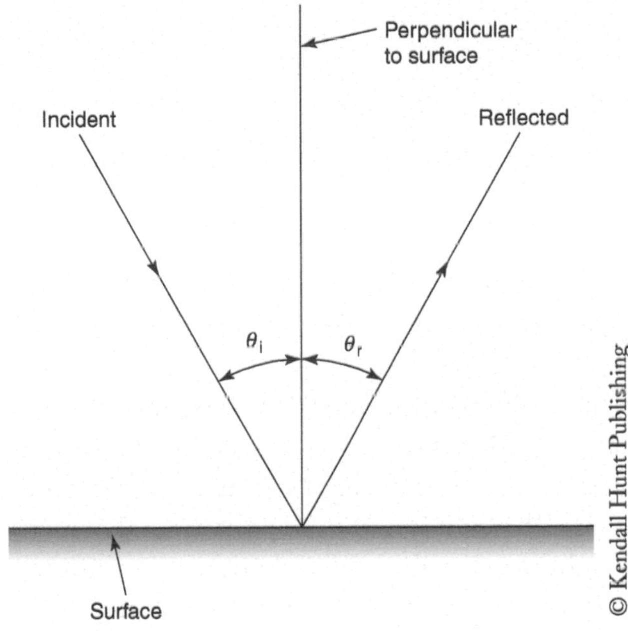

Figure 14.4 Diffuse reflection: When light is reflected from a rough surface, it comes off the surface in many directions.

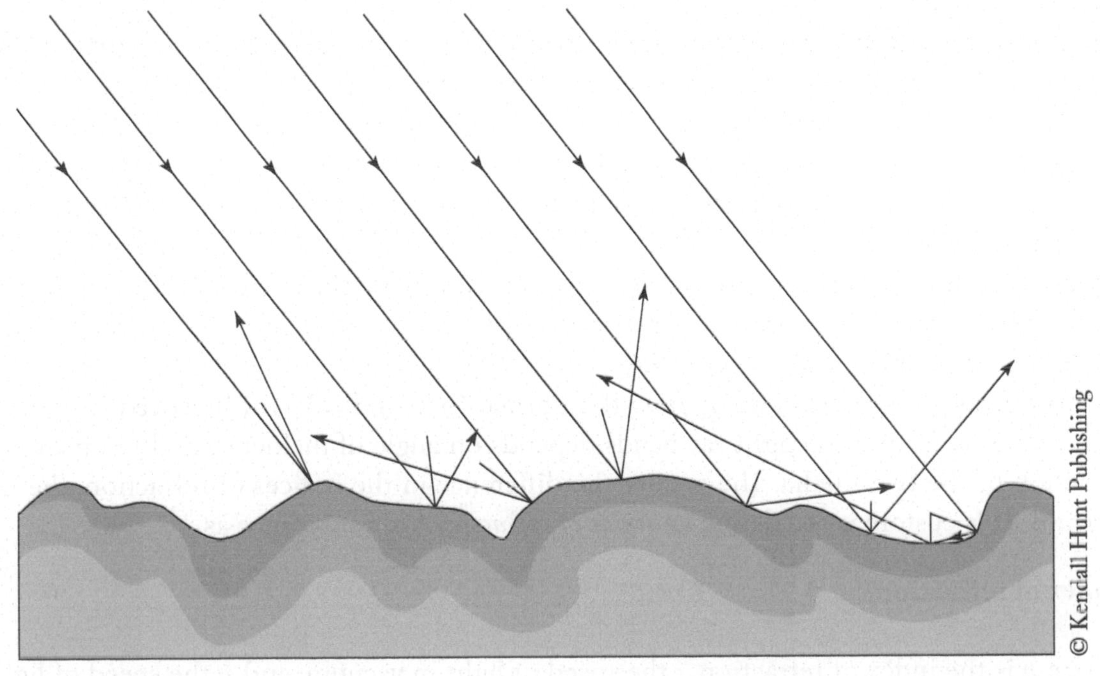

Figure 14.5 The refraction of light as it leaves the water and enters the air on its way to your eye causes this distorted view of a thermometer.

© imstock/Shutterstock.com

14.1.2 Refraction

Light usually changes directions when it goes from one medium to another. This is most easily observed for objects partially submerged in water, such as the thermometer in Figure 14.5. The disjointed appearance of the thermometer is caused by the light's changing direction when it goes from water to air (or from water to glass to air). The light is said to be *refracted* at the interface between each medium.

Which way does light bend at the surface between two substances? How much does the light bend? The answers to these two questions have to do with the speed of light in various substances.

Light travels extremely fast in all substances, but it travels fastest in a vacuum: the speed of light in a vacuum denoted c, is

Speed of light in a vacuum $c = 3.00 \times 10^8$ m/s Know

The speed of light is very large; the earth could be circled seven times in one second at that speed. Nothing can go faster than c, and no object having mass can attain that speed. It is the ultimate speed limit in the universe.

Light travels more slowly in any medium than it does in a vacuum, because it interacts with the atoms and molecules in a substance, sometimes being absorbed and reemitted, but it travels at "c" between atoms and molecules. The net result is that light slows down in a substance by an amount that depends on its molecular structure, density, and similar factors. When light goes from one medium to another, it changes direction, refracts, only if the speed of light is different in the two media (as it usually is).

Figure 14.6 shows light rays going from one medium to another. The law of refraction says that light will bend *toward* the normal, (perpendicular) if the speed of light is smaller in the second medium than in the first, as in Figure 14.6.

Conversely, light will bend *away from* the normal (perpendicular) if its speed is greater in the second. The amount that the light ray bends depends on angle of incidence and the relative indices of refraction of the two media. The greater the difference in the indices of refraction, the more the light bends. It is customary to define an *index of refraction* for a substance as

Index of refraction, $n \equiv \frac{c}{v}$ *Know*

where n is the index of refraction, c the speed of light in vacuum, and v the speed of light in the medium. Note that the index of refraction, n, is always greater than or equal to 1. Table 14.1 lists the

Figure 14.6 The law of refraction: The Speed of light is greater in medium 1 than in medium 2. (a) Light bends toward the perpendicular when it goes from fast to slow. (b) Light bend away from the perpendicular when it goes from slow to fast. Note that the path is precisely reversible.

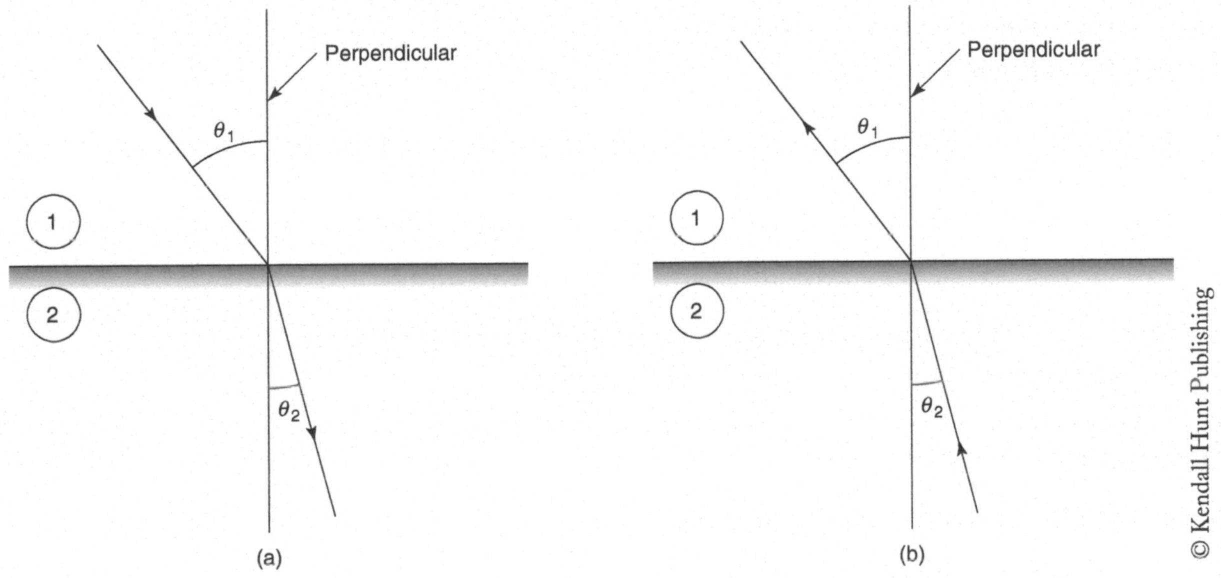

Table 14.1 Indices of Refraction

Material	$n \equiv c/v$
Air (0°C, 1 atm)	1.0003
Ice	1.31
Water	1.33
Ethyl alcohol	1.36
Glasses	
Fused quartz	1.46
Crown glass	1.52
Light flint	1.58
Heavy flint	1.66
Benzene	1.50
Lucite, Plexiglas	1.51
Sodium chloride	1.53
Diamond	2.42

index of refraction for several substances. High index of refraction corrective lenses have an index of refraction of approximately 1.74.

In the next section, the laws of reflection and refraction are used to explain image formation by mirrors and lenses.

The precise mathematical relationship of the angle of incidence, the angle of refraction, and the indices of refraction is known as *Snell's law*:

$$\text{Sin } \theta_1 n_1 = \text{Sin } \theta_2 n$$

θ_1 is known as the angle of incidence. It is the angle measured between the normal and the ray in the first medium.

θ_2 is the angle of refraction. It is the angle between the normal and the ray in the second medium.

This relationship is the quantitative statement of the *law of refraction*.

We will explore Snell's law in more detail in Lab 9.

Things to know in section 14.1

Speed of light in a vacuum $c = 3.0 \times 10^8$ m/s

Index of refraction, by definition: $n \equiv \frac{c}{v}$

Snell's law: $\text{Sin } \theta_1 n_1 = \text{Sin } \theta_2 n_2$

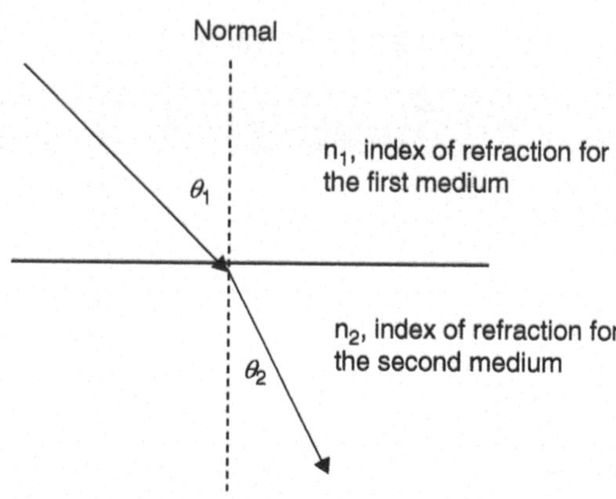

14.2 LENSES AND MIRRORS: IMAGE FORMATION

The laws of reflection and refraction form the basis of geometric optics. They can be used to explain lenses and mirrors and to design lenses, mirrors, and such diverse optical devices as microscopes, telescopes, and eyeglasses.

14.2.1 Lenses: Convex and Concave

Convex Lenses. Let us begin by considering how a simple convex lens, such as that in a magnifying glass, can focus sunlight to a small point, as pictured in Figure 14.7. Light rays reaching the lens

Figure 14.7 The parallel light rays from the distant sun converge at the focal point.

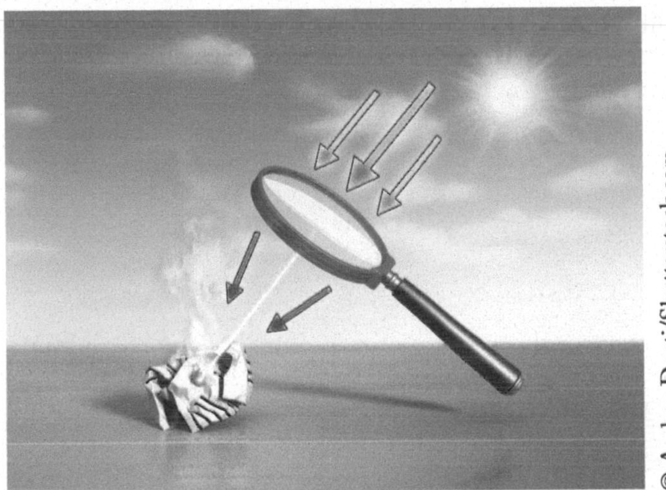

© Andrea Danti/Shutterstock.com

from a distant source, such as the sun, are nearly parallel because they must be nearly parallel at the outset to travel a large distance and reach almost the same point far away.

Figure 14.8 shows how parallel rays of light are refracted by a convex lens. Because the speed of light is slower in glass than in air, the light rays bend toward the perpendicular when they enter the glass. Conversely, light rays bend away from the perpendicular when they leave the glass. Rays passing through the center of the lens are deflected by a negligible amount. The convex shape of the glass is designed so that all parallel rays cross at the same point after passing through the lens.

It is not difficult to shape the glass so that it focuses the light to a very small point. A spherical surface does a reasonably good job and is easy to form. Convex lenses also are called converging lenses because of their effect on light passing through them.

The point at which a convex lens converges parallel light rays, as in Figure 14.8, is called the *focal point F* of the lens. The distance of the focal point from the center of the lens is called the *focal length* of the lens and is given the symbol f.

$f \equiv$ is the distance from the focal point to the center of the lens.

The stronger the lens, the more it refracts the light passing through it. The stronger the lens, the smaller its focal length f is. Consequently, it is convenient to define the *strength* of a lens as

$$S \equiv \frac{1}{f} \qquad \textit{Know} \qquad \text{eq. 14.2}$$

where S is the symbol for *strength*. The units of strength are diopters (abbreviated D) if f is in meters. (Note: f must be in meters for S to be in diopters.) Strength is more commonly called power, but that name is not used here to avoid confusion with ordinary power in watts, a completely different concept. *For clear distance viewing, the strength of a person's eye should be 50 D. For reading, the strength of a person's eye should be approximately 54D.*

EXAMPLE 14.1

(a) What is the strength in diopters of a magnifying glass that converges sunlight to a point a distance 0.100 m from the center of the lens? (b) A farsighted person has a prescription for reading glasses requiring a strength of 3.0 D. What is the focal length of this person's eyeglass lens?

Figure 14.8 Converging lens. Rays of light entering a convex lens parallel to its axis are converged at its focal point *F*. (Ray 2 lies on the axis of this lens.) The expanded view shows the perpendicular and the angle of incidence and refraction for ray 1.

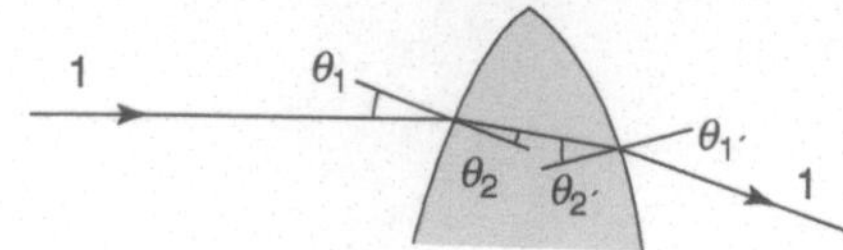

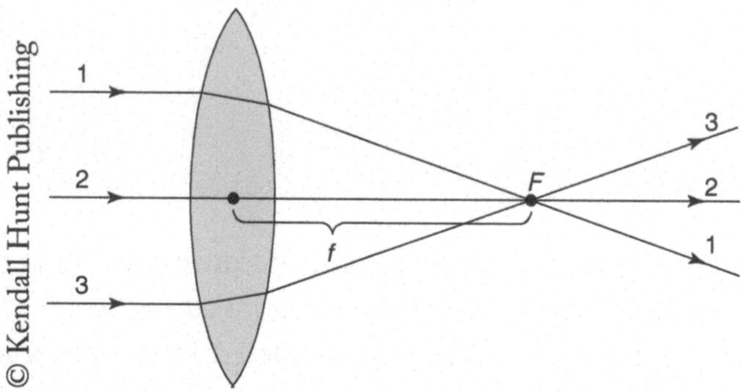

Solution: (a) What is the strength? This magnifying glass has a focal length of 0.100 m. Its *strength* is therefore

$$S \equiv \frac{1}{f} = \frac{1}{0.100} = 10.0\,\mathrm{D}$$

Solution: (b) What is the focal length of a 3.0 D lens? Starting with the definition of strength, we solve for the focal length:

$$S \equiv \frac{1}{f}; \text{therefore } f = \frac{1}{S}$$

The *focal length* of the farsighted person's eyeglass is,

$$f = \frac{1}{S} = \frac{1}{3} = 0.33\,\mathrm{m}$$

Don't forget that when strength is in diopters, the focal length will be in meters.

14.2.2 Concave Lenses

Consider the effect the concave lens in Figure 14.9 has on parallel rays of light. The rays are diverged by the lens, and *consequently concave lenses are called diverging lenses.* Notice that the diverging rays of light all appear to come from the point *F*, defined as the focal point of the lens, and the distance to *F* from the center of the lens is defined as the focal length of the lens, *f*. *The focal length and strength of a concave lens are negative.* The negative focal length indicates that the lens has the opposite effect of a convex lens.

If the person in **Example 14.1(b)** wore concave eyeglasses, the focal length and strength would both be negative rather than positive. *Concave lenses are used to correct nearsightedness.* Nearsightedness means that the person can read without eye correction. They need their glasses in order to clearly see

Figure 14.9 Diverging lens. Rays of light entering a concave lens parallel to its axis all *appear* to originate from its focal point *F*. The dashed lines are not rays; they indicate the directions from which the rays *appear* to come.

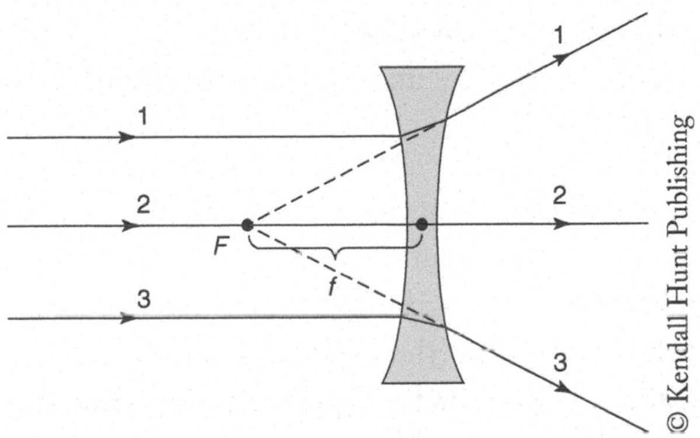

objects that are further away than reading distance. If you need glasses for distant viewing, then you are said to be nearsighted.

14.2.3 Image Formation

The *convex lens* of a magnifying glass can project an image on a piece of paper as shown in Figure 14.10.

The paper has to be held at just the right distance from the lens or the image will be blurry. A little experimenting produces a reasonably good color image, which is inverted. Such an image is called a real image because it can be projected. Slide and movie projectors, cameras, and the eye also

Figure 14.10 A convex lens can be used to project a real image.

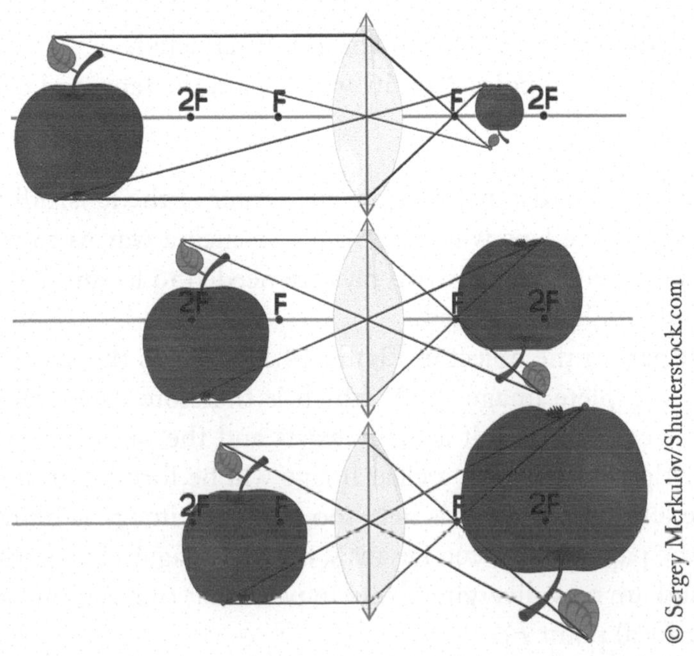

produce real images on screens, film, and the retina, respectively. Notice that the images vary in size and distance from the lens in different applications.

There are *two methods* for predicting the characteristics of images and the general behavior of lenses. Both assume the lens is thin compared to its diameter. The first method is called *ray tracing*, and the second method is the use of the *thin lens equations*.

Ray tracing accurately predicts the size and location of images. It is performed with pencil and paper, using the following rules:

1. A ray entering a *convex lens* parallel to its axis passes through the focal point F of the lens on the other side. (See rays 1 and 3 in Figure 14.8.)
2. The paths of rays are reversible. That is, if a light ray enters a convex lens by passing through its focal point, it exits parallel to the axis (the reverse of paths 1 and 3 in Figure 14.8), and if rays enter a concave lens heading toward its focal point on the other side, they exit parallel to the axis (reverse of paths 1 and 3 in Figure 14.9.)
3. Rays passing through the center of the lens are unaffected by the lens. (See ray 2 in Figures 14.8 and 14.9.)
4. A ray entering a *concave lens* parallel to its axis appears to come from the focal point F. (See rays 1 and 3 in Figure 14.9.)

Now, consider the situation shown in Figure 14.11, where ray tracing is used to predict the size and location of an image produced by a convex lens.

The object on the left is at a distance D_o from a lens of focal length f. D_o is called the object distance. Light is diffused in all directions by the object. In ray tracing, only those rays that fit one of the rules above mentioned are considered. Here, three rays from the top of the object are considered:

Ray 1 enters the lens parallel to the axis and goes through the focal point on the other side of the lens.

Ray 2 goes through the center of the lens and is not refracted.

Ray 3 goes through the focal point F on its way toward the lens, and exits the lens moving parallel to the axis.

The three rays intersect at a distance D_i from the center of the lens. All rays that come from the top of the object and strike the lens will be refracted in such a way as to cross at the point shown. Although three rays are traced here, only two rays are needed to locate the image and determine its size. Anyone of the three could be omitted.

Rays from another part of the object in Figure 14.11, such as the skirt, will also cross at a common point, forming a complete image at D_i, which is therefore called the image distance. A *real image* is formed in this case because it actually exists and the rays can be projected. If a sheet of paper is placed on the right of the lens, a clear image will be formed on it when it is a distance D_i from the center of the lens. If the paper is held too close, the image is blurry because the rays have not yet converged. If the paper is held too far away, the image again becomes blurry because the rays have already crossed and are now diverging. Note in particular that the image is formed at the image distance D_i, not at the focal point F.

Figure 14.11 The formation of a real image by a convex lens, demonstrated by ray tracing. A *real image* is formed whenever all rays from one part of an object actually cross at a common point.

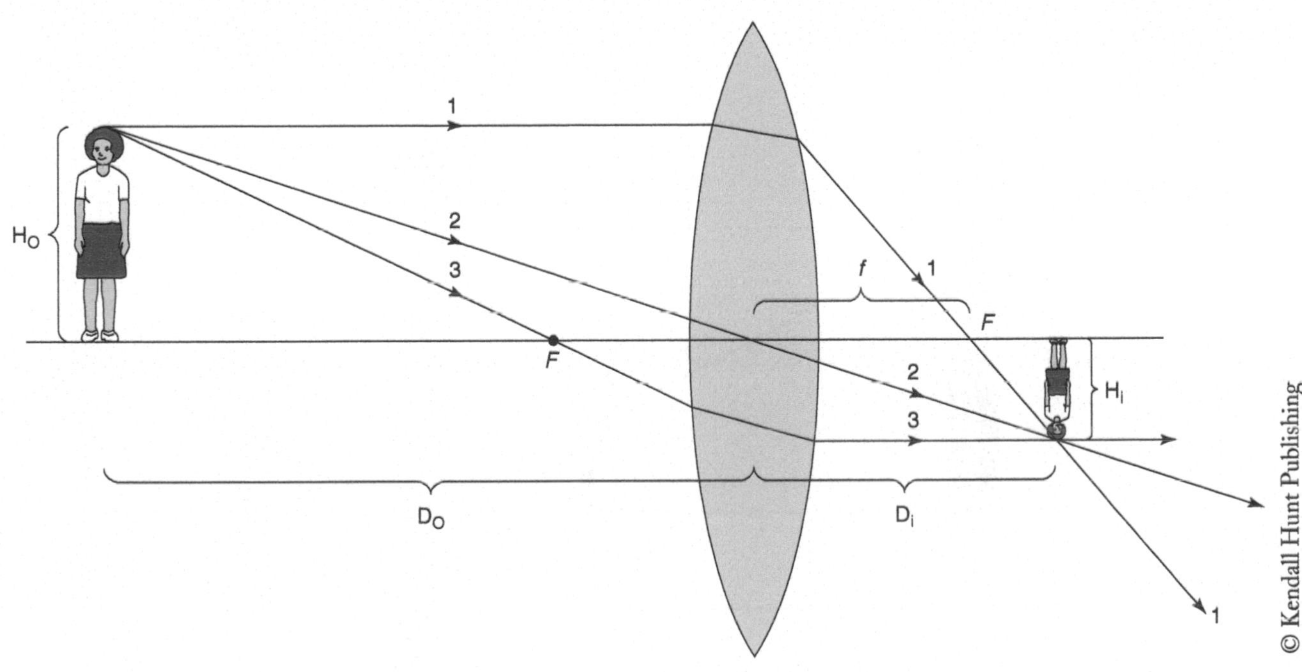

Both the distance to the image, D_i, and the size of the image, H_i, are predicted by ray tracing. If drawings such as that in Figure 14.11 are made to scale, and then D_i and H_i can be measured from the drawing. The accuracy of ray tracing is limited only by the assumption that the lens is thin and the accuracy with which the tracing is drawn.

In Figure 14.12, ray tracing is used to predict the size and location of an image produced by a *concave lens*.

Two rays are traced: *Ray 1* enters the lens parallel to its axis and is refracted upward; *ray 2* goes through the center of the lens and is negligibly deflected. To an observer on the right side of the lens, both rays appear to originate from a common point a distance D_i from the lens, as shown. Furthermore, all rays from the top of the object that strike the lens will be refracted in such a way as to appear to come from this point. * A complete image is produced at D_i in a manner analogous to that described in detail for a convex lens. Both the location D_i and size H_i of the image are correctly predicted by ray tracing. If you look at objects through a concave lens, they appear upright and smaller, consistent with Figure 14.12.

The image produced by the concave lens in Figure 14.12 *cannot be projected*. If a sheet of paper is placed at the location of the image, it simply blocks the rays from reaching the lens, and no image appears on the paper. If a sheet of paper is placed on the right of the lens, no image is formed since rays of light coming from common parts of the object are diverging on the right of the lens. The image produced is therefore called a *virtual image*. A single concave lens always produces a virtual image. Although virtual images cannot be projected, they still can be seen with the eye. This is possible because the convex lens system in the observer's eye converges the rays on the retina.

Figure 14.12 The formation of a *virtual image* by a *concave* lens, demonstrated by ray tracing. A *virtual image* is formed whenever all rays from one part of an object *appear* to originate from a common point.

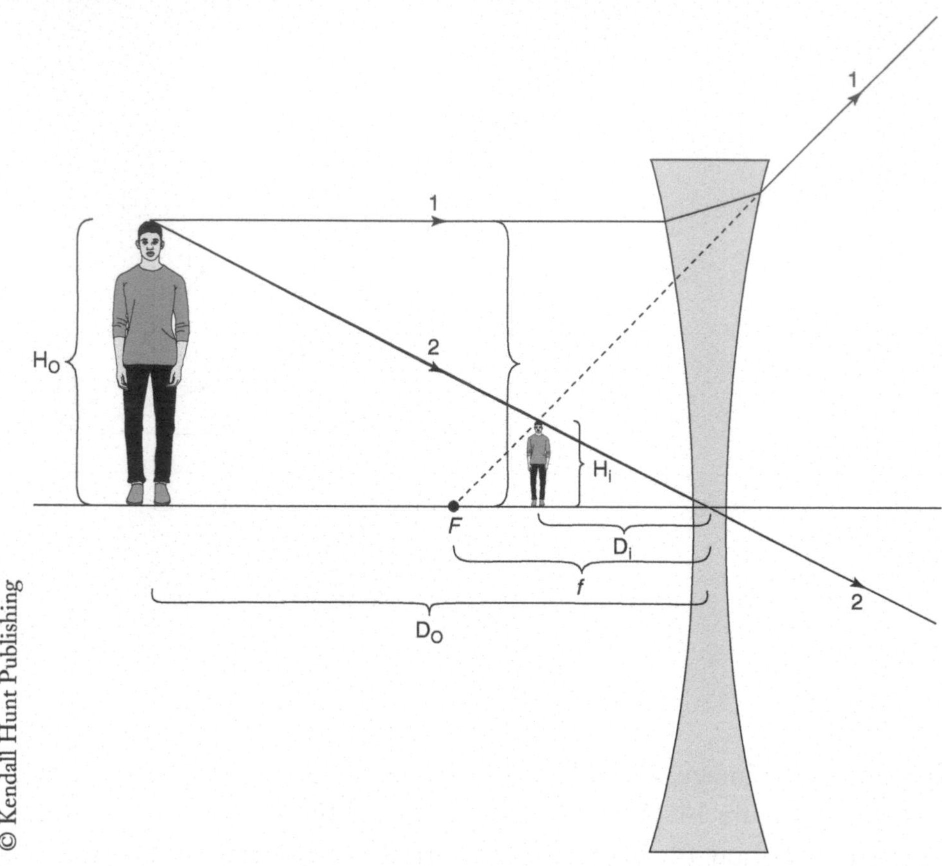

* *In all the figures, solid lines denote the actual paths of rays, and dashed lines denote the path that the ray appears to take if it differs from the actual path. See, for example, Figures 14.9* and *14.12.*

Thin lens equations ray tracing is a geometric technique (consistent with the name geometric optics), but an equivalent algebraic technique for describing optics is available. The *thin lens equations* come from an analysis of the triangles in ray tracing; we simply state them here.

Thin lens equations:

$$\frac{1}{f} = \frac{1}{D_o} + \frac{1}{D_i} \quad \textit{Know} \qquad \text{eq. 14.3}$$

"If I Do, I Di" is a little phrase that might help you remember this rule.

$$\frac{H_i}{H_o} = \frac{D_i}{D_o} \quad \textit{Know} \qquad \text{eq. 14.4}$$

Another little diddi that you might find useful is, "Hi Ho, Di Do, its off to physics we go."
Magnification M,

$$M \equiv \frac{H_i}{H_o} = \frac{D_i}{D_o} \quad \textit{Know} \qquad \text{eq. 14.4b}$$

EXAMPLE 14.2

An object 5.0 cm high is placed 24.0 cm away from a *converging lens* of focal length 8.0 cm. (a) Calculate the location of the image. (b) Calculate the height of the image. (c) Sketch the ray diagram.

Solution: The information above gives the values of H_o (height of object), D_o (distance between the object and the center of the lens, and the focal length *f*, as 5.0, 24.0, and 8.0 cm, respectively.

a. Calculate the location of the image. That means that we must find D_i.

Because we are given H_o, *f*, and D_o, we can use $\frac{1}{f} = \frac{1}{D_o} + \frac{1}{D_i}$ in order to determine D_i.

$$\frac{1}{f} = \frac{1}{D_o} + \frac{1}{D_i}, \text{ therefore } \frac{1}{D_i} = \frac{1}{f} - \frac{1}{D_o}$$

$$\frac{1}{D_i} = \frac{1}{f} - \frac{1}{D_o} = \frac{1}{8} - \frac{1}{24}$$

$$\frac{1}{D_i} = \frac{3}{24} - \frac{1}{24} = \frac{2}{24}$$

Therefore,

$$D_i = \frac{24}{2} = 12\,\text{cm}$$

b. Calculate the height of the image. In other words, find H_i

Come on and sing it with me, Hi Ho, Di Do, its off to physics we go.

$$\frac{H_i}{H_o} = \frac{D_i}{D_o}$$

$$H_i = \frac{H_o D_i}{D_o} = \frac{5\,\text{cm} \cdot 12\,\text{cm}}{24\,\text{cm}} = 2.5\,\text{cm}$$

c. Draw a ray diagram.

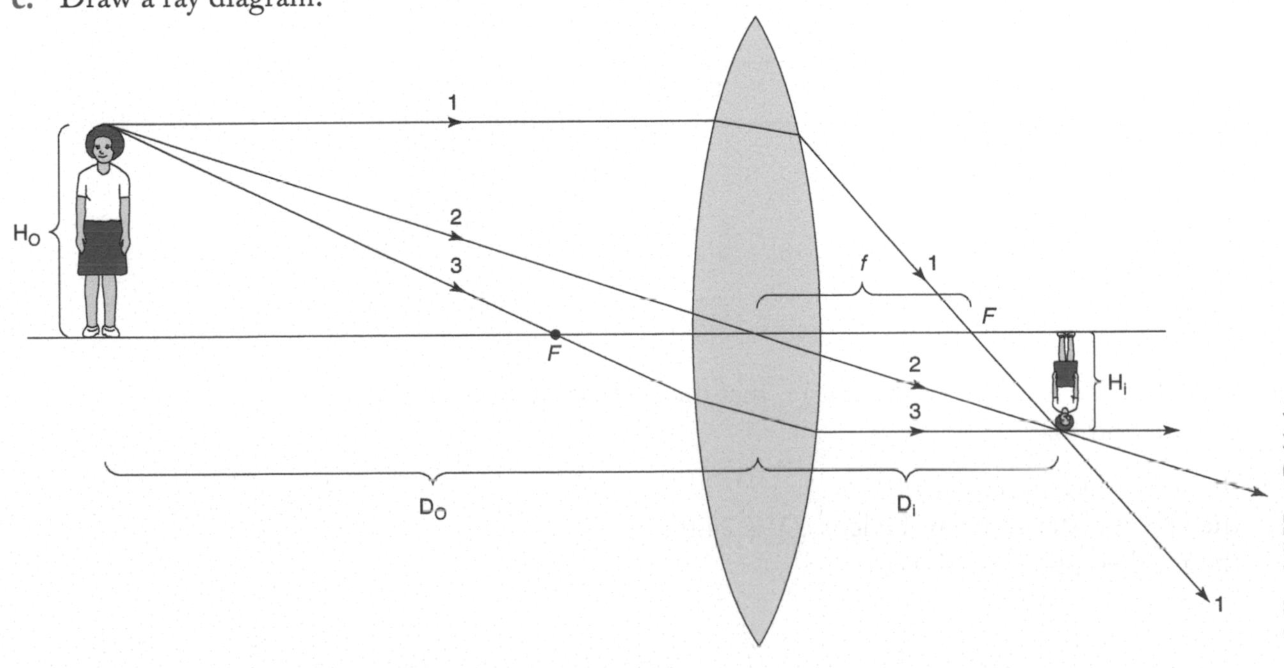

Please notice that the focal length was 8.0 cm and the object was placed 24.0 cm away from the lens. The *image will always be real and inverted when the object is placed outside the focal point of a converging lens.*

Lab 10 will help you to become familiar with these diagrams and equations and should provide you with a better understanding of the relationship between an object and the image produced by a converging lens.

EXAMPLE 14.3

An object 5.0 cm high is placed 24.0 cm away from a *concave lens* of focal length −8.0 cm. (a) Sketch the ray diagram. (b) Calculate the location of the image. (c) Calculate the height of the image.

Solution.

a. Ray 1 appears to come from the focal point. Ray 2 goes through the center of the lens and is not refracted.

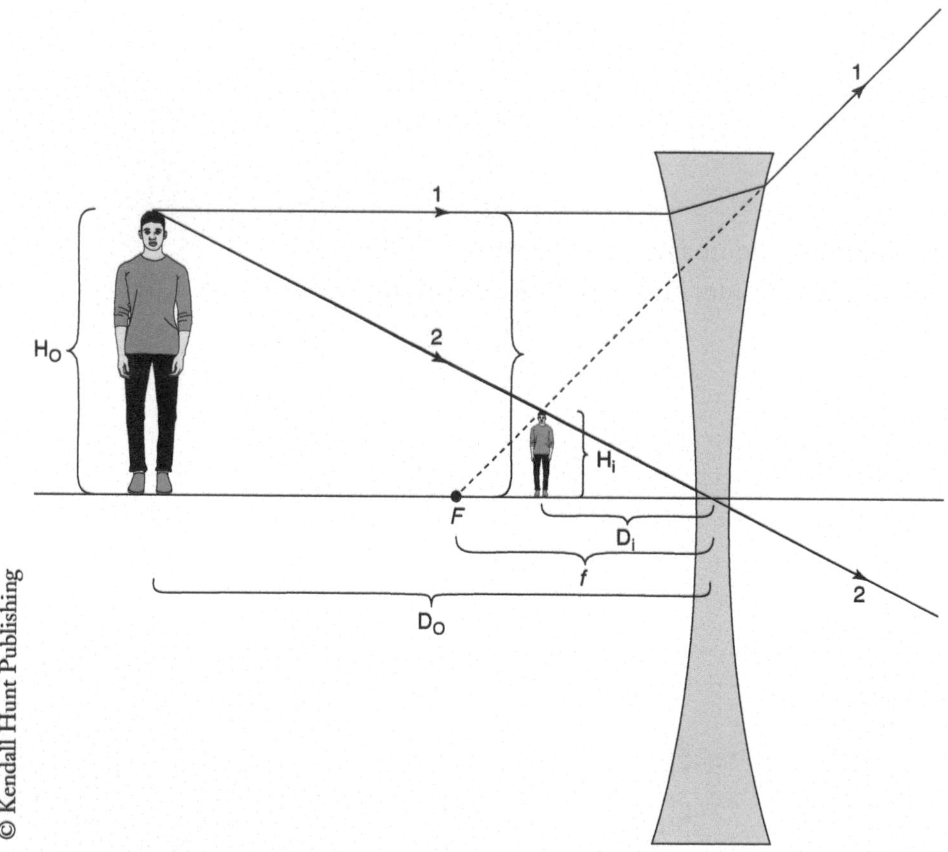

b. Calculate the location of the image. That means that we must find D_i.
We are told that:

the object height, H_o, is 5.0 cm.
the object distance from the lens, D_o is 24.0 cm.
the focal length, f, is −8.0 cm.

Because we are given H_o, f and D_o, we can use $\frac{1}{f}=\frac{1}{D_o}+\frac{1}{D_i}$ in order to determine D_i.

$$\frac{1}{f}=\frac{1}{D_o}+\frac{1}{D_i}\text{, therefore }\frac{1}{D_i}=\frac{1}{f}-\frac{1}{D_o}$$

$$\frac{1}{D_i}=\frac{1}{f}-\frac{1}{D_o}=\frac{1}{-8}-\frac{1}{24}=-\frac{3}{24}-\frac{1}{24}=-\frac{4}{24}$$

Therefore the distance of the image from the lens is $D_i=\frac{-24}{4}=-6\,\text{cm}$

The negative sign for the image distance indicates that the image is on the same side of the lens as the object and that the image is not real. It is a virtual image.

Calculate the height of the image.

$$\frac{H_i}{H_o}=\frac{D_i}{D_o}$$

$$H_i=\frac{H_o D_i}{D_o}=\frac{5\,\text{cm}(-6\,\text{cm})}{24\,\text{cm}}=-1.25\,\text{cm}$$

The negative sign for the image height indicates that the image is on the same side of the lens as the object and that the image is not real. It is a *virtual image*.

Although both the last two examples calculate only D_i and Hi, the thin lens equations can be used to calculate any of the quantities Hi, Ho, Di, Do, f, M as long as sufficient information is known about a particular situation.

Simple Magnifiers. One further example merits discussion: the simple magnifier. Any convex lens acts as a magnifier for any *object closer to the lens than its focal length. Notice that the object in* Figure 14.13 *is inside the focal point.*

Figure 14.13 The formation of a virtual image by a convex lens constitutes a magnifier. The dashed lines are the directions from which the refracted rays *appear* to originate.

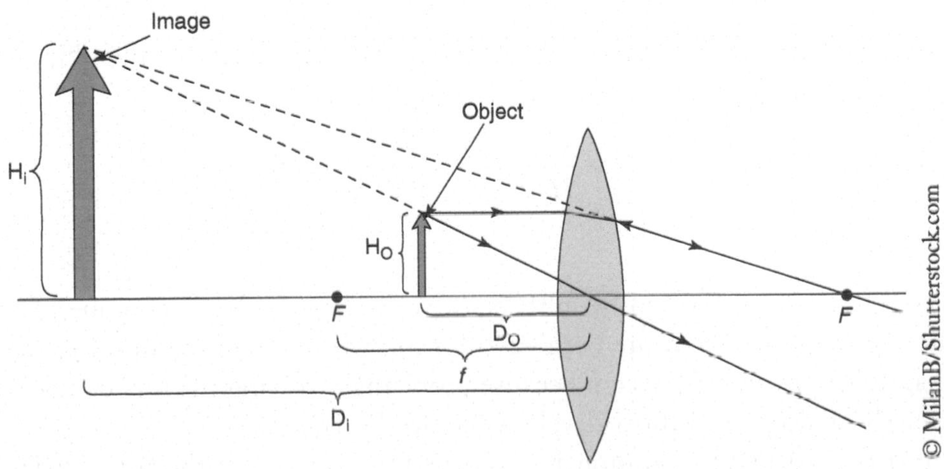

Figure 14.13 uses ray tracing to demonstrate how a magnifier works; a virtual image with a magnification greater than 1 is produced. The interested reader can try the following experiment. Take a magnifying glass and look through it at a small object. Start with the object close to the lens and see what happens to its image as you move it farther and farther away from the lens. Observe that the image becomes larger and larger as you move the object away from the lens until it becomes blurry and then appears again, but inverted. As the object is moved still farther away, the image becomes smaller and smaller and remains inverted. The object distance at which the image flips from upright to inverted is the focal length of the magnifying glass.

EXAMPLE 14.4

An object is placed 6.0 cm away from a lens of focal length 9.0 cm. Calculate its magnification M.

Solution: Magnification is defined in Equation 14.4 as the ratio of the height of the image to the height of the object:

$$M \equiv \frac{H_i}{H_o}$$

We are given the values of the object distance D_o and the focal length f.

$$D_o = 6.0\,\text{cm and } f = 9.0\,\text{cm}$$

Because the ratio of the height of the image to the height of the object is the same as the ratio of the distance of the image from the lens to the distance of the object from the lens. $\frac{H_i}{H_o} = \frac{D_i}{D_o}$, we can determine the magnification by comparing the image distance, D_i to the object distance from the lens, D_o. But there is one little problem. We don't know the value of D_i. So, first let's determine D_i.

$$\frac{1}{f} = \frac{1}{D_o} + \frac{1}{D_i} \text{ therefore } \frac{1}{D_i} = \frac{1}{f} - \frac{1}{D_o}$$

$$\frac{1}{D_i} = \frac{1}{f} - \frac{1}{D_o} = \frac{1}{9} - \frac{1}{6} = \frac{2}{18} - \frac{3}{18} = -\frac{1}{18}$$

$$D_i = -18\,\text{cm}$$

The negative means that the image is on the same side of the lens as the object and that it is a virtual image. Now, finally we can determine the magnification.

$$M \equiv \frac{H_i}{H_o} = \frac{D_i}{D_o} = \frac{18\,\text{cm}}{6.0\,\text{cm}} = 3.0$$

14.2.4 Mirrors

Real and virtual images are formed by mirrors as well as lenses. The technique of ray tracing and the use of the thin lens equations are both valid for mirrors as well as lenses. Less space is devoted to mirrors than lenses in this treatment because problems involving mirrors can be solved by using techniques exactly analogous to those developed above for lenses.

The flat mirror is extremely common. Ray tracing for mirrors uses the law of reflection. It is used in Figure 14.14 to show how an image is formed behind a flat mirror. It is upright and *not real—it is a virtual image*. The image formed by a flat mirror has the same size as the object and is located a

Figure 14.14 The formation of a virtual image by a flat mirror.

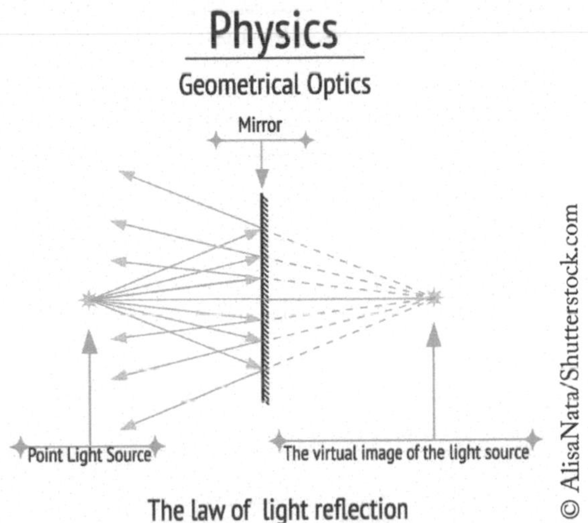

distance behind the mirror equal to the distance of the object from the mirror. Flat mirrors often are used to make small rooms look bigger because of the type of image they form.

Another way to describe the image formed by a flat mirror is to use the thin lens equations (Equations 14.3 and 14.4). The focal length of a flat mirror is infinite since it neither converges nor diverges the rays striking it. If f is infinite, then $1/f$ is zero, and from Equation 14.3, it can be seen that $D_i = -D_o$. That is, the image is the same distance from the mirror as the object. Because $\frac{D_i}{D_o} = 1$, then $\frac{H_i}{H_o} = 1$ must be equal to 1 because we all know that $\frac{H_i}{H_o} = \frac{D_i}{D_o}$, so the image has the same size as the object and is upright. This also implies that it is a virtual image.

Figure 14.15 shows how concave and convex *mirrors* affect parallel rays of light and have focal points and focal lengths analogous to those discussed earlier for lenses. *A major difference between lenses and mirrors is that a concave mirror is converging and a convex mirror is diverging*, just the opposite of lenses. The focal length of a concave mirror is therefore positive and the focal length of a convex mirror is negative. In short, the behavior of mirrors is analogous to that of lenses except that a concave mirror is analogous to a convex lens and a convex mirror is analogous to a concave lens. One example is the use of convex security mirrors in stores. They produce a small upright image similar to the one produced by the concave lens in Figure 14.12. Another example is a concave makeup mirror, which magnifies in a manner analogous to a convex magnifying lens.

Brief summary of converging lens:

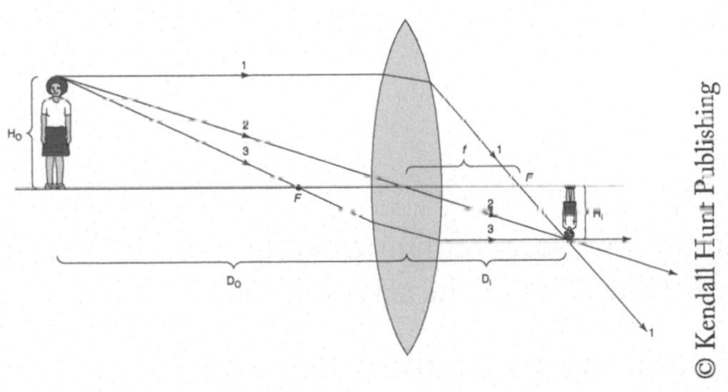

Figure 14.15 Focal points and focal lengths of *curved mirrors* determined by using ray tracing and the law of reflection.

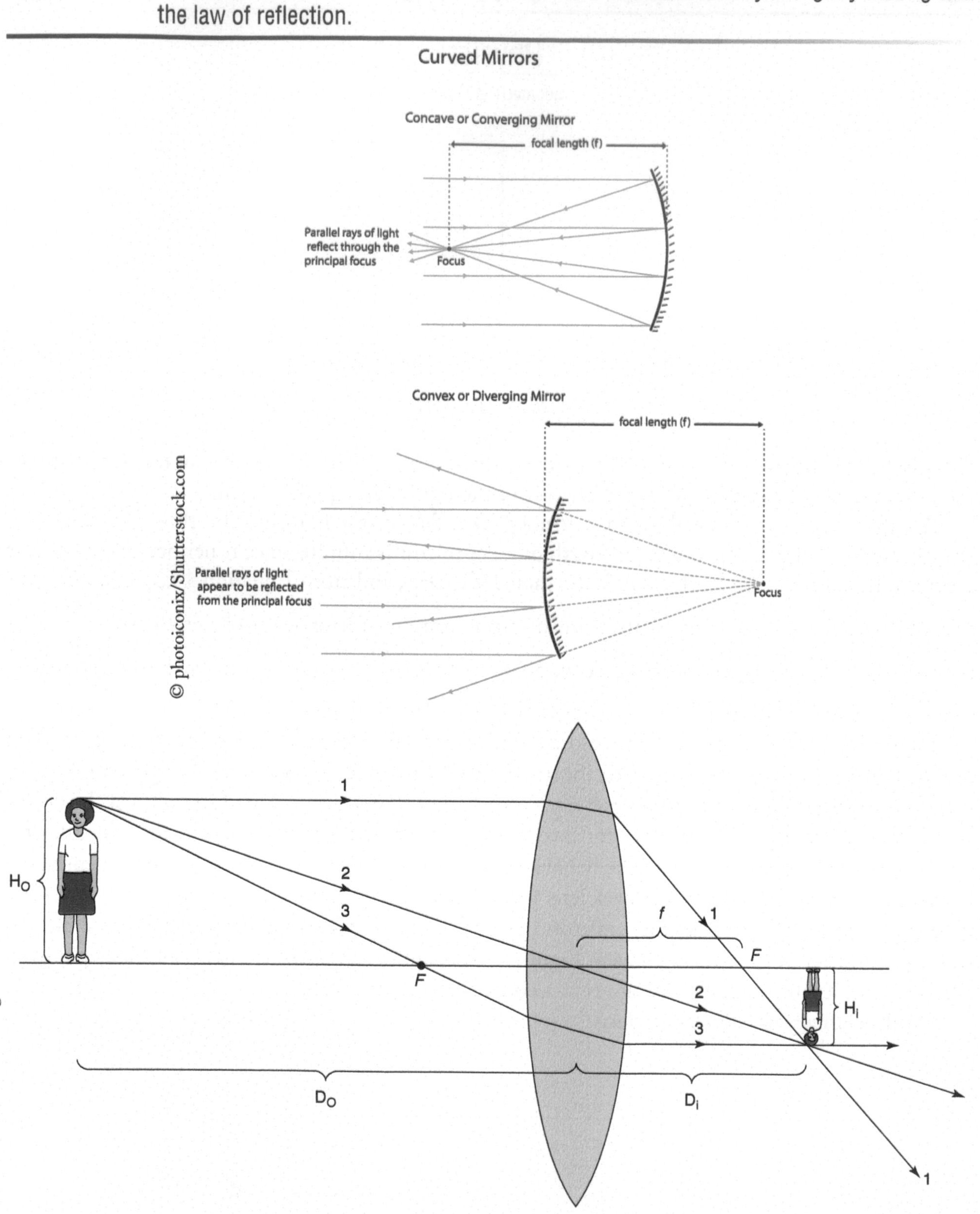

Figure 14.8 Converging lens. Rays of light entering a convex lens parallel to its axis are converged at its focal point *F*. (Ray 2 lies on the axis of this lens.)

$f \equiv$ is the distance from the focal point to the center of the lens.

The strength of a lens is $S \equiv \frac{1}{f}$ when the focal length f is in meters the units of strength are diopters.

Figure 14.11 The formation of a real image by a convex lens, demonstrated by ray tracing. A *real image* is formed whenever all rays from one part of an object actually cross at a common point.

Things to know for section 14.2

$$\frac{1}{f} = \frac{1}{D_o} + \frac{1}{D_i} \quad \frac{H_i}{H_o} = \frac{D_i}{D_o}$$

Magnification:

$$M \equiv \frac{H_i}{H_o} = \frac{D_i}{D_o}$$

Strength,

$S \equiv \frac{1}{f}$ when the focal length "f" is in meters, the units of strength are diopters.

14.3 MULTIPLE-ELEMENT SYSTEMS

Multiple-element optical systems are composed of more than a single lens or mirror and may have both lenses and mirrors in them. Such devices as microscopes, telescopes, cameras, and movie projectors are multiple-element systems. If these devices are composed of thin lenses and mirrors that obey the thin lens equations (Equations 14.3 and 14.4), then it is not difficult to analyze their behavior. The thin lens equations are simply applied to each lens or mirror in the system in succession, from the first to the last. Ray tracing can similarly be applied to multiple-element systems.

Figure 14.16 A simple compound microscope consisting of two thin lenses.

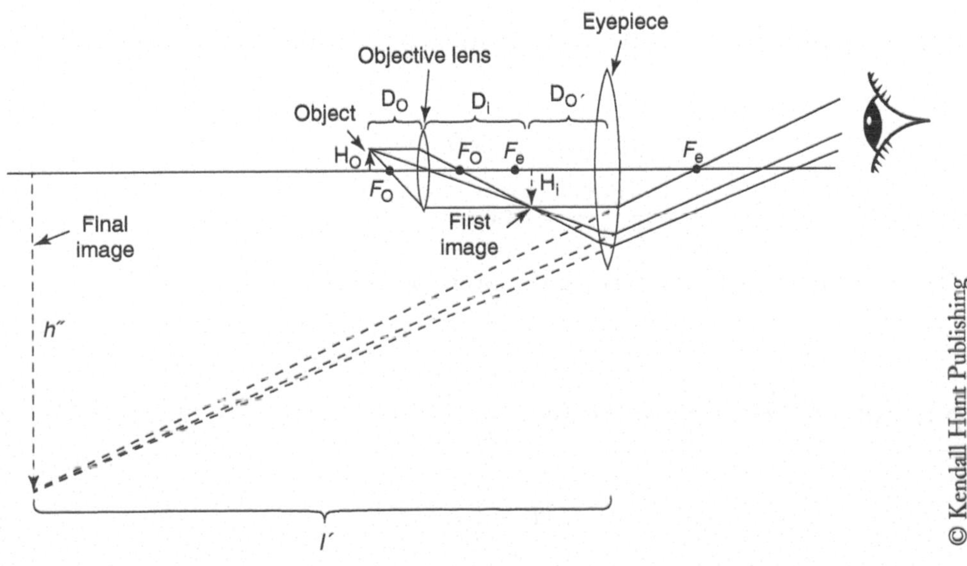

The simplest *compound microscope*, one constructed from two thin lenses, is considered here to illustrate multiple-element systems. Figure 14.16 uses ray tracing to illustrate how the microscope forms a greatly magnified image. The first lens is called an *objective*. The *object* is slightly farther away from the "objective lens" than its focal length. This produces a real image (see Table 14.2 for confirmation) that is larger than the object and is located a distance D_i to the right of the objective in Figure 14.16.

The second lens in the microscope is called the *eyepiece*. The image produced by the first lens becomes the object for the eyepiece. It is a distance D_o from the eyepiece. The eyepiece is positioned so that D_o is less than the focal length of the eyepiece. The eyepiece then acts as a simple magnifier (see Section 14.2 and Table 14.2), producing a greatly magnified virtual image a distance D_i to the left of the eyepiece. This is the final image, and is what is seen by the observer looking through the microscope.

All multiple-element systems are approached the same way. The first element produces an image. That image becomes the object for the second element of the system, which then produces another image (an image of an image). The second image becomes the object of the third element in the system, and so on until the last element of the system is considered. The problems are not so difficult as they are long. The following example considers a microscope similar to the one pictured in Figure 14.16. Because the microscope has two elements, it is really two problems in one.

EXAMPLE 14.5

A microscope similar to the one in Figure 14.16 has an *objective lens of focal length 0.60 cm* and an *eyepiece of focal length 5.0 cm*. The two lenses are 23.0 cm apart and an *object 0.010 cm in size is placed 0.62 cm away from the objective*. (a) Calculate the location, size, and magnification of the image produced by the objective. (b) Using the image produced by the objective as the object of the eyepiece, calculate the location, size, and magnification of the image produced by the eyepiece. (c) What is the overall magnification of this microscope?

Example 14.5(a_1) The location of the image is D_i. Note that the *object height, H_o, is 0.010 cm* and the object's distance from the lens, D_o, *is 0.62 cm*. The *focal length, f, of the objective lens is 0.60 cm*.

Step by step 1: First, solve for $1/D_i$ in Equation 14.3,

$$\frac{1}{f}=\frac{1}{D_o}+\frac{1}{D_i} \quad \text{Therefore} \quad \frac{1}{D_i}=\frac{1}{f}-\frac{1}{D_o}$$

Step 2: *Substitute the values for the focal length and the location of the object.*

$$\frac{1}{D_i}=\frac{1}{f}-\frac{1}{D_0}=\frac{1}{0.60}-\frac{1}{0.62}=\frac{0.0538}{\text{cm}}$$

Step 3: Solve for D_i. $D_i=\dfrac{1\,\text{cm}}{0.0538}=18.6\,\text{cm}$.

We see that the image is 18.6 cm from the objective lens.

(a_2) The size of the image. Note that we now know the image distance from the lens, D_i, equals 18.6 cm.

Step by step 1: Solve for H_i in Equation 14.4. $\dfrac{H_i}{H_o}=\dfrac{D_i}{D_o}$

$$H_i=\frac{D_i}{D_o}\cdot H_o$$

Step 2: Substitute the values.

$$H_i = \frac{D_i}{D_o} \cdot H_o = \frac{18.6\,\text{cm}}{0.62\,\text{cm}} \cdot 0.010\,\text{cm} = 0.30\,\text{cm}$$

The image is 0.30 cm in length.

(a_3) Magnification of the image.

Step by step 1: Equation 14.4b gives us two ways to determine the magnification.

$$M \equiv \frac{H_i}{H_o} = \frac{D_i}{D_o}$$

We will use the definition of magnification:

$$M \equiv \frac{H_i}{H_o} = \frac{0.30\,\text{cm}}{0.010\,\text{cm}} = 30$$

The image is 30 times the size of the object.

Example 14.5(b) Using the image produced by the objective as the object of the eyepiece, calculate the location, size, and magnification of the image produced by the eyepiece.

(b_1) Using the image produced by the objective as the *object* D_o' of the eyepiece, *calculate the location of its image,* D_i'.

We will use Equation 14.3, $\frac{1}{f} = \frac{1}{D_o} + \frac{1}{D_i}$ to determine the distance of the second image from the eyepiece. In Figure 14.16, note that the lenses are separated by 23.0 cm and the image produced by the first lens is 18.6 cm to its right, placing it 4.40 cm to the left of the eyepiece. Therefore, $D_o' = 4.40\,\text{cm}$. ***The eyepiece has a focal length of 5.00 cm.***

Solving $\frac{1}{f} = \frac{1}{D_o} + \frac{1}{D_i}$ for $1/D_i$ yields $\frac{1}{D_i} = \frac{1}{f} - \frac{1}{D_o}$

$$\frac{1}{D_i} = \frac{1}{5.00\,\text{cm}} - \frac{1}{4.40\,\text{cm}} = \frac{0.0273}{\text{cm}} \quad \text{Therefore, } D_i = 36.7\,cm$$

The final image is 36.7 cm from the eyepiece.

(b_2) Using the image produced by the objective as the object of the eyepiece, calculate size H_i', of the final image.

Solve for H_i in Equation 14.4. $\frac{H_i}{H_o} = \frac{D_i}{D_o}$

$$H_i' = \frac{D_i'}{D_o'} \cdot H_o' = \frac{36.7\,\text{cm}}{4.40\,\text{cm}} \cdot 0.30\,\text{cm} = 2.50\,\text{cm}$$

The final image is 2.50 cm in length.

(b_3) What is the magnification of the original object? $M \equiv \frac{H_i}{H_o} = \frac{2.50\,\text{cm}}{0.010\,\text{cm}} = 250$

The magnification is 250.

Not surprisingly, *the overall magnification is the product of the magnifications of the individual lenses,* a result that is true in general.

One further generalization for multiple-element systems is worth noting. If the elements are thin and close together, then they have an effective focal length that can be calculated with the following equation (stated without proof):

$$\frac{1}{f_{\text{effective}}} = \frac{1}{f_1} + \frac{1}{f_2} + \ldots \quad \text{eq. 14.5}$$

where $f_{\text{effective}}$ is the effective focal length of the system and $f_1, f_2 \ldots$ are the focal lengths of the individual elements (lenses and mirrors).

Recall that the definition of lens strength is $S \equiv \frac{1}{f}$, so Equation 14.5 can also be stated as $S_{\text{effective}} = S_1 + S_2 + \ldots$ eq. 14.6

where $S_{\text{effective}}$ is the effective strength of the system and $S_1, S_2, \ldots$ are the strengths of the individual elements. The fact that the overall strength of a system is the sum of the strengths of its elements (provided they are close together) helps in understanding lenses such as those shown in Figure 14.17. For example, even though contact lenses must be shaped to fit the wearer's cornea, they can still be either converging ($S > 0$) or diverging ($S < 0$) as needed for vision correction.

For example, recall that for distance viewing a strength of 50 D is required. Someone with a lens strength is 52.8 D can use a lens with a strength of −2.8 D to correct their vision.

$$S_{\text{effective}} = S_1 + S_2 = 52.8\text{D} + -2.8\text{D} = 50\text{D}$$

14.4 OPTICAL INSTRUMENTS

The purpose of this section is to illustrate how the principles of geometric optics are put to use in a few optical instruments. Among the instruments already discussed are the magnifying glass, flat mirrors, convex security mirrors, concave makeup mirrors, and the microscope. Eyeglasses, also mentioned briefly, will be covered more thoroughly in the chapter15.

Figure 14.17 Lenses without simple convex or concave shapes can often be thought of as combinations of closely spaced thin lenses.

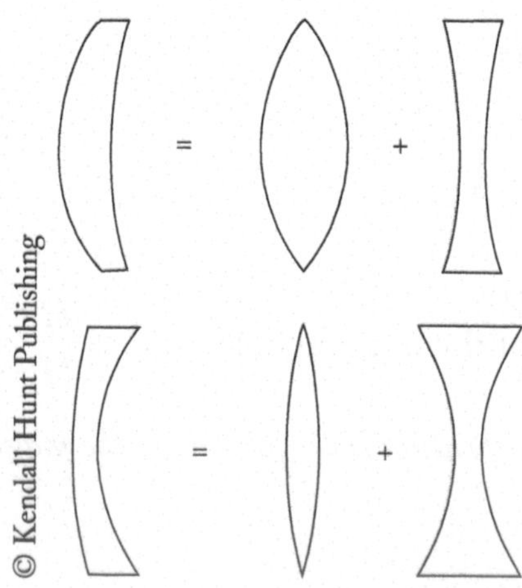

14.4.1 Fiber Optics: Endoscopes to Telephones

Tiny fibers of glass and plastic are used to transmit light for any number of purposes. Sometimes they are used to transmit telephone conversations on light emitted by lasers. In medical devices called endoscopes, one bundle of fibers carries light into the body and another carries the reflected light back out to an observer (see Figure 14.18). Some automobiles use bundles of fibers to carry light from turn signals to the dashboard to give a direct indication that the signals are working. Novelty table lamps also use plastic fibers for special lighting effects.

The fibers used in these and other applications of fiber optics can go around corners or even be tied in knots. Most of the light that enters one end of such fibers comes out of the other end, even though the sides of the fiber are transparent. Why doesn't most of the light simply leak out of the sides of the fiber? Usually, when light strikes the boundary between two substances, part of the light is reflected and part goes into the next substance. It is possible, however, to make 100% of the light reflect from the boundary. Two conditions must be met: The speed of light in the second medium must be greater than in the first, and the light must strike the boundary at an angle greater than a certain critical angle, as we shall see in a moment. These two conditions are met in fiber optics, as illustrated in Figure 14.19.

The law of refraction states that light bends away from the perpendicular when it goes into a medium where its speed increases. Depending on the two substances, there is a critical angle at which the angle of refraction is 90°—that is, light is refracted along the interface between the substances. For any larger incident angles, such as the grazing angles shown in Figure 14.19, the angle of refraction would have to be larger than 90°, meaning that the condition cannot be satisfied. Consequently, all the light is reflected and none is refracted.

14.4.2 Concentrated Light Sources

Concentrated beams, of light, as in surgical and dental lights, flashlights, and headlights, are obviously useful. One of the most common ways of producing a beam of light is to place a light bulb at

Figure 14.18 An endoscope uses fiber optics to inspect the interior of the body. The endoscope may be inserted through a bodily orifice or a small incision. It can also be used to take samples and to perform surgery.

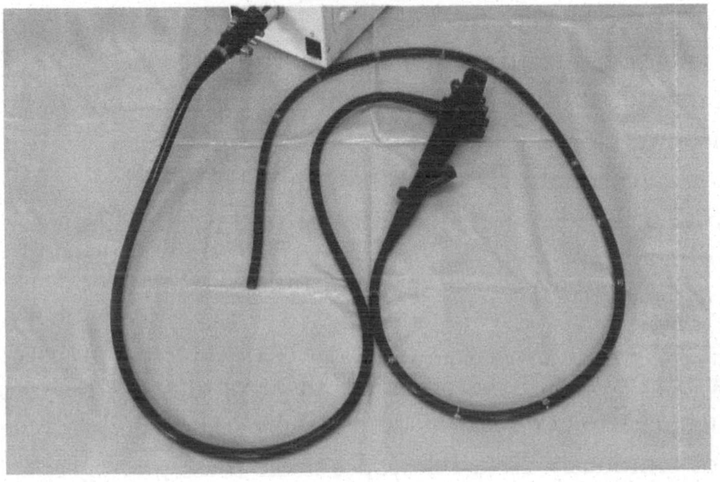

Figure 14.19 Enlarged view of a transparent fiber showing a ray making many grazing reflections as it travels within the fiber. The ray is totally reflected in each reflection.

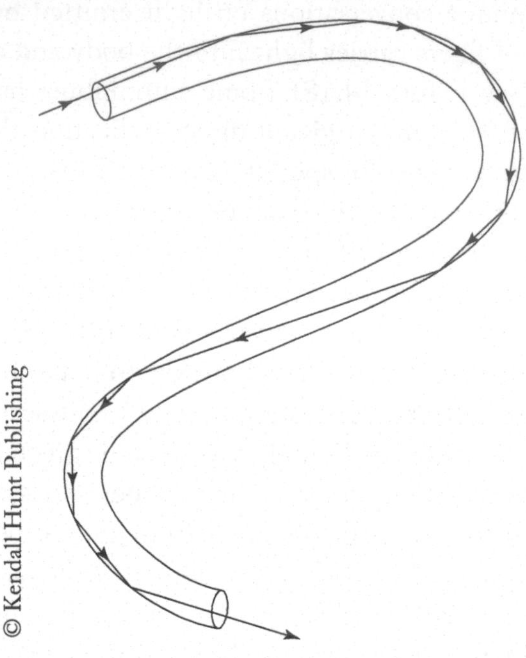

the focal point of a concave mirror, as shown in Figure 14.20. The light rays that strike the mirror follow the reverse paths shown in Figure 14.15(a). The paths of light rays are reversible; therefore, light originating at the focal point goes out parallel to the axis. The same effect can be achieved by placing the light bulb at the focal point of a convex lens. (Remember that concave mirrors and convex lenses behave analogously.)

14.4.3 Measurement of the Curvature of the Cornea

In order to make contact lenses fit properly, the curvature of the cornea must be measured. One method is to measure the size of an image reflected from the cornea, as· shown in Figure 14.21.

When light is reflected from the cornea, it acts as a convex mirror, forming a virtual upright image that is smaller than the object. If the object is held a known distance away from the cornea, then the height of the image is directly related to the curvature of the cornea (the smaller the size of the image, the greater the curvature). Note that the focal length of a mirror is determined by its curvature. If the object distance D_o is known and the height of the image, H_i, is measured, the thin lens equations can be used to calculate the focal length of the cornea and hence its curvature. Instruments designed for this purpose, called keratometers, usually do the calculations for the observer and give the corneal curvature directly.

14.4.4 Telescopes

Among the most beautiful of all photographs are those of star fields and galaxies. Hundreds of thousands of individual stars may be visible in graceful patterns and turbulent clouds of gas may be brilliantly colored or ominously dark. Telescopes make these photographs possible because they

Figure 14.20 A light bulb placed at the focal point of a concave mirror produces a concentrated beam of light. A small mirror is sometimes placed in front of the bulb as shown to get more light into the beam.

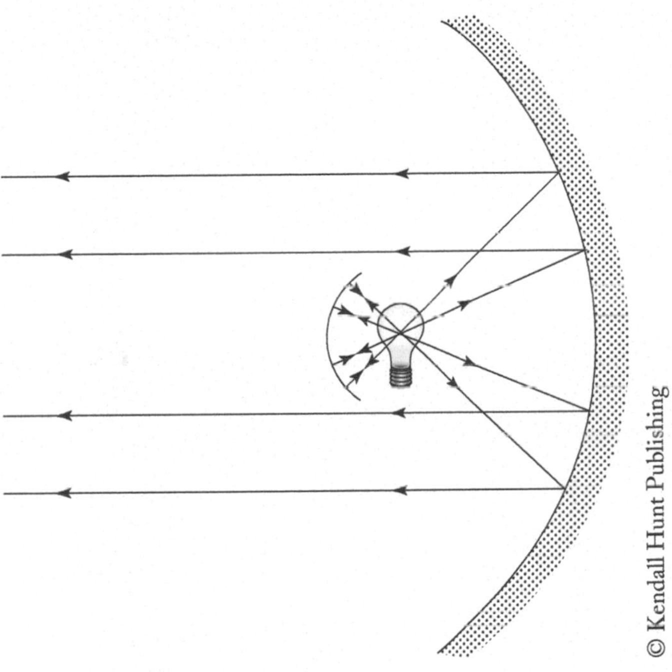

Figure 14.21 The size of the reflected image is related to the curvature of the cornea. Any convex mirror will form a virtual image as shown. F_e is the focal point of the convex mirror formed by the cornea. Instruments used to measure the curvature of the cornea are called keratometers.

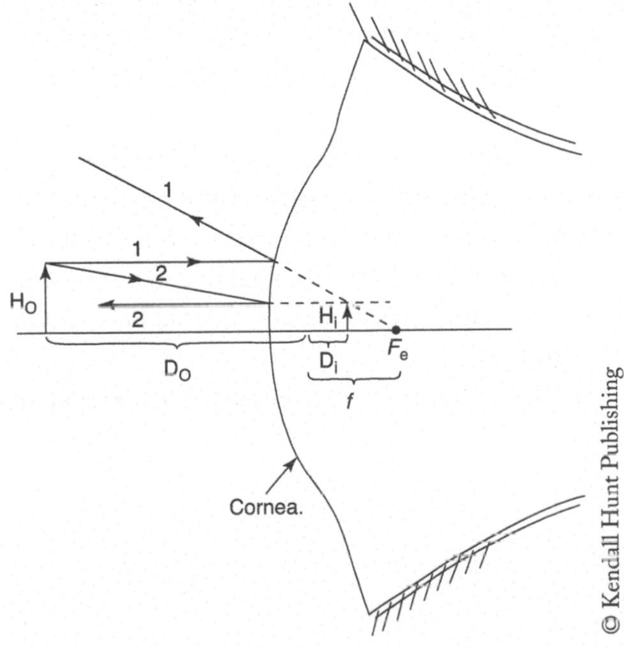

Figure 14.22 Reflecting telescope using a concave mirror as an objective and a convex lens as an eyepiece.

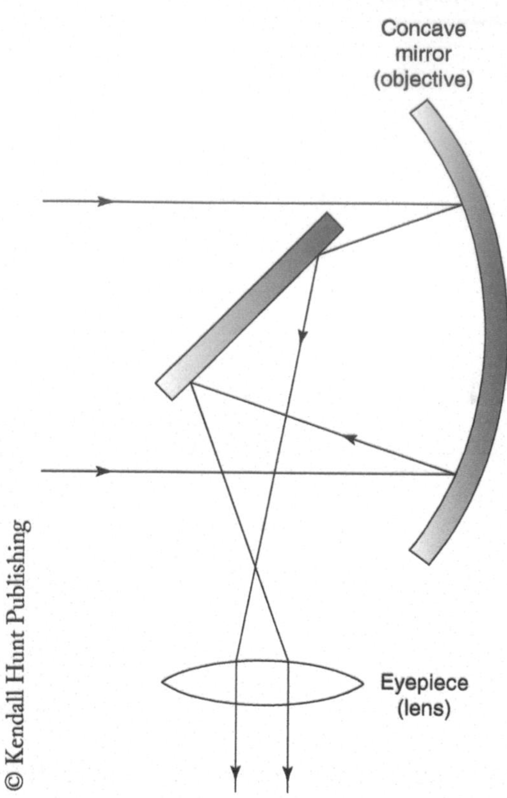

gather millions of times more light than ordinary cameras or the human eye. The largest telescopes use a concave mirror to gather light, as is shown for one type in Figure 14.22. They are called reflecting telescopes. Since light rays from distant objects are nearly parallel, an image is formed at the focal point of the mirror. A small mirror reflects the gathered light to an observer or other optical equipment, such as a camera.

Another type of telescope uses two convex lenses and is actually quite similar to the microscope in design. It is called a refracting telescope. The objective lens of such a telescope is much larger in diameter than a microscope objective so that it can gather more light. The eyepiece of the telescope acts as a magnifier of the image produced by the objective. Many small telescopes, spyglasses, and similar devices, use lenses, but all large telescopes use mirrors because they can be made larger than lenses to gather even more light.

Chapter 15 is devoted to one of the most interesting of all optical systems—the human eye.

Things to know for Ch 14.1

Speed of light in a vacuum $c = 3.0 \times 10^8$ m/s

Index of refraction, by definition: $n \equiv \frac{c}{v}$

Snell's law: $\mathrm{Sin}(\theta_1)n_1 = \mathrm{Sin}(\theta_2)n_2$

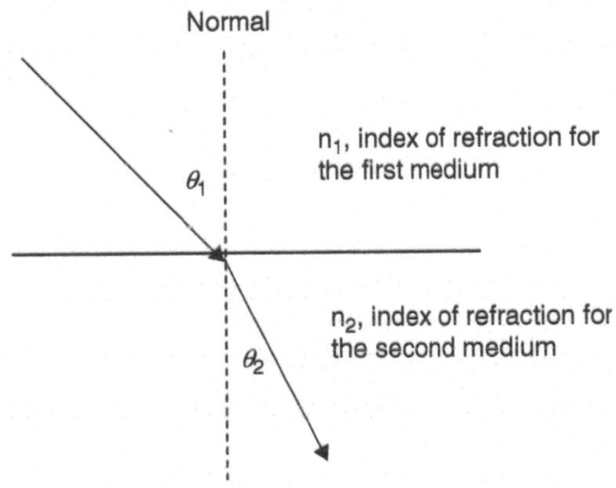

$$\frac{1}{f} = \frac{1}{D_o} + \frac{1}{D_i} \quad \frac{H_i}{H_o} = \frac{D_i}{D_o}$$

Magnification:

$$M \equiv \frac{H_i}{H_o} = \frac{D_i}{D_o}$$

Strength,

$$S = \frac{1}{f} = \frac{1}{D_o} + \frac{1}{D_i}$$

when the focal length "f" is in meters, the units of strength are diopters.

Questions

14.1 Describe an easily observed phenomenon that supports the contention that light moves in straight lines, changing directions only when it interacts with something. (One such observation is that light does not go around corners without a mirror to reflect it.)

14.2 Diffusion of light by reflection from a rough surface is described in this chapter. How might light be diffused by refraction? Describe a situation where diffuse refraction occurs, such as light interacting with crushed ice.

14.3 The speed of light increases when it goes from glass to air. Explain how this is possible when no force is exerted.

14.4 Explain why the index of refraction always is greater than or equal to 1.

14.5 A ring with a clear gemstone is dropped into water.
The gemstone becomes invisible under water. Can it be a diamond? Explain.

14.6 Will light bend toward or away from the perpendicular when it goes from air to glass? Water to glass? Diamond to water?

14.7 In view of the fact that light bends away from the perpendicular when it goes from water to air (see Figure 14.6(b), for example), explain why people's legs look very short when they wade in a pool. Justify your explanation with a ray diagram.

14.8 What happens to the strength of a glass lens when it is immersed in water? Furthermore, could a convex lens be made diverging and a concave lens converging by encasing them in a substance other than air? Explain.

14.9 A thin lens has two focal points, one on either side, equidistant from its center. If such a lens is used in a person's eyeglasses, does it matter which side of the lens is closest to the wearer's eye? Try looking through your glasses backward and comment on whether they are thin lenses. If you don't wear glasses, then experiment with a friend's.

14.10 What is the difference between real and virtual images? How can, you tell if an image formed by a single lens or mirror is real or virtual?

14.11 Can you see a virtual image? Can you photograph it? Can you project it on a screen with additional lenses or mirrors? Explain your response to each question.

14.12 Identify which rule of ray tracing governs each ray in Figures 14.11 and Figure 14.12.

14.13 Why is the front surface of a thermometer curved as shown in Figure 14.23?

14.14 Where is an image always produced at D_o, D_i, f, or F?

14.15 When you focus a camera, you are adjusting the distance between the lens and the film. Why can't the lens be a fixed distance from the film for all pictures, whether of close or distant objects? (Note: A camera lens usually has a fixed focal length.)

14.16 What is meant by a magnification less than 1? What does a negative magnification imply?

Figure 14.23 A clinical thermometer is designed to read the maximum temperature until it is shaken down.

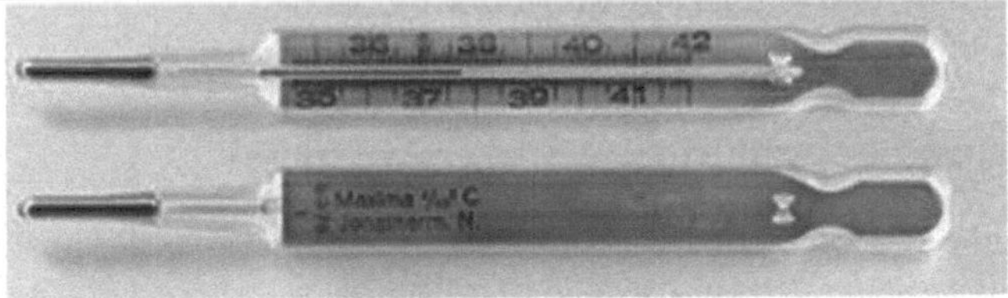

© Hemarat Studio/Shutterstock.com

14.17 Can the image produced by the microscope shown in Figure 14.16 be projected on a screen without *r* using an additional lens or mirror? Explain why or why not.

14.18 One condition that must be satisfied when applying geometric optics is that the light interacts with. macroscopic objects. Why, then, is it correct to use geometric optics to explain the operation of a microscope?

14.19 Which of the lenses in Figure 14.17 have positive strengths? Which have negative strengths?

14.20 Diamonds sparkle because light entering them gets trapped and can exit only at a few points. The trapping is due to a high probability of total internal reflection in diamonds. Why are diamonds more likely than other substances to have total internal reflections?

14.21 Some traffic lights use a lens to aim, a beam of light at the appropriate lane of traffic. How is the light bulb positioned relative to the lens to do this?

14.22 The light bulb in Figure 14.20 is positioned at the focal point of t4e large mirror. Should the bulb also be at the focal point of the small mirror? If so, why? If not, where should it be positioned relative to the small mirror?

14.23 Explain why the image in Figure 14.21 will be smaller if the cornea has a greater curvature.

14.24 Use the rules of ray tracing to explain the behavior of rays 1 and 2 in Figure 14.21.

14.25 The image of a very distant object will be at a distance $D_i = f$ from the lens or mirror. Explain why. (Note: This is the only case in which image distance equals focal length, in all other cases the image is not at a distance f.)

14.26 In view of the dimness of the incoming light from outer space, it is obvious that a large lens or mirror is useful in a telescope. For the same reason a convex lens or a concave mirror is used. Explain why.

PROBLEMS

SECTION 14.1

√14.1 (I) What is the speed of light in water? In crown glass?

√14.2 (I) Calculate the speed of light in a diamond.

√14.3 (I) Calculate the index of refraction of a liquid in which the speed of light is found to be 2.21×10^8 m/s. Identify the most likely substance using Table 14.1.

√14.4 (I) What is the index of refraction of a substance in which the speed of light is 2.0×10^8 m/s? Which of the substances in Table 14.1 is this likely to be?

√14.5 (I) How long does it take light to reach the earth from the sun 1.5×10^8 km away? Light travels at 3.0×10^8 m/s.

√14.6 (I) How long does it take light to cross a room 5.0 m wide?

√14.7 (I) What is the strength in diopters of a camera lens of 50-mm focal length?

√14.8 (I) A zoom lens on a camera has a focal length adjustable from 80 to 200 mm. What are the largest and smallest strengths of this lens in diopters?

√14.9 (I) Calculate the focal length of an eyeglass lens of strength –3.5 D.

√14.10 (I) How far from a lens of strength 25.0 D will sunlight be maximally converged?

√14.11 (I) It is possible to observe the glowing coils of a light bulb filament by projecting its image on a screen. Suppose a light bulb filament is 15.0 cm from a lens of 12.0-cm focal length. (a) Where is its image located? (b) What is its magnification?

√14.12 (I) A camera with a 50-mm focal length lens is being used to photograph a person 1.75 m tall standing 2.0 m away. (a) How far from the lens must the film be in order for the image to be in focus? (b) What is the height of the person's image on the film?

14.13 (I) A slide projector uses a lens of 10.0-cm focal length. A 35-mm slide (actual dimensions 24 mm × 36 mm) is placed 10.3 cm from the lens. (a) How far from the lens must the screen be in order to display a clear image? (b) What will be the dimensions of the picture on the screen?

14.14 (I) A doctor is examining a mole using a magnifying glass of focal length 15.0 cm. The lens is held 13.5 cm from the mole. (a) Calculate the position of the image formed. (b) What is the magnification? (c) What is the height of the image if the mole is 0.50 cm in diameter?

14.15 (II) Using the law of reflection and ray tracing, show that a flat mirror causes rays of light neither to converge nor to diverge. It is easiest to show this if the incoming rays are parallel.

14.16 (II) A lens for taking close-up photographs has a focal length of 40 mm. The farthest the lens can be from the film in the camera is 60 mm. (a) What is the closest object that can be photographed? (b) What will the magnification be for the closest object?

14.17 (II) When focusing a camera, you adjust the distance between the lens and the film so that it equals the image distance. Suppose your camera has a lens of focal length 50.0 mm and the distance between the film and the lens is 51.0 mm. (a) How far away is the object being photographed? (b) If the image has a height of 2.0 cm, what is the height of the object?

14.18 (II) The eye projects a real image on the retina 2.00 cm from its lens system. A person is looking at a poster 2.0 m away. (a) Calculate the focal length of her eye. (b) Calculate its strength in diopters. (c) If the poster is 0.75 m high, what is the height of its image on the retina?

14.19 (II) A magnifier is being used to examine a rare butterfly being held 5.0 cm from the lens. If the butterfly is magnified by 3.0, what is the focal length of the lens? (Note that you must first find the distance to its image.)

14.20 (II) Suppose a person with normal eyes looks through one lens of a nearsighted person's eyeglasses at an object 25 cm on the other side of the lens. If the strength of the lens is –4.0 D, what is the magnification of the object? (Note that the distance to the image must be found before the magnification can be found.)

14.21 (II) A woman using a magnifying makeup mirror holds it 15 cm from her face and gets a magnification of 1.5. Calculate the focal length of the mirror. Note that you must first find the distance to the image.

14.22 (II) A shopper in a drugstore looks into a convex security mirror 3.0 m away and sees an image of himself that is 0.25 his height (that is, $M = 0.25$). (a) Find the location of his image. (b) What is the focal length of the mirror?

14.23 (III) A telephoto lens on a camera has a focal length of 200 mm. Suppose this lens is used to take a picture of a mountain range 10 km away. (a) Where is the image produced by this lens? (b) What is the height of the image of one mountain's 1.0-km-high cliff?

14.24 (III) A camera lens of 100-mm focal length is used to take a picture of the sun and the moon. (a) What is the height of the image of the sun on the film given that it is 1.4×10^6 km in diameter and 1.50×10^8 km away? (b) What is the height of the image of the moon, given that the moon is 3500 km in diameter and 3.84×10^5 km away? (c) What is the relevance of the answers to eclipses of the sun by the moon?

14.25 (III) A dentist would like to use a mirror that gives an upright image of a tooth with a magnification of 6.0 when held 2.0 cm from the tooth. Calculate the focal length of such a mirror.

SECTIONS 14.3 AND 14.4

14.26 (I) A nearsighted person has eyeglasses with a strength of –2.0 D, and is having difficulty seeing distant objects clearly. When a lens of strength –0.50 D is held just in front of his glasses, he finds that his distant vision is clear. What strength should his new eyeglasses have? (This is similar to how eyeglass prescriptions are actually determined.)

14.27 (I) A farsighted person has eyeglasses with a strength of 0.75 D, and is having difficulty reading close up. When a lens of strength 0.50 D is held just in front of her eyeglasses, she finds that reading at a normal distance is possible. What strength should her new eyeglasses have?

14.28 (II) As discussed in Section 14.4 and illustrated in Figure 14.21, an object 1.5 cm high is held 3.0 cm from the cornea and the height of its reflected image is measured to be 0.167 cm. (a) Calculate the magnification. (b) Calculate the location of the image. (You may assume that 3.0 cm is the object distance.) (c) What is the focal length of the convex mirror formed by the cornea? (For a sphere this is half the radius of curvature.)

14.29 (II) An amoeba is 0.305 cm away from the objective lens of a microscope. (a) If the objective lens has a focal length of 0.300 cm, where is the image formed by the objective? (b) What is the magnification of this image? (c) The eyepiece of the microscope has a focal length of 2.00 cm and is 20.0 cm away from the objective. Where is the final image? (d) What is the magnification produced by the eyepiece? (e) What is the overall magnification?

14.30 (III) A small reflecting telescope has a focal length of 1.0 m. (a) What magnification does it produce when photographing the sun, 1.50×10^8 km away? Note that the magnification should be very small since the object being photographed is very large. (b) How large is the image of a sunspot 25,000 km in diameter?

15 Vision

"An investment in knowledge pays the best interest."

—Benjamin Franklin

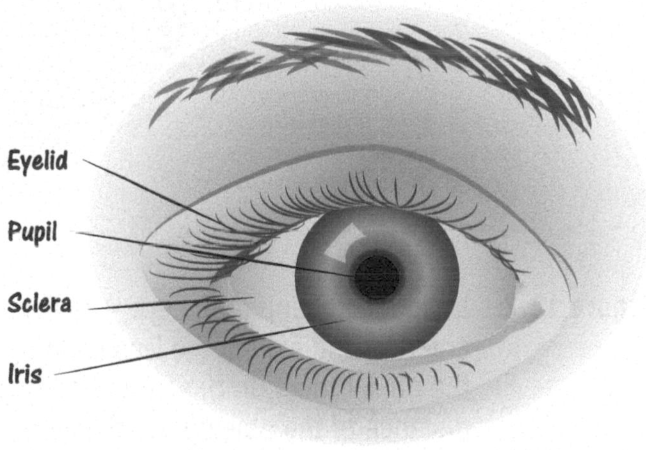

Vision is the perception of light through the eye-brain system—a rather dry definition for the most amazing and complex of our senses. Three major disciplines are involved in any complete description of vision. Physics describes light and how an image is formed on the retina. Physiology describes how the image is processed by the eye-brain system. Perceptual psychology deals with higher-level processing of the information and subsequent formation of a perception. This chapter concentrates on the physics of image formation and vision correction, including some elementary physiology of the eye. The last section of this chapter, on color vision, introduces the subject of color and gives an indication of how complex the sense of vision is.

15.1 IMAGE FORMATION BY THE EYE

The eye uses a convex lens system to project a real image on the retina. Figure 15.1 shows the anatomy of the human eye. Both the cornea, with an index of refraction of 1.38 and the lens of the eye act as convex lenses, projecting a real image on the light-sensitive retina. The center of the image falls on the most sensitive part of the retina, a small region called the fovea. The opening in the iris is the pupil of the eye. It automatically adjusts to the amount of light entering the eye.

The pupil plus chemical adaptation allow the eye to function over an intensity range of 10^{10}:1. The two eyes together provide depth perception. The eyes also sense the direction in which they point, movements, and the color of objects and light sources.

The ray diagram in Figure 15.2 shows image formation by the cornea and lens. The cornea does roughly two-thirds of the bending of rays. The lens does the remaining fine tuning necessary to produce an image on the retina.

Figure 15.1 Cross-sectional anatomy of the human eye seen from above.

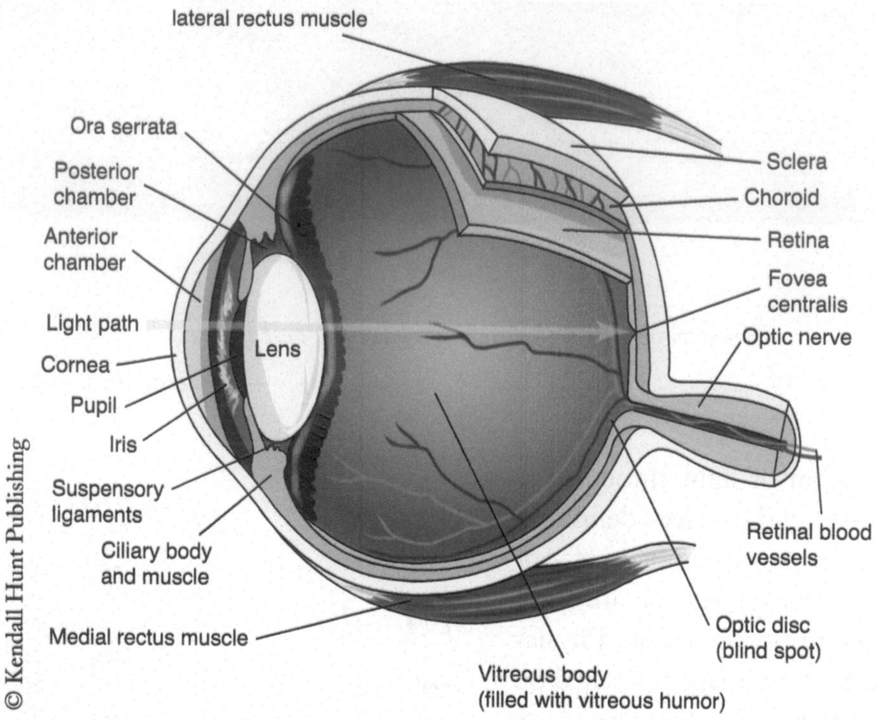

Figure 15.2 Image formation in the eye. The light rays bend at the surface of the cornea and upon entering and exiting the lens an image is formed where the rays from the same part of the object meet. In this case the image lies on the retina, producing clear vision. The distance of the object from the eye is not drawn to scale.

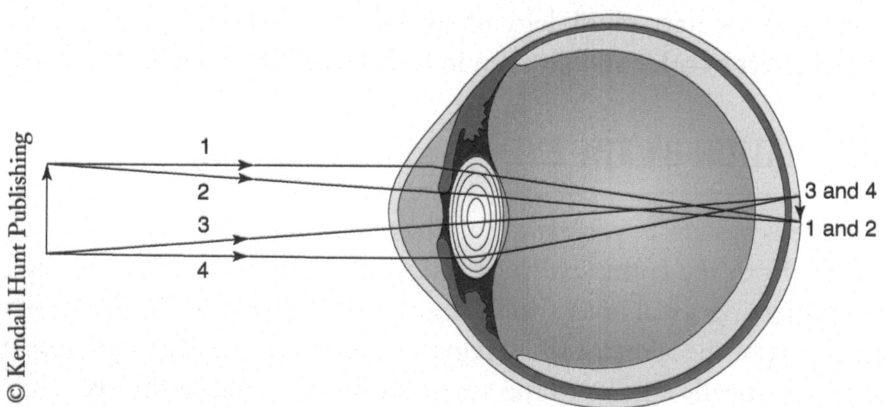

Light rays pass through many layers of materials (air, cornea, aqueous humor, lens periphery, lens core, lens periphery, and vitreous humor) on their way to the retina, changing direction at each interface. However, the indices of refraction of the materials involved are such that significant bending occurs only at the cornea and lens. Furthermore, the cornea and lens can be treated as a single lens with a strength equal to the sum of their individual strengths, a tremendous simplification. (See the discussion at end of Section 14.3.)

15.1.1 Accommodation

For an object to be seen clearly, its image must lie precisely on the fovea in the retina. In slightly more quantitative terms this means that the image distance D_i must equal the lens-to-retina distance for clear vision. Because the lens-to-retina distance in the eye does not change, the image distance D_i must be the same for all objects at all distances. The only way the eye can do this is to vary its focal length, and hence its strength, to accommodate various object distances. The process is thus called *accommodation*. Reminder, we will use D_i = 2 cm for all of our eye problems.

Figure 15.3 illustrates how the strength of the eye varies for distant and close objects. Rays from a close object can diverge and still enter the eye, as seen in Figure 15.3(a). The eye must therefore be more converging to cause the rays to meet on the retina when viewing a close object. Rays of light from a distant object, as shown in Figure 15.3(b), are nearly parallel, and the eye is easily able to converge them on the retina. A more-converging lens has greater strength, as indicated by the thicker lens for close vision. Muscle action in the eye adjusts the thickness and strength of the lens, doing the fine tuning necessary to produce an image precisely on the retina. The muscle is relaxed for a thin lens and tensed for a thick lens. Distant vision is thus termed totally relaxed and close vision is termed *accommodated.*

The fact that the strength of the eye must vary can be more rigorously demonstrated by examining the thin lens equations (14.3 and 14.4), which apply reasonably well to the eye and are repeated here for convenience:

The strength of the eye, S is defined as,

$$S \equiv \frac{1}{f}$$

Figure 15.3 Accommodation of the eye lens for **(a)** Close vision and **(b)** distance vision.

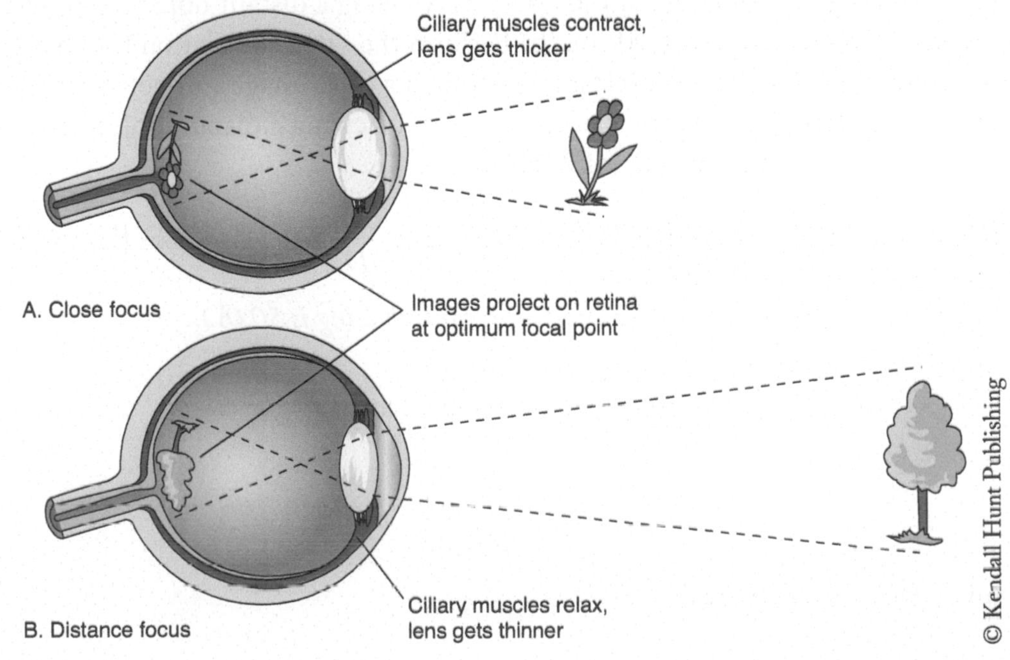

We also know that

$$\frac{1}{f}=\frac{1}{D_o}+\frac{1}{D_i} \quad \textit{(If I do, I die)}$$

And who can forget, hi ho, di do, its off to physics we go:

$$\frac{H_i}{H_o}=\frac{D_i}{D_o}$$

Magnification is defined as,

$$M \equiv \frac{H_i}{H_o}$$

As before, D_o is the object distance, D_i the image distance from the lens, f the focal length, S the strength, H_o the object height, H_i the image height, and M the magnification. In equation 15.1, if D_i is fixed and D_o varies, then f and hence S must vary.

To give some consistency to calculations for vision, certain average values are adopted. *The adult lens-to-retina distance is taken to be 2.00 cm (0.02 m).* When we want strength in diopters, we must enter values for D_o and D_i in meters. Therefore in problems involving the eye, D_i = 0.02 m. The normal range of clear vision is taken to be from a close point of 25 cm for reading, to a far point of infinity—that is, for comfortable reading, D_o = 25 cm, to gazing at stars with D_o = infinity.

In the following example the strength of a normal eye is calculated for distant and close vision.

EXAMPLE 15.1

(a) Calculate the strength in diopters of the eye when viewing a distant object. Suppose the eye has a normal lens-to-retina distance of 2.00 cm. (b) Perform the same calculation for an eye viewing an object 25 cm away, the normal closest distance for which a clear image can be obtained.

(a) Calculate the strength in diopters of the eye when viewing a distant object. Assume that the eye has a normal lens-to-retina distance of 2.00 cm.

Solution. (a): By distant we mean that D_o is so large, that $\frac{1}{D_o}=0$. Also note that we must convert 2.00 cm to meters. 2.00 cm = 0.0200 m.

Therefore, the *strength* of the human eye, for *distant viewing is 50.0*D.

$$S \equiv \frac{1}{f}$$ Now remember that

$$\frac{1}{f}=\frac{1}{D_o}+\frac{1}{D_i}$$

Combining the two equations:

$$S \equiv \frac{1}{f}=\frac{1}{D_o}+\frac{1}{D_i}=0+\frac{1}{.0200}=50.0\,\mathrm{D}$$

Therefore, for distance vision, the normal eye has a strength of approximately 50.0 diopter (D). This is the situation shown in Figure 15.3b.

(b) Calculate the strength of a human eye viewing an object 25 cm away, the average closest distance for which a clear image can be obtained.

Solution. (b): ***Don't forget to convert 25 cm to meters.***

$$S \equiv \frac{1}{f} = \frac{1}{D_o} + \frac{1}{D_i} = \frac{1}{0.250} + \frac{1}{.0200} = 54.0\,\text{D}$$

Therefore, for near vision, the normal eye has a strength of approximately 54.0 D when fully accommodated, (viewing an object only 25 cm from the eye). This is the situation shown in Figure 15.3a.

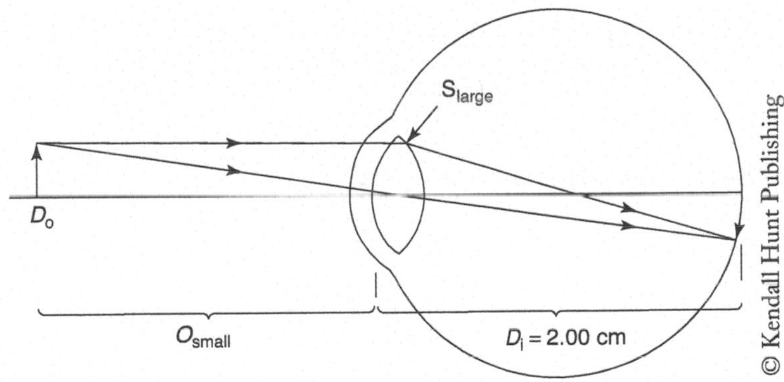

When the eye looks at an object somewhere between infinity and 25 cm, its strength will lie somewhere between 50 and 54 D. The closer the object, the larger the strength.

The maximum strength of 54 D is only 8% greater than the minimum strength of 50 D; this is the fine tuning done by the lens of the eye. The ability to adjust strengths from 50 to 54 D is called a normal accommodation ability of 8%. Accommodation ability is usually better than 8% when a person is younger than 40, but it decreases steadily with age.

15.1.2 Acuity

How fine a detail can the normal eye discern? And what prevents the eye from resolving even smaller detail? The answers to these questions give some insight into how sophisticated an instrument the eye is. Visual acuity is the sharpness of detail detectable with the eye—a property physicists call optical resolution. The human eye can resolve points with a minimum angular separation of 3×10^{-4} radians = 1.0 minute of arc. Vision tests, such as reading lines of letters on a chart, assume a certain acuity is normal. A patient's eyes are tested against this norm to ascertain whether vision correction is called for. Someone with 20/20 is just able to decipher an object (letter) that subtends an angle of 1 minute of arc at the eye. 20/20 is considered the norm but it is not perfect. Young adults have a range of 20/16 to 20/12. It is said that Ted Williams, the famous baseball player, had an acuity of 20/10. By comparison, a hawk has 20/2 vision.

The various methods of measuring visual acuity yield somewhat different results. They are all roughly consistent, however, and correspond to the ability to resolve detail as small as the diameter of a human hair held at arm's length. This is impressive in itself, but it is even more impressive when the sizes of images on the retina are considered, as in the following example.

EXAMPLE 15.2

(a) What is the height of the image on the retina of a dot 0.0120 cm in diameter held 60 cm from an eye with a 2.00-cm lens-to-retina distance? (This dot is the size of a human hair held at arm's length.) (b) What is the height of the image on the retina of a physics instructor 1.8 m tall standing 10 m away from a student in a lecture hall? The lens-to-retina distance of the student is 2.00 cm.

Solution. (a): We are given the following values: $D_i = 2.00$ cm; $D_o = 60$ cm; $H_o = 0.0120$ cm.

$$\frac{H_i}{H_o} = \frac{D_i}{D_o} \text{ Solve for } H_i.\ H_i = \frac{D_i}{D_o} H_o$$

$$H_i = \frac{D_i}{D_o} H_o = \frac{2.00\,\text{cm}}{60.0\,\text{cm}} 0.0120\,\text{cm} = 4.00 \times 10^{-4}\,\text{cm}$$

That is only 1/1000 of 1 mm. It is tiny, tiny, tiny.

Solution. (b):

$$H_i = \frac{D_i}{D_o} H_o = \frac{2.00\,\text{cm}}{10.0\,\text{m}} \cdot 1.8\,\text{m} = 0.36\,\text{cm}$$

That is only 3.6 millimeter. Still pretty small.

As shown in the first part of the preceding example, the smallest discernible detail makes an image about 4.00×10^{-4} m in diameter. This is two or three times the size of retinal cells. Outside the fovea, light-sensitive cells are not as closely spaced, and peripheral visual acuity is consequently less by a factor of about 10. The eye cannot detect detail that creates an image smaller than the spacing of light-sensitive cells. In fact, in order for the eye to resolve two adjacent dots, their images must fall on two nonadjacent cells, as shown in Figure 15.4. If the two dots are images of small light sources, for example, and fall on adjacent cells, then they are interpreted as a single larger dot. The fact that cell B in Figure 15.4 gets less light than its neighbors tells the eye—brain system that there are two images instead of a single larger one. In short, one limit to visual acuity is the size and spacing of light sensitive cells on the retina.

Are there other limits to visual acuity? There is one other limit, called the diffraction limit, caused by the wave nature of light. Diffraction is the tendency of waves to spread out after going through an opening or to bend around an obstacle. The amount of diffraction depends on the relative sizes of the waves and the opening or obstacle: the smaller the opening or obstacle, the greater the tendency of waves to spread out. This is why geometric optics limits itself to the interaction of light with macroscopic objects—so that diffraction effects are negligible.

In the eye, diffraction occurs because light passes through the pupil. The diameter of the pupil varies with light intensity, averaging about 3 mm. Ordinarily 3 mm is large enough that diffraction effects are negligible, but when we are discussing the limits of visual acuity, even small effects become important. The diffraction of light passing through an opening is illustrated in Figure 15.5. The effect in the eye is such that a point source of light will not make a point image on the retina; it will make

Figure 15.4 Position of two just resolvable images on the mosaic of retinal cells. The distance between the centers of the two images must be about two times greater than the size of the cells, so that an intervening cell B receives less light than cells A and C.

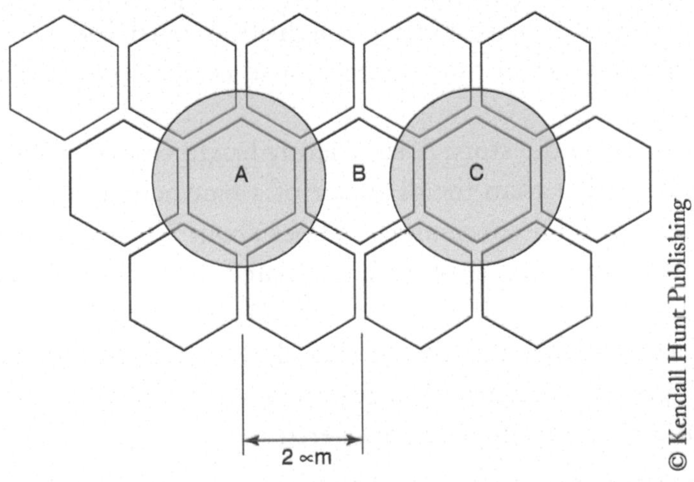

Figure 15.5 Diffraction: **(a)** Light passes through a circular opening much larger than its wavelength, casting a clean circular spot of light on the screen. **(b)** The spot on the screen is smaller for a smaller opening, **(c)** If the opening is very small, diffraction is noticeable. Light spreads out producing a fuzzy spot that is brightest at its center. Diffraction effects from light passing through the pupil of the eye are one limit to visual acuity.

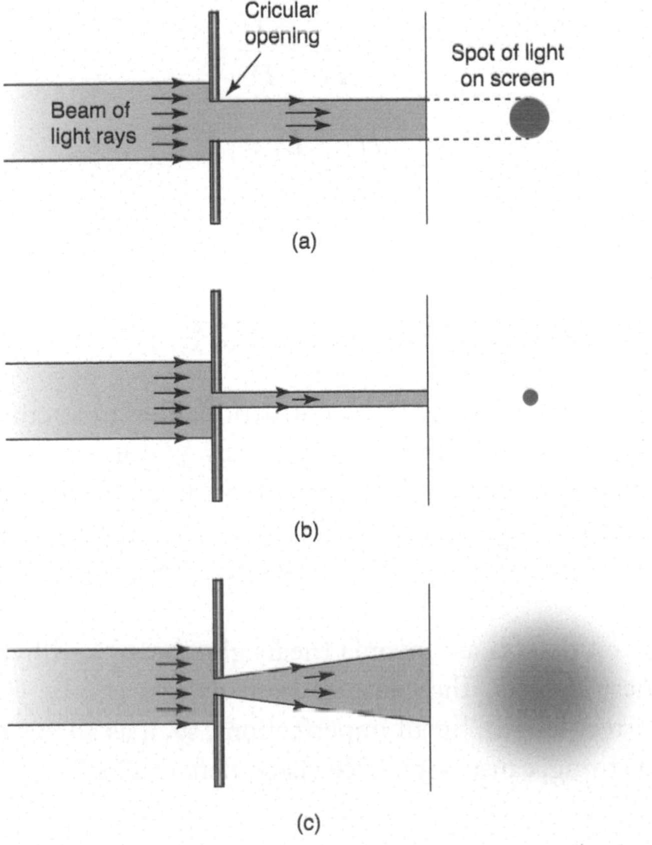

a spot about 2 μm in diameter, very nearly the size of a retinal cell. It is very likely that the evolution of retinal cells to their present size is related to the diffraction limit. The wave nature of light will be discussed in more detail in Chapter 16.

The eye has mechanisms that apparently eliminate other possible limits to visual acuity, such as nonuniformities in the lens and the scattering of light from particles in the eye. None of the other possible limits to visual acuity seem to be important in comparison to the limits imposed by cell spacing and diffraction.

As always, there is more to the story. A sufficiently bright source of light on a dark background can be seen even if it is smaller than the diameter of a hair held at arm's length. It is still able to stimulate a light-sensitive cell. Discontinuities in lines about 30 times smaller than the above limit to visual acuity can be observed because of the coordinated effort of many cells and processing in the retina and brain.

Two Americans, David Hubel and Tosten Weisel, won the, 1981 Nobel prize in medicine for their work on this and related problems of image processing by the eye-brain system, but many questions about the exact mechanisms remain unanswered.

Things to know for section 15.1

Strength of a human eye:

$$S \equiv \frac{1}{f} = \frac{1}{D_o} + \frac{1}{D_i} = \frac{1}{D_o} + \frac{1}{.0200\,\text{m}}$$

Thin lens equations:

$$\frac{1}{f} = \frac{1}{D_o} + \frac{1}{D_i}$$

$$\frac{H_i}{H_o} = \frac{D_i}{D_o}$$

15.2 CORRECTION OF COMMON VISION DEFECTS

A majority of people need some kind of vision correction. The most common vision defects are simple in nature and easily corrected with spectacle lenses. Vision correction has been performed for many centuries. Spectacle lenses are known to have been used as early as the fourteenth century, and cataract removal was practiced by the Romans and possibly in earlier civilizations. Only in the last hundred years or so has vision correction become available to other than the very rich.

The most common vision defects are myopia (nearsightedness) and hyperopia (farsightedness). Figure 15.6 illustrates these defects. The cause of both defects can be either a lens of incorrect strength or an eye of incorrect length. Small imperfections, such as an eye that is 0.01 cm too long or a lens that is 0.5 D too strong, cause noticeable vision deficiencies.

Figure 15.6 The myopic eye converges rays from a distant object before they strike the retina. This person can read without glasses but needs correction for viewing distant object. They are said to be near-sighted. The hyperopic lens is unable to converge rays from a close object onto the retina. This person may be able to view distant objects without correction but needs reading glasses. This person is said to be far-sighted.

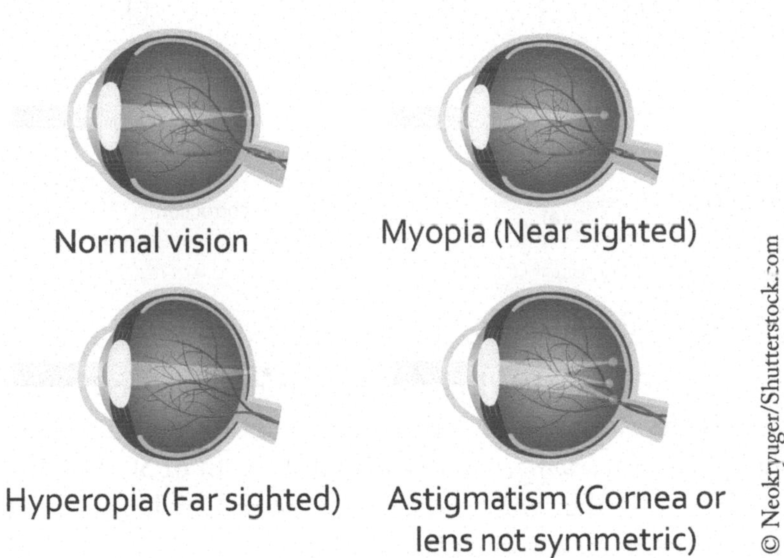

15.2.1 Myopia (Near sighted)

A myopic person can see close objects clearly, but distant vision is blurry—hence the common term nearsightedness. The strength of their lens is greater than 50.0 D. As shown in Figure 15.6, parallel rays from a distant object converge before they meet the retina and are diverging again at the retina. The myopic eye cannot relax its lens enough to avoid over converging rays from a distant object. If the eye is a normal length (lens-to-retina distance of 2.00 cm), this means that the strength of the fully relaxed eye is greater than 50.0 D [see Example 15.1(a)]. To correct for myopia, a diverging spectacle lens of negative strength is used, as shown in Figure 15.7(a).

15.2.2 Hyperopia (far sighted)

A hyperopic person can see distant objects clearly, but close objects are blurry—hence the common term farsightedness. As shown in Figure 15.6(b), rays from a close object do not converge by the time they strike the retina. The lens of the hyperopic eye has insufficient strength to converge rays from a close object on the retina, even when fully accommodated. For an eye of normal length this means that the maximum strength of the eye is less than the needed 54.0 D required for reading. [see Example 15.1(b)]. To correct for hyperopia, a converging spectacle lens of positive strength is used, as shown in Figure 15.7(b).

Figure 15.7 **(a)** Correction for myopia: A diverging lens is used since the eye is too converging. **(b)** Correction for hyperopia: A converging lens is used since the eye is not converging enough. The heavy lines are the paths that the rays take with a spectacle lens in place; the thinner lines are the paths without a spectacle lens.

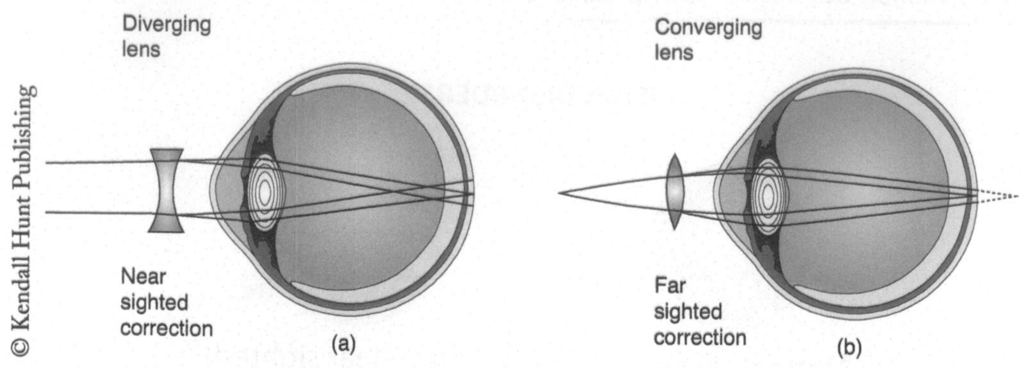

15.2.3 Vision Correction Calculations

To calculate the strength of a spectacle lens for vision correction, recall that the effective strength of several thin lenses is the sum of their individual strengths, provided that they are placed close together. For example, *distant viewing*, requires a total strength of 50.0 D. If the strength of a near sighted person's lens is 52.5 D, they will need a correction of −2.5 D as seen below.

Distant viewing correction for a near sighted person:

Effective strength equals 52.5 D − 2.5 D = **50.0 D.**

On the other hand, as we have shown, reading a book held at 25 cm from the eye requires a strength of 54.0 D. If a far sighted person has a lens strength of 52.5 D, they will require a lens of +1.5 D.

Close viewing and reading correction:

Effective strength equals 52.5 D + 1.5 D = **54.0 D.**

The following example illustrates how data on the severity of a vision defect can be used to calculate an appropriate strength for a spectacle lens.

EXAMPLE 15.3

A myopic person can see objects clearly that are no farther away than 1.0 m and has a lens-to-retina distance of 2.00 cm. What strength spectacle lens will enable him to see distant objects clearly?

Solution: $D_o = 1$ m; $D_i = 0.02$ m

Therefore the strength of this person's eye while viewing an object that is 1.0 m away is:

$$S \equiv \frac{1}{f} = \frac{1}{D_o} + \frac{1}{D_i} = \frac{1}{1} + \frac{1}{.0200} = 51.0\,\text{D}$$

We know that distant viewing requires 50.0 D. Therefore their correcting lens must have a strength of −1.0 D.

Effective strength equals 51.0 D − 1.0 D = 50.0 D.

Note that this and any *spectacle lens for myopic vision has a negative strength*, consistent with preceding discussion and figures.

When calculations similar to that in Example 15.3 are performed for hyperopic eyes, (far sighted), it is found that a positive-strength (convex) lens is needed to correct close vision. Someone with a 2.00-cm lens-to-retina distance needs a strength of 54.0 D to see objects 25 cm away clearly. The accommodated strength of the eye can be calculated from the closest distance at which objects are seen clearly and a spectacle strength can be prescribed to raise the total strength to 54.0 D.

A myopic or hyperopic person with a normal 8% ability to accommodate usually has a normal range of vision with glasses. It is common for hyperopic people, (far sighted), to be told to wear glasses while driving even though they can see distant objects clearly without glasses. The reason is that the hyperopic eye is partially accommodated when distant objects are viewed without glasses. Less eyestrain results if the eye is fully relaxed when viewing distant objects, so it is better to have the glasses on. Both myopic vision and hyperopic vision often are corrected so that the eye is totally relaxed when viewing distant objects.

15.2.4 Loss of Accommodation Ability: Presbyopia

The ability to accommodate decreases with age. The onset of presbyopia, which literally means "old eye," usually is noticed when reading materials must be held at arm's length to be legible. Distant vision usually is not affected, just the ability of the lens of the eye to increase its strength. The loss of accommodation ability is attributed primarily to stiffening of the lens. Presbyopia eventually affects nearly everyone, whether their eyes are normal, myopic, or hyperopic.

Often, a person who could see distant objects clearly for most of their life, will find that sometime after the age of forty, that they have to hold their reading materials further and further from their eyes in order to see clearly. Because the lens in their eye can no longer change shape and adjust its focal length, the eye will have one strength that is unchanging. If for example, it is fixed with a strength of 50.0 D this person will need a correction of + 4 D for reading. (Remember, in order to read at a distance of 25 cm, the eye has to have a strength of 54.0 D). One correction for presbyopia is the use of bifocals. A bifocal lens is ground to two different curvatures. The bottom part of the lens has greater strength than the top since we generally look downward at close objects and need greater strength to see them. Bifocals reputedly were invented by Benjamin Franklin. Trifocals are also used, but they are harder to adapt to. Other possibilities are different glasses for viewing different distances or half-lens reading glasses. Progressive lenses gradually vary their strength from top to bottom, eliminating the line across the lens that identifies the user as over the age of forty.

15.2.5 Astigmatism

In astigmatism, another very common vision defect, the cornea or lens of the eye is not symmetric. Figure 15.8 shows a chart used to detect astigmatism. To a person with astigmatism, some lines appear darker and sharper because they are properly focused while other lines are not. Figure 15.9 shows how an astigmatic lens affects rays from a distant object.

Astigmatism often can be corrected by using a spectacle lens with the opposite asymmetry of the astigmatic eye. For example, if the astigmatic lens is too strong vertically and too weak horizontally a lens that is diverging vertically and converging horizontally correct the problem.

Figure 15.8 Chart for detecting astigmatism. If some of the lines appear darker or clearer to you than others when viewed without corrective lenses, you have an astigmatism. Check each eye separately and look at the center cross.

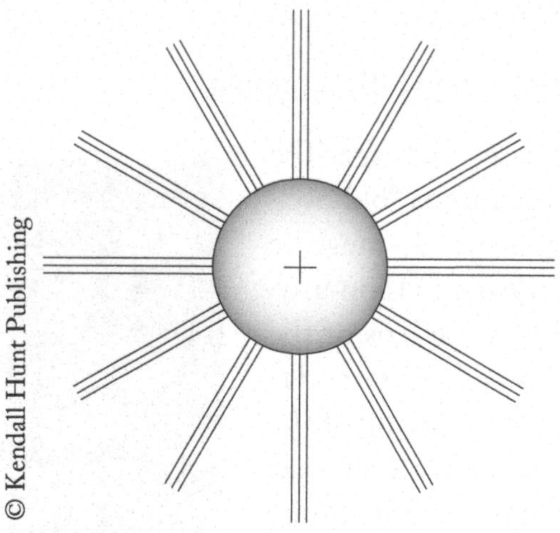

Figure 15.9 Impaired vision with astigmatism.

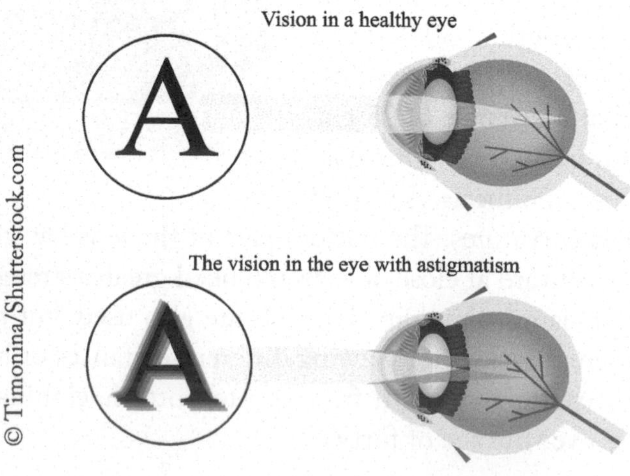

Astigmatism corrections are accomplished by adding a cylindrical lens to the normal lens of the spectacle. Cylindrical lenses look as if they were cut from a can (cylinder) rather than from a sphere. An eyeglass prescription for astigmatism specifies the cylinder strength in diopters and its orientation (determined by using the chart in Figure 15.8). Corrections for myopia or hyperopia are made in the same lens that corrects astigmatism. Corrections for myopia and hyperopia are specified as spherical (symmetric) corrections, also in diopters. Contact lenses ordinarily have no correction for astigmatism because of the difficulty in orienting the lens. Contact lenses can be used to reshape the cornea, and surgery may also be performed on the cornea to correct astigmatism, myopia, or hyperopia.

15.2.6 Cataracts

A cataract is an opacity of the lens of the eye. Cataracts are common in the elderly, but their advent can be hastened by exposure to ultraviolet radiation, microwaves, nuclear radiation, and certain chemicals. Many patients' first symptoms are strong glare from lights and small light sources at night, along with reduced acuity at low light levels. The cloudy lens must be removed, as it is not possible to make it clear again. During cataract surgery, a patient's cloudy natural lens is removed and replaced with a synthetic lens to restore the lens's transparency.

Cataract

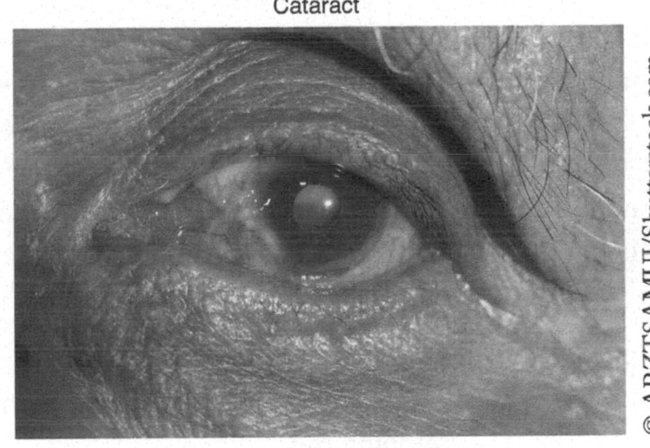

Other recently developed methods of vision correction include the use of lasers to spot-weld detached retinas. The lens of the eye is used to focus the laser beam on the desired spot, and the scar tissue from small burns forms "welds." Electrical devices are even being used to stimulate the optic nerve artificially. This can give a very primitive sense of vision, but it may be improved in the future.

There are many other less-common vision defects not easily corrected by simple spectacle lenses.

Things to know for section 15.2

Strength of a human eye:

$$S \equiv \frac{1}{f} = \frac{1}{D_o} + \frac{1}{D_i} = \frac{1}{D_o} + \frac{1}{.0200} = \frac{1}{D_o} + 50\,\text{D}$$

15.3 COLOR VISION

The gift of vision is made richer by the existence of color. The mechanism by which we sense color is complex and not completely understood. Yet certain characteristics of color vision are well established and give us insight into the physical nature of light and the complexity of the visual process.

15.3.1 Background

The great Isaac Newton was the first to demonstrate that white light is composed of several colors. He did this by shining a beam of sunlight onto a prism and examining the colors obtained. He found that the colors could be recombined into white light with another prism, as shown in the middle of *Figure* 15.10. He also found that the colors were not produced by the first prism, only separated by it. This he demonstrated by observing that a second prism produced no further colors from a "pure" color, as seen in the bottom of *Figure* 15.10.

The sequence and number of colors separated from sunlight by a prism are the same as in the rainbow of color separated from sunlight by water droplets. The processes are similar, resulting in red, orange, yellow, green, blue, and violet light.

Apparently the indices of refraction are greatest for violet light and least for red light since violet light bends most and red light least, as illustrated in Figure 15.10. There must be some physical characteristic of light associated with color, but what is it?

A partial answer was obtained in the nineteenth century, when light was proven to be a wave. It was found that the colors of the rainbow are associated with the frequency of light, with violet having the shortest wavelength, highest frequency and red the longest wavelength, lowest frequency. Visible light is but a small part of the electromagnetic spectrum (which will be discussed at length in Chapter 16, coming between infrared and ultraviolet, as shown in Figure 15.11. Color vision is

Figure 15.10 (Top) Refraction of white light through prism. (Middle) A rainbow of colors refracted by a second prism to produce white light. (Bottom) Red light refracted by a prism remains red.

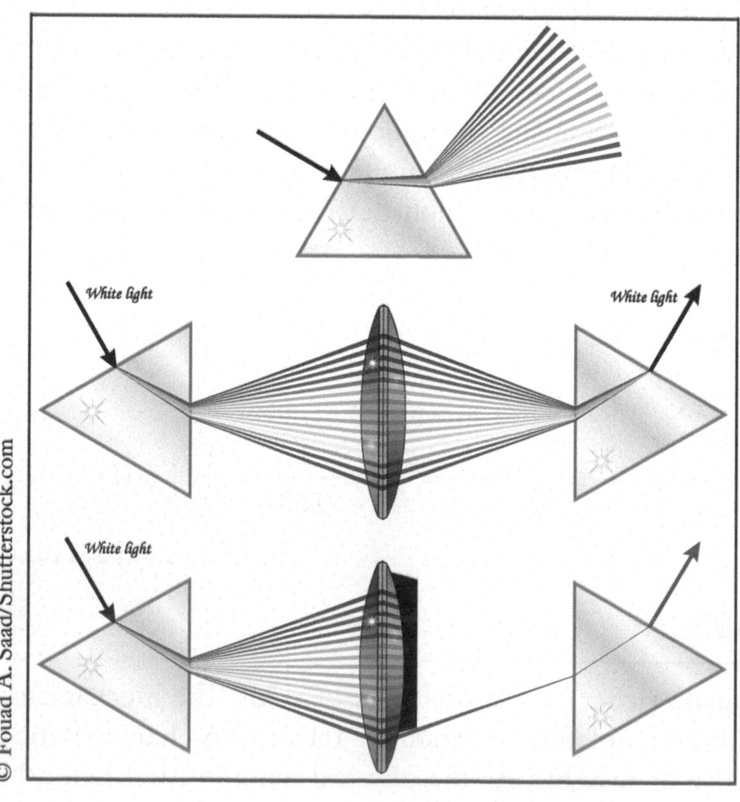

Figure 15.11 Wavelengths of visible light.

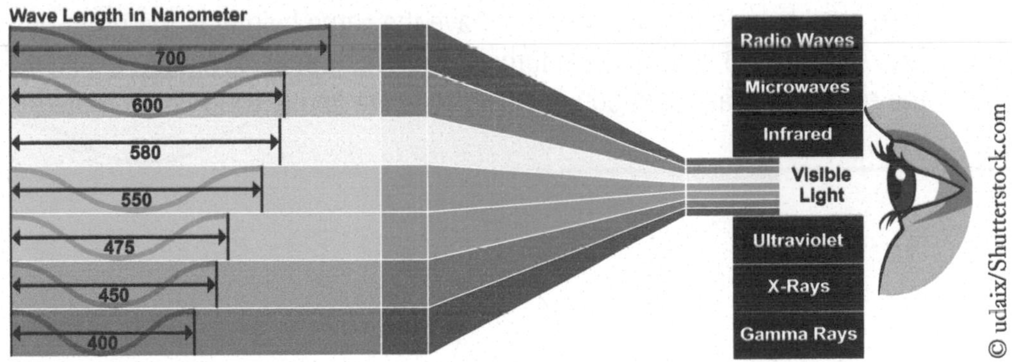

associated with the different frequencies of visible light and how they interact with the cones in our eyes. The different frequencies also correspond to different wavelength of light. It is therefore common practice to separate the different colors by their wavelengths.

15.3.2 Simplified Theory of Color Vision

Rods and Cones are photo-receptors in the eye that convert the energy contained in light photons to electrical signals that are sent to the brain by the optic nerve. There are two major types of light-sensitive cells in the retina: rods and cones. There are 1.3×10^8 rods found throughout the retina, except for a small region in the center of the fovea.

Rods are more sensitive than cones and are solely responsible for vision in very dark environments, and for peripheral vision. Rods do not yield color information. 7×10^6 cones are concentrated in the fovea and are associated with color vision. There are three types of cones, each type is sensitive to a different range of wavelengths. Direct evidence of the three types of cones comes from very difficult experiments. Figure 15.12 shows the absorption characteristics of cone pigments. Indirect evidence includes studies of people with color blindness.

About 8% of men and 1% of women are partially color blind. Most of these are missing only one type of cone, and the various types of color blindness correspond to the absence of different types of cones. There are also rare individuals who lack all three types of cones and are totally color blind. Those individuals who are color blind only in one eye are of particular interest because they can compare normal color vision to their type of color blindness.

Simplified Theory An initial theory of color vision is that there are three primary colors corresponding to the maximum sensitivities of the three types of cones. The hundreds of other hues that the eye can distinguish are created by various stimulations of two or three types of cones. All hues can then be reproduced by adding three primary colors together in various proportions.

A scheme like this is used in color television. The screen of a color TV is covered with equal numbers of red, green, and blue phosphor dots. These three colors correspond to the maximum sensitivities of the three types of cones. A broad range of hues are perceived by the viewer, as various combinations of red, green, and blue dots are lit up. These include yellow, orange, violet, and many hues not in the rainbow, such as flesh tones and brown. It is not necessary to use red,

Figure 15.12 The relative sensitivity of the three types of cones is indicated by their relative absorption of light as a function of frequency. The curve on the left has been increased by a factor of about 8 to have the same height as the other two. The eye is less sensitive to small wavelengths than to long ones. A similar sensitivity curve for the rods peaks at about 500 nm. The rods are about 1000 times more sensitive than the cones.

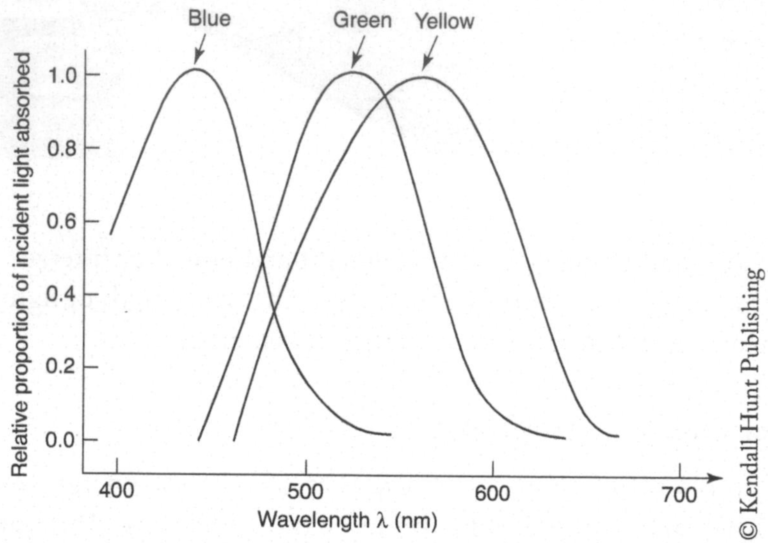

green, and blue dots to stimulate color vision. Because the sensitivities of the three types of cones are broad and overlap, many different sets of three colors can be used to give the impression of hundreds of hues.

The characteristics of cones prove that color vision is associated with the frequency and therefore the wavelength of light, but they also prove that it is not a simple direct association. The three-color aspect is well established; more sophisticated theories do not deny it, but rather expand on it.

15.3.3 Why Do Objects and Light Sources Display Color?

To see how the simplified theory must be modified, it is necessary to consider why various objects and light sources display color; that is, why is a blue jay blue and a neon light red?

Objects The absorptive or reflecting characteristics of various substances differ. Figure 15.13 shows white light falling on four different substances, one pure blue, one pure red, one pure white, and the other jet black. All wavelengths of light except blue are absorbed by the blue substance, and all except red are absorbed by the red substance. More complex hues are created by various degrees of absorption. A white object reflects all wavelengths equally well and a black object absorbs all wavelengths. Gray is a partial absorption of all wavelengths. True color is perhaps more accurately defined as the reflecting characteristics of a substance.

Curve A is average sunlight curve. B is light from a fluorescent lamp. Curve C is light from an incandescent bulb. The eye is able to distinguish the true color of objects when illuminated with any of these sources. The sharp spikes on curve B are atomic spectral lines from mercury (see Chapter 17 for a discussion of atomic spectra).

Lights Similarly, light sources can be of different colors if they emit only certain wavelengths of light. The sun emits a broad spectrum of wavelengths, whereas incandescent and fluorescent lights emit noticeably different spectra (see Figure 15.14).

Figure 15.13 The absorption characteristics determine the true color of an object. Here four objects are illuminated by white light.

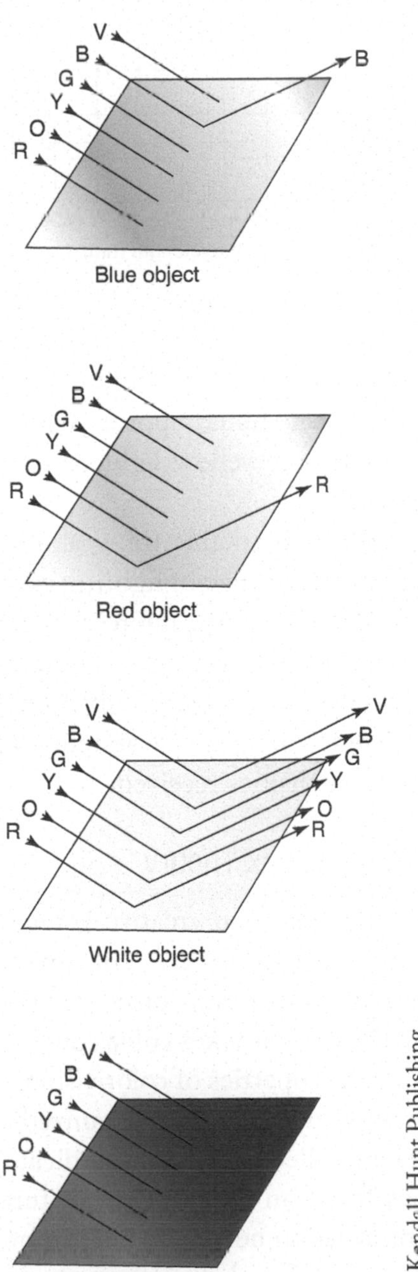

Figure 15.14 A graph of intensity versus wavelength for three different sources of light.

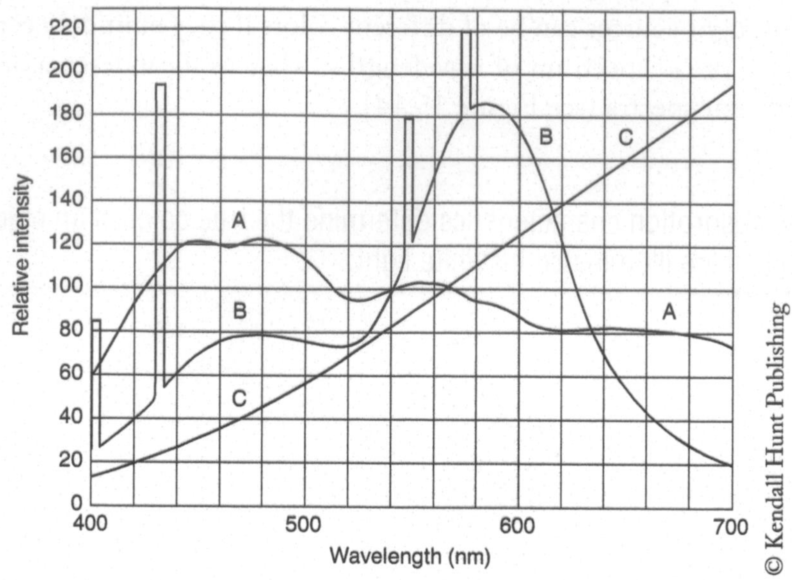

One would then expect objects to appear to be different colors when illuminated by different light sources. For example, a white object should appear yellow when illuminated by yellow light. Under extreme circumstances, such as pure yellow light projected on a white object with a black background, this does in fact happen.

Color Constancy However, a very remarkable thing about color vision is that one is able to discern the true color of an object in a wide range of lighting conditions. For example, a white tablecloth illuminated by the sun, incandescent lights, fluorescent lights, or even yellowish candlelight still appears white. This ability is called color constancy. The eye-brain system does more than just sense the wavelengths of light it receives; it compares light from various objects and arrives at their relative absorption of various wavelengths of light. The system is sensitive to the true colors of objects rather than just the wavelengths and intensities received.

15.3.4 One Modified Theory: The Retinex Theory

Edwin H. Land, (1909–1991), the founder and creative genius behind Polaroid Corporation and instant photography, performed many elegant experiments on color vision. He helped quantify the observations of color constancy, for example. Land considered color constancy to be of fundamental importance and has put forward a theory that takes color constancy into account in an elegant fashion and is consistent with other known properties of color vision. Land's retinex theory proposes that cones do not operate completely independently but are organized into three independent systems, one for each type of cone. The systems are called retinexes to indicate that the systems *involve the retina and cortex of the brain*. Each retinex forms an independent picture of the field of view. Comparisons between retinexes then give the information needed to determine the true color of an object.

A striking experiment performed by Land demonstrates the power of his retinex theory and strains existing theories. Two pictures of a scene are taken on black-and-white film, one with a red filter, the other with a blue filter. The resulting black-and-white slides are then projected and superimposed on a screen, yielding a black-and-white image. Then a red filter is placed in front of the slide taken with a red filter, and again the images from the two are superimposed on the screen. Instead of becoming shades of red, pink, and white, the image suddenly appears to a human observer to be in full color. This demonstration lends credence to the existence of a red retinex system; we can guess that it is this retinex system that is able to make sufficient comparisons from a black-and-white picture superimposed on a red and white picture to discern true colors. If the cones responded only to the wavelengths and intensities they receive, the picture would appear red, pink, and white, because red wavelengths are the most abundant.

Adjacent light receptors do not act independently of one another. The anatomy of the retina is consistent with complex interconnections between light-sensing cells. There are far more cells than nerves to the brain, and there are definite physical links between cells. When we view a picture, only the rods and cones near the edges send large signals to the brain, while those in a region of constant stimulation are inhibited. This is because adjacent cells can affect signals from one another, a phenomenon called lateral inhibition. A number of experiments show this effect and the importance of edges, further reinforcing the conclusion that the light receptors make comparisons and are not sensitive to intensity or wavelength alone.

The retinex theory, though attractive, has not yet been completely verified. Some researchers argue that older theories can explain Land's observations. More experiments are needed to determine the nature of color vision and to distinguish between competing theories.

15.3.5 Summary

Color vision is sensitive to "true color"-that is, to the reflecting properties of an object nearly independently of the wavelengths and intensities used to illuminate it. This ability is indicative of the tremendous sophisticationand beauty of the human senses. Much more is known about vision than has been presented here.

Things to know for Chapter 15

Strength of a human eye:

$$S \equiv \frac{1}{f} = \frac{1}{D_o} + \frac{1}{D_i} = \frac{1}{D_o} + \frac{1}{.0200}$$

Strength of a human eye when viewing objects that are really, really far away:

$$S \equiv \frac{1}{f} = \frac{1}{D_o} + \frac{1}{D_i} = 0 + \frac{1}{.0200} = 50.0\,\text{D}$$

Strength of the human eye *reading* or viewing an object that is 25 cm away:

$$S \equiv \frac{1}{f} = \frac{1}{D_o} + \frac{1}{D_i} = \frac{1}{0.250} + \frac{1}{.0200} = 54.0\,\text{D}$$

Questions

15.1 What aspect of the law of refraction explains the fact that the cornea of the eye does more bending of rays than the lens of the eye?

15.2 If the lens of the eye is removed because of cataracts, why would you expect a spectacle lens approximately 17 D in strength to be prescribed?

15.3 Why do things appear so blurry when viewed while swimming under water? How does a flat face mask restore normal vision to an underwater swimmer? Why must the face mask be flat?

15.4 What is visual accommodation? Is the normal eye totally relaxed when viewing distant or close objects? When is it fully accommodated?

15.5 The lens of the' eye is thinner when viewing distant objects than when viewing close objects, as shown in Figure 15.3. Explain why.

15.6 Why must the strength of the eye change to view objects at various distances?

15.7 Describe the two major factors limiting visual acuity.

15.8 Would visual acuity be improved if the light-sensitive cells on the retina were smaller? Explain.

15.9 Is the strength of a dry contact lens the same as when it is floating on the eye's tear layer? Explain why or why not. If the answer is no, is its strength greater when dry, or when on the eye?

15.10 Give two reasons why Equation 15.3 is less accurate for regular eyeglasses than for contact lenses.

15.11 What happens in a myopic eye to rays from a distant object? Why can a myopic person see close objects clearly? Why is a negative strength spectacle lens necessary to correct myopia?

15.12 What happens in a hyperopic eye to rays from a distant object? Why can a hyperopic person see distant objects clearly? Why is a positive strength spectacle lens necessary to correct hyperopia?

15.13 If the cornea is reshaped to correct myopia, should its curvature be made greater or smaller? Explain. Is the opposite true for hyperopia?

15.14 When light from a laser is shone into a relaxed eye to spot-weld the retina to the back of the eye, the, rays must be parallel as they enter the eye. Why?

15.15 How do we know that colors are contained in white light and not created by the action of a prism?

15.16 Justify the statement that the index of refraction for violet light is greater than that for red light by referring to Figure 15.10 and the law of refraction.

15.17 How does the lack of color vision in the periphery of your field of view support the contention that cones are responsible for color vision?

15.18 In view of the fact that the rods are about 1000 times more light sensitive than the cones, how could one study vision with the rods alone?

15.19 Discuss some evidence for the existence of three types of cones.

15.20 Name and describe three major types of evidence for the coordinated activity of the eye's light-sensitive cells.

15.21 In terms of the wavelengths of light absorbed, reflected, or emitted, what is meant by the "true color" of an object? A light source?

15.22 What is the apparent color of a pure 'red object illuminated by pure green light and located on a black background?

15.23 Why do clear ice cubes appear white when crushed?

15.24 What is the attribute of color constancy? What are its limits?

15.25 What is a retinex, and what is retinex theory?

15.26 The eye-brain system is sensitive to the true color of an object and to edges in its field of view, rather than directly sensitive to wavelength and intensity of light. Give supporting evidence for this.

Problems

Note: In all problems involving the eye, the lens-to-retina distance can be assumed to be 2.00 cm unless specified otherwise.

√**15.1 (1)** Calculate the strength in diopters of a normal eye when viewing an object 0.50 m away. Repeat the calculation for an object 1.0 m away and for one 5.0 m away.

√**15.2 (I)** What is the strength in diopters of a normal eye, when reading a newspaper held at a distance of 45.0 cm?

√**15.3 (I)** Books are often printed using letters averaging 3.5 mm in height. How high is the image of these letters on the retina when the book is held at a distance of 30 cm from the eyes?

15.4 (I) People who do very detailed work close up, such as engravers, often are able to see objects much closer than the normal closest distance of 25 cm. (a) What is the strength in diopters of the eye of a man who can see an object held 8.0 cm away? Is it likely that this person is nearsighted? Explain. (b) What is the size of the image of a 1.0-mm object, such as lettering on a gold heart, held at this distance?

15.5 (II) The strength of a nurse's eye while examining a patient is 53.0 D. How far away is the feature being examined?

15.6 (II) The strength of a student's eye while reading a blackboard is 50.1 D. How far away is the board?

15.7 (II) Neglecting distortion by the atmosphere, what is the smallest detail on the moon observable from earth with the naked eye? The moon is 3.8×10^5 km from the earth. [Hint: Use the data in Example 15.2(a)]

15.8 (II) Calculate how far away a person can be from an airplane and still be able to read the letters on its side using the following information. The letters are 0.75 m high and their image must be 10 times the limit of acuity [see Example 15.2(a)] to be easily read.

15.9 (III) Suppose a woman has normal vision. (a) What is the accommodated strength of her eyes if she has a 10% ability to accommodate? (b) What is the closest object she can see clearly?

15.10 (III) The eyes of a certain myopic administrator have a minimum strength of 52.0 D. (a) What is the accommodated strength of his eyes if he has a normal 8% ability to accommodate? (b) What is the most distant object he can see clearly? (c) What is the closest object he can see clearly?

SECTIONS 15.2 AND 15.3

√**15.11 (I)** The eyes of a farsighted professor have a maximum strength of 53.5 D. What strength lens is required to correct her near vision?

√15.12 (I) A patient's eye has a minimum strength of 50.5 D. What strength spectacle lens is required to correct his distant vision?

15.13 (I) Compare the range of wavelengths of light to which the eye responds and the range of sound wavelengths to which the ear responds. To do this, calculate the ratio of the longest wavelength of light to the shortest and the longest wavelength of sound to the shortest. (The ear responds to sound wavelengths from 1.7 cm to 17 m.)

15.14 (I) Compare the range of light intensities to which the eye responds and the range of sound intensities to which the ear responds. (The ear responds to sounds of intensities from 10^{-12} to 1 W/m^2.)

15.15 (II) (a) What is the most distant object a man can see clearly if the relaxed strength of his eye is 50.5 D? (This is 0.5 D too great.) (b) What strength spectacle lens would give him normal distant vision?

15.16 (II) (a) What is the closest object a woman can see clearly if the accommodated strength of her eye is 53.5 D? (This is 0.5 D too small.) (b) What strength spectacle lens would give her normal close vision (25 cm)?

15.17 (II) (a) What is the accommodated strength of a farsighted man who can see objects clearly that are no closer than 1.50 m? (b) What spectacle lens will allow him to see objects clearly at 25 cm distance?

15.18 (II) (a) A very myopic woman is able to see clearly no farther away than 20 cm. What is the relaxed strength of her eye? (b) What spectacle lens will allow her to see distant objects clearly?

15.19 (II) (a) A mildly nearsighted man can see objects clearly no farther away than 4.0 m. Calculate the relaxed strength of his eye. (b) What spectacle lens will allow him to see objects at a great distance?

15.20 (III) A patient sees in his medical file that his spectacle prescription is −4.0 D. What is the most distant object he can see clearly with glasses off?

15.21 (III) A mother sees that her child's spectacle prescription is +0.75 D. What is the closest object her child can see clearly with glasses off?

15.22 (III) (a) Calculate the strength of a spectacle lens that will correct the distant vision of a woman who can see objects clearly no farther away than 1.0 m. (b) Assuming she has an 8% ability to accommodate, what will be the accommodated strength of her eye? (c) What will be the closest object she can see with glasses on?

15.23 (III) (a) A certain hyperopic man can see objects clearly that are no closer than 1.0 m. What strength spectacle lens will allow him to read at a normal distance of 25 cm? (b) Will this man's eye be totally relaxed when viewing distant objects with glasses on, assuming a 9% ability to accommodate?

15.24 (III) The lens-to-retina distance of a patient is 1.98 cm and the fully accommodated strength of her eye is 54.0 D. (a) What is the closest object she can see clearly? (b) What spectacle lens strength will allow her to read at a normal distance of 25 cm?

15.25 (III) The lens-to-retina distance of a patient is 2.03 cm and the totally relaxed strength of his eye is 50.0 D. (a) What is the most distant object he can see clearly? (b) What spectacle lens strength will correct his distant vision?

15.26 (III) The lens-to-retina distance of a patient is 2.01 cm and the totally relaxed strength of her eye is 50.0 D. (a) What spectacle lens strength will correct her distant vision? (b) If she has an 8% ability to accommodate, what will be the closest object she can see clearly with her glasses on?

16 Electromagnetic Radiation

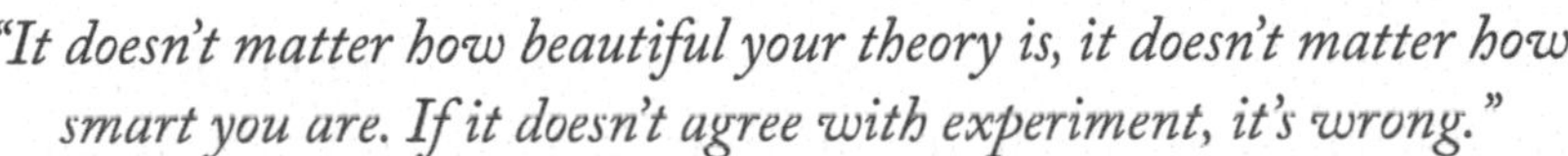

"It doesn't matter how beautiful your theory is, it doesn't matter how smart you are. If it doesn't agree with experiment, it's wrong."

—Richard P. Feynman

INTRODUCTION TO MODERN PHYSICS

A number of types of electromagnetic (EM) radiation have already been mentioned. Chapters 14 and 15 were devoted to just one type; visible light. In Chapter 5, several types of EM radiation-microwaves, infrared radiation, visible light, and ultraviolet (UV) radiation were described as carriers of heat transferred by radiation. Other examples of EM radiation include radio and TV transmission waves, x-rays, and gamma rays.

There are three reasons for further discussion of EM radiation in this text. The *first* is to explore more of its numerous medical applications. The *second* is to examine the physical principles underlying EM radiation and to explain its characteristics in terms of those physical principles. So far, EM radiation has been defined by example only. A more complete definition in terms of underlying physical principles allows more insight. It can explain, for example, how two such apparently different things as radio waves and x-rays are actually members of one large class of closely related entities. The *third* reason for studying EM radiation is that certain of its characteristics cannot be explained by classical physics. The study of EM radiation thus provides an introduction to modern physics.

We begin with the wave aspect of EM radiation, using the most familiar type, visible light, as an example.

16.1 WAVE CHARACTER OF VISIBLE LIGHT

What is meant when something is said to be a wave? As discussed in Chapter 8, all waves—from sound waves to water waves to vibrations in a guitar string exhibit have some common characteristics. One of the unique properties of waves is that they interfere constructively or destructively when superimposed. Because sound exhibits constructive and destructive interference, it is identified as a wave. Experiments show that light also exhibits constructive and destructive interference. Light therefore has wave properties.

Although the wave character of light had been speculated on for centuries, the first definitive proof came in 1801. In that year, the brilliant English physicist and physician Thomas Young (1773–1829) superimposed two light beams and observed constructive and destructive interference. Young obtained his two light beams by passing sunlight through two closely spaced narrow slits. After the light passed through the slits it fell on a screen as shown in Figure 16.1. The pattern formed on the screen was a series of many bright lines instead of just two, as would be expected from the ray model of light. Each bright line in the observed pattern is a region where the two light beams interfere constructively. Each dark space is a region of destructive interference.

A pattern like that on the screen in Figure 16.1 is one type of interference pattern. Such patterns occur when waves combine or a wave encounters an obstacle or goes through an opening. Interference patterns depend strongly on the relative size of the obstacle or opening and, the particular wavelength

Figure 16.1 The wave nature of visible light is demonstrated by interference effects. Light entering two closely spaced narrow slits from the left forms an interference pattern on the screen.

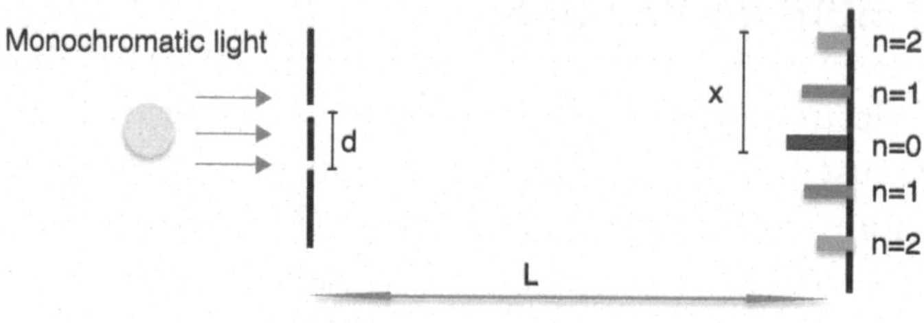

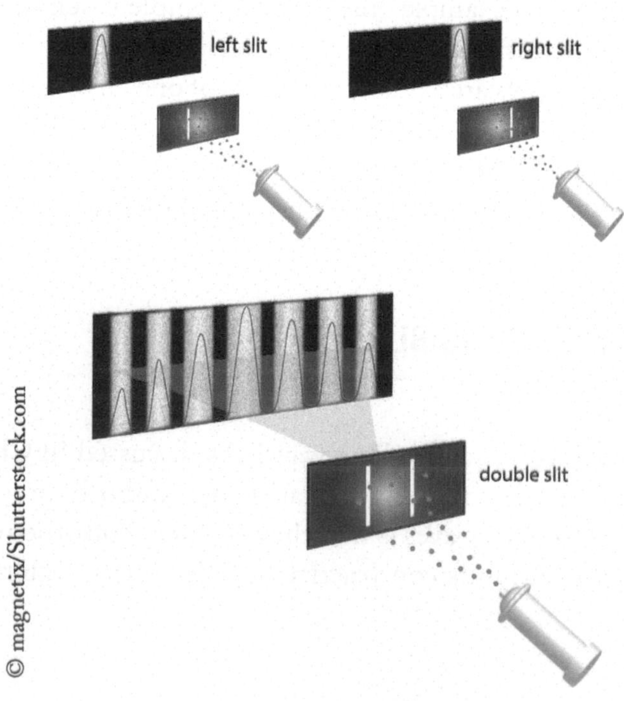

involved. These patterns are easily observed only when the obstacle or opening is roughly the same size or smaller than the wavelength. Young's slits had to be closely spaced and narrow because the wavelength of visible light is very small. If the slits are gradually moved apart and widened then the interference pattern eventually becomes very difficult to observe and the expected pattern of two bright lines is observed. One aspect of the ray model of light (see Section 14.1) is that interference effects usually are negligible when light interacts with objects larger than about a hair's diameter.

Interference patterns are one consequence of the wave character of visible light. Another is the limit to detail observable with a microscope. *A general rule of thumb is that when a wave is used as a probe, no detail smaller than about one wavelength is observable,* because interference and diffraction effects become important when a wave interacts with a small object. These effects blur the image. This is referred to as *the limit of resolution.*

Diffraction is the tendency of waves to bend around objects or spread out after going through an opening. Sound, for example, bends around corners. Diffraction is most prominent when the object or opening are approximately the same size as the wavelength. No matter how well designed a microscope is, diffraction and interference effects prevent observation of details smaller than the wavelength of light. Another related effect of the wave character of light is the limit on the detail observable with the human eye. Diffraction occurs in the eye when light passes through the pupil, resulting in a small blurring of viewed objects, effectively limiting the smallest detail observable with the eye. (See Section 15.1 for more discussion of this effect.)

Experiments such as *Young's double-slit experiment* can be used to measure the wavelength of light.

$$\lambda = \frac{xd}{nL}$$

where,

x is the distance from the central fringe to the fringe being used.
d is the distance between the slits.
n is the fringe number
L is the distance from the slits to the screen.
For example, let
$L = 12.0$ m $\quad n = 2$
$x = 2.00$ cm $\quad d = 0.092$ cm

Determine the wavelength of the light in meters and nm.

$$\lambda = \frac{xd}{nL} = \frac{0.020\,\text{m} \times 0.00092\,\text{m}}{2 \times 12.0\,\text{m}} = 7.67 \times 10^{-7}\,\text{m}$$

The wavelength is 767 nm. This is red, near the infrared part of the electromagnetic spectrum.

Visible light is found to have wavelengths ranging from 380 to 770 nm. The letters nm are used to represent nanometers.

$$1\ \text{nm} \equiv 1 \times 10^{-9}\ \text{m}$$ ***Know***

The smallest detail observable is limited by the wavelength of the light being used. An optical microscope cannot resolve detail smaller than approximately 200 nm. The sense of color is associated with the frequency and therefore wavelength of light. Violet light is the highest frequency of visible light and therefore has the shortest wavelength. Red light has the lowest frequency of visible light and therefore has the longest wavelength of visible light. Infrared radiation has wavelengths longer than red and ultraviolet (UV) has wavelengths shorter than violet. There are specially designed UV microscopes that can detect smaller detail than those using visible light. As other types of EM radiation, such as radio waves, microwaves, and x-rays were discovered and investigated, it was found that they also exhibit interference effects and could be classified according to their wavelengths.

James Clerk Maxwell.

So why is light named electromagnetic (EM) radiation?

In 1865, James Maxwell calculated the speed at which an electric field and a magnetic field would radiate from their source as they exchanged energy.

The result was completely unexpected.

$$c = \frac{1}{\sqrt{\mu_0 E_0}} = 3.00 \times 10^8 \frac{\text{m}}{\text{s}}$$

To his great surprise and delight, the speed was a number that he was familiar with. It was the speed of light. He realized that he had accidentally discovered what light is. Light is a combination of an electric field and magnetic field exchanging energy as they radiate from their source. Therefore the name, electromagnetic radiation?

The next section lists the major types of EM radiation as a function of wavelength and begins the general discussion of their properties.

Things to know for Section 16.1

1 nanometer

$$1 \text{ nm} \equiv 1 \times 10^{-9} \text{ m}$$

16.2 THE ELECTROMAGNETIC SPECTRUM

All types of electromagnetic radiation exhibit wave properties. Figure 16.2 shows the various types of EM radiation organized by wavelength and frequency. What do all EM waves have in common? Answering this question helps in understanding why we say that all types of EM waves are just different manifestations of the same thing.

All EM waves travel at the same speed in a vacuum—namely, at the speed of light, c = 3.0×10^8 m/s. Furthermore, all EM waves obey the same wave equation. This equation, first given in Chapter 8 for sound waves, is written as

$$v = f\lambda$$

Figure 16.2 The electromagnetic spectrum, showing the major classifications of EM radiation as a function of wavelength and frequency.

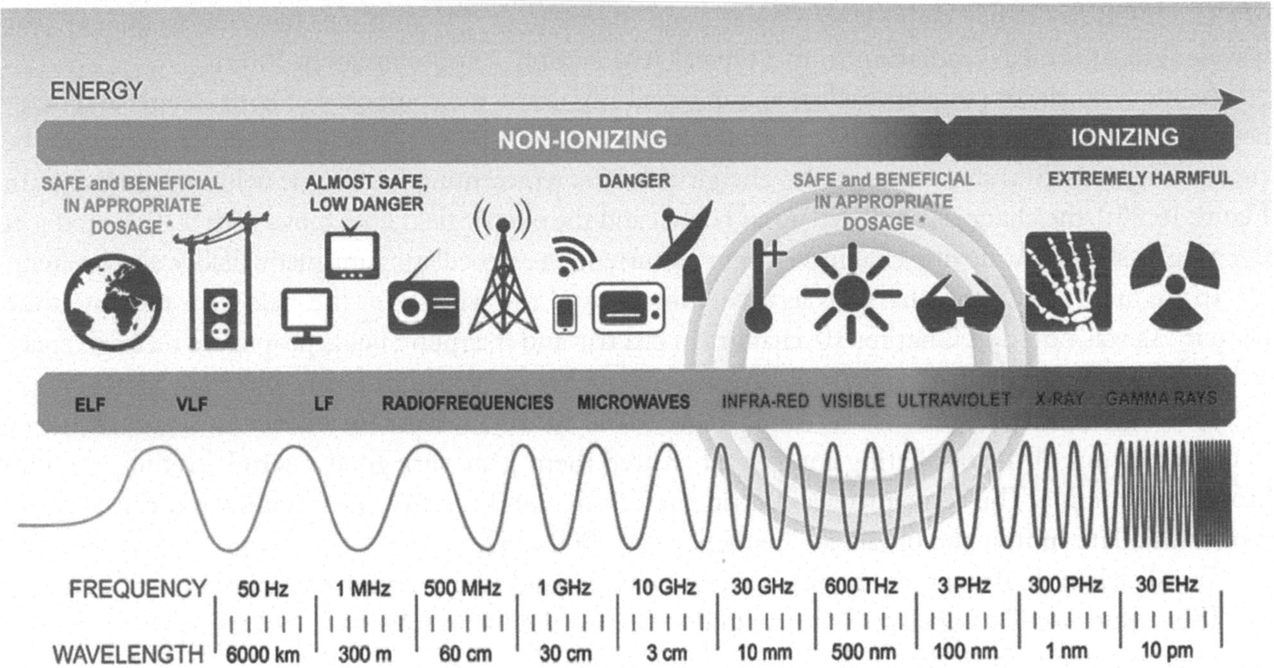

where, v is the speed of the wave, f its frequency, and λ its wavelength. The speed of light is represented by the letter c. Because light travels as a wave, the speed of any EM radiation is;

$$c = f\lambda \quad \textbf{\textit{Know}} \qquad \text{eq.16.1}$$

The frequency of EM radiation is the number of cycles or oscillations per unit time. The frequency and wavelength are inversely proportional to one another: the greater the frequency, the shorter the wavelength. Frequencies of EM waves, in addition to wavelengths, are given in Figure 16.2. The wavelengths are across the top of the diagram.

EXAMPLE 16.1

Cell phones typically operate within a frequency range of 0.500 GHz to 4.00 GHz. Calculate the range of wavelengths of the EM radiation emitted at those frequencies.

Solution: Solving equation 16.1, $c = f\lambda$, for frequency yields

$$\lambda = \frac{c}{f} = \frac{3.0 \times 10^{8}\,\frac{\text{m}}{\text{s}}}{\frac{0.50 \times 10^{9}}{\text{s}}} = 0.60 \text{ m}$$

$$\lambda = \frac{c}{f} = \frac{3.0 \times 10^{8}}{4.0 \times 10^{9}} = 0.075 \text{ m}$$

AM radio dials are often numbered in kilohertz. A frequency of 1000 kHz is near the middle of the AM band ("1000 on your radio dial," as a DJ might say). In tuning a radio or selecting a TV channel you are selecting the frequency at which a tuning circuit resonates. Then only EM waves of the chosen frequency cause it to resonate, and the radio or TV amplifies just that frequency. The wavelength of the EM radiation from a typical AM station is approximately 300 m.

As the term electromagnetic radiation implies, all EM waves are composed of both electric and magnetic fields, which were introduced in Chapter 10. Figure 16.3 shows one way that an EM wave can be created. Figure 16.3(a) shows a stationary charge with lines representing the electric field surrounding it. In Figure 16.3(b), the charge is oscillated up and down and the electric field lines move with it. This produces the wiggles shown. A moving charge constitutes a current, so an oscillating magnetic field is also created.

In Figure 16.3(c), the charge is again stationary and the wiggles in the fields are moving away from it. As was noted in Chapter 10, changes in electric and magnetic fields propagate through space at c, the speed of light. That is why all EM waves travel at the speed of light—they are light.

The wiggles in the field lines shown in Figure 16.3 are EM waves. They have the same frequency as the frequency of the oscillating source that created them. One very good example of this is a radio broadcast antenna. The transmitter's electric circuits are connected to the antenna and cause a current to oscillate in it at the broadcast frequency, say 1000 kHz.

The ideal length of an antenna, with one end in the ground, is one-fourth of the broadcast frequency.

This creates oscillating electric and magnetic fields—that is, an EM wave—of the same frequency that then move out from the antenna (see Figure 16.4).

If EM waves consist of oscillating electric and magnetic fields, then they should be able to exert electric and magnetic forces on objects. They can and do. Consider what happens when a radio broadcast wave reaches an automobile's radio antenna. The oscillating electric and magnetic fields exert forces on electrons in the metal antenna, causing them to oscillate at the same frequency as the EM wave. These oscillating charges constitute a small current that the radio amplifies if the tuning circuit of the radio resonates at the same frequency as the incoming EM wave (see Figure 16.5).

Similar scenarios can be given for all types of EM radiation. Most EM radiation is created by oscillating charged sources. EM radiation of higher frequency than radio waves and microwaves is difficult

Figure 16.3 EM wave created by an oscillating charge. (a) Stationary charge with lines representing electric field. (b) As the charge is oscillated up and down, the electric field lines follow the charge, creating waves. An oscillating magnetic field is also created, but it is omitted for ease of drawing. (c) Charge is again stationary, but EM waves continue to move away from it at the speed of light.

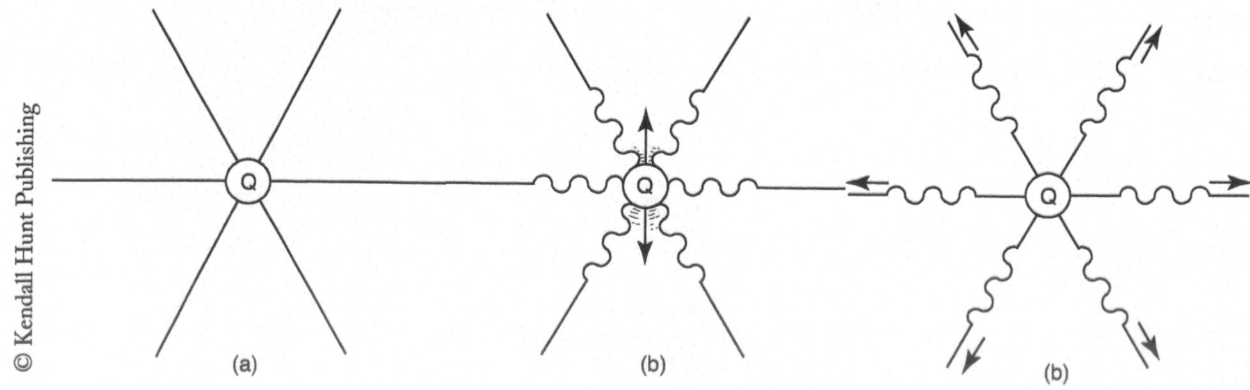

Figure 16.4 A radio transmitter drives an oscillating current up and down an antenna. This creates EM waves of the same frequency as the oscillating current. These waves move out from the antenna as shown.

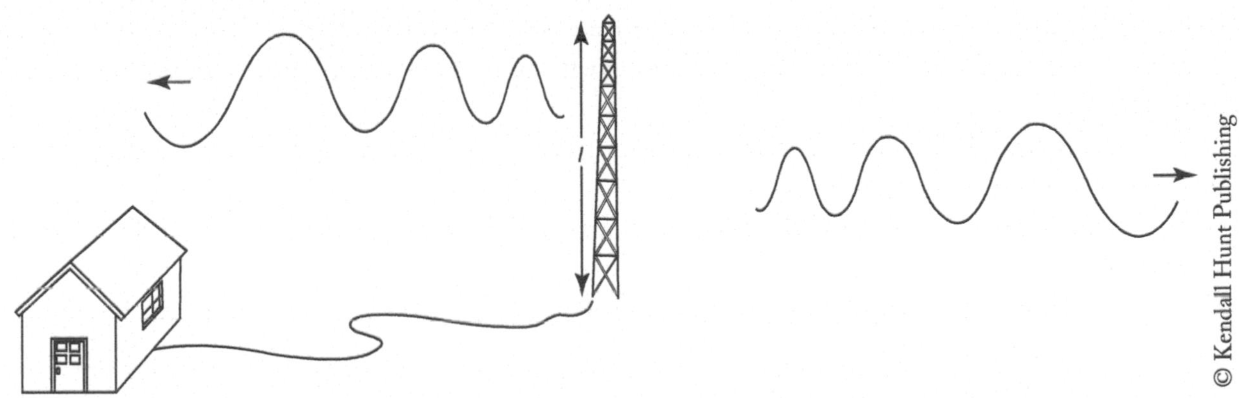

Figure 16.5 The oscillating electric and magnetic fields of which an EM wave is composed exert forces on charges in the automobile's antenna, causing a small alternating current having the same frequency as the EM wave.

to produce with electronic circuits and is generally produced by atoms, molecules, and nuclei, which themselves contain circulating charges. (More will be said about these higher-frequency waves later.) All EM radiation travels at the speed of light in a vacuum and exerts EM forces on objects it encounters, causing oscillations in the object of the same frequency as the radiation. Let us now examine the major divisions of the EM spectrum. (Section 16.4 considers these divisions relative to medical applications.)

16.2.1 Major Divisions of the EM Spectrum

AM Radio. The lowest frequencies for which there are practical applications are used for radio transmission. Easily produced, these were historically among the first developed. Very-low-frequency and very-long-wavelength EM waves are produced by ac power transmission lines and are one cause of energy loss in power transmission.

Shortwave Radio. This type of transmission is so named because the wavelengths are shorter than those of commercial radio. These EM waves are used for a variety of communication purposes (police radio, citizen band, etc.). The effective range of shortwave radio is sometimes less than that of AM radio because the shorter wavelength is more easily blocked by obstacles. Longer wavelengths diffract around large obstacles more easily. However, shortwave radio transmission can sometimes reach long distances by bouncing off the upper atmosphere, avoiding obstacles on the ground.

FM Radio and Television. The frequencies of the entire FM radio band lie between the frequencies of television channels 6 and 7.

The wavelength of a typical FM station is approximately 3 m. The wavelength of AM stations is closer to 300 m.

The broadcast frequency of KC 101 is 101.3×10^6/s

$$\lambda = \frac{c}{f} = \frac{3\times10^8 \ \frac{\text{m}}{\text{s}}}{\frac{101.3\times10^6}{\text{s}}} = 2.96 \text{ m}$$

Microwaves. Microwaves are so named because they are very short compared to those of AM radio. They too are used for communications, such as long-distance telephone transmission. Radar systems use microwaves to determine distances to objects, such as clouds or aircraft, by measuring microwave echo time. The speed of objects, such as cars, can also be measured by observing the Doppler shift of the reflected microwaves. A more recent application is food preparation in microwave ovens. Water molecules in food resonate at frequencies in the microwave region. Food therefore preferentially absorbs microwave energy, resulting in a very efficient method of cooking.

Infrared Radiation. Infrared means "below red," and infrared radiation has frequencies lower than red light. Generally produced by thermal motion it is one of the most important EM waves in heat transfer by radiation. Human skin absorbs almost all infrared falling on it and senses its presence by the warmth produced. A significant fraction of EM radiation from the sun is infrared light.

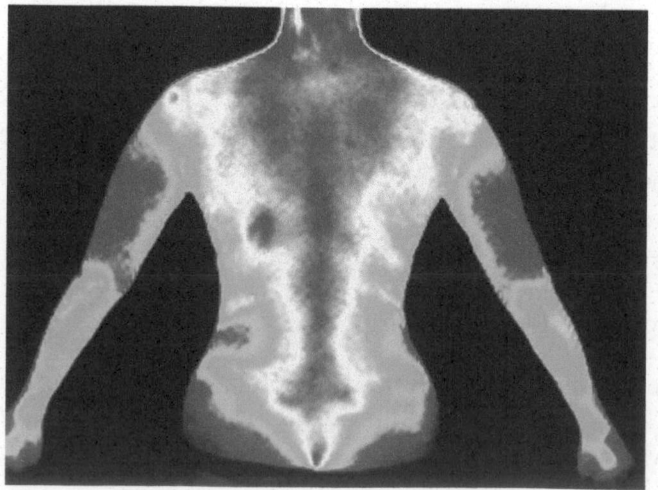

Infrared image

White light

- White light consists of all the colors of the rainbow.
- The light travels as individual bundles of energy called photons
- They travel at 3×10^8 m/s in a vacuum

Visible Light. Visible light is the narrow band of EM waves to which the normal eye responds. The retina actually senses the lowest ultraviolet (UV) frequencies (see below), but these are absorbed by the cornea and lens of the eye before they reach the retina. Red light has the lowest frequencies and the longest wavelengths (770 nm) of visible light, while violet has the highest frequencies and shortest wavelengths (380 nm). The most intense EM waves from the sun are in the middle of the visible spectrum. The wavelength of visible light is so small that light behaves like a ray when interacting with macroscopic objects.

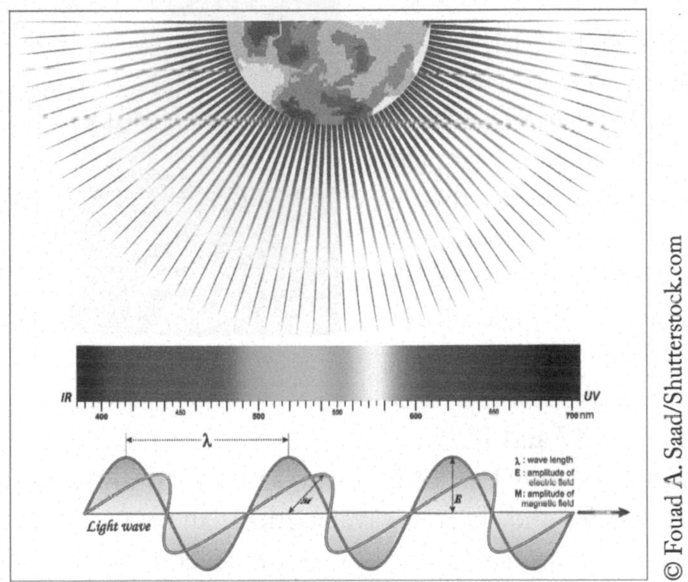

Visible portion of the electromagnetic spectrum

UV Radiation This type of EM radiation causes tanning, sunburn, and nearly all skin cancers. It also causes a chemical reaction in the skin that produces vitamin D. UV radiation can also be an effective sterilizing agent. Most UV radiation from the sun is absorbed by ozone in the upper atmosphere. UV light has a fairly large range of wavelengths. For convenience, UV radiation is separated into three categories or bands. UVA, UVB, and UVC.

UVA is a long wavelength UV with a range of 320–380 nm. It can penetrate glass and clouds and accounts for 97% of the UV radiation that reaches the Earth's surface. It penetrates into the deeper layers of skin and plays a major role in skin aging. It can initiate the development of skin cancers. Tanning booths typically emit two to five times the intensity of UVA that reaches the Earth's surface from sunlight.

UVB wavelengths range from 280 nm to 320 nm. It is responsible for burning, tanning, skin aging, and plays a major role in the development of skin cancer. It cannot penetrate glass.

UVC wavelengths range from 100 nm to 280 nm. Luckily, UVC is filtered out by the Earth's ozone layer and does not reach the Earth's surface.

When selecting a sunscreen and Sun protection factor (SPF), it is important that you select a sunscreen that provides broad spectrum protection in order to filter out both UVA and UVB.

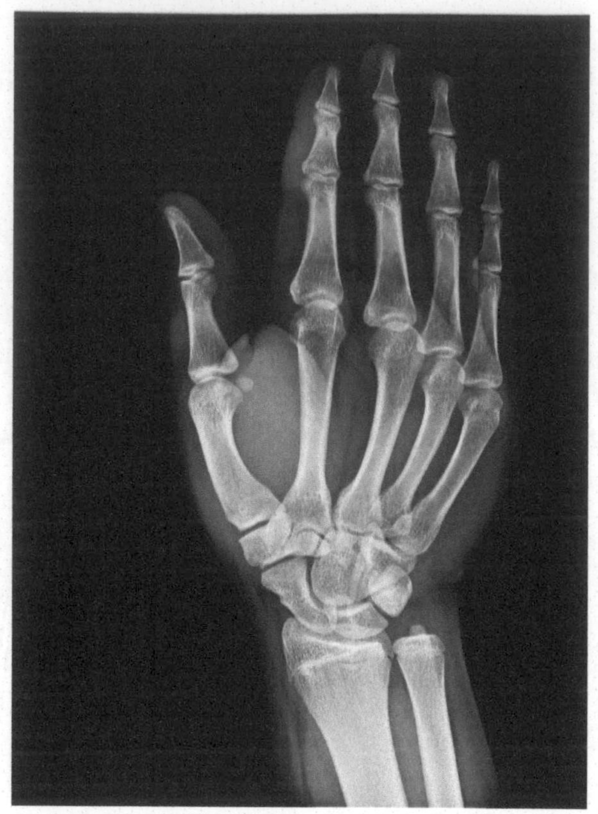
© ChooChin/Shutterstock.com

X-Rays. This type of electromagnetic radiation is formed in extremely hot gases, as on the surface of the sun, and is also formed in cathode ray tubes (CRTs) when the beam of electrons strikes the screen. Because x-rays are biologically hazardous, televisions, oscilloscopes, computer terminals, and so on are constructed so that the user is shielded against them. This is easy in such devices because the x-rays they produce are low in energy and not very penetrating. Higher-energy x-rays are produced in x-ray tubes and can be very penetrating; their greatest use is in medical diagnostics and therapeutics and in materials testing, such as aircraft crack detectors. X-ray tubes are a form of CRT, except that the beam of electrons is accelerated with a much higher voltage and is directed against a metal target rather than a phosphor-coated screen. The wave properties of x-rays are significant only when they interact with objects about the size of atoms or molecules—hence the name x-ray (rather than X-wave). Quite a bit more will be said about x-rays later in this chapter and in Chapters 17 and 18.

Gamma Rays. These are the only type of EM waves produced in the nuclei of atoms. Generally, they can have higher frequencies and shorter wavelengths than x-rays. Their physical characteristics and biological hazards are very similar to those of x-rays since they are identical (except for their source) at the same frequencies. Gamma rays are produced in stars, nuclear reactors, and nuclear bombs and by some radioactive substances. They are sometimes used in medical diagnostics and therapeutics. Gamma rays will be considered in more detail in Chapter 18.

A number of features of EM radiation and what it is have now been covered. The next section introduces the photon and explains more of the particular properties of the various types of EM radiation. The final section of this chapter concentrates on medical applications.

Gamma rays γ-*ray*

- ❑ Gamma rays are high-energy photons of light that are emitted from the nuclei of some atoms.

16.2.2 Home Experiment: Determine the Speed of Light in Your Kitchen!

Microwave ovens use nearly the same frequency (2450 MHz) as microwave diathermy. Hot spots in foods cooked in microwave ovens result from constructive interference and are separated by distances that equal one-half of the wavelengths of the radiation. You can use this information to determine the speed of light. Turn the dish inside the oven upside down. This should prevent it from rotating. Place a bar of chocolate or a stick of butter in the microwave oven. Set the timer for approximately 30 seconds. Constructive interference in the oven will cause two spots on the chocolate bar to begin to melt. The distance between those two spots is equal to half the wavelength of the microwaves. It should be approximately 6.1 cm = 0.061 m. Since half the wavelength is 0.061 m, the full wavelength equals 0.122 m. You can now use equation 16.1 to determine the speed of light. Why not try the experiment yourself and see what you get for the speed of light?

Hint: Remember, $c = f\lambda$

Things to know for Section 16.2

Light travels as a wave, but has particle properties.

The letter '*c*' represents the speed of light.

The speed of any wave is the product of its frequency and wavelength.

Therefore the speed of light is, $c = f\lambda$

$c = 3 \times 10^8$ m/s in a vacuum.

16.3 PHOTONS

By 1900, it was well established that EM radiation has wave properties. It should be remembered, however, that this is a model or mental image based on the experimental results of experiments designed to study the behavior of EM radiation. It is not possible actually to see EM radiation wiggling through space like water waves on a pond, and limits to that analogy were discovered before the turn of the century. The discovery of those limits gave further insight into the nature of EM radiation and helped lead to the revolutionary ideas of modern physics.

16.3.1 Photoelectric Effect and Photons

In 1887, Heinrich Hertz observed what is now called the *photoelectric effect*. When certain frequencies of light land on the surface of some materials, an electrical current is produced. The current is proportional to the light intensity. If the frequency was lowered below a minimum frequency, called the 'threshold frequency,' no electricity was created—no matter how bright the light. It soon became apparent to physicists, that the photoelectric effect could not be completely explained by a simple picture of EM radiation as a wave. The wave picture of light needed a little modification.

Let us begin by summarizing the important features of the photoelectric effect:

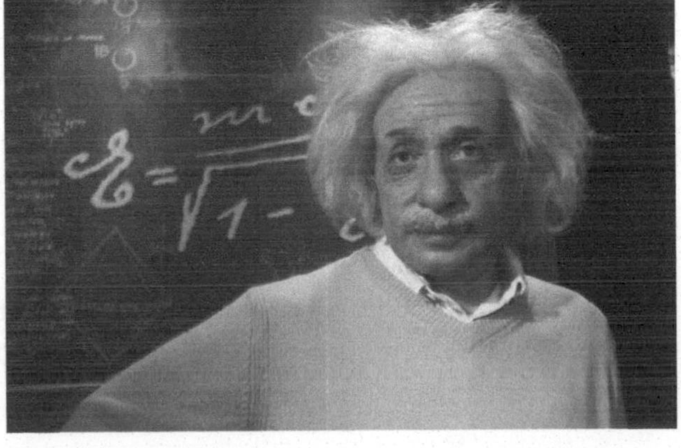

1. Electromagnetic (EM) radiation falling on a material can eject electrons from the material.
2. No electrons will be ejected if the frequency of the EM radiation is less than a certain threshold value (except for extremely high intensity EM radiation). That value depends on the material.

3. The number of electrons ejected per unit time is directly proportional to the intensity of the EM radiation, provided that its frequency is above the threshold value.
4. The energy of the ejected electrons depends on the frequency of the incident EM radiation but not on its intensity.
5. The electrons are ejected as soon as the EM radiation falls on the material, even at very low intensities (provided that the frequency is above threshold).

Now let us consider these points in more detail. This will lead us to the idea of the *photon.*

It is not surprising that EM radiation can eject electrons from a material. EM radiation carries energy and can exert electric and magnetic forces, so it is entirely reasonable that it should be able to exert forces on electrons and transfer energy to them. The second point is surprising. The intensity of a wave is defined as power per unit area, so it would seem that EM radiation of any frequency should be able to transfer enough energy to electrons to eject them. Yet it cannot. For example, blue light may be able to eject electrons from a certain material whereas red light cannot, even if the red light is intense and the blue light very dim. This implies that the energy in EM radiation is distributed in such a way that red light cannot give as much energy to an individual electron as blue light can.

The third point, that the number of electrons ejected is directly proportional to the intensity of the EM radiation, is logical. The greater the intensity, the more energy is available, and so more electrons can be ejected, but again the EM radiation must have a minimum frequency. The shocking thing is that the energy in EM radiation of low frequencies is unable to eject electrons. In 1900, physics could not explain the photoelectric effect using the wave theory of light.

Photons. In 1905, Albert Einstein published a Nobel Prize winning paper that completely explained the photoelectric effect. He did this by adding the concept that light must be traveling as *bundles of energy* called quanta. These bundles of energy were later given the name photons. A *photon* is a massless particle of EM radiation that has an energy proportional to its frequency:

$$E_{\text{photon}} = hf \;\; \textit{Know} \qquad \text{eq. 16.2}$$

where, h is Planck's constant ($h = 6.63 \times 10^{-34}$ J s) and f is the frequency of the EM radiation. It is now known that EM radiation is composed entirely of photons.

We can derive a second equation for the energy of a photon, if we use Equation 16.1,

$c = f\lambda$. Solving for f yields, $f = \dfrac{c}{\lambda}$.

Next we substitute for the frequency in Equation 16.2.

$$E_{\text{photon}} = hf = \frac{hc}{\lambda}$$

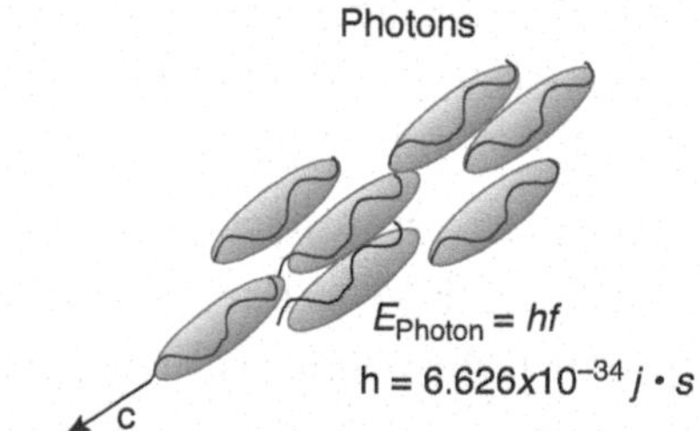

• The energy of a photon equals planck's constant 'h' multiplied by the frequency of the light.

We see that the energy of a photon of light is dependent on its frequency and therefore on its wavelength. High frequency, short wavelength photons are high energy photons.

The minimum amount of energy that EM radiation of frequency f can have is the energy of one photon. Low-frequency EM radiation has a small amount of energy per photon, and high-frequency EM radiation has a large amount of energy per photon (see Figure 16.6).

The threshold frequency in the photoelectric effect is explained by the properties of photons. Low-frequency photons have insufficient energy to eject electrons. More intense EM radiation of low-frequency light has more photons of the same frequency, but the electron would have to absorb two or more photons simultaneously to be ejected. However, there are so many electrons in an object that this is very unlikely (see Figure 16.7). The fact that electrons are ejected immediately when EM radiation falls on a material is also explained by the idea of photons. Each photon, with a high enough frequency, carries enough energy to eject an electron. The energy required to release an electron is called the *work function*, ϕ. If the photon has more energy than is required to release the electron, conservation of energy tells us that it must go somewhere. In fact, the extra energy will become the kinetic energy of the ejected electron. We can summarize this with the following equation:

$$E_{\text{photon}} - \Phi = \text{Kinetic energy of the electron} \quad \textbf{\textit{Know}} \qquad \text{eq. 16.3}$$

In the hypothetical material mentioned above, the energy of red photons was less than the work function, ϕ of the material. The red photons had too low a frequency and therefore not enough energy

Figure 16.6 EM radiation is composed of photons. The spatial extent of photons is limited, like particles, and the photons have a wavelength and frequency, like waves. (a) Several photons of the same wavelength and frequency. (b) A single photon of greater wavelength and lower frequency and energy than those in part (a).

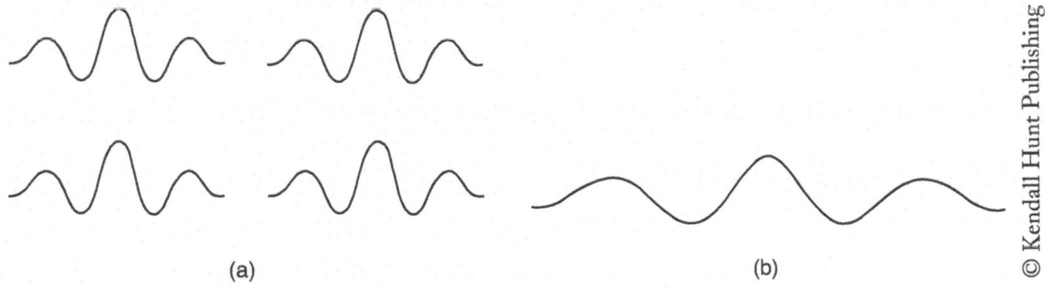

Figure 16.7 The photoelectric effect: A photon of EM radiation strikes a material and gives its energy to an electron. Photons act like particles of pure energy, interacting with individual electrons. It is very unlikely that one electron would absorb two photons simultaneously.

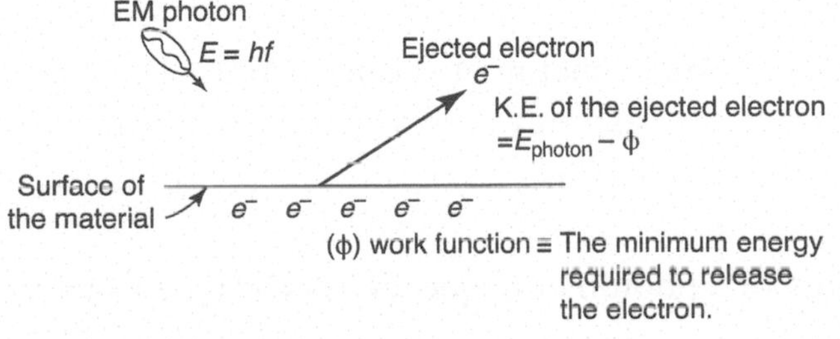

to release an electron. The energy of the higher frequency blue photons was greater. The blue photons could therefore eject electrons from the material, but red photons could not. If we increase the intensity of the blue light, we will have more high-energy blue photons, so it can eject more electrons. On the other hand, increasing the intensity of the red light just delivers more low-energy photons to the surface. None of the red photons have enough energy to release an electron. Their energy is less than the work function. Each photon interacts independently of the others, and the energy it transfers to an electron does not depend on intensity. All the features of the photoelectric effect are elegantly explained in terms of photons. Albert Einstein won the Nobel Prize in physics in 1921 for the explanation of the photoelectric effect in terms of the photon idea (not for his much more famous theory of relativity).

16.3.2 The Photon Picture

The photon picture has far-reaching consequences and explains much more than the photoelectric effect. Photons are an example of quantization. One example of quantization has already been encountered in this text-the basic units of matter: electrons, protons, neutrons, atoms, and molecules. All objects are composed of integral numbers of these basic units. Another example of quantization is the existence of the smallest units of charge carried by electrons and protons. All other charges are built from multiples of electron and proton charges. Other than quarks, no fractional charges are known to exist; that is, it is not possible to have a fraction of an electron or proton charge. Charge is thus said to be quantized.

- *Quarks come in six different varieties know as flavors: up, down, strange, charm, bottom and top. Quarks combine to form composite particles, for example, protons and neutrons are made of quarks. Quarks have a fractional charge but are never directly observed or found in isolation.*

Photons

- The energy of a photon equals Planck's constant "h" multiplied by the frequency of the light.

Similarly to charge, EM radiation is also quantized. The fundamental unit of light is called a photon. Only integral numbers of photons can exist, each photon having an energy determined by its frequency $E_{photon} = hf = \frac{hc}{\lambda}$. As with the quantization of mass and charge, the quantization of EM energy into photons was not detected by the earliest researchers. There are so many photons in normal situations that it is difficult to notice individual ones. Consider the following example.

EXAMPLE 16.2

a. Calculate the energy of a single photon of frequency 6.0×10^{14} Hz (visible).
b. If a 1.0 W flashlight put out light only at this frequency, how many photons per second would it produce?

Solution: (a) The energy of a photon of frequency f is given by Equation 16.2:

$$E_{photon} = hf$$

$$E_{\text{photon}} = 6.63\times10^{-34}\ \text{J}\times\text{sec}\left(\frac{60\times10^{-14}}{\text{s}}\right) = 3.98\times10^{-19}\ \text{J}$$

which is an exceedingly small amount of energy in macroscopic terms.
(b)Number of photons per second required to produce a 1.0-W light?

$$1\ \text{W} \equiv 1\frac{\text{J}}{\text{s}}$$

Think of this problem as one where you are converting units from joules to numbers of photons.

$$1\frac{\text{J}}{\text{s}}\times\left(\frac{1\ \text{photon}}{3.98\times10^{-19}\ \text{J}}\right) = 2.51\times10^{15}\ \frac{\text{photons}}{\text{s}}$$

The energies of individual photons are such small fractions of a joule that it is more convenient to use another unit of energy: the electron volt. One electron volt (eV) is defined as the energy given to one electron accelerated through a potential difference of 1 V. Electrical potential energy. From the definition of Voltage, we can determine the electrical potential energy of one electron that is about to cross a 1 volt potential difference.

$$V \equiv \frac{E}{q};\ \text{therefore } E = Vq$$

$$E = Vq = \frac{1\ \text{J}}{\text{C}}\times\left(1.6\times10^{-19}\text{C}\right) = 1.6\times10^{-19}\ \text{J}$$

Therefore, an *electron volt*, by definition,

$$1\ \text{eV} \equiv 1.6\times10^{-19}\text{J}\ \textit{Know}$$

Small amounts of energies often will be expressed in units of electron volts in the remainder of this text. Electron volts are also used in chemistry, where it is more convenient to express the binding energies in that unit. As an example, we can express the energy of the photons in Example 16.2 in electron volts,

$$3.98\times10^{-19}\text{J}\ \frac{1\ \text{eV}}{1.6\times10^{-19}\ \text{J}} = 2.49\ \text{eV}$$

Photon energies in electron volts are not only convenient in magnitude but can also give insight into why various types of EM radiation have the effects they do. The 2.49-eV photon of Example 16.2 has enough energy to cause changes in organic compounds. Photons may be visible when they have enough energy to alter chemicals in the retina and trigger nerve impulses. Furthermore, typical chemical binding energies are on the order of several electron volts, so individual 2.49-eV photons cannot do much damage to biological organisms. Higher-frequency photons-UV, x-ray, and gamma ray photons have much more energy and can do considerable damage to biological organisms. Radio waves, on the other hand, have such low frequencies that they can be made perfectly safe. Only when low-frequency photons are present in huge numbers, as in a microwave oven, can they deposit enough energy to be hazardous. Without the photon picture it is extremely difficult to explain why radio waves can be harmless and x-rays cannot.

EXAMPLE 16.3

If 25,000 V is used to accelerate electrons in a cathode ray tube (CRT), then the maximum energy x-ray created when the electrons strike the front screen is 25,000 eV, or 25 keV. Calculate (a) the frequency and (b) the wavelength of a 25-keV photon and verify that this is in the x-ray region.

Solution: (a) The frequency—The relationship between photon energy and frequency is given by Equation 16.2,

$$E_{\text{photon}} = hf \,.$$

Remember that Planck's constant $h = 6.63 \times 10^{-34}$ J s. Take a close look at its units. They are joules times seconds. So, let's first convert 25,000 eV to joules.

$$25{,}000 \text{ eV}\left(\frac{1.6\times10^{-19}\,\text{J}}{\text{eV}}\right) = 4\times10^{-15}\,\text{J}$$

Next we solve Equation 16.2 for frequency.

$$E_{\text{photon}} = hf;\ \text{therefore}\ f = \frac{E_{\text{photon}}}{h}$$

$$f = \frac{E_{\text{photon}}}{h} = \frac{4\times10^{-15}\ \text{J}}{6.63\times10^{-34}\ \text{J}\cdot\text{sec}} = \frac{6.03\times10^{18}}{\text{sec}} = 6.03\times10^{18}\,\text{Hz}$$

EX. 16.3B

(b) determine the wavelength. The relationship between frequency and wavelength is given by equation 16.1, $c = f\lambda$. Solving for wavelength,

$$c = f\lambda;\ \text{therefore}\ \lambda = \frac{c}{f}$$

$$\lambda = \frac{c}{f} = \frac{3.0\times10^{8}\,\frac{\text{m}}{\text{s}}}{6.03\times10^{-18}} = 4.97\times10^{-11}\ \text{m}$$

Converting the wavelength to nanometers we get,

$$4.97\times10^{-11}\ \text{m}\left(\frac{1\ \text{nm}}{1\times10^{-9}\text{m}}\right) = 0.0497\ \text{nm}$$

A photon with a wavelength in the range 0–10 nm is considered to be an x-ray. A wavelength of 0.0497 nm is about half the size of an atom, so x-rays are small enough to "observe" individual atoms or crystal structures.

Please try the ***photoelectric self-quiz***.

Summary

EM radiation consists of photons, which behave like bundles or particles of pure energy. Yet photons also have a wavelength. On a macroscopic scale, dirt particles and water waves are quite different things, but photons behave like both particles and waves. This is often called the wave–particle duality. It isn't easy to visualize a particle with a wave nature because we live in the macroscopic world where those things are quite different. We don't really see photons as drawn in Figures 16.6 and 16.7, but we think of them that way because they behave analogously to things we can see. The wave–particle picture of EM radiation is extremely powerful in explaining the characteristics of all types of EM radiation.

Things to know for Section 16.3

The energy of a photon:

$$E_{\text{photon}} = hf$$

Substituting for the frequency f, this equation becomes $E_{\text{photon}} = \dfrac{hc}{\lambda}$

To calculate the kinetic energy of an ejected electron during the photoelectric effect:

$$K.E._{\text{electron}} = E_{\text{photon}} - \Phi$$

Work function, ϕ, represents the minimum energy required to release an electron.

The electron volt is a tiny unit of energy: $1\ \text{eV} \equiv 1.6 \times 10^{-19}$ J of energy

Blast from the past: The definition of power: $Power = \dfrac{Energy}{time}$

Units of a watt: $1\ \text{W} \equiv 1\dfrac{\text{J}}{\text{s}}$

The purpose of this section is to present a small sampling of the many medical applications of EM radiation and to explain them in terms of the wavelengths and photon energies of the radiation. The examples are arranged from longest to shortest wavelength, starting with radio waves and ending with gamma rays.

Radio: Radio is used mostly for communications purposes because of the ease of generating, transmitting, and receiving radio waves. One of the more interesting applications of radio transmission to medical situations is the remote monitoring of electrocardiograms (ECGs) in a hospital. Heart patients are fitted with small radio transmitters that broadcast ECG signals to a central monitor. One person is then able to observe the ECGs of several patients on a single CRT display. Each patient's radio transmitter is set to broadcast at a unique frequency so that there is no mixing of signals. Low-intensity radio waves are harmless because of their low photon energies. (For an application of radio waves in medical diagnostics, see the discussion of nuclear magnetic resonance in Section 10.4.)

Microwaves: The most common medical use of microwaves is for deep heating, called microwave diathermy, previously mentioned in Section 5.5. Water molecules in tissue absorb a range of microwave frequencies. The microwave energy is converted to thermal energy, increasing the temperature of the tissue. The energy per photon of microwaves is small, so very large numbers of them (high intensities) must be used. Microwaves penetrate much deeper into the body than the shorter-wavelength infrared. The beneficial effects of microwave diathermy result from the increased temperature produced (see Figure 16.8).

EXAMPLE 16.4

The most common microwave frequency used in microwave diathermy is 2450 MHz. (a) Calculate the energy in electron volts of a photon of this EM radiation. (b) How many of these photons are needed to break up a compound with a 5.0-eV binding energy? (c) What is the wavelength of this microwave radiation?

Solution: (a) The energy of a photon is given by Equation 16.2:

$E_{\text{photon}} = hf$ where, $h = 6.63 \times 10^{-34}$ J.s

$$E_{\text{photon}} = 6.63 \times 10^{-34} \text{ J.s} \left(\frac{2450 \times 10^{6}}{\text{s}} \right) = 1.62 \times 10^{-24} \text{ J}$$

Next convert joules to electron volts:

$$1.62 \times 10^{-24} \text{ J} \left(\frac{1 eV}{1.6 \times 10^{-19} \text{ J}} \right) = 1.02 \times 10^{-5} \text{ eV}$$

Solution: (b) Notice that each of the microwave photons has only 1.02×10^{-5} eV of energy. In order to determine the number of photons required to break up a compound that has a binding energy of 5.0 eV, we do the following conversion of electron volts to number of photons:

$$5 \text{eV} \left(\frac{1 \text{ photon}}{1.02 \times 10^{-5} \text{eV}} \right) = 4.93 \times 10^{5} \text{ photons}$$

All the photons must arrive at the same molecule at nearly the same time, and all must be absorbed. Because this is exceedingly unlikely, microwaves do not directly destroy compounds. (Thermal agitation resulting from microwaves that cause water molecules to resonate can destroy compounds. It is called cooking.)

Figure 16.8 Diathermy using microwaves.

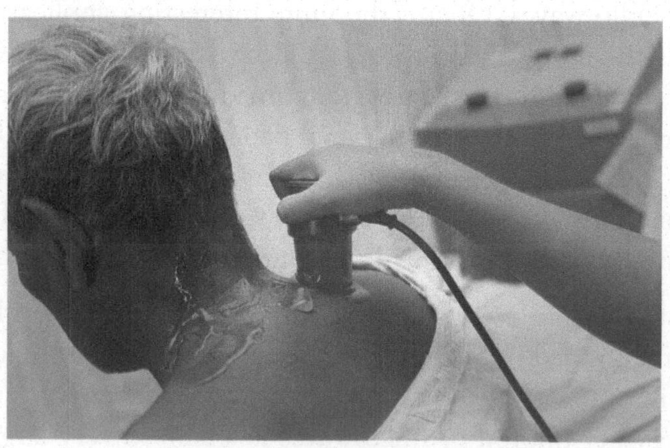

EXAMPLE 16.4

Solution: (c) The wavelength and frequency of EM radiation are related by equation 16.1, $c = f\lambda$. Solve this equation for the wavelength:

$$\lambda = \frac{c}{f} = \frac{3.0 \times 10^{8}\ \frac{\text{m}}{\text{sec}}}{\frac{2450 \times 10^{6}}{s}} = 0.120\ \text{m} = 12.0\ \text{cm}$$

The hazards as well as benefits of microwave diathermy are associated with increased temperature. Because of the wave nature of microwaves, constructive interference can occur at surfaces that reflect microwaves, especially near bones. Furthermore, since bones are curved they can act like mirrors and focus microwaves just as a curved mirror focuses visible light. These effects produce so called bone burns, which those who administer microwave diathermy are taught to avoid. In addition, some structures in the body, such as the lens of the eye, do not self-repair and are thus more susceptible to microwave damage. This destructive aspect of microwaves is now showing promise in cancer therapy, where it sometimes can be used to heat and kill tumors.

Infrared Radiation. Infrared is used for deep heating, first mentioned in Section 5.5. The common heat lamp (Figure 16.9) puts out most of its energy in the infrared region of the EM spectrum and a small amount in the red. Infrared does not penetrate as deeply and does not present the same hazards as microwaves. About 95% of infrared that falls on skin is absorbed in the upper layers of tissue. The interior of the body is heated by blood circulation (forced convection) from the warmer surface. Infrared, like microwave, photon energies are too small to damage organic compounds directly.

Thermography is a diagnostic use of infrared radiation (see Section 5.5). The emission of infrared radiation, like all EM radiation, depends strongly on temperature. Sophisticated methods of photographing the body in the infrared region (thermography) can detect small temperature variations indicative of a variety of medical conditions, such as tumors and impaired blood circulation.

Figure 16.9 Diathermy using infrared lamp.

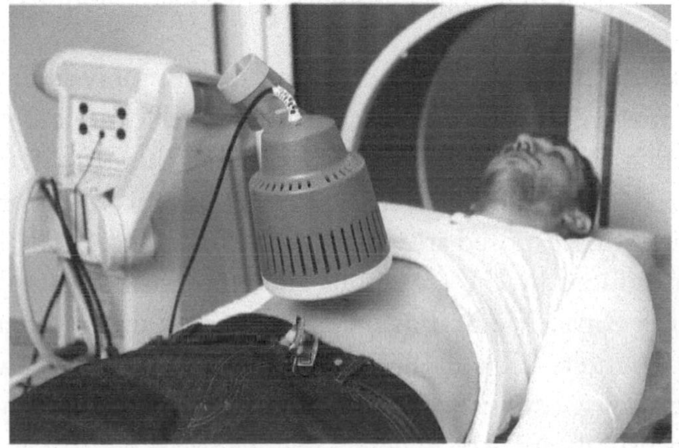

Visible Light. One simple but useful medical application of visible light is to shine it through parts of the body, a process called *transillumination*. With the patient in a darkened room a light source such as a flashlight is shown into the body at the point of interest. If the light is transmitted, a reddish glow appears. You can try this yourself by shining a flashlight through your hand. The amount of light transmitted can indicate whether a lump is a solid or watery mass, giving some indication of the type of tumor. Transillumination can also detect water on the brain (hydrocephalus) in children and collapsed lungs in infants.

Several uses of visible light employ selective absorption of particular wavelengths. The color of an object is related to the wavelengths of visible light that it absorbs and reflects. Strawberry birthmarks absorb much more green light than normal skin. These birthmarks can be removed in some cases by shining intense pure green light on them. The birthmark absorbs more of this light than normal skin, and consequently the temperature of cells in the birthmark can be increased more than that of normal cells. Successive treatments kill off the birthmark cells while doing relatively little damage to normal cells.

Another example of selective absorption is in self-cauterizing surgery using lasers. Lasers emit very pure wavelengths that can be focused to be very intense. (How lasers do this is described in the next chapter.) The type of laser is chosen so that the light it emits is selectively absorbed by blood and therefore causes more heating and cauterizing of vessels. Heat produced by absorption of the light evaporates tissue; thus lasers can be used in a variety of surgical procedures from microsurgery on fallopian tubes to plaque removal in blood vessels using fiber optics.

Newborn infants, particularly if they are premature, sometimes suffer from jaundice, a condition in which the liver excretes excessive bilirubin into the blood. The bilirubin is part of the waste produced as the infant reduces the number of red blood cells in its system. Recovery from jaundice is accelerated by exposing the infant to visible light, especially bluish green.

Fluorescent lights are used because they emit a higher percentage of blue and green than incandescent lights. The infant's eyes are blindfolded during treatment and a UV screen is placed between the light and the infant, to filter out the weak UV emitted by fluorescent lights (see Figure 16.9). The light changes the shape and structure of bilirubin molecules in such a way that they can be excreted in the urine and stool.

Ultraviolet (UV) Radiation. The energy per photon of UV is large enough to destroy organic compounds, and UV thus can be used as a germicide or sterilizer to kill microorganisms. It is most useful in sterilizing the surfaces of instruments such as combs and brushes in barber shops (it does not penetrate deeply into most materials). Similarly, UV causes sunburn but affects only the upper layers of tissue.

UV radiation is used to treat many skin conditions, commonly with careful use of an ordinary sunlamp. The high photon energy of UV apparently initiates chemical processes, such as the production of vitamin D, that alleviate certain skin conditions. There are definite long-term hazards to excessive UV exposure. The incidence of most types of skin cancer is strongly correlated with exposure to UV radiation, affecting exposed areas such as the nose, neck, and backs of hands. Luckily, the varieties of skin cancer induced by UV are among the more easily treatable.

X-Rays and Gamma Rays. Medical uses of both x-rays and gamma rays will be discussed in considerable detail in Chapters 17 and 18. Briefly, their uses fall into two major categories. First, both are used in medical diagnostics; since they are quite penetrating they can be used to probe any part of the body. X-rays usually are used to detect relative tissue density, as in finding cavities in teeth

Figure 16.10 Infant being exposed to fluorescent lighting to treat jaundice. The baby's eyes are masked to protect them from the weak UV emitted by the fluorescent light.

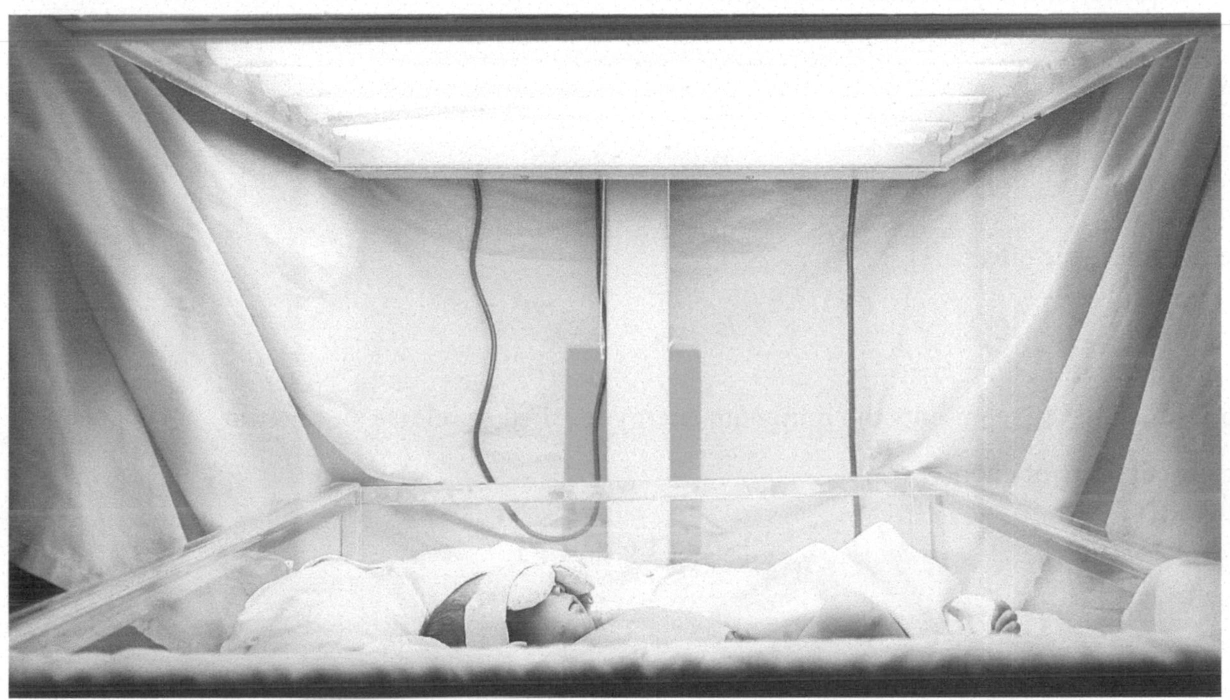

and breaks in bones. Gamma ray–emitting nuclei can be placed in various chemicals that are concentrated in specific organs, such as the thyroid. An image then can be made for medical diagnostic purposes by measuring the pattern of gamma ray emission.

The other major use of x-rays and gamma rays is in cancer therapy.

Because the energy of their photons is quite large, these rays do considerable biological damage. (A typical diagnostic x-ray photon has an energy of 50 keV, enough to disrupt about 10,000 molecules.) Most cancer cells are more sensitive to this damage than are normal cells, and the radiation often can be localized to cancerous tissue. Other therapeutic uses, such as acne treatment, were fairly common as recently as 50 years ago but have been discontinued. Exposure to x-rays and gamma rays increases the long-term risk of cancer. They are used therapeutically when their benefits exceed their risks. Diagnostic uses of x-rays and gamma rays employ much lower intensities (about 10^5 less intense), and the benefits often outweigh the risks.

Things to know for Chapter 16:

1 nanometer (nm);

$$1 \text{ nm} \equiv 1 \times 10^{-9} \text{ m}$$

The speed of light in a vacuum;

$$c = 3.00 \times 10^{8} \frac{\text{m}}{\text{s}}$$

$c = f\lambda$;

f is frequency and λ represents wavelength

Energy of a photon:

$$E_{\text{photon}} = hf = \frac{hc}{\lambda}$$

The photoelectric effect:

$$K.E._{\text{electron}} = E_{\text{photon}} - \phi$$

Work function, ϕ, represents the minimum energy required to release an electron.

Very small units of energy:

$$1 \text{ eV} \equiv 1.6 \times 10^{-19} \text{ J}$$

Questions

16.1. What type of experimental evidence indicates that EM radiation is a wave?

16.2. Explain how the pattern on the screen in Figure 16.1(c) is consistent with that expected from the ray model of light.

16.3. Why do radio waves go around buildings while visible light doesn't?

16.4. In what way do interference and diffraction effects depend upon the relative sizes of the wavelength and the object or opening that a wave encounters? Describe a circumstance under which interference or diffraction is very noticeable.

16.5. Gamma rays behave like rays most of the time and exhibit interference effects only under special circumstances. Why is this?

16.6. Given the nature of EM radiation, why is it expected to travel at the speed of light in a vacuum?

16.7. What types of forces create EM radiation? What types of forces can EM radiation exert on an object? Give an example not discussed in the text.

16.8. Explain how an oscillating charge, like the one in Figure 16.3(b), creates an oscillating magnetic field.

16.9. FM radio signals sometimes overlap with television channel 6. Would stations at the high or low end of the FM band be most likely to overlap channel 6? Explain.

16.10. Do you expect a person who has had the lens of his eye removed because of a cataract to be able to see UV radiation? Why or why not?

16.11. Resonance is the forced oscillation of an object at its natural frequency. Give an example of resonance involving EM radiation.

16.12. What is a photon? Why are effects due to individual photons difficult to observe?

16.13. What aspects of the photoelectric effect cannot be explained with the wave aspect of EM radiation alone (that is, without the photon concept)?

16.14. Exposure to UV rays, x-rays, or gamma rays cannot be made perfectly safe. On the other hand, exposure to all lower-frequency EM radiation can be made perfectly safe by limiting intensity. Explain.

16.15. There is a hazard to wearing cheap sunglasses, particularly those with plastic lenses. Such sunglasses block part of the visible light, to which the pupil responds, while almost all the infrared and UV rays pass through them. What are the hazards of the infrared and UV, respectively, in this circumstance? (Infrared reaches the retina, while UV is absorbed in the lens of the eye.)

16.16. What is a hazard of microwave deep-heat treatments (microwave diathermy)? How does this hazard differ from that presented by UV radiation?

16.17. Give an example of a medical application of EM radiation not described in the text. Explain its effects in terms of the wavelength and photon energy of the type of EM radiation used.

Problems

SECTIONS 16.1–16.4

√16.1 (I) (a) Calculate the frequency range of visible light, given its wavelength range to be 380–770 nm. (b) Calculate the ratio of the highest to lowest frequency the eye can see. (c) Calculate the ratio of the highest to lowest frequency sounds that the human ear can hear. (d) Compare the two ratios. Which of the ratios is larger?

√16.2 (I) What is the wavelength of FM radio transmission waves having a frequency of 102 MHz?

√16.3 (I) Two microwave frequencies are authorized for use in microwave ovens: 900 and 2560 MHz. Calculate the wavelength of each.

√16.4 (I) EM radiation having a wavelength of 10 μm (10,000 nm) is classified as infrared radiation. What is its frequency?

√16.5 (I) What is the smallest possible detail observable (in theory) with a microscope that uses UV light having a frequency of 1.5×10^{15} Hz?

√16.6 (I) Radar systems can be used to detect the shape and size of objects, such as aircraft or geological terrain. If a radar system uses microwaves of frequency 500 MHz, what is the smallest detail that it can detect?

√16.7 (I) The ideal size of a broadcast antenna with one end on the ground is one-fourth the wavelength of the EM radiation being broadcast. Other sizes will work, but an antenna of size about one-fourth the wavelength is most efficient. Suppose you read that a new radio station has an antenna 50 m high. (a) What frequency does this new station broadcast, assuming that it is the ideal height? (b) The frequency of an AM station is approximately 1000 kHz. The broadcast frequency of an FM station is approximately 100 megahertz. Is the station AM or FM? (c) What is the wavelength of the EM radiation being broadcast by the station?

√16.8 (I) Radar is used to determine distances to various objects by measuring the round-trip time of an echo from the object. Calculate the echo times for the following: (a) a car 100 m from the transmitter; (b) an airplane 10 km from the transmitter; and (c) the planet Venus when it is 1.50×10^{8} km from earth.

√16.9 (I) (a) Calculate the energy in electron volts of an infrared photon of frequency 1.0×10^{13} Hz. (b) Calculate the energy in electron volts of an UV photon of frequency 2.0×10^{16} Hz. (c) Compare both of these energies with the 5.0 eV needed to disrupt a certain compound and comment on their likely effects.

√16.10 (I) Repeat problem 16.9 for a radio photon of frequency 8.0×10^{5} Hz and (b) a gamma ray of frequency 2.5×10^{20} Hz.

√16.11 (II) The range of visible EM wavelengths is from 380 to 770 nm. Calculate the range of visible photon energies in electron volts.

√16.12 (II) The lowest frequency EM waves commonly encountered are those from power transmission lines, having a frequency of 60 Hz. The highest frequency gamma rays commonly encountered have a frequency of about 1.0×10^{21} Hz. Lower and higher frequencies exist but are relatively rare. (a) Calculate the energy in electron volts of a photon of each of these. (b) How many of the low-frequency photons are required to damage a DNA molecule? (About 1 eV is needed to damage DNA.) (c) How many DNA molecules could be damaged by one of these gamma ray photons?

√16.13 (II) Suppose you want to see submicroscopic details the size of atoms, about 1×10^{-10} m. This distance is the definition of an angstrom. You need to use EM radiation of wavelength 1×10^{-10} m or less. (a) Calculate the frequency of EM radiation of wavelength 1×10^{-10} m. (b) What is the energy in electron volts of a photon of this radiation? Note that since this energy is far more than is needed to disrupt the atom you observe, the measurement disrupts the object being observed.

√16.14 (II) The smallest details observable using EM radiation as a probe have a size of about one wavelength. (a) What is the smallest detail observable with UV photons of energy 10 eV? (b) What is the smallest detail observable with x-ray photons of energy 100 keV?

√16.15 (II) TV-reception antennas are constructed as shown in Figure 16.10. Each of the cross wires has a length ideal for reception of a single channel. The ideal length for reception when wires are supported at their centers as shown is $1/2\ \lambda$, where λ is the broadcast wavelength of the particular station. Suppose you measure the length of the wires for channels 5 and 11 and find them to be 1.94 and 0.753 m long, respectively. What are the broadcast frequencies of channels 5 and 11?

Figure 16.11 Analog television antenna. Each cross wire has the ideal length to receive one channel.

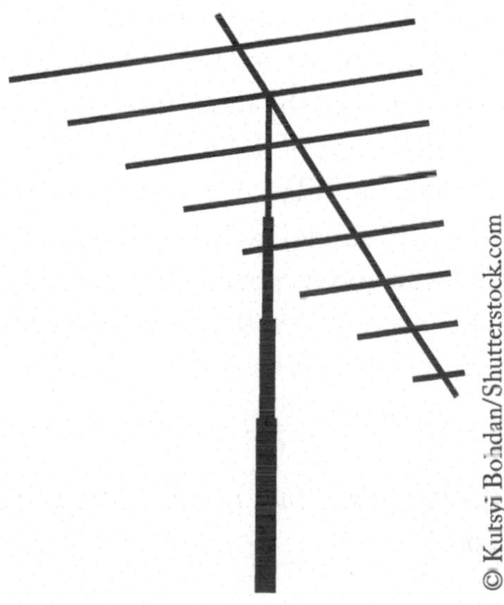

√16.16 (III) Calculate the kinetic energies in electron volts of an electron ejected from a material with a work function $\phi = 1.80$ eV by a photon of wavelength 400 nm. (b) What is the speed of the ejected electron in meters per second given its mass to be 9.11×10^{-31} kg.

√16.17 (III) A certain material has a work function $\phi = 1.50$ eV. (a) What is the frequency of a photon that ejects an electron from this material, giving the electron an energy of 1.6 eV? (b) What type of EM radiation is this?

√16.18 (III) A certain material has a work function, ϕ, of 2.00 eV and is struck by photon with wavelength of 300 nm. (a) What is the frequency of the photon? (b) What is the speed of the photon? (c) What is the energy of the photon in joules? (d) What is the energy of the photon in electron volts (eV)? (e) What is the kinetic energy in eV of the ejected electron? (f) What is the kinetic energy of the ejected electron in joules? (g) What is the velocity of the ejected electron in meter per second? Note that the mass of an electron is 9.11×10^{-31} kg.

√16.19 (III) How many x-ray photons are emitted per second by an x-ray tube that puts out 1.0 W of 75-keV x-rays?

√16.20 (III) Some older police radar units determine the speed of motor vehicles using a method analogous to the ultrasound Doppler shift technique used in medical diagnostics. Radar is bounced from a moving vehicle, and the returning wave is Doppler shifted. The echo is mixed with the original frequency, producing beats. If the police radar uses microwave frequency of 1.5×10^9 Hz and a beat frequency of 150 Hz is obtained, what is the speed of the vehicle? Because of the echo, the Doppler shift is doubled. The equation for the Doppler shift involving an echo is:

$$f_D = \frac{2 f_o V_{scatter} \cos\theta}{V_w - V_{scatter}}$$

where,

f_D is the Doppler shift. In this example it is the beat frequency.
θ is the angle of incidence which we will assume to be 0 degrees for this example,
f_0 is the operating frequency of the radar gun.
V_w in this example is the speed of light.
$V_{scatter}$ is the speed of the car. It is extremely slow compared to the speed of light.

√16.21 (III) A heat lamp emits mostly infrared radiation. Suppose the average wavelength emitted by a heat lamp is 1500 nm and its power output in EM radiation is 200 W. (a) Calculate the average photon energy in joules. (b) How many of these photons are required to raise the temperature of a person's shoulder by 2.0°C if the mass of the shoulder is 4.0 kg and its specific heat is $0.83 \frac{\text{cal}}{\text{g} \cdot {}^\circ\text{C}}$? (c) How long will this take, assuming half of the power emitted by the lamp goes into the shoulder and neglecting any bodily action to keep the shoulder cool?

√16.22 (III) On a medium power setting, a certain microwave oven emits 500 W of 2450-MHz microwaves. (a) Calculate the number of photons emitted per second. (b) How many photons are required to increase the temperature of a dish of spaghetti by 45°C, given the mass of the spaghetti to be 0.40 kg and its specific heat to be $0.90 \frac{\text{cal}}{\text{g} \cdot {}^\circ\text{C}}$? Assume that 90% of the microwaves are absorbed by the spaghetti and neglect other forms of heat transfer. (c) How many seconds are required?

√16.23 (III) One of the dangers of x-rays is that they are not sensed. The temperature rise due to exposure to x-rays is very small, whereas temperature rise is one way they might be felt. Calculate the increase in body temperature due to the energy deposited by a lethal dose of x-rays accidentally received in a short time by a scientist working on a very powerful research x-ray machine. Assume the energy of the photons is 200 keV, 3.13×10^{14} of them are absorbed per kilogram of tissue, and the specific heat of the tissue is $0.83 \frac{\text{cal}}{\text{g} \cdot {}^\circ\text{C}}$.
(This could not happen with a clinical diagnostic x-ray machine because several hours of unshielded exposure would be required.)

17

Atomic Physics

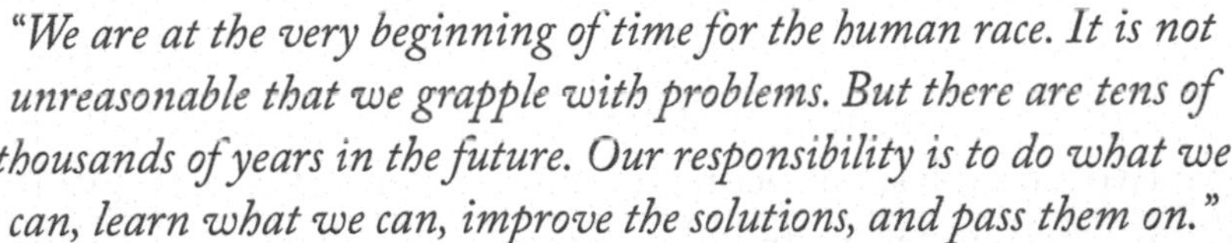

"We are at the very beginning of time for the human race. It is not unreasonable that we grapple with problems. But there are tens of thousands of years in the future. Our responsibility is to do what we can, learn what we can, improve the solutions, and pass them on."

—Richard P. Feynman

Dmitri Mendeleev

Atomic physics is the study of the characteristics of atoms and the physical principles underlying those characteristics. It is firmly in the realm of what is called modern physics. *Modern physics* is more properly referred to as quantum mechanics and relativity. Both theories emerged in the early twentieth century, refining and departing from classical physics. Our major contact with modern physics so far has been in connection with quantization, discussed in some detail for the photon in Section 16.3. We shall cover additional modern physics and its applications in atomic and nuclear physics, the subjects of the last two chapters of this book.

Atoms and molecules were previously defined as the smallest units of elements and compounds, respectively. Molecules are combinations of two or more atoms. Atoms and molecules were postulated by the ancient Greeks, but the first experimental proof of their existence came only at the turn of the 20th century, more than two thousand years later. By the end of the nineteenth century some of the properties of atoms were known in detail. For example, systematics in the chemical characteristics of the elements formed the basis of the periodic table of the elements, developed by Dmitri Mendeleev (1834–1907), a Russian chemist. A recent version of the periodic table of elements is given in Appendix C.

Although many properties of atoms were known, virtually nothing was understood about why they had these properties. Furthermore, nothing was known about what goes on inside atoms, and there were indications that classical physics might not be able to explain the behavior of atoms. This led to intensive investigations of atoms by physicists and chemists alike and culminated in the revolution called modern physics.

The purpose of this chapter is to introduce elementary atomic physics and a few of the more important new physical principles arising from it. We shall also use atomic physics to explain a number of phenomena, such as the color of lights and objects, the operation of lasers, and fluorescence. We begin with a brief description of atomic spectra. Attempts to understand atomic spectra were a strong motivation in the development of modern physics. Section 17.2 discusses the internal structure of the atom and some of its ramifications beyond the explanation of atomic spectra. The final section explains a number of other phenomena in terms of atomic physics.

17.1 ATOMIC SPECTRA

The nature of an object or substance can be studied by measuring the intensity and wavelengths of electromagnetic (EM) radiation coming from it, whether it is the original source or just reflects the EM radiation. To give two examples, hot neon gas emits wavelengths that give it a red appearance, whereas solid gold illuminated under white light appears yellow because it partially absorbs certain wavelengths. EM radiation from an object or substance can be used to identify the substance or to measure such things as its temperature, velocity, and period of rotation. Certain features of the EM radiation emitted or reflected by substances are very difficult to explain, and their study led to the development of modern atomic physic.

How does one measure the intensity of EM radiation as a function of wavelength? One method is shown in Figure 17.1. Light from an object or substance is passed through a prism. The prism

Figure 17.1 Visible light spectrum. The short wavelength, highest frequency light is refracted the most.

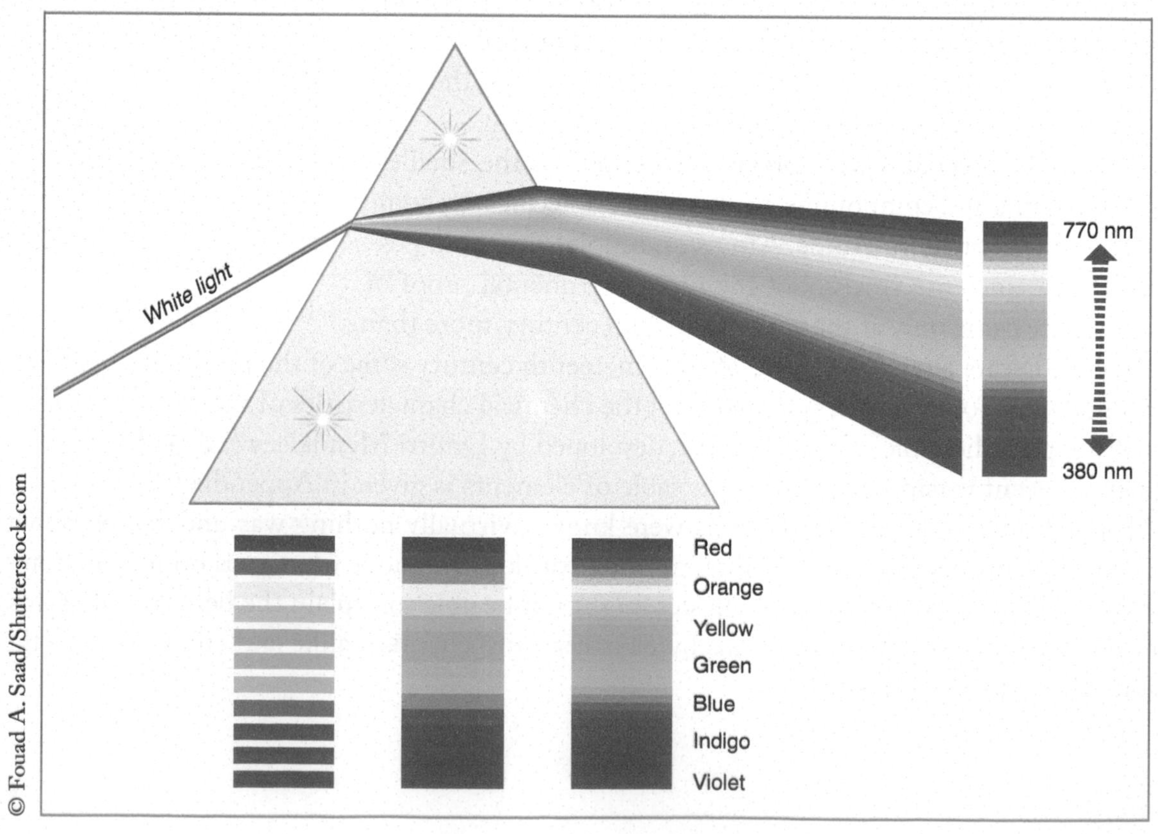

spreads out light according to its wavelength because the angle of refraction depends on wavelength. The resulting array of colors is an orderly arrangement, with larger wavelengths on the top and smaller on the bottom in the setup shown in Figure 17.1. Such as orderly array of EM radiation as a function of wavelength, as we saw in Chapter 16, is known as a spectrum. In the situation shown, intensity information is obtained from photographic film. Spectra can be extended beyond the visible, into the infrared and ultraviolet regions and even further with specialized instruments.

Let us consider the *spectra obtained in three different situations.*

1. *Cool solids* illuminated by white light. "Cool" in this case means low enough in temperature that the object does not emit large amounts of infrared or higher-frequency EM radiation. Room temperature is cool enough. The true color of an object is related to its relative absorption of visible wavelengths of EM radiation. This is exactly what is measured when a spectrum is recorded, since a spectrum gives intensity as a function of wavelength. Suppose a spectrum is taken of a yellow solid illuminated with white light. Typically, the solid reflects wavelengths near yellow very well and absorbs more of the other visible wavelengths. There are usually no sharp spikes or dips in the spectrum, which looks something like that in Figure 17.2. The spectra of cool solids are characterized by absorption of broad bands of wavelengths. The color of the substance can be explained by the response of the eye to the reflected frequencies and wavelengths (color vision is discussed in Section 15.3).
2. **Hot solids.** "Hot" in this case means at least red hot. An interesting thing happens when hot objects begin to emit infrared and visible EM radiation: most having the same temperature emit the same spectrum of EM radiation. The properties of atoms and molecules in such a solid that give it a color when cool are disrupted by thermal agitation. The resulting emissions are called blackbody radiation. (The broad bands emitted by hot solids result from the interaction of atoms with one another.) Their spectra are representative of temperature, as seen in Figure 17.3. The higher the temperature, the higher is the average kinetic energy of atoms and molecules in the solid and the higher average photon energy emitted. The shapes of blackbody spectra cannot be described with classical physics; the existence of photons is necessary to explain them. The spectra of hot solids was a major motivation for the development of modern physics.
3. **Hot gases.** The spectra of hot gases are of particular interest because they are atomic spectra—they represent EM radiation emitted by individual atoms and molecules—provided that the

Figure 17.2 EM spectrum of visible light reflected by a cool yellow object illuminated with white light.

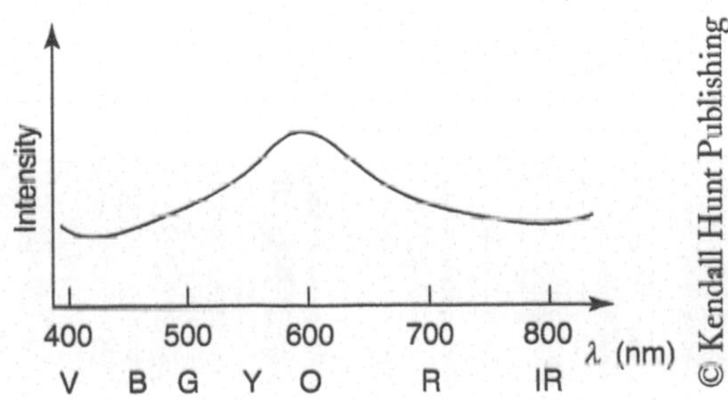

Figure 17.3 Blackbody radiation. The EM spectra of hot solids at three different temperatures. As temperature increases, the intensity of the radiation increases and its maximum shifts to smaller wavelengths. Most solids have the same shape spectrum at the same temperature.

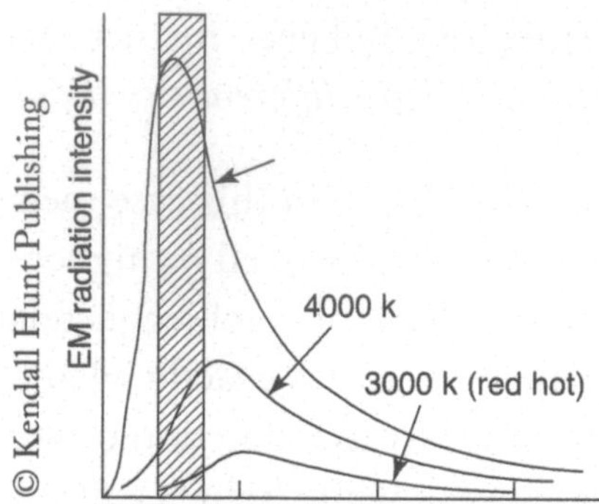

gas is not at high pressure. Atoms and molecules in such gases are separated by several atomic diameters and the forces between them are quite small. Atomic spectra are much different than the EM of spectra solids. They are notable for the presence of spectral lines; that is, a few frequencies are strongly emitted while all others are totally absent (see Figure 17.4). In addition, the frequencies emitted are very systematic; that is, atomic spectra show clear patterns.

Atomic spectra can be used as analytical tools, because of their different electron configurations, each element or compound has a unique spectrum. A sample of an unknown material is placed in an electric arc and heated to an incandescent gas, and the resulting spectrum is analyzed for the presence and relative abundance of elements and compounds. This technique has been used for over one hundred years and has applications in such diverse fields as medicine, air and water pollution studies, and forensics. More important to a physicist is the fact that to understand atomic spectra one must understand the atom.

Figure 17.4(a) Hydrogen emission spectrum.

Figure 17.4(b) Iron Emission Spectrum.

Two major questions arise at the outset: *Why are there sharp emission lines in the spectra*, and *why do the emitted frequencies fall into systematic patterns?* To answer these questions, picture a hot gas in which atoms are colliding. Part of their kinetic energy can be absorbed and stored in the atom following a collision. Eventually this energy is reemitted as photons. The sharp emission lines mean that the atoms emit only certain energies. The regular pattern for a particular element or substance means that there is some pattern in the energies that its atoms or molecules can emit.

Two other major questions now arise. *Why can atoms only emit certain energies*, and *what causes the particular patterns observed?* These questions are explored in the next section.

17.1.1 The Planetary Model of the Atom-Early Model, Useful but Incomplete

At the turn of the century a great deal was known about the chemical interactions of atoms and molecules. The sizes of atoms and molecules were also relatively well established. There was a growing body of experimental data on atomic and molecular spectra, which were used as analytical tools. However, all that was known about the internal structure of atoms and molecules was that they contained negatively charged electrons and positively charged matter of an unknown nature. Nothing else was known about the interiors of atoms until 1911 when the British physicist Ernest Rutherford (1871–1937) discovered that the nucleus of the atom was extremely small and dense. Rutherford's experiment and its implications with respect to the nucleus are discussed in Chapter 18.

Caution, not to scale. The electrons are really much further from the nucleus than depicted here. They are also not revolving around the nucleus.

In 1913, the Danish physicist Niels Bohr (1885–1962) proposed the planetary model of the atom based on the results of Rutherford's experiment. Bohr's model pictures electrons orbiting the nucleus just as planets orbit the sun, as illustrated in Figure 17.5. In addition, Bohr said that only certain electron orbits were allowed. Although we now know that electrons don't actually orbit the nucleus, Bohr's model of the atom explained two important features of atomic spectra: the existence of strong emission lines and a pattern in those lines, as illustrated in *Figure 17.6*.

When an atom absorbs energy, one or more electrons are elevated to higher orbits. Sometime later, the attractive forces between the positive nucleus and the negative electrons cause them to fall back to a lower orbit. The energy they absorbed is emitted, often as a photon of EM radiation.

If only certain orbits are allowed, called quantization of orbits, then only certain frequencies of photons can be emitted. The pattern in the allowed orbits explains the pattern in the emission spectra. In a hot gas, all possible excitations and deexcitations occur because there are many, many atoms involved. The resulting spectrum contains information on all possible energy configurations. Conversely, experimentally determined spectra (many were very well known by 1913) can be used to determine which 'orbits' are allowed. Conservation of energy tells us that the amount of energy absorbed or emitted by an electron changing energy levels is given by

$$\Delta E = E_i - E_f \qquad \text{(eq. 17.1)}$$

where, ΔE is the difference in the energy of the two "orbits" and E_i and E_f are the initial and final energies. The most common units used in this context are electron volts (eV), defined in Chapter 16 as $1 \text{ eV} = 1.6 \times 10^{-19}$ J.

If energy is emitted as a photon of EM radiation, then the energy of the photon will equal the difference between the two energy levels. In Figure 17.7, for example, the energy difference between the second excited state and the first excited state is 1.4 eV. The photon that is emitted when the electron drops to a less excited state will exactly equal the energy difference between the two states.

$$\Delta E = E_i - E_f = hf$$

Therefore, *photon emission information* can easily be calculated using:

$$\Delta E = E_{photon} = hf$$

where, ΔE is the energy difference between the two excited states. See fig. 17.7

If you haven't already done so, please view the lecture on atomic spectra now.

Figure 17.5 The old planetary *model* of the atom pictures electrons orbiting the nucleus as planets orbit the sun. The EM force between the nucleus and its electrons is attractive, as is the gravitational force between the sun and its planets. Most of the mass of the atom is in the nucleus, just as most of the mass of the solar system is in the sun.

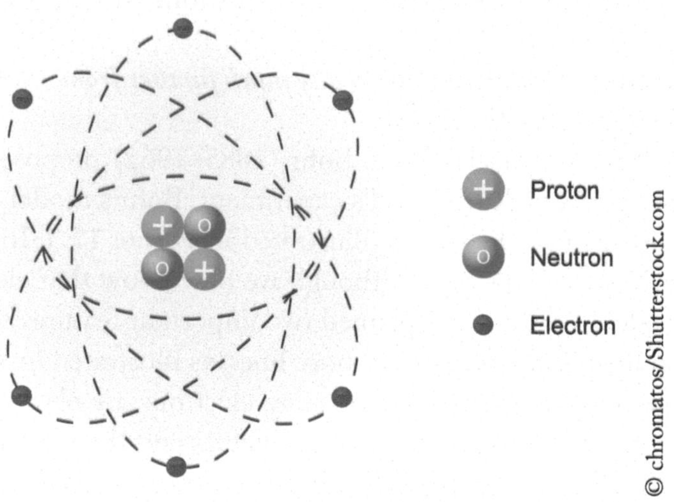

Figure 17.6(a) Energy absorbed by an atom raises one or more electrons into larger energy level. The electrons eventually fall back into lower, less excited states, reemitting the energy as a photon of light.

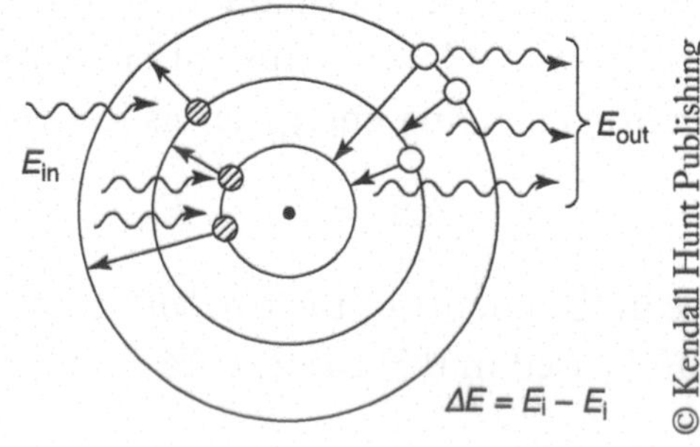

Figure 17.6(b) An energy-level diagram of the same atom's "orbits." Each horizontal line represents the energy of an electron in an "orbit" with the vertical axis proportional to energy. The bottom line is the ground state or the energy of the smallest, lowest energy level. Its energy is taken to be zero as a reference point. The next two lines represent the energies of electrons in the next two "orbits." They are the first and second excited states, respectively.

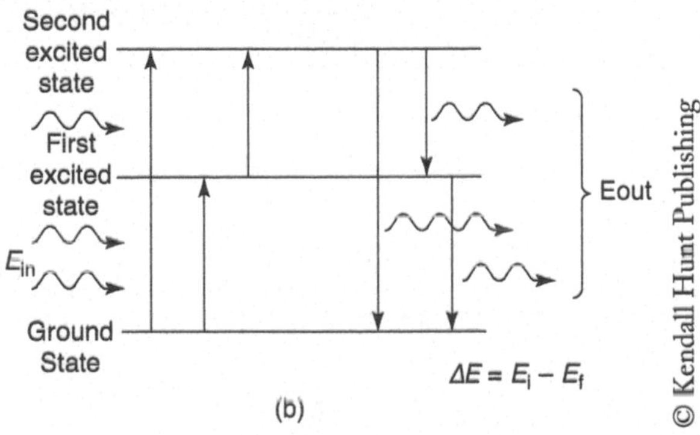

Figure 17.7 Energy diagram for Example 17.1, showing all three possible deexcitations of this atom.

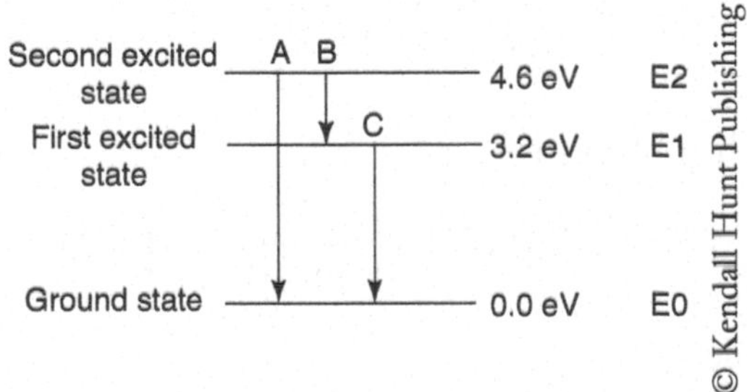

Bohr's planetary model of the atom was a giant stride forward. Now he was able to calculate the emission spectrum of the simplest atom, hydrogen, which no one had been able to do. Bohr received the Nobel Prize in 1922 for his work in this area. Let us consider an example in which the energies of the lowest levels of an atom are known and their emission spectrum can thus be calculated.

We will use the following range of wavelengths to identify types of light: x-ray 0–10 nm; *ultraviolet* 10–380 nm; *visible* 380–770 nm; *infrared* 770 nm to 3×10^6 nm; *radio* any wavelength longer than 3×10^6 nm.

EXAMPLE 17.1

Calculate the energy in joules, the frequencies in hertz and wavelengths in nanometers of transitions A, B, and C for the energy level diagram shown in Figure 17.7. Also state what type of light it is.

Iron emission spectrum

Solution: The emitted energies are given by equation 17.1, $\Delta E = E_i - E_f$, so the energies of transitions A, B, and C are:

Transition A represents an electron dropping from a 4.6 eV excited state, down to the ground state of 0 eV.

The energy in eV is:

$$\Delta E = E_{\mathrm{i}} - E_{\mathrm{f}} = 4.6\ \text{eV} - 0.0\ \text{eV} = 4.6\ \text{eV}$$

Energy in joules:

$$4.6\ \text{eV}\ \frac{1.6\times10^{-19}\ \text{J}}{1\ \text{eV}} = 7.36\times10^{-19}\ \text{J}$$

Frequency:

$$\Delta E = E_{\mathrm{i}} - E_{\mathrm{j}} = hf\text{, therefore } f = \frac{\Delta E}{h}$$

$$f = \frac{\Delta E}{h} = \frac{7.36\ \times\ 10^{-19}\ \text{J}}{6.63\ \times\ 10^{-34}\ \text{J}\times\text{s}} = 1.11\times10^{15}\ \text{Hz}$$

Wavelength:

Let's use the speed of wave equation for EM.

$$c = f\lambda\text{; therefore } \lambda = \frac{c}{f}$$

$$\lambda = \frac{c}{f} = \frac{3\times10^{8}\ \frac{\text{m}}{\text{s}}}{\frac{1.11\times10^{15}}{\text{s}}} = 2.7\times10^{-7}\ \text{m}$$

Convert the wavelength to nanometers.

$$2.7\times10^{-7}\,\text{m}\left(\frac{1\ \text{nm}}{1\times10^{-9}\,\text{m}}\right) = 270\ \text{nm}$$

Type of light: Ultraviolet

Transition B represents an electron dropping from a 4.6 eV excited state, down to a 3.2 eV.

Transition B

The energy in eV is:

$$\Delta E = E_{\mathrm{i}} - E_{\mathrm{f}} = 4.6\ \ \text{eV} - 3.2\ \ \text{eV} = 1.4\ \ \text{eV}$$

Energy in joules:

$$1.4\ \text{eV}\ \frac{1.6\times10^{-19}\ \text{J}}{1\ \text{eV}} = 2.24\times10^{-19}\ \text{J}$$

Frequency:

$$\Delta E = E_{\mathrm{i}} - E_{\mathrm{f}} = hf\text{, therefore } f = \frac{\Delta E}{h}$$

$$f = \frac{\Delta E}{h} = \frac{2.24\times10^{19}\ \text{J}}{6.63\times10^{-34}\ \ \text{J}\times\text{s}} = 3.38\times10^{14}\text{Hz}$$

Wavelength:

Let's use the speed of wave equation for EM.

$$c = f\lambda\text{; therefore } \lambda = \frac{c}{f}$$

$$\lambda = \frac{c}{f} = \frac{3\times10^{8}\ \frac{\text{m}}{\text{s}}}{\frac{3.38\times10^{14}}{\text{s}}} = 8.87\times10^{-7}\,m$$

Convert the wavelength to nanometers.

$$8.87\times10^{-7}\ \text{m}\left(\frac{1\ \text{nm}}{1\times10^{-9}\ \text{m}}\right) = 887\ \text{nm}$$

Type of light: Infrared.
Transition C represents an electron dropping from a 3.2 eV excited state, down to a 0 eV.
Transition C
The energy in eV is:

$$\Delta E = E_{\text{i}} - E_{\text{f}} = 3.2\ \text{eV} - 0.0\ \text{eV} = 3.2\ \text{eV}$$

Energy in joules:

$$3.2\ \text{eV}\ \frac{1.6\times10^{-19}\ \text{J}}{1\ \text{eV}} = 5.12\times10^{-19}\ \text{J}$$

Frequency:

$$\Delta E = E_{\text{i}} - E_{\text{f}} = hf\text{, therefore } f = \frac{\Delta E}{h}$$

$$f = \frac{\Delta E}{h} = \frac{5.12\ \times\ 10^{19}\ \text{J}}{6.63\ \times\ 10^{-34}\ \text{j}\ \times\ \text{s}} = 7.72\times10^{14}\ \text{Hz}$$

Wavelength: Let's use the speed of wave equation for EM.

$$c = f\lambda\text{; therefore } \lambda = \frac{c}{f}$$

$$\lambda = \frac{c}{f} = \frac{3\ \times\ 10^{8}\ \frac{\text{m}}{\text{s}}}{\frac{7.72\ \times\ 10^{14}}{\text{s}}} = 3.89\ \times\ 10^{-7}\ \text{m}$$

Convert the wavelength to nanometers.

$$3.89\times10^{-7}\ \text{m}\left(\frac{1\ \text{nm}}{1\times10^{-9}\ \text{m}}\right) = 389\ \text{nm}$$

Type of light: Visible (violet).

Note that these three frequencies are also ones that these atoms can absorb. This can be thought of as a resonance phenomenon; atoms of a given type can absorb and emit only certain specific frequencies of EM radiation, called the resonant frequencies of the atom. This fact helps explain why the sky is blue. You might now want to try the electron transition practice quiz. Don't worry, it has answers.

Bohr's planetary *model* of the atom is extremely useful in visualizing the nature of electron "orbits" and the emission and absorption spectra of atoms and molecules. However, two questions were not answered by Bohr's model. Bohr did not say why only certain "orbits" are allowed and why they have their observed patterns. Those questions were answered 10 years later when the wave character of matter was proposed.

17.1.2 Wave Character of Matter

In 1923, a French graduate student named Louis de Broglie (1892–1987) proposed that matter is a wave. His proposal was based on the idea that nature should be symmetric. He reasoned that since EM waves have particle properties (EM waves are composed of one or more photons, and photons behave like particles), then particles of matter should have wave properties. If true, then particles of matter, such as electrons, should exhibit constructive and destructive interference, a revolutionary idea. The importance of this idea for atomic orbits is that the "*allowed orbits" of electrons around nuclei are the only places where electrons interfere constructively* (see Figure 17.8). So we see that it was the wave properties of electrons that produced the allowed energy levels.

De Broglie also proposed an equation for the wavelength of particles of matter:

$$\lambda = \frac{h}{mv} \qquad \text{eq. 17.3}$$

where, h is Planck's constant, m is the mass, and v is the speed of the particle. This is called the de Broglie wavelength. Note that this equation is valid only for matter, not for photons. Likewise, the equation for the wavelength of photons $\left(\lambda = \frac{c}{f}\right)$ is not valid for matter.

When electrons "orbit" atoms, an integral number of electron wavelengths must fit into an "orbit" for constructive interference to occur. This is illustrated in Figure 17.8 and is written quantitatively for circular orbits as

$$2\pi r = n\lambda \text{ where, n must be integers } 1,2,3\ldots\ldots \qquad \text{eq. 17.4}$$

Figure 17.8 Electrons represented as waves in circular orbits about nuclei. (a) An allowed orbit: There is constructive interference of the electron with itself because the circumference of the orbit equals an integral number of wavelengths (three in this case). (b) A forbidden orbit. There is destructive interference of the electron with itself because the crests and troughs representing the electron fill one another. The circumference of this orbit is not equal to an integral number of electron wavelengths.

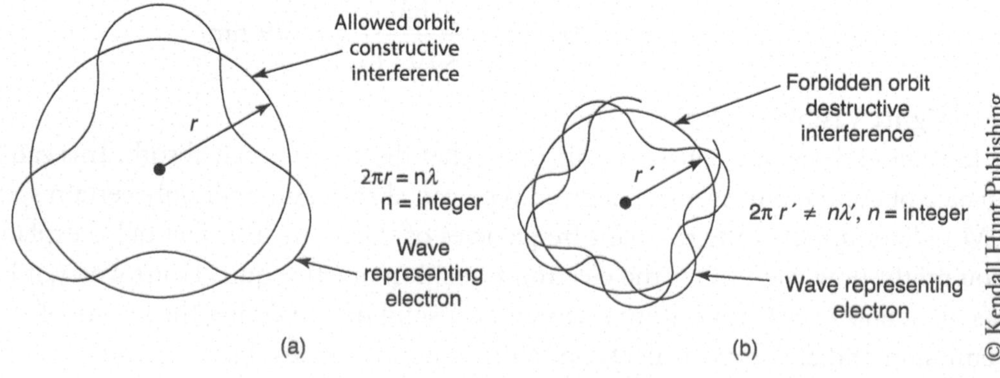

Figure 17.9 Interference pattern of electrons scattered from a crystal. Electrons strike a phosphor; dark areas are regions of destructive interference, and light spots are regions of constructive interference.

r is the radius of the orbit and n, called the principle quantum number, is the number of electron wavelengths in one circumference ($n = 3$ in Figure 17.8(a)). For those familiar with chemical notation, $n = 1$ is the K shell, $n = 2$ the L shell, $n = 3$ the M shell, and so on.

De Broglie's hypothesis of matter having a wavelength supplies the answers to the two questions of why only certain "orbits" are allowed and why the allowed "orbits" have their observed patterns. Only certain 'orbits' can satisfy Equations 17.3 and 17.4, and when physicists worked out the details, they found that those orbits had energies exactly the same as those observed experimentally. This is strong evidence for de Broglie's theory, but it is not quite direct confirmation. Direct confirmation requires actual observation of constructive and destructive interference, as in Young's double slit experiment for light. Such confirmation of the wave character of matter was supplied by C. J. Davisson and L. H. Germer in 1927, when they scattered electrons from a crystal and observed an interference pattern (see Figure 17.9). De Broglie won a Nobel Prize in 1929 for his theory, Davisson in 1937 for the crucial experimental confirmation.

You will recall that interference effects are most noticeable when waves interact with objects about the same size or smaller than their wavelength. The spacing of atoms in a crystal is similar to the wavelength of electrons used by Davisson and Germer, and they were therefore able to observe interference effects similar to those in Figure 17.9. We do not ordinarily observe them in everyday situations because the wavelength of matter is very small when compared to the size of most objects, as the next example shows.

EXAMPLE 17.2

Calculate the wavelength of the following: (a) a 5.0-kg bowling ball moving at 7.5 m/s and (b) an electron accelerated by a potential of 100 V, similar to that used by Davisson and Germer.

Solution: (a) From de Broglie's expression for the wavelength of matter,

$$\lambda = \frac{h}{mv} = \frac{6.63 \times 10^{-34}\ \text{J} \times \text{s}}{5.0\ \text{kg} \times 7.5\ \dfrac{\text{m}}{\text{s}}} = 1.8 \times 10^{-35}\ \text{m}$$

This is such an incredibly small wavelength that the bowling ball will never interact with anything small enough to exhibit noticeable interference.

Solution: (b) In order to calculate the wavelength of the fast moving electron, we see from de Broglie's equation that we must first determine its speed. An electron accelerated through an electrical potential difference of 100 Volts will gain 100 eV of kinetic energy.

Step by step 1: Convert from electron volt to joules,

$$100\ \text{eV} \times \frac{1.6\times10^{-19}\ \text{J}}{1\ \text{eV}} = 1.6\times10^{-17}\ \text{J}$$

step 2: Set the electron's kinetic energy equal to 1.6×10^{-17} J and then solve for the speed of the electron.

$$\text{KE} = \frac{1}{2}\ mv^2 = 1.6\times10^{-17}\,\text{J}$$

The mass of an electron is 9.11×10^{-31} kg
The speed of the electron is:

$$v\ = \sqrt{\frac{2\times1.6\times10^{-17}\ \text{J}}{9.11\times10^{-31}\ \text{kg}}} = 5.9\times10^{6}\ \frac{\text{m}}{\text{s}}$$

step 3: Enter the values into de Broglie's (Equation 17.3) for the wavelength of matter.

$$\lambda = \frac{h}{mv} = \frac{6.63\times10^{-34}\ \text{J}\times\text{s}}{9.11\times10^{-31}\ \text{kg}\times5.9\times10^{6}\ \frac{\text{m}}{\text{s}}} = 1.2\times10^{-10}\ \text{m}$$

This wavelength is about the same as the spacing of atoms in a crystal and therefore these electrons will exhibit interference patterns as they are scattered by a crystal. See Figure 17.9. Electron diffraction is used in solid state physics and chemistry to study the crystalline structure of solids.

There are other implications of the wave character of matter. One is that there is a limit to the smallest detail observable with an electron microscope; that limit is the electron's wavelength. Electron microscopes form an image of an object by focusing scattered electrons with magnets. Its operation is analogous to an optical microscope except that electrons, not light, are scattered from the object and that magnets, rather than glass lenses, focus the electrons. A schematic of one type of electron microscope is shown in Figure 17.10. If the electrons are accelerated by a 100-V potential, then their wavelength (see Example 17.2) is 1.2×10^{-10} m, which will be the size of the smallest detail observable. Higher accelerating potentials yield faster electrons and shorter wavelengths, allowing smaller detail to be observed. An extreme example is the 2-mile-long Stanford Linear Accelerator, which accelerates electrons with an effective potential of 25×10^9 V, giving them wavelengths small enough for the observation of details inside nuclei.

The wave character of matter is universally true. It explains why only certain electron orbits are allowed in atoms, another example of quantization. Furthermore, it appears that all particles in nature, whether they have mass-like electrons or are massless like photons, have wave characteristics. The wave-particle duality is also universal and is one of the underlying connections among

Figure 17.10 The transmission electron microscope: Electrons boiled from a hot filament are accelerated by a high potential. The condensing magnet produces a pencil-like beam, analogous to parallel rays of light. The electrons pass through the object and are differentially absorbed. The objective and projection magnets act like lenses to produce an image with a large magnification. In this type of electron microscope, the object must be thin and in a vacuum.

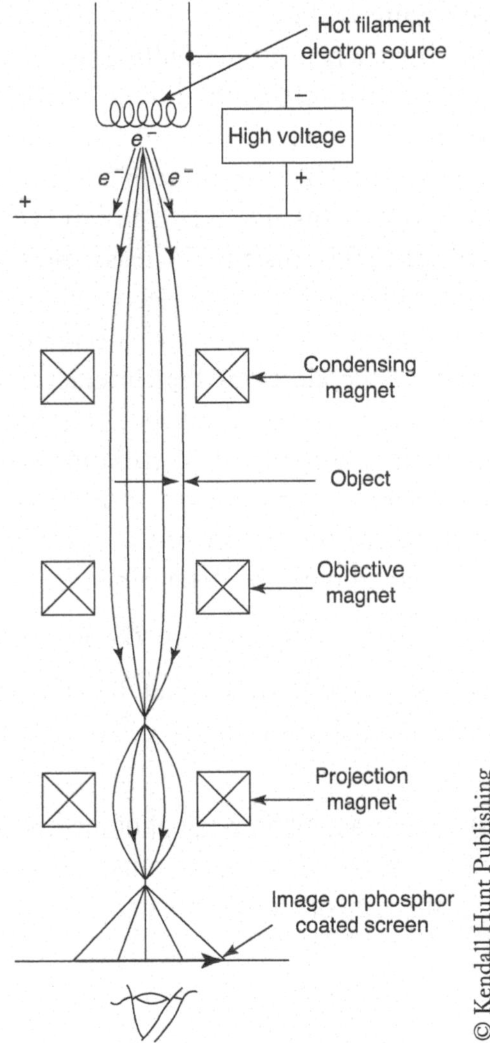

seemingly different entities. This leads to another question: *Why do these things have wave and particle characteristics? This question remains unanswered.*

Let us now use the introductory atomic physics presented in the preceding two sections to explore a number of additional phenomena, devices, and entities that are best understood at the atomic level.

17.1.3 X-Rays

X-rays are very familiar because of their widespread use in medical diagnostics. As defined in Chapter 16, they are high-frequency, short wavelength photons of light. X-rays are created in two ways, one of which is distinctly atomic in nature.

X-rays were discovered in 1895 by W. C. Roentgen (1845–1923), a German physicist. Roentgen found that photographic film wrapped in light-tight paper and placed near a cathode ray tube (CRT) was exposed and that fluorescent materials glowed near operating CRTs. He named the new radiation x-rays because its nature and identity was unknown. We know now that x-rays are energetic photons of light that are produced when electrons are accelerated by a high voltage. High voltage provides enough energy to ionize atoms, usually stripping away one or more electrons. If an inner-shell ($n = 1$, or K shell) electron is removed, an x-ray is created when an electron from a higher orbit falls into the vacant space (see Figure 17.11).

In all but the lightest atoms the energy of the resulting photon is quite large-that is, in the kilo-electron-volt (keV) range. This is the atomic method of producing x-rays and such x-rays are called *characteristic x-rays* since their energy is characteristic of the type of the atom. Characteristic x-rays always have the exact amount of energy that the electron lost as it dropped to a lower energy shell.

The other method for producing x-rays also occurs in CRTs. Most of the electrons striking the target lose their energy in multiple collisions that increase the temperature of the target and also produce a wide range of photon energies. As these electrons interact with the nuclei of the atoms in the target, they experience a change in direction or a decrease in speed and produce photons of a wide range of energies. A few of the electrons that interact with the target create high energy, photons of light as they slow down or change direction. This process is known as bremsstrahlung (*brems*) radiation. These high-energy photons of EM radiation are also x-rays. If you believe in the conservation of energy, as I know you do, then you will realize that the maximum possible energy of the resulting photons will never exceed the energy of the electron that produced it. Recall that the electrical potential energy of the electron can be determined from the definition of voltage.

$$V \equiv \frac{E}{q}; \text{ therefore } E = Vq$$

Machines designed to produce x-rays use tubes that direct a beam of electrons at a metal (tungsten) target. Both processes described above produce x-rays, and the resulting spectrum looks like that in Figure 17.12.

Figure 17.11 The production of characteristic x-ray photons. (a) An energetic electron knocks an electron from the K shell. (b) When an outer shell electron fills the vacancy, an x-ray characteristic of the type of atom is created.

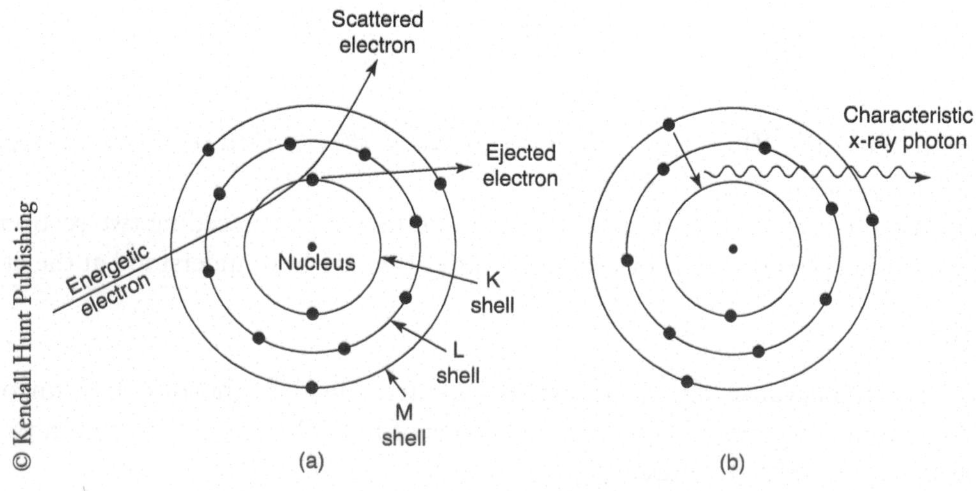

Figure 17.12 The spectrum of x-rays produced by 50-keV electrons striking a molybdenum target. The two sharp peaks are characteristic x-rays created when electrons fill vacancies in the K shell of molybdenum. The two different peaks result from electrons from different orbits falling into the K shell. The smooth, continues part of the curve represents brems radiation.

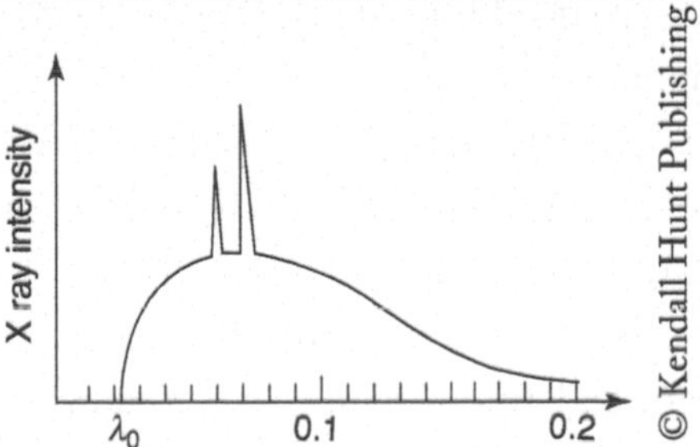

In many medical applications the voltage on the tube can be adjusted; higher voltages produce higher energy and more penetrating x- rays. Often the current in the tube can also be adjusted to control the number of x-ray photons created per unit time.

A few of the applications of x-rays have already been mentioned. Medical applications are discussed in detail in Chapter 18.

17.1.4 Fluorescence and Phosphorescence

When an atom absorbs energy, it is said to be in an excited state. This means that one or more of its electrons have been placed in larger, higher-energy configurations. Atoms can be excited by a variety of processes, such as absorption of photons and thermal collisions with other atoms. An electric current through a material can also excite atoms, since the moving charges transfer some of their energy to atoms in collisions. As discussed in Section 17.2, excited atoms eventually deexcite. Energy that was previously absorbed is reemitted during deexcitation, usually in the form of photons. The atom may return directly to its ground state, or it may deexcite in several smaller steps; both possibilities are shown in Figure 17.13. The second type of deexcitation is called fluorescence, and a different type of energy is emitted than was absorbed.

Figure 17.13 (a) Energy absorbed by an atom moves an electron to a higher energy level. (b) The electron may return to its ground state or (c) drop down in several smaller steps.

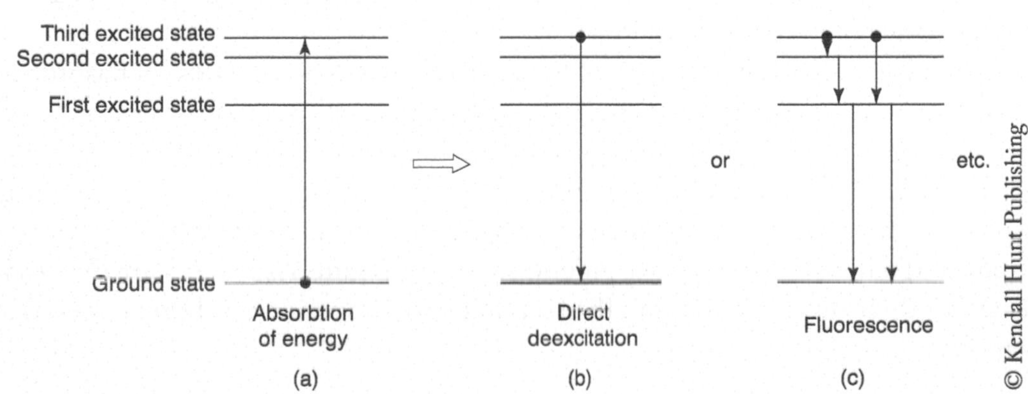

Fluorescence is a common phenomenon. The aptly named fluorescent light bulb uses it to produce light. Such lights are filled with gas through which current is passed, exciting the gas atoms, which then deexcite mostly by emission of ultraviolet radiation. The inside walls of the bulb are coated with a fluorescent material that absorbs ultraviolet and reemits visible light. The spectrum from one type of fluorescent light is shown in Figure 15.14 (notice the sharp atomic emission lines). Different fluorescent materials are used for various applications—for example, grow lights for plants and soft white lights for humans.

Fluorescence is sometimes used to identify substances. When an ultraviolet, or so-called black light, falls on certain minerals they fluoresce, emitting visible light whose color can indicate the type of mineral. X-ray induced fluorescence is a more sophisticated analytical tool that works in the same way. X-rays are shone on a sample, and the resulting fluorescence is analyzed for the type and amount of elements present. X-ray-induced fluorescence can detect very small amounts of most elements and is used in medical sample analyses, air and water pollution monitoring, and other trace element analyses. Televisions, ECG monitors, oscilloscopes, computer terminals, and all other applications of CRTs use electron-induced fluorescence to create light. The type of coating is chosen for the desired color of visible light-white in black-and-white TVs; red, blue, and green in color TVs, and so on.

The coatings in fluorescent lights and CRTs are often phosphors. Phosphorescence is a type of fluorescence in which some of the emitted photons are delayed. The use of phosphors thus prolongs the time span of fluorescence. Most excited states in atoms have short lives, typically about 10^{-8} sec. Some excited states are *metastable states* meaning that electrons in those states remain there significantly longer than the typical 10^{-8} seconds (see Figure 17.14). Lives of metastable states can be as long as minutes, many orders of magnitude longer than typical excited states. Phosphors are used, for example, in luminous clock dials that glow in the dark for a long time after being illuminated by visible light.

17.1.5 Lasers

Lasers are devices that emit EM radiation of a single, very pure wavelength. A few uses of lasers have already been discussed, but it is now possible to explain how lasers work and what makes them so useful.

There are many types of lasers made of various types of materials. Different lasers have been designed to produce microwaves, infrared radiation, visible light, and ultraviolet radiation. The one thing that all laser materials have in common is that their atoms or molecules have a metastable state. Lasers emit EM radiation by phosphorescence but in a special way. The EM radiation produced by lasers is emitted during a chain reaction of deexcitations of many atoms or molecules in *metastable states*. The sequence of events is as follows.

Atoms in a laser are excited by any of a number of possible methods. A flashtube can be used to produce a very intense, very brief flash of photons to excite the laser's atoms. Another method is to pass a current of electricity through the laser material. The excitations are random; that is, electrons are raised into many different orbits. Most excited states then decay rather quickly, with the exception of those electrons in the metastable state. This leaves many atoms in the metastable excited state and all other atoms in the ground state (see Figure 17.15). *If more than half the atoms or molecules* are in their metastable states, then a population inversion has been achieved and the laser is ready to fire.

Figure 17.14 Energy level diagram of an atom that has a metastable excited state. All possible deexcitations of the first three excited states are shown. Transition *F* will be delayed relative to all the rest. This atom will phosphoresce, emitting photons of the energy of transition *F* (pure wavelength).

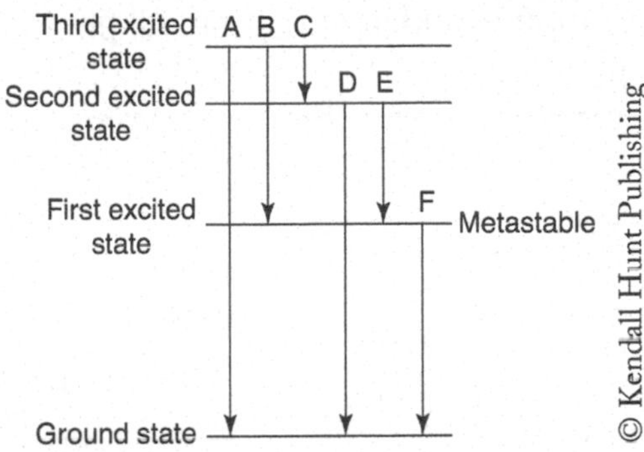

Figure 17.15 First step in the production of EM radiation by a laser. (a) Atoms are excited at random or to specific excited states. (b) Some atoms deexcite to the metastable state as shown.

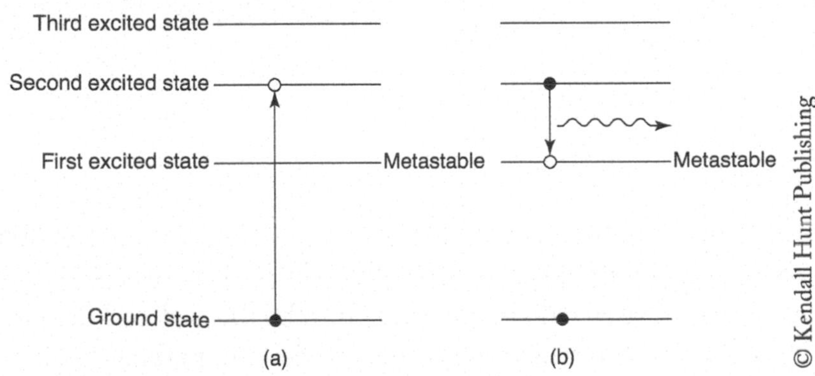

Metastable states deexcite relatively slowly by themselves, but each one emits a photon that can stimulate the deexcitation of others. A chain reaction can be produced as illustrated in Figure 17.14. The word laser is an acronym for this process, standing for light amplification by stimulated emission of radiation.

This may seem like more trouble than it's worth to produce EM radiation, but laser-produced EM radiation has two very special characteristics. First, only photons of one or at most a few, very pure wavelengths are produced. Second, photons of equal wavelengths are nearly perfectly in phase and interfere constructively with one another. They are said to be very coherent. Ordinary sources of EM radiation produce incoherent photons of many, many different wavelengths that interfere constructively and destructively with one another. The pure wavelength and coherent nature of laser-produced radiation is what makes lasers so useful.

The pure wavelengths emitted by lasers can be selectively absorbed. For example, lasers emitting infrared radiation are used to burn away cervical tumors because infrared is strongly absorbed by tissue. Infrared lasers are also used for cutting tissue because the absorbed energy can vaporize cells. Since infrared is invisible, the laser is aimed with the aid of a small visible-light source pointing in

Figure 17.16 The final steps in the production of EM radiation by a laser, (a) the metastable state in an atom spontaneously deexcites producing a photon, (b) the emitted photon stimulates the deexcitation of another atom, producing another photon of the same energy and in phase with the first photon. (c) The chain reaction of stimulated deexcitations is enhanced by mirrors. Note that most of the photons produced move parallel to one another and are in phase. All photons have exactly the same energy and wavelength.

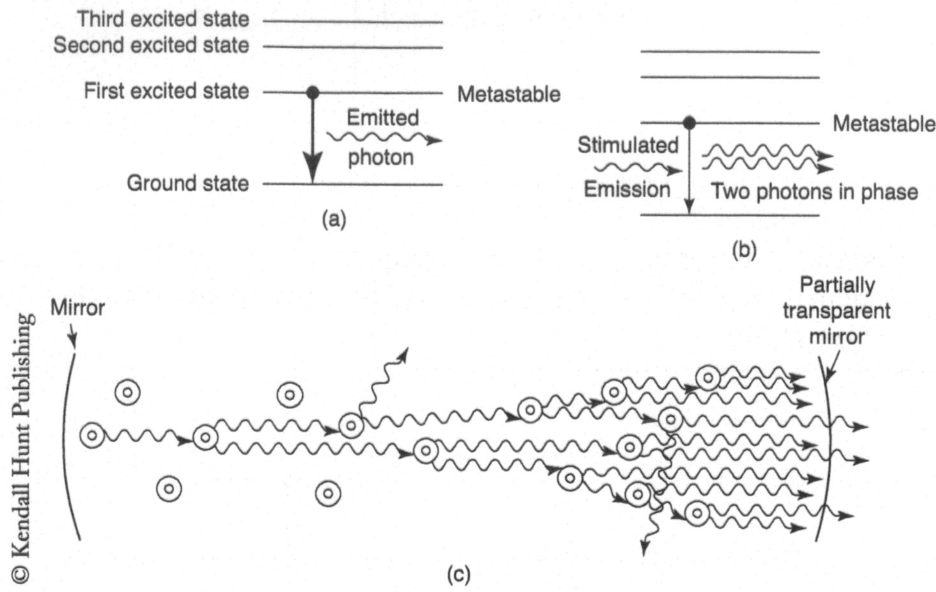

the same direction as the laser beam. A laser emitting green light is used in "bloodless" surgery. Red blood cells absorb more green light than other cells and vessels tend to self-cauterize as they are cut.

The pure wavelength of laser radiation permits it to be focused and directed very accurately. The index of refraction of most materials depends on wavelength, making it difficult to focus different wavelengths at one place. This problem is nonexistent for single-wavelength laser radiation. Thus lasers can be used for microsurgery and to spot-weld detached retinas with more accuracy than ordinary light sources. Lasers are also used to cut tiny holes in metals and to cut patterns from many layers of cloth at one time. The narrow beam of a laser spreads only slightly as it travels. It is now common practice to use lasers with great accuracy in surveying to determine distances, level fields, and layout buildings. Lasers are also used to transmit information, such as in carrying telephone conversations several miles along optical fibers. The coherent nature of laser radiation makes the wave characteristics of EM radiation easier to observe. Constructive and destructive interference are easily observed because only one pattern is produced where incoherent radiation produces many overlapping patterns.

An interesting application of the coherence of laser radiation is holography, a process for recording and reproducing three-dimensional information. One method of producing a hologram is illustrated in Figure 17.17(a).

Ordinary photography produces an image on film, but in holography the film records the interference pattern. An exposed piece of holographic film just looks gray and stippled. When laser radiation of the same type that exposed the film is passed through it, as in Figure 17.17(b), the original interference pattern is reproduced.

Figure 17.17(a) A hologram is produced when film is exposed to the interference pattern between the direct beam and light reflected from the object. The partially silvered mirror passes some light directly to the film and reflects light to illuminate the object.

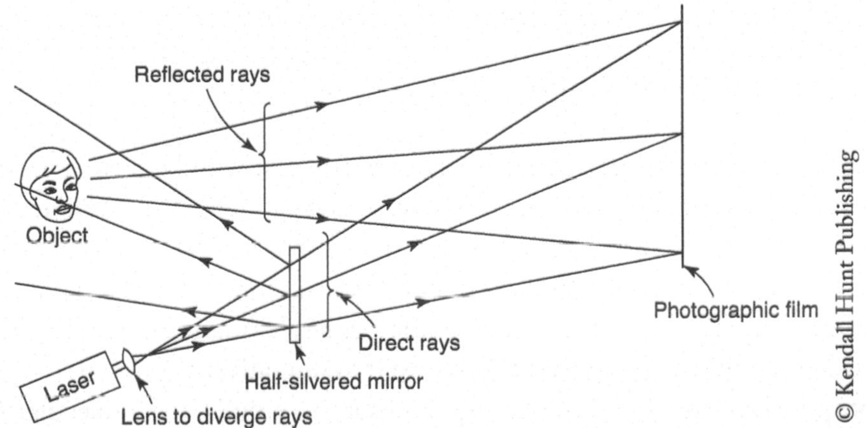

Figure 17.17(b) When laser light is passed through developed film as shown, the film scatters the light and reproduces the same interference pattern as that which exposed it. The observer therefore sees the same pattern he would see if the object were there—a three-dimensional image.

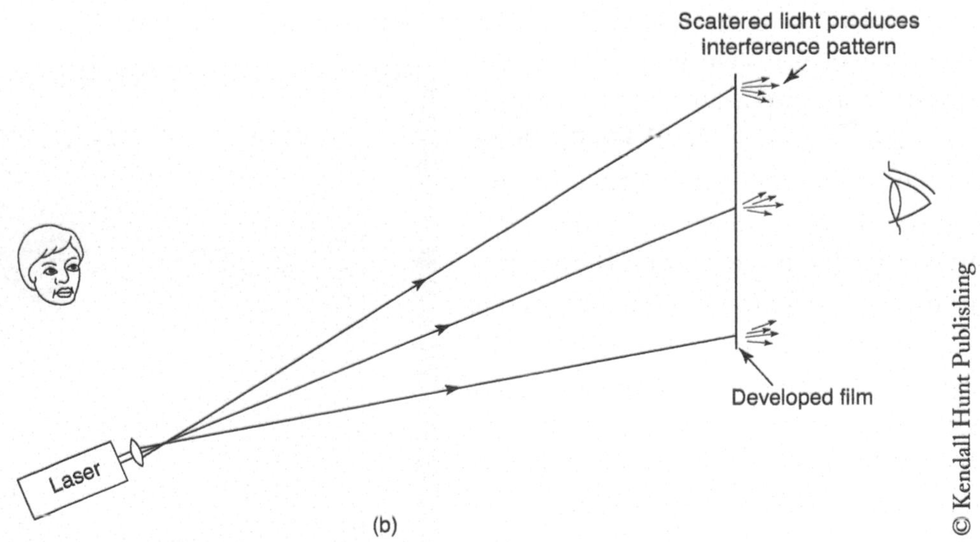

Since this is exactly like what would be produced by the objects, an observer sees a three-dimensional image of them. Holograms are often used for amusement purposes, but they also have such applications as industrial quality control, three-dimensional information storage, and automated reading of prices.

17.1.6 Other Atomic Characteristics

Atomic physics has many physical implications that differ greatly from classical concepts. References at the end of this chapter explore these in more detail than space permits here. The purpose of this subsection is to introduce a few of those atomic characteristics and their implications.

Hydrogen Atom. Hydrogen is the simplest atom, having only one electron. From the condition $2\pi r = n\lambda$ (Equation 17.4), it can be shown that the radius of an electron's orbit is proportional to n^2 and the energy of the orbit is proportional to $\frac{1}{n^2}$, where n is the principle quantum number. The last relationship is used to explain the spectrum of atomic hydrogen, since

$$E_{\text{photon}} = \Delta E = E_{\text{m}} - E_{\text{n}} \propto \left(\frac{1}{n^2} - \frac{1}{m^2} \right)$$

where, n and m are positive integers. These equations apply only to hydrogen or hydrogen-like atoms (with one electron or with one electron outside a closed shell), but they accurately describe the Balmer, Lyman, and other series of lines in such spectra that had baffled scientists for decades. More complex atoms are now well understood, although they cannot be described with equations as simple as the one above. The main point here is that the principle quantum number n not only identifies orbits but also helps explain their energies and their systematics.

More Complex Atoms. As more and more puzzles in atomic and molecular spectra were solved, it was discovered that other characteristics of the atom are also quantized. Among these are angular momentum (roughly orbital rotation) and spin momentum (roughly rotation on an axis), as well as their spatial orientations. These quantized entities are given the symbols l, m, and n, and every atomic "orbit" can be uniquely labeled in terms of n, l, m_1, and m_s. Wolfgang Pauli (1900–1958), an Austrian physicist, introduced the idea that *no two electrons can have the same quantum numbers*. Therefore, only a limited number of electrons can be in the same orbit since there are only a limited number of possible values n, l, m_l, and m_s. Pauli won the 1945 Nobel Prize for this idea, now called the *Pauli exclusion principle*, which successfully explains electron shells around atoms, a feat of major importance to chemistry.

Atoms and Chemistry: The Periodic Table. Chemistry is based on the interactions of atoms. The number of electrons in the outermost orbit of an atom is crucial to its chemistry. Those atoms with only one electron in their outer shell tend to give it up in chemical reactions, while those with one less than allowed tend to capture an electron. Atoms with filled outer orbits do neither very readily and are chemically inert (the noble gases). The periodic table of elements (see Appendix C) was originated based on the chemical properties of various elements, long before much was known about atomic physics. The exclusion principle and knowledge of the quantum numbers n, l, m_l, and m_s can be used to predict the number of electrons in various atomic orbits and hence explain the periodic table. Each grouping of elements—noble gases, halides, metals, and so on—can be explained by electron orbit configurations described by atomic physics.

Wolfgang Pauli

Quantization at the Submicroscopic Level: A measurable property is quantized if it can have only certain values; for example, n can only be a positive integer. Quantization and the rules, symmetries, and quantum numbers discovered in the study of atoms and molecules are applicable in general but are most noticeable at the submicroscopic level. The formal name for the physics developed to describe such phenomena is *quantum mechanics,* which closely approximates classical physics for large, slow-moving objects. Because we live in the macroscopic world, some of the results seem strange, even bizarre. Only a limited number of electrons can occupy the same atomic orbit, but any number of satellites can orbit the earth at the same altitude. Only certain orbits are allowed for electrons around atoms and molecules, but planets face no such restrictions. The term "atomic orbit," is a little misleading. *The electrons do not revolve around the nucleus of an atom, as planets do around the sun.* As physicists continued to explore the submicroscopic world, more quantized entities were discovered, showing that the revolutionary developments in atomic physics were but the beginning. The chain of revelations continues, and the complete picture remains to be constructed.

It is our intention here not to explore or define these concepts fully, but to indicate their existence and importance.

Things to know for chapter 17.

We will use the following range of wavelengths to identify types of light: x-ray 0–10 nm; ultraviolet 10–380 nm; visible 380–770 nm; infrared 770 nm to 3×10^6 nm; radio any wavelength longer than 3×10^6 nm.

Range of Wavelengths of Major Categories of Light

0–10 nm	10–380 nm	380–770 nm	770–3×10^6 nm	>3×10^6 nm
gamma and x-ray	ultraviolet	Visible light Violet-red	infrared	radio waves and microwaves

Questions

17.1 What is a blackbody spectrum? How is the wavelength at which it has its maximum intensity dependent on temperature?

17.2 Incandescent light bulbs usually have tungsten filaments, but they can also be constructed with filaments of other elements, such as carbon. Does the EM spectrum of an incandescent bulb vary with filament substance, temperature, or both?

17.3 Describe how the EM spectrum of a hot sparse gas can be analyzed for the relative abundances of the elements and compounds it contains.

17.4 Describe the planetary model of the atom.

17.5 Explain how the wave nature of electrons leads to the quantization of electron orbits. That is, why are only certain orbits allowed?

17.6 What experimental observations of electron behavior are considered direct evidence that electrons have wave properties?

17.7 Of all particles in nature that have nonzero masses, the electron's mass is the smallest. Explain how the small mass of the electron makes it easier to observe its wave characteristics.

17.8 Why is the wave character of macroscopic objects nearly impossible to observe?

17.9 Transmission electron microscopes are not useful for studies of living tissue. Explain why.

17.10 Redraw the energy level diagram from Figure 17.11, showing another possible mode of deexcitation of the third excited state.

17.11 How are fluorescence and phosphorescence related? Define each term as part of your answer.

17.12 Identify some of the methods by which energy can be transferred into an atom or molecule; that is, what are some of the methods for exciting atoms and molecules?

17.13 What is a metastable excited state in an atom?

17.14 The metastable state in a phosphorescent material that produces white light cannot be the first excited state. Explain why. (Hint: A phosphorescent material which has a first excited state that is metastable will produce a pure color.) You may find it useful to draw an energy level diagram as part of your explanation.

17.15 Describe how a laser produces EM radiation. What are the special characteristics of EM radiation from lasers?

17.16 Give two examples of applications of lasers not described in the text. Explain how the unique characteristics of laser-produced EM radiation relate to your examples.

17.17 The atoms or molecules in a laser material must have a metastable excited state. If the metastable state is the first excited state, the laser will emit a single pure wavelength of EM radiation. If the metastable state is a higher excited state (second, third, etc.), the laser will emit a number of very pure wavelengths. Explain. You may find it useful to draw an energy level diagram as part of your explanation.

17.18 What is unique about a hologram image? What are some uses for holography?

17.19 There are two distinct ways in which x-rays are created on an atomic scale. Describe both.

17.20 Use the conservation of energy principle to explain why the maximum x-ray photon energy produced by an x-ray tube is equal to the tube voltage.

17.21 Are the peaks in the x-ray spectrum in Figure 17.17 representative of the target material or the tube voltage? Explain.

17.22 Limited success in tattoo removal has been achieved using lasers. To remove a green tattoo with as little damage to surrounding tissue as possible, would it be best to use a green, a red, or an infrared laser? Explain.

PROBLEMS

√17.1 (I) (a) Calculate the energies in joules and frequencies in hertz, of the photons emitted as the electrons transition to a lower excited states for transition A, C, and F in Figure 17.18. (b) Identify the type of light of the emitted photons.

√17.2 (I) Calculate the maximum and minimum photon energies produced by a hot gas having the energy level scheme shown in Figure 17.18. What types of radiation are these two extremes?

√17.3 (I) If one of the excited states of the atom having the energy level scheme shown in Figure 17.18 is metastable, then a laser may be constructed from this material. If the 4.0-eV state is metastable, what wavelength EM radiation would the laser produce?

17.4 (I) Calculate the wavelength of a 10-g marble traveling at a speed of 5.0 m/s in a children's marble game. Is it likely to exhibit any wave effects?

√17.5 (II) Calculate the frequencies and wavelengths of the photons emitted in transitions B, D, and E in Figure 17.18. What types of radiation are produced by these three transitions?

√17.6 (II) What wavelengths of EM radiation are produced by a hot gas of atoms having the energy level scheme in Figure 17.18? Which if any of these photons are hazardous to humans?

√17.7 (II) (a) Which transition in an atom with the level scheme shown in *Figure 17.18* produces photons of wavelength 1.38×10^{-6} m? (b) Which produces photons of frequency 1.57×10^{15} Hz?

17.8 (II) Suppose a hot gas composed of atoms having only two excited states is observed to emit photons of energies 1.2, 1.5, and 2.7 eV. Construct an energy level diagram for this atom. See Figure 17.7 for a similar energy level diagram.

Figure 17.18 The energy-level scheme considered in Problems 17.1–17.3, 17.5–17.7, 17.9, and 17.11.

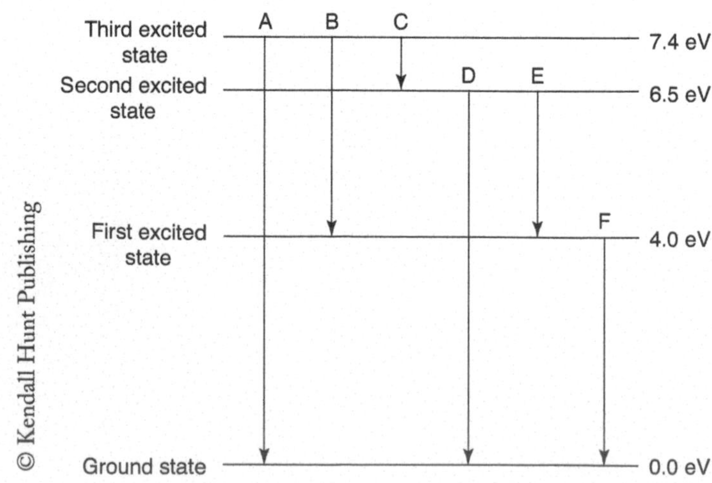

17.9 (II) A laser constructed of an element whose atomic energy level scheme shown in Figure 17.18 is observed to emit EM radiation of wavelengths 191, 311, and 497 nm. Which excited state in this atom is metastable?

17.10 (II) What wavelengths of EM radiation can be absorbed by a gas having the energy level scheme shown in Figure 17. 7?

17.11 (II) What wavelengths of EM radiation can be absorbed by a gas having the energy level scheme shown in Figure 17.18?

17.12 (II) In Example 17.2(b), the wavelength of an electron accelerated by a 100-V potential was found to be 1.2×10^{-10} m. Calculate the energy in electron volts of a photon of the same wavelength. Which do you think is easier to produce?

17.13 (II) What is the shortest-wavelength x-ray that can be created by an x-ray tube operating with a 50-kV potential? Is your answer consistent with Figure 17.17?

17.14 (II) A hot gas is observed to emit photons of six different energies: 1.3, 2.1, 2.2, 3.5, 4.3, and 5.6 eV. Construct an energy-level diagram for the atoms in this gas. (Hint: This atom has three excited states.)

17.15 (III) A fine grain of sand or a particle of dust may have a mass of 1 μg. At what speed does it have a wavelength of 600 nm (equivalent to EM radiation near the middle of the visible range)? In view of your answer, are the wave characteristics of dust particles likely to be observed?

18 Radioactivity and Nuclear Physics

"If you can't explain it simply, then you don't understand it well enough."

—Albert Einstein

Two discoveries made near the beginning of the twentieth century had major implications with respect to the atom's nucleus. In 1896 nuclear radiation was discovered accidentally by Antoine Henri Becquerel (1852–1908), a French physicist. The energy associated with nuclear radiation was large and had no apparent source. About 1911, Ernest Rutherford and his colleagues discovered that nuclei are extremely dense and therefore unlike any macroscopic substance. Classical physics could explain neither discovery. Further investigations of nuclear radiation and the nucleus were part of the revolutionary advances made in physics in the 1920s and 1930s.

The study of radiation and nuclear physics gives further insight into the exotic nature of the submicroscopic world, but applications such as cancer therapy and CT scanners are important in themselves. The chapter begins with the discovery of nuclear radiation and concludes with sections on its biological effects and medical applications.

18.1 DISCOVERY OF RADIOACTIVITY

Becquerel observed in 1896 that pitchblende, now called Uraninite, (a mineral containing uranium) emits some type of invisible, penetrating rays that can darken a photographic plate. The rays therefore carry energy. Amazingly, these rays emanate from pitchblende without energy input. This was an apparent violation of the law of conservation of energy! It was not until 1905, when Einstein published the special theory of relativity, that the source of energy could be explained as the destruction of a small amount of mass. (This explanation was not recognized for several years after 1905). The law of conservation of energy was modified to the conservation of mass plus energy, as will be discussed in *Section 18.2*. The emission of these new types of rays is nuclear radiation, also

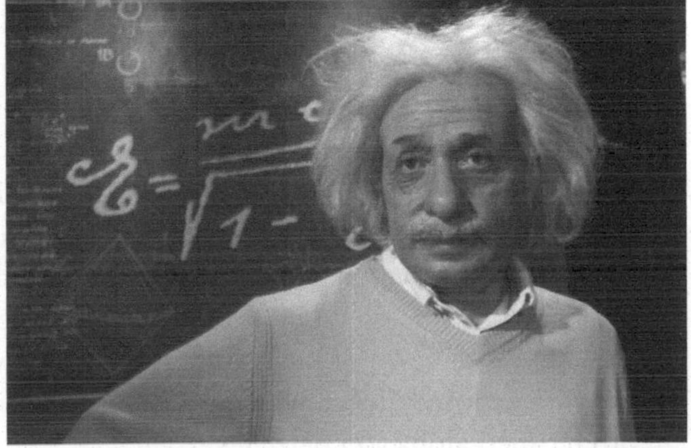

called radioactivity. *Radioactivity* is now known as the emission of particles or photons from the nucleus of an atom.

© Everett Historical/Shutterstock.com

Many scientists quickly turned their attention to Becquerel's discovery of radioactivity. Becquerel handed the study of radioactivity over to a young graduate student by the name of Marie Curie. In 1898, Marie Curie (1867–1934) and her husband Pierre Curie (1859–1906) extracted two previously unknown and highly radioactive elements from pitchblende, a dark, heavy, lustrous mineral. They named them radium and polonium. Gradually a number of elements, including uranium, were found to be radioactive. Certain types of nuclei are unstable and will eventually decay, emitting the penetrating and energetic rays discovered by Becquerel. One unstable nucleus may decay into another unstable nucleus that will also eventually decay, and so on. Such sequences are called decay chains; they end when a stable nucleus is produced. Radium and polonium are two radioactive elements in a decay chain that begins with uranium. Becquerel and the Curies shared the 1903 Nobel Prize in physics for their discoveries. In 1911, Marie also received a Nobel Prize in Chemistry. Marie Skłodowska-Curie was a French–Polish physicist and is still the only scientist to have received the prize in both Physics and Chemistry. It soon became clear that radioactivity originates deep inside the atom—that is, in the nucleus. Radioactivity was found to be independent of the chemical state of the radioactive element. It is also independent of temperature, pressure, and other physical conditions that affect electrons in their orbits. Such independence implies that radioactivity has nothing to do with electrons or their orbits about nuclei. Another indication that radioactivity is nuclear in origin is that radioactive rays have energies ranging from approximately 0.1 to 8.0 MeV (1 MeV = 10^6 eV). Radioactivity thus entails huge amounts of energy compared to typical chemical binding energies (a few electron volts) and even compared to x-rays (perhaps a few hundredths of a MeV).

18.1.1 The Three Major Types of Radioactivity

Three distinct types of nuclear radiation were quickly identified and named alpha, beta, and gamma radiation. One experiment that can distinguish them is illustrated in *Figure 18.1*. In time, alpha radiation was identified as the emission of a helium nucleus, beta radiation as the emission of an energetic electron, and gamma radiation as the emission of a high-frequency, high-energy photon of light. The following three paragraphs briefly summarize certain properties of alpha, beta, and gamma radiation.

Alpha radiation (^4_2He), is the emission of a helium nucleus by a radioactive nucleus. Helium nuclei have a positive charge twice the magnitude of an electron charge. The mass of a helium nucleus is approximately 7000 times that of an electron. Most of the energy associated with alpha radiation is in the form of kinetic energy carried by the helium nucleus. The speeds of these helium nuclei when emitted are a few percent of the speed of light. Helium nuclei emitted in alpha radiation are often called alpha (α) particles.

Beta radiation ($^{\ 0}_{-1}\beta$), is the emission of an electron by a radioactive nucleus. The electron, or beta particle, is created at the time of decay as a neutron changes into a proton. The beta particle (β)

Figure 18.1 Identification of the three major types of radioactivity: alpha, beta, and gamma radiation, A beam of radiation from several radioactive sources in a lead box splits into three parts when passed through a strong magnetic field, The undeflected component must be neutral particles (gammas), while the other two components must be oppositely charged.

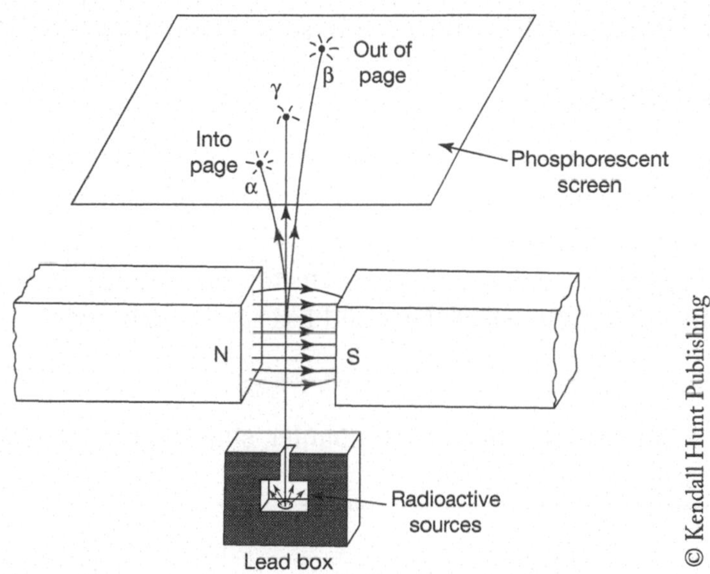

does not exist in the nucleus prior to its decay. Electrons have a negative charge (-1.6×10^{-19} C) and a small mass (9.11×10^{-31} kg). The energy carried by electrons emitted in beta decay is kinetic. The speeds of beta particles when emitted are typically greater than half the speed of light. Because of their smaller mass, beta particles have higher speeds than alpha particles at similar energies.

Gamma radiation $({}^0_0\gamma)$, is the emission of high-frequency photons from excited nuclei, similar to the emission of lower-frequency photons from excited atoms. The term gamma ray (γ) is used only for photons originating in nuclei. All photons, including gamma rays, are chargeless and massless and travel at the speed of light. The energy of a photon is related to its frequency by the familiar equation $E_{photon} = hf$ (see Section 16.3).

18.1.2 Range and Ionization

An important feature of radiation is how far it can travel in a given material before dissipating all its energy. This distance is called the range of the radiation; it has implications in shielding people from radiation and in the biological effects of radiation. At equal energies and in the same material, alpha radiation has the smallest range, beta a larger range, and gamma the largest range. Two factors that affect the range of a particular type of radiation are its charge and its speed. Radiation dissipates its energy mostly by interacting with electrons in the material. The greater the charge of the radiation, the greater are the forces between it and electrons in the material, and the more quickly it will dissipate its energy. The slower the particle, the longer time it spends in the vicinity of an atom and the greater the chance that it will interact with electrons in that atom. Both factors cause alpha radiation to have a shorter range than beta radiation and beta radiation to have a shorter range than gamma radiation. A third factor is the choice of material. Materials with the greatest electron density stop radiation best.

An important effect of nuclear radiation on materials, especially biological tissues, is that it produces ions, which are charged atoms or molecules, as it loses energy. The production of ions by any method is called *ionization*. Because the energy carried by each particle of nuclear radiation is far greater than chemical binding energies, nuclear radiation can easily split molecules into ions and strip electrons from atoms. An appreciable fraction of the energy dissipated by radiation in materials goes into ionization. It is the ionization produced in tissue that is responsible for all of the biological effects of nuclear radiation, as will be discussed in some detail in Section 18.4.

18.1.3 Radiation Detectors

All nuclear radiation detectors are related in one way or another to ionization produced by the radiation. Because x-rays also have enough energy to create significant amounts of ionization, they too may be detected using nuclear radiation detectors.

One of the most common types of radiation detectors is the Geiger counter, shown in *Figure 18.2*. The tube is filled with a specially selected gas and has a wire on its axis.

Figure 18.2 Typical construction of a Geiger counter, showing the ionization created along the path taken by the radiation.

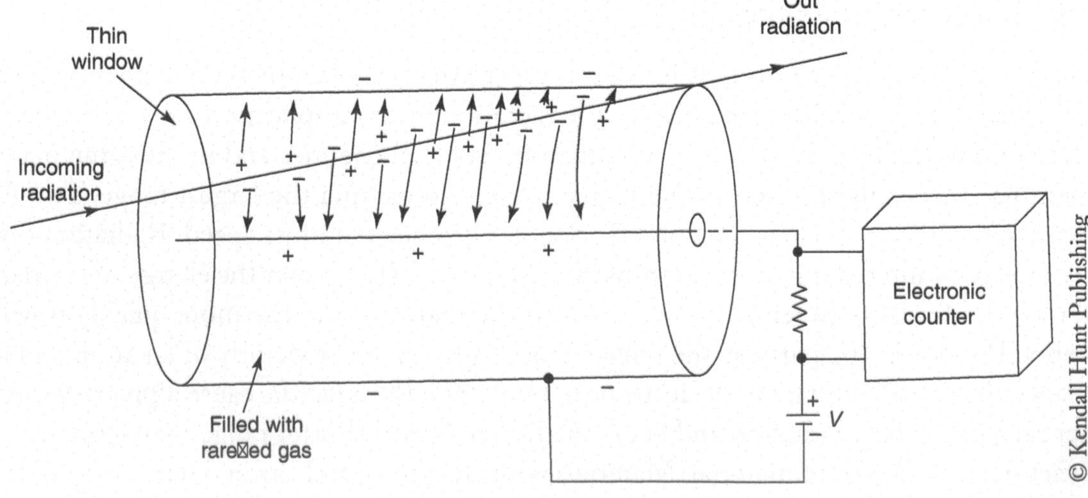

A potential difference of about 1000 V is applied between the wire and the tube. Ions produced along the path of nuclear radiation are free to move and are attracted to charges of the opposite sign. These moving charges constitute a current, which is detected by electronics connected to the Geiger tube. Once all the ions have migrated to either the wire or the tube (about 10^{-5} sec is required), the high voltage is reestablished and the tube can detect another particle. Geiger counters are simple, sturdy, and dependable, but they give no information about the type of radiation detected or its energy, nor are they 100% efficient. Geiger counters only tell how many events, or "counts," have been detected. Similar detectors, filled with different gases and using different voltages, can yield information on the energy and type of radiation detected. These more sophisticated detectors are usually more fragile, more expensive, and less efficient than Geiger counters and are used only where necessary.

Fluorescence and phosphorescence (discussed in *Section 17.3*) are also commonly used to detect radiation. One example is seen in *Figure 18.1*. In many materials some of the energy deposited by radiation causes fluorescence or phosphorescence. A flash of EM radiation called a scintillation is produced, and such materials are called scintillators. EM radiation produced by scintillators can yield information about the number, type, energy, and position of the radiation.

One method for extracting such information utilizes the photoelectric effect in a device called a photomultiplier tube (see Figure 18.3). Fluorescence and photomultiplier tubes are used in x-ray-image intensifiers and in many other applications.

Photographic films are still very useful in detecting radiation. Personnel film badges contain film in a light-tight package protected in places by various absorber materials. The amount of radiation that a person wearing the badge received is estimated by observing the darkening of the film when developed. The type and energy of the radiation can be estimated by observing how well the radiation penetrated the absorbers. Films are also used in medical diagnostic x-rays (see Section 18.5). Nuclear radiation is also detected on film, mostly in experiments.

Many other types of radiation detectors are in use, and dozens of new designs are developed each year. Those listed here are the most commonly encountered in medical and health science applications.

Figure 18.3 A scintillating material coupled to a photomultiplier (PM) tube. Radiation interacting with the scintillator creates a photon, which in turn gives energy to a photoelectron. The photoelectron is attracted to the first dynode by an applied voltage. When it strikes the dynode, it frees several more electrons, which are attracted to the next dynode, and so on, until they reach the final stage, producing an output current pulse of about a milliamp.

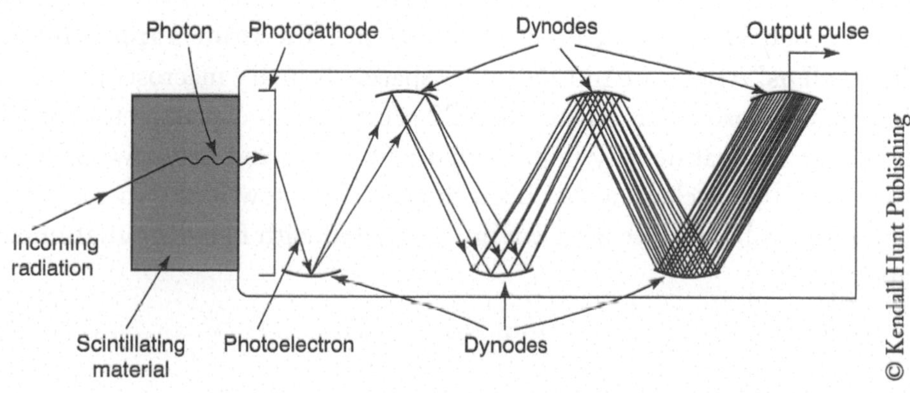

Things to know for 18.1

Marie and Pierre Curie separated three types of radiation with the use of a magnetic field and photographic plates.

The three types were:

Alpha ^4_2He the nucleus of a helium atom

Beta $^0_{-1}\beta$ an electron

Gamma $^0_0\gamma$ an energetic photon of light

18.2 NUCLEAR COMPOSITION AND NUCLEAR DECAY

18.2.1 The Composition of the Nucleus

The existence of the atom had been largely accepted by the end of the nineteenth century. About 1897, J. J. Thomson (1856–1940), a British physicist and Nobel laureate, established the existence of electrons as one basic constituent of atoms. He is credited with discovering electrons and isotopes, and inventing the mass spectrometer. Electrons were found to be negatively charged and to carry only about one part in 1830 of the mass of a proton. In 1911 Earnest Rutherford and his colleagues found that electrons form the outer boundary of the atom, while the interior is mostly empty space with a tiny nucleus at its center. Rutherford was a New Zealand-born British chemist and physicist who became known as the father of nuclear physics. He is considered the greatest experimentalist since Michael Faraday.

The Rutherford experiment is shown in *Figure 18.4(a)*. A beam of alpha particles from a radioactive source was cast upon a very thin gold foil. Most of the alpha particles passed through the foil without being deflected, indicating that the atom was mostly empty space. A few were scattered to very large angles, indicating that the nucleus was a concentrated mass, much smaller than the atom itself. Electrons surrounding the nucleus do not deflect the alpha particles, which are much more massive. (An analogy would be a car running into a moth.) Only two years after Rutherford's experiment, Bohr devised the planetary model of the atom (see Chapter 17). An expanded view of the gold foil with alpha particles passing through it is shown in *Figure 18.4 (b)*.

Detailed experiments have shown that nuclei have diameters of 10^{-14} to 10^{-15} m and atoms have diameters of about 10^{-10} m. The nucleus is very small and contains most of the mass of the atom, so its density is extremely large, on the order of 10^{12} g/cm^3. **The atom is approximately 100,000 times the size of its nucleus!** Atoms are mostly empty space. Ordinary macroscopic substances under normal conditions have densities no greater than 22 g/cm^3, about eleven factors of 10 less than nuclei.. The extremely large nuclear density implied that new forces and new laws of physics would be discovered by studying the nucleus, as has been the case. The new forces in the nucleus are extremely strong and are responsible for the large energy associated with nuclear radiation, nuclear power, and nuclear weapons.

Figure 18.4 (a) Rutherford's experiment: A beam of alpha particles is directed toward a thin gold foil. The wraparound fluorescent screen allows an observer to measure how many alpha particles are deflected and to what angles. (b) An enlarged view of the foil showing individual atoms with their tiny but massive nuclei. Most alpha particles pass through undeflected, but a few are scattered to very large angles.

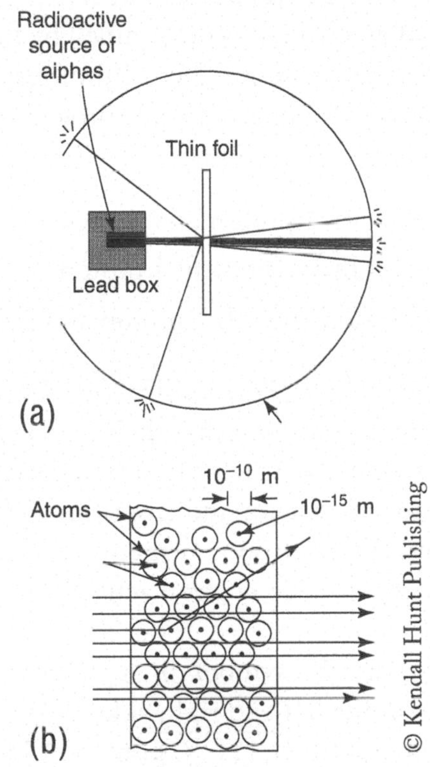

18.2.2 Constituents of Nuclei: ${}^{A}_{Z}X$

The nucleus, like the atom, has a substructure. The simplest atom is hydrogen-1, which has one proton in its nucleus and one electron relatively very far from the nucleus. Most positive charge in nature is carried by protons, which are found in every nucleus. Most nuclei also contain another particle, the neutron. Neutrons are nearly identical to protons but have no charge. They were first observed in 1932 by James Chadwick (1891–1974). Sir James Chadwick was an English Nobel laureate in physics awarded for his discovery of the neutron. Protons and neutrons are referred to as nucleons, because they are usually found in the nuclei of atoms. Each has a mass over 1830 times that of an electron. The neutron is slightly more massive than the proton and helps to glue the nucleus of an atom by adding to the strong nuclear force.

Although many combinations of protons and neutrons are possible, only a fraction of the possible combinations are stable. The nuclei that are unstable, decay and are said to be radioactive. Radioactive nuclei emit photons of light called gamma-rays, or pieces of itself in the form of particles, and are responsible for nuclear radiation. Complex nuclear forces determine which nuclear types are stable. **Appendix D** of this text lists characteristics of some of the more common radioactive nuclei.

The notation $^A_Z X$ is used to identify the type of nucleus. The symbol **X** is the chemical symbol for the atom. The symbol **A** is the total number of protons and neutrons in the nucleus. "A" is called the atomic mass or mass number because most of the mass of an atom is in its nucleus. The symbol **Z** is the number of protons in the nucleus and is called the atomic number. In one sense, Z and X are redundant; the number of protons in a nucleus indirectly determines the chemistry of an atom and therefore determines which element it is. (A nucleus with Z protons will collect Z electrons to form a neutral atom. The chemistry of an element is determined by how many electrons it has.)

An isotope of an element has the same number of protons but a different number of neutrons than other isotopes of that element. As an example, hydrogen has two naturally occurring isotopes, $^1_1H, ^2_1H$ and one man-made isotope, ^3_1H, tritium.

^1_1H is an isotope of hydrogen with 1 proton and no neutrons in its nucleus.

^2_1H, deuterium, is an isotope of hydrogen with 1 proton and 1 neutron in its nucleus.

^3_1H, tritium, is an isotope of hydrogen with 1 proton and 2 neutrons in its nucleus.

^1_1H and ^2_1H are not radioactive but ^3_1H is radioactive. Because they all have one proton, their electron configurations are identical, meaning that their chemical properties are the same.

Hydrogen-3 is produced in a nuclear reactor by a process called neutron activation. This occurs when lithium-6 is placed in a nuclear reactor and is surrounded by a large number of free neutrons. The lithium-6 nucleus captures one neutron and then splits apart, ejecting a helium nucleus and the newly created tritium isotope, ^3_1H. The neutron activation of Li-6 used to create tritium can be written as follows:

$^6_3Li + ^1_0n \rightarrow ^4_2He + ^3_1H$ where the symbol ^1_0n represents the captured neutron. Notice that there is a total charge of plus 3 on the left and right sides of the equation.

As another example of an isotope: carbon-11 $\left(^{11}_{6}C\right)$, carbon-12, $\left(^{12}_{6}C\right)$ and carbon-14, $\left(^{14}_{6}C\right)$, are all isotopes of carbon. Carbon-12 has 6 protons and 6 neutrons in its nucleus and is not radioactive. Carbon-11 has 6 protons and 5 neutrons in its nucleus while carbon-14 has 6 protons and 8 neutrons in its nucleus and both are radioactive. Note that the number of neutrons is determined by subtracting the number of protons, Z, from the mass number A. Remember, different isotopes of an element have different nuclear properties even though they are the same element and have identical chemical properties. All properties of nuclei are determined entirely by the number of protons and neutrons in its nucleus.

The notation $^A_Z X$ is also used for particles other than nuclei, such as neutrons and electrons. The neutron in this notation is written as ^1_0n, and the electron appears as $^0_{-1}\beta$ (although it is sometimes written as $^0_{-1}e$). In these cases, Z has the more general definition of the number of units of charge.

18.2.3 Decay Equations and Conservation Laws

Now that the major types of nuclear radiation and the constituents of nuclei have been discussed, we can use the $^A_Z X$ notation to write rules for nuclear decay. In practice, we usually look up a particular isotope in a table, such as that in Appendix D, to determine its mode of decay, if any. The following decay equations can then be used to determine the decay products.

Alpha Decay (^4_2He) When a nucleus alpha decays, it ejects two protons combined with two neutrons. Because it has two protons, it is a helium nucleus. Only helium has two protons. The symbol used for an alpha particle is ${}_{2}^{4}He$. The original nucleus is left with two fewer protons and neutrons than it had before it underwent decay.

An example of a nucleus that alpha decays is ${}_{92}^{238}U$ is as follows:

$${}_{92}^{238}U \rightarrow {}_{2}^{4}He + {}_{90}^{234}Th^{*} + \text{energy}$$

Total charge is always conserved. Notice that there are 92 positive charges on the left of the arrow and a total of 92 positive charges on the right of the arrow. Say it with me. *I believe in the conservation of charge.* The daughter nucleus (Th) has two less protons, so its atomic number equals 90. By looking at a periodic table of the elements, such as Appendix C, we can determine the element corresponding to 90 protons is thorium (Th). Also notice that there are 238 nucleons on the left hand side of the equation. In order to conserve mass, there must also be a total of 238 nucleons (protons + neutrons) on the right side. The atomic mass of the alpha particle is always 4. Because this is an example of alpha decay, the resulting mass of the thorium atom must be 238 - 4 = 234. Energy is released in any nuclear decay, and it need not be written explicitly in the decay equation.

Beta Decay (${}_{-1}^{0}\beta$) Beta decay follows the same conservation of charge and mass principles. We will use ${}_{-1}^{0}\beta$ to represent beta decay in our equations. Consider the following example of a carbon-14 nucleus undergoing beta decay.

$${}_{6}^{14}C \rightarrow {}_{-1}^{0}\beta + {}_{7}^{14}N + \overline{v}_{e}$$

$\overline{v}_{e}$ is a nearly massless particle known as an antineutrino that travels at almost the speed of light and does not interact with matter very easily. They rarely deposit energy and hence have no biological effects. Except for the antineutrino, the rest of the equations is pretty straight forward. As each carbon-14 atom undergoes beta decay, total mass and charge are conserved.

Let's baby step through the carbon-14 decay equation.

Step by step 1: If you know that Carbon-14 is a beta emitter, you start by writing the following:

$${}_{6}^{14}C \rightarrow {}_{-1}^{0}\beta + ?$$

Step 2: Notice that there are six positive charges on the left side of our equation. Because we believe in the conservation of charge we know that we must have a total of six positive charges on the right side. The beta has a negative one charge. Seven plus minus one equals six. Therefore we need seven more positive charges on the right hand side of the arrow in order to balance our equation and conserve charge:

$${}_{6}^{14}C \rightarrow {}_{-1}^{0}\beta + {}_{7}^{?}X$$

Step 3: Next, look at the periodic table of the elements and find atomic number 7. It has a big capital N on its chest.

$${}_{6}^{14}C \rightarrow {}_{-1}^{0}\beta + {}_{7}^{?}N$$

Step 4: Conservation of mass tells us that we need a mass number of 14 on the right hand side of the arrow:

$$^{14}_{6}C \rightarrow \, ^{0}_{-1}\beta + \, ^{14}_{7}N$$

Now you add in an antineutrino, (conservation law requires an antineutrino to be created when an electron is created in beta decay. It also requires that a neutrino be created when a positron is created in positron decay), and you get your final result of:

$$^{14}_{6}C \rightarrow \, ^{0}_{-1}\beta + \, ^{14}_{7}N + v_e$$

Carbon-14 has undergone beta decay and changed into nitrogen-14.

We can use tritium as yet another example of an isotope that undergoes beta decay.

$$^{3}_{1}H \rightarrow \, ^{0}_{-1}\beta + \, ^{3}_{2}He + \bar{v}_e$$

Please check to see that both charge and mass were conserved. The beta particle, $^{0}_{-1}\beta$, that is emitted as the tritium decays to helium, only travels approximately 6–8 mm in air.

Positron Decay Positron decay is another fairly common mode of nuclear decay. A positron is the antiparticle of electrons. Positrons have the same mass as electrons, however, they have a positive charge. The symbol $^{0}_{+1}\beta$ stands for the positron; it is also written e^+ or β^+. Positrons are rare in nature but are emitted when certain nuclei decay.

An example of a positron emitter is carbon-11:

$$^{11}_{6}C \rightarrow \, ^{0}_{+1}\beta + \, ^{11}_{5}B + v_e$$

Following the rules of conservation of charge and mass, we see that the atomic mass of the new element stays the same in positron decay, but the number of protons decreases by 1, as if one of the protons had turned into a neutron. A neutrino v_e is always emitted in positron decay; the antineutrino is its antiparticle.

When a positron and an electron meet, they annihilate one another, converting all their mass into energy in the form of two gamma-ray photons of light in the following manner:

$$^{0}_{+1}\beta + \, ^{0}_{-1}\beta \rightarrow \, ^{0}_{0}\gamma + \, ^{0}_{0}\gamma$$

Once again we see that both charge and mass are conserved. The two gamma rays photons that are produced when a positron and electron collide always have a characteristic energy of 0.511 MeV and move in opposite directions from each other. See Example 18.0 for a closer look.

EXAMPLE 18.0

To four significant digits, the mass of each electron or positron is 9.109×10^{-31} kg and the speed of light, c, is $2.998 \times 10^{8} \frac{\text{m}}{\text{s}}$. When an electron and its anti-matter equivalent, a positron, collide, their mass is converted to pure energy in the form of two identical gamma-ray photons:

$$^{0}_{+1}\beta + \, ^{0}_{-1}\beta \rightarrow \, ^{0}_{0}\gamma + \, ^{0}_{0}\gamma$$

Use Einstein's famous equation, $E = mc^2$ to determine:

a. The energy in joules of the gamma-ray photons produced when an electron and positron collide.
b. Convert the energy from joules to electron-volts and then to Mega-electron-volts.

$$\text{Note that } 1 \text{ eV} \equiv 1.602 \times 10^{-19}\, joules$$

Ans. a) The energy of each gamma ray photon produced is:

$$E = mc^2 = 9.109 \times 10^{-31}\,\text{kg}\left(2.998 \times 10^8\,\frac{\text{m}}{\text{s}}\right)^2 = 8.198 \times 10^{-14}\ joules$$

Ans. b) Converting to electron-volts and then to MeV:

$$8.198 \times 10^{-14}\ j \bullet \frac{1\text{ eV}}{1.602 \times 10^{-19}\ j} = 511{,}000\text{ eV} = 0.511\text{ MeV}$$

These gamma-ray photons have a characteristic energy of 511 keV and are used in a medical imaging technique known as positron emission tomography (PET).

Isotopes with short half-lives such as carbon-11 (~20 min), nitrogen-13 (~10 min), oxygen-15 (~2 min) are typically used in PET scans. These radionuclides are incorporated either into compounds normally used by the body such as glucose, water ammonia, or into molecules that bind to receptors or other sites of drug action. Such labeled compounds are known as radio-tracers.

The positron travels approximately 1 mm inside a patient before it encounters an electron, creating the two gamma rays that allow the imaging. PET technology can be used to trace the biologic pathway of any compound in living humans (and many other species as well), provided it can be radio-labeled with a PET isotope.

Gamma Decay ($^0_0\gamma$) A gamma ray is an energetic photon of light emitted from the nuclei of some radioactive atoms. It has zero charge and zero mass. The symbol we will use for gamma decay is $^0_0\gamma$. When the nucleons of an atom that are in an excited state, reconfigure into a lower energy state, they must give off the energy difference between the two states. In the process, the nucleus emits one or more gamma rays without changing its identity.

Below is an example of gamma decay.

An excited ^{40}Ca nucleus would gamma decay in the following manner:

$$^{40}_{20}Ca^* \rightarrow {}^{40}_{20}Ca + {}^0_0\gamma$$

The asterisk indicates that the nucleus is in an excited state. Something must excite the nucleus; in radioactive substances gamma decay is preceded by another type of decay that leaves the residual nucleus in an excited state, such as for ^{60}Co, which beta decays to Nickel nuclei whose nucleons are in an energetic configuration.

$$^{60}_{27}Co \rightarrow {}^{60}_{28}Ni^* + {}^{0}_{-1}\beta + \overline{\nu}_e$$

This is followed by the gamma decay of $^{60}Ni^{*}$ where two gamma photons of different energies are emitted. See *figure 18.5*

$$^{60}_{28}Ni^{*} \rightarrow {}^{60}_{28}Ni + {}^{0}_{0}\gamma_1 + {}^{0}_{0}\gamma_2$$

Figure 18.5 shows Cobalt-60 decaying to $^{60}_{28}Ni^{*}$ which in turn emits two gamma rays as its nucleons decay to the ground state.

Although the energy-level diagram for $^{60}_{28}Ni^{*}$ clearly shows that the two gamma photons, $^{0}_{0}\gamma_1$ and $^{0}_{0}\gamma_2$ are emitted by the Nickel nuclei, they are conventionally referred to as cobalt gamma rays. ^{60}Co is sometimes used in cancer therapy.

18.2.4 Conservation Laws

Observations of nature reveal that certain rules are always followed, related to fundamental conservation laws—such as conservation of energy and conservation of charge.

The study of nuclear radiation revealed new conservation laws. One is the conservation of the total number of nucleons. Notice that the total number of nucleons is the same on each side of all nuclear decay equations. Another conservation law is called electron family number. This conservation law requires an antineutrino to be created when an electron is created in beta decay. It also requires that a neutrino be created when a positron is created in positron decay.

There are several other conservation laws in nature (momentum, angular momentum, parity, time-reversal invariance, etc.) but we shall consider only one other here: The law of conservation of energy must be modified to be conservation of mass plus energy. You may have heard that matter cannot be created or destroyed, but that is now known to be false. The source of energy in all nuclear decays is the destruction of a small amount of mass. The decay products have a total mass slightly less than the original nucleus. Einstein was the first to realize that energy and mass are related and that each could be converted into the other. One part of Einstein's special theory of relativity, published in 1905, is his mass-energy equivalence statement:

$$E = mc^2 \qquad \text{eq. 18.1}$$

One meaning of this equation is that if an amount of mass m is "destroyed," then an amount of energy mc^2 is created, where c is the speed of light. Because c is a large number, a large amount of energy is created when a small amount of mass is destroyed. The mass destroyed in nuclear decay is

Figure 18.5 Beta decay of cobalt-60 into meta-stable Ni-60, which subsequently gamma decays to Ni-60

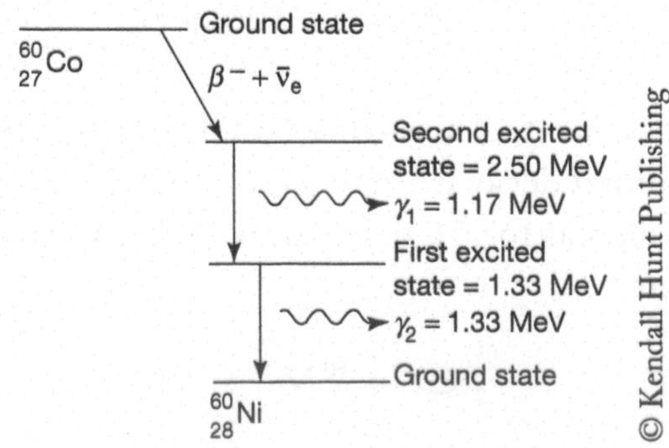

small enough that its loss is difficult to observe. Conversely, mass can be created from energy, but a large amount of energy is required to create a small amount of mass. **Most important is the fact that the total mass plus energy is conserved even though mass and energy are interconvertible**. Every second our star, the sun, converts 4 million tons of its matter into energy in the form of sunlight. The sun is losing mass as it shines. The conservation of mass plus energy is obeyed without exception in all of nature. (See also the discussion in *Section 4.2.*)

EXAMPLE 18.1

(a) Radium-226 undergoes alpha decay. Write the decay equation for ^{226}Ra. (b) The energy released in this decay is 4.8 MeV. Calculate the amount of mass "destroyed" to produce 4.8 MeV. (c) What fraction of the total mass of the ^{226}Ra nucleus was "destroyed" (Converted to energy)?

Solution 18.1: (a) From *Appendix D* we find that *Ra*-226 undergoes alpha decay and has an atomic number of 88. Therefore,

$$^{226}_{88}Ra \rightarrow {}^{4}_{2}He + {}^{222}_{86}Rn + \text{energy}$$

From Appendix C we see that an atomic number of 86 is the element radon.
Radon is also a known alpha decayer, and decays to Polonium.

$$^{222}_{86}Rn \rightarrow {}^{4}_{2}He + {}^{218}_{84}Po + \text{energy}$$

18.1 (b) The amount of mass required to release 4.8 eV of energy can be calculated from Einstein's equation, $E = mc^2$

Let's first convert the energy from electron-volts to joules:

$$4.8 \times 10^{6}\ \text{eV} \times \frac{1.6 \times 10^{-19}\ joules}{1\ \text{eV}} = 7.68 \times 10^{-13}\ joules$$

Solving for mass:

$$\text{m} = \frac{\text{E}}{\text{c}^2} = \frac{7.68 \times 10^{-13}\ \text{J}}{(3 \times 10^{8}\ \frac{\text{m}}{\text{s}})^2} = 8.53 \times 10^{-30}\ \text{kg}$$

(c) The fraction of mass "destroyed" is the ratio of mass destroyed to the mass of Ra-226.

The mass of a R-226 nucleus can be obtained by using the relation between Avogadro's number and atomic mass, namely that 1 mol (6.02×10^{23} atoms) of Ra-226 has a mass of 226 grams. Therefore, the mass of one Ra-226 atom is

$$1\ {}^{226}_{86}Ra\ \text{atom}\left(\frac{226\ \text{grams}}{6.02 \times 10^{23}\ {}^{226}_{86}Ra\ \text{atoms}}\right) = 3.75 \times 10^{-22}\ \text{grams}$$

$$3.75 \times 10^{-22}\ \text{g}\left(\frac{1\ \text{Kg}}{1000\ \text{g}}\right) = 3.75 \times 10^{-25}\ \text{Kg}$$

The fraction destroyed is $\dfrac{\text{mass destroyed}}{\text{mass of 1 atom}} = \dfrac{8.53 \times 10^{-30}\ \text{kg}}{3.75 \times 10^{-25}\ \text{kg}} = 2.27 \times 10^{-5}$

We see that 2.27 parts per 100,000 are "destroyed" That small fraction of the element's mass was converted into energy. The "missing energy" can be calculated using Einstein's famous energy—mass equivalence statement, E = mc^2. It is not surprising that early researchers did not observe these small mass losses. The amount of mass lost in a single nuclear decay is exceedingly small but can be detected by very precise measurements. In a pure macroscopic sample of a radioactive substance, all the nuclei must decay before the sample loses this fraction of its mass, and this does not happen instantaneously. The lifetimes of isotopes vary from infinite to much less than a second. For example, the half-life of ${}^{11}_{6}C$ is approximately 20 minutes, whereas the half-life of ${}^{14}_{6}C$ is 5,730 years. See appendix D for more examples and information. The next section explores these half-lives in more detail.

Things to know for section 18.2

An isotope of an element has the same number of protons but a different number of neutrons than other isotopes of that element.

$${}^{A}_{Z}X$$

Z represents the number of protons or charge.

A represents the number of protons plus the number of neutrons.

X represents the element or particle's symbol.

The symbols used for an alpha particle is ${}^{4}_{2}He$; we will use ${}^{0}_{-1}\beta$ to represent beta decay ${}^{0}_{+1}\beta$ is the symbol that represents a positron. A positron is the antiparticle of an electron. It has the same mass as an electron, but has a positive charge.

The symbol used for gamma decay is ${}^{0}_{0}\gamma$.

Total charge is always conserved

18.3 HALF-LIFE AND ACTIVITY

Some nuclei are unstable and decay; others are stable and last forever. When we say that a nucleus is decays, we mean that it emits a particle or a photon of light. It is said to be radioactive. The most unstable nuclei decay rapidly. That is, they have the shortest lifetimes.

18.3.1 Half-Life: $t_{1/2}$

The half-life of a particular radioactive substance indicates whether or not it will remain in its present form for a long time. If we start with a large number of nuclei of a particular atom ${}^{A}_{Z}X$, then *the half-life* is the time in which half the remaining nuclei will decay. The reason that a half-life is defined in this manner is that nuclear decay is a statistical process. That means that we cannot determine exactly when a nucleus will decay; it decays by chance with a certain probability. If the nucleus makes it through one half-life without decaying, its chances of living another half-life are still 50%. It is very much like tossing a coin; the chance of getting heads or tails does not depend on what happened on the last toss. Some nuclei will get through dozens of half-lives without decaying, but eventually all unstable nuclei will decay. Half-lives of various types of radioactive nuclei range from as short as 10^{-20} sec to as long as billions of years. Stable nuclei have infinite half-lives.

Because nuclear decay is a random, statistical process, more than two half-lives are required for all nuclei in a sample to decay. Table 18.1 and *Figure 18.5* illustrate what happens if we start with 1,000,000 nuclei of type A, the parent nucleus, that decay into nuclei of type B, the daughter nucleus. Fully 25% of the original nuclei are left after two half-lives, and some are still present after ten half-lives.

Radioactive dating is based on knowledge of the half-lives of naturally occurring radioactive substances. In an abstract case, if nucleus A decays producing nucleus B, then the relative amounts of A and B in an object can be used to determine its age.

Figure 18.6 is a graph of the amounts of nuclei A and B in an object as a function of time.

There are two assumptions made in the case shown. First, there were no nuclei of type B in the object when it was formed. Second, nucleus B is stable. (Radioactive dating can also be performed if these assumptions do not apply, but only after more detailed analysis.) The ratio of A to B decreases with time, and if this ratio can be measured, then the age of the object can be determined. The technique becomes inaccurate after most of A has decayed.

The best-known application of radioactive dating is used in archaeology and is called radiocarbon dating. The amount of the isotope Carbon-14 in a once living object is measured to determine the object's age. Carbon-14 has a half-life of 5730 years and although the Earth is much, much older than this (4.54 billion years), carbon-14 exists in nature. It is continually created in the upper atmosphere by energetic neutrons from the sun striking nitrogen nuclei. Among the many reactions that occur is one that produces carbon-14. The entire food chain contains Carbon-14, and it is ingested by every living organism on earth. Once the organism dies, it no longer ingests anything and the ^{14}C decays away.

The fraction of ^{14}C to ^{12}C in living organisms is small (about 1 part in 10^{12}), so only a few half-lives can pass before the remaining ^{14}C is impossible to detect accurately. This limits radiocarbon

Table 18.1 Decay of A to B

Time	A	B
0	1,000,000	0
$t_{1/2}$	500,000	500,000
$2t_{1/2}$	250,000	750,000
$3t_{1/2}$	125,000	875,000
$4t_{1/2}$	62,500	937,500
$5t_{1/2}$	31,250	968,750
$6t_{1/2}$	15,625	984,375
$7t_{1/2}$	7,813	992,187
$8t_{1/2}$	3,906	996,094
$9t_{1/2}$	1,953	998,047
$10t_{1/2}$	977	999,023

Figure 18.6 Graph of the decay of parent nucleus A and growth of daughter nucleus B as a function of time, from the data in Table 18.1.

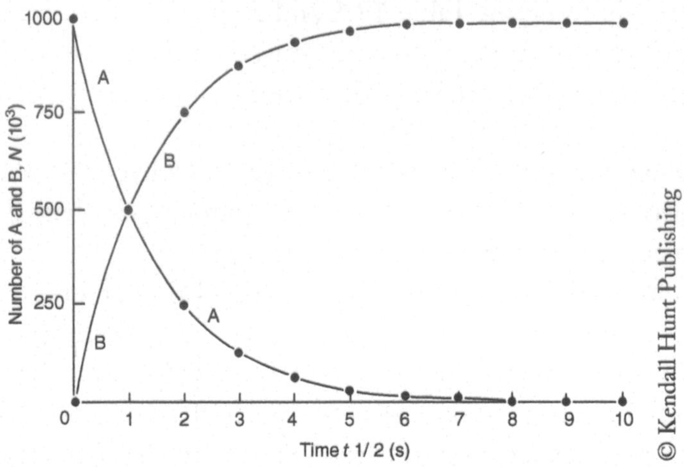

dating to objects younger than about 50,000 years (about ten half-lives). The younger the object, the more accurately it can be dated. Egyptian mummies, for example, can be dated with an accuracy of about 100 years. Radiocarbon dating has had a dramatic impact on certain fields of archaeology and earned the 1960 Nobel Prize in chemistry for its developer, W. F. Libby (1908–1980). Willard Frank Libby was an American physical chemist noted for his role in the 1949 development of radiocarbon dating, a process which revolutionized archaeology. For more on radioactive dating, follow this link.

EXAMPLE 18.2

Iodine-123, ^{123}I, is sometimes used in thyroid scans. Suppose you originally store 50 mg of the radioactive isotope ^{123}I. Approximately (a) how many half-lives and (b) how long will it take until only 1.5 mg of the 50 mg of ^{123}I is left? Note that the residual nucleus ^{123}Te is stable. From Appendix D, we see that the half-life of Iondine-123 is 13 hours..

Solution: (a) The approximate number of half-lives required to reduce the original 50 mg to 1.5 mg is obtained by repeatedly dividing the amount of ^{123}I remaining by 2, until a number close to 1.5 mg is obtained. You should try this yourself. If we divide by 2, five times, we will have approximately 1.5 mg left. Because we cut it in half, 5 times, we say that we would have to wait approximately **5 half-lives** in order for your original 50 mg sample of Iodine-123 to be reduced to 1.5 mg.

Solution: (b) The half-life of ^{123}I is 13 hours. Waiting five half-lives is the same as waiting 5×13 hours = **65 hours..**

(A little later in this section, we will show you a more accurate method to solve part b of this problem).

18.3.2 Activity: The Rate of Decay

Geiger counters often have an audio output of a pop for every count detected. When radiation levels are high, the Geiger counter emits a crackling buzz, almost like static on a radio. The number of counts per unit time is high when radiation levels are high. Activity is defined as the number of decays per unit time occurring. in a radioactive source.

$$\text{Activity} \equiv \frac{\text{Number of decays}}{\text{time}} \qquad \textbf{\textit{Know}} \qquad \text{eq. 18.2a}$$

The activity of a radioactive source depends on the number N of radioactive nuclei present and their half-life $t_{\frac{1}{2}}$. The more nuclei present, the more decays there are per unit time. The shorter the half-life, the more rapidly the nuclei decay. The relation between the activity A, the number of nuclei N, and the half-life is

$$\textit{Activity} = \frac{0.693}{t_{\frac{1}{2}}} \times \text{N} \qquad \textbf{\textit{Know}} \qquad \text{eq. 18.2b}$$

If more than one radioactive isotope is present, the total activity is the sum of the activities of the individual isotopes. (The number of nuclei, *N*, present, can be determined using Avogadro's number and the atomic mass of the isotope as seen in *Example 18.3.*)

The units used for activity are mostly straightforward units of decays per unit time, such as decays per second, decays per hour, or decays per year. Units of *curies* are also very common. The curie (Ci) has the following value:

$$1 \text{ becquerel (Bq)} \equiv 1 \text{ decay/s}$$

$$1 \text{ Ci} \equiv 3.7 \times 10^{10} \text{ decays/s}$$

1 Ci represents a large number of decays per second, and is the activity of 1 gram of radium. The curie is named after Madame and Pierre Curie, who discovered radium. Because 1 curie is so large, activity is often given in micro-curies (μCi) or millicuries (mCi).

EXAMPLE 18.3

Calculate the activity of 1.00×10^{-6} g (barely visible) of ^{198}Au, in units of decays per second and curies. This isotope is sometimes used medically. The half-life of ^{198}Au is listed in *Appendix D*, and is 2.7 days.

Solution: The activity A, can be calculated using equation 18.2. Notice that we need to know both the half-life and number of nuclei.

$$\textit{Activity} = \frac{0.693}{t_{\frac{1}{2}}} \cdot N$$

We are given the half-life of ^{198}Au as 2.7 days. We are not given the number of nuclei but we are told that we have a mass of 1.0 x 10^{-6} g of gold 198.

Step by step 1: The number of nuclei N can be calculated from the mass. One **mole** of an isotope will have a mass in grams, equal to its mass number. A mole is,

$1 \text{ mole} \equiv 6.02 \times 10^{23}$. This is also known as Avogadro's number.

In other words, 198 grams of ^{198}Au would contain 6.02 x 1023 gold nuclei.

Therefore,

$$N = 1.00 \times 10^{-6}\ \text{g} \times \frac{6.02 \times 10^{23}\ \text{nuclei}}{198\ \text{g}} = 3.04 \times 10^{15}\ \text{nuclei are present.}$$

Step 2: Substitute the half-life and the number of nuclei into the activity equation.

$$Activity = \frac{0.693}{t_{\frac{1}{2}}} \cdot N = \frac{0.693}{2.7\ \text{days}} \cdot 3.04 \cdot 10^{15} = 7.8 \times 10^{14}\ \frac{\text{decays}}{\text{day}}$$

Step 3: Convert the activity to decays per second.

$$7.8 \times 10^{14}\ \frac{\text{decays}}{\text{day}} \times \frac{1\ \text{day}}{24\ \text{h}} \times \frac{1\ \text{h}}{60\ \text{min}} \times \frac{1\ \text{min}}{60\ \text{s}} = 9.03 \times 10^{9}\ \frac{\text{decays}}{\text{s}}$$

Step 4: Convert from decays per second to Curies

$$9.03 \times 10^{9}\ \frac{\text{decays}}{\text{s}} \times \frac{1.0\ \text{Ci}}{3.7 \times 10^{10}\ \frac{\text{decays}}{\text{s}}} = 0.24\ \text{Ci}$$

An activity of 0.24 Ci is considered to be a very active or "hot" source. The curie is large unit of activity.

Looking at the equation for activity, $Activity = \frac{0.693}{t_{\frac{1}{2}}} \times N$, notice that the activity is directly proportional to the number of nuclei present. Therefore, the activity decreases in the same way that the number of nuclei N decreases. The activity will decrease by a factor of 2 every half-life. For example, cobalt 60 has a half-life of 5.3 y. After 5.3 y from the date of manufacture, a ^{60}Co source with an original activity of 50 mCi will have an activity of 25 mCi. This fact can be used to estimate when a radioactive source will reach some lower level of activity. A rule of thumb is, that after 10 to 20 half-lives, a radioactive substance decays to background. The source may then no longer be useful or, more positively, may be safe to handle.

18.3.3 How Do We Determine the Half-life of Long Lived Radioactive Substances?

As an example, let us mathematically determine the half-life of radium-226. The activity of radium-226 has been experimentally determined to be equal to $1\ \text{Ci} \equiv 3.7 \times 10^{10}\ \frac{\text{decays}}{\text{s}}$

Step by step 1: Solve equation 18.2 for half-life.

$$Activity = \frac{0.693}{t_{\frac{1}{2}}} \times N$$

$$t_{\frac{1}{2}} = \frac{0.693 \times N}{A}$$

We know that the activity, A, of 1 gram of radium-226 is $1\ \text{Ci} \equiv 3.7 \times 10^{10}$ decays/sec.

Step 2: We can use Avogadro's number to determine the number if nuclei, N, in 1 gram of radium-226 as follows:

$$N = 1\text{ g} \cdot \frac{6.022\times10^{23}}{226\text{ g}} = 2.66\times10^{21}\text{ N}$$

Step 3: Enter these values in our equation.

$$t_{\frac{1}{2}} = \frac{0.693\times N}{A} = \frac{0.693\cdot 2.66\times10^{21}\text{ nuclei}}{3.7\times10^{10}\,\frac{\text{decays}}{\text{s}}} = 4.99\times10^{10}\text{ s} = 1.58\times10^{3}\ y$$

You can compare this value to that of radium-226 listed in appendix D.

18.3.4 The Half-Life Equation

It is now time to introduce one convenient equation that connects activity, time, and half-life.

$$A = A_0 e^{-0.693\frac{T}{t_{\frac{1}{2}}}} \qquad \text{eq. 18.3a}$$

A is the present activity; A_a is the original activity

T is the elapsed time $t_{\frac{1}{2}}$ is the half-life of the isotope of that element and e is a mathematical constant, approximately equal to 2.72, which serves as the base of the natural logarithm. When using your calculator to take the natural log of something, press the ln key.

Because the activity results from the decay of individual nuclei, we can rewrite eq. 18.3 as:

$$N = N_0 e^{-0.693\frac{T}{t_{\frac{1}{2}}}} \qquad \text{eq. 18.3b}$$

N is the number of nuclei remain after time T

N_0 is the original number of nuclei

18.2b We will now use equation 18.3b to solve part b of Example 18.2b

How long will it take until only 1.5 mg of the 50 mg of ^{123}I is left? Note that the residual nucleus ^{123}Te is stable.

Solution: (a) From appendix D, we know that the half-life of ^{123}I is 13.0 h.

The half-life of ^{123}I is 13.0 h

Step by step 1: Using equation 18.3b $N = N_0 e^{-0.693\frac{T}{t_{\frac{1}{2}}}}$, we enter the known values. Because the mass is proportional to the number of nuclei, we can enter masses for N and N_0.

$$N = N_0 e^{-0.693\frac{T}{t_{\frac{1}{2}}}}$$

$$1.50\text{ mg} = 50\text{ mg}\times e^{-0.693\frac{T}{13.0\text{ h}}}$$

Step 2: Divide both sides of the equation by 50 mg.

$$\frac{1.50\text{ mg}}{50\text{ mg}} = 0.0300 = e^{-0.693\frac{T}{13.0\text{ hr}}}$$

Step 3: Note that the natural log of e to a power, is the power. Take the natural log of both sides of this equation.

$$-3.51 = -0.693 \times \frac{T}{13.0\text{ h}}$$

Step 4: Algebraically solve for *T*.

$$T = \frac{-3.51}{-0.693} \times 13.0\text{ hrs} = 65.8\text{ hr}$$

We see that after **65.8 h**, 1.5 mg of the original 50 mg sample of ^{123}I will remain.
$Activity = \frac{0.693}{t_{\frac{1}{2}}} \times N$ to determine the activity of 1.5 mg of Iodine-123.

As more time passes, ^{123}I will continue to decay and after approximately ten half-lives, its radiation will not be distinguishable from background radiation.

The next section describes the effects of ionizing radiation on biological organisms, including you and me.

The biological effects of nuclear radiation and X-rays are due to the ionization they produce. They are therefore considered together under the name ionizing radiation. Ionizing radiation can induce cancer and cause genetic defects. It can also cure cancer and is used routinely for medical diagnostic purposes.

18.3.5 Major Effects of Ionizing Radiation at the Cellular Level

1. It can interfere with cell reproduction by either inhibition or mutation.
2. It can lead to cell death.

Both effects occur for any exposure to ionizing radiation, but there will be more of the first than the second. These two effects of ionizing radiation at the cellular level can be used to explain all the macroscopic effects of radiation, such as cancer, radiation sickness, and cataract inducement, as we shall see below.

18.3.6 Rads and Rems

The biological effects of ionizing radiation are directly proportional to the amount of ionization produced in living tissue. The amount of ionization produced is in turn proportional to the energy deposited. Units of radiation dosage are therefore related to the amount of energy in ionizing radiation deposited per unit mass of tissue.

Radiation absorbed dose (Rad) One of the most common units of radiation dosage is the rad. One rad is defined as:

$$1\text{ rad} \equiv 1.00 \times 10^{-2} \frac{\text{joules of energy}}{\text{kg of exposed tissue}} \tag{18.3}$$

where, the energy in joules is the energy deposited by ionizing radiation. Note that the mass of tissue is only the tissue exposed, not the mass of the entire organism. Rads are in very common use, but a new unit called the Gray (Gy) has been introduced; 1 Gy = 100 rad.

EXAMPLE 18.4 From activity to dose

A technician at a cancer treatment facility is accidentally exposed to gamma radiation from a ^{60}Co source. Calculate the radiation dose received in rads and millirads (mrads), given the following information. The technician's mass is 75.0 kg, his entire body is exposed to 5.00×10^9 gamma rays/sec for 1 min. The gammas have an average energy of 1.25 MeV and are 50% absorbed.

Solution: Anytime a radiation dose in rads is to be calculated, the energy absorbed in joules must be found and then divided by the mass of tissue exposed (see equation 18.3):

Step by Step 1: The total number of the gamma rays in 1 minute:

$$5.00\times10^{9}\,\frac{\gamma \text{ rays}}{\text{s}}\cdot 60.0\text{ s} = 3.00\times10^{11}\ \gamma\text{ ray}$$

Step 2: The total energy of the gamma rays is:

$$3.00\times10^{11}\ \gamma\text{ rays}\cdot\frac{1.25\times10^{6}\text{ eV}}{1\ \gamma\text{ ray}} = 3.75\times10^{17}\text{ eV}$$

Step 3: Convert the energy to joules:

$$3.75\cdot10^{17}\text{ eV}\cdot\frac{1.60\times10^{-19}\,j}{1.00\text{ eV}} = 0.0600\,\textit{Joules}$$

This is the total energy contained in all of the gamma rays. However, we are told that only 50% of the energy is absorbed by the patient.

Step 4: Only 50% of the energy is deposited in the patient's tissue is:

$$0.500\times0.0600\text{ joules} = 0.0300\text{ joules deposited}$$

Step 5: The technician's mass is 75.0 kg. The **dose** in joules per kg is:

$$\frac{0.0300\text{ joules}}{75.0\text{ kg}} = 4.00\times10^{-4}\,\frac{\text{J}}{\text{kg}}$$

Step 6: Convert the dose to rads using

$$1\text{ rad} \equiv 1.00\times10^{-2}\,\frac{\text{joules of energy}}{\text{kg of exposed tissue}}$$

$$4.00\times10^{-4}\,\frac{\text{j}}{\text{kg}}\times\frac{1\text{ rad}}{1.00\times10^{-2}\,\frac{\text{j}}{\text{kg}}} = 0.0400\text{ rad}$$

Step 7: Convert rad to mrad:

$$0.0400\text{ rad}\cdot\frac{1000\text{ mrad}}{1\text{ rad}} = 40.0\text{ mrad}$$

In table 18.3, we shall see that this is a small dose.

The biological effects, such as the risk of cancer, from X-rays, beta radiation, or gamma radiation can be estimated from the number of rads of exposure. For other types of radiation, the same dose in rads may produce different effects. Another unit is needed—the **rem**, or Roentgen equivalent man.

Relative Biological Effectiveness (RBE) is the ratio of two different types of radiation required to produce the same biological effect. X-rays of 250 kV_p are often used as a frame of reference.

X-rays, gamma rays and Beta particles with energy greater than 30 KeV have a RBE of 1.
Beta particles with less than 30 KeV and neutrons with less than 0.02MeV have a RBE of 2-5.
Protons and neutrons with 1-10 MeV of energy have a RBE of 10 for the body and 30 for eyes.
Alpha particles from natural radioactivity has a RBE of 10 to 20.

Roentgen Equivalent Man (rem) Units of rems are valid for exposure to any type of ionizing radiation. The definition is

$$\text{rem} = (\text{RBE})(\text{rad}) \qquad \textbf{\textit{Know}} \qquad (18.4)$$

where, RBE stands for the relative biological effectiveness of the type of radiation. *Table 18.2* lists RBEs for several types of radiation. The larger the RBE, the more severe will be the effect of that radiation on tissue. The advantage of using rems, is that the same dose in rem produces the same effect independent of the type of radiation. [Another unit, the **Sievert** (Sv), has been introduced, but it will not be used here. **1 Sv = 100 rem**. 1 rem = 10 mSv]

The RBE is related to how well the radiation interacts with tissue: the shorter the range, the more likely the radiation will be absorbed by tissue and the larger the RBE. Alphas, for example, have an RBE of 20 because their range is so short. *Figure 18.6* shows how ionization is spread out along the paths of a gamma ray and an alpha particle having the same energies. Because their energies are the same, each creates the same number of ions; but since the alpha has such a short range, the ionization it produces is much more concentrated; that is, it produces a higher ionization density along its path. The same number of cells are adversely affected by the gamma and alpha rays, but the latter produces more-concentrated damage which the body has more difficulty repairing.

EXAMPLE 18.5

Calculate the radiation dose in rems due to exposures of (a) 0.10 rad of alpha radiation given a RBE of 20 for the alphas.

(b) Given a RBE of 1 for gamma radiation, determine the dose in rems due to exposure of 2.0 rad of gamma radiation, and (c) 40 mrad of gamma radiation (see Example 18.4).

Solution: Use the definition of rems from equation 18.4 to make the following calculations:
rem = (RBE)(rad)

a. Multiply the RBE by the exposure in rads:
Dose = 20×0.1 rad = 2.0 rem
b. Similarly,
Dose = 1×2.0 rad = 2.0 rem
c. and

Table 18.2 Relative Biological Effectiveness

Type and Energy of Radiation	RBE
X-rays	1
Gamma rays	1
Beta particles of 30 keV or more	1
Beta particles of less than 30 keV	1.7
Neutrons, thermal to slow (<0.02 MeV)	2–5
Neutrons, 1–10 MeV	10 (body), 30 (eyes)
Protons, 1–10 MeV	10 (body), 30 (eyes)
Alphas from natural radioactivity	10–20

Figure 18.7 Ionization densities created in tissue by a gamma ray and an alpha ray. Because the alpha ray interacts more strongly, its ionization density is higher and the damage it produces is more concentrated.

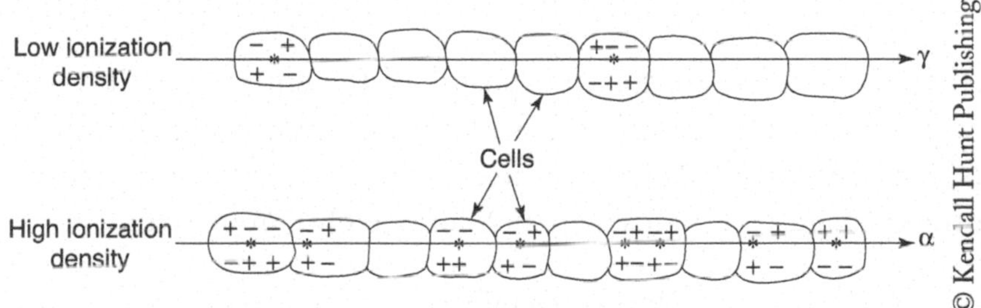

Dose = 1 × 40 mrad = 40 mrem

Notice in examples a and b, the exposure to alpha radiation produced the same biological effects as the exposure to gamma radiation, even though the alphas deposited much less energy per kilogram of tissue (0.10 versus 2.0 rad).

Because X-rays, gamma rays, and energetic beta particles have large ranges, it is not surprising that they have the same RBE. Lower kinetic energy beta particles are moving slower than the higher energy betas, and therefore have more time to interact with the tissue and hence a higher RBE than the more energetic betas. The RBEs for some other forms of radiation—such as neutrons and protons—are also listed. Slow neutrons deposit some of their energy as ionization, but their greatest effect is that they are often absorbed by a nucleus and make that nucleus unstable. It then decays, creating more ionization. The eyes are more sensitive to radiation than many other organs; hence the RBE for radiation deposited in them is higher. The lens of the eye is particularly sensitive because it does not repair damaged cells, so all damage to it is cumulative. Whenever there is a spread in the value of RBE, it is prudent to use the highest value given.

18.3.7 Biological Hazards

The biological effects of radiation exposure are directly related to the number of rems received. The effects can be divided into two categories: immediate (within days of exposure) and long-term effects.

Table 18.3 Immediate biological effects of ionizing radiation are observable only for moderate-to-large doses

Dose (rem)	
0–10 (0 to 100 mSv)	No damage
10–100 (100 mSv to 1,000 mSv)	Decrease in white blood cell count
100–200 (1000 to 2000 mSv)	Nausea is mild to severe, no appetite, considerably higher susceptibility to infection. The patient will most likely recover
200–500 (2,000 to 5,000 mSv)	Nausea much more severe, serious risk of infections, diarrhea, severe blood damage, sterility. 450 rem is lethal to 50% exposed within 30 days if untreated
500–2000 (5,000 to 20,000 mSv)	Malfunction of small intestine and blood system. Central nervous system becomes severely damaged
>2000 (20,000+ mSv)	Fatal with hours

Note that the effects listed are for an exposure of the entire body in a short time. If only part of the body is exposed, the effects will be less severe. If the exposure is spread out over days, weeks, months, or years, the effects are also less severe. This is because most organs in the body are able to repair some of the damage, so the doses are not completely cumulative.

The most easily produced observable effects are changes in blood count. Because radiation interferes with cell reproduction, the systems with the greatest cell division, such as bone marrow, are most sensitive to radiation. At somewhat higher doses, nausea and vomiting occur because the lining of the digestive system is rapidly reproducing and is therefore more sensitive than most body systems.

Death usually occurs because the immune system depends on white blood cells and is badly disrupted; the victim dies of infection or pneumonia. Treatments for very high doses of radiation include bone marrow transplants.

Fetuses and children are much more sensitive than adults. Increases in miscarriages and birth defects are noticeable in women exposed to more than 20 rem. Values of LD50, the dose that is lethal 50% of the time if untreated, vary for different types of plants and animals, the more primitive being the least susceptible. For viruses, for example, LD50 is at least 100,000 rem.

Doses of less than 10 rem are classified as low doses, those between 10 and 100 rem moderate doses, and those above 100 rem high doses.

Long-term biological effects of ionizing radiation are well established for high doses, fairly well established for moderate doses, and poorly known for low doses. The long-term effects are the inducement of cancer and genetic defects. The risk of cancer is greater and far better known than the risk of genetic effects. Both effects are due to the interruption of normal cell reproduction.

Organisms with rapid cell reproduction are again seen to be more susceptible. The only exception is that the risk to children for all cancers other than leukemia is one death per million children who have been exposed to 1 rem of radiation. The risk to adults is five deaths per million. Children are apparently able to repair more of the radiation-induced damage. The latency period for radiation-induced cancer is shortest for leukemia (blood cancer). Again, this is because of the very rapid cell division in the bone marrow. Once the latency period is over, cancer may appear anytime in the following 25–30 y.

It is necessary to consider the relative risks of low, moderate, and high doses of radiation. The values given above are based on statistics gathered for people exposed to large doses of radiation. When they are used to estimate the risk for low doses, the *linear hypothesis* is being assumed. The *linear hypothesis* is that low doses of radiation are proportionally as dangerous as high doses. For several reasons this is an overestimation of risk for low doses. One is that the body has a known ability to repair damage. Two doses of 100 rem each separated by only a week produce significantly milder immediate effects than a single dose of 200 rem. The long-term effects of 100 rem should therefore be less than half those of 200 rem. Some researchers even conclude that there is a threshold below which repair is complete and that there will be no risk from an exposure less than this threshold. The subject is still controversial, and it is perhaps prudent to overestimate risk where safety is involved.

We can estimate the percentage of people who will die of cancer from radiation exposer as follows:

$$\% = \frac{\text{Sum of absolute risk}}{1\times 10^{6}\cdot \text{yr.rem}} \cdot \text{time} \cdot \text{rem}$$

The following example illustrates how to calculate the risk of death from radiation-induced cancer.

EXAMPLE 18.6

Calculate the risk of death from radiation-induced cancer to a Hiroshima adult survivor over a period of 45 years due to a 100-rem exposure (average for survivors). The risk of death from leukemia is one person per million and the risk from all other types of cancer is five per million adults. There is a 15-year latency period.

Solution: The total risk of death is six per 10^6 people per year per rem (one for leukemia plus five for all others). Notice that there is a 15 year latency period. 45 years – 15 years = 30 years.

The percentage risk of death is therefore:

$$\% = \frac{\text{Sum of absolute risk}}{1 \times 10^6 \cdot \text{yr.rem}} \cdot \text{time} \cdot \text{rem}$$

$$\frac{6}{10^6 \text{ yr} \cdot \text{rem}} \cdot 30 \text{ y} \cdot 100 \text{ rem} = 1.8 \cdot 10^{-2} = 1.8\%$$

Note: The 15-y latency period is subtracted from the time of 45 y. The chances of contracting leukemia also diminish 25–30 y after its shorter 2-y. latency period, so the use of 30 y in the calculation is correct for all forms of cancer.

Genetic defects are another long-term effect of ionizing radiation.

They do not occur in humans with as high a frequency as predicted by laboratory studies with animals. It is possible that this is due to a larger incidence of miscarriages in humans exposed to radiation than in animals so exposed. The transmission of recessive genetic defects to subsequent generations is also a genuine concern. No numbers will be quoted in this text because of their large uncertainties, but it can be said with confidence that the genetic effects of radiation occur with less frequency than does radiation-induced cancer.

Maximum permissible doses are not set to protect the individual, because the risk to the individual from such low doses per year is negligible. The philosophy is rather to protect the public as a whole from even a small number of cancers and genetic defects nationwide. Finally, the linear hypothesis is used to set these limits, another prudent overestimation of risk.

On April 26, 1986, a catastrophic nuclear accident occurred at the Chernobyl Nuclear Power Plant in the Ukraine.

Table 18.4 gives the maximum permissible doses in rems per year as set by law in the USA.

Table 18.4 Gives the Maximum Permissible doses in Rems Per Year a Set By Law

Maximum Permissible Doses—Rem/Year	
Occupational[a], medical diagnostics	
Whole body, adults	5.0[b]
Whole body, minors	0.5
Whole body, pregnant women	0.5
Gonads, bone marrow, lens of eye	5.0
Skin, thyroid, bone (external exposure)	30
Skin, thyroid (internal exposure)	15
Hands, Forearms, ankles, feet	75
Soft tissues and most other organs	15

[a]Nonoccupational, 0.1 of occupational. From nuclear power plants, 0.001 of occupational at plant boundary.
[b]Limited to no more than 3 rem/quarter.

We now turn our attention to some of the sources of ionizing radiation in our environment and methods of radiation protection.

18.3.8 Background Radiation

We are routinely exposed to a number of sources of radiation, referred to collectively as **background radiation.** Cosmic rays from the sun and outside the solar system cause more exposure at high altitudes, where there is less atmospheric shielding, than at low altitudes. The amount of natural radioactive isotopes in the environment also varies with location. Medical exposure comes almost entirely from X-rays and is a significant fraction of average annual background radiation in the United States. *Table 18.5* lists the yearly doses from various sources.

The effects of background radiation are so small that no one has been able to measure them. There are places in the world, specifically parts of Brazil and India, where background radiation is far higher than in the United States. Those populations show no measurable increase in cancer. Furthermore, workers in the nuclear weapons industry and people living at high altitudes have

Table 18.5 Background Radiation

Source	mrem/y
Natural radiation—external	
Cosmic rays,	35–70 (from sea level to 5000 ft.)
US. average	44
Radionuclides in soil,	35–70
building materials,	
US average	40
Natural radiation—internal	
^{14}C	1
^{40}K	16
^{226}Ra	1–350 (average close to 1)
Artificial radiation—external and internal	
Medical, dental	50–100 (average 73)
Fallout (e.g., ^{90}Sr)	4
Occupational	1
Nuclear power, to public	0.003
Miscellaneous	2
Total	144–608
Average US total	182

lower than average incidences of cancer. This does not necessarily imply that background radiation is harmless or beneficial, but rather that its effects are so small that they are masked by other effects.

18.3.9 Radiation Protection

Efforts are made to limit exposure to any source of radiation since all radiation causes some damage. If there is no benefit to the exposure, as in medical diagnostics, then there is no reason to sustain even slight damage. The three methods of limiting doses are to limit exposure time, use shielding, and increase distance from the source.

Time of Exposure If the source of radiation is relatively constant, the dose is directly proportional to the time of exposure. One example is the use of sensitive x-ray film so that exposure times can be small.

Shielding Shielding placed between a person and a source will absorb part or all of the radiation. Heavy elements are ideal absorbers because they have many electrons and because radiation loses most of its energy by interacting with electrons in a material. Lead aprons, for example, are used to shield patients being x-rayed.

Distance If the source of radiation is nondirectional, then the radiation levels around it decrease proportional to distance squared in vacuum. If the source is not in vacuum, the radiation levels decrease even faster with distance because of the absorption by materials, including air.

Other precautions, such as using radiation monitors and personnel film badges to record radiation levels and doses, usually amount to common sense. Whenever the risk is great, the rules and procedures for working with radiation necessarily become more restrictive.

Ionizing radiation is used to diagnose numerous medical conditions. The doses in rems from diagnostic procedures are usually small. Large doses of ionizing radiation are applied for therapeutic purposes, almost exclusively for the treatment of cancer. The risk of exposure versus the benefit of the medical procedure is usually carefully considered, as in any medical procedure.

18.3.10 Diagnostic Uses of Ionizing Radiation

Nearly half the average background radiation in the United States comes from medical diagnostics. Most of that exposure is due to x-rays. X-rays were first used for medical diagnostic purposes shortly after their discovery by Roentgen in 1895. X-ray photons travel a significant distance in tissue before being absorbed. Their range depends on the density of the material encountered. The simplest medical x-ray diagnostics (called x-rays for brevity) produce a shadow on film. The darkness of the shadow is representative of tissue density and thus can be used to detect, for example, broken bones or decay in teeth.

In Figure 18.9 CT scanner taking an x-ray "slice" of a patient: The stationary array of x-ray detectors encircles the area examined so that the x-ray tube may make a 360° rotation, centered about the area of interest. The data is fed to a computer for analysis.

Wilhelm Roentgen 1845–1923

Figure 18.8 (a) X-ray showing spiral fracture of the fibula produced while shot putting. (x-ray courtesy of the victum.) (b) the patient has drunk an x-ray absorbing contrast medium to make the upper gastrointestinal track visible.

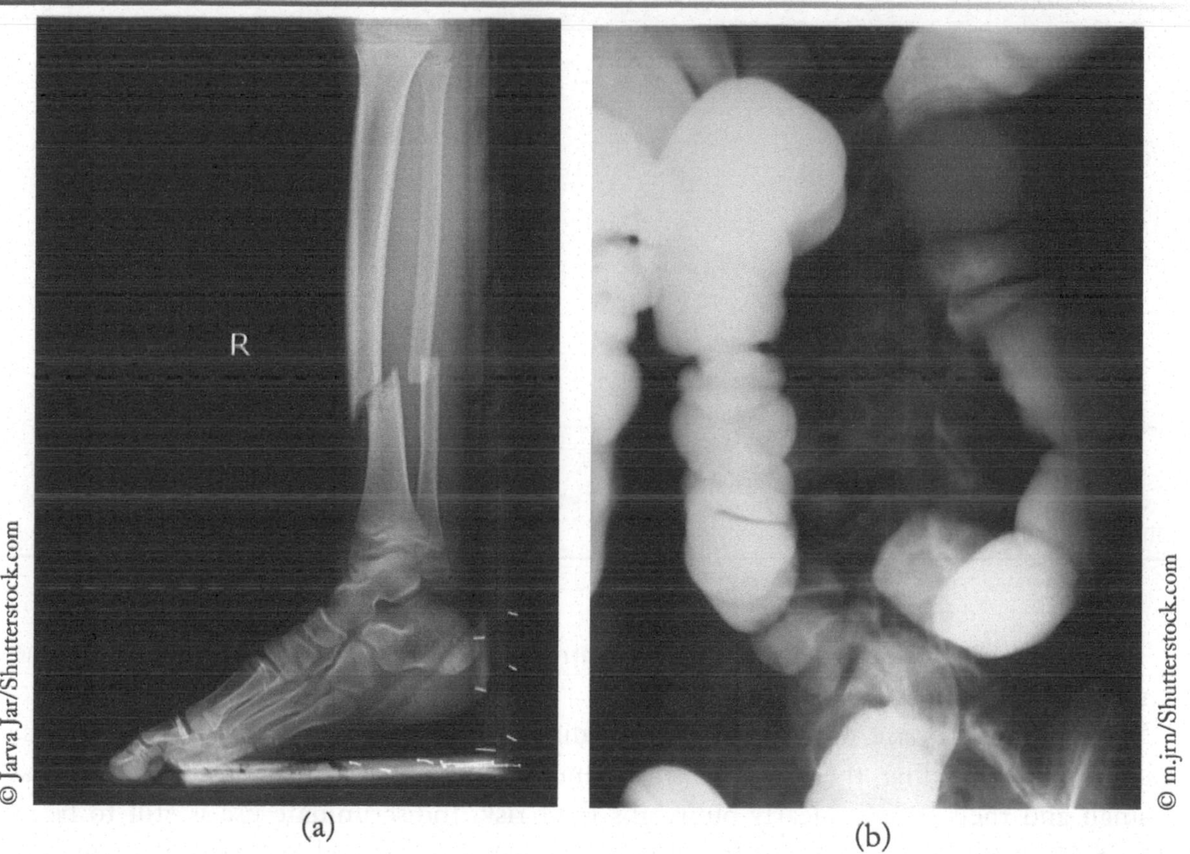

© Jarva Jar/Shutterstock.com

(a)

(b)

© m.jrn/Shutterstock.com

Figure 18.9 CT scanner.

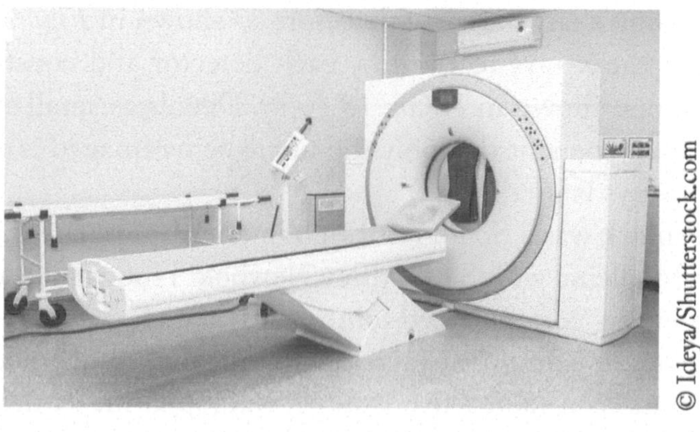

© Ideya/Shutterstock.com

Table 18.6 Typical Doses for X-ray Diagnostic Procedures

Procedure	Dose per Film in Rem to Organ of Region Affected
Chest	0.07
Kidney, intravenous	0.2
Cerebral, intravenous	0.3
Skull	0.4
Dental	0.9
Spine or lower back	1–3
CT scan	1–7
Pneumoenchephalogram (brain)	9
Gastrointestinal fluoroscopic exam with image intensifier	1–4
without image intensifier	2.5–10

The patient has drunk an x-ray absorbing contrast medium containing barium to make the upper gastrointestinal tract visible.

Many techniques are used to minimize radiation doses given in medical diagnostics. These include careful shielding, the use of sensitive films, and fluorescent image intensifiers. Most doses are small and their benefit clearly outweighs their risk. Indiscriminate use is still to be avoided. *Table 18.6* lists representative doses for several x-ray diagnostic procedures. Most doses are small and their benefits outweigh the risks.

Tomography and CT Scanners Tomography is a technique for obtaining a cross-sectional image of high quality and is most often performed using x -rays and machines called computerized tomography (CT) scanners. In standard x-ray images overlapping organs shadow and block one another, but CT scanners eliminate these shadows. The scans are made by rotating the x-ray tube around the patient and using a large array of detectors, as shown in *Figure 18.10(a)*.

A computer analyzes the x-rays received by each detector and constructs an image from this information, such as the one shown in *Figure 18.10(b)*. Details as small as 1 mm are observable in CT scans. There is the added benefit that only the tissue being imaged is exposed to radiation since a very narrow beam of x -rays is used.

Allan MacLeod Cormack was a South African–American physicist who shared the 1979 Nobel Prize for Physiology or Medicine with Sir Godfrey Newbold Hounsfield, an English electrical engineer of Britain, for developing computerized tomography.

Radiopharmaceuticals are being used increasingly in medical diagnostics. A radiopharmaceutical is any drug that contains a radioactive isotope. Radiopharmaceuticals can be designed to be very organ or system specific; that is, the body will concentrate them in specific places. Diagnostic information is then obtained by measuring the nuclear radiation emitted by the selected isotope. Gamma-ray emitters are usually employed because gammas can escape the patient's body. Bone

Figure 18.10a CT scanner taking an x-ray "slice" of a patient: The stationary array of x-ray detectors encircles the area examined so that the x-ray tube may make a 3600 rotation, centered about the area of interest. The data is fed to a computer for analysis.

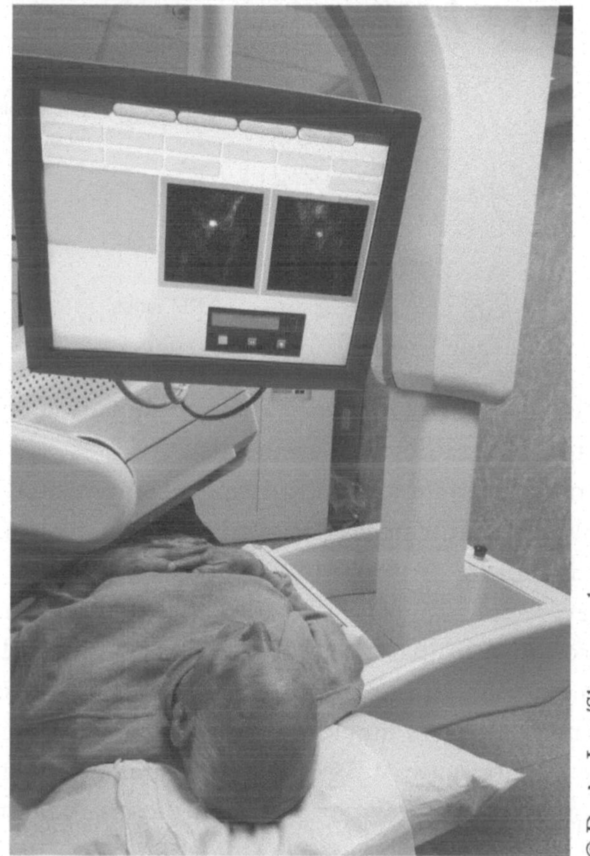

Figure 18.10b Cross-sectional image from a CT scanner. The liver, gallbladder, kidneys, small intestine, backbone, and spinal cord are visible.

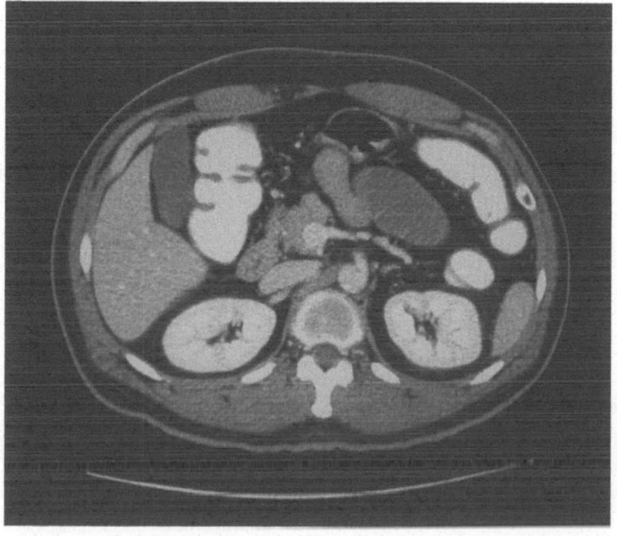

Figure 18.11 Anger camera used to obtain images in radiopharmaceutical diagnostics.

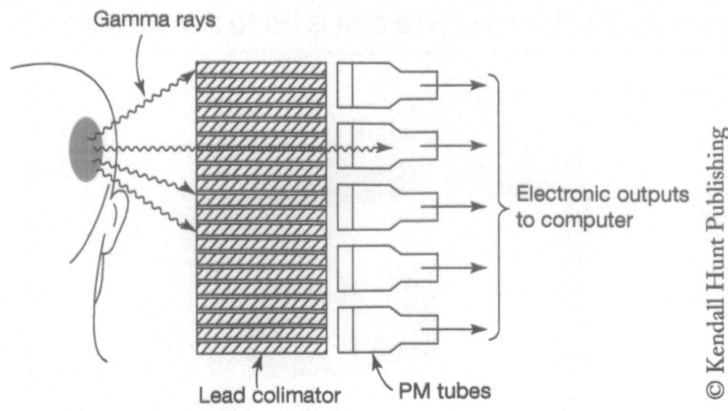

cancer, inflammation of tissues, organ function, blood supply, various tumors, and a variety of other information can be detected with radiopharmaceuticals.

Figure 18.11 shows a detector system, called an Anger camera, used to measure nuclear radiation from radiopharmaceuticals. The collimator is designed to obtain position information. Detector output is sent to a computer for image construction.

A broad range of clever techniques is used to locate radiopharmaceuticals in specific places and to minimize the dose given in diagnostic applications. One of the most effective techniques for minimizing radiation exposures from radiopharmaceuticals is to use isotopes with short half-lives and drugs that the body excretes quickly. About 80% of all radiopharmaceutical procedures performed today employ $^{99m}T_c$ because of its 6.03 h half-life is long enough for imaging but short enough to decay to background in a few days. (The m means that the nucleus is in a metastable excited state.)

Table 18.7 lists some commonly used radiopharmaceuticals and their applications. The applications listed are a small fraction of those in use, and more are being developed every year.

Biological Half-Life and Effective Half-Life The body removes substances by excretion, in an amount of time that depends on the chemical compound. This process is often like the decline of radiation due to nuclear decay; that is, half the substance is excreted in a certain amount of time, half of what remains in a similar amount of time, and so on. In such instances the substance is said to have a *biological half-life*. If a radiopharmaceutical is made from such a substance, then the radioactivity in the patient will decrease faster than it would from nuclear decay alone because part of the radioactivity is being excreted into the sewer system. The radiopharmaceutical has an *effective half-life* given by

$$t_{eff} = \frac{t_{\frac{1}{2}} \times t_B}{t_{\frac{1}{2}} + t_B} \qquad \text{(eq. 18.5)}$$

Where, t_{eff} is the effective half-life and t_B is the biological half-life. Note that the effective half-life is always equal to or shorter than either the nuclear or biological half-lives.

Figure 18.12 Shows the image of a patient obtained by scanning lengthwise with an Anger camera.

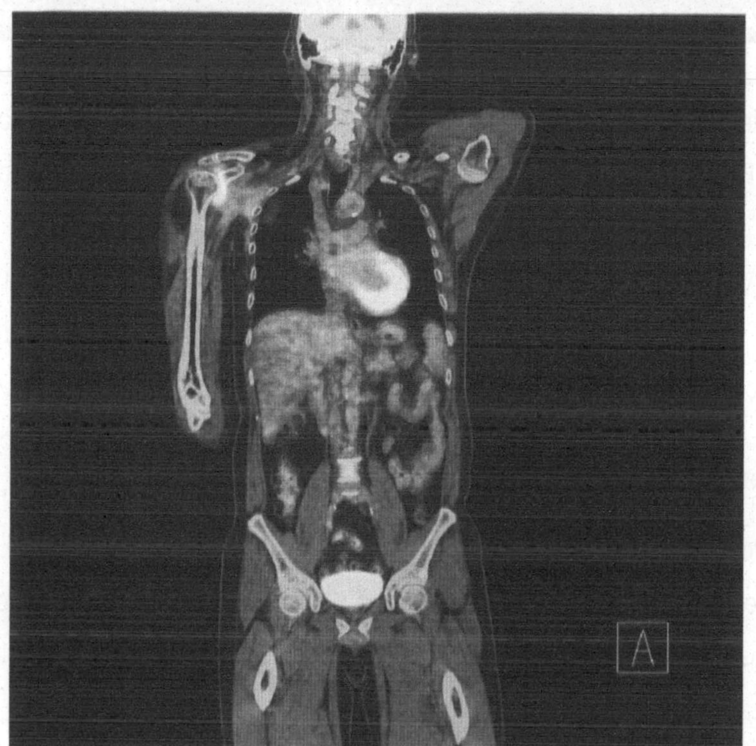

Table 18.8 gives the nuclear, biological and effective half-lives for selected isotopes. Note that the effective half-life is always equal to or less than either the nuclear or biological half-life.

EXAMPLE 18.7

A patient is given an injection containing 1.0 x 10^{-12} g of $^{99m}T_c$. (a) Calculate the effective half-life. (b) Calculate the activity in curies. (c) What activity remains in the patient 3.0 days after injection? Note that the half-life of $^{99m}T_c$ is 6.03 h and its biological half-life is 24.0 h.

Solution: *(a)* Effective half-life can be calculated using equation 18.5:

$$t_{eff} = \frac{t_{\frac{1}{2}} \times t_B}{t_{\frac{1}{2}} + t_B} = \frac{6.03 \text{ h} \times 24 \text{ h}}{(6.03 \text{ h} + 24 \text{ h})} = 4.82 \text{ hr}$$

(b) Calculate the activity in curies. Activity is given by equation 18.2,

$Activity = \dfrac{0.693}{t_{\frac{1}{2}}} \times N$ where, N is the number of nuclei.

Table 18.7 Diagnostic Uses Of Radiopharmaceuticals

Procedure and Agent	Typical Activity (mCi)	Radiation Dose (rem)
Brain scan		
^{99m}Tc-pertechnetate	7.5	1.5 (colon)
^{113m}In-DTPA	7.5	4 (bladder)
Lung scan		
^{99m}Tc-MAA	2	0.7 (lung)
^{133}Xe	7.5	0.4 (lung)
Cardiovascular blood pool		
^{131}I-HSA	0.2	3 (blood)
^{99m}Tc-HAS	2	0.08 (blood)
Placental localization		
^{99m}Tc-pertechnetate	0.7	0.1 (colon)
^{113m}In-transferrin	1	0.1 (blood)
Thyroid scan		
^{131}I	0.05	75 (thyroid)
^{123}I	0.07	1.5 (thyroid)
Liver scan		
^{198}Au-colloid	0.1	5 (liver)
^{99m}Tc-sulfur colloid	2	0.6 (liver)
Bone scan		
^{85}Sr	0.1	4 (bone)
^{99m}Tc-STPP	10	0.5 (bone)
Kidney scan		
^{197}Hg-chlormerodrin	0.1	1.5 (kidney)
^{99m}Tc-iron ascorbate	1.5	0.8 (kidney)

Step by step 1: Determine the number of nuclei in 1.0 x 10^{-12}g of $^{99m}T_c$.

$$1.00 \times 10^{-12}\ \text{g} \bullet \frac{6.022 \times 10^{23}\ \text{nuclei}}{99.0\ \text{g}} = 6.08 \times 10^{9}\ \text{nuclei}$$

Step 2: The nuclear half-life of $^{99m}T_c$ is 6.03 h. Note that we are asked for the activity in curies. The units of curies are decays per second. So, we convert 6.03 h to seconds.

Table 18.8 Nuclear, Biological, And Effective Half-Lives Of Selected Isotopes[a]

	Half-Life (days)		
Isotope	Nuclear	Biological	Effective
^{3}H	4.5×10^{3}	12	12
^{14}C	2.1×10^{6}	40	40
^{22}Na	850	11	11
^{32}P	14.3	1155	14.1
^{35}S	87.4	90	44.3
^{36}Cl	1.1×10^{8}	29	29
^{45}Ca	165	1.8×10^{4}	164
^{59}Fe	45	600	42
^{65}Zn	244	933	193
^{86}Rb	18.8	45	13
^{90}Sr	1.1×10^{4}	1.8×10^{4}	6.8×10^{3}
^{99m}Tc	0.25	1	0.20
^{123}I	0.54	138	0.54
^{131}I	8.0	138	7.6
^{137}Cs	1.1×10^{4}	70	70
^{140}Ba	12.8	65	10.7
^{198}Au	2.7	280	2.7
^{210}Po	138	60	42
^{226}Ra	5.8×10^{5}	1.6×10^{4}	1.5×10^{4}
^{235}U	2.6×10^{11}	15	15
^{239}Pu	8.8×10^{6}	7.3×10^{4}	7.2×10^{4}

[a]See Appendix D for additional nuclear half-lives.

$$6.03\ \text{h} \cdot \frac{3600\ \text{s}}{1\ \text{h}} = 21{,}700\ \text{s}$$

Step 3: Rewrite equation 18.2 and fill in the values.

$$Activity = \frac{0.693}{t_{\frac{1}{2}}} \times N = \frac{0.693}{21{,}700} \times 6.08 \times 10^{9}\ N = 1.94 \times 10^{5}\ \frac{\text{decays}}{\text{s}}$$

Step 4: Convert the activity to curies.

$$1.94\times10^{5}\,\frac{\text{decays}}{\text{s}}\times\frac{1\text{ curie}}{3.7\times10^{10}\,\frac{\text{decays}}{\text{s}}}=5.24\times10^{-6}\text{ curies}$$

Often written in microcuries: 5.24μCi

Example 18.7 (c) What activity remains in the patient 3.0 days, (72 h), after injection? *Note: In order to calculate the activity that remains in the patient, we must use the effective half-life of 4.8 h r.*

Recall equation 18.3 $A = A_0 e^{-0.693\frac{T}{t_{\frac{1}{2}}}}$

$$A = A_0 e^{-0.693\frac{T}{t_{\frac{1}{2}}}} = 5.24\ \mu\text{Ci}\bullet e^{\frac{-0.693\times72\text{ h}}{4.8\text{ h}}} = 1.60\times10^{-4}\ \mu\text{Ci}$$

This activity is undetectable.

The next example is an approximate calculation of the dose received from a radiopharmaceutical used in medical diagnostics. It also illustrates how important the choice of isotope can be in minimizing the dose received.

EXAMPLE 18.8

The standard radiopharmaceuticals for thyroid scans contain ^{131}I, which has a nuclear half-life of 8.0 days and an effective half-life of 7.6 days. Calculate *(a)* the amount of energy delivered to the patient in the first hour, *(b)* the number of joules per kilogram of exposed tissue, (rads), received in the first hour of exposure, and *(c)* the dose in rems received in the first hour of exposure using the following information. The patient is given 0.050 mCi of ^{131}I, which decays by beta and gamma emissions totaling approximately 0.50 MeV of energy per decay. Note that one-half of the energy emitted is absorbed in a total mass of 0.15 kg.

Solution: (a) Since the effective half-life is 7.6 days, the activity will not decrease significantly in the first hour. Therefore, we will assume constant activity.

Step by step 1: Calculate the number of decays per second.

$$0.050\times10^{-3}\text{ Ci}\times\frac{3.7\times10^{10}\,\frac{\text{decays}}{\text{s}}}{1\text{ Ci}}=1.85\times10^{6}\,\frac{\text{decays}}{\text{s}}$$

Step 2: Calculate the amount of energy in eV delivered to the patient in one hour.

$$1.85\times10^{6}\,\frac{\text{decays}}{\text{s}}\bullet\frac{3600\text{ s}}{1\text{ hr}}\bullet 0.50\times10^{6}\,\frac{\text{eV}}{\text{decay}}=3.33\times10^{15}\text{ eV}$$

Step 3: Convert the energy from eV to joules.

$$3.33\times10^{15}\text{ eV}\bullet\frac{1.6\times10^{-19}\text{joules}}{1\text{ eV}}=5.33\times10^{-4}\text{joules}$$

Solution: (b) The number of joules per kilogram of exposed tissue, (rads), received in the first hour of exposure.

Step by step 1: Recall that only 50% of the energy is absorbed in 0.15 kg of tissue.
50% of the energy absorbed is $0.5 \cdot 5.33 \times 10^{-4}$ joules $= 2.66 \times 10^{-4}$ joules

Step 2: Convert to Rads.

$$v \cdot \frac{2.66 \times 10^{-4} \text{ joules}}{0.15 \text{ kg tissue}} \cdot \frac{1 \text{ rad}}{1 \times 10^{-2} \frac{\text{j}}{\text{kg}}} = 0.176 \text{ rads}$$

Solution: (c) The dose in rems. Recall that the relative biological effectiveness, (RBE) of beta and gamma =1.

From equation 18.4, rem = (RBE)(rad)

$$1 \frac{\text{rem}}{\text{rad}} \cdot 0.176 \text{ rad} = 0.176 \text{ rem}$$

The patient in this example will continue to receive radiation for weeks. The dose per hour will decline to half the value found in 7.6 days. In some locations it is possible to obtain ${}^{123}I$ which has a half-life of only 13 h. If it is used instead of ${}^{131}I$ the dose received will be much less because it will be gone in a matter of days instead of months.

18.3.11 Therapeutic Uses of Ionizing Radiation

Therapeutic applications of ionizing radiation, called radiation therapy or radiotherapy, have existed for decades. Radiotherapy was once applied to ailments from acne to rheumatism, often with tragic consequences. It was not until the 1950s that abuses of radiotherapy were prohibited by law. Radiotherapy is now used almost exclusively for the treatment of cancer.

Radiotherapy is appropriate for cancer treatment because cancer cells are rapidly dividing and consequently sensitive to ionizing radiation. Furthermore, there is often no effective risk-free alternative to radiation therapy. Finally, radiotherapy works; it improves survival rates for some types of cancer.

Radiotherapy is sometimes used in combination with the other two major types of cancer treatment, surgery and chemotherapy. Chemotherapy uses chemicals that, like radiation, inhibit cell division. As a result many of the side effects of chemotherapy are similar to those produced by radiation. It is usually possible to localize radiation better than chemicals, so the side effects of radiotherapy tend to be more localized than those of chemotherapy.

The central problem in radiotherapy is to concentrate radiation in abnormal tissue, giving as little dose as possible to normal tissue. The ratio of abnormal cells killed to normal cells killed is called the therapeutic ratio, and all radiotherapy techniques are designed to enhance this ratio. The effects of radiation on cancer cells are generally the same as those for normal cells, with two qualifications. First, rapid cell division in cancer increases its sensitivity to radiation compared to normal tissue. Second, cancer tissue is usually oxygen poor, which decreases its sensitivity to radiation compared to normal tissue; this is known as the oxygen effect. Ionizing radiation produces more toxic

Figure 18.13 ^{60}Co treatment of malignant tumor. The dose to normal tissue is minimized by rotating the source about the patient in a circle centered on the tumor so that the common crossing point for the ^{60}Co gammas is the tumor.

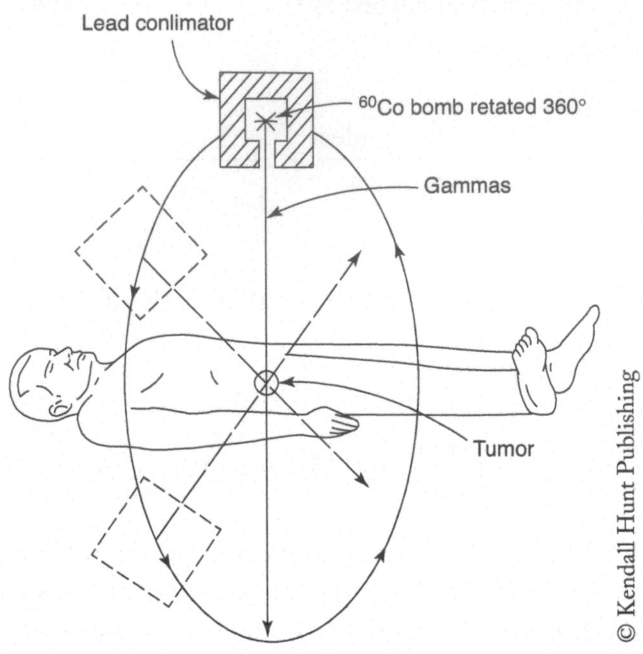

end products when oxygen is plentiful in tissue. It is sometimes possible to oxygenate a tumor to overcome this problem, but in general larger doses of radiation must be used to compensate for the oxygen effect.

Many different forms of radiotherapy are in use. An older form of external beam radiotherapy, cobalt-60, was widely used beginning in the 1950s. As illustrated in *Figure 18.13*, these machines used ^{60}Co gamma rays. Cobalt-60 has been replaced by linear accelerators in many developed countries. However, because the machinery is reliable and simple to use, Cobalt treatment is still in wide use worldwide.

The linear accelerators can generate higher energy radiation than the cobalt-60 units and don't produce radioactive waste. Linear accelerators employ several types of ionizing radiation, such as high-energy nitrogen nuclei, pi mesons, or neutrons. Although more expensive than cobalt-60, these accelerator-produced beams are often more effective than x-rays or gammas.

Radiotherapy can also be administered internally with radiopharmaceuticals or implanted capsules, needles, or pellets. Internally applied radiotherapy has the advantage that radiation having a larger RBE can be used than in externally applied radiotherapy. (Large RBE implies short range, so the radiation could not penetrate from outside the body.) When capsules of radium, radioactive gold needles, or other radioactive isotopes are implanted in tumors, they are removed once a sufficient dose is administered.

Radiopharmaceuticals are used for cancer therapy only if they can be sufficiently localized to produce a favorable therapeutic ratio. One example is the use of radioactive iodine for thyroid cancer, but with larger doses than are used in thyroid imaging. A technique called *immuno-radiotherapy*,

Table 18.9 Cancer Radiotherapy

	Typical Doses[a] (rem)
Lung	1000–2000
Hodgkin's disease	4000–4500
Skin	4500
Ovarian	5000–7500
Breast	5000–8000+
Bladder	7000–7500
Head (brain)	
Neck	
Bone	8000 +
Soft tissue	
Thyroid	

[a]Usually given at 200 rem/treatment, from three to five times per week.

currently being developed, attaches radioactive isotopes to antibodies produced by the patient against his cancer. The antibodies, it is hoped, will locate themselves almost exclusively in cancerous tissue.

Table 18.9 indicates typical doses given to cancerous tissue in radiotherapy. The doses are not fatal because they are localized in one part of the body and spread out over several weeks of time. The upper limit to the radiation given is always determined by the unavoidable exposure of normal tissue. Note that larger doses are given to cancers located in tissues that are relatively resistant to radiation (e.g., the adult brain, since there is no cell reproduction in it).

Radiotherapy is spread over many weeks because it is not possible to give enough radiation in a single dose to kill the cancer but spare the patient. Both the patient and the cancer recover somewhat between treatments, but cancerous tissue recovers less since the radiation is concentrated in it. Complete eradication is sometimes possible. Cures are not always achieved or even attempted. Lung cancer, for example, cannot ordinarily be cured using radiation because lung tissue and blood are too sensitive to radiation to permit doses large enough to eradicate cancer completely. But statistics show that life is prolonged, symptoms are alleviated, and survival chances are improved by radiotherapy for many forms of cancer, including lung cancer.

The decision whether to use radiotherapy in cancer treatment and whether to use it in conjunction with surgery and chemotherapy is complex and contains an element of subjectivity. Physicians rely on an evolving body of statistical data and their own experience. Many factors must be weighed, including the possible inducement of another cancer and the chances of survival without radiotherapy.

Things to know for section 18.

Marie and Pierre Curie separated three types of radiation with the use of a magnetic field and photographic plates.

The three types were:

Alpha	${}_{2}^{4}He$	the nucleus of a helium atom
Beta	${}_{-1}^{0}\beta$	an electron
Gamma	${}_{0}^{0}\gamma$	an energetic photon of light

An isotope of an element has the same number of protons but a different number of neutrons than other isotopes of that element.

${}_{Z}^{A}X$

Z represents the number of protons or charge.
A represents the number of protons plus the number of neutrons.
X represents the element or particle's symbol.

The symbol used for an alpha particle is ${}_{2}^{4}He$.

We will use ${}_{-1}^{0}\beta$ to represent beta decay

${}_{+1}^{0}\beta$ is the symbol that represents a positron. A positron is the antiparticle of an electron. It has the same mass as an electron, but has a positive charge.

The symbol we will use for gamma decay is ${}_{0}^{0}\gamma$

Total charge is always conserved

Half-life is the **time** in which one-half of the remaining nuclei will decay.

$$Activity \equiv \frac{\text{number of decays}}{time} \qquad A = A_0 e^{-0.693\frac{T}{t_{\frac{1}{2}}}}$$

A is the present activity; Ao is the original activity

T is the elapsed time, $t_{\frac{1}{2}}$ is the half-life of the isotope of that element and e is a mathematical constant, approximately equal to 2.72, which serves as the base of the natural logarithm. When using your calculator to take the natural log of something, press the ln key.

An equivalent statement uses N for number of nuclei as follows:

$$N = N_0 e^{-0.693\frac{T}{t_{\frac{1}{2}}}}$$

rem = (RBE)(rad)

Questions

18.1 What characteristics of radioactivity indicate that it is nuclear in origin and not atomic?

18.2 What characteristics of radioactivity eliminate the possibility that it might be a long-term fluorescence process in which stored energy, perhaps from visible light, is slowly being emitted in another form?

18.3 What is the source of energy emitted in radioactive decay?

18.4 Consider *Figure 18.1*. If an electric field were substituted for the magnetic field with a positive voltage instead of a north pole and a negative voltage instead of a south pole, in which directions would the alpha, beta, and gamma rays bend?

18.5 Explain the relative ranges of alpha, beta, and gamma rays in terms of their charges and masses.

18.6 Why do materials composed of heavy elements make the best radiation absorbers?

18.7 An alpha or beta particle that penetrates a thin material always loses some energy, but some· gamma rays that penetrate a thin material lose no energy at all. Explain why arid comment on safety implications for shielding gamma rays.

18.8 What is the difference between x-rays and gamma rays? What are their similarities?

18.9 Alpha, beta, gamma, and x-ray radiation are all classified as ionizing radiation. What does the term ionizing radiation mean, and why isn't visible light ionizing radiation?

18.10 Devise a method for identifying whether radiation is alpha, beta, or gamma using a magnet and a Geiger counter.

18.11 How is the photoelectric effect used in a photomultiplier tube?

18.12 Why can't nuclei be seen—with visible light?

18.13 What are the similarities of protons and neutrons? In what ways are they dissimilar?

18.14 What is an isotope? Why do different isotopes of the same element have nearly identical chemistries?

18.15 What is the conservation of mass plus energy principle, and why is it understandable that it was not discovered earlier in history?

18.16 What is the antimatter counterpart of the electron, and what are, its characteristics?

18.17 Why does biological half-life depend on the chemical compound rather than on the isotope?

18.18 Why is the RBE for alphas so large?

18.19 Why is the RBE, the same for beta and gamma rays?

18.20 Radioactive sources that emit only alpha particles are relatively safe when external to the body but extremely hazardous if ingested. Explain why.

18.21 When stating rems of radiation exposure, it is necessary not only to specify the dose in rems but also whether the exposure was to the whole body or localized. Explain why in terms of the definition of rems and rads.

18.22 What are the major effects of ionizing radiation on biological organisms at the cellular level?

18.23 What characteristic of cancer cells makes them more sensitive to radiation than normal cells? What tends to make them less sensitive?

18.24 The linear hypothesis is generally considered to over estimate the hazards of low-level radiation. Explain why.

18.25 What is the maximum permissible dose of radiation parts of the body?

18.26 A change in blood count is the first immediately observable effect of radiation. Explain why.

18.27 Is radiation the major cause of cancer? What are other causes of cancer, and what similarities do they have to the effects of radiation on biological tissue?

18.28 Why do you think there is a latency period for radiation-induced cancer? Furthermore, why does leukemia have a 2-yr latency period, while other forms have 15-yr latency periods?

18.29 What methods of radiation protection does an x-ray technician use when she retreats behind a lead lined door?

18.30 How does the use of sensitive x-ray film allow for smaller radiation doses?

18.31 What are the advantages of CT scanners over conventional x-rays in diagnostic applications?

18.32 A CT scanner takes pictures in 4-mm slices, giving a dose of 2.5 rem per slice. If a CT scan is made of a patient's entire body, what would the whole body dose be, assuming that no x-rays scatter out of directly exposed tissue? (CT scans produce only about 1% outscatter.)

18.33 Why is ultrasound, rather than x-rays, used for fetal imaging?

18.34 What is the effective half-life of a radioisotope implanted in a patient in a chemically inert capsule?

18.35 Why isn't a 5000-rem radiotherapy program for breast cancer fatal when LD50 is only 450 rem?

18.36 Describe the techniques that can be used to increase the therapeutic ratio in radiotherapy.

Problems

SECTIONS 18.1

√18.1(I) How many ion pairs are created in a Geiger counter by a 5.4-MeV alpha particle if 80% of its energy goes to create ion pairs and 30 eV (average in gases) is required per ion pair?

√18.2(I) How many organic molecules can be disrupted by a 4.8-MeV alpha particle absorbed in living tissue? About 5 eV is required to ionize or disrupt an organic molecule.

SECTIONS 18.2

√18.3 (I) If a model of an atom were created on an enlarged scale with a nucleus 10 cm in diameter, approximately how far away would the outer electrons be?

√18.4 (I) If a scale model of a helium atom has a nucleus with a mass of 1.0 kg, what would the mass of an electron be on that scale?

18.5 (II) If a truck of mass 20,000 kg could be compressed into a cube with the density of nuclear matter, what would be the dimensions of the cube?

18.6 (II) No probe can detect features smaller than its own wavelength. Calculate the energy in MeV of a photon that could detect details of 10^{-16} m (about one-tenth the radius of a nucleon).

18.7 (II) Calculate the frequencies and wavelengths of the two gamma rays emitted by ^{60}Co. They each have an energy of 0.511 MeV.

18.8 (III) Use the approximate size of the nucleus versus the atom to show that the density of the nucleus is about 10^{12} g/cm^3. Assume that the atom has a density of 1.0 g/cm^3.

In each of the following nine problems write the decay equation or equations if the daughter(s) is (are) radioactive. The table of isotopes in Appendix C can be used to determine atomic numbers. Appendix D gives decay modes of some nuclei.

√18.9 (I) The neutron. Neutrons are produced in large numbers in a nuclear reactor. They are stable inside the nucleus but unstable outside.

√18.10 (I) ^{3}H. A part of radioactive fallout due to weapons tests.

√18.11 (I) ^{40}K. A naturally occurring isotope spread uniformly throughout the ecosphere.

√18.12 (I) ^{131}I. The isotope most often used in thyroid scans.

√18.13 (I) ^{137}Cs. One of the major waste products of nuclear reactors.

18.14 (II) ^{52}Mn. The daughter of ^{52}Fe positron decays.

18.15 (II) ^{90}Sr. One of the major waste products of nuclear reactors. Its daughter beta decays.

18.16 (III) ^{238}U, a material contained in pitchblende. Its daughter beta decays, leaving a residual nucleus that beta decays. This is not the end of the decay chain, but you may stop here.

18.17 (III) ^{239}Po, one material used in all nuclear weapons. Also a major waste product of nuclear reactors. Its daughter alpha decays, leaving a residual nucleus that beta decays, leaving another nucleus that alpha decays. This is not the end of the decay chain, but you may stop here.

18.18 (II) Every exothermic process destroys mass. How much mass in grams is destroyed in the digestion of 1.0 g of butter, releasing 7.95 kcal of energy? Can this decrease be easily detected?

18.19 (II) Every endothermic process is really an energy-storage process in which energy is turned into mass. How much will the mass of a storage battery increase if 200 kJ of energy is stored in it? Could this increase be detected?

18.20 (II) A positron emitted in radioactive decay annihilates with an electron in a short time (averaging about 2 nsec), converting all its mass and all the electron's mass into pure energy in the form of photons. Calculate the amount of energy in MeV released in such an annihilation.

18.21 (III) A large nuclear power plant generates 1000 MW of electricity with an efficiency of 35%. Calculate the mass destroyed by the plant in one year of operation if it averages 50% of full power for the entire year.

18.20 (III) A nuclear power plant is refueled after one year of operation, and it is observed that the fuel elements have 2.0 kg less mass than when they were installed in the reactor. If the plant averaged 2000 MW of electric power output for the entire year, calculate (a) the amount of waste heat in calories and (b) the efficiency of the plant.

SECTION 18.3

Additional data necessary for some of these problems can be found in the tables in this chapter and in the appendices.

√18.23 (II) During an archaeological dig an old campfire is discovered. The charcoal in it has one-hundredth the normal amount of ^{14}C. (One part in 10^{14} rather than one part in 10^{12}.) Given that the half-life of carbon-14 is 5,730 y, calculate the approximate age of the charcoal.

√18.24 (II) A 250-mCi (millicurie) radioactive source has a half-life of 12 y and is considered safe if its activity is less than 1.0 μCi (microcurie). How much time must pass before the source is safe?

18.25 (II) A ^{60}Co source is labeled 2.0 mCi, but is found to have a present activity of 1.85×10^7 decays/s. (a) What is the current activity of the source in millicuries? (b) How long ago was it manufactured with 2.0 mCi of activity?

18.26 (II) If a person has 1.00 μg of naturally occurring ^{40}K in his body, calculate the number of ^{40}K decays/s taking place in him.

18.27 (III) What is the mass of ^{60}Co in a 1000 Ci ^{60}Co "bomb" used for cancer therapy?

√18.28 (III) A 48.0 gram sample of carbon is taken from the bones of a skeleton and is found to have a carbon-14 decay activity of 160 decays/min. It is known that carbon from a living person has a decay rate of $15.0 \frac{\text{decays}}{\text{min} \bullet \text{gram}}$. Given that the half-life of carbon-14 is 5,730 y, how old is the bone? Should we call the police or an archeologist?

18.29 (III) The energy produced in nuclear decay eventually ends up as heat after the emitted particles are stopped. Heat produced by a radioactive source can be used to produce electric power, as is done in nuclear reactors and smaller thermoelectrical devices. Such devices are small enough to be carried on satellites. (a) Calculate the activity of a ^{210}Po source that produces 1500 W of heat. (b) What is the mass of the ^{210}Po?

SECTIONS 18.4 AND 18.5

Additional data necessary for some of these problems can be found in the tables in this chapter and in the appendices.

√18.30 (I) What is the dose in rems due to exposures to (a) 7.0 rad of x-rays in an upper GI series? (b) four dental x-rays of 800 mrad each? (c) 0.50 rad of alpha radiation from an ingested ^{226}Ra source? The RBE of x-rays = 1 rem/rad and the RBE of alpha particles = 20 rem/rad.

√18.31 (I) How many rads produced the following exposures to a human body? (a) 5.0 rem of proton radiation. (b) 5.0 rem of fast neutron radiation. (c) 20 rem of energetic beta radiation.

18.32 (II) Calculate the dose in rem to a patient who is given a chest x-ray under the following circumstances. The intensity of the x-ray beam is 2.0 W/m^2, the area of the person's chest is 0.080 m^2, 30% of the x-rays are absorbed by the person's chest, the time of exposure is 0.25 s, and the mass of the tissue exposed is 20 kg.

18.33 (II) A cancer patient is exposed to a 4000-Ci ^{60}Co "bomb" for 45 s. If the gammas from the ^{60}Co source are collimated in such a way that only 1.0% of them strike the patient, and if 15% of those are absorbed in a region having a mass of 2.0 kg, what was the dose to that region in rem? (The average energy per decay is 2.5 MeV.)

18.34 (II) People living in areas of high background radiation may experience higher death rates from radiation-induced cancers. Calculate the increase in deaths from radiation-induced cancer in a city of 1,000,000 persons, such as Denver. Assume the average residence time is 20 y. Compare your result with the 200,000 deaths per 1,000,000 persons from all cancers.

18.35 (II) How many deaths from radiation-induced cancer can be expected in a group of 1,000,000 people exposed to the average medical dose of 73 mrem/y for 40 y? How does this compare with the total risk of dying from cancer, which is about 20%?

18.36 (II) In the 1950s and 1960s one treatment for a rheumatic condition of the spine, called ankylosing spondylitis, was the administration of 400–1800 rad of ^{60}Co gammas to the affected region. Calculate the risk of death from radiation-induced cancer to a patient over 30 y subjected to an exposure of 750 rad in such treatments.

REFERENCE TABLE 18.8 TO SOLVE PROBLEMS 18.37-18.43

18.37 (I) Calculate the effective half-life of ^{3}H.

18.38 (I) Calculate the effective half-life of ^{90}Sr.

18.39 (I) Calculate the effective half-life of ^{131}I.

18.40 (II) A dose of 0.05 mCi of ^{131}I is given to a patient for a thyroid scan. Calculate the activity of the iodine remaining in the patient after 30 days.

18.41 (II) A dose of 0.10 mCi of ^{198}Au is given to a patient for a liver scan. Calculate the activity of the gold remaining in the patient after 19 days.

18.42 (II) Large amounts of ^{65}Zn are produced in copper exposed to accelerator beams. While machining copper contaminated with ^{65}Zn, a scientist ingests 35 μCi of ^{65}Zn. What is the activity of the ^{65}Zn in this person's body after 10 y?

18.43 (III) A radiation worker ingests 1.0-μgm of ^{239}Pu into her lung in an industrial accident. (a) Calculate the effective half-life of ^{239}Pu. (b) ^{239}Pu decays by emitting a 5.5-MeV alpha particle. Assuming that the alphas are absorbed in 50 gram of surrounding tissue, calculate the total dose in rem to that tissue over the next 40 y.

18.44 (III) Naturally occurring ^{40}K gives the average person a radiation dose of 16 mrem/y. Calculate the mass of ^{40}K that must be inside a woman's body to give this dose, assuming that her mass is 50 kg. Each decay of ^{40}K produces a 1.31 MeV beta. Assume that 50% of the total energy is absorbed by her.

18.45 (III) The amount of ^{226}Ra in humans depends on where they live and how much of it is there in the food and water they consume. Calculate the mass of ^{226}Ra in a 75-kg man who gets a dose of 50 mrem/yr from ^{226}Ra. Each ^{226}Ra decay produces a 4.8-MeV alpha and ^{226}Ra has a half-life of 1620 y. Neglect the dose due to the daughters of ^{226}Ra for simplicity. The amount of ^{226}Ra in his body can also be assumed constant because he ingests it continuously. **Figure 18.6** CT scanner taking an x-ray "slice" of a patient: The stationary array of x-ray detectors encircles the area examined so that the x-ray tube may make a 3600 rotation, centered about the area of interest. The data is fed to a computer for analysis.

APPENDIX A

Mathematical Review

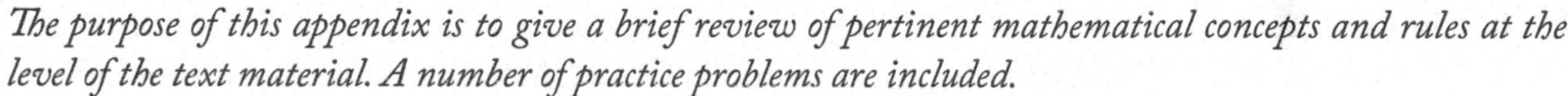

The purpose of this appendix is to give a brief review of pertinent mathematical concepts and rules at the level of the text material. A number of practice problems are included.

A.1 SIMPLE GUIDELINES FOR SOLVING ALGEBRAIC EQUATIONS

Algebra is a generalization of arithmetic in which symbols are used for numbers. It is used extensively in the text to solve problems, illustrate physical principles, derive equations, and so on. The rules of algebra are used to manipulate equations to produce a desired outcome, perhaps to solve for the net force in the equation for Newton's second law: $a = \frac{\Sigma F}{m}$

It is often useful to follow these three steps, in this specific order, when attempting to solve an equation for an unknown variable:

1. Make sure that the variable is in the numerator.
2. Make sure that it is alone.
3. Make sure that it is only raised to the first power.

EXAMPLE 1

Let's say that you want to solve Newton's second law for net force, ΣF.

$$a = \frac{\Sigma F}{m}$$

Following our three steps in order, we ask the following questions:

1. Is ΣF in the numerator? Yes. √
2. Is ΣF alone? No it is not alone. It is being divided by m. What do we do to get ΣF alone?

   Multiply both sides of the equals sign by m.

   $$ma = \frac{\Sigma F \cdot m}{m}$$

   Notice that no matter what the value of m is, when divided by itself, it will equal one. Now we have net force alone.

$$\Sigma F = ma \ \surd$$

3. Is $\sum F$ raised to the first power? Yes. √

EXAMPLE 2

Solve the equation for kinetic energy for velocity:

$$K.E. = \frac{1}{2}mv^2$$

1. Is v in the numerator? Yes. √
2. Is v alone? No. It is being divided by 2 and multiplied by m. What do we do to get v alone? Multiply both sides of the equals sign by 2 and divide both sides of the equation by m. $\frac{2 \cdot K.E.}{m} = \frac{1}{2}\frac{mv^2 \cdot 2}{m}$. This results in $v^2 = \frac{2 \cdot K.E.}{m}$

   Is v alone? Yes. √
3. Is v to the first power? $v^2 = \frac{2 \cdot K.E.}{m}$. No.

   Take the square root of both sides.

$$\left(v^2\right)^{\frac{1}{2}} = \left(\frac{2 \cdot K.E.}{m}\right)^{\frac{1}{2}}$$

$v = \sqrt{\frac{2 \cdot K.E.}{m}}$ Now v is to the first power and we have solved for v.

Addition and Subtraction: An equation remains true if equal quantities are added or subtracted to both sides of the equation.

For example, let's solve the following equation for temperature in Celsius: $T_F = \frac{9}{5}T_C + 32$

1. Is T_C in the numerator? Yes. √
2. Is T_C alone? No.

Subtract 32 from both sides of the equation. $T_F - 32 = \frac{9}{5}T_C + 32 - 32$ Rewrite the equation with T_C on the left.

$\frac{9}{5}T_C = T_F - 32$ Notice that T_C is still not alone.

Multiply both sides by $\frac{5}{9}$.

$$\frac{5}{9}\left(\frac{9}{5}T_C\right) = (T_F - 32)\frac{5}{9}$$

Therefore $T_C = \frac{5}{9}(T_F - 32)$

Is T_C raised to the first power? Yes. √

Basic Right Triangle Trigonometric functions useful for adding vectors.

$$\text{Sin}\ \theta \equiv \frac{\text{opposite}}{\text{hypotenuse}} \quad \text{Cosine}\ \theta \equiv \frac{\text{adjacent}}{\text{hypotenuse}} \quad \text{Tangent}\ \theta \equiv \frac{\text{opposite}}{\text{adjacent}}$$

$$hyp^2 = opp^2 + adj^2 \qquad \text{Therefore, } hyp = \sqrt{opp^2 + adj^2}$$

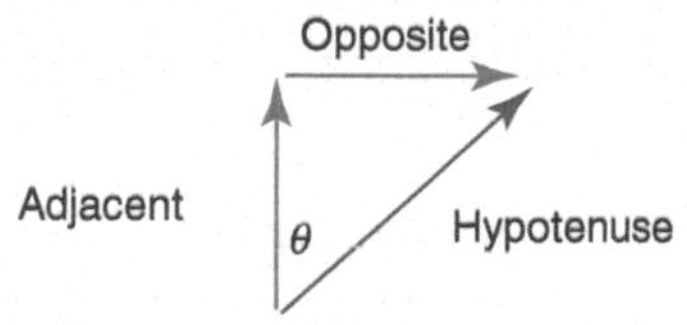

Opposite

Adjacent

θ

Hypotenuse

A.2 MEASUREMENT, ACCURACY, AND SIGNIFICANT FIGURES

Physical laws are based upon the results of many measurements (i.e., experiments). No measurement has perfect accuracy. The result always has some uncertainty, which may be caused by many factors, including the measuring device, the skill of the person making the measurement, and irregularities in the object being measured. Because the validity of models, theories, and laws are judged on their ability to describe nature, it is important to know and state the uncertainty in any measurement. Models, theories, and laws need only describe nature within experimental uncertainties to be considered valid.

If a measured quantity is represented by the symbol A, then its uncertainty is represented by δA, where δ is the Greek letter delta. These are written together as $A \pm \delta A$. For example, a person's temperature might be measured to be 37.1°C with an uncertainty of 0.2°C. This is written as;

$$T \pm \delta T = (37.1 \pm 0.2)°\text{C},$$

indicating that the actual temperature lies between $T - \delta A$ and $T + \delta A$ (36.9°C and 37.3°C). The uncertainty of 0.2°C depends on many factors and is estimated following a careful assessment of the possible sources of uncertainty. (This is a common laboratory activity.) In this example, a more accurate measurement might have an uncertainty of 0.1°C, a less accurate one an uncertainty of 0.5°C.

Percent uncertainties are useful and are defined to be % Uncertainty = ($\delta A/A$) × 100 eq. A.1

For example, the percent uncertainty in the temperature above is

$$\frac{0.2}{37.1} \times 100 = 0.5\%$$

There is an uncertainty in any quantity calculated from measured quantities. For example, if the area of a floor is calculated from measurements of its length and width, there is an uncertainty in that area because the length and width have uncertainties. When measured quantities are used in a calculation, their percent uncertainties are added (this is a good approximation for uncertainties less than about 5%). For example, if a floor has a length of 4.0 m and a width of 3.0 m, with uncertainties of 2% and 1%, respectively, then the area of the floor is 12.0 m^2 and has an uncertainty of 3%.

A.2.1 Significant Figures

An approximate method of keeping track of the accuracy of numbers is to write only those figures that are significant. The last figure in a number then has some significance but is not exact. For example, the temperature of (37.1 ± 0.2)°C quoted above has three significant figures. The 1 is significant but is not exact, and its true significance depends on the accuracy of the experiment. It is incorrect to write this temperature as 37°C, which would imply less accuracy than is present. It is equally incorrect to write this temperature as 37. 10°C; here the 0 indicates that the answer is known to 4 significant figures- and that is not true. In summary, the last figure written down should be the first uncertain figure.

Zeros present special problems as significant figures. The zeros in 0.053 are not significant, while the zero in 1.053 is. The zeros in 1300 may or may not be significant, while those in 1006 are significant. When zeros are "place keepers" for the decimal point, as in 1300 and 0.053, they are usually considered not to be significant. The best solution is to use scientific notation (see Appendix A.3).

When calculations are performed, the final result can have only a limited number of significant figures. (This is equivalent to saying that the calculated result has an uncertainty.) There are two approximate rules for finding the number of significant figures in a calculated result.

1. ***Multiplication and division.*** The result has the same number of significant figures as the least accurate number entering into the calculation. For example, let us find the area of a circle 1.2 m in radius:

$$A = \pi r^2 = (3.14159)(1.2\text{ m})^2 = 4.52389\text{ m}^2 = 4.5\text{ m}^2$$

   Because the radius was only given to two significant figures, the result has only two significant figures, despite the fact that π was quoted to six significant figures.
2. ***Addition and subtraction of linear vectors.*** Suppose there are three horizontal, parallel forces acting on an object. They are 7.56 and 13.7 N to the right and 6.052 N to the left. The net force is then

$$\begin{array}{l} 7.56 \\ 13.7 \\ \underline{-6.052} \\ 15.208 = 15.2\text{ N} \end{array}$$

   In addition and subtraction, then, the last significant figure of the result is in the same column relative to the decimal point as the last significant figure of the least accurate number in the addition or subtraction.

The rules for significant figures are approximate. A more accurate treatment involves mathematics beyond that used in this text.

Significant Figures in This Text Since the rules for significant figures are approximate, numbers in the text are usually assumed to be precise enough to yield a final answer accurate to three significant figures. The exception is in optics, where more accurate numbers are sometimes assumed. Because the purpose of the text material is to illustrate physical principles rather than duplicate laboratory procedures, the simplicity of such a three-significant-figure convention outweighs its lack of purity.

A.3 EXPONENTS

When we write 2^4, we mean 2 multiplied by itself 4 times, or $2^4 = 2 \times 2 \times 2 \times 2 = 16$. The superscript 4 is the exponent, 2 is the base, and 2^4 in words is 2 to the fourth power. More generally, a^n is a times a, taken n times and is stated in words as a to the nth power. Two special cases are $a^1 = a$ and $a^0 = 1$, (the last by definition).

The rule for *multiplying* two numbers expressed as powers is

$$a^n a^m = a^{n+m} \quad \text{For example, } 5^2 5^4 = 5^6$$

Note that this rule is valid only if each number has the same base (5 in this example).

Similarly, the rule for *division* of one power by another of the same base is $\frac{a^n}{a^m} = a^{n-m}$ For example, $\frac{5^2}{5^4} = 5^{-2}$

Also, note that $\frac{1}{a^m} = a^{-m}$ you can even use this rule when writing units. For examples, $\frac{1}{\text{ml}} = \text{ml}^{-1}$; $\frac{1}{\text{cm}^2} = \text{cm}^{-2}$

A.4 SCIENTIFIC NOTATION AND POWERS OF 10

Scientific notation is a convenient way of writing large or small numbers using powers of 10. For example, the distance from the earth to the sun averages approximately 150,000,000 km. In scientific notation, this is 1.50×10^8 km.

Scientific notation has the advantage of being shorter, and it can eliminate ambiguities in significant figures. The average earth-sun distance is accurate to three significant figures when written 1.50×10^8 km but has an unknown number of significant figures when written 150,000,000 km. Similarly, small numbers, such as the smallest detail observable with an optical microscope (0.000001 m or 1×10^{-6} m), are more conveniently written in scientific notation.

Writing numbers in scientific notation, or powers of 10, is based on the concept of exponents (see Appendix A.2) and the fact that our number system is a base-10 system. Thus $1.50 \times 10^8 = 1.50 \times (10)(10)(10)(10)(10)(10)(10)(10) = 1.50 \times 100{,}000{,}000 = 150{,}000{,}000$. Notice that the exponent 8 is the number of places to the right that the decimal point moves. For numbers smaller than 1, such as 1×10^{-6} m, the negative exponent equals the number of places the decimal moves to the left.

Arithmetic operations using scientific notation follow simple rules.

For addition or subtraction both numbers must be expressed in the same power of 10. For example, $(3 \times 10^3) + (4 \times 10^3) = 7 \times 10^3$. The same rule works for subtraction. If two numbers with different powers of 10 are added or subtracted, one must first be converted to the same power of 10 as the other.

The rules for multiplication and division are more general.

For *multiplication*

$$(a \times 10^m) \cdot (b \times 10^n) = ab \times 10^{m+n}$$

For *division*

$$\frac{a \times 10^m}{b \times 10^n} = \frac{a}{b} \times 10^{m-n}$$

It is most common to choose the exponent so that "*a*" is a number between 1 and 10. This is only a convention, but it is very common. For example, $5100 = 5.1 \times 10^3$, the last being the most common in scientific notation.

Entering numbers into your calculator in scientific notation:

Most scientific calculators have an EE key or EXP key. When you press that key, your calculator interprets this as "times 10 raised to a power." The calculator is then waiting for the power.

For example, if you want to enter the number, 9.11×10^{-31} into your calculator, you would proceed as follows:

Step by step 1. Enter 9.11

Step 2. Press the EE or EXP key

Step 3. Press the ± key for the negative power

Step 4. Enter 31

A.5 LOGARITHMS

Log base 10 When you take the $\log_{10}$ of a number, you are asking for the power of 10 that would give you that number. For example, the log of 100 is 2 and the log of 1000 is 3 because 100 is 10^2 and 1000 is 10^3. If we take the log of a number between 100 and 1000, we expect an answer between 2 and 3. Take the log of 500 on our calculator. You should get approximately 2.70.

$$\log_{10} 10^{power} = power$$

The antilog of a number equals 10 raised to that power. For example, the antilog of 2 is $10^2 = 100$. We will use this information in Chapter 8 when calculating sound level and intensity.

Log base e, sometimes called the natural log. Let's now turn our attention to log base e. On the calculator, log base e is the LN key. When you see LN of a number, it means to take the natural log, log base e of the number. In a similar fashion to log base 10, $\ln e^{power} = power$

We will use this rule when solving the half-life equation in Chapter 18.

APPENDIX B
Units and Conversion Factors

BASIC AND DERIVED SI UNITS

Quantity	Unit name (symbol)	In Terms of Basic Units	In Terms of Other SI Units
Basic Units			
Length	meter (m)		
Mass	kilogram (kg)		
Time	second (sec)		
Temp.	kelvin (K)		
Current	ampere (A)		C/s
Derived Units			
Force	newton (N)	$kg{\cdot}m/s^2$	
Energy, work	joule (J)	$kg{\cdot}m^2/s^2$	N·m
Power	watt (W)		J/s
Pressure	pascal (Pa)		N/m^2
Frequency	hertz (Hz)	/second	
Charge	coulomb (C)	A·s	
Electrical potential diff.	volt (V)		J/C
Electric resistance	ohm (Ω)		V/A

CONVERSION FACTORS

The relationships marked with an asterisk are exact.

Length

*1 in. = 2.54 cm
1 cm = 0.394 in.
1 ft = 30.5 cm
1 m = 39.4 in. = 3.28 ft
*1 mi = 5280 ft = 1.61 km
1 km = 0.621 mi
*1 fermi = 1 fm = 10^{-15} m
*1 angstrom (A) = 10^{-10} m = 10 nm
*1 micron = 1 μm = 10^{-6} m

Mass	Time
*1 kg = 1000 g	*1 hr = 60 min = 3600 s
1 slug = 14.6 kg	*1 hr = 60 min = 3600 s
1 kg = 0.0685 slug	*1 day = 24 hr = 8.64×10^4 s
* 1 metric ton = 10^3 kg	1 yr = 365.24 days = 3.16×10^7 s

Speed	Force
1 mile/hr = 1.47 ft/s = 1.61 km/hr	I lb = 4.45 N
1 km/hr = 0.278 m/s = 0.621 mi/hr	1 N = 0.225 lb
1 ft/sec = 0.305 m/s = 0.682 mi/hr	*1 N = 10^5 dyn
1 m/sec = 3.28 ft/s = 3.60 km/hr	*1 ton = 2000 lb

Pressure (see also Table 6.1)

1 atm = 1.013 bar = 1.013×10^5 N/m^2
= 14.7 Ib/in^2 = 760 mm Hg
1 lb/in^2 = 6.90×10^3 N/m^2
1 Pa = 1 N/m^2 = 1.45×10^{-4} lb/in^2

Energy and Work

*1 J = 10^7 ergs = 0.738 ft·lb
1 ft·lb = 1.36 J = 1.29 × 10-3 Btu = 3.25×10^{-4} kcal
1 kcal = 4.186×10^3 J = 3.97 Btu
1 Btu = 252 cal = 778 ft·lb = 1054 J
1 eV = 1.60×10^{-19} J
1 kWh = 3.60×10^6 J = 860 kcal

Power

*1 W = 1 J/s = 0.738 ft·lb/s
*1 hp (US) = 550 ft·lb/s = 746 W
*1 hp (metric) = 750 W
*1 kg has a weight of 2.21 lb where g = 9.80 m/s^2

Radiation related quantities

Activity

1 curie (Ci) = 3.7×10^{10} decays/s
1 becquerel (Bq) = 1 decay/s

Dose

1 rad = 0.01 joule/kg of exposed tissue
1 rem = #rad × RBE
1 rem = 10 mSv (milliSievert)

APPENDIX C

Periodic Table of the Elements

PERIODIC CHART OF THE ELEMENTS

IA	IIA	IIIB	IVB	VB	VIB	VIIB	VIII			IB	IIB	IIIA	IVA	VA	VIA	VII A	INERT GASES
1 H 1.00797																1 H 1.00797	2 He 4.0026
3 Li 6.939	4 Be 9.0122											5 B 10.811	6 C 12.0112	7 N 14.0067	8 O 15.9994	9 F 18.9984	10 Ne 20.183
11 Na 22.9898	12 Mg 24.312											13 Al 26.9815	14 Si 28.086	15 P 30.9738	16 S 32.064	17 Cl 34.453	18 Ar 39.948
19 K 39.102	20 Ca 40.08	21 Sc 44.956	22 Ti 47.90	23 V 50.942	24 Cr 51.996	25 Mn 54.9380	26 Fe 55.847	27 Co 58.9332	28 Ni 58.71	29 Cu 63.54	30 Zn 65.37	31 Ga 69.72	32 Ge 72.59	33 As 74.9216	34 Se 78.96	35 Br 79.909	36 Kr 83.80
37 Rb 85.47	38 Sr 87.62	39 Y 88.905	40 Zr 91.22	41 Nb 92.906	42 Mo 95.94	43 Tc (99)	44 Ru 101.07	45 Rh 102.905	46 Pd 106.4	47 Ag 107.870	48 Cd 112.40	49 In 114.82	50 Sn 118.69	51 Sb 121.75	52 Te 127.60	53 I 126.904	54 Xe 131.30
55 Cs 132.905	56 Ba 137.34	57 La 138.91	72 Hf 178.49	73 Ta 180.948	74 W 183.85	75 Re 186.2	76 Os 190.2	77 Ir 192.2	78 Pt 195.09	79 Au 196.967	80 Hg 200.59	81 Tl 204.37	82 Pb 207.19	83 Bi 208.980	84 Po (210)	85 At (210)	86 Rn (222)
87 Fr (223)	88 Ra (226)	89 ‡ Ac (227)	104 Rf (261)	105 Db (262)	106 Sg (266)	107 Bh (262)	108 Hs (265)	109 Mt (266)	110 ? (271)	111 ? (272)	112 ? (277)						

Numbers in parenthesis are mass numbers of most stable or most common isotope.
Atomic weights corrected to conform to the 1963 values of the commission on Atomic Weights.
The group designations used here are the former Chemical Abstract Service numbers.

* Lanthanide Series

58 Ce 140.12	59 Pr 140.907	60 Nd 144.24	61 Pm (147)	62 Sm 150.35	63 Eu 151.96	64 Gd 157.25	65 Tb 158.924	66 Dy 162.50	67 Ho 164.930	68 Er 167.26	69 Tm 168.934	70 Yb 173.04	71 Lu 174.97

□ Actinide Series

90 Th 232.038	91 Pa (231)	92 U 238.03	93 Np (237)	94 Pu (242)	95 Am (243)	96 Cm (247)	97 Bk (247)	98 Cf (249)	99 Es (254)	100 Fm (253)	101 Md (256)	102 No (256)	103 Lr (257)

APPENDIX D

Selected Radioactive Isotopes*

Decay modes are α, β^-, β^+, and electron capture (EC). EC results in the same daughter nucleus as would β^+ decay. IT stands for isomeric transition from metastable to ground state. Energies quoted for $\beta^\pm$ decays are the maxima; average $\beta^\pm$ energies are roughly one-half the maxima. Percents are approximate in some cases.

Isotope	$t_{1/2}$	Decay Mode and Energy (MeV)					
$^{3}_{1}H$	12.33 yr	β^-	0.0186	100%			
$^{14}_{6}C$	5730 yr	β^-	0.156	100%			
$^{13}_{7}N$	9.96 min	β^+	1.20	100%			
$^{22}_{11}Na$	2.602 yr	β^+	0.55	90%	γ	1.27	100%
$^{32}_{15}P$	14.28 d	β^-	1.71	100%			
$^{35}_{16}S$	87.4 d	β^-	0.167	100%			
$^{36}_{17}Cl$	3.00×10^5 yr	β^-	0.710	100%			
$^{40}_{19}K$	1.28×10^9 yr	β^-	1.31	89%			
$^{43}_{19}K$	22.3 hr	β^-	0.827	87%	γ's	0.373	87%
						0.618	87%
$^{45}_{20}Ca$	165 d	β^-	0.257	100%			
$^{51}_{24}Cr$	27.70 d	EC			γ	0.320	10%
$^{52}_{25}Mn$	5.59 d	β^+	3.69	28%	γ's	1.33	28%
						1.43	28%
$^{52}_{26}Fe$	8.27 hr	β^+	1.80	43%	γ's	0.169	43%
						0.378	43%
$^{59}_{26}Fe$	44.6 d	β^-'s	0.273	45%	γ's	1.10	57%
			0.466	55%		1.29	43%

(Continued)

Isotope	$t_{1/2}$	Decay Mode and Energy (MeV)					
$^{60}_{27}Co$	5.271 yr	β^-	0.318	100%	γ's	1.17	100%
						1.33	100%
$^{65}_{30}Zn$	244.1 d	EC			γ	1.12	51%
$^{67}_{31}Ga$	78.3 hr	EC			γ's	0.0933	70%
						0.185	25%
						0.300	19%
						Others	
$^{75}_{34}Se$	118.5 d	EC			γ's	0.121	20%
						0.136	65%
						0.265	68%
						0.280	20%
						Others	
$^{86}_{37}Rb$	18.8 d	β^-'s	0.69	9%	γ	1.08	9%
			1.77	91%			
$^{85}_{38}Sr$	64.8 d	EC			γ	0.514	100%
$^{90}_{38}Sr$	28.8 yr	β^-	0.546	100%			
$^{90}_{39}Y$	64.1 hr	β^-	2.28	100%			
$^{90m}_{43}Tc$	6.02 hr	IT			γ	0.142	100%
$^{113m}_{49}In$	99.5 min	IT			γ	0.392	100%
$^{123}_{53}I$	13.0 hr	EC			γ	0.159	~100%
$^{131}_{53}I$	8.040 d	β^-'s	0.248	7%	γ's	0.364	85%
			0.607	93%		Others	
			Others				

Isotope	$t_{1/2}$	Decay Mode and Energy (MeV)					
$^{129}_{55}Cs$	32.3 hr	EC			γ's	0.0400	35%
						0.372	32%
						0.411	25
						Others	
$^{137}_{55}Cs$	30.17 yr	β^-'s	0.511	95%	γ	0.662	95%
			1.17	5%			

Isotope	$t_{1/2}$	Decay Mode and Energy (MeV)					
$^{140}_{56}Ba$	12.79 d	β^-	1.035	~100%	γ's	0.030	25%
						0.044	65%
						0.537	24%
						Others	
$^{137}_{56}Ba$	2.55 min				γ	0.662	100%
$^{198}_{79}Au$	2.696 d	β^-	1.161	~100%	γ	0.412	~100%
$^{197}_{80}Hg$	64.1 hr	EC			γ	0.0773	100%
$^{210}_{84}Po$	138.38 d	α	5.41	100%			
$^{226}_{88}Ra$	1.60×10^3 yr	α's	4.68	5%	γ	0.186	5%
$^{222}_{86}Rn$	3.8 d	α	4.87	95%			
$^{235}_{92}U$	7.038×10^8 yr	α	4.68	~100%	γ's	Numerous	<0.400
$^{238}_{92}U$	4.468×10^9 yr	α's	4.22	23%	γ	0.050	23%
			4.27	77%			
$^{237}_{98}Np$	2.14×10^6 yr	α's	Numerous		γ's	Numerous	<0.250
			Max. en.				
			4.96				
$^{239}_{94}Pu$	2.41×10^4 yr	α's	5.19	11%	γ's	75×10^{-5}	73%
			5.23	15%		0.013	15%
			5.24	73%		0.052	10%
						Others	

APPENDIX E
Important Constants and Other Data

The values given are those used in the text. Most are known to much greater precision.

Quantity	Value
Speed of light in a vacuum, c	3.00×10^8 m/s
Gravitational constant, γ	$6.67 \times 10^{-11} \frac{N \cdot m^2}{kg^2}$
Avogadro's number, N_A	6.02×10^{23}/mole
Charge of electron, q_e	-1.60×10^{-19} C
Coulomb (force) constant, k	$9.00 \times 10^9 \frac{N \cdot m^2}{C^2}$
Rest mass of electron, m_e	9.11×10^{-31} kg
Rest mass of proton, m_p	1.67×10^{-27} kg
Rest mass of neutron, m_n	1.67×10^{-27} kg
Stefan-Boltzmann constant, σ	$5.67 \times 10^{-8} \frac{W}{m^2 \cdot K^4}$
Planck's constant, h	6.63×10^{-34} J·s
Electron volt (eV)	1.60×10^{-19} J
Absolute zero (0 K)	–273. 15°C
Kilocalorie (kcal)	4186 J
1 calorie	4.186 J
Earth	
Mass	5.98×10^{24} kg
Radius (average)	6.37×10^6 m
Density (average)	5.50×10^3 kg/m^3

(Continued)

Quantity	Value
Moon	
Mass	7.36×10^{22} kg
Radius (average)	1.74×10^{6} m
Density (average)	3.34×10^{3} kg/m^3
Sun	
Mass	1.99×10^{30} kg
Radius (average)	6.96×10^{8} m
Density (average)	1.46×10^{3} kg/m^3
Earth-sun distance (average)	1.50×10^{11} m
Earth-moon distance (average)	3.84×10^{8} m
Earth orbital period	3.16×10^{7} s
Moon orbital period (average)	2.36×10^{6} s

APPENDIX F

Answers to All Odd and Some Even Problems

CHAPTER 2

1. 25.0 m west
2. (a) speed = 66.7 km/hr (b) 66.7 km/hr North
3. 6.82 m/s
4. 5.00 m/s^2 forward
5. $-4.00\ m/s^2$
6. 61.7 m/s
7. 1.03×10^4 s
9. (a) 30.0 km/s (b) 0
11. (a) 5.10 s (b) 78.0 m
13. 26 m/s
15. 14.5 m/s
16. 4.5 m/s
17. 162 m
19. 3.7 s
21. (a) 600 m (b) 120 s
23. (a) 2.54 s (b) 38.7 m/s
25. (a) Yes (b) 2.84 m/s^2
29. 0.159 m
31. t = 1 s, $\Delta X = -4.9$ m, v = −9.8 m/s
    t = 2 s, $\Delta X = -19.6$ m, v = −19.6 m/s
    t = 3 s, $\Delta X = -44.1$ m, v = −29.4 m/s
    t = 4 s, $\Delta X = -78.4$ m, v = −39.2 m/s
33. (a) $\Delta X = 327$ m (b) 16.3 s
35. 1.11 sec
37. 40.7 m
39. (a) 3.35 s (b) 4.88 s

CHAPTER 3

1. 24.5 N
2. 120 kg
3. 1.67 m/s^2, 0.170 g
4. 490 Newtons
5. 5.10×10^{-7} kg
6. 7.50 Newtons
7. 50 N
8. 0.300
9. 9.8×10^{-4} N
11. 18.5 N
13. (a) 875 N (b) −875 N
15. 7.7 m/s^2
17. (a) 4.2 m/s (b) 29.4 m/s^2 (c) 4.31×10^3 N
19. (a) 0.0469 m/s^2 (b) 6.79×10^4 N (c) 5.83×10^4 N
20. 250 N at 53.1°
21. 250 N, 36.9° north of east
22. F1 = 28.7, N F2 = 41.0 N
23. 1.11 m/s^2, 29.9° to right
25. 2.30×10^3 N
27. 376 N upward
28. 24.5 N
29. 95.4 N, 59° above horizontal
31. (a) 2.0 m/s^2 (b) F > 73.5 N
33. 110 N
35. 25.5 N·m
41. F > 294 N
43. 2.25×10^3 N
45. $1.75 \times 10^4\ m/s^2$
46. (a) 5,820 N (b) 6.13 m/s^2 (c) 0.625g
47. (a) $2.27 \times 10^3\ m/s^2$ (b) $2.30 \times 10^5\ m/s^2$
48. 12.3 m/s^2
49. 27.0 N
50. 18.9 rev/min
51. (a) 17.1 m/s (b) 14.3 m/s
53. 75.7°

CHAPTER 4

1. 7.50×10^7 J
2. 1,080 J
3. 1.08×10^3 J
4. (a) 4,250 J, (b) 28,800 J, (c) 6.78 times
5. 263 kcal
6. 11.4 min.
7. 61.0 min
8. 500 joules
9. 4.90×10^{16} J
11. (a) 19.6 kcal (b) 4.9 kcal
12. 3,220 kcal
13. (a) 3.53×10^5 N, (b) 2.79×10^3 N, (c) 601 W and 4.75 W
15. (a) 935 J (b) 62.3 N 17. 5.1 m
18. (a) 39.6 m/s (b) 39.7 m/s
19. (a) 31.9 m (b) 28 kcal
22. (a) 174 w (b) 0.233 hp 23. 31.0 W
24. 19.9%
25. 16.7%
26. 96.4 cents
27. (a) 144 J (b) 288 W
29. (a) 293 kcal, (b) 10.2 kW, (c) 306 W
31. (a) 2.2×10^3, b) 900 s
32. 20.4 liters/day
33. (a) 25% (b) 0.3 kcal/min 35. 9 g
36. 26.5 %
37. 2.35×10^3 liters

CHAPTER 5

1. (a) 98.6°F (b) 40.0°C
3. −459.67°F
4. 6,170°F
5. −173°F
7. −40.0
8. 6×10^{-3} m
9. 2.02 m
11. 1.8 m less 0.77 mm; No
13. 7.47×10^4 cal, 74.7 food cal
14. 429 cal
15. 24, kcal
17. (a) 330 cal (b) 3030 cal
19. (a) 1.25°C (b) 172 liters
21. (a) 428 kcal (b) 57.1 min
23. 717°C
25. −9.88°C
27. 16. 7°C
29. 19.1°C
30. 87.0%
31. 5.11 g/m^3
33. 74.0%
35. (a) 7.87 g/m^3 (b) 4.56 kcal
36. 40.3 W
37. 1.89 kW
39. 39.0 W
40. 12,900 W
41. 6.60 kW
43. 1.58 kW
45. 30.6°C
47. 129 kcal
48. 725,000 cal
49. 172 g
50. 20,800 cal
51. (a) 3.48×10^5 cal (b) 16.9 W
53. 2.46 kg
55. 1.23°C
57. 293 W

CHAPTER 6

1. 2.45×10^6 N/m^2, 24.1 atm
2. 2.83×10^9 N/m^2
3. 5.44 atm
5. 392 cm^2
7. 1.1×10^8 N/m^2
9. (a) 1.54×10^4 N/m^2, (b) 1.16×10^5 N/m^2
11. 4.08 m
13. (a) 1.43×10^4 N/m^2, (b) 107 mm Hg
15. (a) 100:1, (b) 10:1, (c) 100
17. 2.52×10^4 N
18. (a) 94.7 % submerged, (b) 85.7 % submerged
19. 0. 700 g/cm^3
21. 0.950 g/cm^3
23. (a) 63.5 kg (b) 0.0635 m^3, (c) 1.02×10^3 kg/m^3 (d) Yes
25. (a) 8.67×10^{-2} m^3 (b) 0.112 kg (c) 1.10 N (833 N)
27. 1.7×10^{-3} m^3
28. 83.3 cm^3/s
29. 500 min
30. 2 cm^3/min
31. (a) 33.3 cm^3/s (b) 66.7 cm^3/s (c) 0.160 cm^3/s (d) 200 cm^3/s (e) 98.4 cm^3/s
32. (a) 5.71 cm^3/s (b) 8.2 cm^3/s (c) 32.0 cm^3/s (d) 1.02 cm^3/s
33. 12 m H_2O
34. 1.52ΔP
35. 15.9% or r' = 0.841r
36. 31.7 cm/s

37. 1.20 liters/min 38. (a) 94.2 cm^3/s (b) 5.65 liter/min
39. (a) 20.4 cm^3/s, (b) 12.2 liters/min 41. 3.8×10^{-9} cm^3/s 43. 800 cm/s
47. 37.7 m^2; 0.126 m^2 49. 0.0400 μm
51. (a) 57.6 N/m^2 (b) 18.4 N/m^2 (c) 29.6 N/m^2

CHAPTER 7

1. −13.6 cm H_2O 3. (a) 921 mm Hg (b) 1.88×10^4 mm Hg
5. (a) 3.00×10^4 N/m^2, 226 mm Hg (b) 3.10 m 7. 0.34 m
9. 189 cm^3/min 11. 2.36×10^{-8} cm^3/s 13. 6.5 cm/s
15. (a) 96.8 mm Hg (b) 243 mm Hg 17. 5.83 liters/min
19. 9.00% decrease 21. 39.6% 23. (a) 73.2 N/m (b) 439 N/m
25. 226 mm Hg 27. (a) 1.76 W (b) 12.6 W 29. 75 m^2
31. 2.00% 33. 7.4 mm Hg
35. (a) 0.5 liter (b) 3.10 liter (c) 1.20 liter (d) 4.80 liter
37. (a) 49.0 N/m^2 (b) 0.370 mm Hg

CHAPTER 8

1. 5.00×10^{-7} s 2. 4100 Hz 3. (a) 5.0×10^{-2} s (b) 5.0×10^{-5} s
5. 200 Hz 7. (a) 17.2 m (b) 0.0172 m
8. (a) 15.6 m (b) 3.44×10^{-3}m = 3.44 mm 9. 126 m 10. 102 m
11. 13.9 days 12. 7.20/min 13. 6.16 MHz 14. 1.50 m/s
15. 50 dB 16. 130-dB 17. 1.00×10^{-8} W/m^2 19. 60.0 dB
21. 60.0 dB 23. (a) 1.50 MHz, minimum (b) 0.20 m 25. 1.33×10^{-6} s
27. 3.18×10^{-4} J 29. (a) 0.200 Hz (b) 0.300 Hz (c) 0.500 Hz
31. (a) 0.400 Hz (b) 2.50 s 32. (a) 1,330 Hz
33. (a) 20,500 Hz (b) 2200 Hz 35. 15.0 cm/s
37. (a) 9.63 cm/s (b) no

CHAPTER 9

1. (a) 33 Hz (b) 88 Hz (c) 88 Hz; 176 Hz 3. (a) 0.300 Hz (b) 12.0 Hz (c) 45.0 Hz
5. (a) 64.0 dB (b) 26.0 dB (c) 40.0 dB
7. (a) 50 dB = 36, 50, 53, 38 phons (b) 100 dB = 100, 100, 109, 91 phons
(c) 0 dB = inaudible, inaudible, 7 phons, inaudible
9. (a) 1.00×10^{-11} W/m^2 (b) 1.00×10^{-12} W/m^2 11. 1×10^4
13. (a) No (b) yes (c) 10 dB 15. 20 dB

CHAPTER 10

1. 6.25×10^9 electrons 2. 1.25×10^9 electrons
3. (a) 9.00×10^{-5} N (b_1) 2.25×10^{-5} N (b_2) 1.00×10^{-5} N (c) 9.00×10^{-3} N

5. 4.24 km
7. (a) 27.0 J (b) 1.20×10^4 J
8. 1.00×10^{-4} J
9. (a) 6.49×10^7 m/s (b) 9.37×10^7 m/s
11. 25.6 V
12. (a) 2.00 amps (b) 2.00 milliamp
13. 0.0600 C
15. 1.50×10^{25} electrons
17. (a) 3.00×10^3 C (b) 3.60×10^4 J

CHAPTER 11

1. 29.3 A
2. 2×10^{-4} A
3. 5.00 A
4. (a) 15.0 mA (b) 0.240 mA
5. 8.57 Ω
7. 280 Ω
9. 12.0 V
11. 2.00×10^{12} W
12. 6000 W
13. 43.2 W
14. (a) 144, 000 W (b) 1200 W
15. 1.44 W
16. (a) 4.50 W (b) 0.375 W
17. 67.2 W
19. 12.5 Ω
21. (a) 5.00A (b) 2.40 Ω
23. 1.0×10^5 V
25. $256
26. $402
27. (a) $3.94 billion (b) $39.40
29. 16%
31. (a) 1.50 kW (b) 12.5 A
32. (a) 18.0 Ω (b) 4.00 Ω
33. (a) 14.4 Ω (b) 1440 Ω
34. (a) 120 Ω (b) 16.7 Ω
35. (a) 0.208 A, 576 Ω 0.500 A, 240 Ω (b) 12.4 W; 5.20 W
37. Yes
39. 45 light bulbs
41. (a) 14.4 Ω; 28.8 Ω 1.00 kW; 1.50 kW (b) 0.00 W; 0.330 kW; 0.500 kW;
43. (a) 112 V (b) 874 W (c) 216 V, 959 W
45. 13.0 Ω

CHAPTER 12

1. (a) 0.600 rnA (b) 24.0 rnA
2. 400 V
3. (a) 0.120 MΩ (b) 20 rnA
4. 15.0 mV
5. 6.00 mV
6. (a) 144 W (b) 1.20 mA
7. 8,500 V
8. 1.19 mA
9. (a) 10.0 mV (b) 10.0 μA
11. (a) 37.5 V (b) 0.75 rnA (c) 18.8 rnA (d) 8.00×10^{-4} Ω
13. (a) 2.00 Ω (b) 40.0 V (c) 8.00 kΩ

CHAPTER 14

1. 2.26×10^8 m/s; 1.97×10^8 m/s
2. 1.24×10^8 m/s
3. 1.36; ethyl alcohol
4. 1.50, Benzene
5. 500 s
6. 1.67×10^{-8} s
7. 20 D
8. Largest is 12.5 Diopter, Smallest is 5 Diopter
9. -0.286 m
10. I.0400 m
11. (a) 60.0 cm (b) -4.00
12. (a) 51.3 mm, (b) 44.9 mm
13. (a) 3.43 m (b) 0.800 m by 1.20 m
17. (a) 2.55 m (b) 1.00 m
19. 7.5 cm
21. 45.0 cm
23. (a) 0.200 m (b) -0.0200 m
25. 2.40 cm
27. 1.25 D
29. (a) 18.3 cm (b) -60 (c) -11.3 cm (d) 6.67 (e) -400

CHAPTER 15

1. 52.0 D; 51.0 D; 50.2 D
2. 52.2 D
3. – 0.230 mm
5. 33.3 cm
7. 76.0 km
9. (a) 55.0 D (b) 20.0 cm
11. + 0.500 D
12. – 0.5000 D
13. Light: 2; Sound: 1000
15. (a) 2.00 m (b) – 0.500 D
17. (a) 50.67 D (b) 3.33 D
19. (a) 50.25 D (b) – 0.25 D
21. 30.8 cm
23. (a) 3.00 D (b) No
25. (a) 1.35 m (b) – 0.74 D

CHAPTER 16

1. (a) 7.89×10^{14} Hz to 3.90×10^{14} Hz (b) 2: 1 (c) 1000:1
2. 2.54 m
3. 0.33 m; 0.117 m
4. 3×10^{13} Hz
5. 200 nm
6. 0.600 m
7. 1500 kHz, AM
8. (a) 6.67×10^{-7} s (b) 6.67×10^{-5} s (c) 1000 s
9. (a) 4.14×10^{-2} eV (b) 82.9 eV
10. (a) 3.31×10^{-9} eV (b) 1.04×10^{6} eV
11. 1.61 eV → 3.27 eV
13. (a) 3.00×10^{18} Hz (b) 12.4 keV
14. 1.24×10^{-7} m
15. 77.3 MHz, 199.2 MHz
16. (a) 1.31 eV (b) 6.79×10^{5} m/s
17. (a) 7.5×10^{14} Hz (b) Visible light
19. 8.3×10^{13}/s
20. 15.0 m/s
21. (a) 1.33×10^{-19} J (b) 2.10×10^{23} photons (c) 278 s
23. 0.00300°C

CHAPTER 17

1. A: 7.4 eV; 1.79×10^{15} Hz; ultraviolet C: 0.90 eV; 2.17×10^{14} Hz; infrared
   F: 4.0 eV; 9.65×10^{14} Hz; ultraviolet
2. Max. energy is transition A with 7.4 eV; 1.18×10^{-18} J
3. 311 nm
4. 1.33×10^{-32} m
5. B: 8.21×10^{14} Hz; 366 nm; ultraviolet
   D: 1.57×10^{15} Hz; 191 nm; ultraviolet E: 6.03×10^{14} Hz; 497 nm; blue-green
6. A 168nm; B 365 nm; C 1380 nm; D 191 nm; E 497 nm; F 311 nm
7. (a) C (b) D
9. 2nd
11. 168 nm; 191 nm; 311 nm; 366 nm; 497 nm; 1383 nm
13. 2.49×10^{-11} m; Yes
15. 1.1×10^{-18} m/s

CHAPTER 18

1. 1.44×10^{5}
2. 9.60×10^{5} disruptions
3. 1.00 – 10.0 km
4. 0.139 grams
5. 2.71×10^{-4} m to a side
7. (a) 2.82×10^{20} Hz, (b) 1.06×10^{-12} m; (c) 3.21×10^{20} Hz, (d) 9.35×10^{-13} m
9. ${}^{1}_{0}n \rightarrow {}^{0}_{-1}\beta + {}^{1}_{1}H$
10. ${}^{3}_{1}H \rightarrow {}^{0}_{-1}\beta + {}^{3}_{2}He$
11. ${}^{40}_{19}K \rightarrow {}^{0}_{-1}\beta + {}^{40}_{20}Ca$
12. ${}^{132}_{50}I \rightarrow {}^{0}_{-1}\beta + {}^{131}_{54}Xe$
13. ${}^{137}_{55}Cs \rightarrow {}^{0}_{-1}\beta + {}^{137}_{56}\mathrm{Ba}$
16. (a) ${}^{238}_{92}U \rightarrow {}^{4}_{2}He + {}^{234}_{90}Th$ (b) ${}^{234}_{90}Th \rightarrow {}^{0}_{-1}\beta + {}^{234}_{91}Pa$ (c) ${}^{234}_{91}Pa \rightarrow {}^{0}_{-1}\beta + {}^{234}_{92}U$

19. 2.22×10^{-12} kg **21.** 0.502 kg **23.** 38,100 yr **24.** 215 yr
25. (a) 0.500 mCi (b) 10.5 yr **27.** 8.86×10^{-4} kg **28.** 12,400 yr
29. (a) 4.68×10^{4} Ci (b) 1.04×10^{-2} kg
30. (a) 7.00 REM (b) 3.20 REM (c) 10.0 REM
31. (a) 0.500 RAD (b) 0.500 RAD (c) 20.0 RAD
33. 200 REM **35.** Approximately 440 deaths, or 0.044% versus 20%
37. 12.0 days **39.** 7.60 days **41.** 0.780 μCi
43. (a) 7.2×10^{4} days (b) 1.0×10^{5} REM **45.** 2.14×10^{-12} kg